ST/ESA/STAT/SER.S/35

Department of Economic and Social Affairs
Statistics Division

Département des affaires économiques et sociales
Division de statistique

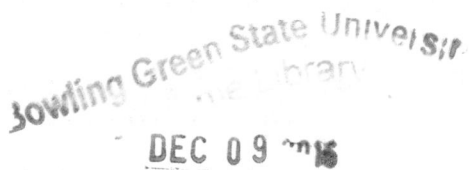

Statistical Yearbook
2016 edition
Fifty-ninth issue

Annuaire statistique
2016 édition
Cinquante-neuvième édition

United Nations | Nations Unies
New York, 2016

Department of Economic and Social Affairs

The Department of Economic and Social Affairs of the United Nations Secretariat is a vital interface between global policies in the economic, social and environment spheres and national action. The Department works in three main inter-linked areas: (i) it compiles, generates and analyses a wide range of economic, social and environmental data and information on which Member States of the United Nations draw to review common problems and to take stock of policy options; (ii) it facilitates the negotiations of Member States in many intergovernmental bodies on joint courses of action to address ongoing or emerging global challenges; and (iii) it advises interested Governments on the ways and means of translating policy frameworks developed in United Nations conferences and summits into programmes at the country level and, through technical assistance, helps build national capacities.

Département des affaires économiques et sociales

Le Département des affaires économiques et sociales du Secrétariat de l'Organisation des Nations Unies assure le lien essentiel entre les politiques adoptées au plan international dans les domaines économique, social et écologique et les mesures prises au plan national. Il mène ses activités dans trois grands domaines interdépendants : i) il compile, produit et analyse une grand variété de données et d'informations économiques, sociales et écologiques dont les États Membres de l'ONU tirent parti pour examiner les problèmes communs et faire le point sur les possibilités d'action; (ii) il facilite les négociations que les États Membres mènent dans un grand nombre d'organes intergouvernementaux sur les moyens d'action à employer conjointement pour faire face aux problèmes mondiaux existants ou naissants; et (iii) il aide les gouvernements intéressés à traduire les orientations politiques établies lors des conférences et sommets de l'ONU en programmes nationaux et contribue à renforcer les capacités des pays en leur apportant une assistance technique.

Note

The designations employed and the presentation of the material in this publication do not imply the expression of any opinion whatsoever on the part of the Secretariat of the United Nations concerning the legal status of any country, city or area, or of its authorities, or concerning the delimitation of its frontiers or boundaries.

In general, statistics contained in the present publication cover a period up to 2016 and as available to the United Nations Secretariat as of 31 July 2016. They reflect the country nomenclature currently in use.

The term "country" as used in the text of this publication also refers, as appropriate, to territories or areas.

The designations "developed" and "developing" which appear in some tables are intended for statistical convenience and do not necessarily express a judgement about the stage reached by a particular country or area in the development process.

Symbols of the United Nations documents are composed of capital letters combined with figures.

Note

Les appellations employées dans la présente publication et la présentation de données qui y figurent n'impliquent, de la part du Secrétariat de l'Organisation des Nations Unies, aucune prise de position quant au statut juridique des pays, territoires, villes ou zones, ou de leurs autorités, ni quant au tracé de leurs frontières ou limites.

En règle générale, les statistiques contenues dans la présente publication couvrent pour la période jusqu'en 2016 et sont celles dont disposait le Secrétariat de l'Organisation des Nations Unies au 31 juillet 2016. Elles reflètent donc la nomenclature de pays en vigueur à l'époque.

Le terme « pays », tel qu'il est utilisé dans la présente publication peut également désigner des territoires ou des zones.

Les appellations « développées » et « en développement » qui figurent dans certains tableaux sont employées à des fins exclusivement statistiques et n'expriment pas nécessairement un jugement quant au niveau de développement atteint par tel pays ou telle région.

Les cotes des documents de l'Organisation des Nations Unies se composent de lettres majuscules et de chiffres.

ST/ESA/STAT/SER.S/35

UNITED NATIONS PUBLICATION
Sales No. B.16.XVII.1.H

ISBN 978-92-1-061387-3
e-ISBN 978-92-1-058368-8
ISSN 0082-8459

ST/ESA/STAT/SER.S/35

PUBLICATION DES NATIONS UNIES
Numéro de vente : B.16.XVII.1.H

ISBN 978-92-1-061387-3
e-ISBN 978-92-1-058368-8
ISSN 0082-8459

Preface

The 2016 edition of the United Nations *Statistical Yearbook* is the fifty-ninth issue of the publication, prepared by the Statistics Division of the Department of Economic and Social Affairs. Ever since the compilation of data for the *Statistical Yearbook* series was initiated in 1948, it has consistently provided a wide range of internationally available statistics on social, economic and environmental conditions and activities at the national, regional and world levels.

The contents of the *Yearbook* are currently under review and the number of tables has been reduced as a result. Additional tables are expected to be introduced in future editions. Please send any comments or views to our email address, statistics@un.org. The tables include series covering an appropriate historical period, depending upon data availability and space constraints, for as many countries, territories and statistical areas of the world as available. The tables cover a period up to 2016, with some of the data being estimated.

The *Yearbook* tables are based on data which have been compiled by the Statistics Division mainly from official national and international sources as these are more authoritative and comprehensive, more generally available as time series and more comparable among countries than other sources. These sources include the United Nations Statistics Division in the fields of national accounts, industry, energy and international trade, the United Nations Statistics Division and the Population Division in the field of demographic statistics, and over 20 offices of the United Nations system and international organizations in other specialized fields. In some cases, official sources have been supplemented by other sources and estimates, where these have been subjected to professional scrutiny and debate and are consistent with other independent sources.

The United Nations agencies and other international, national and specialized organizations which furnished data are listed under "Source" at the end of each table. Acknowledgement is gratefully made for their generous and valuable cooperation in continually providing data.

After the first table which presents key world aggregates and totals, the *Yearbook* is organized in three parts as follows; part one, relating to population and social topics; part two, relating to economic activity; and part three relating to energy, environment and infrastructure. The tables in the three parts are presented mainly by countries or areas, and world and regional aggregates are shown where available.

The four annexes contain information on country and area nomenclature (annex I), summary technical notes on statistical definitions, methods and sources for all tables of the *Yearbook* (annex II), the conversion coefficients and factors used in certain tables (annex III), and a list of those tables which were added, omitted or discontinued since the last issue of the *Yearbook* (annex IV).

The *Statistical Yearbook* is prepared by the Statistical Dissemination Section, Statistical Services Branch of the Statistics Division, Department of Economic and Social Affairs of the United Nations Secretariat. The programme manager is Matthias Reister and the chief editor is Ian Rutherford. They are assisted by David Carter, Anuradha Chimata and Aida Diawara. Bogdan Dragovic provided IT support.

Comments on the present *Yearbook* and its future evolution are welcome. They may be sent via e-mail to statistics@un.org or to the United Nations Statistics Division, Statistical Dissemination Section, New York, NY 10017, USA.

Préface

L'Annuaire statistique des Nations Unies 2016 est la cinquante-neuvième édition de cette publication, préparée par la Division de statistique du Département des affaires économiques et sociales. Depuis son instauration en 1948 comme outil de compilation des données statistiques internationales, l'*Annuaire statistique* s'efforce de constamment diffuser un large éventail de statistiques disponibles sur les activités et conditions économiques et sociales, aux niveaux national, régional et mondial.

Le contenu de l'Annuaire est actuellement en cours de révision, de ce fait le nombre de tableaux a été réduit. Des tableaux seront inclus dans les prochaines éditions. Pour tout commentaire, veuillez envoyer un courriel à notre adresse électronique : statistics@un.org. Les tableaux présentent des séries qui couvrent une période historique appropriée, en fonction de la disponibilité des données et des contraintes d'espace, pour autant de pays, territoires et zones statistiques du monde comme disponibles. Les tableaux couvrent généralement la période allant jusqu'en 2016.

Les tableaux de l'*Annuaire* sont construits essentiellement à partir des données compilées par la Division de statistique et provenant de sources officielles, nationales et internationales; c'est en effet la meilleure source si l'on veut des données fiables, complètes et comparables, et si l'on a besoin de séries chronologiques. Ces sources sont: la Division de statistique du Secrétariat de l'Organisation des Nations Unies pour ce qui concerne la comptabilité nationale, l'industrie, l'énergie et le commerce extérieur, la Division de statistique et la Division de la population du Secrétariat de l'Organisation des Nations Unies pour les statistiques démographiques; et plus de 20 bureaux du système des Nations Unies et d'organisations internationales pour les autres domaines spécialisés. Dans quelques cas, les données officielles sont complétées par des informations et des estimations provenant d'autres sources qui ont été examinées par des spécialistes et confirmées par des sources indépendantes.

Les institutions spécialisées des Nations Unies et les autres organisations internationales, nationales et spécialisées qui ont fourni des données sont énumérées dans la "Source" sur chaque tableau. Les auteurs de l'*Annuaire statistique* les remercient de leur précieuse et généreuse collaboration.

Après le premier tableau qui fournit les principaux agrégats et totaux au niveau mondial, l'*Annuaire* est groupés en trois parties. La première partie est consacrée à la population et aux questions sociales, la deuxième est consacrée à l'activité économique, et la dernière est consacrée à l'énergie, environnement et infrastructures. Dans ces trois parties, les tableaux sont généralement présentés par pays ou régions, mais les agrégats mondiaux ou régionaux sont indiqués si disponibles.

Les quatre annexes donnent des renseignements sur la nomenclature des pays et des zones (annexe I), des notes récapitulatives sur les définitions ,méthodes statistiques et techniques utilisées dans les sources de chaque tableau de l'*Annuaire* (annexe II), ainsi que sur les coefficients et facteurs de conversion employés dans les différents tableaux (annexe III), enfin une liste de nouveaux tableaux, de tableaux omis ou supprimés depuis la dernière édition est disponible (annexe IV).

L'Annuaire statistique est préparé par la Section de la diffusion statistique, Service des statistiques de services de la Division de statistique, Département des affaires économiques et sociales du Secrétariat de l'Organisation des Nations Unies. La responsable du programme est Matthias Reister, et le rédacteur en chef est Ian Rutherford. Ils sont secondés par David Carter, Anuradha Chimata et Aida Diawara. Bogdan Dragovic est responsable du support en informatique.

Les observations sur la présente édition de l'*Annuaire* et les suggestions de modification pour l'avenir seront reçues avec intérêt. Elles peuvent être envoyées par message électronique à statistics@un.org, ou adressées à la Division de statistique des Nations Unies, Section de la diffusion statistique, New York, NY 10017 (États-Unis d'Amérique).

Explanatory notes

Symbols and conventions used in the tables

. A point is used to indicate decimals.

- A hyphen between years, for example, 2010-2015, indicates the full period involved, including the beginning and end

/ A slash indicates a financial year, school year or crop year, for example 2014/15.

... Data are not available or not applicable.

* Data are provisional, estimated or include a major revision.

Marked break in the time series.

<0 Not zero, but less than half of the unit employed.

−<0 Not zero, but negative and less than half of the unit employed.

A space is used as a thousands separator, for example 1 000 is one thousand. Subtotals and percentages in the tables do not necessarily add to totals because of rounding.

Country notes and nomenclature

As a general rule, the data presented in the *Yearbook* relate to a given country or area as described in the complete list of countries and territories, see annex I.

References to statistical sources and methods in tables

For brevity the *Yearbook* omits specific information on the source or methodology for individual data points, and when a data point is estimated no distinction is made between whether the estimation was done by the international or national organization. See the technical notes to the tables in annex II for the primary source which may provide this information.

Units of measurement

The metric system of weights and measures has been employed throughout the *Yearbook*. For conversion coefficients and factors, see annex III.

Notes explicatives

Signes et conventions employés dans les tableaux

.   Les décimales sont précédées d'un point.

-     Un tiret entre des années, par exemple "2013-2014", indique que la période est embrassée dans sa totalité, y compris la première et la dernière année.

/   Une barre oblique renvoie à un exercice financier, à une année scolaire ou à une campagne agricole, par exemple

...   Données non disponibles ou non applicables.

*   Données provisoire, estimatif ou avec une révision majeure.

\#   Discontinuité notable dans la série chronologique.

<0   Non nul mais inférieur à la moitié de l'unité employée

−<0   Non nul mais négatif et inférieur à la moitié de l'unité employée

Le séparateur utilisé pour les milliers est l'espace : par exemple, 1 000 correspond à un millier. Les chiffres étant arrondis, les sous-totaux ou pourcentages ne correspondent pas toujours à la somme exacte des éléments figurant dans les tableaux.

Notes sur les pays et nomenclature

En règle générale, les données renvoient au pays ou zone en question que décrite dans la liste complète des pays et territoires figure à l'annexe I.

Références à des sources et des méthodes statistiques dans les tableaux

Par souci de brièveté, l'*Annuaire* omet les informations relatives aux sources et méthodologies employées pour les points de données individuels. De plus, quand un point de données est estimé aucune distinction n'est faite entre une estimation provenant d'une organisation internationale ou d'une organisation nationale. Voir les notes techniques relatives aux tableaux dans l'annexe II ou se trouvent les références aux différentes sources contentant ces informations.

Unités de mesure

Le système métrique de poids et mesures a été utilisé dans tout l'*Annuaire*. On trouvera à l'annexe III les coefficients et facteurs de conversion.

Contents

Part Three: Energy, environment and infrastructure

Annexes

Note: See **Annex IV** for table names presented in previous issues of the *Statistical Yearbook* which are not contained in the present issue.

Table des matières

Troisième partie: Énergie, environnement et infrastructures

Voir Annexe IV pour les tableaux publiés dans les éditions précédentes de l'Annuaire statistique mais qui n'ont pas été repris dans la présente edition.

Introduction

The 2016 edition of the United Nations *Statistical Yearbook* is the fifty-ninth issue of this publication, prepared by the Statistics Division, Department of Economic and Social Affairs, of the United Nations Secretariat. The contents of the *Yearbook* are currently under review and the number of tables has been reduced as a result. Additional tables are expected to be introduced in future editions. Please send any comments or views to our email address, statistics@un.org. The tables include series covering an appropriate historical period, depending upon data availability (as of 31 July 2016) and space constraints, for as many countries, territories and statistical areas of the world as available. The tables cover a period up to 2016, with some of the data being estimated.

Objective and content of the Statistical Yearbook

The main purpose of the *Statistical Yearbook* is to provide in a single volume a comprehensive compilation of internationally available statistics on social, economic and environmental conditions and activities, at world, regional and national levels, for an appropriate historical period.

Most of the statistics presented in the *Yearbook* are extracted from more detailed, specialized databases prepared by the Statistics Division and by many other international statistical services. Thus, while the specialized databases concentrate on monitoring topics and trends in particular social, economic and environmental fields, the *Statistical Yearbook* tables aim to provide data for a comprehensive, overall description of social, economic and environmental structures, conditions, changes and activities. The objective has been to collect, systematize, coordinate and present in a consistent way the most essential components of comparable statistical information which can give a broad picture of social, economic and environmental processes.

The content of the *Statistical Yearbook* is planned to serve a general readership. The *Yearbook* endeavours to provide information for various bodies of the United Nations system as well as for other international organizations, governments and non-governmental organizations, national statistical, economic and social policy bodies, scientific and educational institutions, libraries and the public. Data published in the *Statistical Yearbook* may also be of interest to companies and enterprises and to agencies engaged in market research. The *Statistical Yearbook* thus provides information on a wide range of social, economic and environmental issues which are of concern in the United Nations system and among the governments and peoples of the world. A particular value of the *Yearbook* is that it facilitates meaningful analysis of issues by systematizing and coordinating the data across many fields and shedding light on such interrelated issues as:

- General economic growth and related economic conditions;
- Gender equality;
- Population by sex and rate of increase, surface area and density;
- Unemployment, inflation and prices;
- Energy production and consumption;
- Expansion of trade;
- The financial situation of countries;
- Education;
- Improvement in general living conditions;
- Pollution and protection of the environment;
- Assistance provided to developing countries for social, economic and environmental development purposes.

Organization of the Yearbook

After the first table which presents key world aggregates and totals, the tables of the *Yearbook* are grouped into three parts as follows:

- Part One: Population and Social Statistics (chapters II-VI: tables 2-9);
- Part Two: Economic Activity (chapters VII-XI: tables 10-17);
- Part Three: Energy, environment and infrastructure (chapters XII-XVII: tables 18-26).

The first table provides a summary picture of development at the global level. More specific and detailed information for analysis concerning regions, individual countries or areas are presented in the three thematic parts.

Part One, Population and Social Statistics, comprises 8 tables which contain more detailed statistical series on population and migration, gender, education, health and crime.

Part Two, Economic Activity, comprises 8 tables on national accounts, finance, labour market, price and production indices and international merchandise trade.

Part Three, Energy, environment and infrastructure, comprises 9 tables on energy, environment, science and technology, communication, international tourism and transport and development assistance.

Annexes and regional groupings of countries or areas

The annexes to the *Statistical Yearbook*, and the section "Explanatory notes" preceding the Introduction, provide additional essential information on the *Yearbook*'s contents and presentation of data.

Annex I provides information on countries or areas covered in the *Yearbook* tables and on their arrangement in geographical regions and economic or other groupings. The geographical groupings shown in the *Yearbook* are generally based on continental regions unless otherwise indicated. However, strict consistency in this regard is impossible. A wide range of classifications is used for different purposes in the various international agencies and other sources of statistics for the *Yearbook*. These classifications vary in response to administrative and analytical requirements.

Annex II provide brief descriptions of major statistical concepts, definitions and classifications required for interpretation and analysis of the data.

Annex III provides detailed information on conversion coefficients and factors used in various tables, and Annex IV provides a list of tables added, omitted and discontinued in the present edition of the *Yearbook*. Some Tables do not feature in this edition of the *Yearbook* due to space limitations or an insufficient amount of new data being available. Their titles nevertheless are still listed in this Annex since it is planned that they will be published in a later issue as new data are compiled by the collecting agency and where space permits.

Data comparability, quality and relevance

The major challenge continuously facing the *Statistical Yearbook* is to present series which are as comparable across countries as the available statistics permit. Considerable efforts have already been made among the international suppliers of data and by the staff of the *Statistical Yearbook* to ensure the compatibility of various series by aligning time periods, base years, prices chosen for valuation, and so on. This is indispensable in relating various bodies of data to each other and in facilitating analysis across different sectors. In general, the data presented reflect the methodological recommendations of the United Nations Statistical Commission issued in various United Nations publications, and of other international bodies concerned with statistics. The use of international recommendations promotes international comparability of the data and ensures a degree of compatibility regarding the underlying concepts, definitions and classifications relating to different series. However, much work remains to be done in this area and, for this reason, some tables serve only as a first source of data and require further adjustment before being used for more in-depth analytical studies. While on the whole, a significant degree of comparability has been achieved in international statistics, there will remain some limitations, for a variety of reasons.

One common cause of non-comparability of economic data is different valuations of statistical aggregates such as national income, wages and salaries, output of industries and so forth. Conversion of these and similar series originally expressed in national prices into a common currency, for example into United States dollars, through the use of exchange rates, is not always satisfactory owing to frequent wide fluctuations in market rates and differences between official rates and rates which would be indicated by unofficial markets or purchasing power parities. The use of different kinds of sources for obtaining data is another cause of incomparability. This is true, for example, in the case of employment and unemployment, where data are obtained from different sources, namely household and labour force sample surveys, establishment censuses or surveys, official estimates, social insurance statistics and employment office statistics, which are not fully comparable in many cases. Non-comparability of data may also result from differences in the institutional patterns of countries. Certain variations in social, economic and environmental organization and institutions may have an impact on the comparability of the data even if the underlying concepts and definitions are identical. These and other causes of non-comparability of the data are briefly explained in the technical note associated with each table (see Annex II).

A further set of challenges relate to timeliness, quality and relevance of the data contained in the *Yearbook*. Users generally demand the most up-to-date statistics. However, due to the different development stages of statistical capacity in different countries, data for the most recent years may only be available for a small number of countries. For a global print publication, therefore, a balance has to be struck between presenting the most updated information and satisfactory country coverage. Of course the United Nations Statistics Division's website offers greater flexibility in presenting continuously updated information and is therefore a useful complement to the annual print publication. Furthermore, as most of the information presented in this *Yearbook* is collected through specialized United Nations agencies and partners, the timeliness is continuously enhanced by improving the communication and data flow between countries and the specialized agencies on the one hand, and between the United Nations Statistics Division and the specialized agencies on the other. The development of new XML-based data transfer protocols will address this issue and is expected to make international data flows more efficient in the future.

Data quality at the international level is a function of the data quality at the national level. The United Nations Statistics Division in close cooperation with its partners among the UN agencies and the international statistical system continues to support countries' efforts to improve both the coverage and the quality of their data. Metadata, as for example reflected in the footnotes and technical notes of this publication, are an important service to the user to allow an informed assessment of the quality of the data. Given the wide variety of sources for the *Yearbook*, there is of course an equally wide variety of data formats and accompanying metadata. An important challenge for the United Nations Statistics Division and its partners for the future is to work further towards the standardization, or at least harmonization, of metadata.

A crucial challenge is to maintain the relevance of the series included in the *Yearbook*. As new policy concerns enter the developmental debate, the United Nations Statistics Division will need to introduce new series that describe concerns that have gained prominence as well as to prune data as they become outdated and continue to update the recurrent *Yearbook* series that still address those issues which are most pertinent. Often choosing the appropriate moment when the statistical information on new topics has matured sufficiently so as to be able to disseminate meaningful global data can be challenging. Furthermore, a balance has to continuously be found between the ever-increasing amount of information available for dissemination and the space limitations of the print version of the *Statistical Yearbook*. International comparability, data availability, data quality and relevance will remain the key criteria to guide the United Nations Statistics Division in its selection.

Needless to say, more can always be done to improve the *Statistical Yearbook*'s scope, coverage, design, metadata and timeliness. The *Yearbook* team continually strives to improve upon each of these aspects and to make its publication as responsive as possible to its users' needs and expectations, while at the same time focusing on a manageable body of data and metadata. Since data disseminated in digital form have clear advantages over those in print, as much of the *Yearbook* information as possible will continue to be included in the Statistics Division's online databases. Please feel welcome to provide feedback and suggestions to statistics@un.org.

Introduction

L'*Annuaire statistique* des Nations Unies 2016 est la cinquante-neuvième édition de cette publication, établi par la Division de statistique du Département des affaires économiques et sociales du Secrétariat de l'Organisation des Nations Unies. Le contenu de l'*Annuaire* est actuellement en cours de révision, de ce fait le nombre de tableaux a été réduit. Des tableaux seront inclus dans les prochaines éditions. Pour tout commentaire, veuillez envoyer un courriel à notre adresse électronique : statistics@un.org. Les tableaux présentent des séries qui couvrent une période historique appropriée, en fonction de la disponibilité des données (à la date du 31 juillet 2016) et des contraintes d'espace, pour autant de pays, territoires et zones statistiques du monde comme disponibles. Les tableaux couvrent généralement pour la période jusqu'à 2016.

Objectif et contenu de l'Annuaire statistique

Le principal objectif de *l'Annuaire statistique* est de fournir en un seul volume un inventaire complet de statistiques internationales concernant la situation et les activités sociales, économiques et environnementales aux niveaux mondial, régional et national, sur une période adéquate.

La plupart des données qui figurent dans *l'Annuaire statistique* proviennent de bases de données spécialisées davantage détaillées, préparées par la Division de statistique et par bien d'autres services statistiques internationaux. Tandis que les bases de données spécialisées se concentrent sur le suivi de domaines socioéconomiques et environnementaux particuliers, les données de *l'Annuaire* sont présentées de telle sorte qu'elles fournissent une description globale et exhaustive des structures, conditions, transformations et activités socioéconomiques et environnementaux. On a cherché à recueillir, systématiser, coordonner et présenter de manière cohérente les principales informations statistiques comparables, de manière à dresser un tableau général des processus socioéconomiques et environnementaux.

Le contenu de *l'Annuaire statistique* a été élaboré en vue d'un lectorat large. Les renseignements fournis devraient ainsi pouvoir être utilisés par les divers organismes du système des Nations Unies, mais aussi par d'autres organisations internationales, les gouvernements et les organisations non gouvernementales, les organismes nationaux de statistique et de politique économique et sociale, les institutions scientifiques et les établissements d'enseignement, les bibliothèques et les particuliers. Les données publiées dans *l'Annuaire* peuvent également intéresser les sociétés et entreprises, et les organismes spécialisés dans les études de marché. L'*Annuaire* présente des informations sur un large éventail de questions socioéconomiques et environnementales liées aux préoccupations actuelles du système des Nations Unies, des gouvernements et des peuples du monde entier. Une qualité particulière de l'*Annuaire* est de faciliter une analyse approfondie de ces questions en systématisant et en articulant les données d'un domaine/secteur à l'autre, et en apportant un éclairage sur des sujets interdépendants, tels que :

- La croissance économique générale, et les conditions économiques qui lui sont liées;
- La situation de femmes;
- La population, taux d'accroissement, superficie et densité;
- Le chômage, l'inflation et les prix;
- La production et la consommation d'énergie;
- L'expansion des échanges;
- La situation financière des pays;
- L'éducation;
- L'amélioration des conditions de vie;
- La pollution et la protection de l'environnement;
- L'assistance aux pays en développement à des fins socioéconomiques et environnementaux.

Présentation de l'Annuaire

Après le premier tableau qui fournit les principaux agrégats et totaux au niveau mondial, l'Annuaire est groupés en trois parties comme suit:

- Première partie: Statistiques démographiques et sociales (chapitres II à VI, tableaux 2 à 9)
- Deuxième partie: Activité économique (chapitres VII à XI, tableaux 10 à 17)
- Troisième partie: Energie, environnement et infrastructures (chapitres XII à XVII, tableaux 18 à 26)

Le premier tableau donne un aperçu du développement à l'échelon mondial, tandis que les trois autres parties contiennent des renseignements plus précis et détaillés qui se prêtent mieux à une analyse par régions, pays ou par zones. Chacune de ces trois parties est subdivisée en unités thématiques.

La première partie, intitulée "Population et statistiques sociales", comporte 8 tableaux où figurent des séries plus détaillées concernant la population et la migration, la situation des femmes, l'éducation, la santé et la criminalité.

La deuxième partie, intitulée "Activité économique", comporte 8 tableaux qui présentent des statistiques concernant les comptes nationaux, le finances, le marché du travail, les indices des prix et de la production, et au commence international des marchandises. Les tableaux traitant de certains produits de base associent autant que possible les données relatives à la consommation aux valeurs concernant la production.

La troisième partie, intitulée "Energie, environnement et infrastructures", comprend 9 tableaux relatifs à l'énergie, l'environnement, la science et technologie, à la communication, au tourisme et transport internationaux et à l'aide au développement.

Annexes et groupements régionaux des pays et zones

Les annexes à l'*Annuaire statistique*, et la section intitulée "Notes explicatives" qui précède l'introduction, offrent d'importantes informations complémentaires quant à la teneur et à la présentation des données figurant dans le présent ouvrage.

L'annexe I donne des renseignements sur les pays ou zones couverts par les tableaux de l'*Annuaire* et sur leur regroupement en régions géographiques et groupements économiques ou autres. Sauf indication contraire, les groupements géographiques figurant dans l'*Annuaire* sont généralement fondés sur les régions continentales, mais une présentation absolument systématique est impossible à cet égard car les diverses institutions internationales et autres sources de statistiques employées pour la confection de l'*Annuaire* emploient, selon l'objet de l'exercice, des classifications fort différentes en réponse à diverses exigences d'ordre administratif ou analytique.

L'annexe II contient les notes techniques de chaque chapitre apportent une brève description des principales notions, définitions et classifications statistiques nécessaires pour interpréter et analyser les données.

L'annexe III fournit des renseignements sur les coefficients et facteurs de conversion employés dans les différents tableaux, et l'annexe IV contient la líste de tableaux qui ont été ajoutés omis ou supprimés dans la présente édition de l'*Annuaire*. Certains tableaux ne figurent pas dans cet *Annuaire* en raison du manque d'espace ou de données nouvelles. Comme ils seront repris dans une prochaine édition à mesure que des données nouvelles seront dépouillées par l'office statistique d'origine, ses titres figurent toujours dans la table des matières.

Comparabilité, qualité et pertinence des statistiques

Le défi majeur auquel l'*Annuaire Statistique* fait continuellement face est de présenter des séries aussi comparables entre les pays que la disponibilité des statistiques le permettent. Les sources internationales de données et les auteurs de l'*Annuaire* ont réalisé des efforts considérables pour faire en sorte que diverses séries soient compatibles, en harmonisant les périodes de référence, les années de base, les prix utilisés pour les évaluations, etc. Cette démarche est indispensable si l'on veut rapprocher divers ensembles de données, et faciliter l'analyse intersectorielle de l'économie. De façon générale, les données sont présentées selon les recommandations méthodologiques formulées par la Commission de statistique des Nations Unies, et par les autres entités internationales impliquées dans les statistiques. Le respect des recommandations internationales tend non seulement à promouvoir la comparabilité internationale des données, mais elle assure également une certaine comparabilité entre les concepts, les définitions et classifications utilisés. Mais comme il reste encore beaucoup à faire dans ce domaine, les données présentées dans certains tableaux n'ont qu'une valeur indicative, et nécessiteront des ajustements plus poussés avant de pouvoir servir à des analyses approfondies. Bien que l'on soit parvenu, dans l'ensemble, à un degré de comparabilité appréciable en matière de statistiques internationales, diverses raisons expliquent que subsistent encore de nombreuses limitations.

Une cause commune de non comparabilité des données économiques réside dans la diversité des méthodes d'évaluation employées pour comptabiliser des agrégats tels que le revenu national, les salaires et traitements, la production des différentes branches d'activité industrielle, etc. Il n'est pas toujours satisfaisant de ramener la valeur des séries de ce type—exprimée à l'origine en prix nationaux—à une monnaie commune (par exemple le dollar des États-Unis) car les taux de change du marché connaissent fréquemment de fortes fluctuations, et parce que les taux officiels ne coïncident pas avec ceux des marchés officiels ni avec les parités réelles de pouvoir d'achat. Le recours à des sources diverses pour la collecte des données est un autre facteur qui limite la comparabilité. C'est le cas, par exemple, des données d'emploi et de chômage, obtenues par des moyens aussi peu comparables que les sondages, le dépouillement des registres d'assurances sociales et les enquêtes auprès des entreprises. Dans certains cas, les données ne sont pas comparables en raison de différences entre les structures institutionnelles des pays. Des changements dans l'organisation et les institutions sociales, économiques et environnementales peuvent affecter la comparabilité des données, même si les concepts et définitions sont fondamentalement identiques. Ces causes, et d'autres, de non comparabilité des données sont brièvement expliquées dans les notes techniques associée à chaque tableau (voir annexe II).

Un autre ensemble de défis à relever concerne la fraîcheur, la qualité et la pertinence des données présentées dans l'*Annuaire*. Les utilisateurs exigent généralement des données les plus récentes possibles. Toutefois, selon le niveau de développement de la capacité statistique des pays, les données pour les dernières années peuvent n'être disponibles que pour un nombre limité de pays. Dans le cadre d'une publication mondiale, un équilibre doit être trouvé entre la présentation de l'information la plus récente et une couverture géographique satisfaisante. Bien entendu, le site Internet de la Division de statistique des Nations Unies offre une plus grande flexibilité, puisqu'il propose une information actualisée au fil de l'eau, et constitue ainsi un complément utile à la publication papier annuelle. Par ailleurs, étant donné que la plupart des informations présentées dans cet *Annuaire* sont collectées parmi les agences spécialisées des Nations Unies et autres partenaires, la fraîcheur des données est continuellement améliorée, grâce à une meilleure communication et un meilleur échange de données entre les pays et les agences spécialisées d'une part, et entre la Division de statistique des Nations Unies et les agences spécialisées d'autre part. Le développement de nouveaux protocoles de transfert de données basés sur le langage XML devrait contribuer à rendre, à l'avenir, les échanges de données internationales encore plus efficaces.

La qualité des données au niveau international est fonction de la qualité des données au niveau national. La Division de statistique des Nations Unies, en étroite collaboration avec ses partenaires dans les agences de l'ONU et dans le système statistique international, continue de soutenir les efforts des pays pour améliorer à la fois la couverture et la qualité de leurs données. Des métadonnées, comme l'illustrent les notes de bas de page et les notes techniques de cette publication, constituent un important service fourni à l'utilisateur pour lui permettre d'évaluer de manière avisée la qualité des données. Etant donné la grande variété des sources de l'*Annuaire*, il y a bien entendu une non moins grande variété de formats de données et de métadonnées associées. Un important défi que la Division de statistique des Nations Unies et ses partenaires doivent relever dans le futur est d'aboutir à la standardisation, ou au moins l'harmonisation, des métadonnées.

Le défi majeur reste la constance de la pertinence des séries présentées dans l'*Annuaire*. Au fur et à mesure que de nouvelles préoccupations politiques pénètrent le débat lié au développement, la Division de statistique des Nations Unies doit introduire dans l'*Annuaire* de nouvelles séries qui leur sont liées, et, ce faisant, effectuer une coupe sombre parmi les données qui lui semblent dépassées, tout en s'assurant de continuer à actualiser les séries récurrentes de qui paraissent encore pertinentes. Souvent, choisir le moment idoine auquel les données statistiques sur de nouveaux thèmes sont suffisamment matures pour qu'elles puissent, au niveau mondial, être diffusées sans hésitation, est un défi en soi. Par ailleurs, un équilibre doit continuellement être trouvé entre le volume toujours croissant d'informations disponibles à la diffusion, et les contraintes d'espace de la version papier de l'*Annuaire* statistique. La comparabilité internationale, la disponibilité, la qualité et la pertinence des données devront rester les principaux critères à considérer par la Division de statistique des Nations Unies dans sa sélection.

Inutile de dire qu'il est toujours possible d'améliorer l'*Annuaire* statistique en ce qui concerne son champ, sa couverture, sa conception générale, ses métadonnées et sa mise à jour. L'équipe en charge de l'*Annuaire* s'évertue en permanence à améliorer chacun de ces aspects, et de faire en sorte que cette publication réponde au plus près aux besoins et aux attentes de ses utilisateurs, sans toutefois oublier de mettre l'accent sur un corpus gérable de données et de métadonnées. Puisqu'il est avéré que les données diffusées de manière digitale ont des avantages comparés à celles diffusées sur papier, autant d'informations de l'*Annuaire* que possible continueront d'être inclues dans les bases de données électroniques de la Division de statistique. L'*Annuaire* statistique garde toujours une place de choix parmi les produits de la Division de statistique comme une ressource utile pour une compréhension général de la situation sociale, économique et environnementale globale. N'hésitez pas à nous faire part de vos commentaires et suggestions à statistics@un.org

World statistics: selected series
Population and social statistics, economic activity and energy, environment and infrastructure

Statistiques mondiales : séries principales
Population et statistiques sociales, activité économique et énergie, environnement et infrastructures

Series Séries	Unit or base Unité ou base	1985	1995	2005	2010	2012	2013	2014	2015	2016
Population • Population										
World population [1] Population mondial [1]	million million	4 852.54	5 735.12	6 519.64	6 929.73	7 097.5	7 181.71	7 265.79	7 349.47	7 432.66
International migrants and refugees • Les migrants et les réfugiés internationaux										
International migrant stock - Total Stock de migrants internationaux - Total	thousand millier	...	160 802	191 269	221 714	...	...	...	243 700	...
International migrant stock - Total Stock de migrants internationaux - Total	percentage pourcentage	...	2.8	2.9	3.2	...	...	...	3.3	...
International migrant stock - Male Stock de migrants internationaux - Hommes	percentage pourcentage	...	2.8	3.0	3.3	...	...	...	3.4	...
International migrant stock - Female Stock de migrants internationaux - Fémmes	percentage pourcentage	...	2.8	2.9	3.1	...	...	...	3.2	...
Refugees Réfugiés	thousand millier	...	...	...	...	...	...	...	15 098	...
Asylum seekers Demandeurs d'asile	thousand millier	...	...	...	...	...	...	...	2 344	...
Other of concern to UNHCR Autres personnes relevant de la compétence du HCR	thousand millier	...	...	...	...	...	...	...	40 518	...
Total population of concern to UNHCR Population totale de préoccupation pour le HCR	thousand millier	...	...	...	...	...	...	...	57 960	...
Education • Enseignement										
Students enrolled in primary education Les étudiants inscrits dans l'enseignement primaire	thousand millier	551 115	623 179	678 999	697 216	708 940	715 906	719 059	...	...
Gross enrollment ratio - Primary, Male Taux brut de scolarisation - Primaire, Hommes	percentage pourcentage	106.6	103.5	104.6	106.7	106.9	105.6	105.0	...	...
Gross enrollment ratio - Primary, Female Taux brut de scolarisation - Primaire, Fémmes	percentage pourcentage	90.8	93.6	99.7	103.8	104.4	104.8	104.1	...	...
Students enrolled in secondary education Les étudiants inscrits dans l'enseignement secondaire	thousand millier	305 502	395 032	509 073	546 230	561 296	571 116	568 019	...	...
Gross enrollment ratio - Secondary, Male Taux brut de scolarisation - Secondaire, Hommes	percentage pourcentage	51.8	60.9	65.6	72.2	74.8	75.8	75.6	...	...
Gross enrollment ratio - Secondary, Female Taux brut de scolarisation - Secondaire, Fémmes	percentage pourcentage	42.0	53.4	62.1	69.7	72.6	74.5	74.5	...	...
Students enrolled in tertiary education Les étudiants inscrits dans l'enseignement supérieur	thousand millier	60 929	79 978	139 293	181 531	197 031	198 892	207 516	...	...
Gross enrollment ratio - Tertiary, Male Taux brut de scolarisation - Tertiaire, Hommes	percentage pourcentage	14.5	15.9	23.7	28.3	30.8	31.2	32.8	...	...
Gross enrollment ratio - Tertiary, Female Taux brut de scolarisation - Tertiaire, Féminin	percentage pourcentage	12.2	15.1	24.8	30.4	33.7	34.4	36.2	...	...
Gross domestic product • Produit intérieur brut										
GDP in current prices PIB aux prix courants	billion US $ milliard $ E.-U.	13 502	30 860	47 265	65 645	74 222	76 176	78 037	...	...
GDP per capita PIB par habitant	US $ $ E.-U.	2 784	5 384	7 251	9 475	10 460	10 609	10 742	...	...
GDP in constant 2005 prices PIB aux prix constants	billion US $ milliard $ E.-U.	26 013	34 592	47 265	52 895	55 618	56 813	58 254	...	...
GDP real rates of growth Taux de croissance	percentage pourcentage	3.7	2.9	3.6	4.1	2.2	2.1	2.5	...	...

1

World statistics: selected series *(continued)*
Population and social statistics, economic activity and energy, environment and infrastructure

Statistiques mondiales : séries principales *(suite)*
Population et statistiques sociales, activité économique et énergie, environnement et infrastructures

Series Séries	Unit or base Unité ou base	1985	1995	2005	2010	2012	2013	2014	2015	2016
Labour market • Marché du travail										
Labour force - Total Force de travail - Total	percentage pourcentage	...	66.0	64.7	63.2	62.9	62.9	62.9	62.9	62.8
Labour force - Male Force de travail - Hommes	percentage pourcentage	...	79.7	77.6	76.5	76.3	76.2	76.1	76.1	76.1
Labour force - Female Force de travail - Femmes	percentage pourcentage	...	52.3	51.9	50.0	49.7	49.6	49.6	49.6	49.5
Unemployment rate - Total Chômage - Total	percentage pourcentage	...	6.2	6.2	6.1	6.0	6.0	5.8	5.8	5.8
Unemployment rate - Male Chômage - Mâle	percentage pourcentage	...	5.9	5.8	5.8	5.7	5.7	5.5	5.5	5.5
Unemployment rate - Female Chômage - Feminin	percentage pourcentage	...	6.5	6.7	6.5	6.4	6.4	6.3	6.2	6.3
Employment in Agriculture - Total L'emploi dans Agriculture - Total	percentage pourcentage	...	41.2	36.7	32.7	30.9	30.6	29.8	29.3	28.9
Employment in Agriculture - Male L'emploi dans Agriculture - Hommes	percentage pourcentage	...	40.5	35.3	32.2	30.4	30.1	29.4	29.1	28.7
Employment in Agriculture - Female L'emploi dans Agriculture - Femmes	percentage pourcentage	...	42.3	38.8	33.5	31.6	31.4	30.3	29.7	29.4
Employment in Industry - Total L'emploi dans industrie - Total	percentage pourcentage	...	20.9	20.7	21.6	22.1	21.8	22.0	22.0	21.9
Employment in Industry - Male L'emploi dans industrie - Hommes	percentage pourcentage	...	21.6	23.7	25.0	26.0	26.0	26.0	26.3	26.3
Employment in Industry - Female L'emploi in industrie - Femmes	percentage pourcentage	...	19.8	16.2	16.4	16.1	15.5	15.7	15.3	15.1
Employment in Services - Total L'emploi dans Services - Total	percentage pourcentage	...	37.9	42.6	45.7	47.0	47.6	48.2	48.7	49.1
Employment in Services - Male L'emploi dans Services - Hommes	percentage pourcentage	...	37.9	41.0	42.9	43.6	43.9	44.6	44.6	45.0
Employment in Services - Female L'emploi dans Services - Femmes	percentage pourcentage	...	37.9	45.0	50.1	52.3	53.1	53.9	55.0	55.5
Agricultural production indices • Indices de la production agricole										
Agriculture (gross) Agriculture (brut)	2004-2006 = 100	65.5	78.3	100.0	112.9	118.2	121.9	...	...	...
Food (gross) Ailmentaires (brut)	2004-2006 = 100	65.0	78.2	100.0	113.4	118.4	122.4	...	...	...
International merchandise trade • Commerce international des marchandises										
Imports CIF Importations CIF	billion US $ milliard $ E.-U.	2 037	5 158	10 602	15 157	18 125	18 423	18 597	16 449	...
Exports FOB Exportations FOB	billion US $ milliard $ E.-U.	1 985	5 128	10 356	15 106	18 081	18 468	18 642	16 462	...
Balance Balance	billion US $ milliard $ E.-U.	-51	-30	-246	-51	-44	45	45	12	...
Energy • Ènergie										
Primary energy production Production d'énergie primaire	petajoules pétajoules	378 635	474 857	532 611	556 304	567 050	...	...	...	...
Net imports Importations nettes	petajoules pétajoules	-8 893	-14 084	-14 795	-14 871	-16 621	...	...	...	...
Changes in stocks Variations des stocks	petajoules pétajoules	128	1 131	-454	630	228	...	...	...	...
Total supply Approvisionnement total	petajoules pétajoules	369 614	459 642	518 270	540 803	550 201	...	...	...	...
Supply per capita Approvisionnement par habitant	gigajoules gigajoules	65	71	75	76	77	...	...	...	...

1

World statistics: selected series *(continued)*
Population and social statistics, economic activity and energy, environment and infrastructure

Statistiques mondiales : séries principales *(suite)*
Population et statistiques sociales, activité économique et énergie, environnement et infrastructures

Series Séries	Unit or base Unité ou base	1985	1995	2005	2010	2012	2013	2014	2015	2016
CO_2 emissions estimates • Estimations des émissions de CO_2										
Emissions Émissions	000 metric ton millier de tonnes	19 864	23 263	29 427	33 505	35 463	35 848	...	...	...
Emissions per capita Émissions par habitant	metric ton tonne	4.1	4.1	4.5	4.8	5.0	5.0	...	...	...
Science and technology • Science et technologie										
Patents Brevets	thousand millier	398	428	631	912	1 136	1 173	1 177	...	...
Development assistance • Aide au developpement										
Net ODA received - Bilateral APD nette reçue - Bilatérale	million US $ million $ E.-U.	26 356	48 142	83 033	94 207	96 003	111 066	119 123	...	...
Net ODA received - Multilateral APD nette reçue - Multilatérale	million US $ million $ E.-U.	...	11 004	25 620	37 133	36 735	39 735	41 952	...	...
Net ODA received - Total APD nette reçue - Total	million US $ million $ E.-U.	32 277	59 145	108 652	131 340	132 738	150 800	161 075	...	...
Net ODA received - Total APD nette reçue - Total	% of GNI % du RNB	1.3	1.0	1.2	0.7	0.6	0.6	0.6	...	...

Source:
These data are presented in other tables of this Yearbook; please refer to the relevant table for source notes and last access date.

Source:
Ces données sont présentées dans d'autres tableaux du présent Annuaire, veuillez donc vous reporter au tableau correspondant pour les notes de source et la date du dernier accès.

1   Mid-year estimates.

1   Les estimations au milieu de l'année.

Part One

Population and social statistics

Première partie

Population et statistiques sociales

2

Population, surface area and density
Population, superficie et densité

Region, country or area	Mid-year population estimates (millions) Estimations de population au milieu de l'année (millions)						Surface area Superficie (000 km²)	Density Densité	Région, pays ou zone
	1995	2005	2010	2014	2015	2016	2014	2016	
Total, all countries or areas	5 735.12	6 519.64	6 929.73	7 265.79	7 349.47	7 432.66	136 162.0	57.1	Total, tous pays ou zones
Africa	720.42	920.24	1 044.11	1 156.65	1 186.18	1 216.13	30 311.0	41.0	Afrique
Northern America	295.70	328.52	344.13	355.16	357.84	360.53	21 776.0	19.3	Amérique septentrionale
Latin Amer. and the Caribb.	487.33	563.83	599.82	627.64	634.39	641.03	20 546.0	31.8	Amérique latine et Caraïbes
Asia	3 474.85	3 944.67	4 169.86	4 349.56	4 393.30	4 436.22	31 915.0	143.0	Asie
Europe	727.78	729.01	735.39	738.02	738.44	738.85	23 049.0	33.4	Europe
Oceania	29.05	33.37	36.41	38.76	39.33	39.90	8 564.0	4.7	Océanie
Afghanistan	16.77	24.40	27.96	31.63	32.53	33.37	652.9	51.1	Afghanistan
Albania	3.11	3.08	2.90	2.89	2.90	2.90	28.7	106.0	Albanie
Algeria	28.90	33.27	36.04	38.93	39.67	40.38	2 381.7	17.0	Algérie
American Samoa	0.05	0.06	0.06	0.06	0.06	0.06	0.2	278.0	Samoa américaines
Andorra	0.06	0.08	0.08	0.07	0.07	0.07	0.5	147.2	Andorre
Angola	13.04	17.91	21.22	24.23	25.02	25.83	1 246.7	20.7	Angola
Anguilla	0.01	0.01	0.01	0.01	0.01	0.01	0.1	164.0	Anguilla
Antigua and Barbuda	0.07	0.08	0.09	0.09	0.09	0.09	0.4	210.8	Antigua-et-Barbuda
Argentina	34.99	39.15	41.22	42.98	43.42	43.85	2 780.4	16.0	Argentine
Armenia	3.22	3.01	2.96	3.01	3.02	3.03	29.7	106.3	Arménie
Aruba	0.08	0.10	0.10	0.10	0.10	0.10	0.2	579.2	Aruba
Australia [1]	18.12	20.27	22.16	23.62	23.97	24.31	7 692.1	3.2	Australie [1]
Austria	7.97	8.23	8.39	8.52	8.54	8.57	83.9	104.0	Autriche
Azerbaijan	7.77[2]	8.56[2]	9.10[2]	9.63[2]	9.75[2]	9.87[2]	86.6	119.4[2]	Azerbaïdjan
Bahamas	0.28	0.33	0.36	0.38	0.39	0.39	13.9	39.2	Bahamas
Bahrain	0.56	0.87	1.26	1.36	1.38	1.40	0.8	1 837.9	Bahreïn
Bangladesh	118.43	142.93	151.62	159.08	161.00	162.91	147.6	1 251.5	Bangladesh
Barbados	0.27	0.27	0.28	0.28	0.28	0.29	0.4	662.8	Barbade
Belarus	10.16	9.64	9.49	9.50	9.50	9.48	207.6	46.7	Bélarus
Belgium	10.16	10.56	10.93	11.23	11.30	11.37	30.5	375.6	Belgique
Belize	0.21	0.28	0.32	0.35	0.36	0.37	23.0	16.1	Belize
Benin	5.99	8.18	9.51	10.60	10.88	11.17	114.8	99.0	Bénin
Bermuda	0.06	0.07	0.06	0.06	0.06	0.06	0.1	1 233.2	Bermudes
Bhutan	0.51	0.65	0.72	0.77	0.77	0.78	38.4	20.6	Bhoutan
Bolivia (Plurinational State of)	7.57	9.13	9.92	10.56	10.72	10.89	1 098.6[3]	10.1	Bolivie (État plurinational de)
Bonaire, St Eust. and Saba	0.02	0.01	0.02	0.02	0.02	0.03	...	77.2	Bonaire, Saint-Eustache et Saba
Bosnia and Herzegovina	3.88	3.83	3.84	3.82	3.81	3.80	51.2	74.6	Bosnie-Herzégovine
Botswana	1.58	1.86	2.05	2.22	2.26	2.30	582.0	4.1	Botswana
Brazil	162.76	188.48	198.61	206.08	207.85	209.57	8 514.9	25.1	Brésil
British Virgin Islands	0.02	0.02	0.03	0.03	0.03	0.03	0.2	204.4	Îles Vierges britanniques
Brunei Darussalam	0.30	0.36	0.39	0.42	0.42	0.43	5.8	81.4	Brunéi Darussalam
Bulgaria	8.36	7.68	7.41	7.20	7.15	7.10	111.0	65.4	Bulgarie
Burkina Faso	10.09	13.42	15.63	17.59	18.11	18.63	273.0	68.1	Burkina Faso
Burundi	6.24	7.93	9.46	10.82	11.18	11.55	27.8	449.9	Burundi
Cabo Verde	0.39	0.47	0.49	0.51	0.52	0.53	4.0	130.8	Cabo Verde
Cambodia	10.69	13.32	14.36	15.33	15.58	15.83	181.0	89.7	Cambodge

Region, country or area	Mid-year population estimates (millions) Estimations de population au milieu de l'année (millions)						Surface area Superficie (000 km²)	Density Densité	Région, pays ou zone
	1995	2005	2010	2014	2015	2016	2014	2016	
Cameroon	13.93	18.13	20.59	22.77	23.34	23.92	475.7	50.6	Cameroun
Canada	29.30	32.26	34.13	35.59	35.94	36.29	9 984.7	4.0	Canada
Cayman Islands	0.03	0.05	0.06	0.06	0.06	0.06	0.3	253.2	Îles Caïmanes
Central African Rep.	3.34	4.06	4.45	4.80	4.90	5.00	623.0	8.0	Rép. centrafricaine
Chad	7.00	10.07	11.90	13.59	14.04	14.50	1 284.0	11.5	Tchad
Channel Islands[4]	0.14	0.15	0.16	0.16	0.16	0.16	0.2	865.6	Îles Anglo-Normandes[4]
Chile	14.19	16.10	17.02	17.76	17.95	18.13	756.1	24.4	Chili
China[5]	1 227.84	1 305.60	1 340.97	1 369.44	1 376.05	1 382.32	9 597.0	147.2	Chine[5]
China, Hong Kong SAR	6.14	6.84	6.99	7.23	7.29	7.35	1.1	6 996.4	Chine, Hong Kong RAS
China, Macao SAR	0.40	0.47	0.53	0.58	0.59	0.60	<0.0	19 970.8	Chine, Macao RAS
Colombia	37.44	43.29	45.92	47.79	48.23	48.65	1 141.7	43.9	Colombie
Comoros	0.48	0.62	0.70	0.77	0.79	0.81	2.2	433.7	Comores
Congo	2.72	3.50	4.07	4.51	4.62	4.74	342.0	13.9	Congo
Cook Islands	0.02	0.02	0.02	0.02	0.02	0.02	0.2[6]	87.3	Îles Cook
Costa Rica	3.51	4.25	4.55	4.76	4.81	4.86	51.1	95.1	Costa Rica
Côte d'Ivoire	14.40	18.13	20.13	22.16	22.70	23.25	322.5	73.1	Côte d'Ivoire
Croatia	4.62	4.38	4.32	4.26	4.24	4.23	56.6	75.5	Croatie
Cuba	10.91	11.26	11.31	11.38	11.39	11.39	109.9	107.0	Cuba
Curaçao	0.14	0.13	0.15	0.16	0.16	0.16	0.4	357.3	Curaçao
Cyprus	0.86[7]	1.03[7]	1.10[7]	1.15[7]	1.17[7]	1.18[7]	9.3	127.3[7]	Chypre
Czech Republic	10.34	10.23	10.51	10.54	10.54	10.55	78.9	136.6	République tchèque
Dem. P. R. Korea	21.76	23.81	24.50	25.03	25.16	25.28	120.5	210.0	R. p. dém. de Corée
Dem. Rep. of the Congo	42.18	56.09	65.94	74.88	77.27	79.72	2 344.9	35.2	Rép. dém. du Congo
Denmark	5.23	5.42	5.55	5.65	5.67	5.69	42.9	134.1	Danemark
Djibouti	0.66	0.78	0.83	0.88	0.89	0.90	23.2	38.8	Djibouti
Dominica	0.07	0.07	0.07	0.07	0.07	0.07	0.8	97.4	Dominique
Dominican Republic	7.89	9.24	9.90	10.41	10.53	10.65	48.2	220.4	Rép. dominicaine
Ecuador	11.44	13.74	14.93	15.90	16.14	16.39	257.2	66.0	Équateur
Egypt	62.43	74.94	82.04	89.58	91.51	93.38	1 002.0	93.8	Égypte
El Salvador	5.59	5.95	6.04	6.11	6.13	6.15	21.0[8]	296.6	El Salvador
Equatorial Guinea	0.45	0.63	0.73	0.82	0.85	0.87	28.1	31.0	Guinée équatoriale
Eritrea	3.16	4.19	4.69	5.11	5.23	5.35	117.6	53.0	Érythrée
Estonia	1.43	1.36	1.33	1.32	1.31	1.31	45.2	30.9	Estonie
Ethiopia	57.24	76.61	87.56	96.96	99.39	101.85	1 104.3	101.9	Éthiopie
Falkland Is. (Malvinas)	<0.00	<0.00	<0.00	<0.00	<0.00	<0.00	12.2	0.2	Îles Falkland (Malvinas)
Faroe Islands	0.04	0.05	0.05	0.05	0.05	0.05	1.4	34.6	Îles Féroé
Fiji	0.78	0.82	0.86	0.89	0.89	0.90	18.3	49.1	Fidji
Finland[9]	5.11	5.25	5.37	5.48	5.50	5.52	338.4	18.2	Finlande[9]
France	58.22	61.24	62.96	64.12	64.40	64.67	551.5	118.1	France
French Guiana	0.14	0.20	0.23	0.26	0.27	0.28	83.5	3.4	Guyane française
French Polynesia	0.22	0.25	0.27	0.28	0.28	0.29	4.0	78.1	Polynésie française
Gabon	1.09	1.38	1.54	1.69	1.73	1.76	267.7	6.8	Gabon
Gambia	1.07	1.44	1.69	1.93	1.99	2.06	11.3	203.1	Gambie
Georgia	5.07[10]	4.48[10]	4.25[10]	4.03[10]	4.00[10]	3.98[10]	69.7	57.3[10]	Géorgie

Region, country or area	Mid-year population estimates (millions) Estimations de population au milieu de l'année (millions)						Surface area Superficie (000 km²)	Density Densité	Région, pays ou zone
	1995	2005	2010	2014	2015	2016	2014	2016	
Germany	81.61	81.25	80.44	80.65	80.69	80.68	357.3	231.5	Allemagne
Ghana	16.76	21.39	24.32	26.79	27.41	28.03	238.5	123.2	Ghana
Gibraltar	0.03	0.03	0.03	0.03	0.03	0.03	<0.0	3 237.3	Gibraltar
Greece	10.64	11.07	11.18	11.00	10.95	10.92	132.0	84.7	Grèce
Greenland	0.06	0.06	0.06	0.06	0.06	0.06	2 166.1	0.1	Groenland
Grenada	0.10	0.10	0.10	0.11	0.11	0.11	0.3	315.7	Grenade
Guadeloupe	0.41[11]	0.45[11]	0.46[11]	0.47[11]	0.47[11]	0.47[11]	1.7	278.4[11]	Guadeloupe
Guam	0.15	0.16	0.16	0.17	0.17	0.17	0.5	318.7	Guam
Guatemala	10.36	13.18	14.73	16.02	16.34	16.67	108.9	155.6	Guatemala
Guinea	7.86	9.67	11.01	12.28	12.61	12.95	245.9	52.7	Guinée
Guinea-Bissau	1.18	1.46	1.63	1.80	1.84	1.89	36.1	67.2	Guinée-Bissau
Guyana	0.73	0.74	0.75	0.76	0.77	0.77	215.0	3.9	Guyana
Haiti	7.82	9.26	10.00	10.57	10.71	10.85	27.8	393.6	Haïti
Holy See [12]	<0.00	<0.00	<0.00	<0.00	<0.00	<0.00	<0.0[13]	1 820.5	Saint-Siège [12]
Honduras	5.59	6.88	7.50	7.96	8.08	8.19	112.5	73.2	Honduras
Hungary	10.35	10.10	10.01	9.89	9.86	9.82	93.0	108.5	Hongrie
Iceland	0.27	0.30	0.32	0.33	0.33	0.33	103.0	3.3	Islande
India	960.88	1 144.33	1 230.98	1 295.29	1 311.05	1 326.80	3 287.3	446.3	Inde
Indonesia	196.96	226.25	241.61	254.45	257.56	260.58	1 910.9	143.8	Indonésie
Iran (Islamic Rep. of)	60.32	70.12	74.25	78.14	79.11	80.04	1 628.8[14]	49.1	Iran (Rép. islamique d')
Iraq	20.22	27.02	30.87	35.27	36.42	37.55	435.1	86.5	Iraq
Ireland	3.65	4.20	4.62	4.68	4.69	4.71	69.8	68.4	Irlande
Isle of Man	0.07	0.08	0.08	0.09	0.09	0.09	0.6	155.1	Île de Man
Israel	5.33	6.60	7.42	7.94	8.06	8.19	22.1	378.6	Israël
Italy	57.12	58.66	59.59	59.79	59.80	59.80	302.1	203.3	Italie
Jamaica	2.49	2.68	2.74	2.78	2.79	2.80	11.0	258.9	Jamaïque
Japan	124.48	126.98	127.32	126.79	126.57	126.32	377.9[15]	346.5	Japon
Jordan	4.32	5.33	6.52	7.42	7.59	7.75	89.3	87.3	Jordanie
Kazakhstan	15.93	15.45	16.31	17.37	17.63	17.86	2 724.9	6.6	Kazakhstan
Kenya	27.37	35.35	40.33	44.86	46.05	47.25	592.0	83.0	Kenya
Kiribati	0.08	0.09	0.10	0.11	0.11	0.11	0.7[16]	141.2	Kiribati
Kuwait	1.64	2.26	3.06	3.75	3.89	4.01	17.8	224.9	Koweït
Kyrgyzstan	4.59	5.12	5.46	5.84	5.94	6.03	199.9	31.5	Kirghizistan
Lao People's Dem. Rep.	4.86	5.75	6.26	6.69	6.80	6.92	236.8	30.0	Rép. dém. pop. lao
Latvia	2.49	2.23	2.09	1.99	1.97	1.96	64.6	31.4	Lettonie
Lebanon	3.03	3.99	4.34	5.61	5.85	5.99	10.5	585.4	Liban
Lesotho	1.75	1.93	2.01	2.11	2.14	2.16	30.4	71.2	Lesotho
Liberia	2.08	3.27	3.96	4.40	4.50	4.62	111.4	47.9	Libéria
Libya	4.88	5.80	6.27	6.26	6.28	6.33	1 676.2	3.6	Libye
Liechtenstein	0.03	0.03	0.04	0.04	0.04	0.04	0.2	236.1	Liechtenstein
Lithuania	3.63	3.34	3.12	2.92	2.88	2.85	65.3	45.5	Lituanie
Luxembourg	0.41	0.46	0.51	0.56	0.57	0.58	2.6	222.5	Luxembourg
Madagascar	13.45	18.29	21.08	23.57	24.24	24.92	587.3	42.8	Madagascar
Malawi	9.82	12.75	14.77	16.70	17.22	17.75	118.5	188.3	Malawi

Region, country or area	Mid-year population estimates (millions) Estimations de population au milieu de l'année (millions)						Surface area Superficie (000 km²)	Density Densité	Région, pays ou zone
	1995	2005	2010	2014	2015	2016	2014	2016	
Malaysia	20.73[17]	25.80[17]	28.12[17]	29.90[17]	30.33[17]	30.75[17]	330.4	93.6[17]	Malaisie
Maldives	0.25	0.31	0.33	0.36	0.36	0.37	0.3	1 232.7	Maldives
Mali	9.64	12.88	15.17	17.09	17.60	18.13	1 240.2	14.9	Mali
Malta	0.37	0.40	0.41	0.42	0.42	0.42	0.3	1 311.3	Malte
Marshall Islands	0.05	0.05	0.05	0.05	0.05	0.05	0.2	294.8	Îles Marshall
Martinique	0.37	0.40	0.39	0.40	0.40	0.40	1.1	373.9	Martinique
Mauritania	2.33	3.15	3.59	3.97	4.07	4.17	1 030.7	4.0	Mauritanie
Mauritius	1.13[18]	1.22[18]	1.25[18]	1.27[18]	1.27[18]	1.28[18]	2.0[19]	629.3[18]	Maurice
Mayotte	0.12	0.18	0.21	0.23	0.24	0.25	...	657.3	Mayotte
Mexico	94.43	109.75	118.62	125.39	127.02	128.63	1 964.4	66.2	Mexique
Micronesia (Fed. States of)	0.11	0.11	0.10	0.10	0.10	0.11	0.7	150.0	Micronésie (États féd. de)
Monaco	0.03	0.03	0.04	0.04	0.04	0.04	<0.0	25 411.4	Monaco
Mongolia	2.30	2.53	2.71	2.91	2.96	3.01	1 564.1	1.9	Mongolie
Montenegro	0.62	0.62	0.62	0.63	0.63	0.63	13.8	46.6	Monténégro
Montserrat	0.01	<0.00	0.01	0.01	0.01	0.01	0.1	51.5	Montserrat
Morocco	27.16	30.39	32.11	33.92	34.38	34.82	446.6	78.0	Maroc
Mozambique	15.91	21.13	24.32	27.22	27.98	28.75	799.4	36.6	Mozambique
Myanmar	44.71	49.98	51.73	53.44	53.90	54.36	676.6	83.2	Myanmar
Namibia	1.65	2.03	2.19	2.40	2.46	2.51	824.3	3.1	Namibie
Nauru	0.01	0.01	0.01	0.01	0.01	0.01	<0.0	513.2	Nauru
Nepal	21.39	25.51	26.88	28.17	28.51	28.85	147.2	201.3	Népal
Netherlands	15.45	16.33	16.63	16.87	16.92	16.98	37.4	503.6	Pays-Bas
New Caledonia	0.19	0.23	0.25	0.26	0.26	0.27	18.6	14.6	Nouvelle-Calédonie
New Zealand	3.67	4.13	4.37	4.50	4.53	4.57	268.1	17.3	Nouvelle-Zélande
Nicaragua	4.61	5.38	5.74	6.01	6.08	6.15	130.4	51.1	Nicaragua
Niger	9.36	13.49	16.29	19.11	19.90	20.72	1 267.0	16.4	Niger
Nigeria	108.42	139.61	159.42	177.48	182.20	186.99	923.8	205.3	Nigéria
Niue	<0.00	<0.00	<0.00	<0.00	<0.00	<0.00	0.3	6.2	Nioué
Northern Mariana Islands	0.06	0.06	0.05	0.05	0.06	0.06	0.5	120.4	Îles Mariannes du Nord
Norway[20]	4.36	4.62	4.89	5.15	5.21	5.27	386.2	14.4	Norvège[20]
Oman	2.19	2.51	2.94	4.24	4.49	4.65	309.5	15.0	Oman
Other non-specified areas	21.16	22.70	23.20	23.36	23.38	23.40	...	660.7	Autres zones non-spécifiées
Pakistan	122.60	153.36	170.04	185.04	188.92	192.83	796.1	250.1	Pakistan
Palau	0.02	0.02	0.02	0.02	0.02	0.02	0.5	46.7	Palaos
Panama	2.74	3.32	3.62	3.87	3.93	3.99	75.3	53.7	Panama
Papua New Guinea	4.72	6.09	6.85	7.46	7.62	7.78	462.8	17.2	Papouasie-Nvl-Guinée
Paraguay	4.76	5.80	6.21	6.55	6.64	6.73	406.8	16.9	Paraguay
Peru	24.04	27.61	29.37	30.97	31.38	31.77	1 285.2	24.8	Pérou
Philippines	69.84	86.14	93.04	99.14	100.70	102.25	300.0	342.9	Philippines
Poland	38.59	38.46	38.57	38.62	38.61	38.59	311.9	126.0	Pologne
Portugal	10.08	10.48	10.58	10.40	10.35	10.30	92.2	112.5	Portugal
Puerto Rico	3.69	3.76	3.71	3.69	3.68	3.68	8.9	415.0	Porto Rico
Qatar	0.50	0.84	1.77	2.17	2.24	2.29	11.6	197.4	Qatar
Republic of Korea	44.65	47.61	49.09	50.07	50.29	50.50	100.3	519.4	République de Corée

Region, country or area	Mid-year population estimates (millions) Estimations de population au milieu de l'année (millions)						Surface area Superficie (000 km²)	Density Densité	Région, pays ou zone
	1995	2005	2010	2014	2015	2016	2014	2016	
Republic of Moldova	4.34[21]	4.16[21]	4.08[21]	4.07[21]	4.07[21]	4.06[21]	33.8	123.7[21]	République de Moldova
Réunion	0.67	0.79	0.83	0.86	0.86	0.87	2.5	346.9	Réunion
Romania	22.97	21.41	20.30	19.65	19.51	19.37	238.4	84.2	Roumanie
Russian Federation	148.29	143.62	143.16	143.43	143.46	143.44	17 098.2	8.8	Fédération de Russie
Rwanda	5.91	9.01	10.29	11.34	11.61	11.88	26.3	481.7	Rwanda
Saint Helena [22]	0.01	<0.00	<0.00	<0.00	<0.00	<0.00	0.3	10.1	Sainte-Hélène [22]
Saint Kitts and Nevis	0.04	0.05	0.05	0.05	0.06	0.06	0.3	216.1	Saint-Kitts-et-Nevis
Saint Lucia	0.15	0.17	0.18	0.18	0.19	0.19	0.5[23]	305.5	Sainte-Lucie
Saint Pierre and Miquelon	0.01	0.01	0.01	0.01	0.01	0.01	0.2	27.4	Saint-Pierre-et-Miquelon
Saint Vincent-Grenadines	0.11	0.11	0.11	0.11	0.11	0.11	0.4	281.1	Saint-Vincent-Grenadines
Samoa	0.17	0.18	0.19	0.19	0.19	0.19	2.8	68.7	Samoa
San Marino	0.03	0.03	0.03	0.03	0.03	0.03	0.1	532.5	Saint-Marin
Sao Tome and Principe	0.13	0.15	0.17	0.19	0.19	0.19	1.0	202.5	Sao Tomé-et-Principe
Saudi Arabia	18.85	24.75	28.09	30.89	31.54	32.16	2 206.7	15.0	Arabie saoudite
Senegal	8.71	11.27	12.96	14.67	15.13	15.59	196.7[24]	81.0	Sénégal
Serbia	9.88[25]	9.19[25]	9.06[25]	8.89[25]	8.85[25]	8.81[25]	88.4	100.8[25]	Serbie
Seychelles	0.08	0.09	0.09	0.10	0.10	0.10	0.5	210.9	Seychelles
Sierra Leone	3.84	5.07	5.78	6.32	6.45	6.59	72.3	91.3	Sierra Leone
Singapore	3.48	4.50	5.08	5.51	5.60	5.70	0.7[26]	8 137.9	Singapour
Sint Maarten (Dutch part)	0.03	0.03	0.03	0.04	0.04	0.04	<0.0	1 162.9	Saint-Martin (partie néerlandaise)
Slovakia	5.36	5.39	5.41	5.42	5.43	5.43	49.0[27]	112.9	Slovaquie
Slovenia	1.99	2.00	2.05	2.07	2.07	2.07	20.3	102.7	Slovénie
Solomon Islands	0.36	0.47	0.53	0.57	0.58	0.59	28.9	21.3	Îles Salomon
Somalia	6.35	8.47	9.58	10.52	10.79	11.08	637.7	17.7	Somalie
South Africa	41.43	48.35	51.62	53.97	54.49	54.98	1 221.0	45.3	Afrique du Sud
South Sudan	5.45	8.10	10.06	11.91	12.34	12.73	658.8	20.8	Soudan du sud
Spain	39.76[28]	43.85[28]	46.60[28]	46.26[28]	46.12[28]	46.06[28]	506.0	92.4[28]	Espagne
Sri Lanka	18.25	19.53	20.20	20.62	20.72	20.81	65.6	331.9	Sri Lanka
State of Palestine	2.62[29]	3.58[29]	4.07[29]	4.54[29]	4.67[29]	4.80[29]	6.0	796.9[29]	État de Palestine
Sudan	24.69	31.99	36.11	39.35	40.23	41.18	...	23.3	Soudan
Suriname	0.45	0.49	0.52	0.54	0.54	0.55	163.8	3.5	Suriname
Swaziland	0.96	1.10	1.19	1.27	1.29	1.30	17.4	75.8	Swaziland
Sweden	8.83	9.03	9.38	9.70	9.78	9.85	450.3	24.0	Suède
Switzerland	7.02	7.41	7.83	8.21	8.30	8.38	41.3	212.1	Suisse
Syrian Arab Republic	14.33	18.13	20.72	18.77	18.50	18.56	185.2	101.1	Rép. arabe syrienne
Tajikistan	5.78	6.81	7.58	8.30	8.48	8.67	142.6	61.9	Tadjikistan
TFYR of Macedonia	1.95	2.04	2.06	2.08	2.08	2.08	25.7	82.5	ex-R.Y. de Macédoine
Thailand	59.27	65.86	66.69	67.73	67.96	68.15	513.1	133.4	Thaïlande
Timor-Leste	0.86	0.99	1.06	1.16	1.18	1.21	14.9	81.5	Timor-Leste
Togo	4.28	5.58	6.39	7.12	7.30	7.50	56.8	137.8	Togo
Tokelau	<0.00	<0.00	<0.00	<0.00	<0.00	<0.00	<0.0	127.6	Tokélaou
Tonga	0.10	0.10	0.10	0.11	0.11	0.11	0.7	148.5	Tonga
Trinidad and Tobago	1.26	1.30	1.33	1.35	1.36	1.37	5.1	266.1	Trinité-et-Tobago
Tunisia	9.11	10.10	10.64	11.13	11.25	11.38	163.6	73.2	Tunisie

Region, country or area	Mid-year population estimates (millions) Estimations de population au milieu de l'année (millions)						Surface area Superficie (000 km²)	Density Densité	Région, pays ou zone
	1995	2005	2010	2014	2015	2016	2014	2016	
Turkey	58.52	67.86	72.31	77.52	78.67	79.62	783.6	103.5	Turquie
Turkmenistan	4.19	4.75	5.04	5.31	5.37	5.44	488.1	11.6	Turkménistan
Turks and Caicos Islands	0.02	0.03	0.03	0.03	0.03	0.03	0.9[30]	36.7	Îles Turques-et-Caïques
Tuvalu	0.01	0.01	0.01	0.01	0.01	0.01	<0.0	331.4	Tuvalu
Uganda	20.41	28.04	33.15	37.78	39.03	40.32	241.6	201.8	Ouganda
Ukraine [31]	50.81	46.80	45.65	45.00	44.82	44.62	603.5	77.0	Ukraine [31]
United Arab Emirates	2.35	4.48	8.33	9.09	9.16	9.27	83.6	110.8	Émirats arabes unis
United Kingdom	57.90	60.21	62.72	64.33	64.72	65.11	242.5	269.1	Royaume-Uni
United Rep. of Tanzania [32]	29.90	39.07	45.65	51.82	53.47	55.16	947.3	62.3	Rép.-Unie de Tanzanie [32]
United States	266.28	296.14	309.88	319.45	321.77	324.12	9 833.5	35.4	États-Unis
United States Virgin Is.	0.11	0.11	0.11	0.11	0.11	0.11	0.3	304.0	Îles Vierges américaines
Uruguay	3.22	3.33	3.37	3.42	3.43	3.44	173.6	19.7	Uruguay
Uzbekistan	22.69	25.92	27.74	29.47	29.89	30.30	447.4	71.2	Ouzbékistan
Vanuatu	0.17	0.21	0.24	0.26	0.26	0.27	12.2	22.2	Vanuatu
Venezuela (Boliv. Rep. of)	22.19	26.77	29.00	30.69	31.11	31.52	912.1	35.7	Venezuela (Rép. boliv. du)
Viet Nam	75.20	84.20	88.36	92.42	93.45	94.44	331.0	304.6	Viet Nam
Wallis and Futuna Islands	0.01	0.01	0.01	0.01	0.01	0.01	0.1	93.7	Îles Wallis-et-Futuna
Western Sahara	0.25	0.43	0.51	0.56	0.57	0.58	266.0[33]	2.2	Sahara occidental
Yemen	15.27	20.50	23.59	26.18	26.83	27.48	528.0	52.0	Yémen
Zambia	9.25	12.04	13.92	15.72	16.21	16.72	752.6	22.5	Zambie
Zimbabwe	11.68	12.98	13.97	15.25	15.60	15.97	390.8	41.3	Zimbabwe

Source:
Mid-year population estimates, surface area and density: United Nations Population Division, New York, *World Population Prospects: The 2015 Revision.* Surface area: United Nations Statistics Division, New York, *Demographic Yearbook 2014.*

Source:
Estimations de population au milieu de l'année, superficie et densité : Organisation des Nations Unies, Division pour la population, New York, *Perspectives de la population mondiale: révision de 2015.* Superficie: Organisation des Nations Unies, Division de statistique, New York, *Annuaire démographique 2014.*

1   Including Christmas Island, Cocos (Keeling) Islands and Norfolk Island.
2   Including Nagorno-Karabakh.
3   Data updated according to "Superintendencia Agraria". Interior waters correspond to natural or artificial bodies of water or snow.
4   Refers to Guernsey and Jersey.
5   For statistical purposes, the data for China do not include those for the Hong Kong Special Administrative Region (Hong Kong SAR), Macao Special Administrative Region (Macao SAR) and Taiwan Province of China.
6   Excluding Niue, shown separately, which is part of Cook Islands, but because of remoteness is administered separately.
7   Refers to the whole country.
8   The total surface is 21 040.79 square kilometres, without taking into account the last ruling of The Hague.
9   Including Åland Islands.
10   Including Abkhazia and South Ossetia.
11   Including Saint-Barthélemy and Saint-Martin (French part).
12   Refers to the Vatican City State.

1   Y compris les îles Christmas, Cocos (Keeling) et Norfolk.
2   Y compris le Haut-Karabakh.
3   Données actualisées d'après la « Superintendencia Agraria ». Les eaux intérieures correspondent aux étendues d'eau naturelles ou artificielles et aux étendues neigeuses.
4   Se rapporte à Guernsey et Jersey.
5   Pour la présentation des statistiques, les données pour la Chine ne comprennent pas la Région Administrative Spéciale de Hong Kong (Hong Kong RAS), la Région Administrative Spéciale de Macao (Macao RAS) et la province de Taiwan.
6   Non compris Niue, qui fait l'objet d'une rubrique distincte et qui fait partie des îles Cook, mais qui, en raison de son éloignement, est administrée séparément.
7   Ensemble du pays.
8   La superficie totale est égale à 21 040.79 km2, sans tenir compte de la dernière décision de la Haye.
9   Y compris les Îles d'Åland.
10   Y compris l'Abkhazie et l'Ossétie du Sud.
11   Y compris Saint Barthélémy et Saint Martin.
12   Se réfère à l'État de la Cité du Vatican.

13	Surface area is 0.44 Km2.	13	Superficie: 0,44 Km2
14	Land area only.	14	La superficie des terres seulement.
15	Data refer to 1 October 2007.	15	Les données se réfèrent au 1er octobre 2007.
16	Land area only. Excluding 84 square km of uninhabited islands.	16	La superficie des terres seulement. Exclut des îles inhabitées d'une superficie de 84 kilomètres carrés.
17	Including Sabah and Sarawak.	17	Y compris Sabah et Sarawak.
18	Including Agalega, Rodrigues and Saint Brandon.	18	Y compris Agalega, Rodrigues et Saint Brandon.
19	Excludes the islands of St. Brandon and Agalega.	19	Non compris les îles St. Brandon et Agalega.
20	Including Svalbard and Jan Mayen Islands.	20	Y compris îles Svalbard et Jan Mayen.
21	Including Transnistria.	21	Y compris la Transnistrie.
22	Including Ascension and Tristan da Cunha.	22	Y compris Ascension et Tristan da Cunha.
23	Refers to habitable area. Excludes St. Lucia's Forest Reserve.	23	S'applique à la zone habitable. Exclut la réserve forestière de Sainte-Lucie.
24	Surface area is based on the 2002 population and housing census.	24	La superficie est fondée sur les données provenant du recensement de la population et du logement de 2002.
25	Including Kosovo.	25	Y compris Kosovo.
26	The land area of Singapore comprises the mainland and other islands.	26	La superficie terrestre de Singapour comprend l'île principale et les autres îles.
27	Excluding inland water.	27	Exception faite des eaux intérieures.
28	Including Canary Islands, Ceuta and Melilla.	28	Y compris les Iles Canaries, Ceuta et Melilla.
29	Including East Jerusalem.	29	Y compris Jérusalem-Est.
30	Including low water level for all islands (area to shoreline).	30	Incluent le niveau de basses eaux pour toutes les iles.
31	Including Crimea.	31	Y compris Crimea
32	Including Zanzibar.	32	Y compris Zanzibar.
33	Comprising the Northern Region (former Saguia el Hamra) and Southern Region (former Rio de Oro).	33	Comprend la région septentrionale (ancien Saguia el Hamra) et la région méridionale (ancien Rio de Oro).

International migrants and refugees
International migrant stock (number and percentage) and refugees and others of concern to UNHCR

Migrants internationaux et réfugiés
Stock de migrants internationaux (nombre et pourcentage) et réfugiés et autres personnes relevant de la compétence du HCR

Region, country or area Région, pays ou zone	Year Année	International Migrant Stock (mid-year) Stock de migrants internationaux (milieu de l'année)				Refugees and others of concern to UNHCR (end-year) Réfugiés et autres personnes relevant de la compétence du HCR (fin d'année)			
		Total total	% of Total Population % de la population totale			Refugees[&] Réfugiés[&]	Asylum seekers Demandeurs d'asile	Other[&&] Autres[&&]	Total population Population totale
		MF	MF	M	F				
Total, all countries or areas	2005	191 269 100	2.9	3.0	2.9	...	...	...	...
Total, tous pays ou zones	2010	221 714 243	3.2	3.3	3.1	...	...	...	...
	2015	243 700 236	3.3	3.4	3.2	15 097 633[1]	2 343 919[1]	40 518 150[1]	57 959 702[1]
Africa	2005	15 191 146	1.7	1.8	1.5	...	...	...	...
Afrique	2010	16 840 014	1.6	1.7	1.5	...	...	...	...
	2015	20 649 557	1.7	1.9	1.6	4 493 139[1]	1 044 031[1]	11 530 138[1]	17 067 308[1]
Northern America	2005	45 363 387	13.8	13.9	13.7	...	...	...	...
Amérique septentrionale	2010	51 220 996	14.9	14.7	15.0	...	...	...	...
	2015	54 488 725	15.2	15.0	15.5	416 385[1]	238 989[1]	...	655 374[1]
Latin America and the	2005	7 233 098	1.3	1.3	1.3	...	...	...	...
Caribbean	2010	8 238 795	1.4	1.4	1.4	...	...	...	...
Amérique latine et Caraïbes	2015	9 233 989	1.5	1.5	1.4	336 552[1]	37 378[1]	6 697 290[1]	7 071 220[1]
Asia	2005	53 371 224	1.4	1.5	1.2	...	...	...	...
Asie	2010	65 914 319	1.6	1.8	1.4	...	...	...	...
	2015	75 081 125	1.7	1.9	1.5	8 178 380[1]	320 437[1]	19 921 907[1]	28 420 724[1]
Europe	2005	64 086 824	8.8	8.8	8.8	...	...	...	...
Europe	2010	72 374 755	9.8	9.8	9.9	...	...	...	...
	2015	76 145 954	10.3	10.2	10.4	1 626 214[1]	678 737[1]	2 368 815[1]	4 673 766[1]
Oceania	2005	6 023 421	18.1	17.9	18.2	...	...	...	...
Océanie	2010	7 125 364	19.6	19.4	19.8	...	...	...	...
	2015	8 100 886	20.6	20.3	20.9	46 963[1]	24 347[1]	...	71 310[1]
Afghanistan	2005	87 300	0.4	0.4	0.3	32	14	159 551	159 597
Afghanistan	2010	102 246	0.4	0.4	0.3	6 434	30	1 193 523	1 199 987
	2015	382 365	1.2	1.2	1.2	225 714[1]	101[1]	1 195 604[1]	1 421 419[1]
Albania	2005	64 739	2.1	2.1	2.1	56	35	1	92
Albanie	2010	52 784	1.8	1.8	1.8	76	23	...	99
	2015	57 616	2.0	2.0	1.9	154[1]	501[1]	7 443[1]	8 098[1]
Algeria	2005	247 537[2,3]	0.7[2,3]	0.8[2,3]	0.7[2,3]	94 101	306	...	94 407
Algérie	2010	244 964[2,3]	0.7[2,3]	0.7[2,3]	0.6[2,3]	94 144	304	...	94 448
	2015	242 391[2,3]	0.6[2,3]	0.7[2,3]	0.6[2,3]	94 144[1,4]	5 892[1,4]	...	100 036[1,4]
American Samoa	2005	24 233	41.0	...	...	...	...	...	...
Samoa américaines	2010	23 555	42.3	...	...	...	...	...	...
	2015	23 216	41.8	...	...	...	...	...	...
Andorra [3]	2005	50 298	61.9	...	...	...	...	...	...
Andorre [3]	2010	52 053	61.7	...	...	...	...	...	...
	2015	42 082	59.7	...	...	...	...	...	...
Angola	2005	61 329[2]	0.3[2]	0.3[2]	0.3[2]	13 984	885	45	14 914
Angola	2010	76 549[2]	0.4[2]	0.4[2]	0.4[2]	15 155	4 241	...	19 396
	2015	106 845[2]	0.4[2]	0.4[2]	0.4[2]	15 572[1]	30 086[1]	2 887[1]	48 545[1]
Anguilla	2005	4 684	37.1	...	...	...	...	...	...
Anguilla	2010	5 103	37.1	...	...	...	...	...	...
	2015	5 470	37.4	...	...	...	...	...	...
Antigua and Barbuda	2005	24 741	30.0	28.0	31.7	...	...	...	...
Antigua-et-Barbuda	2010	26 412	30.3	28.3	32.1	...	...	...	...
	2015	28 083	30.6	28.5	32.5	...	10[1]	...	10[1]
Argentina	2005	1 673 088	4.3	4.0	4.5	3 074	825	2	3 901
Argentine	2010	1 805 957	4.4	4.1	4.6	3 276	947	...	4 223
	2015	2 086 302	4.8	4.5	5.1	3 523[1]	897[1]	...	4 420[1]
Armenia	2005	469 119[2]	15.6[2]	12.9[2]	18.1[2]	219 550	70	...	219 620
Arménie	2010	221 560[2]	7.5[2]	5.9[2]	9.1[2]	3 296	23	82 525	85 844
	2015	191 199[2]	6.3[2]	5.5[2]	7.0[2]	15 690[1]	114[1]	238[1]	16 042[1]
Aruba	2005	32 540	32.5	30.6	34.3	...	...	...	...
Aruba	2010	34 327	33.8	31.6	35.8	...	1	...	1
	2015	36 114	34.8	32.6	36.7	...	2[1]	...	2[1]

International migrants and refugees *(continued)*
International migrant stock (number and percentage) and refugees and others of concern to UNHCR

3

Migrants internationaux et réfugiés *(suite)*
Stock de migrants internationaux (nombre et pourcentage) et réfugiés et autres personnes relevant de la compétence du HCR

Region, country or area Région, pays ou zone	Year Année	International Migrant Stock (mid-year) Stock de migrants internationaux (milieu de l'année)				Refugees and others of concern to UNHCR (end-year) Réfugiés et autres personnes relevant de la compétence du HCR (fin d'année)			
		Total total	% of Total Population % de la population totale			Refugees[&] Réfugiés[&]	Asylum seekers Demandeurs d'asile	Other[&&] Autres[&&]	Total population Population totale
		MF	MF	M	F				
Australia Australie	2005	4 878 030[5]	24.1[5]	23.9[5]	24.2[5]	64 964	1 822	8	66 794
	2010	5 882 980[5]	26.5[5]	26.4[5]	26.7[5]	21 805	3 760	15	25 580
	2015	6 763 663[5]	28.2[5]	27.9[5]	28.6[5]	35 582[1,6]	22 837[1,6]	...	58 419[1,6]
Austria Autriche	2005	1 136 270	13.8	13.6	14.0	21 230	40 710	830	62 770
	2010	1 275 992	15.2	14.9	15.5	42 630	25 625	470	68 725
	2015	1 492 374	17.5	17.1	17.8	60 747[1,7]	30 900[1,7]	570[1,7]	92 217[1,7]
Azerbaijan Azerbaïdjan	2005	302 220[2,8]	3.5[2,8]	3.2[2,8]	3.8[2,8]	3 004	115	581 194	584 313
	2010	276 901[2,8]	3.0[2,8]	2.9[2,8]	3.2[2,8]	1 891	17	594 969	596 877
	2015	264 241[2,8]	2.7[2,8]	2.6[2,8]	2.8[2,8]	1 357[1]	262[1]	626 477[1]	628 096[1]
Bahamas Bahamas	2005	45 595	13.8	14.7	13.0	...	...	...	...
	2010	54 736	15.2	15.8	14.6	28	9	...	37
	2015	59 306	15.3	15.8	14.8	7[1]	19[1]	30[1]	56[1]
Bahrain Bahreïn	2005	404 018[3]	46.6[3]	55.4[3]	33.3[3]	...	15	...	15
	2010	657 856[3]	52.2[3]	60.5[3]	38.4[3]	165	69	...	234
	2015	704 137[3]	51.1[3]	59.6[3]	37.3[3]	277[1]	78[1]	...	355[1]
Bangladesh Bangladesh	2005	1 166 700[2]	0.8[2]	1.4[2]	0.2[2]	21 098	58	250 094	271 250
	2010	1 345 546[2]	0.9[2]	1.5[2]	0.2[2]	229 253	...	...	229 253
	2015	1 422 805[2]	0.9[2]	1.5[2]	0.2[2]	232 975[1,9]	11[1,9]	...	232 986[1,9]
Barbados Barbade	2005	30 624	11.2	10.3	12.0	...	...	...	...
	2010	32 825	11.7	11.0	12.4	...	...	...	...
	2015	34 475	12.1	11.4	12.8	1[1]	...	...	1[1]
Belarus Bélarus	2005	1 106 982	11.5	11.3	11.7	725	56	12 421	13 202
	2010	1 090 378	11.5	11.3	11.6	589	66	7 734	8 389
	2015	1 082 905	11.4	11.2	11.5	1 369[1]	257[1]	6 302[1]	7 928[1]
Belgium Belgique	2005	870 862[3]	8.2[3]	8.6[3]	7.9[3]	15 282	18 913	398	34 593
	2010	1 052 844[3]	9.6[3]	10.0[3]	9.3[3]	17 892	18 288	696	36 876
	2015	1 387 940[3]	12.3[3]	12.9[3]	11.7[3]	31 115[1]	9 396[1]	5 267[1]	45 778[1]
Belize Belize	2005	41 424[2]	14.6[2]	14.8[2]	14.5[2]	624	14	...	638
	2010	46 360[2]	14.4[2]	14.6[2]	14.2[2]	134	30	2	166
	2015	53 860[2]	15.0[2]	15.1[2]	14.9[2]	...	146[1]	...	146[1]
Benin Bénin	2005	171 499[2,3]	2.1[2,3]	2.3[2,3]	1.9[2,3]	30 294	1 695	77	32 066
	2010	209 267[2,3]	2.2[2,3]	2.5[2,3]	1.9[2,3]	7 139	101	47	7 287
	2015	245 399[2,3]	2.3[2,3]	2.5[2,3]	2.0[2,3]	488[1]	84[1]	...	572[1]
Bermuda Bermudes	2005	18 276	28.1	...	...	...	...	...	...
	2010	18 884	29.5	...	...	...	...	...	...
	2015	19 126	30.8	...	...	...	...	...	...
Bhutan Bhoutan	2005	40 279	6.2	9.4	2.5	...	...	...	...
	2010	48 420	6.7	10.2	2.7	...	...	...	...
	2015	51 106	6.6	10.0	2.7	...	...	...	...
Bolivia (Plurin.l State of) Bolivie (État plurin. de)	2005	107 745	1.2	1.2	1.1	535	3	2	540
	2010	122 846	1.2	1.3	1.2	695	41	3	739
	2015	142 989	1.3	1.4	1.3	767[1]	8[1]	...	775[1]
Bonaire, St. Eustatius and Saba Bonaire, St.-Eustache et Saba	2005[10]	49 105	...	...	...	...	...	...	...
	2010	11 445	54.7	...	...	...	1[11]	...	1[11]
	2015	13 002	52.3	...	...	...	...	...	...
Bosnia and Herzegovina Bosnie-Herzégovine	2005	* 47 272[2]	* 1.2[2]	* 1.2[2]	* 1.3[2]	10 568	215	188 735	199 518
	2010	* 38 792[2]	* 1.0[2]	* 1.0[2]	* 1.1[2]	7 016	153	171 790	178 959
	2015	* 34 803[2]	* 0.9[2]	* 0.9[2]	* 1.0[2]	6 805[1]	11[1]	137 035[1]	143 851[1]
Botswana Botswana	2005	88 829[3]	4.8[3]	5.3[3]	4.2[3]	3 109	47	592	3 748
	2010	120 912[3]	5.9[3]	6.5[3]	5.3[3]	2 986	249	111	3 346
	2015	160 644[3]	7.1[3]	7.8[3]	6.4[3]	2 164[1]	248[1]	...	2 412[1]
Brazil Brésil	2005	638 582	0.3	0.4	0.3	3 458	195	4 093	7 746
	2010	592 568	0.3	0.3	0.3	4 357	872	8	5 237
	2015	713 568	0.3	0.4	0.3	7 762[1]	17 902[1]	40 338[1]	66 002[1]

3

International migrants and refugees *(continued)*
International migrant stock (number and percentage) and refugees and others of concern to UNHCR

Migrants internationaux et réfugiés *(suite)*
Stock de migrants internationaux (nombre et pourcentage) et réfugiés et autres personnes relevant de la compétence du HCR

Region, country or area / Région, pays ou zone	Year / Année	International Migrant Stock (mid-year) / Stock de migrants internationaux (milieu de l'année) Total total MF	% of Total Population % de la population totale MF	M	F	Refugees and others of concern to UNHCR (end-year) / Réfugiés et autres personnes relevant de la compétence du HCR (fin d'année) Refugees[&] / Réfugiés[&]	Asylum seekers Demandeurs d'asile	Other[&&] Autres[&&]	Total population population Population totale
British Virgin Islands Îles Vierges britanniques	2005	13 808	59.6	…	…	…	…	…	…
	2010	15 558	57.2	…	…	2	…	…	2
	2015	17 308	57.5	…	…	…	…	…	…
Brunei Darussalam Brunéi Darussalam	2005	98 441	27.2	29.8	24.5	…	…	…	…
	2010	100 587	25.6	27.9	23.1	…	…	20 992	20 992
	2015	102 733	24.3	26.7	21.7	…	…	20 524[1]	20 524[1]
Bulgaria Bulgarie	2005	61 074	0.8	0.7	0.9	4 413	805	…	5 218
	2010	76 287	1.0	0.9	1.1	5 530	1 412	…	6 942
	2015	102 113	1.4	1.4	1.5	11 046[1,12]	7 840[1,12]	67[1,12]	18 953[1,12]
Burkina Faso Burkina Faso	2005	596 972[2]	4.4[2]	4.3[2]	4.6[2]	511	784	7	1 302
	2010	673 904[2]	4.3[2]	4.1[2]	4.5[2]	531	534	9	1 074
	2015	704 676[2]	3.9[2]	3.7[2]	4.0[2]	34 027[1]	180[1]	…	34 207[1]
Burundi Burundi	2005	172 874[2]	2.2[2]	2.2[2]	2.2[2]	20 681	19 900	12 988	53 569
	2010	235 259[2]	2.5[2]	2.5[2]	2.5[2]	29 365	12 062	159 338	200 765
	2015	286 810[2]	2.6[2]	2.6[2]	2.6[2]	54 126[1]	2 733[1]	80 851[1]	137 710[1]
Cabo Verde Cabo Verde	2005	12 700	2.7	2.8	2.5	…	…	1	1
	2010	14 373	2.9	3.0	2.8	…	…	…	…
	2015	14 924	2.9	2.9	2.8	…	…	115[1]	115[1]
Cambodia Cambodge	2005	114 031	0.9	0.9	0.8	127	68	207	402
	2010	81 977	0.6	0.6	0.5	129	51	…	180
	2015	73 963	0.5	0.5	0.4	80[1]	33[1]	131[1]	244[1]
Cameroon Cameroun	2005	258 737	1.4	1.6	1.3	52 042	6 766	7 444	66 252
	2010	289 091	1.4	1.6	1.3	104 275	2 383	66	106 724
	2015	381 984	1.6	1.6	1.7	302 293[1]	7 835[1]	81 693[1]	391 821[1]
Canada Canada	2005	6 078 985	18.8	18.2	19.4	147 171	20 552	12	167 735
	2010	7 011 226	20.5	19.8	21.3	165 549	51 025	7	216 581
	2015	7 835 502	21.8	21.0	22.6	149 163[1,12]	14 481[1,12]	…	163 644[1,12]
Cayman Islands Îles Caïmanes	2005[3]	21 655	44.5	…	…	…	…	…	…
	2010	24 057[3]	43.3[3]	…	…	1	4	…	5
	2015	23 726[3]	39.6[3]	…	…	6[1]	1[1]	100[1]	107[1]
Central African Rep. Rép. centrafricaine	2005	94 449[3]	2.3[3]	2.5[3]	2.1[3]	24 569	1 960	2 167	28 696
	2010	93 466[3]	2.1[3]	2.3[3]	1.9[3]	21 574	1 219	192 531	215 324
	2015	81 598[3]	1.7[3]	1.8[3]	1.5[3]	7 906[1]	394[1]	508 904[1]	517 204[1]
Chad Tchad	2005	352 062[2]	3.5[2]	3.2[2]	3.8[2]	275 412	68	…	275 480
	2010	416 924[2]	3.5[2]	3.1[2]	3.9[2]	347 939	110	185 000	533 049
	2015	516 968[2]	3.7[2]	3.4[2]	4.0[2]	420 774[1]	2 749[1]	50 000[1]	473 523[1]
Channel Islands[13] Îles Anglo-Normandes[13]	2005	70 941	46.0	44.3	47.6	…	…	…	…
	2010	77 581	48.6	46.8	50.4	…	…	…	…
	2015	82 307	50.3	48.5	52.0	…	…	…	…
Chile Chili	2005	273 384	1.7	1.6	1.8	806	107	…	913
	2010	369 436	2.2	2.1	2.3	1 621	274	…	1 895
	2015	469 436	2.6	2.5	2.7	1 798[1]	719[1]	…	2 517[1]
China Chine	2005	678 947[3,14]	0.1[3,14]	0.1[3,14]	<0.0[3,14]	301 041	84	1	301 126
	2010	849 861[3,14]	0.1[3,14]	0.1[3,14]	0.1[3,14]	300 986	122	…	301 108
	2015	978 046[3,14]	0.1[3,14]	0.1[3,14]	0.1[3,14]	301 057[1,15]	564[1,15]	…	301 621[1,15]
China, Hong Kong SAR Chine, Hong Kong RAS	2005	2 721 235	39.8	36.2	43.0	1 934	1 097	…	3 031
	2010	2 779 950	39.8	35.0	44.0	154	486	3	643
	2015	2 838 665	38.9	32.7	44.5	151[1]	9 940[1]	1[1]	10 092[1]
China, Macao SAR Chine, Macao RAS	2005	279 308	59.7	58.1	61.1	…	…	…	…
	2010	318 506	59.6	57.1	61.9	…	9	1	10
	2015	342 703	58.3	55.0	61.4	…	6[1]	…	6[1]
Colombia Colombie	2005	107 612	0.2	0.3	0.2	155	41	2 000 009	2 000 205
	2010	124 271	0.3	0.3	0.3	212	167	3 672 065	3 672 444
	2015	133 134	0.3	0.3	0.3	219[1]	56[1]	6 520 304[1]	6 520 579[1]

International migrants and refugees *(continued)*
International migrant stock (number and percentage) and refugees and others of concern to UNHCR

Migrants internationaux et réfugiés *(suite)*
Stock de migrants internationaux (nombre et pourcentage) et réfugiés et autres personnes relevant de la compétence du HCR

Region, country or area Région, pays ou zone	Year Année	International Migrant Stock (mid-year) Stock de migrants internationaux (milieu de l'année)				Refugees and others of concern to UNHCR (end-year) Réfugiés et autres personnes relevant de la compétence du HCR (fin d'année)			
		Total total	% of Total Population % de la population totale			Refugees[&] Réfugiés[&]	Asylum seekers Demandeurs d'asile	Other[&&] Autres[&&]	Total population Population totale
		MF	MF	M	F				
Comoros Comores	2005	13 209	2.1	2.0	2.3	1	...	...	1
	2010	12 618	1.8	1.7	1.9	...	...	...	...
	2015	12 555	1.6	1.5	1.7	...	...	...	...
Congo Congo	2005	315 238	9.0	9.8	8.2	66 075	3 486	8 079	77 640
	2010	419 649	10.3	11.2	9.4	133 112	5 524	59	138 695
	2015	392 996	8.5	9.3	7.7	61 492[1]	3 248[1]	1 070[1]	65 810[1]
Cook Islands Îles Cook	2005	3 277	16.9	...	...	...	...	...	...
	2010	3 769	18.6	...	...	...	...	...	...
	2015	4 152	19.9	...	...	...	...	...	...
Costa Rica Costa Rica	2005	358 175[2]	8.4[2]	8.3[2]	8.6[2]	11 253	223	...	11 476
	2010	405 404[2]	8.9[2]	8.6[2]	9.2[2]	19 505	375	40	19 920
	2015	421 697[2]	8.8[2]	8.4[2]	9.1[2]	3 475[1]	1 819[1]	2 613[1]	7 907[1]
Côte d'Ivoire Côte d'Ivoire	2005	2 010 824[3]	11.1[3]	12.0[3]	10.2[3]	41 627	2 443	71 050	115 120
	2010	2 095 185[3]	10.4[3]	11.3[3]	9.5[3]	26 218	256	538 068	564 542
	2015	2 175 399[3]	9.6[3]	10.4[3]	8.7[3]	1 972[1]	667[1]	724 131[1]	726 770[1]
Croatia Croatie	2005	579 273[2]	13.2[2]	12.9[2]	13.6[2]	2 927	8	7 932	10 867
	2010	573 248[2]	13.3[2]	12.9[2]	13.6[2]	936	81	24 463	25 480
	2015	576 883[2]	13.6[2]	13.2[2]	14.0[2]	710[1]	90[1]	16 684[1]	17 484[1]
Cuba Cuba	2005	17 023	0.2	0.1	0.2	706	32	...	738
	2010	14 818	0.1	0.1	0.1	411	11	1	423
	2015	13 336	0.1	0.1	0.1	313[1]	12[1]	...	325[1]
Curaçao Curaçao	2010	34 627	23.5	21.0	25.6	7	2	...	9
	2015	37 611	23.9	21.3	26.1	44[1]	41[1]	...	85[1]
Cyprus Chypre	2005	117 165[16]	11.3[16]	9.7[16]	13.1[16]	701	13 067	1	13 769
	2010	187 923[16]	17.0[16]	14.6[16]	19.5[16]	3 394	5 396	...	8 790
	2015	196 167[16]	16.8[16]	14.6[16]	19.1[16]	5 763[1]	2 339[1]	6 000[1]	14 102[1]
Czech Republic République tchèque	2005	322 540[3]	3.2[3]	3.6[3]	2.7[3]	1 802	924	...	2 726
	2010	397 785[3]	3.8[3]	4.6[3]	3.0[3]	2 449	1 065	1	3 515
	2015	405 093[3]	3.8[3]	4.5[3]	3.2[3]	3 137[1,12]	409[1,12]	1 502[1,12]	5 048[1,12]
Dem. P. R. Korea * R. p. dém. de Corée *	2005	40 097	0.2	0.2	0.2	...	...	...	...
	2010	44 010	0.2	0.2	0.2	...	...	...	...
	2015	48 458	0.2	0.2	0.2	...	...	...	...
Dem. Rep. of the Congo Rép. dém. du Congo	2005	622 869[2]	1.1[2]	1.1[2]	1.1[2]	204 341	140	47 435	251 916
	2010	588 950[2]	0.9[2]	0.9[2]	0.9[2]	166 336	932	2 196 591	2 363 859
	2015	545 694[2]	0.7[2]	0.7[2]	0.7[2]	* 160 271[1]	* 1 124[1]	* 1 839 611[1]	* 2 001 006[1]
Denmark Danemark	2005	440 383	8.1	7.9	8.3	44 374	510	573	45 457
	2010	509 740	9.2	8.9	9.4	17 922	3 363	3 238	24 523
	2015	572 520	10.1	9.8	10.3	17 785[1,12]	4 566[1,12]	4 984[1,12]	27 335[1,12]
Djibouti Djibouti	2005	92 091[2]	11.8[2]	13.1[2]	10.5[2]	10 456	19	7 668	18 143
	2010	101 575[2]	12.2[2]	12.8[2]	11.7[2]	15 104	732	7	15 843
	2015	112 351[2]	12.7[2]	13.3[2]	12.1[2]	14 787[1]	2 586[1]	...	17 373[1]
Dominica Dominique	2005	4 744	6.7	...	...	...	...	...	...
	2010	5 765	8.1	...	...	...	...	...	...
	2015	6 720	9.2	...	...	...	...	...	...
Dominican Republic Rép. dominicaine	2005	376 001	4.1	4.9	3.2	...	...	...	...
	2010	393 720	4.0	4.8	3.1	599	1 759	...	2 358
	2015	415 564	3.9	4.8	3.1	609[1,17]	752[1,17]	133 770[1,17]	135 131[1,17]
Ecuador Équateur	2005	187 404[2]	1.4[2]	1.4[2]	1.3[2]	10 063	2 489	250 001	262 553
	2010	325 366[2]	2.2[2]	2.2[2]	2.1[2]	121 249	49 887	...	171 136
	2015	387 513[2]	2.4[2]	2.5[2]	2.3[2]	121 722[1,7]	11 583[1,7]	...	133 305[1,7]
Egypt Égypte	2005	274 001[2]	0.4[2]	0.4[2]	0.3[2]	88 946	11 005	203	100 154
	2010	295 714[2]	0.4[2]	0.4[2]	0.3[2]	95 056	14 303	570	109 929
	2015	491 643[2]	0.5[2]	0.6[2]	0.5[2]	226 344[1]	30 019[1]	21[1]	256 384[1]
El Salvador El Salvador	2005	36 019[2]	0.6[2]	0.6[2]	0.6[2]	49	1	44	94
	2010	40 324[2]	0.7[2]	0.7[2]	0.7[2]	38	18	...	56
	2015	42 045[2]	0.7[2]	0.7[2]	0.7[2]	48[1]	...	...	48[1]

International migrants and refugees *(continued)*
International migrant stock (number and percentage) and refugees and others of concern to UNHCR

Migrants internationaux et réfugiés *(suite)*
Stock de migrants internationaux (nombre et pourcentage) et réfugiés et autres personnes relevant de la compétence du HCR

Region, country or area Région, pays ou zone	Year Année	International Migrant Stock (mid-year) Stock de migrants internationaux (milieu de l'année)				Refugees and others of concern to UNHCR (end-year) Réfugiés et autres personnes relevant de la compétence du HCR (fin d'année)			
		Total total	% of Total Population % de la population totale			Refugees[&] Réfugiés[&]	Asylum seekers Demandeurs d'asile	Other[&&] Autres[&&]	Total population Population totale
		MF	MF	M	F				
Equatorial Guinea [3] Guinée équatoriale [3]	2005	6 588	1.1	1.1	1.0	...	...	...	...
	2010	8 658	1.2	1.3	1.1	...	...	...	...
	2015	10 825	1.3	1.4	1.2	...	...	...	...
Eritrea Érythrée	2005	* 14 314	* 0.3	* 0.4	* 0.3	4 418	1 591	31	6 040
	2010	* 15 676	* 0.3	* 0.4	* 0.3	4 809	137	13	4 959
	2015	* 15 941	* 0.3	* 0.3	* 0.3	2 944[1]	1[1]	22[1]	2 967[1]
Estonia Estonie	2005	233 701	17.2	15.0	19.2	7	8	136 000	136 015
	2010	217 890	16.4	14.1	18.3	39	10	100 983	101 032
	2015	202 348	15.4	13.2	17.3	117[1,18]	117[1,18]	86 522[1,18]	86 756[1,18]
Ethiopia Éthiopie	2005	514 242[2]	0.7[2]	0.7[2]	0.6[2]	100 817	209	4 110	105 136
	2010	567 720[2]	0.6[2]	0.7[2]	0.6[2]	154 295	1 028	47	155 370
	2015	1 072 949[2]	1.1[2]	1.1[2]	1.1[2]	702 467[1]	2 871[1]	348[1]	705 686[1]
Falkland Is. (Malvinas) Îles Falkland (Malvinas)	2005	1 166	39.6	...	...	...	...	...	...
	2010	1 436	50.3	...	...	...	...	...	...
	2015	1 571	54.1	...	...	...	...	...	...
Faroe Islands Îles Féroé	2005	4 583	9.5	...	...	...	...	...	...
	2010	5 096	10.5	...	...	...	...	...	...
	2015	5 517	11.4	...	...	...	...	...	...
Fiji Fidji	2005	12 435	1.5	1.6	1.5	...	...	...	...
	2010	13 351	1.6	1.6	1.5	1	6	...	7
	2015	13 751	1.5	1.6	1.4	12[1]	8[1]	...	20[1]
Finland Finlande	2005	192 169[19]	3.7[19]	3.8[19]	3.6[19]	11 809	...	849	12 658
	2010	248 135[19]	4.6[19]	4.8[19]	4.5[19]	8 724	2 097	3 133	13 954
	2015	315 881[19]	5.7[19]	5.9[19]	5.6[19]	11 798[1,12]	2 622[1,12]	1 928[1,12]	16 348[1,12]
France France	2005	6 737 600	11.0	11.0	11.0	137 316	41 279	946	179 541
	2010	7 196 481	11.4	11.4	11.4	200 687	48 576	1 171	250 434
	2015	7 784 418	12.1	12.1	12.1	264 972[1]	53 827[1]	1 290[1]	320 089[1]
French Guiana Guyane française	2005	86 468	42.5	42.4	42.6	...	...	...	...
	2010	96 288	41.2	39.9	42.5	...	...	...	...
	2015	106 108	39.5	37.5	41.5	...	...	...	...
French Polynesia Polynésie française	2005	32 286	12.7	14.1	11.2	...	...	...	...
	2010	31 640	11.8	13.1	10.4	...	...	...	...
	2015	30 058	10.6	11.9	9.3	...	...	...	...
Gabon Gabon	2005	214 123[3]	15.5[3]	17.7[3]	13.3[3]	8 545	4 843	291	13 679
	2010	243 992[3]	15.8[3]	17.9[3]	13.7[3]	9 015	4 132	84	13 231
	2015	268 384[3]	15.6[3]	17.6[3]	13.5[3]	1 008[1]	1 886[1]	...	2 894[1]
Gambia Gambie	2005	181 905	12.6	13.5	11.8	7 330	602	42	7 974
	2010	185 763	11.0	11.7	10.3	8 378	74	31	8 483
	2015	192 540	9.7	10.3	9.1	11 773[1]	2[1]	...	11 775[1]
Georgia Géorgie	2005	199 805[20]	4.5[20]	4.1[20]	4.8[20]	2 497	8	236 097	238 602
	2010	182 202[20]	4.3[20]	3.9[20]	4.6[20]	639	44	361 547	362 230
	2015	168 802[20]	4.2[20]	3.8[20]	4.6[20]	1 659[1]	587[1]	266 060[1]	268 306[1]
Germany Allemagne	2005	10 299 160	12.7	12.6	12.8	700 016	71 624	12 340	783 980
	2010	11 605 690	14.4	14.0	14.8	594 269	51 991	24 378	670 638
	2015	12 005 690	14.9	14.4	15.3	250 299[1]	311 551[1]	11 978[1]	573 828[1]
Ghana Ghana	2005	304 436	1.4	1.5	1.4	53 537	5 496	1 212	60 245
	2010	337 017	1.4	1.5	1.3	13 828	749	184	14 761
	2015	399 471	1.5	1.6	1.3	18 476[1]	2 855[1]	...	21 331[1]
Gibraltar Gibraltar	2005	9 211	31.7	...	...	...	...	...	...
	2010	10 369	33.7	...	...	...	...	...	...
	2015	11 065	34.3	...	...	...	...	...	...
Greece Grèce	2005	1 190 707	10.8	10.8	10.7	2 390	8 867	3 000	14 257
	2010	1 269 749	11.4	11.2	11.5	1 444	55 724	260	57 428
	2015	1 242 514	11.3	10.8	11.9	8 231[1]	29 157[1]	214[1]	37 602[1]

International migrants and refugees *(continued)*
International migrant stock (number and percentage) and refugees and others of concern to UNHCR

Migrants internationaux et réfugiés *(suite)*
Stock de migrants internationaux (nombre et pourcentage) et réfugiés et autres personnes relevant de la compétence du HCR

Region, country or area Région, pays ou zone	Year Année	International Migrant Stock (mid-year) Stock de migrants internationaux (milieu de l'année)				Refugees and others of concern to UNHCR (end-year) Réfugiés et autres personnes relevant de la compétence du HCR (fin d'année)			
		Total total	% of Total Population % de la population totale			Refugees[&] Réfugiés[&]	Asylum seekers Demandeurs d'asile	Other[&&] Autres[&&]	Total population Population totale
		MF	MF	M	F				
Greenland	2005	6 686	11.7	...	...	...	...	...	...
Groenland	2010	6 226	11.0	...	...	...	...	...	...
	2015	6 009	10.7	...	...	...	...	...	...
Grenada	2005	6 902	6.7	6.7	6.7	...	...	...	...
Grenade	2010	6 980	6.7	6.6	6.7	...	3	...	3
	2015	7 057	6.6	6.5	6.7	...	...	...	...
Guadeloupe [21]	2005	89 065	19.8	19.3	20.1	...	...	...	...
Guadeloupe [21]	2010	94 942	20.8	20.0	21.4	...	...	...	...
	2015	98 507	21.0	20.2	21.7	...	...	...	...
Guam	2005	74 743	47.2	47.8	46.5	...	...	...	...
Guam	2010	75 416	47.3	48.0	46.5	...	...	...	...
	2015	76 089	44.8	45.6	44.0	...	...	...	...
Guatemala	2005	57 252[2]	0.4[2]	0.4[2]	0.5[2]	391	3	...	394
Guatemala	2010	66 384[2]	0.5[2]	0.4[2]	0.5[2]	138	2	...	140
	2015	76 352[2]	0.5[2]	0.5[2]	0.5[2]	202[1]	73[1]	...	275[1]
Guinea	2005	229 611[2,3]	2.4[2,3]	2.3[2,3]	2.4[2,3]	63 525	3 808	29 645	96 978
Guinée	2010	205 111[2,3]	1.9[2,3]	1.8[2,3]	1.9[2,3]	14 113	764	117	14 994
	2015	228 413[2,3]	1.8[2,3]	1.8[2,3]	1.8[2,3]	8 704[1]	293[1]	...	8 997[1]
Guinea-Bissau	2005	20 736[2]	1.4[2]	1.4[2]	1.4[2]	7 616	166	...	7 782
Guinée-Bissau	2010	21 061[2]	1.3[2]	1.3[2]	1.3[2]	7 679	330	5	8 014
	2015	22 333[2]	1.2[2]	1.2[2]	1.2[2]	8 684[1]	123[1]	...	8 807[1]
Guyana	2005	10 868	1.5	1.6	1.3	...	...	...	...
Guyana	2010	13 126	1.7	1.9	1.6	7	...	...	7
	2015	15 384	2.0	2.1	1.9	11[1]	1[1]	...	12[1]
Haiti	2005	30 468	0.3	0.4	0.3	...	...	...	...
Haïti	2010	35 104	0.4	0.4	0.3	...	4	...	4
	2015	39 529	0.4	0.4	0.3	5[1]	5[1]	...	10[1]
Holy See *	2005	793	99.4	...	...	...	...	...	...
Saint-Siège *	2010	799	100.0	...	...	...	...	...	...
	2015	800	100.0	...	...	...	...	...	...
Honduras	2005	27 875[2]	0.4[2]	0.4[2]	0.4[2]	22	50	...	72
Honduras	2010	27 288[2]	0.4[2]	0.4[2]	0.3[2]	14	...	1	15
	2015	28 070[2]	0.3[2]	0.4[2]	0.3[2]	23[1]	19[1]	...	42[1]
Hungary	2005	366 787[2]	3.6[2]	3.6[2]	3.6[2]	8 046	684	192	8 922
Hongrie	2010	436 616[2]	4.4[2]	4.4[2]	4.3[2]	5 414	367	62	5 843
	2015	449 632[2]	4.6[2]	4.7[2]	4.4[2]	4 192[1]	24 431[1]	128[1]	28 751[1]
Iceland	2005	25 492	8.6	8.3	8.9	293	29	53	375
Islande	2010	35 091	11.0	11.0	11.1	83	39	113	235
	2015	37 522	11.4	11.0	11.8	104[1]	225[1]	119[1]	448[1]
India	2005	5 923 642[2]	0.5[2]	0.5[2]	0.5[2]	139 283	303	2 792	142 378
Inde	2010	5 436 012[2]	0.4[2]	0.4[2]	0.4[2]	184 821	3 746	5 109	193 676
	2015	5 240 960[2]	0.4[2]	0.4[2]	0.4[2]	200 383[1]	5 381[1]	...	205 764[1]
Indonesia	2005	289 568[2]	0.1[2]	0.1[2]	0.1[2]	89	58	263	410
Indonésie	2010	305 416[2]	0.1[2]	0.1[2]	0.1[2]	811	2 071	12	2 894
	2015	328 846[2]	0.1[2]	0.1[2]	0.1[2]	5 277[1]	7 911[1]	...	13 188[1]
Iran (Islamic Rep. of)	2005	2 568 930[2]	3.7[2]	3.9[2]	3.4[2]	974 302	140	344 914	1 319 356
Iran (Rép. islamique d')	2010	2 761 561[2]	3.7[2]	4.2[2]	3.2[2]	1 073 366	1 775	10 168	1 085 309
	2015	2 726 420[2]	3.4[2]	3.6[2]	3.3[2]	979 441[1]	42[1]	8[1]	979 491[1]
Iraq	2005	132 915[2,3]	0.5[2,3]	0.6[2,3]	0.4[2,3]	50 177	1 948	1 526 104	1 578 229
Iraq	2010	117 389[2,3]	0.4[2,3]	0.5[2,3]	0.3[2,3]	34 655	3 073	1 758 603	1 796 331
	2015	353 881[2,3]	1.0[2,3]	1.1[2,3]	0.8[2,3]	288 035[1,22]	7 420[1,22]	4 016 205[1,22]	4 311 660[1,22]
Ireland	2005	589 046	14.0	14.8	13.3	7 113	2 414	4	9 531
Irlande	2010	730 542	15.8	15.7	16.0	9 107	5 129	1	14 237
	2015	746 260	15.9	15.5	16.3	5 853[1,12]	4 300[1,12]	99[1,12]	10 252[1,12]

International migrants and refugees *(continued)*
International migrant stock (number and percentage) and refugees and others of concern to UNHCR

Migrants internationaux et réfugiés *(suite)*
Stock de migrants internationaux (nombre et pourcentage) et réfugiés et autres personnes relevant de la compétence du HCR

Region, country or area Région, pays ou zone	Year Année	International Migrant Stock (mid-year) Stock de migrants internationaux (milieu de l'année)				Refugees and others of concern to UNHCR (end-year) Réfugiés et autres personnes relevant de la compétence du HCR (fin d'année)			
		Total total	% of Total Population % de la population totale			Refugees[&] Réfugiés[&]	Asylum seekers Demandeurs d'asile	Other[&&] Autres[&&]	Total population Population totale
		MF	MF	M	F				
Isle of Man	2005	41 475	51.6	...	...	...	...	...	...
Île de Man	2010	43 447	51.5	...	...	...	...	...	...
	2015	45 221	51.5	...	...	...	...	...	...
Israel	2005	1 889 503[2]	28.6[2]	26.6[2]	30.5[2]	609	939	...	1 548
Israël	2010	1 950 615[2]	26.3[2]	24.3[2]	28.2[2]	25 471	5 575	9	31 055
	2015	2 011 727[2]	24.9[2]	22.8[2]	27.0[2]	38 500[1]	6 591[1]	88[1]	45 179[1]
Italy	2005	3 954 790	6.7	6.4	7.1	20 675	...	940	21 615
Italie	2010	5 787 893	9.7	9.2	10.2	56 397	4 076	858	61 331
	2015	5 788 875	9.7	9.0	10.4	93 715[1,12]	48 307[1,12]	606[1,12]	142 628[1,12]
Jamaica	2005	24 314	0.9	0.9	0.9	...	...	...	...
Jamaïque	2010	23 677	0.9	0.9	0.8	21	...	...	21
	2015	23 167	0.8	0.8	0.8	15[1]	3[1]	...	18[1]
Japan	2005	2 012 916[3]	1.6[3]	1.5[3]	1.7[3]	1 941	533	1 770	4 244
Japon	2010	2 134 151[3]	1.7[3]	1.5[3]	1.8[3]	2 586	3 047	1 397	7 030
	2015	2 043 877[3]	1.6[3]	1.5[3]	1.7[3]	* 2 419[1]	* 10 705[1]	* 631[1]	* 13 755[1]
Jordan	2005	2 325 414[2,3]	43.6[2,3]	42.8[2,3]	44.4[2,3]	965	16 570	329	17 864
Jordanie	2010	2 722 983[2,3]	41.8[2,3]	41.2[2,3]	42.4[2,3]	450 915	2 159	107	453 181
	2015	3 112 026[2,3]	41.0[2,3]	40.3[2,3]	41.7[2,3]	664 102[1,23]	20 693[1,23]	...	684 795[1,23]
Kazakhstan	2005	3 102 962	20.1	19.9	20.3	7 265	65	50 598	57 928
Kazakhstan	2010	3 334 623	20.4	20.7	20.2	4 406	314	7 969	12 689
	2015	3 546 778	20.1	20.7	19.6	662[1]	149[1]	7 038[1]	7 849[1]
Kenya	2005	756 894[2]	2.1[2]	2.2[2]	2.1[2]	251 271	16 460	481	268 212
Kenya	2010	926 959[2]	2.3[2]	2.3[2]	2.3[2]	402 905	27 966	320 083	750 954
	2015	1 084 357[2]	2.4[2]	2.4[2]	2.4[2]	552 272[1]	40 341[1]	21 231[1]	613 844[1]
Kiribati	2005	2 487	2.7	2.9	2.5	...	...	...	...
Kiribati	2010	2 868	2.8	3.0	2.6	...	...	...	...
	2015	3 153	2.8	3.0	2.6	...	...	...	...
Kuwait	2005	1 333 327[2,3]	58.9[2,3]	69.2[2,3]	44.3[2,3]	1 523	203	101 000	102 726
Koweït	2010	1 871 537[2,3]	61.2[2,3]	76.0[2,3]	42.1[2,3]	184	3 275	93 000	96 459
	2015	2 866 136[2,3]	73.6[2,3]	86.4[2,3]	57.3[2,3]	593[1]	1 040[1]	93 000[1]	94 633[1]
Kyrgyzstan	2005	312 897	6.1	5.1	7.1	2 598	498	100 004	103 100
Kirghizistan	2010	231 511	4.2	3.5	5.0	2 458	554	301 164	304 176
	2015	204 382	3.4	2.8	4.1	433[1]	168[1]	13 678[1]	14 279[1]
Lao People's Dem. Rep. [2,3]	2005	20 371	0.4	0.4	0.3	...	...	...	...
Rép. dém. pop. lao [2,3]	2010	21 185	0.3	0.4	0.3	...	...	...	...
	2015	22 244	0.3	0.4	0.3	...	...	...	...
Latvia	2005	376 725	16.9	15.2	18.4	11	9	418 638	418 658
Lettonie	2010	313 786	15.0	13.1	16.7	68	53	326 906	327 027
	2015	263 126	13.4	11.4	15.0	195[1,24]	171[1,24]	262 802[1,24]	263 168[1,24]
Lebanon	2005	756 784[2]	19.0[2]	19.3[2]	18.7[2]	1 078	1 450	451	2 979
Liban	2010	820 655[2]	18.9[2]	19.3[2]	18.5[2]	8 063	1 417	40	9 520
	2015	1 997 776[2]	34.1[2]	32.7[2]	35.6[2]	1 172 388[1]	10 851[1]	5 813[1]	1 189 052[1]
Lesotho	2005[2,3]	6 290	0.3	0.4	0.3	...	...	...	...
Lesotho	2010[2,3]	6 414	0.3	0.4	0.3	...	...	...	...
	2015	6 572[2,3]	0.3[2,3]	0.3[2,3]	0.3[2,3]	44[1]	1[1]	...	45[1]
Liberia	2005	87 188	2.7	3.1	2.2	10 168	29	498 604	508 801
Libéria	2010	99 129	2.5	2.9	2.1	24 743	28	1 855	26 626
	2015	113 779	2.5	2.9	2.2	38 904[1]	18[1]	1 486[1]	40 408[1]
Libya	2005	625 212[3]	10.8[3]	14.9[3]	6.4[3]	12 166	200	35	12 401
Libye	2010	683 998[3]	10.9[3]	15.3[3]	6.4[3]	7 923	3 194	37	11 154
	2015	771 146[3]	12.3[3]	17.4[3]	7.1[3]	27 948[1]	8 904[1]	434 869[1]	471 721[1]
Liechtenstein	2005	18 898	54.2	...	...	150	60	...	210
Liechtenstein	2010	22 342	61.6	...	...	92	44	6	142
	2015	23 493	62.6	...	...	107[1]	75[1]	2[1]	184[1]

International migrants and refugees *(continued)*
International migrant stock (number and percentage) and refugees and others of concern to UNHCR

Migrants internationaux et réfugiés *(suite)*
Stock de migrants internationaux (nombre et pourcentage) et réfugiés et autres personnes relevant de la compétence du HCR

Region, country or area Région, pays ou zone	Year Année	International Migrant Stock (mid-year) Stock de migrants internationaux (milieu de l'année)				Refugees and others of concern to UNHCR (end-year) Réfugiés et autres personnes relevant de la compétence du HCR (fin d'année)			
		Total total	% of Total Population % de la population totale			Refugees[&] Réfugiés[&]	Asylum seekers Demandeurs d'asile	Other[&&] Autres[&&]	Total population Population totale
		MF	MF	M	F				
Lithuania Lituanie	2005	201 209	6.0	6.1	6.0	531	55	8 709	9 295
	2010	160 772	5.1	4.8	5.4	803	71	3 679	4 553
	2015	136 036	4.7	4.3	5.1	1 055[1]	54[1]	3 583[1]	4 692[1]
Luxembourg Luxembourg	2005	150 618	32.9	33.2	32.6	1 822	...	74	1 896
	2010	163 142	32.1	32.3	32.0	3 254	697	173	4 124
	2015	249 325	44.0	44.3	43.6	1 192[1]	831[1]	81[1]	2 104[1]
Madagascar Madagascar	2005[3]	26 058	0.1	0.2	0.1	...	...	...	...
	2010	28 905[3]	0.1[3]	0.2[3]	0.1[3]	...	...	1	1
	2015	32 075[3]	0.1[3]	0.2[3]	0.1[3]	10[1]	9[1]	...	19[1]
Malawi Malawi	2005	221 661[2]	1.7[2]	1.7[2]	1.8[2]	4 240	5 331	49	9 620
	2010	217 722[2]	1.5[2]	1.4[2]	1.5[2]	5 740	9 362	131	15 233
	2015	215 158[2]	1.2[2]	1.2[2]	1.3[2]	8 963[1]	13 669[1]	...	22 632[1]
Malaysia Malaisie	2005	1 722 344[2,3,25]	6.7[2,3,25]	7.7[2,3,25]	5.7[2,3,25]	33 693	10 838	61 555	106 086
	2010	2 406 011[2,3,25]	8.6[2,3,25]	10.3[2,3,25]	6.8[2,3,25]	81 516	11 339	120 015	212 870
	2015	2 514 243[2,3,25]	8.3[2,3,25]	10.2[2,3,25]	6.4[2,3,25]	97 573[1]	54 400[1]	120 000[1]	271 973[1]
Maldives [2,3] Maldives [2,3]	2005	45 045	14.8	17.7	11.8	...	...	...	...
	2010	73 604	22.1	28.8	15.4	...	...	...	...
	2015	94 086	25.9	36.3	15.4	...	...	...	...
Mali Mali	2005	256 797[2]	2.0[2]	2.0[2]	1.9[2]	11 233	1 833	...	13 066
	2010	336 607[2]	2.2[2]	2.3[2]	2.2[2]	13 558	1 703	...	15 261
	2015	363 145[2]	2.1[2]	2.1[2]	2.0[2]	14 970[1]	386[1]	132 953[1]	148 309[1]
Malta Malte	2005	24 560	6.2	6.1	6.3	1 939	149	...	2 088
	2010	33 084	8.0	8.2	7.9	6 136	1 295	...	7 431
	2015	41 442	9.9	10.5	9.3	6 095[1]	425[1]	...	6 520[1]
Marshall Islands Îles Marshall	2005	2 417	4.6	...	...	...	...	...	...
	2010	3 089	5.9	...	...	...	...	...	...
	2015	3 284	6.2	...	...	...	...	...	...
Martinique Martinique	2005	57 034	14.4	14.0	14.7	...	...	...	...
	2010	59 575	15.1	14.4	15.6	...	...	...	...
	2015	61 731	15.6	14.9	16.2	...	...	...	...
Mauritania Mauritanie	2005	58 119[2,3]	1.8[2,3]	2.1[2,3]	1.6[2,3]	632	92	29 500	30 224
	2010	84 679[2,3]	2.4[2,3]	2.7[2,3]	2.0[2,3]	26 717	241	9	26 967
	2015	138 162[2,3]	3.4[2,3]	3.9[2,3]	2.9[2,3]	76 851[1]	407[1]	...	77 258[1]
Mauritius [3,26] Maurice [3,26]	2005	19 647	1.6	1.5	1.7	...	...	...	...
	2010	24 836	2.0	2.1	1.9	...	...	...	...
	2015	28 585	2.2	2.5	2.0	...	...	...	...
Mayotte Mayotte	2005	63 176	35.5	35.7	35.2	...	...	...	...
	2010	72 757	34.9	33.8	35.9	...	...	...	...
	2015	76 992	32.1	30.2	33.9	...	...	...	...
Mexico Mexique	2005	712 487[2]	0.6[2]	0.7[2]	0.6[2]	3 229	161	...	3 390
	2010	969 538[2]	0.8[2]	0.8[2]	0.8[2]	1 395	172	6	1 573
	2015	1 193 155[2]	0.9[2]	1.0[2]	0.9[2]	2 158[1]	...	68[1]	2 226[1]
Micronesia (Fed. States of) Micronésie (États féd. de)	2005	2 905	2.7	2.9	2.6	...	...	...	...
	2010	2 805	2.7	2.8	2.6	...	...	...	...
	2015	2 756	2.6	2.8	2.5	...	34[1]	...	34[1]
Monaco Monaco	2005	21 312	63.0	...	...	...	...	...	...
	2010	21 132	57.4	...	...	...	1	...	1
	2015	21 042	55.8	...	...	33[1]	...	...	33[1]
Mongolia Mongolie	2005	11 475[3]	0.5[3]	0.6[3]	0.3[3]	...	2	581	583
	2010	16 061[3]	0.6[3]	0.9[3]	0.3[3]	12	1	260	273
	2015	17 620[3]	0.6[3]	0.9[3]	0.3[3]	11[1]	5[1]	16[1]	32[1]
Montenegro Monténégro	2010	78 507	12.6	10.3	14.9	16 364	5	1 886	18 255
	2015	82 541	13.2	10.8	15.5	6 203[1]	7[1]	13 602[1]	19 812[1]
Montserrat Montserrat	2005	1 244	26.0	...	...	...	...	...	...
	2010	1 277	25.8	...	...	...	14	...	14
	2015	1 351	26.4	...	...	...	...	...	...

International migrants and refugees *(continued)*
International migrant stock (number and percentage) and refugees and others of concern to UNHCR

Migrants internationaux et réfugiés *(suite)*
Stock de migrants internationaux (nombre et pourcentage) et réfugiés et autres personnes relevant de la compétence du HCR

Region, country or area Région, pays ou zone	Year Année	International Migrant Stock (mid-year) Stock de migrants internationaux (milieu de l'année)				Refugees and others of concern to UNHCR (end-year) Réfugiés et autres personnes relevant de la compétence du HCR (fin d'année)			
		Total total	% of Total Population % de la population totale			Refugees[&] Réfugiés[&]	Asylum seekers Demandeurs d'asile	Other[&&] Autres[&&]	Total population Population totale
		MF	MF	M	F				
Morocco Maroc	2005	54 379[3]	0.2[3]	0.2[3]	0.2[3]	219	1 843	4	2 066
	2010	70 909[3]	0.2[3]	0.2[3]	0.2[3]	792	280	...	1 072
	2015	88 511[3]	0.3[3]	0.3[3]	0.3[3]	2 144[1]	2 216[1]	...	4 360[1]
Mozambique Mozambique	2005	204 830[2]	1.0[2]	1.0[2]	0.9[2]	1 954	4 015	40	6 009
	2010	214 612[2]	0.9[2]	0.9[2]	0.9[2]	4 077	5 914	1	9 992
	2015	222 928[2]	0.8[2]	0.8[2]	0.8[2]	4 552[1]	14 257[1]	7[1]	18 816[1]
Myanmar Myanmar	2005	83 025[3]	0.2[3]	0.2[3]	0.2[3]	...	...	236 495	236 495
	2010	76 414[3]	0.1[3]	0.2[3]	0.1[3]	...	...	859 403	859 403
	2015	73 308[3]	0.1[3]	0.2[3]	0.1[3]	...	...	1 466 501[1,27]	1 466 501[1,27]
Namibia Namibie	2005	106 274	5.2	5.8	4.7	5 307	1 073	2 752	9 132
	2010	102 405	4.7	5.2	4.2	7 254	1 421	93	8 768
	2015	93 888	3.8	4.2	3.4	1 659[1]	1 100[1]	1 684[1]	4 443[1]
Nauru Nauru	2005[3]	2 253	22.3	...	...	...	...	...	...
	2010[3]	2 112	21.1	...	...	...	...	...	...
	2015	3 178[3]	31.1[3]	...	...	506[1]	816[1]	...	1 322[1]
Nepal Népal	2005	679 457[2]	2.7[2]	1.9[2]	3.5[2]	126 436	1 272	410 929	538 637
	2010	578 657[2]	2.2[2]	1.4[2]	2.8[2]	89 808	938	800 571	891 317
	2015	518 278[2]	1.8[2]	1.2[2]	2.4[2]	36 287[1,28]	57[1,28]	409[1,28]	36 753[1,28]
Netherlands Pays-Bas	2005	1 736 127	10.6	10.4	10.8	118 189	14 664	6 872	139 725
	2010	1 832 510	11.0	10.7	11.3	74 961	13 053	2 095	90 109
	2015	1 979 486	11.7	11.2	12.2	82 494[1,12]	8 097[1,12]	1 951[1,12]	92 542[1,12]
New Caledonia Nouvelle-Calédonie	2005	55 405	24.2	26.0	22.4	...	...	...	...
	2010	61 158	24.8	26.3	23.3	...	...	...	...
	2015	64 290	24.4	26.1	22.8	...	...	...	...
New Zealand Nouvelle-Zélande	2005	839 952	20.3	20.1	20.5	5 307	396	6	5 709
	2010	947 443	21.7	21.4	22.0	2 307	216	1	2 524
	2015	1 039 736	23.0	22.7	23.2	1 349[1,12]	251[1,12]	...	1 600[1,12]
Nicaragua Nicaragua	2005	34 918[2]	0.6[2]	0.7[2]	0.6[2]	227	1	...	228
	2010	37 333[2]	0.7[2]	0.7[2]	0.6[2]	64	12	...	76
	2015	40 262[2]	0.7[2]	0.7[2]	0.6[2]	361[1]	25[1]	5[1]	391[1]
Niger Niger	2005	124 461[2]	0.9[2]	0.9[2]	1.0[2]	301	48	37	386
	2010	126 464[2]	0.8[2]	0.7[2]	0.8[2]	314	18	5	337
	2015	189 255[2]	1.0[2]	0.9[2]	1.0[2]	82 064[1]	122[1]	120 000[1]	202 186[1]
Nigeria Nigéria	2005	648 019[2,3]	0.5[2,3]	0.5[2,3]	0.4[2,3]	9 019	420	3 290	12 729
	2010	920 118[2,3]	0.6[2,3]	0.6[2,3]	0.5[2,3]	8 747	1 815	34	10 596
	2015	1 199 115[2,3]	0.7[2,3]	0.7[2,3]	0.6[2,3]	1 279[1]	909[1]	1 508 017[1]	1 510 205[1]
Niue Nioué	2005	522	31.0	...	...	...	...	...	...
	2010	545	33.6	...	...	...	...	...	...
	2015	557	34.6	...	...	...	...	...	...
Northern Mariana Islands Îles Mariannes du Nord	2005	37 542	58.3	...	...	...	...	...	...
	2010	24 168	44.9	...	...	...	...	...	...
	2015	21 648	39.3	...	...	...	...	...	...
Norway Norvège	2005	361 144[29]	7.8[29]	7.7[29]	7.9[29]	43 034	...	1 143	44 177
	2010	526 799[29]	10.8[29]	11.0[29]	10.6[29]	40 260	12 473	3 150	55 883
	2015	741 813[29]	14.2[29]	14.7[29]	13.7[29]	47 043[1,12]	5 885[1,12]	1 997[1,12]	54 925[1,12]
Oman Oman	2005	666 160[3]	26.6[3]	37.7[3]	12.5[3]	7	4	...	11
	2010	816 221[3]	27.7[3]	37.5[3]	13.3[3]	78	13	...	91
	2015	1 844 978[3]	41.1[3]	50.2[3]	23.1[3]	122[1]	268[1]	...	390[1]
Pakistan Pakistan	2005	3 171 132[2]	2.1[2]	2.2[2]	2.0[2]	1 084 694	3 426	461 123	1 549 243
	2010	3 941 586[2]	2.3[2]	2.4[2]	2.3[2]	1 900 621	2 095	2 248 308	4 151 024
	2015	3 628 956[2]	1.9[2]	1.9[2]	1.9[2]	1 540 854[1]	6 103[1]	1 893 008[1]	3 439 965[1]
Palau Palaos	2005	6 043	30.4	...	...	...	...	...	...
	2010	5 787	28.3	...	...	...	...	...	...
	2015	5 664	26.6	...	...	1[1]	...	...	1[1]

3

International migrants and refugees *(continued)*
International migrant stock (number and percentage) and refugees and others of concern to UNHCR

Migrants internationaux et réfugiés *(suite)*
Stock de migrants internationaux (nombre et pourcentage) et réfugiés et autres personnes relevant de la compétence du HCR

Region, country or area Région, pays ou zone	Year Année	International Migrant Stock (mid-year) Stock de migrants internationaux (milieu de l'année)				Refugees and others of concern to UNHCR (end-year) Réfugiés et autres personnes relevant de la compétence du HCR (fin d'année)			
		Total total	% of Total Population % de la population totale			Refugees[&] Réfugiés[&]	Asylum seekers Demandeurs d'asile	Other[&&] Autres[&&]	Total population Population totale
		MF	MF	M	F				
Panama	2005	117 563	3.5	3.6	3.5	1 730	433	10 273	12 436
Panama	2010	157 309	4.3	4.5	4.2	17 073	479	6	17 558
	2015	184 710	4.7	4.8	4.6	17 303[1]	2 038[1]	2[1]	19 343[1]
Papua New Guinea	2005	29 967[2,3]	0.5[2,3]	0.6[2,3]	0.4[2,3]	9 999	4	135	10 138
Papouasie-Nvl-Guinée	2010	25 424[2,3]	0.4[2,3]	0.5[2,3]	0.3[2,3]	9 698	1	...	9 699
	2015	25 782[2,3]	0.3[2,3]	0.4[2,3]	0.2[2,3]	9 510[1]	400[1]	...	9 910[1]
Paraguay	2005	168 243	2.9	3.0	2.8	50	8	...	58
Paraguay	2010	160 299	2.6	2.6	2.5	107	8	...	115
	2015	156 462	2.4	2.4	2.3	161[1]	39[1]	...	200[1]
Peru	2005	77 541	0.3	0.3	0.3	848	336	1	1 185
Pérou	2010	84 066	0.3	0.3	0.3	1 146	264	1	1 411
	2015	90 881	0.3	0.3	0.3	1 407[1]	366[1]	...	1 773[1]
Philippines	2005	257 468[2,3]	0.3[2,3]	0.3[2,3]	0.3[2,3]	96	42	775	913
Philippines	2010	208 599[2,3]	0.2[2,3]	0.2[2,3]	0.2[2,3]	243	73	139 577	139 893
	2015	211 862[2,3]	0.2[2,3]	0.2[2,3]	0.2[2,3]	254[1]	163[1]	385 746[1]	386 163[1]
Poland	2005	722 509	1.9	1.6	2.1	4 604	1 627	75	6 306
Pologne	2010	642 417	1.7	1.4	1.9	15 555	2 126	763	18 444
	2015	619 403	1.6	1.4	1.8	15 741[1,12]	2 470[1,12]	10 825[1,12]	29 036[1,12]
Portugal	2005	771 184	7.4	7.4	7.3	363	...	...	363
Portugal	2010	762 825	7.2	7.3	7.1	384	72	31	487
	2015	837 257	8.1	7.9	8.2	699[1,12]	641[1,12]	14[1,12]	1 354[1,12]
Puerto Rico	2005	352 144	9.4	9.2	9.5	...	...	...	...
Porto Rico	2010	304 969	8.2	8.0	8.4	...	...	...	...
	2015	274 972	7.5	7.2	7.7	...	...	...	...
Qatar	2005	646 026[3]	77.2[3]	89.3[3]	52.9[3]	46	28	...	74
Qatar	2010	1 456 413[3]	82.5[3]	90.7[3]	57.4[3]	51	16	1 200	1 267
	2015	1 687 640[3]	75.5[3]	87.2[3]	44.4[3]	133[1]	100[1]	1 200[1]	1 433[1]
Republic of Korea	2005	485 546[3]	1.0[3]	1.2[3]	0.8[3]	69	519	...	588
République de Corée	2010	919 275[3]	1.9[3]	2.1[3]	1.6[3]	358	712	179	1 249
	2015	1 327 324[3]	2.6[3]	3.0[3]	2.3[3]	1 313[1]	5 102[1]	200[1]	6 615[1]
Republic of Moldova	2005	173 957[30]	4.2[30]	3.8[30]	4.5[30]	84	148	1 543	1 775
République de Moldova	2010	157 668[30]	3.9[30]	3.1[30]	4.6[30]	148	81	2 032	2 261
	2015	142 904[30]	3.5[30]	2.6[30]	4.4[30]	389[1]	164[1]	6 233[1]	6 786[1]
Réunion	2005	115 076	14.5	15.2	13.9	...	...	...	...
Réunion	2010	123 029	14.8	15.6	14.1	...	...	...	...
	2015	127 209	14.8	15.5	14.1	...	...	...	...
Romania	2005	145 162	0.7	0.6	0.7	2 056	264	400	2 720
Roumanie	2010	155 982	0.8	0.8	0.8	1 021	388	321	1 730
	2015	226 943	1.2	1.3	1.1	2 426[1]	138[1]	294[1]	2 858[1]
Russian Federation	2005	11 667 588	8.1	8.7	7.6	1 523	292	481 282	483 097
Fédération de Russie	2010	11 194 710	7.8	8.3	7.4	4 922	1 463	126 170	132 555
	2015	11 643 276	8.1	8.6	7.7	315 313[1,31]	2 423[1,31]	113 474[1,31]	431 210[1,31]
Rwanda	2005	432 797[2]	4.8[2]	5.1[2]	4.6[2]	45 206	4 301	14 849	64 356
Rwanda	2010	436 787[2]	4.2[2]	4.4[2]	4.1[2]	55 398	290	2	55 690
	2015	441 525[2]	3.8[2]	4.0[2]	3.7[2]	132 743[1]	253[1]	2 557[1]	135 553[1]
Saint Helena [32]	2005	487	11.4	...	...	...	...	...	...
Sainte-Hélène [32]	2010	569	13.6	...	...	...	...	...	...
	2015	604	15.2	...	...	...	...	...	...
Saint Kitts and Nevis	2005	6 682	13.6	...	...	...	...	...	...
Saint-Kitts-et-Nevis	2010	7 245	13.8	...	...	...	...	...	...
	2015	7 443	13.4	...	...	1[1]	...	...	1[1]
Saint Lucia	2005	11 468	6.9	7.1	6.7	...	...	...	...
Sainte-Lucie	2010	12 100	6.8	7.0	6.6	...	6	1	7
	2015	12 771	6.9	7.1	6.7	2[1]	2[1]	60[1]	64[1]

3

International migrants and refugees *(continued)*
International migrant stock (number and percentage) and refugees and others of concern to UNHCR

Migrants internationaux et réfugiés *(suite)*
Stock de migrants internationaux (nombre et pourcentage) et réfugiés et autres personnes relevant de la compétence du HCR

Region, country or area Région, pays ou zone	Year Année	International Migrant Stock (mid-year) Stock de migrants internationaux (milieu de l'année)				Refugees and others of concern to UNHCR (end-year) Réfugiés et autres personnes relevant de la compétence du HCR (fin d'année)			
		Total total	% of Total Population % de la population totale			Refugees[&] Réfugiés[&]	Asylum seekers Demandeurs d'asile	Other[&&] Autres[&&]	Total population Population totale
		MF	MF	M	F				
Saint Pierre and Miquelon	2005	1 147	18.3	...	...	...	...	...	...
Saint-Pierre-et-Miquelon	2010	1 017	16.2	...	...	...	...	...	...
	2015	986	15.7	...	...	...	...	...	...
Saint Vincent-Grenadines	2005	4 395	4.0	3.9	4.2	...	...	...	...
Saint-Vincent-Grenadines	2010	4 485	4.1	4.0	4.2	...	...	...	...
	2015	4 577	4.2	4.1	4.3	...	...	...	...
Samoa	2005	5 746	3.2	3.2	3.2	...	...	...	...
Samoa	2010	5 122	2.8	2.7	2.8	...	...	...	...
	2015	4 929	2.6	2.5	2.6	...	...	...	...
San Marino	2005	4 218	14.4	...	...	...	...	...	...
Saint-Marin	2010	4 399	14.3	...	...	...	...	...	...
	2015	4 717	14.8	...	...	...	...	...	...
Sao Tome and Principe [3]	2005	3 433	2.2	2.3	2.2	...	...	...	...
Sao Tomé-et-Principe [3]	2010	2 700	1.6	1.6	1.6	...	...	...	...
	2015	2 394	1.3	1.3	1.3	...	...	...	...
Saudi Arabia	2005	6 501 819[2,3]	26.3[2,3]	32.4[2,3]	18.6[2,3]	240 701	212	70 084	310 997
Arabie saoudite	2010	8 429 956[2,3]	30.0[2,3]	37.5[2,3]	20.3[2,3]	582	87	70 000	70 669
	2015	10 185 945[2,3]	32.3[2,3]	38.9[2,3]	23.7[2,3]	211[1]	93[1]	70 000[1]	70 304[1]
Senegal	2005	238 298[2]	2.1[2]	2.3[2]	2.0[2]	20 712	2 629	12	23 353
Sénégal	2010	256 092[2]	2.0[2]	2.1[2]	1.8[2]	20 672	2 177	1 401	24 250
	2015	263 242[2]	1.7[2]	1.9[2]	1.6[2]	14 304[1]	2 956[1]	...	17 260[1]
Serbia [33]	2005	845 120	9.2	8.3	10.0	148 264	33	338 641	486 938
Serbie [33]	2010	826 066	9.1	8.2	10.0	73 608	209	238 810	312 627
	2015	807 441	9.1	8.2	10.0	35 309[1]	464[1]	223 949[1]	259 722[1]
Seychelles	2005	8 997	10.1	13.1	7.2	...	...	...	...
Seychelles	2010	11 420	12.3	16.4	7.9	...	...	...	...
	2015	12 791	13.3	18.3	8.1	...	...	...	...
Sierra Leone	2005	149 615[2]	3.0[2]	3.2[2]	2.7[2]	59 965	177	6 202	66 344
Sierra Leone	2010	97 452[2]	1.7[2]	1.9[2]	1.5[2]	8 363	210	38	8 611
	2015	91 213[2]	1.4[2]	1.6[2]	1.3[2]	1 371[1]	16[1]	...	1 387[1]
Singapore	2005	1 710 594	38.1	34.0	42.0	3	1	...	4
Singapour	2010	2 164 794	42.6	38.2	47.0	7	...	...	7
	2015	2 543 638	45.4	40.6	50.0	...	...	1[1]	1[1]
Sint Maarten (Dutch part)	2005	13 100	40.3	...	...	...	...	...	...
Saint-Martin (partie	2010	26 200	79.1	...	...	1	3	...	4
néerlandaise)	2015	27 295	70.4	...	...	3[1]	5[1]	...	8[1]
Slovakia	2005	130 491	2.4	2.2	2.6	368	2 707	...	3 075
Slovaquie	2010	146 319	2.7	2.6	2.8	461	267	911	1 639
	2015	177 190	3.3	3.2	3.3	799[1,12]	61[1,12]	1 671[1,12]	2 531[1,12]
Slovenia	2005	197 276	9.9	11.0	8.8	251	185	715	1 151
Slovénie	2010	253 786	12.4	13.9	10.9	314	121	4 119	4 554
	2015	235 966	11.4	13.1	9.7	283[1]	43[1]	4[1]	330[1]
Solomon Islands	2005	3 271	0.7	0.8	0.6	...	...	...	...
Îles Salomon	2010	2 760	0.5	0.6	0.5	...	...	...	...
	2015	2 585	0.4	0.5	0.4	3[1]	...	...	3[1]
Somalia	2005	* 20 670[2]	* 0.2[2]	* 0.3[2]	* 0.2[2]	493	98	400 000	400 591
Somalie	2010	* 23 995[2]	* 0.3[2]	* 0.3[2]	* 0.2[2]	1 937	24 111	1 463 780	1 489 828
	2015	* 25 291[2]	* 0.2[2]	* 0.3[2]	* 0.2[2]	3 582[1]	9 320[1]	1 152 073[1]	1 164 975[1]
South Africa	2005	1 210 936[2]	2.5[2]	3.0[2]	2.0[2]	29 714	140 095	100	169 909
Afrique du Sud	2010	1 943 099[2]	3.8[2]	4.5[2]	3.1[2]	57 899	171 702	134	229 735
	2015	3 142 511[2]	5.8[2]	7.0[2]	4.6[2]	114 512[1]	798 080[1]	...	912 592[1]
South Sudan	2010[2]	257 905	2.6	2.6	2.5	...	...	...	...
Soudan du sud	2015	824 122[2]	6.7[2]	6.8[2]	6.5[2]	265 887[1,34]	632[1,34]	1 792 014[1,34]	2 058 533[1,34]
Spain	2005	4 107 226[35]	9.4[35]	9.9[35]	8.8[35]	5 374	...	27	5 401
Espagne	2010	6 280 065[35]	13.5[35]	14.1[35]	12.9[35]	3 820	2 712	31	6 563
	2015	5 852 953[35]	12.7[35]	12.6[35]	12.8[35]	5 798[1,12]	11 020[1,12]	440[1,12]	17 258[1,12]

International migrants and refugees *(continued)*
International migrant stock (number and percentage) and refugees and others of concern to UNHCR

Migrants internationaux et réfugiés *(suite)*
Stock de migrants internationaux (nombre et pourcentage) et réfugiés et autres personnes relevant de la compétence du HCR

Region, country or area Région, pays ou zone	Year Année	International Migrant Stock (mid-year) Stock de migrants internationaux (milieu de l'année)				Refugees and others of concern to UNHCR (end-year) Réfugiés et autres personnes relevant de la compétence du HCR (fin d'année)			
		Total total	% of Total Population % de la population totale			Refugees[&] Réfugiés[&]	Asylum seekers Demandeurs d'asile	Other[&&] Autres[&&]	Total population Population totale
		MF	MF	M	F				
Sri Lanka Sri Lanka	2005	39 526[2]	0.2[2]	0.2[2]	0.2[2]	106	121	351 884	352 111
	2010	38 959[2]	0.2[2]	0.2[2]	0.2[2]	223	138	434 903	435 264
	2015	38 706[2]	0.2[2]	0.2[2]	0.2[2]	848[1,36]	461[1,36]	50 499[1,36]	51 808[1,36]
State of Palestine État de Palestine	2005[37,38]	266 617	7.4	6.6	8.3	...	...	...	...
	2010[37,38]	258 032	6.3	5.6	7.1	...	...	...	...
	2015	255 507[37,38]	5.5[37,38]	4.8[37,38]	6.2[37,38]	...	...	3[1]	3[1]
Sudan Soudan	2005	541 994[2,39]	1.7[2,39]	1.7[2,39]	1.7[2,39]	147 256	4 425	878 067	1 029 748
	2010	578 363[2]	1.6[2]	1.6[2]	1.6[2]	178 308	6 046	1 767 167	1 951 521
	2015	503 477[2]	1.3[2]	1.3[2]	1.2[2]	356 191[1,40]	11 448[1,40]	2 400 270[1,40]	2 767 909[1,40]
Suriname Suriname	2005[3]	33 664	6.8	7.4	6.3	...	...	...	...
	2010	39 713[3]	7.7[3]	8.3[3]	7.0[3]	1	7	...	8
	2015	46 836[3]	8.6[3]	9.4[3]	7.8[3]	1[1]	...	...	1[1]
Swaziland Swaziland	2005	27 097[2]	2.5[2]	2.7[2]	2.2[2]	760	256	...	1 016
	2010	30 476[2]	2.6[2]	2.7[2]	2.4[2]	759	...	...	759
	2015	31 579[2]	2.5[2]	2.6[2]	2.3[2]	539[1]	321[1]	4[1]	864[1]
Sweden Suède	2005	1 125 790	12.5	12.0	12.9	74 915	15 702	5 785	96 402
	2010	1 384 929	14.8	14.4	15.1	82 629	18 635	9 545	110 809
	2015	1 639 771	16.8	16.6	17.0	142 207[1,12]	56 135[1,12]	27 167[1,12]	225 509[1,12]
Switzerland Suisse	2005	1 805 437	24.4	25.4	23.3	48 030	14 428	990	63 448
	2010	2 075 182	26.5	24.3	28.6	48 813	12 916	127	61 856
	2015	2 438 702	29.4	29.1	29.7	69 390[1]	17 085[1]	76[1]	86 551[1]
Syrian Arab Republic Rép. arabe syrienne	2005	876 410[2,3]	4.8[2,3]	4.8[2,3]	4.8[2,3]	26 089	1 898	300 001	327 988
	2010	1 661 922[2,3]	8.0[2,3]	8.1[2,3]	7.9[2,3]	1 005 472	2 446	300 194	1 308 112
	2015	875 189[2,3]	4.7[2,3]	4.8[2,3]	4.7[2,3]	149 200[1,41]	4 839[1,41]	7 792 500[1,41]	7 946 539[1,41]
Tajikistan Tadjikistan	2005	280 444	4.1	3.5	4.7	1 018	22	25	1 065
	2010	278 152	3.7	3.2	4.2	3 131	1 656	2 338	7 125
	2015	275 059	3.2	2.8	3.7	1 782[1,42]	79[1,42]	10 082[1,42]	11 943[1,42]
Thailand Thaïlande	2005	2 163 447[2]	3.3[2]	3.3[2]	3.2[2]	117 053	32 163	136	149 352
	2010	3 224 131[2]	4.8[2]	4.9[2]	4.7[2]	96 675	10 250	542 505	649 430
	2015	3 913 258[2]	5.8[2]	5.9[2]	5.6[2]	110 372[1,43]	8 166[1,43]	506 718[1,43]	625 256[1,43]
TFYR of Macedonia ex-R.Y. de Macédoine	2005	127 667	6.2	5.2	7.3	1 274	723	2 451	4 448
	2010	129 701	6.3	5.3	7.3	1 398	161	1 731	3 290
	2015	130 730	6.3	5.3	7.3	828[1]	43[1]	717[1]	1 588[1]
Timor-Leste Timor-Leste	2005	11 286	1.1	1.1	1.2	3	10	...	13
	2010	10 983	1.0	1.2	0.9	1	4	...	5
	2015	10 834	0.9	1.1	0.8	...	...	5[1]	5[1]
Togo Togo	2005	203 379[2,3]	3.6[2,3]	3.8[2,3]	3.5[2,3]	9 287	420	9 012	18 719
	2010	255 262[2,3]	4.0[2,3]	4.1[2,3]	3.9[2,3]	14 051	151	5	14 207
	2015	276 844[2,3]	3.8[2,3]	3.9[2,3]	3.7[2,3]	21 877[1]	687[1]	...	22 564[1]
Tokelau Tokélaou	2005	258	21.3	...	...	...	...	...	...
	2010	429	37.8	...	...	...	...	...	...
	2015	487	39.0	...	...	...	...	...	...
Tonga Tonga	2005	4 301	4.3	4.6	3.9	...	...	...	...
	2010	5 022	4.8	5.2	4.4	...	3	...	3
	2015	5 731	5.4	5.9	4.9	...	...	...	...
Trinidad and Tobago Trinité-et-Tobago	2005	44 812	3.5	3.2	3.7	...	...	...	...
	2010	48 226	3.6	3.5	3.7	29	102	...	131
	2015	49 883	3.7	3.6	3.7	121[1]	59[1]	...	180[1]
Tunisia Tunisie	2005	35 040[3]	0.3[3]	0.4[3]	0.3[3]	87	26	...	113
	2010	43 172[3]	0.4[3]	0.4[3]	0.4[3]	89	23	3	115
	2015	56 701[3]	0.5[3]	0.5[3]	0.5[3]	824[1]	156[1]	3[1]	983[1]
Turkey Turquie	2005	1 319 236[2]	1.9[2]	1.9[2]	2.0[2]	2 399	4 872	1 434	8 705
	2010	1 367 034[2]	1.9[2]	1.8[2]	2.0[2]	10 032	6 715	1 086	17 833
	2015	2 964 916[2]	3.8[2]	3.9[2]	3.6[2]	1 838 848[1,44]	145 335[1,44]	1 086[1,44]	1 985 269[1,44]

3

International migrants and refugees *(continued)*
International migrant stock (number and percentage) and refugees and others of concern to UNHCR

Migrants internationaux et réfugiés *(suite)*
Stock de migrants internationaux (nombre et pourcentage) et réfugiés et autres personnes relevant de la compétence du HCR

Region, country or area / Région, pays ou zone	Year / Année	International Migrant Stock (mid-year) / Stock de migrants internationaux (milieu de l'année)				Refugees and others of concern to UNHCR (end-year) / Réfugiés et autres personnes relevant de la compétence du HCR (fin d'année)			
		Total / total	% of Total Population / % de la population totale			Refugees[&] / Réfugiés[&]	Asylum seekers / Demandeurs d'asile	Other[&&] / Autres[&&]	Total population / Population totale
		MF	MF	M	F				
Turkmenistan / Turkménistan	2005	213 051	4.5	4.0	4.9	11 963	2	45	12 010
	2010	197 979	3.9	3.6	4.2	62	...	20 000	20 062
	2015	196 386	3.7	3.5	3.8	27[1]	...	7 144[1]	7 171[1]
Turks and Caicos Islands / Îles Turques-et-Caïques	2005	9 945	37.6	...	...	...	...	...	...
	2010	10 875	35.1	...	...	...	...	...	...
	2015	11 688	34.0	...	...	4[1]	4[1]	...	8[1]
Tuvalu [3] / Tuvalu [3]	2005	183	1.9	...	...	...	...	...	...
	2010	154	1.6	...	...	...	...	...	...
	2015	141	1.4	...	...	...	...	...	...
Uganda / Ouganda	2005	652 968[2]	2.3[2]	2.3[2]	2.3[2]	257 256	1 809	1 636	260 701
	2010	529 160[2]	1.6[2]	1.6[2]	1.6[2]	135 801	20 804	437 775	594 380
	2015	749 471[2]	1.9[2]	1.9[2]	1.9[2]	428 397[1]	38 068[1]	180 000[1]	646 465[1]
Ukraine / Ukraine	2005	5 050 302	10.8	10.0	11.5	2 346	1 618	72 896	76 860
	2010	4 818 767	10.6	9.8	11.2	3 022	2 981	40 357	46 360
	2015	4 834 898	10.8	10.0	11.4	3 232[1]	6 169[1]	1 417 179[1]	1 426 580[1]
United Arab Emirates / Émirats arabes unis	2005	3 281 036[2,3]	73.2[2,3]	75.0[2,3]	68.9[2,3]	104	79	...	183
	2010	7 316 611[2,3]	87.8[2,3]	87.7[2,3]	88.3[2,3]	538	86	...	624
	2015	8 095 126[2,3]	88.4[2,3]	90.1[2,3]	83.7[2,3]	424[1]	378[1]	...	802[1]
United Kingdom / Royaume-Uni	2005	5 926 156	9.8	9.6	10.1	303 181	12 500	909	316 590
	2010	7 604 583	12.1	11.9	12.3	238 150	14 880	229	253 259
	2015	8 543 120	13.2	12.7	13.7	117 234[1,12]	37 829[1,12]	16[1,12]	155 079[1,12]
United Rep. of Tanzania / Rép.-Unie de Tanzanie	2005	770 846[2]	2.0[2]	2.5[2]	1.4[2]	548 824	307	81 519	630 650
	2010	308 600[2]	0.7[2]	0.7[2]	0.7[2]	109 286	1 247	163 269	273 802
	2015	261 222[2]	0.5[2]	0.5[2]	0.5[2]	159 014[1]	1 150[1]	168 019[1]	328 183[1]
United States / États-Unis	2005	39 258 293	13.3	13.4	13.1	379 340	169 743	76	549 159
	2010	44 183 643	14.3	14.2	14.4	264 569	6 285	89	270 943
	2015	46 627 102	14.5	14.3	14.7	267 222[1,45]	224 508[1,45]	...	491 730[1,45]
United States Virgin Is. / Îles Vierges américaines	2005	56 647	52.6	51.8	53.3	...	...	...	...
	2010	56 684	53.3	52.6	53.9	...	...	...	...
	2015	56 721	53.4	52.8	53.9	...	...	...	...
Uruguay / Uruguay	2005	82 318	2.5	2.3	2.6	121	9	...	130
	2010	76 263	2.3	2.1	2.4	189	40	...	229
	2015	71 799	2.1	2.0	2.2	289[1]	68[1]	...	357[1]
Uzbekistan / Ouzbékistan	2005	1 329 345	5.1	4.6	5.6	43 950	587	5	44 542
	2010	1 220 149	4.4	4.1	4.7	311	...	2	313
	2015	1 170 899	3.9	3.7	4.1	118[1,46]	...	86 703[1,46]	86 821[1,46]
Vanuatu / Vanuatu	2005	2 800	1.3	1.3	1.4	...	...	...	...
	2010	2 991	1.3	1.2	1.3	4	...	...	4
	2015	3 187	1.2	1.2	1.2	...	1[1]	...	1[1]
Venezuela (Boliv. Rep. of) / Venezuela (Rép. boliv. du)	2005	1 070 562	4.0	4.0	4.0	408	5 912	200 001	206 321
	2010	1 331 488	4.6	4.6	4.6	201 547	15 859	...	217 406
	2015	1 404 448	4.5	4.5	4.5	174 191[1]	704[1]	...	174 895[1]
Viet Nam / Viet Nam	2005	51 768[2,3]	0.1[2,3]	0.1[2,3]	<0.0[2,3]	2 357	...	15 000	17 357
	2010	61 756[2,3]	0.1[2,3]	0.1[2,3]	0.1[2,3]	1 928	...	10 200	12 128
	2015	72 793[2,3]	0.1[2,3]	0.1[2,3]	0.1[2,3]	...	...	11 000[1]	11 000[1]
Wallis and Futuna Islands / Îles Wallis-et-Futuna	2005	2 365	16.6	...	...	...	...	...	...
	2010	2 776	20.5	...	...	...	...	...	...
	2015	2 849	21.7	...	...	...	...	...	...
Western Sahara * / Sahara occidental *	2005	3 891	0.9	1.0	0.8	...	...	...	...
	2010	4 493	0.9	1.0	0.8	...	...	...	...
	2015	5 179	0.9	1.0	0.8	...	...	...	...
Yemen / Yémen	2005	171 073[2,3]	0.8[2,3]	1.0[2,3]	0.7[2,3]	81 937	798	48	82 783
	2010	285 837[2,3]	1.2[2,3]	1.3[2,3]	1.1[2,3]	190 092	2 557	315 929	508 578
	2015	344 131[2,3]	1.3[2,3]	1.3[2,3]	1.2[2,3]	263 047[1]	9 902[1]	1 267 590[1]	1 540 539[1]

International migrants and refugees *(continued)*
International migrant stock (number and percentage) and refugees and others of concern to UNHCR

Migrants internationaux et réfugiés *(suite)*
Stock de migrants internationaux (nombre et pourcentage) et réfugiés et autres personnes relevant de la compétence du HCR

Region, country or area Région, pays ou zone	Year Année	International Migrant Stock (mid-year) Stock de migrants internationaux (milieu de l'année)				Refugees and others of concern to UNHCR (end-year) Réfugiés et autres personnes relevant de la compétence du HCR (fin d'année)			
		Total total	% of Total Population % de la population totale			Refugees[&] Réfugiés[&]	Asylum seekers Demandeurs d'asile	Other[&&] Autres[&&]	Total population Population totale
		MF	MF	M	F				
Zambia Zambie	2005	252 749[2]	2.1[2]	2.1[2]	2.1[2]	155 718	146	29 833	185 697
	2010	149 637[2]	1.1[2]	1.1[2]	1.1[2]	47 857	325	9 687	57 869
	2015	127 915[2]	0.8[2]	0.8[2]	0.8[2]	25 737[1]	2 606[1]	23 415[1]	51 758[1]
Zimbabwe Zimbabwe	2005	392 693[2]	3.0[2]	3.5[2]	2.6[2]	13 850	118	7	13 975
	2010	397 891[2]	2.8[2]	3.3[2]	2.4[2]	4 435	416	…	4 851
	2015	398 866[2]	2.6[2]	3.0[2]	2.2[2]	* 6 085[1]	* 123[1]	* 301 883[1]	* 308 091[1]

Source:

United Nations Population Division, New York, International migrant stock: The 2015 Revision, last accessed June 2016.
United Nations High Commissioner for Refugees (UNHCR), Geneva, UNHCR Population Statistics Database, last accessed June 2016.

& Number of refugees or persons in refugee-like situations as reported by the Office of the United Nations High Commissioner for Refugees (UNHCR).
&& Figure includes sum of returned refugees, internally displaced persons (IDPs) protected/assisted by UNHCR, including people in IDP like situations, returned IDPs, Persons under UNHCR's stateless mandate, and others of concern to UNHCR categories.

Source:

Organisation des Nations Unies, Division pour la population, New York, Stock de migrants internationaux: La révision 2015, derniér accéss juin 2016.
L'Office du Haut Commisariat des Nations Unies pour les réfugiés (UNHCR), Genève, base de données d'UNHCR, dernier accès juin 2016.

& Nombre de réfugiés ou de personnes en situation analogue à celle de réfugiés, tel que donné par le Bureau des Nations Unies.
&& Les chiffres correspondent au total des réfugiés rapatriés, des personnes déplacées protégées ou assistées par le HCR, notamment celles se trouvant dans une situation analogue à celle des personnes déplacées, des personnes déplacées de retour, des apatrides relevant du mandat du HCR et d'autres catégories de personnes relevant de la compétence du HCR.

1	Refers to mid-year.
2	Including refugees.
3	Refers to foreign citizens.
4	According to the Government of Algeria, there are an estimated 165,000 Sahrawi refugees in the Tindouf camps.
5	Including Christmas Island, Cocos (Keeling) Islands and Norfolk Island.
6	Asylum-seekers are based on the number of applications lodged for protection visas. Refugee figure refers to the end of 2014.
7	All figures relate to the end of 2014.
8	Including Nagorno-Karabakh.
9	The refugee population includes 200,000 persons originating from Myanmar in a refugee-like situation. The Government of Bangladesh estimates the population to be between 300,000 and 500,000.
10	The estimates for 2005 refer to the former Netherlands Antilles.
11	Bonaire only.
12	Refugee population relates to the end of 2014.
13	Refers to Guernsey and Jersey.
14	For statistical purposes, the data for China do not include those for the Hong Kong Special Administrative Region (Hong Kong SAR), Macao Special Administrative Region (Macao SAR) and Taiwan Province of China.
15	The 300,000 Vietnamese refugees are well integrated and in practice receive protection from the Government of China.
16	Including northern Cyprus.

1	Données en milieu d'année.
2	Y compris les réfugiés.
3	Se rapportent aux citoyens étrangers.
4	Selon le Gouvernement algérien, les camps de Tindouf accueillent environ 165 000 réfugiés sahraouis.
5	Y compris les îles Christmas, Cocos (Keeling) et Norfolk.
6	Les chiffres de l'Australie concernant les demandeurs d'asile sont établis sur la base du nombre de demandes de visa de protection présentées. Les chiffres relatifs aux réfugiés se rapportent à fin 2014.
7	Tous les chiffres se rapportent à fin 2014.
8	Y compris le Haut-Karabakh.
9	Au nombre des réfugiés figurent 200 000 personnes originaires du Myanmar se trouvant dans une situation analogue à celle des réfugiés. Le Gouvernement bangladais estime que le nombre de réfugiés se situe entre 300 000 et 500 000.
10	Les estimations pour 2005 se rapportent aux Antilles néerlandaises.
11	Bonaire seulement.
12	Le nombre de réfugiés se rapporte à fin 2014.
13	Se rapporte à Guernsey et Jersey.
14	Pour la présentation des statistiques, les données pour la Chine ne comprennent pas la Région Administrative Spéciale de Hong Kong (Hong Kong RAS), la Région Administrative Spéciale de Macao (Macao RAS) et la province de Taiwan.
15	Les 300 000 réfugiés vietnamiens sont bien intégrés et, dans la pratique, reçoivent la protection du Gouvernement chinois.
16	Y compris la partie nord de Chypre.

3

International migrants and refugees *(continued)*
International migrant stock (number and percentage) and refugees and others of concern to UNHCR

Migrants internationaux et réfugiés *(suite)*
Stock de migrants internationaux (nombre et pourcentage) et réfugiés et autres personnes relevant de la compétence du HCR

17	Revised estimate includes only individuals born in the country where both parents were born abroad. This estimate does not include subsequent generations of individuals of foreign descent as such it does not include all persons without nationality.	17	Cette réestimation concerne uniquement les personnes nées dans le pays dont les deux parents sont nés à l'étranger. Cette estimation ne comprend pas les générations futures de personnes d'origine étrangère tels que les apatrides.
18	Almost all people recorded as being stateless have permanent residence and enjoy more rights than foreseen in the 1954 Convention relating to the Status of Stateless Persons.	18	La quasi totalité des personnes enregistrées comme apatrides ont une résidence permanente et jouissent de davantage de droits que ceux prévus par la Convention de 1954 relative au statut des apatrides.
19	Including Åland Islands.	19	Y compris les Îles d'Åland.
20	Including Abkhazia and South Ossetia.	20	Y compris l'Abkhazie et l'Ossétie du Sud.
21	Including Saint Barthélemy and Saint Martin (French part).	21	Y compris Saint-Barthélémy et Saint-Martin (partie français).
22	Including an estimate for stateless persons populations in line with Law 26 of 2006, which allows stateless persons to apply for nationality in certain circumstances.	22	Y compris une estimation des populations apatrides conforme à la loi 26 de 2006,qui permet aux apatrides de demander la nationalité dans certaines circonstances.
23	Includes 32,800 Iraqi refugees registered with UNHCR in Jordan. The Government estimated the number of Iraqis at 400,000 individuals at the end of March 2015. This included refugees and other categories of Iraqis.	23	Y compris 32 800 réfugiés iraquiens enregistrés par le HCR en Jordanie. À la fin de mars 2015, le Gouvernement estimait le nombre d'Iraquiens à 400 000 (réfugiés et autres).
24	The figure of stateless persons includes persons covered by two separate Laws; Law on Stateless Persons dated 17 February 2004 and the Law on the Status of Those Former USSR Citizens who are not Citizens of Latvia or of Any Other State.	24	Le nombres d'apatrides comprend les personnes couvertes par deux lois distinctes: La loi sur les apatrides du 17 février 2004 et la loi relative au Statut des citoyens de l'ex-URSS qui ne sont pas citoyens de la Lettonie ou d'un autre État.
25	Including Sabah and Sarawak.	25	Y compris Sabah et Sarawak.
26	Including Agalega, Rodrigues and Saint Brandon.	26	Y compris Agalega, Rodrigues et Saint Brandon.
27	Stateless persons population refers to persons without citizenship in Rakhine State only.	27	Les apatrides s'entendent des personnes dépourvues de nationalité, dans l'État de Rakhine uniquement.
28	Various studies estimate that a large number of individuals lack citizenship certificates in Nepal. While these individuals are not all necessarily stateless, UNHCR has been working closely with the Government of Nepal and partners to address this situation.	28	Selon différentes études, un grand nombre de personnes ne disposeraient pas de certificat de nationalité au Népal. Elles ne sont pas nécessairement toutes apatrides, mais le HCR travaille en coopération étroite avec les autorités népalaises et des partenaires pour régler la situation.
29	Including Svalbard and Jan Mayen Islands.	29	Y compris îles Svalbard et Jan Mayen.
30	Including Transnistria.	30	Y compris la Transnistrie.
31	Stateless persons refers to census figure from 2010 adjusted to reflect the number of people who acquired nationality in 2011-2014.	31	Le nombre d'apatrides se rapporte aux données du recensement effectué en 2010, ajustées pour tenir compte du nombre de personnes ayant acquis la nationalité entre 2011 et 2014.
32	Including Ascension and Tristan da Cunha.	32	Y compris Ascension et Tristan da Cunha.
33	Including Kosovo.	33	Y compris Kosovo.
34	IDP figure in South Sudan includes 105,000 people who are in an IDP-like situation.	34	Au nombre des déplacés au Soudan du Sud figurent 105 000 personnes dans une situation analogue à celle des déplacés.
35	Including Canary Islands, Ceuta and Melilla.	35	Y compris les Iles Canaries, Ceuta et Melilla.
36	The statistics of the remaining IDPs as at mid-2015, while provided by the Government authorities at the district level, are being reviewed by the central authorities. Once this review has been concluded, the statistics will be changed accordingly.	36	Les statistiques relatives au reste des déplacés à la mi-2015 ont été fournies par les autorités locales au niveau des districts, mais examinées par les autorités centrales, et seront ajustées en fonction des résultats de l'examen.
37	Including East Jerusalem.	37	Y compris Jérusalem-Est.
38	Refugees are not part of the foreign-born migrant stock in the State of Palestine.	38	Les réfugiés ne sont pas comptabilisés parmi les migrants nés à l'étranger qui se trouvent dans l'État de Palestine.
39	The estimates for 2005 refer to Sudan and South Sudan.	39	Les estimations de 2005 se rapportent au Soudan et au Soudan du Sud.
40	IDP figure in Sudan includes 77,300 people who are in an IDP-like situation.	40	Au nombre des déplacés au Soudan figurent 77 300 personnes dans une situation analogue à celles des déplacés.
41	Refugee figure for Iraqis in the Syrian Arab Republic is a Government estimate. UNHCR has registered and is assisting 23,500 Iraqis at mid-2015.	41	Le nombre de réfugiés iraquiens en République arabe syrienne est une estimation du Gouvernement. À la mi-2015, 23 500 Iraquiens avaient été enregistrés par le HCR et bénéficiaient de son aide.
42	Figure refers to a registration exercise in three regions and 637 persons registered as stateless by the Ministry of Internal Affairs of Tajikistan.	42	Les chiffres se rapportent aux enregistrements effectués dans trois régions; 637 personnes ont été enregistrées comme apatrides par le Ministère de l'intérieur tadjik.
43	Figure of stateless persons in Thailand refers to 2011.	43	Les chiffres relatifs au nombre d'apatrides en Thaïlande remontent à 2011.
44	Refugee figure for Syrians in Turkey is a Government estimate.	44	Le nombre de réfugiés syriens en Turquie est une estimation du Gouvernement.
45	The refugee figure is under review. Refugee figure relates to the end of 2014.	45	Le chiffre des réfugiés est en cours d'étude et se rapporte à la fin de l'année 2014.
46	Figure of stateless persons refers to those with permanent residence reported in 2010 by the Government. Information on other categories of stateless persons is not available.	46	Le nombre d'apatrides renvoie au nombre de résidents permanents recensés en 2010 par le Gouvernement. On ne dispose d'aucune information sur d'autres catégories d'apatrides.

Proportion of seats held by women in national parliament
Percentage, as of January/February each year

Proportion de sièges occupés par les femmes au parlement national
Pourcentage, données disponibles en janvier/février de chaque année

Country or area[&]	Last Election date Dernière date de l'élection	1990	2000	2005	2010	2012	2013	2014	2015	2016	Pays ou zone[&]
Afghanistan	2010-09	3.7	...	...	27.3	27.7	27.7	27.7	27.7	27.7	Afghanistan
Albania	2013-06	28.8	5.2	6.4	16.4	15.7	15.7	20.0	20.7	20.7	Albanie
Algeria	2012-05	2.4	3.2	6.2	7.7	8.0	31.6	31.6	31.6	31.6	Algérie
Andorra	2015-03	...	7.1	14.3	35.7	50.0	50.0	50.0	50.0	39.3	Andorre
Angola	2012-08	14.5	15.5	15.0	38.6	38.2	34.1	36.8	36.8	36.8	Angola
Antigua and Barbuda	2014-06	0.0	...	10.5	10.5	10.5	10.5	10.5	11.1	11.1	Antigua-et-Barbuda
Argentina	2015-10	6.3	28.0	33.7	38.5	37.4	37.4	36.6	36.2	35.8	Argentine
Armenia	2012-05	35.6	3.1	5.3	9.2	8.4	10.7	10.7	10.7	10.7	Arménie
Australia	2013-09	6.1	22.4	24.7	27.3	24.7	24.7	26.0	26.7	26.7	Australie
Austria	2013-09	11.5	26.8	33.9	27.9	27.9	27.9	32.2	30.6	30.6	Autriche
Azerbaijan	2015-11	...	12.0	10.5	11.4	16.0	16.0	15.6	15.6	16.9	Azerbaïdjan
Bahamas	2012-05	4.1	15.0	20.0	12.2	12.2	13.2	13.2	13.2	13.2	Bahamas
Bahrain	2014-11	...	...	0.0	2.5	10.0	10.0	10.0	7.5	7.5	Bahreïn
Bangladesh	2014-01	10.3	9.1	2.0	18.6	19.7	19.7	6.4	20.0	20.0	Bangladesh
Barbados	2013-02	3.7	10.7	13.3	10.0	10.0	10.0	16.7	16.7	16.7	Barbade
Belarus	2012-09	...	4.5	29.4	31.8	31.8	26.6	26.6	27.3	27.3	Bélarus
Belgium	2014-05	8.5	23.3	34.7	38.0	38.0	38.0	41.3	39.3	39.3	Belgique
Belize	2012-03	0.0	6.9	6.7	0.0	0.0	3.1	3.1	3.1	3.1	Belize
Benin	2015-04	2.9	6.0	7.2	10.8	8.4	8.4	8.4	8.4	7.2	Bénin
Bhutan	2013-07	2.0	2.0	9.3	8.5	8.5	8.5	8.5	8.5	8.5	Bhoutan
Bolivia (Plurin. State of)	2014-10	9.2	11.5	19.2	22.3	25.4	25.4	25.4	53.1	53.1	Bolivie (État plurin. de)
Bosnia and Herzegovina	2014-10	...	28.6	16.7	19.0	21.4	21.4	21.4	21.4	21.4	Bosnie-Herzégovine
Botswana	2014-10	5.0	...	11.1	7.9	7.9	7.9	9.5	9.5	9.5	Botswana
Brazil	2014-10	5.3	5.7	8.6	8.8	8.6	8.6	8.6	9.0	9.9	Brésil
Bulgaria	2014-10	21.0	10.8	26.3	20.8	20.8	22.9	24.6	20.4	20.4	Bulgarie
Burkina Faso	2015-11	...	8.1	11.7	15.3	15.3	15.7	18.9	13.3	9.4	Burkina Faso
Burundi	2015-06	...	6.0	18.4	31.4	30.5	30.5	30.5	30.5	36.4	Burundi
Cabo Verde	2011-02	12.0	11.1	11.1	18.1	20.8	20.8	20.8	20.8	20.8	Cabo Verde
Cambodia	2013-07	...	8.2	9.8	21.1	20.3	20.3	20.3	20.3	20.3	Cambodge
Cameroon	2013-09	14.4	5.6	8.9	13.9	13.9	13.9	31.1	31.1	31.1	Cameroun
Canada	2015-10	13.3	20.6	21.1	22.1	24.8	24.7	25.1	25.2	26.0	Canada
Central African Rep.	2011-01	3.8	7.3	...	9.6	12.5	12.5	...	...	...	Rép. centrafricaine
Chad	2011-02	...	2.4	6.5	5.2	12.8	14.9	14.9	14.9	14.9	Tchad
Chile	2013-11	...	10.8	12.5	14.2	14.2	14.2	15.8	15.8	15.8	Chili
China	2013-03	21.3	21.8	20.2	21.3	21.3	21.3	23.4	23.6	23.6	Chine
Colombia	2014-03	4.5	11.8	12.0	8.4	12.1	12.1	12.1	19.9	19.9	Colombie
Comoros	2015-01	0.0	...	3.0	3.0	3.0	3.0	3.0	...	3.0	Comores
Congo	2012-07	14.3	12.0	8.5	7.3	7.3	7.4	7.4	7.4	7.4	Congo
Costa Rica	2014-02	10.5	19.3	35.1	36.8	38.6	38.6	38.6	33.3	33.3	Costa Rica
Côte d'Ivoire	2011-12	5.7	...	8.5	8.9	11.0	10.4	9.4	9.2	9.2	Côte d'Ivoire
Croatia	2015-11	...	...	21.7	23.5	23.8	23.8	23.8	25.8	15.2	Croatie
Cuba	2013-02	33.9	27.6	36.0	43.2	45.2	45.2	48.9	48.9	48.9	Cuba
Cyprus	2011-05	1.8	5.4	16.1	12.5	10.7	10.7	12.5	12.5	12.5	Chypre
Czech Republic	2013-10	...	15.0	17.0	15.5	22.0	22.0	19.5	19.0	20.0	République tchèque
Dem. P. R. Korea	2014-03	21.1	20.1	20.1	15.6	15.6	15.6	15.6	16.3	16.3	R. p. dém. de Corée
Dem. Rep. of the Congo	2011-11	5.4	...	12.0	8.4	8.9	8.9	10.6	8.9	8.9	Rép. dém. du Congo
Denmark	2015-06	30.7	37.4	38.0	38.0	39.1	39.1	39.1	38.0	37.4	Danemark
Djibouti	2013-02	0.0	0.0	10.8	13.8	13.8	13.8	12.7	12.7	12.7	Djibouti
Dominica	2014-12	10.0	9.4	19.4	14.3	12.5	12.5	12.9	21.9	21.9	Dominique
Dominican Republic	2010-05	7.5	16.1	17.3	19.7	20.8	20.8	20.8	20.8	20.8	Rép. dominicaine
Ecuador	2013-02	4.5	17.4	16.0	32.3	32.3	32.3	41.6	41.6	41.6	Équateur
Egypt	2015-10	3.9	2.0	2.9	1.8	2.0	2.0	...	...	14.9	Égypte
El Salvador	2015-03	11.7	16.7	10.7	19.0	19.0	26.2	26.2	27.4	32.1	El Salvador
Equatorial Guinea	2013-05	13.3	5.0	18.0	10.0	10.0	10.0	24.0	24.0	24.0	Guinée équatoriale
Eritrea	1994-02	...	14.7	22.0	22.0	22.0	22.0	22.0	22.0	22.0	Érythrée
Estonia	2015-03	...	17.8	18.8	22.8	19.8	20.8	19.0	19.8	23.8	Estonie
Ethiopia	2015-05	...	2.0	7.7	21.9	27.8	27.8	27.8	27.8	38.8	Éthiopie
Fiji	2014-09	...	11.3	8.5	...	...	...	...	14.0	16.0	Fidji
Finland	2015-04	31.5	37.0	37.5	40.0	42.5	42.5	42.5	42.5	41.5	Finlande
France	2012-06	6.9	10.9	12.2	18.9	18.9	26.9	26.2	26.2	26.2	France

4

Proportion of seats held by women in national parliament *(continued)*
Percentage, as of January/February each year

Proportion de sièges occupés par les femmes au parlement national *(suite)*
Pourcentage, données disponibles en janvier/février de chaque année

Country or area[&]	Last Election date Dernière date de l'élection	1990	2000	2005	2010	2012	2013	2014	2015	2016	Pays ou zone[&]
Gabon	2011-12	13.3	8.3	9.2	14.7	15.8	15.8	15.0	14.2	14.2	Gabon
Gambia	2012-03	7.8	2.0	13.2	7.5	7.5	7.5	9.4	9.4	9.4	Gambie
Georgia	2012-10	...	7.2	9.4	5.1	6.6	12.0	12.0	11.3	11.3	Géorgie
Germany	2013-09	...	30.9	32.8	32.8	32.9	32.9	36.5	36.5	36.5	Allemagne
Ghana	2012-12	...	9.0	10.9	8.3	8.3	10.3	10.9	10.9	10.9	Ghana
Greece	2015-09	6.7	6.3	14.0	17.3	18.7	21.0	21.0	23.0	19.7	Grèce
Grenada	2013-02	...	...	26.7	13.3	13.3	13.3	33.3	33.3	33.3	Grenade
Guatemala	2015-09	7.0	7.1	8.2	12.0	13.3	13.3	13.3	13.3	13.9	Guatemala
Guinea	2013-09	...	8.8	19.3	...	...	...	21.9	21.9	21.9	Guinée
Guinea-Bissau	2014-04	20.0	...	14.0	10.0	10.0	14.0	11.0	13.7	13.7	Guinée-Bissau
Guyana	2015-05	36.9	18.5	30.8	30.0	31.3	31.3	31.3	31.3	30.4	Guyana
Haiti	2015-08	...	3.6	3.6	4.1	4.2	4.2	4.2	4.2	0.0	Haïti
Honduras	2013-11	10.2	9.4	5.5	18.0	19.5	19.5	25.8	25.8	25.8	Honduras
Hungary	2014-04	20.7	8.3	9.1	11.1	8.8	8.8	9.4	10.1	10.1	Hongrie
Iceland	2013-04	20.6	34.9	30.2	42.9	39.7	39.7	39.7	41.3	41.3	Islande
India	2014-04	5.0	9.0	8.3	10.8	11.0	11.0	11.4	12.0	12.0	Inde
Indonesia	2014-04	12.4	...	11.3	18.0	18.2	18.6	18.6	17.1	17.1	Indonésie
Iran (Islamic Rep. of)	2012-05	1.5	4.9	4.1	2.8	2.8	3.1	3.1	3.1	3.1	Iran (Rép. islamique d')
Iraq	2014-04	10.8	6.4	...	25.5	25.2	25.2	25.2	26.5	26.5	Iraq
Ireland	2011-02	7.8	12.0	13.3	13.9	15.1	15.1	15.7	16.3	16.3	Irlande
Israel	2015-03	6.7	11.7	15.0	19.2	20.0	21.7	22.5	22.5	26.7	Israël
Italy	2013-02	12.9	11.1	11.5	21.3	21.6	21.4	31.4	31.0	31.0	Italie
Jamaica	2011-12	5.0	13.3	11.7	13.3	12.7	12.7	12.7	12.7	12.7	Jamaïque
Japan	2014-12	1.4	4.6	7.1	11.3	10.8	7.9	8.1	9.5	9.5	Japon
Jordan	2013-01	0.0	0.0	5.5	6.4	10.8	12.0	12.0	12.0	12.0	Jordanie
Kazakhstan	2012-01	...	10.4	10.4	17.8	24.3	24.3	25.2	26.2	26.2	Kazakhstan
Kenya	2013-03	1.1	3.6	7.1	9.8	9.8	9.8	19.1	19.7	19.7	Kenya
Kiribati	2015-12	0.0	4.9	4.8	4.3	8.7	8.7	8.7	8.7	6.5	Kiribati
Kuwait	2013-07	...	0.0	0.0	7.7	7.7	6.2	4.6	1.5	1.5	Koweït
Kyrgyzstan	2015-10	...	1.4	10.0	25.6	23.3	23.3	23.3	23.3	19.2	Kirghizistan
Lao People's Dem. Rep.	2011-04	6.3	21.2	22.9	25.2	25.0	25.0	25.0	25.0	25.0	Rép. dém. pop. lao
Latvia	2014-10	...	17.0	21.0	22.0	23.0	23.0	25.0	18.0	18.0	Lettonie
Lebanon	2009-06	0.0	2.3	2.3	3.1	3.1	3.1	3.1	3.1	3.1	Liban
Lesotho	2015-02	...	3.8	11.7	24.2	24.2	26.7	26.7	26.7	25.0	Lesotho
Liberia	2011-10	...	...	5.3	12.5	9.6	11.0	11.0	11.0	11.0	Libéria
Libya	2014-06	...	...	...	7.7	...	16.5	16.5	16.0	16.0	Libye
Liechtenstein	2013-02	4.0	4.0	12.0	24.0	24.0	24.0	20.0	20.0	20.0	Liechtenstein
Lithuania	2012-10	...	17.5	22.0	19.1	19.1	24.5	24.1	23.4	23.4	Lituanie
Luxembourg	2013-10	13.3	16.7	23.3	20.0	25.0	21.7	28.3	28.3	28.3	Luxembourg
Madagascar	2013-12	6.5	8.0	6.9	...	17.5	17.5	23.1	20.5	20.5	Madagascar
Malawi	2014-05	9.8	8.3	14.0	20.8	22.3	22.3	22.3	16.7	16.7	Malawi
Malaysia	2013-05	5.1	...	9.1	9.9	10.4	10.4	10.4	10.4	10.4	Malaisie
Maldives	2014-03	6.3	...	12.0	6.5	6.5	6.5	6.8	5.9	5.9	Maldives
Mali	2013-11	...	12.2	10.2	10.2	10.2	10.2	9.5	9.5	8.8	Mali
Malta	2013-03	2.9	9.2	9.2	8.7	8.7	8.7	14.3	12.9	12.9	Malte
Marshall Islands	2015-11	...	...	3.0	3.0	3.0	3.0	3.0	3.0	9.1	Îles Marshall
Mauritania	2013-11	...	3.8	3.7	22.1	22.1	22.1	25.2	25.2	25.2	Mauritanie
Mauritius	2014-12	7.1	7.6	5.7	17.1	18.8	18.8	18.8	11.6	11.6	Maurice
Mexico	2015-06	12.0	18.2	22.6	27.6	26.2	36.8	37.4	38.0	42.4	Mexique
Micronesia (Fed. States of)	2015-03	...	0.0	0.0	0.0	0.0	0.0	0.0	0.0	0.0	Micronésie (États féd. de)
Monaco	2013-02	11.1	22.2	20.8	26.1	19.0	19.0	20.8	20.8	20.8	Monaco
Mongolia	2012-06	24.9	7.9	6.8	3.9	3.9	14.9	14.9	14.9	14.5	Mongolie
Montenegro	2012-10	...	...	...	11.1	12.3	17.3	14.8	17.3	17.3	Monténégro
Morocco	2011-11	0.0	0.6	10.8	10.5	17.0	17.0	17.0	17.0	17.0	Maroc
Mozambique	2014-10	15.7	...	34.8	39.2	39.2	39.2	39.2	39.6	39.6	Mozambique
Myanmar	2015-11	...	...	...	...	3.5	6.0	5.6	6.2	9.9	Myanmar
Namibia	2014-11	6.9	22.2	25.0	26.9[1]	24.4	24.4	25.6	41.3	41.3	Namibie
Nauru	2013-06	5.6	0.0	0.0	0.0	0.0	0.0	5.3	5.3	5.3	Nauru
Nepal	2013-11	6.1	5.9	...	33.2	33.2	33.2	29.9	29.5	29.5	Népal
Netherlands	2012-09	21.3	36.0	36.7	42.0	40.7	38.7	38.7	37.3	37.3	Pays-Bas
New Zealand	2014-09	14.4	29.2	28.3	33.6	32.2	32.2	33.9	31.4	31.4	Nouvelle-Zélande

4

Proportion of seats held by women in national parliament *(continued)*
Percentage, as of January/February each year

Proportion de sièges occupés par les femmes au parlement national *(suite)*
Pourcentage, données disponibles en janvier/février de chaque année

Country or area[&]	Last Election date Dernière date de l'élection	1990	2000	2005	2010	2012	2013	2014	2015	2016	Pays ou zone[&]
Nicaragua	2011-11	14.8	9.7	20.7	20.7	40.2	40.2	40.2	39.1	41.3	Nicaragua
Niger	2011-01	5.4	1.2	12.4	9.7	13.3	13.3	13.3	13.3	13.3	Niger
Nigeria	2015-03	...	...	4.7	7.0	6.8	6.7	6.7	6.7	5.6	Nigéria
Norway	2013-09	35.8	36.4	38.2	39.6	39.6	39.6	39.6	39.6	39.6	Norvège
Oman	2015-10	...	...	2.4	0.0	1.2	1.2	1.2	1.2	1.2	Oman
Pakistan	2013-05	10.1	...	21.3	22.2	22.5	22.5	20.7	20.7	20.6	Pakistan
Palau	2012-11	...	0.0	0.0	0.0	0.0	0.0	0.0	0.0	0.0	Palaos
Panama	2014-05	7.5	...	16.7	8.5	8.5	8.5	8.5	19.3	18.3	Panama
Papua New Guinea	2012-06	0.0	1.8	0.9	0.9	0.9	2.7	2.7	2.7	2.7	Papouasie-Nvl-Guinée
Paraguay	2013-04	5.6	2.5	10.0	12.5	12.5	12.5	15.0	15.0	15.0	Paraguay
Peru	2011-04	5.6	10.8	18.3	27.5	21.5	21.5	22.3	22.3	22.3	Pérou
Philippines	2013-05	9.1	12.4	15.3	21.0	22.9	22.9	27.3	27.2	27.2	Philippines
Poland	2015-10	13.5	13.0	20.2	20.0	23.7	23.7	24.3	24.1	27.4	Pologne
Portugal	2015-10	7.6	18.7	19.1	27.4	28.7	28.7	31.3	31.3	34.8	Portugal
Qatar	2013-07	...	...	...	0.0	0.0	0.0	0.0	0.0	0.0	Qatar
Republic of Korea	2012-04	2.0	3.7	13.0	14.7	14.7	15.7	15.7	16.3	16.3	République de Corée
Republic of Moldova	2014-11	...	8.9	15.8	23.8	19.8	19.8	18.8	20.8	21.8	République de Moldova
Romania	2012-12	34.4	7.3	11.4	11.4	11.2	13.3	13.5	13.7	13.7	Roumanie
Russian Federation	2011-12	...	7.7	9.8	14.0	13.6	13.6	13.6	13.6	13.6	Fédération de Russie
Rwanda	2013-09	17.1	17.1	48.8	56.3	56.3	56.3	63.8	63.8	63.8	Rwanda
Saint Kitts and Nevis	2015-02	6.7	13.3	0.0	6.7	6.7	6.7	6.7	6.7	13.3	Saint-Kitts-et-Nevis
Saint Lucia	2011-11	0.0	11.1	11.1	11.1	16.7	16.7	16.7	16.7	16.7	Sainte-Lucie
Saint Vincent-Grenadines	2015-12	9.5	4.8	22.7	21.7	17.4	17.4	13.0	13.0	13.0	Saint-Vincent-Grenadines
Samoa	2011-03	0.0	8.2	6.1	8.2	4.1	4.1	4.1	6.1	6.1	Samoa
San Marino	2012-11	11.7	13.3	16.7	16.7	18.3	16.7	18.3	16.7	16.7	Saint-Marin
Sao Tome and Principe	2014-10	11.8	9.1	9.1	7.3	18.2	18.2	18.2	18.2	18.2	Sao Tomé-et-Principe
Saudi Arabia	2013-01	...	...	0.0	0.0	0.0	19.9	19.9	19.9	19.9	Arabie saoudite
Senegal	2012-07	12.5	12.1	19.2	22.7	22.7	42.7	43.3	42.7	42.7	Sénégal
Serbia	2014-03	...	...	...	21.6	22.0	33.2	33.6	34.0	34.0	Serbie
Serbia and Monten. [former]		...	5.1	7.9	...	...	...	...	...	...	Serbie-et-Monténégro [anc.]
Seychelles	2011-09	16.0	23.5	29.4	23.5	43.8	43.8	43.8	43.8	43.8	Seychelles
Sierra Leone	2012-11	...	8.8	14.5	13.2	12.9	12.4	12.1	12.4	12.4	Sierra Leone
Singapore	2015-09	4.9	4.3	16.0	23.4	22.2	24.2	25.3	25.3	23.1	Singapour
Slovakia	2012-03	...	12.7	16.7	18.0	16.0	18.7	18.7	18.7	18.7	Slovaquie
Slovenia	2014-07	...	7.8	12.2	14.4	32.2	32.2	33.3	36.7	36.7	Slovénie
Solomon Islands	2014-11	0.0	2.0	0.0	0.0	0.0	2.0	2.0	2.0	2.0	Îles Salomon
Somalia	2012-08	4.0	...	...	6.9	6.8	13.8	13.8	13.8	13.8	Somalie
South Africa	2014-05	2.8	30.0	32.8	44.5	42.3	42.3	44.8	41.5	42.0	Afrique du Sud
South Sudan	2011-08	...	...	...	...	26.5	26.5	26.5	26.5	26.5	Soudan du sud
Spain	2015-12	14.6	21.6	36.0	36.6	36.0	36.0	39.7	41.1	40.0	Espagne
Sri Lanka	2015-08	4.9	4.9	4.9	5.8	5.8	5.8	5.8	5.8	5.8	Sri Lanka
Sudan	2015-04	...	...	...	...	24.6	24.6	24.3	24.3	30.5	Soudan
Sudan [former]		...	...	9.7	18.9	...	...	...	...	...	Soudan [anc.]
Suriname	2015-05	7.8	15.7	19.6	25.5	11.8	11.8	11.8	11.8	25.5	Suriname
Swaziland	2013-09	3.6	3.1	10.8	13.6	13.6	13.6	6.2	6.2	6.2	Swaziland
Sweden	2014-09	38.4	42.7	45.3	46.4	44.7	44.7	45.0	43.6	43.6	Suède
Switzerland	2015-10	14.0	22.5	25.0	29.0	28.5	29.0	31.0	30.5	32.0	Suisse
Syrian Arab Republic	2012-05	9.2	10.4	12.0	12.4	12.4	12.0	12.0	12.4	12.4	Rép. arabe syrienne
Tajikistan	2015-03	...	2.8	12.7	17.5	19.0	19.0	15.9	16.9	19.0	Tadjikistan
Thailand	2014-08	2.8	5.6	8.8	13.3	15.8	15.8	15.8	6.1	6.1	Thaïlande
TFYR of Macedonia	2014-04	...	7.5	19.2	32.5	30.9	32.5	34.1	33.3	33.3	ex-R.Y. de Macédoine
Timor-Leste	2012-07	...	...	25.3	29.2	32.3	38.5	38.5	38.5	38.5	Timor-Leste
Togo	2013-07	5.2	...	6.2	11.1	11.1	11.1	16.5	17.6	17.6	Togo
Tonga	2014-11	0.0	...	0.0	3.1	3.6	3.6	3.6	0.0	0.0	Tonga
Trinidad and Tobago	2015-09	16.7	11.1	19.4	26.8	28.6	28.6	28.6	28.6	31.0	Trinité-et-Tobago
Tunisia	2014-10	4.3	11.5	22.8	27.6	26.7	26.7	28.1	31.3	31.3	Tunisie
Turkey	2015-11	1.3	4.2	4.4	9.1	14.2	14.2	14.4	14.4	14.9	Turquie
Turkmenistan	2013-12	26.0	26.0	...	16.8	16.8	16.8	26.4	25.8	25.8	Turkménistan
Tuvalu	2015-03	7.7	0.0	0.0	0.0	0.0	6.7	6.7	6.7	6.7	Tuvalu
Uganda	2011-02	12.2	17.9	23.9	31.5	35.0	35.0	35.0	35.0	35.0	Ouganda
Ukraine	2014-10	...	7.8	5.3	8.0	8.0	9.4	9.7	11.8	12.1	Ukraine

Proportion de sièges occupés par les femmes au parlement national *(suite)*
Pourcentage, données disponibles en janvier/février de chaque année

Country or area[&]	Last Election date Dernière date de l'élection	1990	2000	2005	2010	2012	2013	2014	2015	2016	Pays ou zone[&]
United Arab Emirates	2011-09	0.0	0.0	0.0	22.5	17.5	17.5	17.5	17.5	22.5	Émirats arabes unis
United Kingdom	2015-05	6.3	18.4	18.1	19.5	22.3	22.5	22.6	22.8	29.4	Royaume-Uni
United Rep. of Tanzania	2015-10	...	16.4	21.4	30.7	36.0	36.0	36.0	36.0	36.6	Rép.-Unie de Tanzanie
United States	2014-11	6.6	13.3	14.9	16.8	16.8	17.8	18.3	19.4	19.4	États-Unis
Uruguay	2014-10	6.1	12.1	12.1	14.1	12.1	12.1	13.1	13.1	16.2	Uruguay
Uzbekistan	2014-12	...	6.8	17.5	22.0	22.0	22.0	22.0	16.0	16.0	Ouzbékistan
Vanuatu	2016-01	4.3	0.0	3.8	3.8	1.9	0.0	0.0	0.0	0.0	Vanuatu
Venezuela (Boliv. Rep. of)	2015-12	10.0	12.1	9.7	17.5	17.0	17.0	17.0	17.0	14.4	Venezuela (Rép. boliv. du)
Viet Nam	2011-05	17.7	26.0	27.3	25.8	24.4	24.4	24.3	24.3	24.3	Viet Nam
Yemen	2003-04	4.1	0.7	0.3	0.3	0.3	0.3	0.3	0.3	0.0	Yémen
Zambia	2011-09	6.6	10.1	12.0	14.0	11.5	11.5	10.8	12.7	12.7	Zambie
Zimbabwe	2013-07	11.0	14.0	10.0	15.0	15.0	15.0	31.5	31.5	31.5	Zimbabwe

Source:

Inter-Parliamentary Union (IPU), Geneva, Women in National Parliament dataset and the Millennium Development Goals Indicators database, last accessed April 2016.

Source:

Union interparlementaire, Genève, données des femmes dans les parlements et la base de données des Objectifs du Millénaire pour le développement, dernier accès avril 2016.

& The data are as at 1 February for 2013 – 2016, as at 31 January for 2005, 2010 and 2012, and as at 25 January for 2000.

& Les données sont au 1er février en 2013 – 2016, au 31 janvier en 2005, 2010 et 2012, et au 25 janvier en 2000.

1 Figure excludes 11 members yet to be sworn in.

1 Ce chiffre ne tient pas compte de 11 membres qui n'avaient pas encore été assermentés.

Education at the primary, secondary and tertiary levels

Number of students enrolled (thousands) and gross enrolment ratio by sex

Enseignement primaire, secondaire et supérieur

Nombre d'élèves inscrits (milliers) et taux brut de scolarisation par sexe

Region, country or area Région, pays ou zone	Year [t] Année [t]	Primary education Enseignement primaire			Secondary education Enseignement secondaire			Tertiary education Enseignement supérieur		
		Total (000)	Gross enrolment ratio Taux brut de scolarisation		Total (000)	Gross enrolment ratio Taux brut de scolarisation		Total (000)	Gross enrolment ratio Taux brut de scolarisation	
			M	F		M	F		M	F
Total, all countries or areas	2005	678 999	104.6	99.7	509 073	65.6	62.1	139 293	23.7	24.8
Total, tous pays ou zones	2010	697 216	106.7	103.8	546 230	72.2	69.7	181 531	28.3	30.4
	*2014	719 059	105.0	104.1	568 019	75.6	74.5	207 516	32.8	36.2
Africa	2005	136 433	99.5	89.1	49 218	42.6	36.1	8 611	10.7	8.6
Afrique	2010	159 501	101.3	94.7	63 020	49.0	42.4	*11 453	*12.5	*10.5
	*2014	179 138	102.9	96.1	72 670	51.1	45.4	13 011	13.1	11.2
America, North	2005	52 783	103.9	102.3	43 698	84.7	88.1	22 881	47.8	64.3
Amérique du Nord	2010	53 217	104.9	103.6	45 193	86.0	89.2	27 279	54.6	73.6
	2014	52 937	104.2	103.2	46 727	90.0	93.1	26 812	52.4	68.3
America, South	2005	43 066	124.6	120.4	42 241	89.5	97.0	11 813	30.5	37.3
Amérique du Sud	2010	40 569	117.4	112.9	42 796	91.9	99.8	16 336	39.9	53.2
	*2014	39 172	109.9	106.8	44 313	96.5	103.4	18 372	45.5	61.3
Asia	2005	*405 137	*105.0	*101.3	305 614	*62.7	*57.9	62 612	18.9	16.3
Asie	2010	403 584	108.5	107.3	333 624	71.7	68.8	91 201	24.3	22.5
	*2014	405 269	105.8	108.0	342 889	75.7	75.2	116 833	31.1	32.1
Europe	2005	38 438	102.6	101.9	64 839	97.8	97.3	32 082	55.3	70.9
Europe	2010	36 749	103.4	102.8	58 232	101.5	100.2	33 688	60.1	77.3
	2014	38 199	103.0	103.0	57 917	109.4	108.7	30 741	63.6	77.8
Oceania	2005	3 142	91.9	90.0	3 463	110.2	107.0	1 295	45.6	58.6
Océanie	2010	3 597	101.8	99.1	3 365	102.3	98.2	1 574	48.9	67.1
	2014	4 344	*110.3	*106.5	3 504	*104.1	*98.5	1 748	*51.9	*72.0
Afghanistan	2004	4 430	153.4	66.7	594	30.2	6.3	28	1.9	0.5
Afghanistan	2005	4 319	130.4	76.5	651	29.3	9.6	...	...	...
	2009	4 946	122.3	81.0	1 716	62.4	30.1	95	6.2	1.5
	2010	5 279	124.6	85.3	2 044	70.3	35.1	...	...	...
	2014	6 218	130.7	91.8	2 603	70.7	39.7	263	13.3	3.7
Albania	2005	238	101.9	101.0	407	78.7	75.0	63	19.1	27.3
Albanie	2010	225	100.2	97.7	356	88.8	88.0	122	37.5	52.4
	2014	196	113.7	111.2	333	100.0	92.6	174	52.4	73.8
Algeria	2005	4 362	111.9	103.4	3 654	75.3	82.5	792	18.3	23.3
Algérie	2010	3 312	119.4	111.3	4 616	95.6	98.9	1 144	24.5	35.3
	2011	3 363	120.8	112.6	4 573	98.1	101.7	1 189	25.5	37.0
	2014	3 765	122.3	115.1	...	...	...	1 245	27.5	41.9
American Samoa Samoa américaines	2007	...	...	...	...	...	...	2	...	...
Andorra	2005	4	...	...	4	...	...	<0	...	...
Andorre	2008	4	...	...	4	...	...	<0	...	...
	2010	4	...	...	4	...	...	...	...	...
	2014	4	...	...	4	...	...	<0	...	...
Angola	2002	...	...	...	*462	*18.4	*15.2	13	*1.0	*0.6
Angola	2005	...	...	...	...	...	...	48	...	...
	2006	...	...	...	...	...	...	49	...	...
	2010	4 273	125.2	101.8	850	34.2	23.4	...	...	...
	2011	5 027	156.9	100.4	885	35.1	22.7	143	10.2	3.8
	2013	...	...	...	...	...	...	219	11.0	8.8
Anguilla	2005	1	...	...	1	...	...	<0	...	...
Anguilla	2008	2	...	...	1	...	...	<0	...	...
	2010	2	...	...	1	...	...	...	...	...
	2011	2	...	...	1	...	...	...	...	...
Antigua and Barbuda	2000	13	...	...	5	*82.8	*73.6	...	...	...
Antigua-et-Barbuda	2010	11	107.9	98.8	8	105.7	106.7	1	9.1	22.7
	2012	10	101.3	94.5	8	97.9	112.9	2	15.1	31.1
	2014	10	100.9	93.3	8	101.8	102.8	...	...	...

5

Education at the primary, secondary and tertiary levels *(continued)*
Number of students enrolled (thousands) and gross enrolment ratio by sex

Enseignement primaire, secondaire et supérieur *(suite)*
Nombre d'élèves inscrits (milliers) et taux brut de scolarisation par sexe

Region, country or area Région, pays ou zone	Year[t] Année[t]	Primary education Enseignement primaire			Secondary education Enseignement secondaire			Tertiary education Enseignement supérieur		
		Total (000)	Gross enrolment ratio Taux brut de scolarisation		Total (000)	Gross enrolment ratio Taux brut de scolarisation		Total (000)	Gross enrolment ratio Taux brut de scolarisation	
			M	F		M	F		M	F
Argentina	2005	4 873	117.7	116.6	3 884	89.8	98.4	2 083	52.1	75.8
Argentine	2010	4 947	117.5	115.9	4 213	97.1	106.6	2 521	59.1	89.2
	2013	4 792	111.1	110.0	4 406	102.5	110.3	2 768	61.5	99.0
Armenia	2000	180	98.7	98.3	409	87.0	94.5	93	30.7	38.7
Arménie	2005	125	91.0	98.2	370	...	...	115	30.5	46.7
	2009	115	95.8	111.2	313	89.9	105.1	146	39.3	60.7
	2010	102	...	...	306	...	...	146	40.3	63.1
	2014	143	...	...	245	...	...	113	40.1	48.5
Aruba	2005	10	115.0	107.8	7	95.7	96.1	2	25.9	37.3
Aruba	2010	10	114.5	113.0	7	93.5	98.3	2	31.1	43.9
	2012	9	102.8	105.4	8	110.3	112.1	3	30.5	45.6
	2014	10	119.1	115.4	...	...	...	1	10.4	23.8
Australia	2005	1 935	101.6	103.2	2 497	150.1	146.5	1 025	64.2	80.8
Australie	2010	2 015	105.8	105.6	2 282	134.4	130.3	1 276	68.4	94.5
	2013	2 128	106.9	106.2	2 384	141.2	133.7	1 390	72.5	101.7
	2014	2 169	...	...	2 371	...	...	1 454	...	...
Austria	2005	363	101.1	100.9	781	102.9	98.0	244	43.5	52.2
Autriche	2010	328	100.5	99.1	744	100.8	96.8	350	63.4	74.2
	2014	327	102.8	101.7	697	101.6	96.9	421	72.8	87.6
Azerbaijan	2005	568	*98.2	*92.8	...	...	...	...	...	...
Azerbaïdjan	2010	482	*94.2	*93.2	1 063	*99.5	*98.0	181	*19.4	*19.2
	2014	518	*106.6	*105.5	949	*103.1	*102.4	195	*21.7	*24.8
Bahamas	2005	37	109.9	108.4	32	89.5	89.1	...	...	...
Bahamas	2010	34	106.9	108.9	34	90.2	95.1	...	...	...
Bahrain	2005	83	...	...	72	96.0	104.3	19	13.5	44.3
Bahreïn	2006	90	...	...	74	97.5	101.4	18	12.7	42.9
	2010	91	...	...	80	...	...	...	...	...
	2014	104	...	...	90	...	...	38	24.2	56.5
Bangladesh	2005	16 219	96.5	100.9	10 109	44.1	47.0	912	8.2	4.2
Bangladesh	2009	*16 539	*97.5	*102.2	10 907	46.7	50.4	1 582	13.1	7.9
	2010	*16 987	*99.6	*105.9	11 334	47.3	53.1	...	...	...
	2011	*18 432	*108.7	*115.1	11 543	47.3	54.5	2 008	15.7	10.9
	2013	...	...	...	13 314	56.1	60.7	...	...	...
	2014	...	...	...	...	...	...	2 068	15.4	11.4
Barbados	2001	24	99.9	98.7	21	103.3	103.7	8	23.1	57.0
Barbade	2005	22	96.3	95.9	21	109.1	108.0	...	...	...
	2010	*23	*98.9	*100.6	*19	*101.7	*102.2	13	43.9	95.9
	2011	*23	*98.4	*98.5	20	101.7	105.5	12	40.3	90.6
	2014	21	93.1	94.2	21	107.9	110.7	...	...	...
Belarus	2005	380	98.9	95.6	...	...	...	529	56.8	77.8
Bélarus	2010	358	103.5	103.4	763	108.7	105.3	569	*65.2	*94.4
	2014	369	98.9	99.1	649	107.9	106.1	518	76.5	102.0
Belgium	2005	739	101.4	101.0	815	110.1	106.5	390	55.2	67.7
Belgique	2010	732	103.3	103.2	806	107.3	104.2	445	59.9	75.8
	2014	774	104.8	104.9	1 210	154.7	175.4	496	63.6	83.4
Belize	2005	50	115.8	110.8	29	75.3	77.7	5	12.5	19.7
Belize	2010	53	114.8	110.3	33	72.5	78.2	7	16.8	26.9
	2013	53	114.6	109.4	37	78.5	81.8	8	18.3	30.0
	2014	52	114.4	109.0	37	78.1	81.7	...	...	...
Benin	2001	1 055	106.3	72.0	*257	*32.1	*14.8	28	*7.3	*1.7
Bénin	2005	1 318	111.3	85.8	*435	*44.8	*24.4	42	...	...
	2010	1 788	123.9	108.6	...	...	...	114	19.6	6.9
	2013	2 064	130.5	118.1	869	65.4	42.9	145	22.4	8.4
	2014	2 133	131.1	119.9	897	64.8	43.9	...	...	...

5

Education at the primary, secondary and tertiary levels *(continued)*
Number of students enrolled (thousands) and gross enrolment ratio by sex
Enseignement primaire, secondaire et supérieur *(suite)*
Nombre d'élèves inscrits (milliers) et taux brut de scolarisation par sexe

Region, country or area Région, pays ou zone	Year[t] Année[t]	Primary education Enseignement primaire			Secondary education Enseignement secondaire			Tertiary education Enseignement supérieur		
		Total (000)	Gross enrolment ratio Taux brut de scolarisation		Total (000)	Gross enrolment ratio Taux brut de scolarisation		Total (000)	Gross enrolment ratio Taux brut de scolarisation	
			M	F		M	F		M	F
Bermuda	2005	5	96.8	101.5	5	76.0	85.9	...	...	...
Bermudes	2006	5	93.7	100.7	5	74.4	81.0	1	18.7	41.6
	2010	4	...	...	4	72.3	85.4	1	19.3	40.6
	2014	4	88.8	86.1	4	68.0	77.4	1	19.1	36.1
Bhutan	2005	99	95.7	93.4	42	48.5	43.0	4	5.6	3.7
Bhoutan	2010	110	108.2	110.7	60	65.6	66.7	5	8.6	5.3
	2013	104	104.3	105.4	69	74.6	79.9	9	12.6	9.2
	2014	102	101.4	102.9	74	81.4	87.1	...	...	...
Bolivia (Plurinational State of)	2003	1 532	120.0	118.2	1 049	92.0	88.4	338	...	...
Bolivie (État plurinational de)	2004	1 504	116.2	114.9	...	...	...	*346	...	...
	2007	1 512	113.2	112.1	1 052	87.7	84.4	*353	*41.8	*34.9
	2008	1 508	112.3	110.7	1 060	86.8	84.3	...	...	...
	2010	1 429	...	...	1 058	83.6	82.6	...	...	...
	2013	1 349	...	...	1 113	85.0	84.4	...	...	...
Bosnia and Herzegovina	2005	...	...	...	...	...	...	84	23.5	31.3
Bosnie-Herzégovine	2010	175	...	...	323	...	...	105	31.5	42.5
	2014	161	100.0	100.4	297	87.4	90.0	112	40.9	54.7
Botswana	2005	329	107.9	106.5	173	77.3	79.7	20	9.7	8.6
Botswana	2006	330	108.7	106.5	177	78.4	82.7	22	11.0	9.0
	2007	328	108.1	105.3	178	79.1	83.6	22	...	...
	2009	331	109.3	105.7	...	...	...	48	...	...
	2010	...	...	...	...	...	...	42	...	...
	2013	340	110.1	107.0	184	81.6	86.2	55	22.3	27.8
	2014	...	...	...	...	...	...	61	23.3	31.8
Brazil	2005	18 661	*137.2	*129.2	24 863	*96.5	*106.2	4 572	*22.8	*29.3
Brésil	2009	17 452	*133.9	*128.2	23 617	*91.5	*102.1	6 115	*31.9	*42.2
	2010	16 893	...	...	23 539	...	...	6 553	...	...
	2013	16 761	*112.4	*107.0	24 881	*97.8	*106.1	7 541	*39.6	*53.4
British Virgin Islands	2004	3	...	...	2	...	...	1	...	...
Îles Vierges britanniques	2005	3	...	...	2	...	...	...	...	...
	2009	3	...	...	2	...	...	1	...	...
	2010	3	...	...	2	...	...	...	...	...
	2014	3	...	...	2	...	...	...	...	...
Brunei Darussalam	2005	46	113.1	111.0	44	97.2	98.0	5	9.9	19.7
Brunéi Darussalam	2010	44	107.3	106.9	49	98.9	99.7	6	11.1	20.6
	2014	41	107.3	107.6	49	99.1	99.1	11	23.8	40.1
Bulgaria	2005	290	101.8	100.6	686	92.4	88.4	238	41.1	47.6
Bulgarie	2010	260	103.6	103.7	532	92.4	88.3	287	50.1	66.4
	2014	259	99.4	98.7	519	102.6	99.0	283	63.0	79.0
Burkina Faso	2005	1 271	64.7	51.8	295	16.7	11.8	28	3.2	1.4
Burkina Faso	2010	2 048	81.6	73.9	538	24.8	18.9	51	4.8	2.3
	2013	2 466	87.2	83.3	762	30.7	26.0	74	6.4	3.1
	2014	2 594	88.7	85.1	842	32.4	28.2	...	...	...
Burundi	2005	1 037	85.8	72.4	*171	*14.8	*10.7	17	*3.4	*1.3
Burundi	2010	1 850	133.7	129.2	338	27.2	18.7	29	4.2	2.2
	2013	2 010	130.6	130.3	491	36.3	28.4	45	6.3	2.7
	2014	2 047	126.8	128.4	583	41.1	34.8	...	...	...
Cabo Verde	2005	83	112.4	108.0	53	64.5	73.0	4	7.2	7.7
Cabo Verde	2010	71	114.7	108.6	62	79.5	93.4	10	15.3	19.4
	2014	67	116.3	109.9	60	86.8	98.5	13	19.1	26.8
Cambodia	2004	2 763	136.8	125.9	*630	*36.6	*25.6	45	3.9	1.7
Cambodge	2005	2 695	135.1	125.6	...	...	...	57	4.6	2.1
	2008	2 341	127.2	119.3	*930	*48.5	*41.4	137	12.0	6.2
	2010	2 273	126.9	119.8	...	...	...	195	17.6	10.5
	2011	2 224	125.9	117.4	...	...	...	223	19.6	12.1
	2014	2 129	119.6	113.1	...	...	...	...	...	...

Education at the primary, secondary and tertiary levels *(continued)*
Number of students enrolled (thousands) and gross enrolment ratio by sex
Enseignement primaire, secondaire et supérieur *(suite)*
Nombre d'élèves inscrits (milliers) et taux brut de scolarisation par sexe

Region, country or area Région, pays ou zone	Year[t] Année[t]	Primary education Enseignement primaire			Secondary education Enseignement secondaire			Tertiary education Enseignement supérieur		
		Total (000)	Gross enrolment ratio Taux brut de scolarisation		Total (000)	Gross enrolment ratio Taux brut de scolarisation		Total (000)	Gross enrolment ratio Taux brut de scolarisation	
			M	F		M	F		M	F
Cameroon	2005	2 978	*109.2	*91.4	784	29.8	23.5	*100	*7.0	*4.6
Cameroun	2009	3 351	111.3	96.3	1 269	43.2	36.0	174	10.0	7.9
	2010	3 510	113.9	98.4	...	...	...	220	12.2	9.9
	2011	3 585	113.3	98.4	1 574	51.2	43.3	244	13.7	10.1
	2014	4 143	120.1	106.9	2 000	60.9	51.9	...	...	...
Canada	2000	2 456	100.3	100.4	2 519	100.6	102.5	1 212	50.5	67.7
Canada	2005	2 321	97.5	96.6	2 602	102.2	99.9	...	...	...
	2010	2 168	98.5	98.9	2 612	103.6	101.1	...	...	...
	2013	2 206	100.1	101.1	2 698	109.8	110.0	...	...	...
Cayman Islands	2003	...	...	...	...	...	...	1	...	...
Îles Caïmanes	2005	3	...	...	3	...	...	...	...	...
	2008	4	...	...	3	...	...	1	...	...
	2013	4	...	...	3	...	...	...	...	...
Central African Rep.	2000	...	...	...	...	...	...	6	3.2	0.6
Rép. centrafricaine	*2001	459	92.7	63.3	70	15.6	8.1	...	...	...
	*2002	411	82.4	55.1	72	...	...	...	...	...
	2004	363	70.4	46.7	...	...	...	6	...	...
	2005	412	77.1	53.1	...	...	...	...	...	...
	2009	608	104.1	73.8	93	17.7	9.8	10	3.4	1.5
	2010	637	107.4	76.4	...	...	...	11	3.9	1.2
	2012	662	107.3	79.8	126	23.0	11.8	13	4.1	1.5
Chad	2005	1 262	85.6	58.0	*245	*23.2	*8.1	12	2.7	0.2
Tchad	2010	1 727	95.1	69.9	430	31.8	13.4	*22	*3.6	*0.6
	2012	2 091	106.6	81.2	458	30.7	14.0	...	...	...
	2013	2 331	114.6	88.0	...	...	...	...	...	...
	*2014	...	...	...	...	...	...	42	5.7	1.1
Chile	2005	1 721	112.3	106.6	1 630	97.5	97.8	664	51.3	48.6
Chili	2010	1 547	104.9	101.6	1 518	93.2	96.0	988	67.1	72.3
	2014	1 469	102.1	98.9	1 556	99.6	101.2	1 205	81.3	92.3
China	2003	121 662	113.7	113.2	95 625	63.7	60.7	15 186	17.1	14.1
Chine	2005	...	...	...	...	...	...	20 601	20.3	18.3
	2010	101 019	113.3	111.6	99 218	85.0	84.8	31 047	23.1	24.8
	2014	95 107	103.8	104.0	88 692	93.4	95.4	41 924	36.6	42.5
China, Hong Kong SAR	2000	497	100.3	97.2	...	...	...	...	...	...
Chine, Hong Kong RAS	2005	451	...	...	498	80.5	80.3	152	32.7	31.9
	2010	349	...	...	508	88.4	87.6	265	*58.9	*58.1
	2014	324	111.7	110.3	416	102.5	98.6	305	64.0	73.8
China, Macao SAR	2005	37	...	...	47	103.8	106.7	23	79.0	50.4
Chine, Macao RAS	2010	25	...	...	38	94.2	93.3	29	62.7	60.9
	2014	23	...	...	32	97.0	95.0	30	60.2	78.6
Colombia	2005	5 298	128.7	126.2	4 297	79.3	87.7	1 224	28.0	30.4
Colombie	2010	5 085	125.4	123.1	5 080	97.3	106.9	1 674	37.6	41.3
	2014	4 543	115.7	111.7	4 828	95.1	102.6	2 138	47.9	54.8
Comoros	2003	104	120.0	98.8	38	42.8	35.9	2	3.4	2.6
Comores	2004	104	113.7	100.9	43	49.1	37.5	...	...	...
	2008	111	111.6	102.8	...	...	...	...	...	...
	2010	...	...	...	...	...	...	4	6.5	4.9
	2013	120	108.3	102.0	68	58.3	60.4	6	9.3	8.1
	2014	...	...	...	...	...	...	6	9.8	8.0
Congo	2003	510	103.9	97.6	204	47.9	32.9	*12	*5.5	*2.2
Congo	2004	584	116.2	108.9	*235	*49.7	*42.0	...	...	...
	2005	597	115.9	107.8	...	...	...	...	...	...
	2009	672	115.0	107.1	...	...	...	23	10.8	2.3
	2010	705	116.3	109.8	...	...	...	...	...	...
	2012	734	107.0	114.8	339	58.4	50.6	39	12.8	8.0
	2013	...	...	...	...	...	...	37	11.1	8.3

Education at the primary, secondary and tertiary levels *(continued)*
Number of students enrolled (thousands) and gross enrolment ratio by sex

Enseignement primaire, secondaire et supérieur *(suite)*
Nombre d'élèves inscrits (milliers) et taux brut de scolarisation par sexe

Region, country or area Région, pays ou zone	Year[t] Année[t]	Primary education Enseignement primaire			Secondary education Enseignement secondaire			Tertiary education Enseignement supérieur		
		Total (000)	Gross enrolment ratio Taux brut de scolarisation		Total (000)	Gross enrolment ratio Taux brut de scolarisation		Total (000)	Gross enrolment ratio Taux brut de scolarisation	
			M	F		M	F		M	F
Cook Islands	2005	2	*110.2	*113.0	2	*75.1	*84.9	...	...	...
Îles Cook	2010	2	*104.6	*103.6	2	*80.7	*88.1	...	...	...
	2014	2	*107.7	*104.2	2	*82.5	*90.6	1	*28.5	*96.2
Costa Rica	2004	558	116.7	114.0	340	80.8	84.3	109	24.5	30.2
Costa Rica	2005	542	114.4	111.5	347	82.6	86.2	...	...	...
	2010	521	118.1	115.8	414	100.1	104.3	...	...	...
	2014	476	111.3	110.3	460	117.2	123.6	217	47.5	58.8
Côte d'Ivoire	*2003	2 046	80.4	64.4	...	...	...	...	...	...
Côte d'Ivoire	2009	2 383	80.3	65.6	...	...	...	*153	*11.5	*6.1
	2010	...	...	...	...	...	...	144	10.6	5.5
	2014	3 177	95.6	83.6	1 418	47.0	33.2	177	10.9	6.4
Croatia	2005	196	103.3	103.3	400	91.6	94.7	135	40.5	49.2
Croatie	2010	167	91.7	91.9	389	97.3	104.2	150	46.8	62.8
	2014	161	99.0	98.7	370	97.2	100.9	166	58.8	80.7
Cuba	2005	895	100.1	97.5	937	91.8	93.6	472	*46.8	*79.6
Cuba	2010	853	103.0	101.8	809	91.3	92.5	801	73.5	122.1
	2014	763	100.3	95.8	830	98.0	101.4	302	32.0	50.8
Curaçao										
Curaçao	2013	21	179.1	171.3	11	86.1	90.6	2	12.2	28.4
Cyprus	2005	61	*100.9	*100.8	64	*95.7	*97.5	20	*31.2	*35.3
Chypre	2010	55	*101.6	*101.6	64	*90.9	*92.0	32	*50.9	*45.6
	2014	53	*98.7	*99.6	59	*99.4	*99.5	34	*44.1	*62.6
Czech Republic	2005	503	100.1	98.7	975	95.0	96.7	336	44.9	52.1
République tchèque	2010	463	104.2	103.8	837	94.6	95.1	437	53.7	75.0
	2014	511	98.9	98.9	781	104.9	105.4	419	55.2	77.4
Dem. P. R. Korea	2009	1 547	99.8	99.8	2 474	102.1	102.0	593	40.8	20.6
R. p. dém. de Corée	2011	...	...	...	...	...	...	600	...	...
Dem. Rep. of the Congo	2002	5 455	71.6	56.5	...	...	...	...	...	...
Rép. dém. du Congo	2009	10 244	103.4	88.7	3 399	50.7	28.5	378	9.9	3.1
	2010	10 572	102.3	89.1	3 484	49.9	28.8	...	...	...
	2013	12 601	107.9	97.8	3 996	50.6	31.4	443	9.1	4.2
	2014	13 535	112.0	101.8	4 388	53.6	33.3	...	...	...
Denmark	2005	414	98.9	98.7	465	122.0	126.5	232	67.7	93.4
Danemark	2010	403	99.5	99.7	504	119.0	120.1	241	60.4	87.5
	2014	467	102.0	100.7	554	127.7	132.2	301	68.2	95.4
Djibouti	2005	51	47.6	39.2	30	26.9	18.0	2	2.7	2.0
Djibouti	2009	56	62.8	56.6	44	38.5	28.4	3	4.0	2.8
	2010	...	...	...	...	...	...	3	4.1	2.8
	2011	61	68.6	62.4	51	43.6	35.0	5	5.9	4.0
	2014	64	72.5	63.3	59	51.1	41.5	...	...	...
Dominica	2005	9	99.4	99.4	7	103.3	111.7	...	...	...
Dominique	2010	8	110.0	107.6	7	92.8	101.4	...	...	...
	2011	8	114.6	113.5	7	93.5	100.1	...	...	...
	2014	8	117.9	118.0	...	...	...	...	...	...
Dominican Republic	2003	1 375	*112.4	*113.4	...	...	...	287	25.9	40.7
Rép. dominicaine	2005	1 290	108.6	103.3	808	63.6	76.0	...	...	...
	2010	1 318	114.1	100.5	905	71.5	80.6	...	...	...
	2014	1 268	105.4	95.7	931	74.4	82.4	456	36.3	58.8
Ecuador	2005	1 998	112.0	111.5	1 053	61.5	62.2	...	...	...
Équateur	2008	...	...	...	...	...	...	535	36.0	41.5
	2010	2 095	115.6	115.6	1 479	82.9	84.1	...	...	...
	2012	2 118	117.0	117.0	1 531	84.8	86.2	573	34.6	45.3
	2013	2 103	115.5	116.2	1 835	99.7	104.5	586	...	...
	2014	2 068	113.1	113.4	1 878	102.2	106.3	...	...	...

Education at the primary, secondary and tertiary levels *(continued)*
Number of students enrolled (thousands) and gross enrolment ratio by sex

Enseignement primaire, secondaire et supérieur *(suite)*
Nombre d'élèves inscrits (milliers) et taux brut de scolarisation par sexe

Region, country or area Région, pays ou zone	Year[t] Année[t]	Primary education Enseignement primaire			Secondary education Enseignement secondaire			Tertiary education Enseignement supérieur		
		Total (000)	Gross enrolment ratio Taux brut de scolarisation M	F	Total (000)	Gross enrolment ratio Taux brut de scolarisation M	F	Total (000)	Gross enrolment ratio Taux brut de scolarisation M	F
Egypt Égypte	2004	*7 928	*101.3	*98.0	*8 330	*82.9	*78.6	2 261	31.5	25.5
	2005	9 564	102.9	97.5	...	...	...	2 352	32.2	27.1
	2010	10 542	108.6	105.5	6 846	72.5	70.5	2 646	32.2	29.5
	2014	11 128	104.1	103.8	8 208	86.3	85.9	2 544	33.4	29.9
El Salvador El Salvador	2005	1 045	125.4	121.2	524	68.5	68.5	122	21.7	24.3
	2010	940	123.6	118.0	577	72.3	71.8	150	25.0	27.6
	2014	777	114.6	109.7	625	80.9	81.3	176	27.5	30.1
Equatorial Guinea Guinée équatoriale	2000	73	115.3	94.5	21	42.4	18.7	1	4.4	2.0
	2005	76	98.3	94.0	26	31.8	22.9	...	...	...
	2010	85	82.6	80.6	...	...	...	...	...	...
	2012	92	85.2	83.7	...	...	...	...	...	...
Eritrea Érythrée	2004	375	77.3	62.9	194	32.6	18.7	5	1.8	0.3
	2005	378	77.5	63.6	217	35.4	21.4	...	...	...
	2010	286	52.6	45.1	248	38.4	29.7	12	3.4	1.3
	2013	350	55.2	47.1	269	39.3	31.6	...	...	...
	2014	...	...	...	...	...	...	13	3.4	1.7
Estonia Estonie	2005	86	100.2	97.6	124	101.0	104.7	68	50.9	85.8
	2010	73	103.8	102.2	95	105.3	105.3	69	51.6	85.8
	2013	76	101.0	100.4	81	109.0	108.1	65	58.7	88.1
Ethiopia Éthiopie	2005	10 020	86.1	71.9	2 488	30.8	18.6	191	4.2	1.3
	2010	13 635	95.6	88.3	4 207	38.0	31.6	578	10.2	4.4
	2012	14 532	98.3	91.2	4 736	37.9	34.5	693	11.1	5.2
	2014	15 733	104.3	95.8	...	...	...	757	10.9	5.3
Fiji Fidji	2004	113	113.4	111.8	102	88.0	93.9	13	14.6	17.4
	*2005	...	...	...	...	...	...	13	14.7	17.6
	2009	101	105.9	104.3	98	83.1	90.9	...	...	...
	2012	103	104.5	105.4	97	84.3	93.4	...	...	...
	2013	105	105.0	106.1	...	...	...	...	...	...
Finland Finlande	2005	382	98.9	98.2	431	109.5	114.5	306	83.4	100.6
	2010	347	99.5	98.8	427	104.9	109.8	304	85.0	103.7
	2014	352	101.5	101.4	537	139.1	152.1	306	80.5	97.2
France France	2005	4 015	109.5	108.2	6 036	112.2	112.4	2 187	49.1	62.0
	2010	4 159	108.9	107.6	5 873	111.0	111.7	2 245	50.6	63.9
	2014	4 189	105.7	105.0	5 947	110.1	111.2	2 389	57.9	71.0
Gabon Gabon	2002	282	142.8	142.1	*105	...	...	...	...	...
	2003	280	139.3	138.7	...	...	...	10	10.6	6.2
	2011	318	144.0	139.9	...	...	...	...	...	...
Gambia Gambie	2004	205	93.3	94.0	...	...	...	2	1.9	0.4
	2005	205	89.3	92.1	...	...	...	...	...	...
	2010	229	83.2	85.6	*124	*58.9	*56.0	3	2.6	1.8
	2012	244	82.2	85.9	...	...	...	5	3.7	2.5
	2014	275	83.5	88.0	...	...	...	...	...	...
Georgia Géorgie	2005	338	96.5	93.6	316	83.0	80.3	174	45.9	47.2
	2008	311	110.6	109.6	305	92.8	88.5	130	*31.1	*37.5
	2009	299	110.5	110.5	342	...	...	95	22.8	28.4
	2010	289	109.3	110.7	...	...	...	106	25.6	32.3
	2014	285	116.3	117.5	282	99.3	99.6	121	34.8	43.7
Germany Allemagne	2005	3 306	104.7	104.7	8 268	103.8	101.1	...	...	...
	2010	3 068	103.8	103.6	7 664	106.6	101.2	...	...	...
	2014	2 863	103.7	103.0	7 201	105.2	99.5	2 912	67.6	63.3
Ghana Ghana	2005	2 930	92.1	88.6	*1 370	*51.6	*42.9	120	7.4	4.0
	2009	3 659	106.5	104.5	1 836	61.7	54.7	203	11.0	6.5
	2014	4 117	106.5	106.4	2 266	69.2	64.9	402	18.6	12.4
Gibraltar Gibraltar	2001	2	...	...	2	...	...	...	...	...
	2009	3	...	...	2	...	...	...	...	...

Education at the primary, secondary and tertiary levels *(continued)*
Number of students enrolled (thousands) and gross enrolment ratio by sex

Enseignement primaire, secondaire et supérieur *(suite)*
Nombre d'élèves inscrits (milliers) et taux brut de scolarisation par sexe

Region, country or area Région, pays ou zone	Year [t] Année [t]	Primary education Enseignement primaire			Secondary education Enseignement secondaire			Tertiary education Enseignement supérieur		
		Total (000)	Gross enrolment ratio Taux brut de scolarisation		Total (000)	Gross enrolment ratio Taux brut de scolarisation		Total (000)	Gross enrolment ratio Taux brut de scolarisation	
			M	F		M	F		M	F
Greece	2005	650	100.3	97.8	716	101.7	97.4	647	81.9	92.7
Grèce	2010	643	101.2	99.3	717	108.3	101.5	642	100.6	104.9
	2013	634	99.2	97.9	695	110.4	105.9	659	110.1	110.2
Grenada	2005	14	98.4	95.5	*13	*96.9	*98.7	...	...	...
Grenade	2009	14	107.6	101.5	11	100.7	100.8	7	44.8	60.9
	2010	14	105.1	101.5	12	106.2	109.4	...	...	...
	2014	13	104.0	102.3	10	101.2	100.5	...	...	...
Guatemala	2002	2 076	109.7	100.5	608	45.2	40.4	112	10.2	7.5
Guatemala	2005	2 345	115.4	107.5	754	51.9	47.6	...	...	...
	2007	2 449	115.6	109.2	864	55.9	52.3	234	16.7	17.1
	2010	2 653	118.7	115.2	1 082	65.2	61.0	...	...	...
	2013	2 499	109.5	105.6	1 160	66.5	61.8	294	17.8	18.9
	2014	2 417	105.5	101.6	1 166	65.5	61.5	...	...	...
Guinea	2005	1 207	87.8	70.4	*420	*38.9	*19.4	24	4.7	1.1
Guinée	2008	1 364	90.9	76.2	531	43.1	24.9	80	13.4	4.4
	2010	1 453	92.6	76.4	...	...	...	99	15.4	5.1
	2014	1 730	98.6	83.8	716	46.8	30.7	118	14.9	6.7
Guinea-Bissau	2000	150	86.1	57.6	26	22.3	12.1	...	...	...
Guinée-Bissau	2005	252	...	...	51	...	...	3	...	...
	2006	269	...	...	55	...	...	4	...	...
	2010	279	117.5	109.8	...	...	...	...	...	...
Guyana	2005	117	102.6	96.1	71	99.3	94.5	7	8.5	17.2
Guyana	2010	99	87.0	83.7	81	88.7	88.6	8	7.5	18.9
	2012	94	86.9	83.9	86	89.7	89.0	9	8.2	16.8
Honduras	2004	1 257	115.0	114.2	...	...	...	*123	*14.1	*20.0
Honduras	2005	1 232	111.8	111.4	...	...	...	...	...	...
	2010	1 275	116.8	116.3	655	65.5	80.0	170	19.3	22.0
	2014	1 150	110.2	108.0	620	63.2	73.7	186	18.0	24.4
Hungary	2005	431	99.1	97.1	960	96.7	95.7	436	53.1	77.6
Hongrie	2010	388	102.6	101.6	905	100.7	99.1	389	51.4	69.7
	2014	393	102.1	100.9	858	106.9	107.1	329	46.6	60.0
Iceland	2005	31	99.1	96.7	33	108.0	110.3	15	48.6	92.9
Islande	2010	30	98.4	99.0	36	108.3	110.4	18	56.2	101.3
	2012	29	98.5	98.9	36	111.6	110.8	19	60.7	104.6
India	2003	125 569	104.3	101.5	81 050	55.0	44.2	11 295	12.6	8.5
Inde	2005	...	...	...	89 462	*59.2	*48.7	11 777	12.5	8.8
	2010	138 414	*108.1	*110.4	107 687	65.5	60.9	20 741	20.6	15.0
	2013	141 155	104.9	117.0	119 401	68.6	69.2	28 175	24.6	23.1
Indonesia	2004	29 142	109.6	108.0	16 354	62.1	61.6	3 551	18.5	14.7
Indonésie	2005	29 150	*109.4	*106.1	15 993	*60.4	*59.8	*3 662	...	...
	2010	30 342	106.9	110.5	19 976	76.3	76.8	5 001	25.8	22.6
	2014	29 838	106.9	104.5	22 587	82.8	82.2	6 463	29.4	32.8
Iran (Islamic Rep. of)	2005	6 207	101.3	98.8	9 066	77.2	74.4	2 126	22.2	23.5
Iran (Rép. islamique d')	2010	5 630	106.2	105.3	7 347	82.7	81.1	3 791	42.9	42.3
	2014	7 441	106.9	111.6	5 795	88.9	87.9	4 685	68.2	63.6
Iraq	2004	4 335	113.1	95.2	1 706	56.5	38.2	413	19.8	11.8
Iraq	*2005	...	...	...	...	...	...	425	20.0	11.9
	2007	4 864	117.0	98.7	2 038	60.9	45.6	...	...	...
Ireland	2005	454	107.0	106.3	317	104.6	114.7	187	48.0	60.7
Irlande	2010	506	106.6	107.0	336	120.8	126.2	194	59.6	66.5
	2012	518	103.3	104.0	341	124.8	127.5	193	65.6	69.7
	2013	528	102.6	103.1	...	...	...	199	71.2	75.2
Israel	2005	722	103.3	103.8	673	105.0	104.4	311	50.1	66.4
Israël	2009	786	102.6	103.2	694	100.9	103.1	343	54.5	70.7
	2010	807	104.0	104.5	708	100.9	103.3	360	...	...
	2014	862	104.0	104.4	768	101.1	102.7	377	56.7	76.2

5

Education at the primary, secondary and tertiary levels *(continued)*
Number of students enrolled (thousands) and gross enrolment ratio by sex
Enseignement primaire, secondaire et supérieur *(suite)*
Nombre d'élèves inscrits (milliers) et taux brut de scolarisation par sexe

Region, country or area Région, pays ou zone	Year[t] Année[t]	Primary education Enseignement primaire			Secondary education Enseignement secondaire			Tertiary education Enseignement supérieur		
		Total (000)	Gross enrolment ratio Taux brut de scolarisation		Total (000)	Gross enrolment ratio Taux brut de scolarisation		Total (000)	Gross enrolment ratio Taux brut de scolarisation	
			M	F		M	F		M	F
Italy Italie	2005	2 771	102.0	100.9	4 507	98.6	97.9	2 015	54.3	74.3
	2010	2 822	102.7	101.6	4 626	102.9	101.5	1 980	54.8	78.1
	2013	2 861	102.3	101.4	4 594	103.4	101.4	1 873	53.2	74.2
Jamaica Jamaïque	2002	330	97.1	96.9	228	84.1	86.9	45	12.0	26.0
	2004	331	97.1	97.4	246	88.9	92.9	52	...	...
	2005	326	...	...	246	87.9	93.4	48	...	...
	2010	294	...	...	260	89.2	96.6	71	16.6	38.3
	2013	273	...	...	229	66.9	71.9	74	17.1	39.0
	2014	266	...	...	224	66.3	71.5	...	...	...
Japan Japon	2005	7 232	101.9	101.9	7 710	100.9	101.1	4 038	58.1	51.7
	2010	7 099	102.2	102.2	7 296	101.5	101.7	3 836	61.3	54.7
	2013	6 802	101.7	101.5	7 281	101.7	102.1	3 863	65.2	59.5
Jordan Jordanie	2005	805	104.8	106.3	626	89.7	92.4	218	37.3	40.6
	2010	820	91.1	90.9	710	86.1	91.1	247	37.7	43.5
	2012	849	89.3	88.2	724	82.3	86.4	307	43.8	51.6
Kazakhstan Kazakhstan	2000	1 208	96.1	96.7	1 994	91.5	95.4	418	29.4	34.1
	2005	1 024	101.8	101.7	...	...	...	...	...	...
	2010	958	107.3	107.9	1 818	97.5	98.0	757	40.5	51.5
	2014	1 122	111.4	111.3	1 662	104.2	106.8	727	42.9	54.2
Kenya Kenya	2005	6 076	109.8	105.5	*2 468	*49.0	*46.8	*114	*3.7	*2.2
	2009	7 150	114.5	112.0	3 204	63.5	57.4	168	4.8	3.3
	2012	8 053	115.8	116.5	3 833	70.1	65.2	...	...	...
	2014	8 158	111.2	111.6	...	...	...	...	...	...
Kiribati Kiribati	2005	16	111.2	112.8	12	83.1	95.3	...	...	...
	2008	16	108.4	113.1	12	82.9	91.6	...	...	...
	2009	16	108.0	113.8	...	...	...	...	...	...
	2014	16	111.2	115.2	...	...	...	...	...	...
Kuwait Koweït	2004	158	109.2	108.7	267	99.4	111.3	*37	*13.6	*28.1
	2005	203	109.8	107.6	*244	*103.4	*118.3	...	...	...
	2010	214	100.9	104.9	264	96.5	90.6	...	...	...
	2013	239	103.1	105.3	279	89.2	96.2	72	20.4	33.1
	2014	253	102.4	103.0	*283	*88.9	*98.9	...	...	...
Kyrgyzstan Kirghizistan	2005	434	98.4	97.0	721	86.2	86.3	220	37.7	47.1
	2010	391	98.4	97.5	*664	*83.3	*82.5	261	36.6	47.5
	2014	435	108.3	107.0	651	90.4	91.2	268	39.9	52.1
Lao People's Dem. Rep. Rép. dém. pop. lao	2005	891	119.1	104.1	394	49.6	37.4	47	9.2	6.5
	2010	916	127.8	117.9	435	50.4	41.7	118	18.5	14.2
	2014	871	119.1	113.5	601	59.8	54.6	132	17.9	16.7
Latvia Lettonie	2005	84	95.5	91.7	272	103.9	104.2	131	56.9	101.7
	2010	114	106.2	105.4	147	99.2	97.5	113	51.4	90.3
	2014	115	100.8	100.0	122	116.6	114.0	90	55.5	79.2
Lebanon Liban	2005	*476	*107.8	*99.0	*378	*80.2	*80.8	166	43.0	45.2
	2010	462	109.6	100.0	383	74.5	76.1	202	47.4	49.3
	2013	472	101.6	92.7	389	68.0	68.4	230	43.9	47.8
	2014	...	...	...	...	...	...	229	39.5	45.7
Lesotho Lesotho	2005	422	116.8	117.5	94	34.5	44.9	8	3.1	4.3
	2006	425	117.2	117.7	96	34.6	45.2	9	3.4	4.4
	2010	389	111.5	109.4	126	41.7	59.3	...	...	...
	2014	366	108.3	105.8	131	44.1	60.4	24	8.0	11.7
Liberia Libéria	2000	496	129.6	95.4	136	40.7	29.6	52	24.9	13.8
	2009	605	105.0	94.1	...	...	...	...	...	...
	2010	...	...	...	...	...	...	33	12.1	6.4
	2012	...	...	...	...	...	...	44	14.2	9.0
	2014	684	99.5	91.6	223	42.5	33.1	...	...	...

5

Education at the primary, secondary and tertiary levels *(continued)*
Number of students enrolled (thousands) and gross enrolment ratio by sex

Enseignement primaire, secondaire et supérieur *(suite)*
Nombre d'élèves inscrits (milliers) et taux brut de scolarisation par sexe

Region, country or area Région, pays ou zone	Year[t] Année[t]	Primary education Enseignement primaire			Secondary education Enseignement secondaire			Tertiary education Enseignement supérieur		
		Total (000)	Gross enrolment ratio Taux brut de scolarisation		Total (000)	Gross enrolment ratio Taux brut de scolarisation		Total (000)	Gross enrolment ratio Taux brut de scolarisation	
			M	F		M	F		M	F
Libya	2003	739	111.1	107.9	*798	*101.7	*108.4	*375	*58.1	*64.3
Libye	2005	714	108.7	107.6	702	*86.7	*103.3	...	...	...
	2006	755	116.8	112.1	733	93.6	109.9	...	...	...
Liechtenstein	2004	2	*109.9	*108.5	3	*117.2	*103.9	1	*35.7	*13.3
Liechtenstein	2005	2	...	...	3	...	...	1	...	...
	2010	2	*108.6	*102.3	3	*117.6	*100.0	1	*44.3	*27.4
	2014	2	*103.1	*102.3	3	*127.8	*103.3	1	*50.8	*24.1
Lithuania	2005	158	94.9	94.4	424	104.6	103.7	195	62.9	98.5
Lituanie	2010	122	100.6	99.5	343	101.9	100.0	201	68.7	103.1
	2014	108	102.3	102.6	277	108.9	104.6	148	55.8	82.0
Luxembourg	2003	34	99.5	99.3	35	93.6	98.7	3	11.1	13.3
Luxembourg	2005	35	100.7	100.9	36	92.9	97.9	...	...	...
	2010	35	97.1	98.1	43	99.9	102.9	5	17.3	19.2
	2012	35	96.9	97.2	44	98.5	102.0	6	18.2	20.7
	2013	35	96.2	96.9	46	101.1	103.8	...	...	...
Madagascar	2005	3 598	141.4	135.3	*621	*21.6	*20.7	45	3.0	2.7
Madagascar	2009	4 324	151.6	148.2	*1 022	*31.3	29.4	68	3.6	3.3
	2010	4 242	145.9	143.3	...	...	...	74	3.8	3.4
	2013	4 483	145.8	144.7	...	...	...	97	4.4	4.1
	2014	4 611	146.9	146.5	1 494	38.8	38.1	...	...	...
Malawi	2005	2 868	126.3	128.0	516	30.3	24.3	6	*0.6	*0.3
Malawi	2010	3 417	135.1	138.4	692	34.9	31.2	10	0.9	0.5
	2011	3 564	137.1	140.9	736	36.0	32.4	12	1.0	0.6
	2014	4 097	145.1	148.0	920	41.4	37.5	...	...	...
Malaysia	2005	3 202	...	...	2 489	...	...	697	...	...
Malaisie	2010	3 234	...	...	2 616	...	...	1 061	...	...
	2014	3 178	...	...	3 038	...	...	860	...	...
Maldives	2003	66	125.1	122.4	29	66.1	72.9	<0	0.1	0.3
Maldives	2004	63	124.1	120.5	*29	*65.8	*74.0	...	...	...
	2005	58	117.6	114.5	...	...	...	...	...	...
	2008	47	107.3	103.2	...	...	...	5	12.0	13.5
	2009	45	104.0	100.8	...	...	...	...	...	...
	2010	42	...	...	...	...	...	...	...	...
	2014	40	...	...	...	...	...	...	...	...
Mali	2002	1 227	74.3	56.7	...	...	...	23	2.7	1.4
Mali	2005	1 506	81.3	64.8	430	*31.2	*19.6	...	...	...
	2010	2 019	88.7	77.1	758	45.4	31.9	81	8.4	3.6
	2012	2 114	85.7	75.8	...	...	...	97	9.6	4.1
	2014	2 182	81.2	73.0	961	49.4	37.4	...	...	...
Malta	2005	30	97.1	96.3	40	101.8	102.9	9	28.5	41.9
Malte	2010	25	96.1	86.5	37	110.2	87.5	11	32.3	54.1
	2014	24	100.0	95.0	30	91.0	80.6	13	41.9	48.0
Marshall Islands	2002	9	124.6	116.1	6	71.7	74.4	1	14.2	18.3
Îles Marshall	2005	8	103.0	129.9	5	75.6	75.7	...	...	...
	2009	8	107.4	107.9	5	101.4	104.4	...	...	...
	2011	9	105.7	105.0	...	...	...	...	...	...
	2012	...	...	...	...	...	...	1	44.6	41.2
Mauritania	2005	444	89.3	92.2	93	23.7	20.8	9	4.3	1.5
Mauritanie	2010	531	94.3	98.4	*110	*21.9	*18.7	15	6.3	2.5
	2013	569	94.0	98.6	171	30.3	28.4	19	7.5	3.3
	2014	592	95.3	100.7	179	31.2	28.6	...	...	...
Mauritius	2005	124	103.3	103.1	*129	*89.7	*87.6	*21	*21.0	*21.7
Maurice	2010	117	102.5	103.1	*127	*87.2	*91.3	33	30.5	36.9
	2014	105	101.7	103.7	133	96.9	99.0	40	34.7	42.7

5

Education at the primary, secondary and tertiary levels *(continued)*
Number of students enrolled (thousands) and gross enrolment ratio by sex

Enseignement primaire, secondaire et supérieur *(suite)*
Nombre d'élèves inscrits (milliers) et taux brut de scolarisation par sexe

Region, country or area Région, pays ou zone	Year [t] Année [t]	Primary education Enseignement primaire			Secondary education Enseignement secondaire			Tertiary education Enseignement supérieur		
		Total (000)	Gross enrolment ratio Taux brut de scolarisation		Total (000)	Gross enrolment ratio Taux brut de scolarisation		Total (000)	Gross enrolment ratio Taux brut de scolarisation	
			M	F		M	F		M	F
Mexico	2005	14 700	103.8	102.2	10 564	76.0	82.6	2 385	22.9	23.3
Mexique	2010	14 906	104.0	102.8	11 682	79.7	86.8	2 847	25.7	26.6
	2014	14 627	103.6	103.2	12 993	87.7	93.5	3 419	29.9	30.0
Micronesia (Fed. States of)	*2000	...	...	...	...	...	...	2	...	...
Micronésie (États féd. de)	2005	19	112.8	109.8	14	80.1	86.7	...	...	...
	2007	19	110.9	112.4	...	...	...	...	...	...
	2014	14	98.1	97.2	...	...	...	...	...	...
Monaco	2004	2	...	...	3	...	...	...	...	...
Monaco	2010	2	...	...	3	...	...	...	...	...
	2014	2	...	...	3	...	...	...	...	...
Mongolia	2005	251	98.0	98.0	339	85.6	95.0	124	33.8	55.9
Mongolie	2010	274	127.1	124.2	276	88.4	94.8	166	42.5	65.3
	2014	239	102.7	100.6	286	89.6	91.9	175	52.9	75.9
Montenegro	2005	38	112.7	112.6	68	93.2	95.8	11	17.2	27.2
Monténégro	2010	35	107.7	105.8	70	100.1	101.4	24	49.2	61.8
	2012	38	95.2	95.4	63	91.0	91.8	...	...	...
Montserrat	2005	1	*114.6	*118.7	<0	*111.0	*122.5	...	...	...
Montserrat	2007	<0	*101.2	*112.9	<0	*101.1	*103.2	...	...	...
	2009	<0	...	...	<0	...	...	<0	...	...
	2010	...	...	...	...	...	...	<0	...	...
	2014	<0	...	...	<0	...	...	...	...	...
Morocco	2005	4 023	110.0	99.8	1 948	53.7	45.8	368	13.0	10.5
Maroc	2010	3 945	115.6	108.8	2 393	66.9	58.3	447	15.2	13.7
	2012	4 017	119.2	113.7	2 554	74.4	63.5	606	20.0	18.5
	2014	4 030	118.7	113.4	...	...	...	774	25.0	24.1
Mozambique	2005	3 943	104.4	87.8	306	15.5	10.7	28	1.9	0.9
Mozambique	2010	5 278	114.7	103.0	672	26.6	21.7	104	5.5	3.6
	2014	5 670	108.6	99.6	784	25.5	23.5	157	7.0	5.0
Myanmar	2001	4 782	98.4	96.3	2 302	38.6	36.2	553	...	...
Myanmar	2005	4 948	98.2	98.4	2 589	45.0	43.6	...	...	...
	2007	5 014	...	...	2 686	46.3	46.2	508	8.9	12.2
	2010	5 126	97.1	96.3	2 852	47.0	49.3	...	...	...
	2012	...	...	...	...	...	...	634	12.1	14.9
	2014	5 177	101.0	98.3	3 191	50.6	52.0	...	...	...
Namibia	2005	404	109.2	108.9	148	59.8	67.2	14	7.4	6.3
Namibie	2007	410	110.4	108.9	158	60.1	69.6	...	...	...
	2008	407	109.7	107.8	...	...	...	20	8.2	10.4
	2010	407	109.0	106.3	...	...	...	...	...	...
	2013	425	113.3	109.5	...	...	...	...	...	...
Nauru	2005	2	*122.3	*128.5	1	*44.4	*50.3	...	...	...
Nauru	2008	1	*90.1	*96.0	1	*57.6	*68.9	...	...	...
	2014	2	*109.5	*100.4	1	*81.9	*83.4	...	...	...
Nepal	2005	4 030	120.3	110.1	2 054	*53.9	*43.6	187	10.6	5.6
Népal	2010	4 901	137.8	147.1	*2 694	*60.3	*57.3	377	18.2	11.1
	2013	4 577	133.5	143.8	*3 111	*64.5	*67.4	477	18.7	15.3
	2014	4 402	130.1	140.6	*3 164	*64.9	*68.9	459	...	...
Netherlands	2005	1 278	108.2	105.5	1 411	120.5	118.2	565	57.5	62.1
Pays-Bas	2010	1 294	109.8	108.5	1 475	124.0	122.2	651	61.6	68.9
	2012	1 277	108.0	107.1	1 550	131.4	128.9	794	74.7	82.5
	*2013	1 255	106.6	105.7	1 572	134.4	129.8	...	...	...
	2014	1 223	104.8	104.1	*1 574	...	...	...	...	...
Netherlands Antilles [former] Antilles néerlandaises [anc.]	2002	23	...	...	15	...	...	2	...	...
New Zealand	2005	353	100.0	99.2	526	116.9	123.4	240	65.8	95.8
Nouvelle-Zélande	2010	348	100.9	101.3	512	116.3	121.7	266	67.3	98.6
	2014	360	98.8	98.5	492	114.2	121.0	261	68.2	94.0

Education at the primary, secondary and tertiary levels *(continued)*
Number of students enrolled (thousands) and gross enrolment ratio by sex
Enseignement primaire, secondaire et supérieur *(suite)*
Nombre d'élèves inscrits (milliers) et taux brut de scolarisation par sexe

Region, country or area Région, pays ou zone	Year[t] Année[t]	Primary education Enseignement primaire			Secondary education Enseignement secondaire			Tertiary education Enseignement supérieur		
		Total (000)	Gross enrolment ratio Taux brut de scolarisation		Total (000)	Gross enrolment ratio Taux brut de scolarisation		Total (000)	Gross enrolment ratio Taux brut de scolarisation	
			M	F		M	F		M	F
Nicaragua	2002	923	116.4	116.6	383	54.9	65.4	100	16.3	18.2
Nicaragua	2005	945	121.9	120.2	438	63.4	73.8	...	...	...
	2010	924	123.8	122.7	465	69.8	78.8	...	...	...
Niger	2005	1 064	56.7	40.8	182	12.2	7.7	11	1.8	0.6
Niger	2010	1 726	69.6	56.8	307	16.2	10.9	17	2.2	0.8
	2012	2 051	75.3	63.3	389	18.7	12.5	22	2.6	0.9
	2014	2 277	75.9	65.0	515	22.1	15.6	...	...	...
Nigeria	2005	22 115	109.0	92.6	6 398	37.6	31.7	1 392	12.1	8.7
Nigéria	2010	*21 558	*88.3	*80.9	9 057	46.4	41.2	...	...	...
Niue	2005	<0	*118.9	*105.9	<0	*71.3	*116.5	...	...	...
Nioué	2014	<0	*127.4	*125.3	...	...	...	...	...	...
Norway	2005	430	98.4	98.7	403	113.4	114.4	214	62.5	95.2
Norvège	2010	424	98.9	99.1	435	114.0	112.4	225	56.1	90.4
	2014	426	100.2	100.1	439	114.5	110.6	264	63.0	91.4
Oman	2005	312	89.7	87.7	302	91.0	86.0	48	18.6	19.2
Oman	2009	302	106.7	100.0	322	107.1	95.3	76	21.7	28.6
	2010	...	...	...	...	...	...	78	21.8	29.6
	2011	296	104.1	105.0	301	105.5	98.3	89	24.7	33.8
	2014	233	105.8	114.9	391	...	...	85	...	...
Pakistan	2005	17 258	99.2	76.3	6 852	...	...	783	5.3	4.6
Pakistan	2009	18 468	101.1	86.3	9 433	39.2	31.1	*1 226	*7.5	*6.3
	2010	18 756	103.1	87.8	9 655	40.3	31.1	...	...	...
	2014	19 432	100.7	85.8	11 287	46.3	36.6	1 932	10.1	10.7
Palau	*2002	...	...	...	...	...	...	<0	25.6	52.3
Palaos	2004	2	*104.6	*101.1	2	*97.7	*99.9	...	...	...
	2005	2	...	...	2	...	...	...	...	...
	2007	2	...	...	2	...	...	...	...	...
	2013	2	*115.3	*113.4	2	*113.9	*117.1	1	*49.1	*76.1
	2014	2	*116.5	*111.8	2	*110.5	*116.9	...	...	...
Panama	2005	430	109.1	105.4	256	65.9	70.6	126	32.1	52.5
Panama	2010	440	109.0	104.9	284	69.2	73.5	139	35.2	53.8
	2013	436	106.9	103.7	310	73.2	77.9	124	31.2	46.5
Papua New Guinea	2005	532	62.7	53.3	...	...	...	...	...	...
Papouasie-Nvl-Guinée	2008	601	63.8	56.9	...	...	...	...	...	...
	2012	1 427	119.8	109.3	378	45.8	34.6	...	...	...
Paraguay	2005	934	113.5	110.3	529	66.0	67.8	156	24.3	27.5
Paraguay	2010	839	105.9	101.8	561	66.5	70.2	225	29.1	41.2
	2012	838	107.6	104.3	631	74.2	79.1	...	...	...
Peru	2005	4 077	117.0	118.0	2 470	86.3	84.6	908	32.8	33.6
Pérou	2010	3 763	109.8	111.0	2 675	94.5	94.9	1 151	38.6	42.5
	2014	3 496	101.4	101.4	2 671	95.7	95.5	...	...	...
Philippines	2005	13 084	106.6	105.4	6 352	78.3	87.6	2 403	24.7	30.4
Philippines	2009	13 687	109.1	108.6	6 767	81.0	87.6	2 625	25.7	31.8
	2010	...	...	...	...	...	...	2 774	26.5	33.1
	2013	14 460	116.8	116.9	7 220	84.4	92.7	3 317	29.8	37.6
	2014	...	...	...	...	...	...	3 563	31.4	40.3
Poland	2005	2 724	96.5	96.0	3 445	98.9	98.1	2 118	52.4	74.1
Pologne	2010	2 235	98.9	98.3	2 842	97.2	96.2	2 149	58.6	88.3
	2013	2 161	101.2	101.4	2 778	110.6	106.7	1 903	56.1	86.9
Portugal	2005	753	121.2	115.4	670	93.5	102.8	381	48.2	63.1
Portugal	2010	734	113.6	110.2	721	105.5	108.4	384	60.3	71.2
	2014	674	110.8	106.1	769	117.8	115.0	362	60.9	70.4
Puerto Rico	2010	300	93.2	96.5	291	80.7	85.3	249	69.9	103.2
Porto Rico	2013	264	88.7	90.5	269	78.4	83.3	245	70.4	100.4

5

Education at the primary, secondary and tertiary levels *(continued)*
Number of students enrolled (thousands) and gross enrolment ratio by sex

Enseignement primaire, secondaire et supérieur *(suite)*
Nombre d'élèves inscrits (milliers) et taux brut de scolarisation par sexe

| Region, country or area Région, pays ou zone | Year [t] Année [t] | Primary education Enseignement primaire | | | Secondary education Enseignement secondaire | | | Tertiary education Enseignement supérieur | | |
| | | Total (000) | Gross enrolment ratio Taux brut de scolarisation | | Total (000) | Gross enrolment ratio Taux brut de scolarisation | | Total (000) | Gross enrolment ratio Taux brut de scolarisation | |
			M	F		M	F		M	F
Qatar	2005	70	108.1	97.3	56	113.7	108.7	*10	*8.9	*32.3
Qatar	2010	89	98.7	96.1	69	98.1	104.4	14	4.8	26.2
	2011	95	103.7	99.0	73	116.1	103.2	15	5.5	30.5
	2014	117	...	...	88	...	...	25	7.3	45.8
Republic of Korea	2005	4 031	112.0	111.2	3 786	93.9	91.9	3 210	107.4	71.1
République de Corée	2010	3 306	102.5	101.3	3 951	96.4	95.3	3 270	113.5	83.9
	2013	2 791	99.5	98.6	3 720	98.3	97.2	3 342	107.8	81.3
Republic of Moldova	2005	184	*98.6	*97.2	394	*86.2	*90.0	130	*29.5	*42.9
République de Moldova	2010	141	*93.7	*93.3	308	*87.0	*89.0	130	*32.7	*43.7
	2013	138	*93.7	*93.8	260	*87.8	*88.9	122	*36.1	*46.6
	2014	138	*93.3	*92.9	246	*86.8	*87.9	...	...	...
Romania	2005	970	109.4	107.9	2 090	82.9	84.5	739	40.2	51.1
Roumanie	2010	842	98.0	96.5	1 822	100.7	99.8	1 000	57.4	78.9
	2012	807	96.4	94.6	1 714	100.0	98.7	705	48.3	61.1
	2014	...	...	...	1 609	95.2	94.3	579	47.7	59.1
Russian Federation	2005	5 309	95.2	95.4	12 433	83.4	82.5	9 003	61.3	84.2
Fédération de Russie	2009	5 015	99.0	99.4	9 614	85.6	84.2	9 330	64.3	86.9
	2014	5 726	98.2	98.9	9 061	101.6	99.5	6 996	71.5	86.2
Rwanda	2004	1 753	134.9	133.8	204	15.2	13.8	25	3.5	2.0
Rwanda	2005	1 858	136.1	137.1	219	16.9	15.1	28	...	...
	2010	2 299	140.5	144.1	426	32.9	32.6	63	6.3	4.9
	2013	2 402	136.2	139.2	582	38.7	41.6	77	8.5	6.6
	2014	2 399	132.0	135.1	587	37.3	40.9	...	...	...
Saint Kitts and Nevis	2005	6	98.2	100.1	*4	*84.1	*89.9	...	...	...
Saint-Kitts-et-Nevis	2008	6	95.9	97.3	4	99.5	104.9	1	11.8	24.7
	2010	6	93.4	93.5	*4	*97.9	*97.1	...	...	...
	2014	6	83.0	84.5	4	90.3	92.8	3	78.5	79.7
Saint Lucia	2005	24	105.4	100.7	14	71.4	84.0	2	7.0	19.0
Sainte-Lucie	2007	22	102.9	97.7	15	83.4	92.7	1	5.2	12.0
	2010	19	...	...	16	95.8	94.7	2	7.0	18.0
	2014	17	...	...	14	86.7	86.2	3	10.9	22.9
Saint Vincent-Grenadines	2005	18	123.9	112.3	10	79.6	99.5	...	...	...
Saint-Vincent-Grenadines	2010	14	108.8	101.2	11	106.3	108.8	...	...	...
	2014	13	106.3	103.0	10	106.2	103.2	...	...	...
Samoa	2000	28	95.9	97.3	22	73.7	83.9	1	7.8	7.2
Samoa	2005	31	109.0	109.8	24	78.2	88.2	...	...	...
	2010	31	111.5	109.9	26	82.3	93.6	...	...	...
	2014	31	106.1	106.3	26	81.9	92.2	...	...	...
San Marino	2000	1	...	...	1	...	...	1	...	...
Saint-Marin	2004	1	...	...	...	...	...	...	...	...
	2010	2	*88.8	*100.6	2	*96.3	*98.2	1	*53.0	*77.3
	2012	2	*93.9	*92.6	2	*93.5	*95.9	1	*50.5	*69.9
Sao Tome and Principe	2005	30	132.3	127.2	8	41.9	44.4	...	...	...
Sao Tomé-et-Principe	2010	34	122.6	121.4	10	52.5	53.4	1	4.4	4.2
	2013	35	116.0	113.1	16	62.7	69.2	...	...	...
	2014	34	114.0	109.4	19	...	...	2	10.4	9.0
Saudi Arabia	2005	*3 098	*93.7	*92.3	*2 610	*88.4	*83.8	604	25.2	33.9
Arabie saoudite	2009	3 255	98.4	96.2	*2 997	*101.8	*89.4	758	27.1	35.1
	2010	3 321	99.2	99.1	...	...	...	904	33.7	39.6
	2014	3 737	109.2	108.3	*3 419	*122.6	*93.7	1 497	62.4	59.9
Senegal	2005	1 444	80.2	77.2	406	25.2	18.8	*59	...	...
Sénégal	2010	1 695	79.7	84.2	725	38.0	33.1	*92	*9.3	*5.5
	2011	1 726	78.5	83.5	*834	*41.9	*38.2	...	...	...
	2014	1 888	77.5	84.3	...	...	...	...	...	...

Education at the primary, secondary and tertiary levels *(continued)*
Number of students enrolled (thousands) and gross enrolment ratio by sex

Enseignement primaire, secondaire et supérieur *(suite)*
Nombre d'élèves inscrits (milliers) et taux brut de scolarisation par sexe

Region, country or area Région, pays ou zone	Year[t] Année[t]	Primary education Enseignement primaire			Secondary education Enseignement secondaire			Tertiary education Enseignement supérieur		
		Total (000)	Gross enrolment ratio Taux brut de scolarisation		Total (000)	Gross enrolment ratio Taux brut de scolarisation		Total (000)	Gross enrolment ratio Taux brut de scolarisation	
			M	F		M	F		M	F
Serbia	2005	325	*102.5	*103.0	632	*87.3	*90.0	225	*38.6	*50.1
Serbie	2010	283	*96.1	*95.6	591	*90.5	*92.4	227	*42.8	*55.6
	2014	285	*100.9	*101.4	548	*93.3	*95.4	243	*50.5	*66.0
Seychelles	2005	9	110.1	108.0	9	79.0	84.8	...	...	...
Seychelles	2010	9	108.9	112.5	8	*72.4	*75.9	...	...	...
	2014	9	103.6	104.7	7	73.9	75.3	<0	3.8	9.4
Sierra Leone	2001	554	98.3	68.6	156	*31.8	*22.5	9	3.2	1.3
Sierra Leone	*2002	...	...	...	...	...	...	9	3.1	1.2
	2013	1 300	130.3	129.8	417	46.9	40.0	...	...	...
Singapore	2009	295	...	...	232	...	...	199	...	...
Singapour	2010	...	...	...	...	...	...	213	...	...
	2013	...	...	...	...	...	...	255	...	...
Sint Maarten (Dutch part) Saint-Martin (partie néerlandaise)	2014	4	...	...	3	...	...	<0	...	...
Slovakia	2005	242	99.2	98.1	663	93.3	94.2	181	35.4	45.7
Slovaquie	2010	212	102.9	101.8	550	92.2	93.1	235	44.9	69.3
	2014	214	101.6	100.4	465	91.5	92.3	198	41.8	64.6
Slovenia	2005	93	101.0	99.3	181	97.4	96.8	112	65.6	93.4
Slovénie	2010	107	98.5	98.0	138	98.3	97.4	115	71.2	107.0
	2014	112	99.2	99.4	145	110.8	110.5	91	68.2	98.5
Solomon Islands	2005	75	103.6	97.8	22	33.2	27.4	...	...	...
Îles Salomon	2010	95	116.6	113.3	40	52.1	44.9	...	...	...
	2012	98	114.6	111.9	42	49.8	47.0	...	...	...
	2014	102	115.5	112.1	...	...	...	...	...	...
Somalia Somalie	2007	457	37.6	20.8	87	10.1	4.6	...	...	...
South Africa	2005	7 314	96.5	103.0	4 658	84.3	93.7	...	...	...
Afrique du Sud	2010	7 024	93.6	99.3	4 690	84.4	96.7	...	...	...
	2013	7 064	100.5	96.5	4 953	84.9	103.7	1 036	16.0	23.5
	2014	7 195	102.2	97.3	*4 956	*85.3	*103.8	...	...	...
South Sudan Soudan du sud	2011	1 451	101.1	67.0	...	...	...	...	...	...
Spain	2005	2 485	103.8	102.4	3 108	113.9	121.3	1 809	60.6	74.4
Espagne	2010	2 721	106.0	105.5	3 185	123.0	125.8	1 879	70.5	87.4
	2014	2 961	104.1	105.3	3 288	129.9	130.4	1 982	81.5	96.9
Sri Lanka	2005	1 611	99.9	99.3	...	...	...	...	...	...
Sri Lanka	2010	1 721	101.0	98.4	2 525	96.3	97.5	262	11.7	20.9
	2013	1 767	101.9	99.4	2 606	97.5	102.0	298	14.8	22.9
	2014	1 778	102.4	100.1	...	...	...	324	16.6	24.7
State of Palestine	2005	387	88.5	87.9	657	87.7	91.7	138	40.6	41.2
État de Palestine	2010	403	91.6	89.8	711	82.4	89.0	197	41.1	54.9
	2014	442	95.2	94.6	709	78.5	86.2	214	34.6	53.8
Sudan	2005	3 177	66.0	57.8	1 344	38.9	36.1	364	11.0	12.4
Soudan	2010	4 024	73.6	66.1	1 687	44.1	38.2	523	13.8	16.2
	2013	4 292	74.1	66.6	1 871	43.9	41.5	640	16.0	17.9
Sudan [former]	2000	2 567	...	...	980	...	...	*204	...	...
Soudan [anc.]	2005	3 278	...	...	*1 370	...	...	...	...	...
	2009	4 744	...	...	1 837	...	...	...	...	...
Suriname	2002	64	100.7	100.3	42	59.0	81.8	5	9.4	16.1
Suriname	2005	66	113.7	111.2	46	59.7	76.9	...	...	...
	2009	71	121.0	115.9	47	67.2	83.2	...	...	...
	2013	69	120.7	114.6	52	68.2	89.2	...	...	...
	2014	70	121.6	118.2	...	...	...	...	...	...

Education at the primary, secondary and tertiary levels *(continued)*
Number of students enrolled (thousands) and gross enrolment ratio by sex

Enseignement primaire, secondaire et supérieur *(suite)*
Nombre d'élèves inscrits (milliers) et taux brut de scolarisation par sexe

Region, country or area Région, pays ou zone	Year[t] Année[t]	Primary education Enseignement primaire			Secondary education Enseignement secondaire			Tertiary education Enseignement supérieur		
		Total (000)	Gross enrolment ratio Taux brut de scolarisation		Total (000)	Gross enrolment ratio Taux brut de scolarisation		Total (000)	Gross enrolment ratio Taux brut de scolarisation	
			M	F		M	F		M	F
Swaziland	2005	222	106.3	99.9	71	46.2	47.1	6	4.6	4.9
Swaziland	2006	230	111.8	104.5	77	49.7	51.0	6	4.5	4.4
	2010	241	120.7	110.8	89	58.0	58.0	...	...	...
	2013	239	118.2	108.3	93	63.6	62.4	8	5.2	5.5
Sweden	2005	658	96.0	95.7	735	104.3	103.9	427	64.8	100.0
Suède	2010	576	101.8	101.2	731	98.7	97.7	455	59.3	90.8
	2014	757	117.7	124.3	827	124.8	141.6	429	49.4	76.0
Switzerland	2005	524	102.2	101.9	575	98.0	92.2	200	49.0	42.4
Suisse	2010	493	102.8	102.3	605	97.3	94.4	249	53.1	52.5
	2014	484	103.2	103.3	616	101.3	98.2	290	56.8	57.7
Syrian Arab Republic	2005	2 252	123.0	119.5	2 389	71.5	68.4	361	19.2	17.2
Rép. arabe syrienne	2010	2 429	124.4	120.0	2 732	72.8	72.9	574	27.9	24.1
	2013	1 547	81.4	78.7	1 857	50.5	50.5	630	32.4	33.6
Tajikistan	2005	693	100.6	97.2	984	87.7	72.9	149	28.3	13.5
Tadjikistan	2010	682	101.5	98.5	1 032	90.2	78.5	196	29.7	15.6
	2013	665	99.9	98.0	1 063	92.5	83.1	195	28.9	16.2
	2014	662	97.5	97.1	...	...	...	209	30.4	18.4
Thailand	2005	5 975	99.7	96.9	4 533	*69.6	*73.6	2 359	41.4	47.2
Thaïlande	2010	5 147	97.1	95.0	4 807	81.2	86.1	2 427	44.1	56.3
	2013	4 955	98.9	96.9	4 655	83.2	89.3	2 405	44.0	58.9
	2014	5 182	103.3	104.1	...	...	...	2 433	45.0	60.0
TFYR of Macedonia	2005	110	95.2	93.3	214	86.1	83.6	49	24.8	34.1
ex-R.Y. de Macédoine	2010	111	86.3	84.9	197	84.6	82.4	62	36.2	42.3
	2012	107	86.4	85.1	186	82.9	81.1	63	37.2	44.7
	2013	...	...	...	...	...	...	61	34.9	44.1
Timor-Leste	2002	184	...	...	53	...	...	*6	*8.3	*9.5
Timor-Leste	2005	178	98.4	90.1	75	54.3	52.8	...	...	...
	2010	230	133.8	128.4	102	67.2	67.7	19	21.1	15.2
	2014	246	137.5	136.0	119	70.3	76.0	...	...	...
Togo	2005	997	121.9	103.5	404	58.9	31.2	...	...	...
Togo	2007	1 022	118.7	102.7	409	*57.6	*30.5	33	...	...
	2010	1 287	134.9	121.8	...	...	...	56	...	...
	2011	1 300	131.8	120.0	546	...	...	63	...	...
	2014	1 413	128.8	121.4	...	...	...	67	14.3	5.9
Tokelau Tokélaou	2003	<0	*101.8	*132.6	<0	*83.3	*75.8	...	...	...
Tonga	2002	17	113.2	108.1	15	105.7	116.6	1	*4.2	*7.1
Tonga	2003	18	119.5	113.1	16	...	...	1	4.8	8.0
	2004	17	114.9	109.2	14	...	...	...	...	...
	2005	17	113.6	110.3	...	...	...	...	...	...
	2010	17	109.3	107.2	15	100.9	107.7	...	...	...
	2014	17	108.6	107.5	15	86.4	94.2	...	...	...
Trinidad and Tobago	2004	*137	*103.4	*100.0	*105	*82.6	*88.5	17	10.6	13.4
Trinité-et-Tobago	*2005	130	100.7	97.8	...	...	...	...	...	...
	2010	131	108.0	104.3	...	...	...	...	...	...
Tunisia	2005	1 184	114.0	109.3	1 239	81.4	88.3	327	28.6	35.1
Tunisie	2010	1 030	109.0	105.4	1 164	87.6	93.4	370	27.8	42.6
	2011	1 028	109.4	106.0	1 152	90.0	94.2	362	26.7	43.1
	2014	1 089	114.7	111.4	1 020	...	...	332	26.3	43.1
Turkey	2005	6 678	105.8	100.5	6 345	91.0	75.6	2 106	37.8	27.6
Turquie	2010	6 635	102.3	100.8	7 531	87.8	80.5	3 529	61.7	50.2
	2013	5 594	107.2	106.5	10 563	101.8	98.7	4 976	84.9	72.9
Turkmenistan Turkménistan	2014	359	90.1	88.6	651	86.9	83.7	44	9.7	6.2

Education at the primary, secondary and tertiary levels *(continued)*
Number of students enrolled (thousands) and gross enrolment ratio by sex

Enseignement primaire, secondaire et supérieur *(suite)*
Nombre d'élèves inscrits (milliers) et taux brut de scolarisation par sexe

Region, country or area Région, pays ou zone	Year [t] Année [t]	Primary education Enseignement primaire			Secondary education Enseignement secondaire			Tertiary education Enseignement supérieur		
		Total (000)	Gross enrolment ratio Taux brut de scolarisation		Total (000)	Gross enrolment ratio Taux brut de scolarisation		Total (000)	Gross enrolment ratio Taux brut de scolarisation	
			M	F		M	F		M	F
Turks and Caicos Islands	2005	2	...	...	2	...	...	...	...	...
Îles Turques-et-Caïques	2009	3	...	...	2	...	...	...	...	...
	2010	...	...	...	...	...	...	<0	...	...
	2014	3	...	...	2	...	...	...	...	...
Tuvalu	2001	1	*94.9	*106.9	1	*67.4	*74.1	...	...	...
Tuvalu	2005	1	*101.4	*97.6	...	...	...	...	...	...
	2006	1	*102.4	*97.6	...	...	...	...	...	...
	2014	1	*100.7	*102.1	1	*72.3	*90.2	...	...	...
Uganda	2004	7 377	130.3	128.4	733	21.5	17.3	88	4.4	2.8
Ouganda	2005	7 224	123.2	122.2	*767	*21.7	*17.4	...	...	...
	2010	8 375	119.3	121.2	1 265	29.2	25.1	121	4.5	3.5
	2011	8 098	111.4	113.3	1 254	28.1	24.0	140	5.0	3.9
	2013	8 459	108.9	110.9	1 421	29.5	25.7	...	...	...
Ukraine	2005	1 946	106.1	105.7	4 043	*100.6	*93.2	2 605	63.8	78.8
Ukraine	2010	1 540	102.4	103.1	3 133	*96.2	*93.9	2 635	72.8	91.5
	2014	1 685	102.8	105.1	2 714	100.2	98.2	2 146	76.5	88.4
United Arab Emirates	2005	263	105.8	105.2	285	...	...	...	...	...
Émirats arabes unis	2010	327	98.1	101.5	*338	...	...	102	8.2	24.3
	2014	410	106.3	107.1	411	...	...	143	15.4	34.6
United Kingdom	2005	4 635	106.3	106.2	5 761	103.9	107.0	2 288	49.5	68.9
Royaume-Uni	2010	4 422	106.1	105.6	5 538	102.1	101.7	2 479	50.4	68.1
	2014	4 737	108.4	108.1	6 557	125.3	130.4	2 353	49.0	64.1
United Rep. of Tanzania	2005	7 541	106.2	101.0	...	...	...	*52	*2.0	*0.9
Rép.-Unie de Tanzanie	2010	8 419	98.6	99.3	1 826	34.8	27.5	85	2.4	1.9
	2013	8 232	86.2	87.4	2 052	33.7	30.8	158	4.9	2.5
United States	2005	24 455	101.1	100.3	24 432	94.4	97.1	17 272	68.2	96.8
États-Unis	2010	24 393	101.0	100.2	24 193	93.5	95.2	20 428	78.6	110.7
	2014	24 538	99.8	99.2	24 230	96.7	98.5	19 700	73.5	100.7
United States Virgin Is. Îles Vierges américaines	2007	...	...	...	...	...	...	2	...	...
Uruguay	2005	366	115.3	112.4	323	94.2	108.5	111	33.2	57.9
Uruguay	2010	342	113.9	110.2	287	84.6	96.2	163	46.5	80.3
	2013	327	111.1	108.3	291	88.5	99.8	...	...	...
Uzbekistan	2005	2 383	102.7	102.6	4 516	91.2	88.3	266	12.0	8.2
Ouzbékistan	2010	1 971	97.7	95.8	4 449	95.0	94.8	289	11.2	7.6
	2011	1 948	98.0	95.7	4 370	95.9	94.9	277	10.9	6.9
Vanuatu	2004	39	122.4	120.0	14	45.8	39.5	*1	*5.9	*3.5
Vanuatu	2005	39	119.2	116.1	...	...	...	...	...	...
	2010	42	123.0	121.9	20	59.6	59.5	...	...	...
	2013	44	125.2	122.2	...	...	...	...	...	...
Venezuela (Boliv. Rep. of)	2003	3 450	104.0	101.9	1 866	63.4	72.9	*983	*38.4	*41.2
Venezuela (Rép. boliv. du)	2004	3 453	104.1	101.9	1 954	66.1	75.3	*1 050	...	...
	2005	3 449	103.8	101.8	2 028	68.6	77.5	...	...	...
	2008	3 439	103.3	100.5	2 224	76.1	83.8	2 109	*57.8	*97.7
	2009	3 462	103.6	100.8	2 252	77.4	84.6	2 123	...	...
	2010	3 458	103.1	100.1	2 255	77.3	84.7	...	...	...
	2014	3 493	101.9	99.7	2 567	88.3	95.0	...	...	...
Viet Nam	2005	7 773	99.6	94.6	...	...	...	1 355	18.7	13.3
Viet Nam	2010	6 923	107.7	102.3	...	...	...	2 020	22.7	22.7
	2014	7 435	110.0	108.6	...	...	...	2 692	29.8	31.2
Yemen	2005	3 220	100.7	74.5	1 455	61.0	29.8	200	14.0	5.1
Yémen	2010	3 427	99.8	81.5	1 561	53.3	33.1	272	14.6	6.4
	2011	3 641	104.3	85.1	1 643	55.4	35.1	267	13.7	6.1
	2013	3 875	105.7	88.9	1 768	57.4	39.5	...	...	...

5

Education at the primary, secondary and tertiary levels *(continued)*
Number of students enrolled (thousands) and gross enrolment ratio by sex

Enseignement primaire, secondaire et supérieur *(suite)*
Nombre d'élèves inscrits (milliers) et taux brut de scolarisation par sexe

Region, country or area Région, pays ou zone	Year [t] Année [t]	Primary education Enseignement primaire			Secondary education Enseignement secondaire			Tertiary education Enseignement supérieur		
		Total (000)	Gross enrolment ratio Taux brut de scolarisation		Total (000)	Gross enrolment ratio Taux brut de scolarisation		Total (000)	Gross enrolment ratio Taux brut de scolarisation	
			M	F		M	F		M	F
Zambia	2005	2 573	113.1	108.0	...	...	...	...	...	...
Zambie	2010	2 899	106.9	108.3	...	...	...	...	...	...
	2013	3 075	103.3	104.0	...	...	...	...	...	...
Zimbabwe	2003	2 362	96.8	95.4	758	39.6	35.6	...	...	...
Zimbabwe	2010	...	...	...	...	...	...	95	6.6	5.2
	2013	2 663	100.8	99.1	957	48.1	47.1	94	6.4	5.4

Source:
United Nations Educational, Scientific and Cultural Organization (UNESCO) Institute for Statistics, Montreal, the UNESCO Institute for Statistics (UIS) database, last accessed June 2016.

[t] Data relate to the calendar year in which the academic year ends.

Source :
L'Institut de statistique de l'Organisation des Nations Unies pour l'éducation, la science et la culture (UNESCO), Montréal, la base de données de l'institut de statistique de l'UNESCO (ISU), dernier accès juin 2016.

[t] Les données se réfèrent à l'année civile durant laquelle l'année scolaire se termine.

6

Public expenditure on education
By expenditure type, level of education, total government expenditure and GDP

Dépenses publiques afférentes à l'éducation
Par type de dépenses, niveau de scolarité, dépenses publiques totales et PIB

Country or area Pays ou zone	Year Année	Percentage of total expenditure in public institutions Pourcentage des dépenses publiques totales en faveur de l'éducation			Percentage of government expenditure on education by level Pourcentage des dépenses publiques selon le niveau d'enseignement				As % of GDP En % du PIB	As % of Government exp. En % des dépenses du gouvernement
		Current expenditure Dépenses courantes	Staff Compensation Rémunération du personnel	Capital expenditure Dépenses en Capital	Pre-primary Préprimaire	Primary Primaire	Secondary Secondaire	Tertiary Tertiaire		
Afghanistan *	2010	...	...	...	...	...	...	...	22.2	4.5
Afghanistan *	2014	...	...	...	...	...	...	...	18.4	4.8
Albania	2005	...	...	...	...	...	...	...	11.4	3.2
Albanie	2007	...	...	...	...	...	...	...	11.2	3.3
	2013	11.5	79.9	8.6	...	56.9	19.8	21.9	12.1	3.5
Algeria										
Algérie	2008	...	...	...	...	...	...	27.0	11.4	4.3
American Samoa										
Samoa américaines	2006	...	...	6.2	...	...	...	11.2	...	...
Andorra	2005	...	...	...	16.9	24.1	20.6	3.9	...	1.6
Andorre	2010	30.9	36.7	1.7	14.6	28.9	20.9	3.9	...	3.1
	2013	59.4	40.0	0.6	18.2	32.1	35.2	4.6	...	2.5
	2014	...	...	1.1	13.8	22.3	20.5	6.1	...	3.1
Angola	2005	...	...	...	...	36.5	55.2	8.3	8.0	2.1
Angola	2006	...	...	24.3	13.5	31.4	42.4	8.7	7.5	2.3
	2010	...	...	...	...	...	...	...	8.7	3.4
Anguilla	2005	6.7	80.9	...	1.9	28.5	51.3	18.3	...	...
Anguilla	2008	* 9.8	* 69.4	* 20.8	1.6	39.4	56.1	2.8	...	2.8
Antigua and Barbuda	2002	25.7	...	3.9	2.1	28.6	35.2	6.7	8.9	3.4
Antigua-et-Barbuda	2009	21.6	73.2	5.2	0.3	41.6	48.4	7.4	6.9	2.6
Argentina	2004	11.6	87.6	0.8	8.3	36.7	37.8	17.2	15.2	3.2
Argentine	2005 `	...	...	...	7.9	34.2	41.7	16.2	15.8	3.5
	2010	9.9	85.9	4.2	7.4	33.1	39.7	19.9	15.6	4.6
	2013	6.0	88.2	5.9	8.2	30.0	41.5	20.2	15.1	5.3
Armenia	2005	...	...	...	...	...	...	...	13.6	2.7
Arménie	2009	...	...	...	11.3	...	...	9.4	13.5	3.8
	2010	...	...	...	...	...	...	11.7	12.4	3.2
	2013	10.1	67.2	22.6	11.0	20.1	48.9	10.3	11.1	2.7
	2014	8.5	65.5	26.0	...	...	54.5	13.6	9.4	2.2
Aruba	2005	6.1	90.7	3.2	6.8	26.7	26.8	11.5	18.9	4.7
Aruba	2009	5.7	87.0	...	...	...	...	...	21.6	5.9
	2010	...	...	...	...	...	...	...	22.2	6.7
	2011	5.5	94.5	...	...	...	...	...	21.8	6.0
Australia	2000	24.7	68.3	7.0	1.2	34.1	38.6	23.4	13.6	4.9
Australie	2005	...	...	...	1.0	34.6	39.7	22.2	13.6	4.9
	2010	...	...	...	1.2	36.8	36.8	22.3	14.3	5.6
	2013	...	...	...	2.7	32.8	33.1	25.9	14.0	5.3
Austria	2005	23.2	72.1	4.7	7.4	19.0	45.9	27.3	10.3	5.3
Autriche	2010	24.9	70.2	4.9	10.3	17.1	44.4	27.7	10.9	5.7
	2013	27.5	68.0	4.5	8.4	16.3	41.2	32.3	11.0	5.6
Azerbaijan	2005	31.6	66.8	1.6	6.9	...	...	6.9	13.1	3.0
Azerbaïdjan	2010	32.6	64.8	2.5	7.1	...	...	14.0	8.8	2.8
	2011	34.6	62.7	2.8	7.0	...	...	14.6	7.3	2.4
	2013	31.1	64.5	4.4	...	...	...	...	6.5	2.5
Bahamas *										
Bahamas *	2000	...	...	...	...	...	...	...	19.5	2.9
Bahrain	2008	...	...	...	...	...	...	...	10.6	2.5
Bahreïn	2012	13.8	82.3	3.8	...	...	...	...	8.9	2.6
Bangladesh	2004	...	...	...	...	40.0	48.5	11.5	16.0	1.9
Bangladesh	2009	21.2	59.8	19.0	...	44.7	40.2	13.5	14.0	1.9
	*2013	...	...	...	...	...	...	...	13.8	2.0
Barbados	2005	16.2	79.9	4.0	* 6.3	* 27.9	31.7	31.6	15.4	5.6
Barbade	2008	14.5	78.6	6.9	0.4	36.9	29.4	30.1	13.4	5.1
	2010	10.4	83.4	6.2	...	...	29.6	32.5	15.6	5.9
	2014	47.8	48.6	3.6	...	...	22.5	39.3	14.2	6.7

Public expenditure on education *(continued)*
By expenditure type, level of education, total government expenditure and GDP

Dépenses publiques afférentes à l'éducation *(suite)*
Par type de dépenses, niveau de scolarité, dépenses publiques totales et PIB

Country or area Pays ou zone	Year Année	Percentage of total expenditure in public institutions Pourcentage des dépenses publiques totales en faveur de l'éducation			Percentage of government expenditure on education by level Pourcentage des dépenses publiques selon le niveau d'enseignement				As % of GDP En % du PIB	As % of Government exp. En % des dépenses du gouvernement
		Current expenditure Dépenses courantes	Staff Compensation Rémunération du personnel	Capital expenditure Dépenses en Capital	Pre-primary Préprimaire	Primary Primaire	Secondary Secondaire	Tertiary Tertiaire		
Belarus	2005	30.1	65.2	4.8	17.4	...	...	25.5	12.9	5.9
Bélarus	2010	26.5	62.6	10.9	19.0	...	...	17.1	12.9	5.4
	2011	26.8	65.4	7.8	21.0	...	...	17.3	14.0	4.8
	2014	25.7	65.6	8.7	...	...	...	16.9	12.4	5.0
Belgium	2005	11.7	85.5	2.8	9.6	23.6	43.1	21.7	11.4	5.8
Belgique	2010	11.9	84.3	3.8	9.5	23.5	42.5	22.2	12.2	6.4
	2011	11.9	84.5	3.6	9.6	23.3	42.8	22.0	11.9	6.4
Belize	2003	* 18.3	* 62.3	* 19.3	...	...	...	...	16.2	5.2
Belize	2004	...	...	...	0.6	42.6	39.2	10.8	16.8	5.3
	2010	...	...	...	1.5	46.1	41.1	8.7	23.1	6.6
	2013	...	...	...	1.4	40.0	38.1	13.6	22.9	6.2
Benin	2001	15.2	56.8	28.0	* 1.2	* 55.4	* 25.1	18.3	15.3	3.2
Bénin	2005	13.3	73.0	13.7	...	* 46.0	* 32.8	21.2	18.8	3.6
	2010	6.2	85.7	8.2	2.6	52.8	28.0	15.6	26.1	5.0
	2014	19.9	73.5	6.6	4.2	48.9	26.3	20.4	22.2	4.4
Bermuda	2005	...	...	...	* 7.0	* 40.7	* 52.3	...	12.1	2.0
Bermudes	2010	...	...	...	8.9	30.8	45.4	13.5	12.9	2.6
	2014	...	...	...	7.5	29.9	44.3	18.2	7.8	1.8
Bhutan	2005	...	...	...	...	22.3	50.9	14.1	22.8	7.1
Bhoutan	2010	...	...	...	...	29.9	53.9	15.1	12.8	4.0
	2014	...	...	...	...	32.0	55.7	10.3	17.8	5.9
Bolivia (Plurin.State of)	2000	...	...	14.5	6.3	* 40.5	* 18.5	28.9	18.7	5.5
Bolivie (État plurin. de)	2003	...	...	...	3.3	44.1	24.3	21.6	19.9	6.4
	2009	31.8	68.2	...	2.6	40.5	26.0	29.7	22.6	8.1
	2010	...	...	...	2.8	39.4	26.3	29.9	24.1	7.6
	2014	12.9	68.8	18.3	5.0	32.4	36.3	26.2	16.9	7.3
Botswana	2005			...	...	30.0	39.6	27.9	25.8	10.7
Botswana	2009	...	...	9.8	...	17.8	32.7	41.5	20.5	9.6
Brazil	2005	23.6	70.3	6.1	8.2	34.0	38.8	19.0	11.3	4.5
Brésil	2010	24.3	68.5	7.2	7.6	31.3	44.7	16.4	14.6	5.6
	2012	23.7	68.3	8.0	10.1	28.9	44.6	16.4	15.6	5.9
	2013	24.0	69.7	6.2	...	...	...	...	...	...
British Virgin Islands	*2004	2.3	90.3	7.4	...	...	...	...	...	...
Îles Vierges britan.	2005	0.7	99.3	...	0.1	33.7	31.2	34.2	...	...
	2007	15.1	77.8	7.1	0.1	28.4	37.7	33.1	...	3.6
	2010	...	...	...	0.1	34.1	34.0	...	...	4.4
Brunei Darussalam	2000	...	...	...	...	...	...	...	8.9	3.7
Brunéi Darussalam	2010	...	...	...	...	28.5	46.8	24.4	5.3	2.0
	2014	...	...	...	...	...	38.7	31.9	10.0	3.8
Bulgaria	2005	26.7	63.2	10.1	16.9	20.3	45.9	16.9	12.1	4.1
Bulgarie	2010	28.4	66.6	5.1	22.5	19.6	43.0	14.8	11.2	3.9
	2012	22.7	71.4	5.9	24.3	18.6	39.2	17.8	10.7	3.5
Burkina Faso	2005	...	...	...	0.2	71.2	10.3	9.6	19.5	4.4
Burkina Faso	2007	8.0	56.7	35.3	0.6	67.0	15.7	15.2	17.9	4.6
	2010	...	...	...	...	60.3	18.0	18.8	17.3	3.9
	2014	...	...	...	<0.0	65.4	16.9	12.3	19.4	4.5
Burundi	2004	15.7	57.8	26.6	<0.0	50.7	29.5	18.3	13.0	3.7
Burundi	2005	...	...	...	...	51.6	33.1	15.3	13.9	3.6
	2010	23.7	68.5	7.9	<0.0	45.4	27.9	18.0	16.6	6.8
	2012	20.5	77.6	1.9	<0.0	44.0	24.2	20.6	16.4	5.8
	2013	...	...	...	<0.0	45.4	26.8	24.2	17.2	5.4

6

Public expenditure on education *(continued)*
By expenditure type, level of education, total government expenditure and GDP

Dépenses publiques afférentes à l'éducation *(suite)*
Par type de dépenses, niveau de scolarité, dépenses publiques totales et PIB

Country or area / Pays ou zone	Year / Année	Percentage of total expenditure in public institutions / Pourcentage des dépenses publiques totales en faveur de l'éducation			Percentage of government expenditure on education by level / Pourcentage des dépenses publiques selon le niveau d'enseignement				As % of GDP / En % du PIB	As % of Government exp. / En % des dépenses du gouvernement
		Current expenditure / Dépenses courantes	Staff Compensation / Rémunération du personnel	Capital expenditure / Dépenses en Capital	Pre-primary / Préprimaire	Primary / Primaire	Secondary / Secondaire	Tertiary / Tertiaire		
Cabo Verde	2002	...	...	...	...	43.8	29.8	17.5	19.8	7.9
Cabo Verde	2004	3.3	80.0	16.7	...	...	...	...	20.8	7.5
	*2005	2.1	82.4	16.7	...	...	...	...	...	...
	2008	...	...	15.9	0.3	36.3	36.6	11.3	18.6	5.5
	2009	...	...	...	...	44.1	33.0	14.0	16.2	5.3
	2010	15.6	77.1	7.3	...	...	...	...	14.4	5.6
	2013	...	...	...	0.7	38.4	40.3	15.8	15.0	5.0
Cambodia	2001	...	...	...	1.1	74.4	11.2	5.0	11.4	1.7
Cambodge	2002	...	...	...	1.0	64.6	...	...	10.1	1.7
	2004	...	...	...	...	...	...	...	12.4	1.7
	2010	...	...	...	2.2	41.8	17.6	14.5	13.1	2.6
	2013	...	...	...	2.7	49.7	41.5	6.1	9.9	2.0
Cameroon	2005	...	...	...	...	35.1	52.0	12.6	21.4	3.1
Cameroun	2010	6.5	80.2	13.4	3.9	34.2	52.7	9.0	18.8	3.3
	2012	...	...	...	3.5	36.2	52.6	7.8	15.2	3.0
	2013	10.7	75.7	13.6	4.0	33.9	...	10.2	13.8	3.0
Canada	2000	25.9	70.5	3.6	3.9	...	...	...	12.7	5.5
Canada	2005	25.4	70.0	4.6	...	...	...	...	12.3	4.8
	2010	24.5	65.8	9.7	...	...	...	35.4	12.3	5.4
	2011	24.1	65.5	10.4	...	...	26.4	35.6	12.2	5.3
Cayman Islands	2005	...	...	...	1.2	49.3	49.5	...	...	...
Îles Caïmanes	2006	2.4	97.6	...	...	46.0	54.0	...	...	...
Central African Rep.	2005	...	79.2	2.2	...	...	...	...	9.7	1.6
Rép. centrafricaine	2008	13.5	84.2	2.3	...	...	25.9	17.5	7.9	1.3
	2010	...	...	1.1	2.3	53.3	24.0	20.5	6.5	1.2
	2011	...	...	...	...	...	...	27.3	7.8	1.2
Chad	2005	...	...	...	...	45.8	35.5	18.7	14.7	1.7
Tchad	2010	...	...	31.3	0.4	51.6	27.4	19.8	8.1	2.0
	2011	...	...	20.8	0.3	40.7	34.9	16.3	10.1	2.3
	2013	...	...	...	...	...	24.7	31.7	12.5	2.9
Chile	2005	17.5	78.1	4.4	9.6	37.4	39.0	14.0	16.1	3.2
Chili	2009	18.1	78.9	3.0	12.8	36.4	35.2	15.6	17.2	4.2
	2010	...	...	...	13.4	31.7	32.5	22.3	17.5	4.2
	2011	...	14.9	...	13.0	32.8	32.5	21.8	17.5	4.1
	2013	...	...	...	15.5	29.2	29.7	25.6	19.1	4.6
China, Hong Kong SAR	2005	...	...	...	2.3	23.1	33.8	28.4	22.5	4.1
Chine, Hong Kong RAS	2010	...	...	12.7	3.8	20.9	35.5	27.8	19.9	3.5
	2012	...	...	18.2	3.9	18.7	33.6	32.8	18.6	3.5
	2014	...	...	...	4.2	18.9	31.3	29.2	17.6	3.6
China, Macao SAR	2000	...	...	...	*7.6	*24.8	*25.0	28.4	...	3.7
Chine, Macao RAS	2003	...	...	...	...	*24.8	...	41.1	...	2.9
	2005	...	...	...	...	...	...	45.6	...	2.3
	2010	19.5	39.0	41.5	...	...	...	48.4	15.7	2.6
	2013	23.5	49.8	26.7	...	...	55.1	44.9	14.3	2.1
Colombia	2004	23.3	71.0	5.7	*2.7	*48.3	*35.7	13.3	15.4	4.1
Colombie	2005	...	...	...	2.6	47.4	36.1	13.8	15.5	4.0
	2010	6.8	71.0	22.3	5.8	35.9	34.8	22.1	16.4	4.8
	2014	5.5	68.5	25.9	5.7	35.6	37.5	20.8	15.9	4.7
Comoros	2002	...	...	...	*0.4	*45.2	*40.1	*7.7	15.8	3.9
Comores	2008	...	...	...	...	61.7	23.7	14.6	29.2	7.8
	2012	...	...	...	7.4	56.6	27.8	8.2	18.5	5.1
Congo	2005	...	...	...	*2.9	*27.3	*41.2	*25.9	7.6	1.8
Congo	2010	...	...	...	4.6	31.0	53.3	10.9	29.0	6.2
Cook Islands	2000	...	...	...	7.1	50.6	36.5	...	6.7	...
Îles Cook	2013	...	...	...	11.4	40.4	40.6	7.6	...	2.9
	2014	15.7	84.3	...	...	...	...	7.3	...	3.9

Public expenditure on education *(continued)*
By expenditure type, level of education, total government expenditure and GDP

Dépenses publiques afférentes à l'éducation *(suite)*
Par type de dépenses, niveau de scolarité, dépenses publiques totales et PIB

| Country or area
Pays ou zone | Year
Année | Percentage of total expenditure in public institutions
Pourcentage des dépenses publiques totales en faveur de l'éducation | | | Percentage of government expenditure on education by level
Pourcentage des dépenses publiques selon le niveau d'enseignement | | | | As % of GDP
En % du PIB | As % of Government exp.
En % des dépenses du gouvernement |
		Current expenditure Dépenses courantes	Staff Compensation Rémunération du personnel	Capital expenditure Dépenses en Capital	Pre-primary Préprimaire	Primary Primaire	Secondary Secondaire	Tertiary Tertiaire		
Costa Rica	2004	...	...	...	7.9	45.5	27.8	18.8	20.4	4.9
Costa Rica	2010	...	...	...	5.4	41.9	31.5	18.4	22.9	6.8
	2014	...	...	...	6.1	36.3	33.4	21.7	23.1	7.0
Côte d'Ivoire	2000	...	...	...	* <0.0	* 41.6	* 32.7	22.0	20.8	3.7
Côte d'Ivoire	2001	...	...	...	...	* 42.2	...	19.4	23.2	3.7
	2005	...	...	...	...	...	...	20.0	21.8	4.1
	2007	...	...	...	...	...	...	21.0	21.9	4.3
	2008	...	...	...	...	...	...	...	21.9	4.4
	2014	14.8	77.5	7.7	2.7	39.1	37.2	21.0	20.7	4.7
Croatia	2004	30.0	65.6	4.4	9.2	...	...	18.1	8.2	3.8
Croatie	2010	23.0	70.6	6.4	13.8	...	...	18.3	9.1	4.3
	2011	21.3	74.4	4.3	14.1	...	...	22.2	8.6	4.2
Cuba	2005	34.3	52.6	13.2	8.3	30.3	37.7	22.1	...	10.6
Cuba	2007	46.7	50.2	3.1	8.1	30.1	36.0	25.1	...	11.9
	2010	40.5	58.6	0.9	6.8	29.0	29.1	...	...	12.8
Curaçao										
Curaçao	2013	13.9	86.1	...	9.2	27.5	33.9	5.3	...	4.9
Cyprus	2005	9.8	76.9	13.4	4.9	27.3	44.9	22.9	15.7	6.3
Chypre	2010	11.5	77.5	11.0	5.5	31.0	43.0	20.5	15.7	6.7
	2011	10.7	79.0	10.3	5.6	31.3	42.6	20.4	15.5	6.6
Czech Republic	2005	34.4	56.0	9.7	9.8	14.5	51.3	21.0	9.3	3.9
République tchèque	2010	38.4	51.5	10.1	11.4	16.2	45.7	22.5	9.5	4.1
	2012	42.3	48.4	9.3	12.1	16.8	44.0	23.4	9.7	4.3
Dem. Rep. of the Congo	2010	2.6	87.8	9.6	0.6	33.3	33.7	24.0	9.0	1.6
Rép. dém. du Congo	2013	1.7	81.4	16.9	1.0	61.6	14.4	22.0	16.8	2.2
Denmark	2005	20.7	74.1	5.3	8.6	23.3	36.3	28.7	15.8	8.1
Danemark	2010	18.0	74.9	7.1	11.5	23.7	34.5	27.3	15.1	8.6
	2011	25.2	71.1	3.7	15.7	21.6	32.1	27.9	15.0	8.5
Djibouti	2005	...	42.2	2.7	...	...	...	...	22.7	8.4
Djibouti	2007	...	38.8	4.6	...	19.0	...	...	22.3	8.4
	2010	...	...	...	...	34.1	37.3	16.5	12.3	4.5
Dominica										
Dominique	2012	...	100.0	...	...	...	...	...	...	...
Dominican Republic	2002	16.6	76.7	6.7	2.1	62.8	...	...	12.1	2.0
Rép. dominicaine	*2003	...	...	...	...	...	...	...	9.7	1.9
	2004	24.1	72.6	3.4	...	...	...	...	...	...
	2007	15.5	66.9	17.6	4.2	58.7	18.0	14.5	12.6	2.1
	2013	24.1	40.1	35.8	...	...	...	...	...	...
Ecuador	2000	16.7	78.1	5.2	...	38.3	34.6	5.2	5.0	1.2
Équateur	2010	22.1	72.2	5.7	3.0	30.1	35.0	26.6	11.8	4.1
	2012	10.8	81.8	7.4	4.8	25.4	41.2	26.7	10.3	4.2
Egypt	2005	...	...	...	...	...	...	...	14.4	4.8
Égypte	2008	...	...	...	...	...	...	...	10.4	3.8
El Salvador	2005	27.3	66.9	5.8	* 8.5	* 51.0	* 29.1	* 11.1	14.7	2.7
El Salvador	2008	37.8	51.7	10.5	9.1	38.0	22.0	...	19.4	3.7
	2010	...	...	...	7.9	42.7	30.1	12.2	16.2	3.5
	2011	26.7	63.0	10.2	7.6	...	...	8.4	15.9	3.4
Eritrea	2002	8.6	31.0	60.4	...	26.0	35.4	14.9	5.6	3.5
Érythrée	2004	...	...	41.7	...	23.1	21.1	31.9	5.7	3.1
	2006	22.7	57.7	19.6	...	...	...	...	5.2	2.1
Estonia	2002	28.8	56.3	15.0	5.2	28.0	40.9	19.7	15.3	5.5
Estonie	2004	...	...	9.5	6.4	25.8	46.3	17.4	14.5	4.9
	2005	...	...	...	7.4	25.3	43.7	18.9	14.5	4.8
	2010	...	...	...	7.8	24.6	40.1	21.7	13.7	5.5
	2012	...	...	...	9.1	27.0	36.4	21.8	12.1	4.7

Public expenditure on education *(continued)*
By expenditure type, level of education, total government expenditure and GDP

Dépenses publiques afférentes à l'éducation *(suite)*
Par type de dépenses, niveau de scolarité, dépenses publiques totales et PIB

Country or area Pays ou zone	Year Année	Percentage of total expenditure in public institutions Pourcentage des dépenses publiques totales en faveur de l'éducation			Percentage of government expenditure on education by level Pourcentage des dépenses publiques selon le niveau d'enseignement				As % of GDP En % du PIB	As % of Government exp. En % des dépenses du gouvernement
		Current expenditure Dépenses courantes	Staff Compensation Rémunération du personnel	Capital expenditure Dépenses en Capital	Pre-primary Préprimaire	Primary Primaire	Secondary Secondaire	Tertiary Tertiaire		
Ethiopia	*2002	...	...	...	...	...	...	...	16.2	3.6
Éthiopie	2010	21.0	44.8	34.1	<0.0	28.4	25.0	43.9	26.3	4.5
	2013	19.7	44.2	36.1	0.4	24.0	29.3	42.7	27.0	4.5
Fiji	2004	...	...	...	0.2	39.1	32.6	15.9	22.0	6.2
Fidji	2005	...	...	...	...	...	...	...	18.8	5.1
	2009	...	...	...	...	...	...	...	15.4	4.5
	2011	...	...	...	0.4	44.4	15.9	13.0	14.9	4.2
	2013	1.2	98.7	0.1	...	38.6	...	22.6	14.0	3.9
Finland	2005	32.1	60.9	7.0	5.5	20.8	41.9	31.8	12.2	6.0
Finlande	2010	33.6	60.3	6.1	5.8	19.9	42.4	31.8	11.9	6.5
	2013	34.0	60.2	5.8	10.8	18.6	15.0	28.0	12.5	7.2
France	2005	15.6	74.5	9.9	11.4	20.2	47.1	21.1	10.4	5.5
France	2010	17.4	73.2	9.4	11.6	20.4	45.0	22.6	10.1	5.7
	2012	17.7	73.8	8.5	12.3	21.1	43.8	22.3	9.7	5.5
Gabon *										
Gabon *	2000	...	...	...	...	...	...	...	17.7	3.8
Gambia	2005	...	...	...	...	61.4	23.1	14.3	5.3	1.1
Gambie	2010	10.8	46.4	42.8	...	65.3	23.4	9.9	17.6	4.2
	2012	7.4	50.4	42.2	...	60.0	31.4	7.4	13.8	4.1
	2013	...	...	...	...	51.8	35.5	10.7	10.3	2.8
Georgia	2005	...	...	...	...	...	...	...	11.2	2.5
Géorgie	2008	...	...	...	10.7	36.0	37.0	11.6	8.9	2.9
	*2009	...	...	...	12.3	...	...	...	9.0	3.2
	2012	...	...	...	14.4	34.3	32.1	19.2	6.7	2.0
Germany	2010	19.8	70.6	9.6	9.1	13.5	45.5	27.2	10.4	4.9
Allemagne	2012	19.7	72.8	7.6	8.6	12.8	43.2	26.8	11.2	4.9
Ghana	2005	22.3	63.0	14.7	5.9	39.5	32.5	22.1	23.4	7.4
Ghana	2010	14.3	68.8	16.9	6.1	30.9	37.1	25.9	21.2	5.5
	2013	18.7	76.8	4.5	6.6	25.1	48.5	19.8	21.7	6.0
Greece										
Grèce	2005	11.0	67.3	21.7	...	...	33.9	36.1	9.2	4.0
Grenada										
Grenade	2003	8.3	82.2	9.5	5.7	36.4	35.8	9.8	10.8	3.9
Guatemala	2008	9.9	65.2	11.5	10.0	59.8	12.8	10.8	23.3	3.2
Guatemala	2010	...	...	0.1	11.2	55.6	14.8	11.3	19.3	2.8
	2012	28.4	68.2	3.4	11.0	55.5	13.1	12.8	21.1	3.0
	2013	...	...	3.3	10.6	52.7	14.3	12.3	20.6	2.8
	2014	...	...	5.5	...	...	...	...	...	...
Guinea	2005	...	...	...	...	...	...	30.6	10.9	1.8
Guinée	2010	...	...	7.7	...	38.9	24.9	36.2	12.4	3.7
	2013	18.7	58.7	22.6	0.1	40.1	25.2	34.6	14.1	3.5
Guinea-Bissau	2010	12.5	61.5	26.0	0.7	46.7	46.7	5.0	10.0	1.9
Guinée-Bissau	2013	0.9	48.3	50.8	0.8	64.4	30.4	3.9	16.2	2.4
Guyana	2004	37.6	51.8	10.6	11.6	32.1	26.5	6.2	11.1	5.5
Guyana	2005	23.1	62.3	14.6	...	...	...	...	13.8	8.1
	2010	22.8	68.4	8.7	11.5	29.1	33.4	5.3	12.0	3.7
	2012	22.5	62.9	14.6	11.6	30.6	33.3	5.1	10.3	3.2
Honduras										
Honduras	2013	15.6	79.7	4.7	7.3	48.9	25.3	18.5	19.2	5.9
Hungary	2005	21.0	71.6	7.4	14.2	20.1	40.4	18.9	10.8	5.4
Hongrie	2010	27.1	62.5	10.4	14.4	17.8	40.3	20.1	9.7	4.8
	2011	28.6	62.3	9.1	13.3	16.1	39.3	23.4	9.3	4.6
Iceland	2005	23.6	67.9	8.5	6.7	35.0	33.6	19.1	17.8	7.4
Islande	2010	20.8	72.6	6.6	9.6	32.4	31.2	21.5	14.6	7.2
	2011	23.4	71.1	5.5	9.9	33.7	31.7	19.4	15.4	7.0

6

Public expenditure on education *(continued)*
By expenditure type, level of education, total government expenditure and GDP

Dépenses publiques afférentes à l'éducation *(suite)*
Par type de dépenses, niveau de scolarité, dépenses publiques totales et PIB

| Country or area
Pays ou zone | Year
Année | Percentage of total expenditure in public institutions
Pourcentage des dépenses publiques totales en faveur de l'éducation | | | Percentage of government expenditure on education by level
Pourcentage des dépenses publiques selon le niveau d'enseignement | | | | As % of GDP
En % du PIB | As % of Government exp.
En % des dépenses du gouvernement |
		Current expenditure Dépenses courantes	Staff Compensation Rémunération du personnel	Capital expenditure Dépenses en Capital	Pre-primary Préprimaire	Primary Primaire	Secondary Secondaire	Tertiary Tertiaire		
India	2005	9.0	86.2	4.8	1.3	35.6	42.9	19.6	11.5	3.1
Inde	2010	...	...	...	1.1	25.2	37.0	36.1	11.7	3.3
	2012	...	...	...	1.2	27.2	38.7	32.2	14.1	3.8
Indonesia	*2005	...	...	...	...	...	...	...	15.1	2.9
Indonésie	2010	33.6	54.5	11.9	0.7	44.4	24.3	16.1	16.7	2.8
	2014	25.0	63.3	11.7	1.8	43.7	26.8	15.1	17.5	3.3
Iran (Islamic Rep. of)	2002	...	...	8.2	0.9	25.9	36.3	18.5	26.3	4.5
Iran (Rép. islamique d')	2005	...	...	...	1.0	22.4	34.6	15.4	22.3	4.2
	2010	...	...	...	1.0	24.7	49.2	21.4	18.8	3.9
	2014	...	...	6.6	0.2	26.8	39.5	28.3	19.7	3.0
Ireland	2005	17.3	74.2	8.4	0.1	33.6	34.8	23.3	13.6	4.5
Irlande	2010	19.7	71.6	8.7	1.6	35.3	33.8	22.2	9.3	6.1
	2012	19.0	76.0	5.1	1.9	36.9	33.4	21.4	13.8	5.8
Israel	2005	25.1	68.8	6.1	10.7	37.5	28.9	16.6	12.5	5.7
Israël	2010	18.7	72.9	8.4	11.0	41.2	25.7	16.9	13.3	5.5
	2011	18.0	73.6	8.4	10.7	42.6	25.5	16.2	13.5	5.6
	2013	16.2	74.2	9.6	12.6	39.8	25.7	15.8	...	5.9
Italy	2005	20.9	71.9	7.2	10.3	24.6	46.5	17.2	9.0	4.2
Italie	2010	20.5	74.7	4.8	10.0	25.0	42.9	18.8	8.7	4.3
	2011	21.6	73.7	4.7	10.3	24.7	42.5	19.4	8.4	4.1
Jamaica	2004	7.5	88.1	4.5	5.6	32.6	41.9	18.8	11.7	3.9
Jamaïque	2005	...	...	...	4.9	* 34.2	* 38.0	22.1	14.3	4.6
	2008	4.9	85.6	9.4	5.5	30.2	42.6	15.7	19.1	6.2
	2009	7.0	93.0	...	8.2	33.2	37.1	20.3	17.6	6.2
	2010	...	...	...	4.0	36.3	36.6	21.8	16.1	6.4
	2014	19.5	78.5	2.0	3.4	36.4	31.4	17.5	21.8	6.0
Japan	2005	20.2	70.8	9.1	2.7	35.7	38.4	17.3	10.3	3.5
Japon	2010	20.3	66.4	13.3	2.7	34.9	36.8	20.1	9.5	3.8
	2014	21.3	64.8	13.9	2.8	33.1	37.8	20.8	9.3	3.8
Kazakhstan	2005	40.6	53.2	6.2	3.8	...	...	12.3	10.2	2.3
Kazakhstan	2009	34.2	56.8	9.0	6.8	...	...	13.1	13.0	3.1
	2010	32.8	57.8	9.4	...	...	...	...	...	...
	2014	37.2	57.4	5.4	...	...	...	...	...	...
Kenya	2004	...	...	...	1.6	62.1	23.4	12.9	26.7	6.8
Kenya	2005	...	...	...	...	...	...	...	27.3	7.3
	2006	...	...	...	0.1	54.1	21.7	15.4	25.0	7.0
	2010	...	...	...	...	...	...	...	20.6	5.5
Kiribati										
Kiribati	2001	...	...	...	...	34.3	...	...	11.5	12.0
Kuwait	2005	17.4	74.9	7.7	9.8	19.8	36.2	32.6	13.9	4.7
Koweït	*2006	...	...	...	9.6	20.4	36.2	32.6	13.4	3.8
Kyrgyzstan	2005	28.4	66.2	5.4	6.2	...	...	19.2	16.5	4.9
Kirghizistan	2010	29.1	64.5	6.5	8.1	...	15.5	15.7	5.8
	2013	19.9	70.2	9.9	10.7	...	54.4	12.8	17.8	6.8
Lao People's Dem. Rep.	2002	...	...	...	1.9	45.0	19.0	12.6	16.2	2.8
Rép. dém. pop. lao	2005	...	...	...	1.6	62.5	...	...	13.7	2.4
	2010	...	...	...	...	...	...	...	11.9	2.8
	2014	...	...	...	...	...	...	...	15.4	4.2
Latvia	2004	21.9	69.7	8.4	13.0	16.3	56.4	13.4	14.5	4.9
Lettonie	2010	21.6	65.7	12.7	16.8	28.5	38.4	15.9	11.8	5.1
	2013	25.1	62.3	12.6	16.8	30.0	32.7	19.5	13.1	4.9
Lebanon	2005	1.6	94.7	3.7	...	...	...	28.8	8.4	2.7
Liban	2010	0.1	88.0	11.9	...	...	...	27.2	5.5	1.6
	2011	0.1	86.8	13.1	...	...	...	26.5	5.7	1.6
	2013	2.4	97.6	...	...	...	17.1	28.7	8.6	2.6

6

Public expenditure on education *(continued)*
By expenditure type, level of education, total government expenditure and GDP

Dépenses publiques afférentes à l'éducation *(suite)*
Par type de dépenses, niveau de scolarité, dépenses publiques totales et PIB

		Percentage of total expenditure in public institutions Pourcentage des dépenses publiques totales en faveur de l'éducation			Percentage of government expenditure on education by level Pourcentage des dépenses publiques selon le niveau d'enseignement				As % of GDP En % du PIB	As % of Government exp. En % des dépenses du gouvernement
Country or area Pays ou zone	Year Année	Current expenditure Dépenses courantes	Staff Compensation Rémunération du personnel	Capital expenditure Dépenses en Capital	Pre-primary Préprimaire	Primary Primaire	Secondary Secondaire	Tertiary Tertiaire		
Lesotho	2001	...	...	8.7	...	53.2	23.8	18.6	23.8	10.8
Lesotho	2005	...	...	...	...	* 43.0	* 18.3	* 36.4	32.4	14.8
	2008	...	...	* 1.7	0.1	36.0	20.5	36.4	24.7	13.0
Liberia	2008	47.4	41.4	11.2	...	...	...	...	7.3	3.2
Libéria	2012	...	...	...	...	30.7	34.5	* 31.6	8.1	2.8
Liechtenstein	2004	...	...	6.5	29.0	42.6	13.9	...	2.4
Liechtenstein	2007	19.3	80.7	...	6.9	30.6	47.2	9.1	...	1.9
	2008	30.2	69.8	...	7.3	31.1	52.4	...	...	2.0
	2010	18.9	68.4	12.6	...	...	...	...	...	...
	2011	20.7	73.9	5.4	7.1	38.6	47.4	...	...	2.6
Lithuania	2005	20.1	73.0	6.8	12.1	14.9	50.6	21.0	14.6	4.9
Lituanie	2010	19.5	72.9	7.6	12.9	16.7	44.9	23.6	12.9	5.3
	2012	20.1	67.4	12.5	13.1	15.3	39.7	29.0	13.5	4.8
Luxembourg Luxembourg	2001	11.2	71.3	17.5	...	...	...	...	9.7	3.6
Madagascar	2005	* 24.4	* 43.8	31.8	...	57.3	18.9	10.1	18.0	3.8
Madagascar	2008	31.3	58.7	10.0	0.4	53.6	22.1	15.4	16.3	2.9
	2009	...	...	...	0.3	52.3	18.2	15.4	22.6	3.2
	2012	...	...	* 4.6	0.2	47.4	19.3	15.2	20.3	2.7
	2013	29.5	67.4	3.1	...	...	...	...	14.0	2.1
Malawi	2001	...	...	...	...	45.2	22.5	15.3	...	4.5
Malawi	2003	...	...	...	...	54.9	21.1	...	16.6	4.2
	2010	...	...	4.9	<0.0	34.8	24.1	29.8	12.5	4.4
	2013	14.5	71.7	13.8	<0.0	36.7	24.4	28.4	20.4	7.7
	2014	...	...	2.9	...	49.5	24.7	21.5	16.3	6.9
Malaysia	2004	30.2	56.8	13.1	1.1	29.6	35.1	33.4	21.0	5.9
Malaisie	2009	15.7	68.8	15.5	1.2	27.7	32.8	35.9	18.5	6.0
	2010	...	...	...	1.2	28.6	33.7	34.5	18.4	5.0
	2011	19.3	75.1	5.6	1.7	29.3	30.6	37.0	21.0	5.8
	2013	...	...	...	1.6	30.3	34.1	34.0	21.5	6.1
Maldives	2005	...	...	...	...	54.1	...	...	13.0	5.2
Maldives	2008	...	...	...	9.7	50.4	39.9	...	12.6	4.8
	2009	...	...	...	...	...	...	...	17.5	7.0
	2012	...	...	4.1	...	...	...	9.4	15.3	5.2
Mali	2005	...	...	...	...	...	...	...	16.3	4.0
Mali	2010	40.9	54.1	5.0	0.2	25.3	53.8	20.7	16.5	3.9
	2012	34.9	62.7	2.4	0.3	38.3	39.9	20.3	22.4	4.2
	2014	...	...	2.9	0.2	43.2	34.1	21.6	18.2	4.3
Malta	2004	14.5	79.9	5.6	5.2	21.1	39.8	11.0	10.6	4.8
Malte	2010	17.5	72.4	10.1	7.4	21.6	45.7	22.8	15.7	6.9
	2012	16.4	70.9	12.8	7.1	22.7	46.6	22.2	15.3	6.8
Marshall Islands	2002	* 22.6	* 75.0	* 2.4	...	45.5	38.9	14.7	14.8	8.7
Îles Marshall	2003	...	...	...	...	...	...	...	22.5	12.2
	2006	42.8	35.8	21.4	...	...	...	...	...	...
Mauritania	*2004	...	...	...	...	...	...	...	8.2	2.5
Mauritanie	2010	...	...	...	0.5	47.5	23.2	18.8	16.0	3.6
	2013	...	...	...	...	43.2	23.3	11.6	11.4	3.3
Mauritius	2005	...	...	...	1.5	27.2	44.2	11.3	15.8	4.2
Maurice	2010	...	...	13.5	1.4	27.0	52.6	9.4	14.6	3.7
	2013	...	...	...	1.7	27.5	54.5	8.0	14.8	3.7
	2014	...	...	...	...	22.7	61.6	7.1	20.9	5.0
Mexico	2005	11.2	85.7	3.1	10.4	39.1	30.3	17.5	22.2	4.9
Mexique	2010	10.3	85.7	3.9	10.2	36.0	30.3	19.7	19.4	5.2
	2011	10.6	85.5	3.9	10.4	35.5	30.7	18.1	19.0	5.2
Micronesia (Fed. St. of) * Micronésie (États féd. de) *	2000	...	...	...	...	...	...	...	10.0	6.7

Public expenditure on education *(continued)*
By expenditure type, level of education, total government expenditure and GDP

Dépenses publiques afférentes à l'éducation *(suite)*
Par type de dépenses, niveau de scolarité, dépenses publiques totales et PIB

Country or area Pays ou zone	Year Année	Percentage of total expenditure in public institutions Pourcentage des dépenses publiques totales en faveur de l'éducation			Percentage of government expenditure on education by level Pourcentage des dépenses publiques selon le niveau d'enseignement				As % of GDP En % du PIB	As % of Government exp. En % des dépenses du gouvernement
		Current expenditure Dépenses courantes	Staff Compensation Rémunération du personnel	Capital expenditure Dépenses en Capital	Pre-primary Préprimaire	Primary Primaire	Secondary Secondaire	Tertiary Tertiaire		
Monaco	2004	2.1	90.9	7.0	4.3	15.4	45.8	4.4	5.9	1.2
Monaco	2010	8.4	90.7	0.9	3.2	15.3	37.7	...	6.4	1.3
	2014	4.3	87.3	8.4	3.6	16.0	41.1	...	5.0	1.0
Mongolia	2000	55.8	43.4	0.8	...	...	...	...	16.1	5.6
Mongolie	2002	...	...	4.9	14.1	...	...	15.5	20.3	7.2
	2004	...	...	...	18.5	25.1	32.8	18.4	13.6	4.3
	2010	...	...	...	22.0	26.9	33.7	6.7	14.7	4.6
	2011	...	...	...	23.8	32.7	30.4	3.8	12.2	4.6
Montserrat	2001	* 27.5	* 41.4	* 31.2	13.6	20.6	29.5	5.5	...	...
Montserrat	2004	...	...	...	...	20.5	...	...	...	...
	2009	...	...	...	...	...	...	...	...	5.1
Morocco	2008	...	...	...	<0.0	35.5	43.1	...	17.5	5.3
Maroc	2009	...	...	...	...	37.7	41.9	20.2	17.3	5.3
Mozambique	2005	...	...	22.6	...	...	...	...	22.7	4.4
Mozambique	2006	...	...	26.8	...	57.7	29.3	12.1	18.6	4.3
	2013	22.2	58.5	19.3	...	49.2	30.6	13.7	19.0	6.5
Namibia	2000	19.3	73.6	7.2	0.2	58.8	27.2	12.0	21.9	7.0
Namibie	2003	...	...	...	0.5	63.1	25.8	9.1	20.3	6.1
	2008	...	* 79.9	...	0.4	47.6	18.7	9.9	25.1	6.5
	2010	22.1	77.0	0.8	...	40.0	23.5	23.1	26.2	8.3
Nauru	2002	...	...	...	...	67.5	...	...	...	...
Nauru	*2007	20.8	56.9	...	...	...	...	...	...	...
Nepal	2003	...	...	...	...	* 56.5	22.1	10.3	21.3	3.1
Népal	2005	...	...	...	...	...	...	...	22.3	3.4
	2009	...	...	...	1.5	63.0	23.5	12.0	19.8	4.7
	2010	...	...	...	...	...	...	10.7	20.8	4.7
	2014	...	...	...	2.7	50.6	35.3	11.4	22.1	4.7
Netherlands	2005	17.5	69.8	12.7	7.7	25.9	39.4	26.8	12.2	5.2
Pays-Bas	2010	18.5	69.6	11.9	6.9	24.3	40.7	28.0	11.5	5.6
	2013	18.3	69.5	12.2	6.9	23.5	40.7	28.9	12.1	5.6
New Zealand	2005	...	...	...	3.3	25.7	42.6	23.1	17.8	6.3
Nouvelle-Zélande	2010	...	...	...	6.7	24.4	38.7	26.8	16.6	7.1
	2014	...	...	...	6.9	23.0	38.3	24.9	17.8	6.4
Nicaragua	2002	10.7	80.1	9.2	...	...	...	...	15.7	2.4
Nicaragua	*2003	...	...	...	...	...	...	...	14.9	2.4
	2004	...	...	11.7	...	...	...	...	...	...
	2010	10.7	46.2	11.4	3.6	39.8	13.2	26.0	22.8	4.5
Niger	2003	...	...	...	...	72.6	...	...	13.8	2.4
Niger	2010	14.2	77.9	7.9	2.4	61.0	21.3	12.6	18.1	3.7
	2014	15.0	59.1	25.9	4.5	52.2	29.1	13.9	21.7	6.8
Niue										
Nioué	2002	...	...	...	...	28.8	50.7	...	...	...
Norway	2005	22.4	66.9	10.7	4.1	25.0	35.7	32.4	16.8	6.9
Norvège	2010	21.3	67.4	11.3	4.9	26.7	35.3	29.6	15.3	6.8
	2011	20.9	69.0	10.1	5.5	26.0	35.2	29.9	15.0	6.5
	2012	19.1	71.6	9.3	9.7	22.0	...	...	17.5	7.4
Oman	2002	0.9	84.7	14.4	...	46.8	43.2	10.0	11.1	4.3
Oman	2005	...	...	...	...	50.5	41.8	7.6	10.1	3.5
	2009	...	...	...	...	33.0	40.2	26.9	10.9	4.2
	2013	20.7	67.6	11.7	...	...	32.1	...	11.1	5.0
Pakistan	2005	12.3	49.2	38.5	...	...	...	...	13.8	2.3
Pakistan	2007	14.8	59.1	26.1	...	...	...	...	15.4	2.6
	2010	...	...	...	...	...	...	...	11.9	2.3
	2014	...	...	13.6	...	35.1	33.4	22.3	11.3	2.5
Palau	2001	...	...	...	* 14.6	* 44.9	9.6	20.7	15.1	7.5
Palaos	*2002	...	...	...	...	...	...	...	15.3	7.5

6 Public expenditure on education *(continued)*
By expenditure type, level of education, total government expenditure and GDP

Dépenses publiques afférentes à l'éducation *(suite)*
Par type de dépenses, niveau de scolarité, dépenses publiques totales et PIB

Country or area Pays ou zone	Year Année	Percentage of total expenditure in public institutions Pourcentage des dépenses publiques totales en faveur de l'éducation			Percentage of government expenditure on education by level Pourcentage des dépenses publiques selon le niveau d'enseignement				As % of GDP En % du PIB	As % of Government exp. En % des dépenses du gouvernement
		Current expenditure Dépenses courantes	Staff Compensation Rémunération du personnel	Capital expenditure Dépenses en Capital	Pre-primary Préprimaire	Primary Primaire	Secondary Secondaire	Tertiary Tertiaire		
Panama Panama	2002	11.1	67.9	21.0	2.2	32.1	29.2	28.1	17.0	4.4
	*2004	...	...	...	4.0	34.5	25.9	28.4	14.6	3.8
	2008	...	...	14.1	...	...	...	...	14.9	3.5
	2011	...	...	...	3.0	23.5	22.4	22.2	13.0	3.3
Paraguay Paraguay	2004	6.2	89.4	4.4	7.6	46.6	29.8	15.9	18.2	3.4
	2010	9.5	65.5	6.5	6.7	38.5	36.1	18.5	18.7	3.8
	2012	11.7	79.9	8.4	6.1	38.3	33.0	22.4	19.6	5.0
Peru Pérou	2005	9.3	81.3	9.4	8.1	35.6	31.4	10.7	14.3	2.9
	2010	14.5	64.6	20.9	12.0	39.9	34.2	13.8	13.6	2.8
	2014	24.5	55.0	20.5	16.6	38.5	30.7	14.2	16.2	3.7
Philippines Philippines	2005	11.1	85.5	3.4	0.1	52.0	26.8	13.3	12.4	2.4
	2009	18.4	73.4	8.2	1.7	55.0	29.7	12.0	13.2	2.7
Poland Pologne	2005	32.6	60.5	6.9	9.9	31.0	36.6	21.8	12.2	5.5
	2010	29.3	61.4	9.3	10.0	31.0	35.5	22.8	11.1	5.1
	2012	29.5	62.7	7.7	11.4	30.5	34.0	23.4	11.3	4.8
Portugal Portugal	2005	10.2	86.2	3.6	7.4	30.6	40.8	18.1	10.9	5.1
	2010	12.1	84.1	3.8	7.2	27.0	44.0	20.2	10.4	5.4
	2011	12.2	85.0	2.8	7.4	27.7	43.7	19.7	10.2	5.1
Puerto Rico Porto Rico	2013	37.9	57.5	4.6	7.2	21.6	23.6	38.6	21.6	6.4
Qatar Qatar	2005	...	...	...	...	...	...	...	12.7	4.0
	2009	...	...	...	4.4	17.6	14.7	...	14.3	3.4
	2010	...	...	...	...	...	...	...	13.2	4.5
	2014	45.5	22.7	31.9	...	...	...	...	11.1	3.5
Republic of Korea République de Corée	2005	24.0	58.7	17.4	1.6	35.4	41.5	14.0	...	3.9
	2009	27.2	55.8	17.0	2.2	32.6	38.2	17.1	...	4.7
	2012	27.8	58.3	13.9	3.1	30.0	37.9	15.6	...	4.6
Republic of Moldova République de Moldova	2005	...	...	...	...	...	...	...	19.4	7.2
	2010	29.1	62.6	8.2	19.5	18.0	37.3	18.0	22.3	9.1
	2014	28.6	60.8	10.6	21.0	22.7	35.7	17.2	18.8	7.5
Romania Roumanie	2005	34.4	59.1	6.5	9.1	* 13.8	* 44.3	23.2	10.8	3.5
	2010	27.1	62.2	10.8	10.0	16.4	35.8	28.5	9.1	3.5
	2012	34.8	57.3	7.9	9.6	15.9	36.0	26.2	8.4	2.9
Russian Federation Fédération de Russie	2005	...	...	...	13.9	...	...	21.1	12.0	3.8
	2008	...	...	...	15.0	...	...	23.1	12.0	4.1
	2012	24.5	66.4	9.1	...	...	...	21.2	11.1	4.2
Rwanda Rwanda	2000	...	47.3	20.3	* 0.5	* 48.2	* 16.7	* 34.7	21.1	4.1
	2001	24.4	33.8	41.8	...	...	...	...	26.7	5.7
	2010	25.1	58.7	16.1	0.2	36.7	30.7	22.7	17.4	5.0
	2011	29.6	54.8	15.6	0.1	31.8	37.6	18.4	15.4	4.7
	2013	66.7	...	9.9	0.2	30.1	41.7	14.0	16.6	5.0
Saint Kitts and Nevis Saint-Kitts-et-Nevis	2005	...	...	...	...	...	...	...	11.1	3.9
	2007	...	...	...	...	...	...	...	12.8	4.2
Saint Lucia Sainte-Lucie	2004	26.1	73.2	0.7	0.4	47.2	33.4	...	15.0	4.1
	2005	...	...	...	...	...	...	...	16.3	5.1
	2006	28.4	45.3	26.3	...	...	...	...	19.0	5.5
	*2010	...	...	...	1.8	42.3	45.6	4.6	14.2	4.1
	2011	...	...	...	1.5	42.3	45.1	* 5.0	13.8	4.3
	2012	...	...	...	1.6	40.3	45.7	...	12.2	4.0
	2013	...	...	8.7	...	37.1	38.9	...	13.8	4.7
	2014	...	...	...	...	35.5	40.3	...	15.3	4.8
Saint Vincent-Grenadines Saint-Vincent-Grenadines	2005	14.6	51.9	33.6	0.3	41.0	31.7	3.7	22.6	6.4
	2010	5.6	94.4	...	2.0	41.4	36.3	7.0	15.5	5.1
Samoa Samoa	2001	...	...	...	* 0.7	* 36.2	* 29.8	33.2	12.8	3.8
	*2002	...	...	...	...	45.8	...	...	12.8	3.8
	2008	...	...	...	...	...	...	...	16.1	5.1

Public expenditure on education *(continued)*
By expenditure type, level of education, total government expenditure and GDP

Dépenses publiques afférentes à l'éducation *(suite)*
Par type de dépenses, niveau de scolarité, dépenses publiques totales et PIB

Country or area Pays ou zone	Year Année	Percentage of total expenditure in public institutions Pourcentage des dépenses publiques totales en faveur de l'éducation			Percentage of government expenditure on education by level Pourcentage des dépenses publiques selon le niveau d'enseignement				As % of GDP En % du PIB	As % of Government exp. En % des dépenses du gouvernement
		Current expenditure Dépenses courantes	Staff Compensation Rémunération du personnel	Capital expenditure Dépenses en Capital	Pre-primary Préprimaire	Primary Primaire	Secondary Secondaire	Tertiary Tertiaire		
San Marino	2010	4.2	95.8	...	19.0	31.7	39.5	9.8	10.8	2.3
Saint-Marin	2011	5.0	95.0	...	20.2	33.3	35.8	10.8	10.6	2.4
Sao Tome and Principe	2005	...	...	...	...	...	...	...	12.1	5.3
Sao Tomé-et-Principe	2010	...	...	...	...	...	...	...	19.3	9.8
	2014	...	...	...	12.0	55.8	22.6	9.6	12.3	3.9
Saudi Arabia	2005	...	...	...	...	...	...	...	19.3	5.4
Arabie saoudite	2008	...	...	...	...	...	...	...	19.3	5.1
Senegal	2005	...	...	...	0.9	41.8	22.4	22.0	21.8	5.1
Sénégal	2010	33.5	* 58.1	8.4	0.5	39.6	29.0	24.6	20.7	5.6
Serbia	2010	22.4	75.1	2.5	0.5	45.9	23.3	29.1	10.5	4.6
Serbie	2011	22.4	75.0	2.6	1.1	45.3	23.1	29.2	10.6	4.5
	2012	22.4	75.1	2.5	...	...	...	29.1	9.6	4.4
	2013	19.3	76.5	4.2	...	...	...	...	...	...
Seychelles	2002	23.9	66.4	9.7	* 10.5	* 32.0	* 26.1	17.4	9.2	5.2
Seychelles	2003	31.8	59.3	8.9	* 8.6	* 32.8	* 28.9	...	12.0	5.4
	2006	29.8	56.7	13.5	4.1	...	...	17.9	11.0	4.8
	2011	18.7	71.3	10.0	...	24.0	16.2	32.5	10.4	3.6
Sierra Leone	*2005	...	...	...	...	...	...	...	15.5	2.8
Sierra Leone	2010	28.3	69.1	2.5	...	50.7	26.4	19.6	12.8	2.6
	2011	24.5	74.3	1.2	...	52.6	26.3	17.9	12.4	2.7
	2014	...	...	0.5	...	43.1	23.4	30.7	15.1	2.8
Singapore	2005	...	...	...	...	...	...	...	23.9	3.2
Singapour	2010	...	...	9.2	...	20.8	25.0	35.1	16.6	3.1
	2013	...	...	8.2	...	22.0	23.0	35.3	19.9	2.9
Slovakia	2005	34.1	60.2	5.6	10.1	17.1	48.9	21.0	9.6	3.8
Slovaquie	2010	30.8	57.0	12.2	9.4	20.7	46.9	19.6	9.9	4.1
	2013	34.7	58.2	7.2	10.8	19.7	42.7	23.7	10.0	4.1
Slovenia	2003	21.0	69.4	9.6	9.2	* 21.8	* 46.6	22.3	13.7	5.7
Slovénie	2005	20.6	69.4	10.0	10.4	...	...	22.0	13.3	5.6
	2010	21.3	70.0	8.7	10.3	28.5	37.1	24.0	12.1	5.6
	2011	21.5	69.9	8.6	11.3	28.1	36.4	24.2	12.1	5.6
	2012	21.7	69.8	8.5	11.8	27.8	...	21.2	12.6	5.7
Solomon Islands										
Îles Salomon	2010	...	...	...	...	...	...	...	17.5	10.0
South Africa	2005	20.9	66.3	2.9	0.6	42.9	32.6	15.1	19.9	5.1
Afrique du Sud	2006	...	69.8	* 3.5	0.5	45.1	31.1	12.8	19.7	5.1
	2009	...	...	* 3.8	0.9	41.1	30.9	12.5	18.3	5.2
	2010	...	...	...	1.1	42.5	31.4	11.9	18.0	5.7
	2014	29.5	67.2	...	1.5	38.8	30.7	12.2	19.1	6.1
South Sudan										
Soudan du sud	2011	...	...	...	0.3	62.6	11.8	25.3	3.9	0.8
Spain	2005	16.7	73.5	9.8	12.4	25.8	39.5	22.4	10.8	4.1
Espagne	2010	17.7	71.6	10.8	14.1	25.5	37.0	23.4	10.6	4.8
	2013	19.5	74.6	5.8	11.2	26.1	36.9	22.4	9.6	4.3
Sri Lanka	2010	9.9	75.5	14.6	...	25.0	53.2	16.4	8.6	1.7
Sri Lanka	2013	8.5	69.7	21.7	...	23.0	49.1	21.1	9.8	1.6
State of Palestine										
État de Palestine	2013	7.4	87.9	4.7	...	...	...	...	...	...
Sudan	2005	...	...	...	...	...	...	...	6.0	1.6
Soudan	2009	...	...	...	...	...	...	...	10.8	2.2
Swaziland	2005	...	...	...	<0.0	39.9	34.9	23.9	20.9	7.4
Swaziland	2010	...	...	...	0.2	47.2	35.5	16.6	18.3	7.0
	2011	...	...	...	0.2	48.7	36.7	12.8	22.4	8.6
Sweden	2005	28.4	65.0	6.6	7.8	26.2	38.2	27.5	12.8	6.6
Suède	2010	31.3	63.1	5.6	10.2	24.3	35.6	29.0	13.3	6.6
	2013	30.0	65.0	5.1	16.7	22.4	27.9	25.4	15.2	7.7

Public expenditure on education *(continued)*
By expenditure type, level of education, total government expenditure and GDP

Dépenses publiques afférentes à l'éducation *(suite)*
Par type de dépenses, niveau de scolarité, dépenses publiques totales et PIB

Country or area Pays ou zone	Year Année	Percentage of total expenditure in public institutions Pourcentage des dépenses publiques totales en faveur de l'éducation			Percentage of government expenditure on education by level Pourcentage des dépenses publiques selon le niveau d'enseignement				As % of GDP En % du PIB	As % of Government exp. En % des dépenses du gouvernement
		Current expenditure Dépenses courantes	Staff Compensation Rémunération du personnel	Capital expenditure Dépenses en Capital	Pre-primary Préprimaire	Primary Primaire	Secondary Secondaire	Tertiary Tertiaire		
Switzerland	2005	15.5	75.3	9.1	3.6	29.7	37.7	26.0	15.2	5.2
Suisse	2010	16.2	75.2	8.6	3.7	28.0	41.3	25.3	15.9	4.9
	2012	16.0	75.3	8.7	3.6	29.3	39.1	26.3	16.1	5.1
Syrian Arab Republic	2002	...	...	...	...	44.4	33.4	21.0	16.8	5.0
Rép. arabe syrienne	2004	...	...	...	...	...	...	24.8	17.1	5.4
	2009	...	...	...	...	38.9	36.9	24.2	19.2	5.1
Tajikistan	2005	...	...	...	3.8	...	...	7.4	15.3	3.5
Tadjikistan	2010	...	...	...	5.0	...	...	10.1	15.3	4.0
	2012	...	...	...	5.2	...	...	11.2	16.4	4.0
Thailand	2004	...	...	...	10.4	31.0	24.0	20.6	21.5	4.0
Thaïlande	2005	...	...	...	13.6	...	...	21.7	20.5	3.9
	2010	45.8	50.9	3.3	5.7	40.1	28.6	16.5	16.2	3.5
	2013	45.8	49.2	5.0	5.4	41.1	29.8	15.6	18.9	4.1
TFYR of Macedonia ex-R.Y. de Macédoine	2002	10.8	87.1	2.1	...	...	...	15.0	8.6	3.3
Timor-Leste	2010	52.5	37.8	9.7	...	...	...	10.4	12.9	10.5
Timor-Leste	2014	46.4	49.5	4.0	1.2	64.9	29.8	4.1	7.7	7.7
Togo	2000	20.9	69.1	10.0	...	48.6	29.3	17.4	24.4	4.5
Togo	2005	...	...	...	...	41.8	...	...	17.7	3.4
	2010	14.7	74.7	10.6	2.6	51.7	25.6	16.7	19.6	4.4
	2014	11.1	86.6	2.3	1.4	58.9	17.5	21.7	19.4	4.8
Tonga Tonga	2004	...	...	...	...	44.4	30.1	21.7	18.1	3.9
Trinidad and Tobago Trinité-et-Tobago	2003	...	...	...	0.6	39.2	36.8	...	13.3	3.1
Tunisia	2000	8.2	77.3	14.5	...	* 33.3	* 45.0	* 21.7	* 24.6	* 6.2
Tunisie	2002	...	...	11.6	...	* 32.9	* 44.4	22.8	22.9	5.8
	2005	...	...	...	...	* 34.3	* 41.6	24.1	26.7	6.5
	2008	...	...	10.0	...	27.7	47.3	25.0	25.3	6.3
	2010	...	...	10.4	...	...	...	28.2	26.1	6.3
	2012	...	...	...	...	...	...	28.1	21.6	6.2
Turkey	2004	12.6	66.3	21.1	0.6	...	...	27.9	8.8	3.1
Turquie	2006	14.9	74.6	10.6	...	...	...	31.9	8.5	2.9
	2011	21.0	68.0	11.0	...	...	...	...	...	...
Turkmenistan Turkménistan	2012	...	...	33.7	28.1	...	...	9.2	20.8	3.0
Turks and Caicos Islands	2002	9.8	61.1	29.2	* 5.0	* 21.1	31.7	22.9	...	...
Îles Turques-et-Caïques	2003	...	...	...	...	* 21.6	28.0	37.2	...	...
	2005	13.7	69.5	16.8	...	* 30.0	47.7	...	...	...
Uganda	2004	12.1	62.7	25.2	...	61.2	17.3	11.9	20.3	5.0
Ouganda	2009	16.2	70.8	12.9	...	55.4	23.5	11.3	13.9	3.3
	2010	...	...	8.5	...	57.7	24.2	11.3	9.4	2.3
	2013	21.4	66.5	12.1	...	52.7	29.2	13.8	11.8	2.2
Ukraine	2000	...	...	8.3	11.1	...	...	32.3	11.4	4.2
Ukraine	2005	...	...	...	11.0	...	...	29.5	13.7	6.1
	2009	...	...	...	12.5	...	...	32.4	15.1	7.3
	2013	...	...	2.5	16.0	17.5	28.1	31.9	13.9	6.7
United Kingdom	2005	18.5	73.5	8.0	5.9	26.5	45.4	22.3	13.2	5.2
Royaume-Uni	2010	29.8	58.8	11.4	5.2	30.1	48.3	16.4	13.0	5.9
	2013	22.3	75.0	2.7	5.4	29.8	40.1	24.0	13.0	5.7
United Rep. of Tanzania	2005	...	...	...	5.8	56.0	12.9	21.7	18.1	4.6
Rép.-Unie de Tanzanie	2009	...	...	...	4.4	41.9	16.7	33.8	17.4	4.0
	2010	...	...	...	...	...	11.3	28.3	19.6	4.6
	2014	...	...	...	6.0	49.2	18.3	21.4	17.3	3.5
United States	2005	21.0	67.6	11.3	5.8	32.3	36.0	25.8	15.1	5.1
États-Unis	2010	20.5	69.4	10.1	6.5	32.4	35.4	25.7	13.2	5.4
	2011	21.1	69.4	9.6	6.4	31.6	35.9	26.1	13.1	5.2

Public expenditure on education *(continued)*
By expenditure type, level of education, total government expenditure and GDP

Dépenses publiques afférentes à l'éducation *(suite)*
Par type de dépenses, niveau de scolarité, dépenses publiques totales et PIB

Country or area Pays ou zone	Year Année	Percentage of total expenditure in public institutions Pourcentage des dépenses publiques totales en faveur de l'éducation			Percentage of government expenditure on education by level Pourcentage des dépenses publiques selon le niveau d'enseignement				As % of GDP En % du PIB	As % of Government exp. En % des dépenses du gouvernement
		Current expenditure Dépenses courantes	Staff Compensation Rémunération du personnel	Capital expenditure Dépenses en Capital	Pre-primary Préprimaire	Primary Primaire	Secondary Secondaire	Tertiary Tertiaire		
Uruguay	2005	14.5	80.8	4.7	8.9	33.8	35.5	21.8	9.5	2.7
Uruguay	2006	14.1	80.2	5.7	8.7	33.3	36.0	22.0	9.9	2.9
	2011	14.9	80.5	4.6	10.2	27.9	33.2	26.8	14.9	4.4
Vanuatu	2001	1.1	51.7	47.2	0.1	27.9	57.4	10.5	40.1	9.0
Vanuatu	2003	...	...	...	...	...	...	...	44.8	8.4
	2008	11.4	87.4	1.2	0.5	49.7	32.9	5.9	21.4	5.8
	2009	8.8	89.5	1.7	...	54.3	29.7	...	18.7	5.0
	2014	...	...	...	0.1	47.1	32.5	...	21.8	4.9
Venezuela (Boliv. Rep. of) Venezuela (Rép. boliv. du)	2009	20.2	76.9	2.9	11.6	31.5	19.0	22.6	20.7	6.9
Viet Nam	2010	* 17.7	* 64.4	17.9	10.8	32.1	38.1	14.7	20.9	6.3
Viet Nam	2012	...	...	18.2	9.9	30.2	43.3	16.7	21.4	6.3
Yemen	*2001	...	...	...	...	...	...	...	30.3	9.2
Yémen	2008	...	...	25.7	...	...	...	...	12.5	4.6
Zambia	2004	25.2	73.0	1.8	...	63.9	13.3	18.0	10.6	2.5
Zambie	2005	...	...	...	...	59.8	14.5	25.8	7.7	1.7
	2008	...	...	...	...	...	...	...	5.7	1.1
Zimbabwe Zimbabwe	2010	...	...	...	...	51.6	25.6	22.8	8.7	2.0

Source:

United Nations Educational, Scientific and Cultural Organization (UNESCO) Institute for Statistics, Montreal, the UNESCO Institute for Statistics (UIS) statistics database, last accessed June 2016.

Source:

L'Institut de statistique de l'Organisation des Nations Unies pour l'éducation, la science et la culture (UNESCO), Montréal, la base de données de l'institut de statistique de l'UNESCO (ISU), dernier accès juin 2016.

Health Personnel
Number of health workers and workforce density (per 1 000 population)

Le personnel de santé
Personnel de santé et densité (pour 1 000 habitants)

Country or area Pays ou zone	Year Année	Physicians Médecins Number Nombre	Per 1 000 Pour 1 000	Dentists Dentistes Number Nombre	Per 1 000 Pour 1 000	Pharmacists Pharmaciens Number Nombre	Per 1 000 Pour 1 000	Nurses and midwives Infirmières et Sages-femmes Number Nombre	Per 1 000 Pour 1 000
Afghanistan	2001	4 104.0	0.2	630.0	<0.0	525.0	<0.0	4 752.0	0.2
Afghanistan	2005	...	...	900.0	<0.0	900.0	<0.0	14 930.0	0.5
	2009	6 037.0	0.2	107.0	<0.0	847.0	<0.0	17 257.0	0.5
	2010	6 901.0	0.2	103.0	<0.0	814.0	<0.0	...	...
	2013	9 184.0	0.3	124.0	<0.0	1 607.0	<0.0	...	...
Albania	2009	3 656.0	1.1	...	...	1 206.0	0.4	12 455.0	3.9
Albanie	2010	3 640.0	1.1	...	...	1 324.0	0.4	...	...
	2013	3 709.0	1.1	...	...	2 441.0	0.8	12 455.0	3.8
Algeria	2002	35 368.0	1.1	9 553.0	0.3	6 333.0	0.2	69 749.0	2.2
Algérie	2005	33 952.0	1.0	9 022.0	0.3	6 104.0	0.2	...	...
	2007	40 857.0	1.2	11 010.0	0.3	8 232.0	0.2	65 919.0	1.9
Andorra	2003	244.0	3.7	44.0	0.7	68.0	1.0	215.0	3.3
Andorre	2010	272.0	4.0	59.0	0.9	78.0	1.1	324.0	4.8
Angola	2004	1 165.0	0.1	222.0	<0.0	919.0	0.1	18 485.0	1.3
Angola	2009	2 956.0	0.2	...	...	...	...	29 592.0	1.7
Argentina	2004	122 623.0	3.2	35 592.0	0.9	19 510.0	0.5	18 685.0	0.5
Argentine	2013	160 041.0	3.9	...	...	...	...	...	...
Armenia	2010	8 177.0	2.7	1 257.0	0.4	123.0	<0.0	15 076.0	5.1
Arménie	2013	8 412.0	2.7	1 215.0	0.4	148.0	<0.0	15 070.0	4.8
Australia	2001	47 875.0	2.5	21 296.0	1.1	13 956.0	0.7	187 837.0	9.7
Australie	2009	62 800.0	3.0	14 500.0	0.7	21 800.0	1.0	201 300.0	9.6
	2010	81 639.0	3.9	...	...	19 237.0	0.9	...	...
	2011	73 980.0	3.3	12 154.0	0.5	24 636.0	1.1	240 716.0	10.6
	2012	...	...	...	...	23 294.0	1.0	...	...
Austria Autriche	2003	27 413.0	3.4	4 037.0	0.5	4 869.0	0.6	76 161.0	9.4
Azerbaijan	2010	33 085.0	3.8	2 467.0	0.3	1 831.0	0.2	66 401.0	7.6
Azerbaïdjan	2013	32 434.0	3.4	2 452.0	0.3	1 951.0	0.2	62 336.0	6.5
Bahamas Bahamas	2008	947.0	2.8	115.0	0.3	160.0	0.5	1 391.0	4.1
Bahrain	2005	928.0	1.3	243.0	0.3	164.0	0.2	2 389.0	3.3
Bahreïn	2010	1 178.0	1.5	294.0	0.4	196.0	0.2	3 052.0	3.9
	2012	1 244.0	0.9	330.0	0.2	205.0	0.2	3 224.0	2.4
Bangladesh	2004	38 485.0	0.3	2 537.0	<0.0	9 411.0	0.1	46 794.0	0.3
Bangladesh	2005	42 881.0	0.3	2 344.0	<0.0	...	...	39 471.0	0.3
	2007	43 315.0	0.3	2 742.0	<0.0	9 411.0	0.1	39 992.0	0.3
	2011	53 603.0	0.4	4 165.0	<0.0	...	...	32 839.0	0.2
Barbados Barbade	2005	489.0	1.8	94.0	0.3	251.0	0.9	1 311.0	4.9
Belarus	2010	33 325.0	3.5	5 087.0	0.5	2 731.0	0.3	99 995.0	10.5
Bélarus	2013	37 281.0	3.9	5 196.0	0.5	3 264.0	0.3	101 041.0	10.6
Belgium	2002	...	...	...	...	11 775.0	1.1	...	...
Belgique	2003	...	...	...	...	...	...	60 142.0	5.8
	2005	21 599.0	2.1	7 731.0	0.7	...	...	...	...
	2010	39 690.0	3.8	8 313.0	0.8	16 368.0	1.6	165 650.0	15.8
	2013	52 858.0	4.9	8 095.0	0.7	27 681.0	2.6	181 314.0	16.8
Belize	2000	251.0	1.0	32.0	0.1	...	...	303.0	1.3
Belize	2009	241.0	0.8	12.0	<0.0	112.0	0.4	570.0	2.0
Benin	2004	311.0	<0.0	12.0	<0.0	11.0	<0.0	5 789.0	0.8
Bénin	2008	542.0	0.1	37.0	<0.0	20.0	<0.0	7 129.0	0.8
Bhutan	2004	118.0	0.1	58.0	<0.0	79.0	<0.0	515.0	0.2
Bhoutan	2007	144.0	0.1	65.0	<0.0	87.0	<0.0	545.0	0.2
	2008	171.0	0.1	...	...	89.0	<0.0	666.0	0.3
	2012	194.0	0.3	...	...	11.0	<0.0	736.0	1.0
Bolivia (Plurinational State of)	2001	10 329.0	1.2	5 997.0	0.7	4 670.0	0.6	18 091.0	2.1
Bolivie (État plurinational de)	2010	4 406.0	0.4	976.0	0.1	567.0	0.1	8 366.0	0.8
	2011	4 771.0	0.5	1 130.0	0.1	660.0	0.1	10 139.0	1.0
Bosnia and Herzegovina	2010	6 665.0	1.7	797.0	0.2	371.0	0.1	20 331.0	5.2
Bosnie-Herzégovine	2013	7 211.0	1.9	809.0	0.2	429.0	0.1	20 903.0	5.6

Country or area Pays ou zone	Year Année	Physicians Médecins		Dentists Dentistes		Pharmacists Pharmaciens		Nurses and midwives Infirmières et Sages-femmes	
		Number Nombre	Per 1 000 Pour 1 000	Number Nombre	Per 1 000 Pour 1 000	Number Nombre	Per 1 000 Pour 1 000	Number Nombre	Per 1 000 Pour 1 000
Botswana	2004	715.0	0.4	38.0	<0.0	333.0	0.2	4 753.0	2.6
Botswana	2005	466.0	0.3	...	...	...	...	4 468.0	2.5
	2009	693.0	0.4	145.0	0.1	365.0	0.2	5 816.0	3.3
Brazil	2005	310 138.0	1.7	207 904.0	1.1	99 696.0	0.5	703 813.0	3.8
Brésil	2010	355 006.0	1.8	242 266.0	1.2	142 841.0	0.7	1 446 403.0	7.3
	2013	378 354.0	1.9	...	...	...	...	1 520 533.0	7.6
Brunei Darussalam	2005	390.0	1.0	73.0	0.2	41.0	0.1	2 006.0	5.4
Brunéi Darussalam	2010	563.0	1.4	86.0	0.2	43.0	0.1	2 907.0	7.0
	2012	596.0	1.4	174.0	0.4	167.0	0.4	3 323.0	8.0
Bulgaria	2001	...	...	...	...	1 020.0	0.1	...	...
Bulgarie	2010	27 997.0	3.8	6 355.0	0.9	...	...	...	...
	2012	28 599.0	3.9	6 750.0	0.9	...	...	35 350.0	4.8
Burkina Faso	2004	708.0	0.1	58.0	<0.0	343.0	<0.0	6 557.0	0.5
Burkina Faso	2010	713.0	<0.0	32.0	<0.0	339.0	<0.0	8 645.0	0.6
Burundi	2004	200.0	<0.0	14.0	<0.0	76.0	<0.0	1 348.0	0.2
Burundi	2011	...	...	...	...	172.0	<0.0	...	...
Cabo Verde	2004	231.0	0.5	11.0	<0.0	43.0	0.1	410.0	0.9
Cabo Verde	2010	167.0	0.3	4.0	<0.0	4.0	<0.0	257.0	0.5
	2011	153.0	0.3	3.0	<0.0	5.0	<0.0	279.0	0.6
Cambodia	2000	2 047.0	0.2	209.0	<0.0	564.0	<0.0	11 125.0	0.8
Cambodge	2010	3 294.0	0.2	242.0	<0.0	548.0	<0.0	12 251.0	0.8
	2012	2 440.0	0.2	284.0	<0.0	566.0	<0.0	11 454.0	0.8
Cameroon	2005	1 049.0	0.1	26.0	<0.0	27.0	<0.0	6 705.0	0.4
Cameroun	2009	1 346.0	0.1	32.0	<0.0	27.0	<0.0	7 626.0	0.4
Canada	2004	60 612.0	1.8	35 834.0	1.1	28 537.0	0.9	316 512.0	9.5
Canada	2005	...	...	...	...	...	...	321 159.0	10.0
	2008	65 440.0	2.0	41 798.0	1.3	29 010.0	0.9	343 699.0	10.4
	2010	69 648.0	2.1	...	...	25 597.0	0.8	318 565.0	9.5
	2011	...	...	...	...	...	...	319 026.0	9.3
	2012	...	...	...	...	35 555.0	1.0	...	...
Central African Rep.	2004	331.0	0.1	13.0	<0.0	17.0	<0.0	1 613.0	0.4
Rép. centrafricaine	2009	205.0	<0.0	12.0	<0.0	12.0	<0.0	1 097.0	0.3
Chad	2004	345.0	<0.0	15.0	<0.0	37.0	<0.0	2 499.0	0.3
Tchad	2006	368.0	<0.0	...	...	...	...	1 891.0	0.2
Chile	2005	15 865.0	1.0	12.0	<0.0	36.0	<0.0	2 103.0	0.1
Chili	2009	17 382.0	1.0	15.0	<0.0	39.0	<0.0	2 443.0	0.1
China	2002	1 463 573.0	1.1	...	...	357 659.0	0.3	1 246 545.0	1.0
Chine	2005	...	...	51 012.0	<0.0	349 533.0	0.3	1 349 589.0	1.0
	2010	1 972 840.0	1.5	...	...	353 916.0	0.3	2 048 071.0	1.5
	2011	2 020 154.0	1.5	...	...	363 993.0	0.3	2 244 020.0	1.7
Colombia	2005	62 807.0	1.4	37 371.0	0.8	...	...	25 845.0	0.6
Colombie	2010	71 980.0	1.5	44 858.0	0.9	...	...	30 119.0	0.6
Comoros									
Comores	2004	115.0	0.1	29.0	<0.0	41.0	0.1	588.0	0.7
Congo	2004	756.0	0.2	12.0	<0.0	99.0	<0.0	3 672.0	1.0
Congo	2007	401.0	0.1	...	...	63.0	<0.0	3 492.0	0.8
Cook Islands	2004	20.0	1.2	10.0	0.6	2.0	0.1	80.0	4.7
Îles Cook	2009	24.0	1.3	19.0	1.1	8.0	0.4	116.0	6.4
Costa Rica	2000	5 204.0	1.3	1 905.0	0.5	2 101.0	0.5	3 653.0	0.9
Costa Rica	2013	5 411.0	1.1	599.0	0.1	891.0	0.2	3 745.0	0.8
Côte d'Ivoire	2005	1 698.0	0.1	204.0	<0.0	128.0	<0.0	8 988.0	0.5
Côte d'Ivoire	2008	2 746.0	0.1	274.0	<0.0	413.0	<0.0	9 231.0	0.5
Croatia									
Croatie	2012	...	...	...	...	3 070.0	0.7	...	...
Cuba	2002	67 079.0	6.0	9 955.0	0.9	...	...	83 880.0	7.4
Cuba	2005	70 594.0	6.3	10 554.0	0.9	...	...	...	...
	2010	76 506.0	6.7	12 144.0	1.1	4 656.0	0.4	103 014.0	9.1
Cyprus	2002	...	...	...	...	144.0	0.2	2 994.0	3.8
Chypre	2008	2 218.0	2.6	792.0	0.9	210.0	0.2	3 710.0	4.3
	2010	2 442.0	2.8	772.0	0.9	...	...	3 930.0	4.5
	2012	2 630.0	2.3	816.0	0.7	195.0	0.2	...	...

Health Personnel *(continued)*
Number of health workers and workforce density (per 1 000 population)

Le personnel de santé *(suite)*
Personnel de santé et densité (pour 1 000 habitants)

Country or area Pays ou zone	Year Année	Physicians Médecins		Dentists Dentistes		Pharmacists Pharmaciens		Nurses and midwives Infirmières et Sages-femmes	
		Number Nombre	Per 1 000 Pour 1 000	Number Nombre	Per 1 000 Pour 1 000	Number Nombre	Per 1 000 Pour 1 000	Number Nombre	Per 1 000 Pour 1 000
Czech Republic	2003	35 960.0	3.5	6 737.0	0.7	5 610.0	0.5	...	...
République tchèque	2012	...	...	...	...	8 041.0	0.8	...	...
Dem. P. R. Korea									
R. p. dém. de Corée	2003	74 597.0	3.3	8 315.0	0.4	13 497.0	0.6	93 414.0	4.1
Dem. Rep. of the Congo	2004	5 827.0	0.1	159.0	<0.0	1 200.0	<0.0	28 789.0	0.5
Rép. dém. du Congo	2012	...	...	...	...	308.0	<0.0	...	...
Denmark	2004	16 439.0	3.1	4 616.0	0.9	...	...	78 349.0	14.6
Danemark	2005	...	...	4 634.0	0.9	...	...	81 219.0	15.0
	2010	19 173.0	3.5	4 333.0	0.8	...	...	92 351.0	16.8
Djibouti	2005	140.0	0.2	60.0	0.1	41.0	0.1	450.0	0.6
Djibouti	2006	185.0	0.2	96.0	0.1	16.0	<0.0	322.0	0.4
	2008	...	...	99.0	0.1	266.0	0.3	666.0	0.8
Dominica									
Dominique	2001	124.0	1.6	21.0	0.3	18.0	0.2	438.0	5.6
Dominican Republic	2000	15 670.0	1.9	7 000.0	0.8	3 330.0	0.4	15 352.0	1.8
Rép. dominicaine	2008	10 385.0	1.1	1 205.0	0.1	...	...	7 603.0	0.8
	2009	...	...	1 785.0	0.2	...	...	12 686.0	1.4
	2011	14 983.0	1.5	1 910.0	0.2	...	...	13 374.0	1.3
Ecuador	2003	20 020.0	1.5	2 213.0	0.2	497.0	<0.0	20 372.0	1.6
Équateur	2009	23 614.0	1.7	3 363.0	0.2	664.0	<0.0	27 764.0	2.0
	2011	25 277.0	1.7	4 183.0	0.3	810.0	0.1	31 635.0	2.2
Egypt	2005	179 900.0	2.4	25 170.0	0.3	92 540.0	1.3	248 010.0	3.4
Égypte	2009	225 565.0	2.8	33 476.0	0.4	133 107.0	1.7	280 561.0	3.5
El Salvador	2005	10 355.0	1.5	4 255.0	0.6	2 280.0	0.3	2 084.0	0.3
El Salvador	2008	11 542.0	1.6	4 669.0	0.6	2 316.0	0.3	2 929.0	0.4
Equatorial Guinea									
Guinée équatoriale	2004	153.0	0.3	15.0	<0.0	121.0	0.2	271.0	0.5
Eritrea									
Érythrée	2004	215.0	0.1	16.0	<0.0	107.0	<0.0	2 505.0	0.6
Estonia	2000	6 118.0	4.5	1 747.0	1.3	580.0	0.4	12 087.0	8.8
Estonie	2010	4 376.0	3.3	1 218.0	0.9	1 462.0	1.1	8 589.0	6.6
	2012	4 343.0	3.2	1 196.0	0.9	847.0	0.6	8 545.0	6.4
Ethiopia	2003	1 936.0	<0.0	93.0	<0.0	1 343.0	<0.0	15 544.0	0.2
Éthiopie	2005	2 453.0	<0.0	...	...	1 619.0	<0.0	18 809.0	0.2
	2009	2 152.0	<0.0	...	...	2 661.0	<0.0	21 488.0	0.3
Fiji	2003	380.0	0.5	60.0	0.1	90.0	0.1	1 660.0	2.0
Fidji	2009	372.0	0.4	171.0	0.2	76.0	0.1	1 957.0	2.2
Finland	2002	16 446.0	3.2	6 674.0	1.3	5 829.0	1.1	78 402.0	15.1
Finlande	2009	15 384.0	2.9	4 014.0	0.8	...	...	126 869.0	24.0
	2011	...	...	3 905.0	0.7	5 965.0	1.1	58 484.0	10.9
France	2003	...	...	40 648.0	0.7	63 909.0	1.1	439 115.0	7.3
France	2005	...	...	...	...	...	...	469 036.0	7.8
	2007	227 683.0	3.7	42 508.0	0.7	70 498.0	1.2	500 863.0	8.2
	2008	...	...	41 422.0	0.7	72 160.0	1.2	494 895.0	8.1
	2013	203 490.0	3.2	42 051.0	0.7	69 489.0	1.1	#20 752.0	0.3
Gabon									
Gabon	2004	395.0	0.3	66.0	<0.0	63.0	<0.0	6 778.0	5.0
Gambia	2005	166.0	0.1	10.0	<0.0	11.0	<0.0	732.0	0.5
Gambie	2008	175.0	0.1	47.0	<0.0	75.0	<0.0	1 411.0	0.9
Georgia	2009	20 609.0	4.8	...	...	...	...	13 925.0	3.2
Géorgie	2013	18 278.0	4.3	...	...	...	...	#594.0	0.1
Germany	2003	277 885.0	3.4	64 609.0	0.8	...	...	801 677.0	9.7
Allemagne	2012	318 887.0	3.9	66 157.0	0.8	50 456.0	0.6	...	...
Ghana	2004	3 240.0	0.2	393.0	<0.0	1 388.0	0.1	19 707.0	0.9
Ghana	2008	2 587.0	0.1	148.0	<0.0	1 673.0	0.1	22 834.0	1.0
	2010	2 325.0	0.1	...	...	...	...	22 507.0	0.9
Greece	2000	...	...	...	...	8 977.0	0.8	42 129.0	3.9
Grèce	2001	47 944.0	4.4	12 394.0	1.1	...	...	#1 916.0	0.2
Grenada	2003	58.0	0.7	13.0	0.2	22.0	0.3	366.0	4.6
Grenade	2006	69.0	0.7	19.0	0.2	21.0	0.2	398.0	3.8
	2011	...	...	...	...	89.0	0.8	...	...

Health Personnel *(continued)*
Number of health workers and workforce density (per 1 000 population)

Le personnel de santé *(suite)*
Personnel de santé et densité (pour 1 000 habitants)

Country or area Pays ou zone	Year Année	Physicians Médecins		Dentists Dentistes		Pharmacists Pharmaciens		Nurses and midwives Infirmières et Sages-femmes	
		Number Nombre	Per 1 000 Pour 1 000	Number Nombre	Per 1 000 Pour 1 000	Number Nombre	Per 1 000 Pour 1 000	Number Nombre	Per 1 000 Pour 1 000
Guatemala	2007	...	...	2 376.0	0.2	...	...	...	...
Guatemala	2009	12 940.0	0.9	...	...	...	...	12 452.0	0.9
Guinea									
Guinée	2005	940.0	0.1	33.0	<0.0	199.0	<0.0	401.0	<0.0
Guinea-Bissau	2004	188.0	0.1	22.0	<0.0	40.0	<0.0	1 072.0	0.7
Guinée-Bissau	2009	124.0	0.1	13.0	<0.0	21.0	<0.0	1 042.0	0.6
Guyana	2000	366.0	0.5	30.0	<0.0	...	...	1 738.0	2.3
Guyana	2009	161.0	0.2	46.0	0.1	87.0	0.1	989.0	1.3
	2010	161.0	0.2	...	...	...	...	399.0	0.5
Honduras	2000	3 676.0	0.6	1 371.0	0.2	926.0	0.1	8 528.0	1.3
Honduras	2005	2 680.0	0.4	...	...	...	...	7 796.0	1.1
Hungary	2003	38 241.0	3.9	5 364.0	0.5	5 125.0	0.5	87 381.0	8.8
Hongrie	2005	32 563.0	3.2	...	...	...	...	...	...
	2010	33 943.0	3.4	...	...	...	...	...	...
	2012	30 641.0	3.1	5 609.0	0.6	5 683.0	0.6	64 458.0	6.5
Iceland	2005	1 070.0	3.6	314.0	1.1	388.0	1.3	4 378.0	14.8
Islande	2010	1 146.0	3.7	313.0	1.0	595.0	1.9	4 875.0	15.9
	2012	1 141.0	3.5	270.0	0.8	358.0	1.1	5 119.0	15.6
India	2003	625 400.0	0.6	47 318.0	<0.0	559 408.0	0.5	1 382 901.0	1.3
Inde	2005	660 801.0	0.6	55 344.0	0.1	...	...	1 481 270.0	1.3
	2009	757 377.0	0.6	104 603.0	0.1	655 801.0	0.6	1 702 555.0	1.5
	2010	816 629.0	0.7	...	...	656 101.0	0.6	1 894 968.0	1.6
	2011	922 177.0	0.7	117 825.0	0.1	657 230.0	0.5	2 124 667.0	1.7
	2012	883 812.0	0.7	120 897.0	0.1	630 766.0	0.5	...	...
Indonesia	2003	29 499.0	0.1	7 093.0	<0.0	7 580.0	<0.0	179 959.0	0.8
Indonésie	2010	33 736.0	0.1	18 454.0	0.1	18 022.0	0.1	256 625.0	1.1
	2012	49 853.0	0.2	24 147.0	0.1	25 502.0	0.1	338 501.0	1.4
Iran (Islamic Rep. of)	2005	61 870.0	0.9	13 210.0	0.2	13 900.0	0.2	98 020.0	1.4
Iran (Rép. islamique d')	2011	...	...	...	...	40 000.0	0.5	...	...
Iraq									
Iraq	2010	19 738.0	0.6	4 799.0	0.1	5 675.0	0.2	...	...
Ireland	2004	11 141.0	2.8	2 237.0	0.6	3 898.0	1.0	60 774.0	15.2
Irlande	2011	...	...	...	...	5 303.0	1.2	...	...
	2013	12 367.0	2.7	...	...	...	...	...	...
Israel									
Israël	2012	25 733.0	3.3	5 367.0	0.7	5 967.0	0.8	38 133.0	5.0
Italy	2003	...	...	...	...	66 119.0	1.2	312 377.0	5.4
Italie	2004	241 000.0	4.2	33 000.0	0.6	...	...	...	...
	2012	229 445.0	3.8	...	...	...	...	...	...
Jamaica	2003	2 253.0	0.9	212.0	0.1	...	...	4 374.0	1.7
Jamaïque	2008	1 103.0	0.4	244.0	0.1	172.0	0.1	2 930.0	1.1
Japan	2004	270 371.0	2.1	95 197.0	0.7	241 369.0	1.9	1 210 633.0	9.5
Japon	2010	295 049.0	2.3	101 576.0	0.8	276 517.0	2.2	1 395 571.0	10.9
	2012	...	...	...	...	...	...	1 452 635.0	11.5
Jordan	2005	12 909.0	2.3	4 194.0	0.7	7 100.0	1.2	17 428.0	3.1
Jordanie	2010	16 212.0	2.6	5 691.0	0.9	9 151.0	1.4	25 661.0	4.0
	2012	...	...	...	...	13 840.0	2.1	...	...
Kazakhstan	2010	57 179.0	3.9	6 657.0	0.5	11 265.0	0.8	125 492.0	8.5
Kazakhstan	2013	59 872.0	3.6	6 406.0	0.4	13 828.0	0.8	136 640.0	8.3
Kenya	2002	4 506.0	0.1	1 340.0	<0.0	...	...	37 113.0	1.2
Kenya	2004	...	...	...	...	3 094.0	0.1	16 146.0	0.5
	2010	7 129.0	0.2	898.0	<0.0	3 097.0	0.1	29 678.0	0.8
	2013	8 682.0	0.2	1 045.0	<0.0	2 202.0	0.1	37 907.0	0.9
Kiribati	2004	...	...	3.0	<0.0	2.0	<0.0	260.0	3.0
Kiribati	2008	25.0	0.2	18.0	0.2	22.0	0.2	361.0	3.4
	2010	41.0	0.4	...	...	...	...	404.0	3.7
Kuwait	2001	...	...	673.0	0.3	722.0	0.3	9 197.0	3.9
Koweït	2009	5 340.0	1.8	1 054.0	0.4	888.0	0.3	13 554.0	4.6
Kyrgyzstan	2010	13 349.0	2.4	...	...	...	...	29 703.0	5.3
Kirghizistan	2013	10 838.0	2.0	981.0	0.2	317.0	0.1	34 237.0	6.2
Lao People's Dem. Rep.	2005	1 614.0	0.3	...	...	...	...	5 724.0	1.0
Rép. dém. pop. lao	2009	1 211.0	0.2	324.0	0.1	...	...	5 322.0	0.8
	2012	1 160.0	0.2	225.0	<0.0	735.0	0.1	5 581.0	0.9

Le personnel de santé *(suite)*
Personnel de santé et densité (pour 1 000 habitants)

Country or area Pays ou zone	Year Année	Physicians Médecins		Dentists Dentistes		Pharmacists Pharmaciens		Nurses and midwives Infirmières et Sages-femmes	
		Number Nombre	Per 1 000 Pour 1 000	Number Nombre	Per 1 000 Pour 1 000	Number Nombre	Per 1 000 Pour 1 000	Number Nombre	Per 1 000 Pour 1 000
Latvia	2005	8 207.0	3.6	1 450.0	0.6	...	...	10 350.0	4.5
Lettonie	2010	7 951.0	3.5	1 505.0	0.7	...	...	8 936.0	4.0
	2012	7 998.0	3.6	1 463.0	0.7	...	...	7 676.0	3.4
Lebanon	2001	11 505.0	3.3	4 283.0	1.2	3 359.0	1.0	4 157.0	1.2
Liban	2004	...	...	3 260.0	0.9	3 000.0	0.8	...	...
	2005	...	...	...	...	4 105.0	1.1	4 720.0	1.3
	2010	11 583.0	3.1	5 395.0	1.4	5 508.0	1.5	8 791.0	2.3
	2011	13 630.0	3.2	6 261.0	1.5	6 687.0	1.6	11 585.0	2.7
Lesotho									
Lesotho	2003	89.0	<0.0	16.0	<0.0	62.0	<0.0	1 123.0	0.6
Liberia	2004	103.0	<0.0	13.0	<0.0	35.0	<0.0	1 035.0	0.3
Libéria	2008	51.0	<0.0	4.0	<0.0	269.0	0.1	978.0	0.3
Libya	2004	7 070.0	1.2	850.0	0.2	1 130.0	0.2	27 160.0	4.8
Libye	2009	12 009.0	1.9	3 792.0	0.6	2 275.0	0.4	42 982.0	6.8
Lithuania	2010	12 226.0	3.6	2 456.0	0.7	...	...	...	...
Lituanie	2012	13 552.0	4.1	...	...	2 440.0	0.7	...	...
Luxembourg	2003	...	...	323.0	0.7	...	...	...	...
Luxembourg	2013	1 537.0	2.9	472.0	0.9	...	...	6 684.0	12.6
Madagascar	2004	5 201.0	0.3	410.0	<0.0	175.0	<0.0	5 661.0	0.3
Madagascar	2007	3 150.0	0.2	57.0	<0.0	...	...	...	...
Malawi	2004	266.0	<0.0	...	...	...	...	7 264.0	0.6
Malawi	2009	265.0	<0.0	180.0	<0.0	221.0	<0.0	4 812.0	0.3
Malaysia	2002	17 020.0	0.7	2 160.0	0.1	2 880.0	0.1	43 380.0	1.8
Malaisie	2010	32 979.0	1.2	9 995.0	0.4	11 077.0	0.4	90 199.0	3.3
	2011	...	...	...	...	12 402.0	0.4	...	...
Maldives	2004	302.0	0.9	14.0	<0.0	241.0	0.7	886.0	2.7
Maldives	2010	525.0	1.4	32.0	0.1	247.0	0.7	1 868.0	5.0
Mali	2004	1 053.0	0.1	84.0	<0.0	351.0	<0.0	8 338.0	0.6
Mali	2010	1 291.0	0.1	103.0	<0.0	135.0	<0.0	6 715.0	0.4
Malta	2010	1 279.0	3.1	184.0	0.4	301.0	0.7	2 838.0	6.9
Malte	2013	1 466.0	3.5	197.0	0.5	469.0	1.1	3 151.0	7.5
Marshall Islands	2000	24.0	0.5	4.0	0.1	2.0	<0.0	152.0	3.0
Îles Marshall	2008	31.0	0.5	11.0	0.2	10.0	0.1	177.0	2.6
	2010	32.0	0.4	...	...	...	...	127.0	1.7
Mauritania	2004	313.0	0.1	64.0	<0.0	81.0	<0.0	1 893.0	0.6
Mauritanie	2009	445.0	0.1	93.0	<0.0	123.0	<0.0	2 303.0	0.7
Mauritius									
Maurice	2004	1 303.0	1.1	233.0	0.2	1 428.0	1.2	4 604.0	3.7
Mexico	2004	303 519.0	2.9	148 456.0	1.4	79 925.0	0.8	417 665.0	4.0
Mexique	2005	187 144.0	1.7	...	...	...	...	...	...
	2010	225 559.0	2.0	13 225.0	0.1	...	...	275 171.0	2.4
	2011	240 525.0	2.1	13 451.0	0.1	...	...	290 317.0	2.5
Micronesia (Fed. States of)	2005	62.0	0.6	13.0	0.1	16.0	0.1	249.0	2.3
Micronésie (États féd. de)	2008	64.0	0.6	40.0	0.4	16.0	0.1	280.0	2.5
	2009	20.0	0.2	...	...	...	...	375.0	3.3
Monaco	2011	250.0	7.1	38.0	1.1	93.0	2.6	610.0	17.2
Monaco	2012	254.0	7.2	37.0	1.0	96.0	2.7	...	...
Mongolia	2002	6 732.0	2.6	337.0	0.1	1 093.0	0.4	8 826.0	3.4
Mongolie	2010	7 497.0	2.7	533.0	0.2	1 176.0	0.4	9 876.0	3.5
	2011	7 943.0	2.8	652.0	0.2	1 284.0	0.5	10 143.0	3.6
Montenegro	2010	1 268.0	2.0	26.0	<0.0	91.0	0.1	3 410.0	5.4
Monténégro	2013	1 338.0	2.1	25.0	<0.0	105.0	0.2	3 428.0	5.4
Morocco	2004	15 991.0	0.5	3 091.0	0.1	7 366.0	0.2	24 328.0	0.8
Maroc	2009	20 682.0	0.6	2 668.0	0.1	9 006.0	0.3	29 689.0	0.9
Mozambique	2004	514.0	<0.0	159.0	<0.0	618.0	<0.0	6 183.0	0.3
Mozambique	2010	811.0	<0.0	...	...	1 196.0	0.1	8 816.0	0.4
	2012	971.0	<0.0	...	...	1 387.0	0.1	10 081.0	0.4
Myanmar	2004	17 791.0	0.4	1 396.0	<0.0	127.0	<0.0	49 341.0	1.0
Myanmar	2005	18 584.0	0.4	1 756.0	<0.0	...	...	36 521.0	0.7
	2010	26 435.0	0.5	2 849.0	0.1	...	...	45 200.0	0.9
	2012	29 832.0	0.6	3 355.0	0.1	...	...	48 871.0	1.0

Health Personnel *(continued)*
Number of health workers and workforce density (per 1 000 population)

Le personnel de santé *(suite)*
Personnel de santé et densité (pour 1 000 habitants)

Country or area Pays ou zone	Year Année	Physicians Médecins		Dentists Dentistes		Pharmacists Pharmaciens		Nurses and midwives Infirmières et Sages-femmes	
		Number Nombre	Per 1 000 Pour 1 000	Number Nombre	Per 1 000 Pour 1 000	Number Nombre	Per 1 000 Pour 1 000	Number Nombre	Per 1 000 Pour 1 000
Namibia	2004	598.0	0.3	113.0	0.1	288.0	0.1	6 145.0	3.1
Namibie	2007	774.0	0.4	90.0	<0.0	376.0	0.2	5 750.0	2.8
Nauru	2004	10.0	0.8	1.0	0.1	10.0	0.8	63.0	4.8
Nauru	2008	10.0	0.7	1.0	0.1	1.0	0.1	69.0	4.9
	2009	10.0	0.7	3.0	0.2	7.0	0.5	...	...
	2010	10.0	0.7	...	...	...	...	...	...
Nepal	2004	5 384.0	0.2	359.0	<0.0	358.0	<0.0	11 825.0	0.5
Népal	2012	...	...	...	...	4 200.0	0.1	...	...
Netherlands	2003	50 854.0	3.1	7 759.0	0.5	3 134.0	0.2	...	...
Pays-Bas	2008	...	...	...	...	...	...	138 200.0	8.4
	2011	...	...	...	...	3 446.0	0.2	2 692.0	0.2
New Zealand	2001	9 027.0	2.4	2 586.0	0.7	3 495.0	0.9	33 249.0	8.7
Nouvelle-Zélande	2002	8 190.0	2.1	1 620.0	0.4	...	...	...	...
	2007	9 757.0	2.4	1 877.0	0.5	2 889.0	0.7	44 491.0	10.9
	2010	11 412.0	2.7	...	...	...	...	...	...
	2011	...	...	...	...	4 440.0	1.0	...	...
Nicaragua	2005	2 717.0	0.5	246.0	<0.0	...	...	6 294.0	1.1
Nicaragua	2010	4 239.0	0.7	258.0	<0.0	...	...	7 366.0	1.2
	2014	5 495.0	0.9	260.0	<0.0	...	...	8 323.0	1.4
Niger	2004	296.0	<0.0	15.0	<0.0	20.0	<0.0	2 818.0	0.2
Niger	2008	288.0	<0.0	16.0	<0.0	21.0	<0.0	2 115.0	0.1
Nigeria	2005	39 210.0	0.3	2 113.0	<0.0	12 072.0	0.1	213 425.0	1.6
Nigéria	2008	56 526.0	0.4	3 781.0	<0.0	18 682.0	0.1	224 943.0	1.6
	2009	58 363.0	0.4	2 464.0	<0.0	...	...	...	...
	2011	...	...	...	...	17 022.0	0.1	...	...
Niue	2004	4.0	2.0	2.0	1.0	1.0	0.5	22.0	11.0
Nioué	2008	3.0	3.0	4.0	4.0	1.0	1.0	16.0	16.0
Norway	2003	...	...	...	...	1 675.0	0.4	...	...
Norvège	2009	19 579.0	4.2	4 192.0	0.9	...	...	150 334.0	31.9
	2012	21 238.0	4.3	4 340.0	0.9	3 367.0	0.7	85 661.0	17.3
Oman	2005	4 182.0	1.6	448.0	0.2	912.0	0.4	9 277.0	3.6
Oman	2010	5 862.0	2.0	654.0	0.2	2 784.0	1.0	12 865.0	4.5
	2012	7 055.0	2.4	805.0	0.3	5 455.0	1.9	15 627.0	5.4
Pakistan	2004	116 298.0	0.7	7 862.0	0.1	8 102.0	0.1	71 764.0	0.5
Pakistan	2005	126 350.0	0.8	...	...	...	...	47 380.0	0.3
	2010	144 901.0	0.8	10 508.0	0.1	...	...	100 397.0	0.6
Palau	2004	...	...	...	...	...	...	120.0	6.0
Palaos	2007	...	...	5.0	0.3	1.0	0.1	...	...
	2010	29.0	1.4	...	...	...	...	120.0	5.7
Panama	2000	4 431.0	1.5	2 231.0	0.8	2 526.0	0.9	8 158.0	2.8
Panama	2005	4 488.0	1.4	938.0	0.3	...	...	6 675.0	2.1
	2010	5 121.0	1.5	1 091.0	0.3	...	...	9 021.0	2.6
	2013	6 068.0	1.7	1 037.0	0.3	...	...	5 158.0	1.4
Papua New Guinea	2000	275.0	0.1	90.0	<0.0	...	...	2 841.0	0.5
Papouasie-Nvl-Guinée	2009	...	...	121.0	<0.0	...	...	...	...
	2010	376.0	0.1	...	...	...	...	3 643.0	0.6
	2012	...	...	...	...	354.0	<0.0	...	...
Paraguay	2002	6 355.0	1.1	3 182.0	0.6	1 868.0	0.3	10 261.0	1.8
Paraguay	2012	8 203.0	1.2	1 054.0	0.2	...	...	6 689.0	1.0
Peru	2009	27 272.0	0.9	3 570.0	0.1	1 822.0	0.1	37 672.0	1.3
Pérou	2012	33 669.0	1.1	4 471.0	0.2	1 528.0	0.1	45 024.0	1.5
Philippines	2004	93 862.0	1.2	45 903.0	0.6	49 667.0	0.6	488 434.0	6.0
Philippines	2011	...	...	...	...	84 000.0	0.9	...	...
Poland	2003	95 272.0	2.5	11 451.0	0.3	25 397.0	0.7	188 898.0	4.9
Pologne	2010	79 337.0	2.1	12 326.0	0.3	25 120.0	0.7	206 941.0	5.4
	2012	85 025.0	2.2	12 491.0	0.3	26 843.0	0.7	236 006.0	6.2
Portugal	2005	36 138.0	3.4	5 056.0	0.5	9 494.0	0.9	48 155.0	4.6
Portugal	2010	41 431.0	3.9	6 972.0	0.7	10 895.0	1.0	62 433.0	5.8
	2012	43 863.0	4.1	7 533.0	0.7	10 890.0	1.0	65 404.0	6.1
	2013	...	...	8 133.0	0.8	...	...	...	...
Qatar	2005	2 150.0	2.6	690.0	0.8	1 100.0	1.4	4 880.0	6.0
Qatar	2006	2 313.0	2.8	486.0	0.6	1 056.0	1.3	6 185.0	7.4
	2010	6 919.0	7.7	...	...	...	...	10 615.0	11.9

Country or area Pays ou zone	Year Année	Physicians Médecins Number Nombre	Per 1 000 Pour 1 000	Dentists Dentistes Number Nombre	Per 1 000 Pour 1 000	Pharmacists Pharmaciens Number Nombre	Per 1 000 Pour 1 000	Nurses and midwives Infirmières et Sages-femmes Number Nombre	Per 1 000 Pour 1 000
Republic of Korea	2005	85 369.0	1.8	21 581.0	0.5	54 829.0	1.1	222 301.0	4.6
République de Corée	2010	98 293.0	2.0	20 936.0	0.4	32 152.0	0.7	229 819.0	4.7
	2012	104 114.0	2.1	21 888.0	0.5	32 560.0	0.7	243 402.0	5.0
Republic of Moldova	2010	12 780.0	3.1	...	...	...	...	...	...
République de Moldova	2013	10 432.0	3.0	1 751.0	0.5	1 901.0	0.5	22 358.0	6.4
Romania	2002	43 093.0	1.9	13 205.0	0.6	15 121.0	0.7	#4 014.0	0.2
Roumanie	2012	52 362.0	2.4	13 779.0	0.6	15 359.0	0.7	120 084.0	5.6
Rwanda	2004	432.0	0.1	21.0	<0.0	278.0	<0.0	3 647.0	0.4
Rwanda	2010	568.0	0.1	122.0	<0.0	56.0	<0.0	6 975.0	0.7
Saint Kitts and Nevis Saint-Kitts-et-Nevis	2001	49.0	1.2	15.0	0.4	...	...	294.0	7.0
Saint Lucia	2001	130.0	0.9	27.0	0.2	23.0	0.2	371.0	2.5
Sainte-Lucie	2004	68.0	0.5	7.0	<0.0	...	...	375.0	2.5
	2005	83.0	0.5	...	...	...	...	326.0	2.0
	2009	18.0	0.1	5.0	<0.0	...	...	...	...
	2010	...	...	28.0	0.2	...	...	...	...
Saint Vincent-Grenadines Saint-Vincent-Grenadines	2001	62.0	0.5	21.0	0.2	23.0	0.2	476.0	4.0
Samoa	2003	50.0	0.3	10.0	0.1	20.0	0.1	310.0	1.7
Samoa	2005	50.0	0.3	...	...	3.0	<0.0	173.0	0.9
	2008	85.0	0.5	63.0	0.3	59.0	0.3	348.0	1.9
San Marino Saint-Marin	2013	164.0	5.1	...	...	23.0	0.7	284.0	8.8
Sao Tome and Principe Sao Tomé-et-Principe	2004	81.0	0.5	11.0	0.1	24.0	0.1	308.0	1.9
Saudi Arabia	2000	14 950.0	0.7	...	...	863.0	<0.0	36 495.0	1.6
Arabie saoudite	2001	14 464.0	0.6	1 581.0	0.1	...	...	...	...
	2010	66 014.0	2.4	...	...	14 928.0	0.5	129 792.0	4.7
	2012	71 518.0	2.5	...	...	15 590.0	0.5	139 701.0	4.9
Senegal	2004	594.0	0.1	97.0	<0.0	85.0	<0.0	3 287.0	0.3
Sénégal	2008	741.0	0.1	105.0	<0.0	127.0	<0.0	5 254.0	0.4
Serbia Serbie	2009	20 825.0	2.1	2 273.0	0.2	2 034.0	0.2	...	...
Seychelles	2005	104.0	1.3	10.0	0.1	7.0	0.1	390.0	4.8
Seychelles	2010	100.0	1.2	19.0	0.2	4.0	<0.0	412.0	4.9
	2012	93.0	1.1	14.0	0.2	4.0	<0.0	419.0	4.8
Sierra Leone	2004	168.0	<0.0	5.0	<0.0	340.0	0.1	2 510.0	0.5
Sierra Leone	2010	136.0	<0.0	6.0	<0.0	114.0	<0.0	1 017.0	0.2
Singapore	2005	6 748.0	1.6	1 277.0	0.3	1 330.0	0.3	20 167.0	4.7
Singapour	2010	8 819.0	1.9	1 506.0	0.3	1 814.0	0.4	29 340.0	6.4
	2013	10 339.0	2.0	2 149.0	0.4	2 186.0	0.4	30 533.0	5.8
Slovakia	2003	17 172.0	3.2	...	...	...	...	...	...
Slovaquie	2010	18 110.0	3.4	2 663.0	0.5	3 267.0	0.6	34 619.0	6.4
	2012	18 193.0	3.3	2 665.0	0.5	3 522.0	0.6	33 243.0	6.1
Solomon Islands	2005	89.0	0.2	52.0	0.1	53.0	0.1	694.0	1.5
Îles Salomon	2009	118.0	0.2	...	...	...	...	1 080.0	2.1
Somalia	2005	...	...	...	...	...	...	429.0	0.1
Somalie	2006	300.0	<0.0	...	...	50.0	<0.0	965.0	0.1
South Africa	2004	34 829.0	0.8	5 995.0	0.1	12 521.0	0.3	184 459.0	4.1
Afrique du Sud	2005	...	...	...	...	...	...	191 269.0	4.0
	2010	...	...	...	...	...	...	231 086.0	4.8
	2012	...	...	...	...	20 974.0	0.4	248 736.0	4.9
	2013	39 541.0	0.8	10 175.0	0.2	...	...	260 698.0	5.1
Spain	2005	199 123.0	4.6	22 150.0	0.5	59 498.0	1.4	231 001.0	5.4
Espagne	2010	223 484.0	5.1	27 826.0	0.6	64 203.0	1.5	262 915.0	6.0
	2013	232 816.0	4.9	38 684.0	0.8	66 657.0	1.4	266 495.0	5.7
Sri Lanka	2004	10 479.0	0.5	1 245.0	0.1	990.0	0.1	33 233.0	1.7
Sri Lanka	2005	10 198.0	0.5	954.0	<0.0	...	...	27 514.0	1.3
	2007	11 023.0	0.5	1 743.0	0.1	886.0	<0.0	40 678.0	1.9
	2010	14 668.0	0.7	1 046.0	<0.0	...	...	35 367.0	1.6
	2012	...	...	1 224.0	0.1	753.0	<0.0	...	...

7

Health Personnel *(continued)*
Number of health workers and workforce density (per 1 000 population)

Le personnel de santé *(suite)*
Personnel de santé et densité (pour 1 000 habitants)

Country or area Pays ou zone	Year Année	Physicians Médecins		Dentists Dentistes		Pharmacists Pharmaciens		Nurses and midwives Infirmières et Sages-femmes	
		Number Nombre	Per 1 000 Pour 1 000	Number Nombre	Per 1 000 Pour 1 000	Number Nombre	Per 1 000 Pour 1 000	Number Nombre	Per 1 000 Pour 1 000
Sudan [former]	2004	7 552.0	0.2	1 082.0	<0.0	3 558.0	0.1	31 496.0	0.9
Soudan [anc.]	2008	10 813.0	0.3	772.0	<0.0	386.0	<0.0	32 439.0	0.8
Suriname									
Suriname	2004	400.0	0.9	42.0	0.1	...	...	2 580.0	5.9
Swaziland	2004	171.0	0.2	32.0	<0.0	70.0	0.1	6 828.0	6.3
Swaziland	2009	173.0	0.2	43.0	<0.0	51.0	0.1	1 626.0	1.6
Sweden	2002	29 122.0	3.3	7 270.0	0.8	5 885.0	0.7	90 758.0	10.2
Suède	2007	29 693.0	3.3	7 541.0	0.8	...	...	100 480.0	11.0
	2011	37 063.0	3.9	7 604.0	0.8	7 304.0	0.8	#7 001.0	0.7
Switzerland	2000	...	...	...	...	...	...	77 120.0	10.8
Suisse	2002	25 921.0	3.6	...	...	...	...	...	...
	2012	31 313.0	4.0	4 181.0	0.5	4 350.0	0.6	#2 446.0	0.3
Syrian Arab Republic	2005	28 247.0	1.5	15 725.0	0.8	13 218.0	0.7	34 604.0	1.8
Rép. arabe syrienne	2010	31 194.0	1.5	15 984.0	0.7	16 554.0	0.8	40 053.0	1.9
Tajikistan	2005	13 272.0	2.0	...	...	...	...	...	...
Tadjikistan	2008	13 909.0	2.1	...	...	...	...	...	...
	2013	13 778.0	1.9	1 250.0	0.2	...	...	36 044.0	5.0
Thailand	2004	18 918.0	0.3	4 129.0	0.1	7 413.0	0.1	96 704.0	1.5
Thaïlande	2010	26 244.0	0.4	17 222.0	0.3	8 700.0	0.1	138 710.0	2.1
TFYR of Macedonia	2009	5 364.0	2.6	1 425.0	0.7	...	...	...	...
ex-R.Y. de Macédoine	2011	...	...	...	...	3 271.0	1.6	...	...
Timor-Leste	2004	79.0	0.1	45.0	0.1	14.0	<0.0	1 795.0	2.2
Timor-Leste	2010	84.0	0.1	44.0	<0.0	130.0	0.1	1 255.0	1.0
	2011	84.0	0.1	46.0	<0.0	131.0	0.1	1 283.0	1.1
Togo	2004	225.0	<0.0	19.0	<0.0	134.0	<0.0	1 937.0	0.4
Togo	2008	349.0	0.1	19.0	<0.0	11.0	<0.0	1 816.0	0.3
Tonga	2001	35.0	0.3	33.0	0.3	17.0	0.2	341.0	3.3
Tonga	2002	30.0	0.3	...	...	...	...	350.0	3.4
	2003	...	...	23.0	0.2	...	...	...	...
	2009	62.0	0.6	37.0	0.4	15.0	0.1	378.0	3.7
	2010	58.0	0.6	...	...	...	...	400.0	3.9
	2012	...	...	...	...	9.0	0.1	...	...
Trinidad and Tobago	2003	1 038.0	0.8	249.0	0.2	525.0	0.4	3 980.0	3.1
Trinité-et-Tobago	2007	1 543.0	1.2	294.0	0.2	641.0	0.5	4 677.0	3.6
Tunisia	2004	13 330.0	1.3	2 452.0	0.2	2 909.0	0.3	28 537.0	2.9
Tunisie	2005	9 422.0	0.9	1 850.0	0.2	2 114.0	0.2	...	...
	2009	12 535.0	1.2	2 528.0	0.2	2 106.0	0.2	34 551.0	3.3
	2010	12 996.0	1.2	3 130.0	0.3	3 236.0	0.3	...	...
Turkey	2005	100 853.0	1.4	18 149.0	0.2	22 756.0	0.3	121 723.0	1.7
Turquie	2010	123 447.0	1.6	21 432.0	0.3	26 506.0	0.3	165 115.0	2.1
	2011	126 029.0	1.7	21 099.0	0.3	26 089.0	0.4	176 887.0	2.4
Turkmenistan	2002	20 032.0	4.2	876.0	0.2	1 626.0	0.3	43 359.0	9.0
Turkménistan	2013	...	...	610.0	0.1	906.0	0.2	...	...
Tuvalu	2003	10.0	0.9	2.0	0.2	2.0	0.2	50.0	4.5
Tuvalu	2008	10.0	0.9	4.0	0.4	2.0	0.2	64.0	5.8
	2009	12.0	1.1	...	...	...	...	...	...
Uganda	2005	3 361.0	0.1	440.0	<0.0	762.0	<0.0	37 625.0	1.3
Ouganda	2012	...	...	...	...	1 200.0	<0.0	...	...
Ukraine	2001	...	...	...	...	23 576.0	0.5	...	...
Ukraine	2013	158 344.0	3.5	30 934.0	0.7	1 870.0	<0.0	342 833.0	7.7
United Arab Emirates	2004	...	...	1 368.0	0.4	2 006.0	0.7	14 362.0	4.7
Émirats arabes unis	2005	6 946.0	1.5	...	...	...	...	14 844.0	3.3
	2007	9 215.0	1.9	2 053.0	0.4	2 817.0	0.6	17 336.0	3.6
	2010	12 752.0	2.5	...	...	...	...	24 362.0	4.8
	2011	...	...	...	...	...	...	24 936.0	3.2
United Kingdom									
Royaume-Uni	2013	177 449.0	2.8	33 968.0	0.5	51 315.0	0.8	556 051.0	8.8
United Rep. of Tanzania	2002	822.0	<0.0	267.0	<0.0	365.0	<0.0	13 292.0	0.4
Rép.-Unie de Tanzanie	2006	300.0	<0.0	230.0	<0.0	81.0	<0.0	9 440.0	0.2
	2012	1 481.0	<0.0	509.0	<0.0	656.0	<0.0	20 800.0	0.4

Country or area Pays ou zone	Year Année	Physicians Médecins Number Nombre	Per 1 000 Pour 1 000	Dentists Dentistes Number Nombre	Per 1 000 Pour 1 000	Pharmacists Pharmaciens Number Nombre	Per 1 000 Pour 1 000	Nurses and midwives Infirmières et Sages-femmes Number Nombre	Per 1 000 Pour 1 000
United States	2000	730 801.0	2.6	463 663.0	1.6	249 642.0	0.9	2 669 603.0	9.4
États-Unis	2005	718 473.0	2.4	...	...	...	...	2 927 000.0	9.8
	2010	752 572.0	2.4	...	...	275 000.0	0.9	...	...
	2011	767 782.0	2.5	...	...	...	...	...	...
Uruguay	2002	12 384.0	3.7	3 936.0	1.2	...	...	2 880.0	0.8
Uruguay	2008	13 197.0	3.7	2 476.0	0.7	1 877.0	0.5	19 595.0	5.5
Uzbekistan	2010	72 522.0	2.5	4 693.0	0.2	1 074.0	<0.0	324 493.0	11.4
Ouzbékistan	2013	71 971.0	2.5	4 749.0	0.2	1 127.0	<0.0	339 015.0	11.9
Vanuatu	2004	30.0	0.1	...	...	...	...	360.0	1.7
Vanuatu	2008	26.0	0.1	3.0	<0.0	2.0	<0.0	380.0	1.7
Venezuela (Boliv. Rep. of)									
Venezuela (Rép. boliv. du)	2001	48 000.0	1.9	13 680.0	0.6	...	...	28 000.0	1.1
Viet Nam	2005	101 117.0	1.2	...	...	23 218.0	0.3	69 665.0	0.8
Viet Nam	2008	107 131.0	1.2	...	...	28 370.0	0.3	88 025.0	1.0
	2010	99 621.0	1.1	...	...	...	...	94 049.0	1.0
	2011	102 925.0	1.2	...	...	27 734.0	0.3	100 972.0	1.1
	2013	107 867.0	1.2	...	...	...	...	112 029.0	1.2
Yemen	2004	6 739.0	0.3	850.0	<0.0	2 638.0	0.1	13 746.0	0.7
Yémen	2010	4 834.0	0.2	897.0	<0.0	2 295.0	0.1	16 590.0	0.7
Zambia	2004	1 499.0	0.1	491.0	<0.0	804.0	0.1	22 010.0	2.0
Zambie	2005	646.0	0.1	...	...	...	...	8 369.0	0.7
	2010	836.0	0.1	246.0	<0.0	317.0	<0.0	9 932.0	0.8
	2012	2 399.0	0.2	401.0	<0.0	1 223.0	0.1	...	...
Zimbabwe	2004	2 086.0	0.2	310.0	<0.0	883.0	0.1	9 357.0	0.7
Zimbabwe	2010	916.0	0.1	239.0	<0.0	472.0	<0.0	17 029.0	1.3
	2011	1 054.0	0.1	255.0	<0.0	467.0	<0.0	17 022.0	1.3

Source:

World Health Organisation (WHO), Geneva, WHO Global Health
Workforce statistics database, last accessed July 2016.

Source:

Organisation mondiale de la santé (OMS), Genève, base de données OMS
sur les statistiques relatives aux personnels de santé, dernier accès juillet
2016.

Expenditure on health
Percentage of GDP and of government expenditure

Dépenses de santé
Pourcentage du PIB et des dépenses du gouvernement

Region, country or area	1995	2000	2005	2010	2012	2013	2014	Région, pays ou zone
Afghanistan * [1,2]								**Afghanistan * [1,2]**
Total expenditure (% GDP)	...	...	8.1	9.2	8.5	8.1	8.2	Dépenses totales (% du PIB)
General govt.(% expend. total)	...	...	5.5	14.4	11.7	10.6	12.0	Dépenses publiques (en % du totale)
Albania								**Albanie**
Total expenditure (% GDP)	6.6	6.3	6.1	5.3	5.6	5.7	5.9	Dépenses totales (% du PIB)
General govt.(% expend. total)	5.5	7.1	9.7	8.5	9.7	9.8	9.4	Dépenses publiques (en % du totale)
Algeria								**Algérie**
Total expenditure (% GDP)	3.7	3.5	3.2	5.1	6.1	7.1	7.2	Dépenses totales (% du PIB)
General govt.(% expend. total)	8.5	8.8	8.2	9.6	10.0	9.9	9.9	Dépenses publiques (en % du totale)
Andorra								**Andorre**
Total expenditure (% GDP)	5.8	5.8	5.2	8.0	7.5	11.5	8.1	Dépenses totales (% du PIB)
General govt.(% expend. total)	23.6	19.1	19.5	24.0	18.9	22.3	27.9	Dépenses publiques (en % du totale)
Angola								**Angola**
Total expenditure (% GDP)	6.5	2.8	4.1	3.4	3.3	4.3	3.3	Dépenses totales (% du PIB)
General govt.(% expend. total)	5.2	2.9	6.1	5.4	5.6	7.4	5.0	Dépenses publiques (en % du totale)
Antigua and Barbuda *								**Antigua-et-Barbuda ***
Total expenditure (% GDP)	4.1	4.1	4.4	5.6	5.4	5.3	5.5	Dépenses totales (% du PIB)
General govt.(% expend. total)	12.9	11.4	12.1	16.4	17.2	16.3	18.1	Dépenses publiques (en % du totale)
Argentina *								**Argentine ***
Total expenditure (% GDP)	8.3	9.2	6.8	6.6	5.0	5.0	4.8	Dépenses totales (% du PIB)
General govt.(% expend. total)	19.4	17.6	16.7	14.1	8.7	7.7	6.9	Dépenses publiques (en % du totale)
Armenia								**Arménie**
Total expenditure (% GDP)	6.4	6.3	5.3	4.6	4.5	4.6	4.5	Dépenses totales (% du PIB)
General govt.(% expend. total)	8.3	5.3	10.2	7.0	7.4	7.3	7.0	Dépenses publiques (en % du totale)
Australia								**Australie**
Total expenditure (% GDP)	7.3	8.1	8.5	9.0	9.4	9.4	9.4	Dépenses totales (% du PIB)
General govt.(% expend. total)	14.9	16.0	16.9	17.1	17.3	17.3	17.3	Dépenses publiques (en % du totale)
Austria								**Autriche**
Total expenditure (% GDP)	9.5	10.1	10.5	11.2	11.2	11.1	11.2	Dépenses totales (% du PIB)
General govt.(% expend. total)	12.7	14.5	15.7	15.9	16.3	16.3	16.3	Dépenses publiques (en % du totale)
Azerbaijan [3]								**Azerbaïdjan [3]**
Total expenditure (% GDP)	5.8	4.7	7.9	5.3	5.4	5.5	6.0	Dépenses totales (% du PIB)
General govt.(% expend. total)	6.9	5.4	5.2	4.2	3.8	3.5	3.9	Dépenses publiques (en % du totale)
Bahamas								**Bahamas**
Total expenditure (% GDP)	6.9	5.2	6.0	7.4	7.4	7.0	7.7	Dépenses totales (% du PIB)
General govt.(% expend. total)	14.2	14.8	14.7	17.6	15.1	13.4	14.8	Dépenses publiques (en % du totale)
Bahrain *								**Bahreïn ***
Total expenditure (% GDP)	4.1	3.5	3.2	3.6	4.4	4.7	5.0	Dépenses totales (% du PIB)
General govt.(% expend. total)	10.8	10.2	8.6	9.1	9.3	10.6	10.5	Dépenses publiques (en % du totale)
Bangladesh								**Bangladesh**
Total expenditure (% GDP)	3.2	2.3	2.7	3.1	3.1	2.9	2.8	Dépenses totales (% du PIB)
General govt.(% expend. total)	8.4	8.1	7.9	8.3	6.9	5.5	5.7	Dépenses publiques (en % du totale)
Barbados								**Barbade**
Total expenditure (% GDP)	5.2	5.2	5.4	6.2	7.4	7.6	7.5	Dépenses totales (% du PIB)
General govt.(% expend. total)	11.4	9.9	9.2	9.7	10.7	10.7	10.9	Dépenses publiques (en % du totale)
Belarus								**Bélarus**
Total expenditure (% GDP)	6.7	6.1	6.9	5.6	5.0	6.1	5.7	Dépenses totales (% du PIB)
General govt.(% expend. total)	11.5	10.1	10.5	13.4	13.2	13.9	13.8	Dépenses publiques (en % du totale)
Belgium								**Belgique**
Total expenditure (% GDP)	7.6	8.1	9.2	10.2	10.5	10.6	10.6	Dépenses totales (% du PIB)
General govt.(% expend. total)	11.2	12.3	13.7	15.0	14.9	15.1	15.1	Dépenses publiques (en % du totale)
Belize *								**Belize ***
Total expenditure (% GDP)	4.2	4.0	4.5	5.8	5.5	5.8	5.8	Dépenses totales (% du PIB)
General govt.(% expend. total)	10.5	6.5	9.3	13.1	12.5	13.4	13.8	Dépenses publiques (en % du totale)
Benin								**Bénin**
Total expenditure (% GDP)	4.7	4.3	4.7	4.9	4.9	4.6	4.6	Dépenses totales (% du PIB)
General govt.(% expend. total)	10.4	10.0	11.1	12.4	10.2	9.6	9.6	Dépenses publiques (en % du totale)
Bhutan								**Bhoutan**
Total expenditure (% GDP)	4.0	6.9	5.3	5.2	3.7	3.8	3.6	Dépenses totales (% du PIB)
General govt.(% expend. total)	6.9	12.2	11.7	11.0	7.8	8.4	8.0	Dépenses publiques (en % du totale)
Bolivia (Plurinational State of) * [4]								**Bolivie (État plurinational de) * [4]**
Total expenditure (% GDP)	3.7	5.7	5.7	5.4	5.6	6.0	6.3	Dépenses totales (% du PIB)
General govt.(% expend. total)	7.9	8.7	11.8	11.9	11.3	12.0	11.8	Dépenses publiques (en % du totale)

Expenditure on health *(continued)*
Percentage of GDP and of government expenditure

Dépenses de santé *(suite)*
Pourcentage du PIB et des dépenses du gouvernement

Region, country or area	1995	2000	2005	2010	2012	2013	2014	Région, pays ou zone
Bosnia and Herzegovina								**Bosnie-Herzégovine**
Total expenditure (% GDP)	9.0	7.1	8.5	9.6	9.9	9.5	9.6	Dépenses totales (% du PIB)
General govt.(% expend. total)	9.1	11.3	10.5	13.7	14.7	14.1	14.1	Dépenses publiques (en % du totale)
Botswana								**Botswana**
Total expenditure (% GDP)	4.1	4.6	5.6	5.6	6.3	5.8	5.4	Dépenses totales (% du PIB)
General govt.(% expend. total)	5.5	7.5	11.8	8.6	10.8	10.4	8.8	Dépenses publiques (en % du totale)
Brazil								**Brésil**
Total expenditure (% GDP)	6.5	7.0	8.3	8.3	8.3	8.5	8.3	Dépenses totales (% du PIB)
General govt.(% expend. total)	8.4	4.1	5.0	9.9	6.9	7.1	6.8	Dépenses publiques (en % du totale)
Brunei Darussalam								**Brunéi Darussalam**
Total expenditure (% GDP)	3.0	3.1	2.6	2.7	2.3	2.6	2.6	Dépenses totales (% du PIB)
General govt.(% expend. total)	3.5	6.3	6.9	6.3	6.1	6.4	6.5	Dépenses publiques (en % du totale)
Bulgaria								**Bulgarie**
Total expenditure (% GDP)	4.8	6.1	7.1	7.2	7.1	7.9	8.4	Dépenses totales (% du PIB)
General govt.(% expend. total)	8.5	9.0	11.7	11.0	11.5	11.0	11.0	Dépenses publiques (en % du totale)
Burkina Faso								**Burkina Faso**
Total expenditure (% GDP)	4.9	5.1	6.9	7.2	5.3	5.9	5.0	Dépenses totales (% du PIB)
General govt.(% expend. total)	9.1	8.8	18.7	15.7	11.7	12.7	11.2	Dépenses publiques (en % du totale)
Burundi * [5]								**Burundi * [5]**
Total expenditure (% GDP)	4.4	5.0	9.8	8.8	8.2	8.0	7.5	Dépenses totales (% du PIB)
General govt.(% expend. total)	6.8	7.1	11.2	13.5	14.1	14.0	13.2	Dépenses publiques (en % du totale)
Cabo Verde								**Cabo Verde**
Total expenditure (% GDP)	5.3	4.8	4.9	4.8	4.6	4.3	4.8	Dépenses totales (% du PIB)
General govt.(% expend. total)	8.2	9.9	10.4	8.7	9.6	9.2	11.7	Dépenses publiques (en % du totale)
Cambodia								**Cambodge**
Total expenditure (% GDP)	5.4	5.9	5.8	6.0	6.2	5.9	5.7	Dépenses totales (% du PIB)
General govt.(% expend. total)	7.0	8.9	13.1	7.2	7.3	6.8	6.1	Dépenses publiques (en % du totale)
Cameroon *								**Cameroun ***
Total expenditure (% GDP)	3.9	4.5	4.8	5.3	4.3	4.3	4.1	Dépenses totales (% du PIB)
General govt.(% expend. total)	6.1	6.1	7.7	8.5	5.5	4.6	4.3	Dépenses publiques (en % du totale)
Canada								**Canada**
Total expenditure (% GDP)	8.9	8.7	9.6	11.2	10.8	10.7	10.4	Dépenses totales (% du PIB)
General govt.(% expend. total)	13.2	15.1	17.5	18.2	18.6	18.6	18.8	Dépenses publiques (en % du totale)
Central African Rep. [5]								**Rép. centrafricaine [5]**
Total expenditure (% GDP)	3.6	4.2	4.3	3.9	3.6	3.8	4.2	Dépenses totales (% du PIB)
General govt.(% expend. total)	8.0	12.6	12.5	9.7	10.5	15.5	14.2	Dépenses publiques (en % du totale)
Chad [5,6]								**Tchad [5,6]**
Total expenditure (% GDP)	5.8	6.3	3.9	2.9	3.0	3.4	3.6	Dépenses totales (% du PIB)
General govt.(% expend. total)	11.1	12.9	13.5	4.9	5.9	7.8	9.0	Dépenses publiques (en % du totale)
Chile								**Chili**
Total expenditure (% GDP)	5.2	6.4	6.7	7.0	7.2	7.5	7.8	Dépenses totales (% du PIB)
General govt.(% expend. total)	12.7	14.5	12.5	13.8	14.7	15.3	15.9	Dépenses publiques (en % du totale)
China								**Chine**
Total expenditure (% GDP)	3.5	4.6	4.7	4.9	5.3	5.4	5.5	Dépenses totales (% du PIB)
General govt.(% expend. total)	15.9	10.8	9.8	10.2	10.5	10.3	10.4	Dépenses publiques (en % du totale)
Colombia *								**Colombie ***
Total expenditure (% GDP)	6.8	5.9	5.8	6.8	6.9	6.8	7.2	Dépenses totales (% du PIB)
General govt.(% expend. total)	17.2	19.3	15.7	18.1	18.9	18.1	18.1	Dépenses publiques (en % du totale)
Comoros								**Comores**
Total expenditure (% GDP)	4.6	3.6	4.3	5.8	7.1	6.5	6.7	Dépenses totales (% du PIB)
General govt.(% expend. total)	8.2	9.3	11.2	6.2	9.9	7.4	8.7	Dépenses publiques (en % du totale)
Congo								**Congo**
Total expenditure (% GDP)	3.2	2.1	2.4	2.3	4.0	5.1	5.2	Dépenses totales (% du PIB)
General govt.(% expend. total)	5.7	4.8	6.2	6.5	8.7	8.7	8.7	Dépenses publiques (en % du totale)
Cook Islands								**Îles Cook**
Total expenditure (% GDP)	5.2	3.5	4.7	3.6	3.4	3.6	3.4	Dépenses totales (% du PIB)
General govt.(% expend. total)	9.9	10.1	13.2	10.3	8.4	7.9	6.1	Dépenses publiques (en % du totale)
Costa Rica *								**Costa Rica ***
Total expenditure (% GDP)	6.5	7.1	7.7	9.7	9.6	9.5	9.3	Dépenses totales (% du PIB)
General govt.(% expend. total)	20.9	24.3	21.3	29.0	25.7	24.1	23.3	Dépenses publiques (en % du totale)
Côte d'Ivoire								**Côte d'Ivoire**
Total expenditure (% GDP)	6.4	6.0	5.4	6.3	6.1	5.8	5.7	Dépenses totales (% du PIB)
General govt.(% expend. total)	6.8	10.0	7.0	8.2	8.5	8.2	7.3	Dépenses publiques (en % du totale)
Croatia								**Croatie**
Total expenditure (% GDP)	6.7	7.7	6.9	8.2	7.8	7.8	7.8	Dépenses totales (% du PIB)
General govt.(% expend. total)	12.8	13.0	13.7	15.3	14.0	14.0	14.0	Dépenses publiques (en % du totale)

Region, country or area	1995	2000	2005	2010	2012	2013	2014	Région, pays ou zone
Cuba								**Cuba**
Total expenditure (% GDP)	5.2	6.1	9.4	10.2	8.6	9.5	11.1	Dépenses totales (% du PIB)
General govt.(% expend. total)	8.0	10.8	13.7	13.9	11.5	14.1	18.0	Dépenses publiques (en % du totale)
Cyprus [7]								**Chypre** [7]
Total expenditure (% GDP)	4.7	5.8	6.4	# 7.2	7.4	7.5	7.4	Dépenses totales (% du PIB)
General govt.(% expend. total)	5.1	6.5	6.2	# 7.4	7.5	7.6	7.6	Dépenses publiques (en % du totale)
Czech Republic								**République tchèque**
Total expenditure (% GDP)	6.7	6.3	6.9	7.4	7.5	7.5	7.4	Dépenses totales (% du PIB)
General govt.(% expend. total)	11.5	13.7	14.1	14.2	14.2	14.9	14.9	Dépenses publiques (en % du totale)
Dem. Rep. of the Congo *								**Rép. dém. du Congo** *
Total expenditure (% GDP)	3.3	1.4	3.1	4.0	3.8	3.9	4.3	Dépenses totales (% du PIB)
General govt.(% expend. total)	2.2	1.8	6.2	7.8	11.6	9.3	11.1	Dépenses publiques (en % du totale)
Denmark								**Danemark**
Total expenditure (% GDP)	8.1	8.7	9.8	11.1	11.0	11.2	10.8	Dépenses totales (% du PIB)
General govt.(% expend. total)	11.3	13.6	15.6	16.3	15.9	16.8	16.8	Dépenses publiques (en % du totale)
Djibouti								**Djibouti**
Total expenditure (% GDP)	4.0	5.8	7.2	8.8	8.9	9.1	10.6	Dépenses totales (% du PIB)
General govt.(% expend. total)	6.2	12.0	13.4	14.1	14.1	14.1	14.1	Dépenses publiques (en % du totale)
Dominica								**Dominique**
Total expenditure (% GDP)	5.9	4.7	4.6	5.7	5.4	5.6	5.5	Dépenses totales (% du PIB)
General govt.(% expend. total)	11.9	6.6	7.9	10.6	9.6	10.6	10.5	Dépenses publiques (en % du totale)
Dominican Republic *								**Rép. dominicaine** *
Total expenditure (% GDP)	5.2	5.9	4.3	4.1	4.3	4.1	4.4	Dépenses totales (% du PIB)
General govt.(% expend. total)	10.7	15.9	7.9	12.2	14.4	15.5	17.4	Dépenses publiques (en % du totale)
Ecuador *								**Équateur** *
Total expenditure (% GDP)	3.4	3.4	5.9	5.9	6.5	7.3	9.2	Dépenses totales (% du PIB)
General govt.(% expend. total)	9.6	5.1	5.6	7.1	7.3	8.8	10.2	Dépenses publiques (en % du totale)
Egypt								**Égypte**
Total expenditure (% GDP)	3.5	5.6	5.1	4.8	5.3	5.5	5.6	Dépenses totales (% du PIB)
General govt.(% expend. total)	5.3	7.3	6.1	5.6	5.8	5.5	5.6	Dépenses publiques (en % du totale)
El Salvador *								**El Salvador** *
Total expenditure (% GDP)	6.4	8.2	7.2	6.9	6.7	6.9	6.8	Dépenses totales (% du PIB)
General govt.(% expend. total)	13.5	15.1	15.5	13.5	15.5	16.8	16.7	Dépenses publiques (en % du totale)
Equatorial Guinea [5]								**Guinée équatoriale** [5]
Total expenditure (% GDP)	5.4	2.7	1.6	3.8	4.0	3.7	3.8	Dépenses totales (% du PIB)
General govt.(% expend. total)	15.3	8.7	6.5	7.0	7.0	7.0	7.0	Dépenses publiques (en % du totale)
Eritrea [5]								**Érythrée** [5]
Total expenditure (% GDP)	4.5	4.4	3.0	3.2	3.0	3.0	3.3	Dépenses totales (% du PIB)
General govt.(% expend. total)	3.0	2.6	1.8	3.6	3.6	3.6	3.6	Dépenses publiques (en % du totale)
Estonia								**Estonie**
Total expenditure (% GDP)	6.3	5.3	5.0	6.2	6.4	6.5	6.4	Dépenses totales (% du PIB)
General govt.(% expend. total)	13.7	11.3	11.5	12.3	13.0	13.5	13.5	Dépenses publiques (en % du totale)
Ethiopia								**Éthiopie**
Total expenditure (% GDP)	3.0	4.4	4.2	6.9	5.8	5.2	4.9	Dépenses totales (% du PIB)
General govt.(% expend. total)	7.0	9.4	11.2	19.8	18.5	15.9	15.7	Dépenses publiques (en % du totale)
Fiji								**Fidji**
Total expenditure (% GDP)	3.1	3.9	3.6	4.2	4.2	4.3	4.5	Dépenses totales (% du PIB)
General govt.(% expend. total)	9.4	11.3	10.8	10.7	9.5	9.7	9.2	Dépenses publiques (en % du totale)
Finland								**Finlande**
Total expenditure (% GDP)	7.8	7.2	8.4	9.0	9.3	9.5	9.7	Dépenses totales (% du PIB)
General govt.(% expend. total)	9.1	10.6	12.4	12.1	12.4	12.3	12.4	Dépenses publiques (en % du totale)
France [8]								**France** [8]
Total expenditure (% GDP)	10.1	9.8	# 10.6	11.2	# 11.4	11.6	11.5	Dépenses totales (% du PIB)
General govt.(% expend. total)	14.9	15.2	# 15.6	15.4	# 15.5	15.6	15.7	Dépenses publiques (en % du totale)
Gabon								**Gabon**
Total expenditure (% GDP)	3.5	2.9	2.8	3.4	3.1	4.0	3.4	Dépenses totales (% du PIB)
General govt.(% expend. total)	5.2	5.3	5.2	10.7	8.2	10.1	7.4	Dépenses publiques (en % du totale)
Gambia								**Gambie**
Total expenditure (% GDP)	3.3	3.6	5.0	5.8	6.1	6.5	7.3	Dépenses totales (% du PIB)
General govt.(% expend. total)	6.9	10.4	13.0	14.5	12.4	15.3	15.3	Dépenses publiques (en % du totale)
Georgia [9]								**Géorgie** [9]
Total expenditure (% GDP)	5.1	6.9	8.6	10.1	8.6	7.2	7.4	Dépenses totales (% du PIB)
General govt.(% expend. total)	2.5	6.9	6.2	6.6	3.8	4.2	5.0	Dépenses publiques (en % du totale)
Germany								**Allemagne**
Total expenditure (% GDP)	9.4	10.1	10.5	11.3	11.0	11.2	11.3	Dépenses totales (% du PIB)
General govt.(% expend. total)	14.0	17.9	17.3	18.1	18.8	19.3	19.6	Dépenses publiques (en % du totale)

Expenditure on health *(continued)*
Percentage of GDP and of government expenditure

Dépenses de santé *(suite)*
Pourcentage du PIB et des dépenses du gouvernement

Region, country or area	1995	2000	2005	2010	2012	2013	2014	Région, pays ou zone
Ghana [10]								**Ghana** [10]
Total expenditure (% GDP)	3.1	3.0	4.5	5.3	4.8	4.6	3.6	Dépenses totales (% du PIB)
General govt.(% expend. total)	10.8	7.8	15.1	14.9	9.3	10.6	6.8	Dépenses publiques (en % du totale)
Greece								**Grèce**
Total expenditure (% GDP)	8.3	7.6	9.4	9.2	9.2	9.3	8.1	Dépenses totales (% du PIB)
General govt.(% expend. total)	9.4	9.8	12.4	11.8	11.4	10.0	10.0	Dépenses publiques (en % du totale)
Grenada								**Grenade**
Total expenditure (% GDP)	6.9	6.6	5.5	6.4	6.4	6.2	6.1	Dépenses totales (% du PIB)
General govt.(% expend. total)	10.4	13.2	10.8	11.0	11.4	9.8	9.2	Dépenses publiques (en % du totale)
Guatemala *								**Guatemala ***
Total expenditure (% GDP)	3.9	5.3	6.8	6.6	6.3	6.3	6.2	Dépenses totales (% du PIB)
General govt.(% expend. total)	15.0	14.4	14.9	15.7	14.7	16.9	17.8	Dépenses publiques (en % du totale)
Guinea								**Guinée**
Total expenditure (% GDP)	3.5	3.5	2.8	4.5	5.4	5.5	5.6	Dépenses totales (% du PIB)
General govt.(% expend. total)	7.0	6.4	3.3	6.8	8.0	9.5	9.0	Dépenses publiques (en % du totale)
Guinea-Bissau * [5,11]								**Guinée-Bissau * [5,11]**
Total expenditure (% GDP)	6.4	4.9	5.7	6.7	6.0	6.1	5.6	Dépenses totales (% du PIB)
General govt.(% expend. total)	9.2	2.3	5.0	9.1	10.4	13.3	7.8	Dépenses publiques (en % du totale)
Guyana								**Guyana**
Total expenditure (% GDP)	5.1	5.8	5.8	6.6	6.6	5.1	5.2	Dépenses totales (% du PIB)
General govt.(% expend. total)	10.3	10.8	8.1	14.7	13.1	8.8	9.4	Dépenses publiques (en % du totale)
Haiti *								**Haïti ***
Total expenditure (% GDP)	6.6	6.1	4.4	8.1	9.9	8.1	7.6	Dépenses totales (% du PIB)
General govt.(% expend. total)	23.6	16.0	8.2	5.5	3.4	6.1	6.1	Dépenses publiques (en % du totale)
Honduras								**Honduras**
Total expenditure (% GDP)	5.3	6.6	7.8	8.5	9.8	9.2	8.7	Dépenses totales (% du PIB)
General govt.(% expend. total)	11.4	15.3	15.5	15.6	16.7	14.6	15.4	Dépenses publiques (en % du totale)
Hungary								**Hongrie**
Total expenditure (% GDP)	7.2	7.1	8.3	7.9	7.7	7.5	7.4	Dépenses totales (% du PIB)
General govt.(% expend. total)	11.0	10.6	11.8	10.5	10.3	10.1	10.1	Dépenses publiques (en % du totale)
Iceland								**Islande**
Total expenditure (% GDP)	8.2	9.3	9.2	8.9	8.7	8.8	8.9	Dépenses totales (% du PIB)
General govt.(% expend. total)	16.1	18.3	18.1	14.4	15.4	16.0	15.7	Dépenses publiques (en % du totale)
India								**Inde**
Total expenditure (% GDP)	4.0	4.3	4.3	4.3	4.4	4.5	4.7	Dépenses totales (% du PIB)
General govt.(% expend. total)	4.5	4.4	4.5	4.3	4.5	4.7	5.0	Dépenses publiques (en % du totale)
Indonesia								**Indonésie**
Total expenditure (% GDP)	2.0	2.0	2.8	2.7	2.9	2.9	2.8	Dépenses totales (% du PIB)
General govt.(% expend. total)	4.9	4.4	4.2	6.1	6.1	6.0	5.7	Dépenses publiques (en % du totale)
Iran (Islamic Rep. of) [12]								**Iran (Rép. islamique d') [12]**
Total expenditure (% GDP)	3.7	4.5	6.1	8.0	7.0	6.5	6.9	Dépenses totales (% du PIB)
General govt.(% expend. total)	6.2	10.6	9.6	12.7	17.5	17.5	17.5	Dépenses publiques (en % du totale)
Iraq # [5,13]								**Iraq # [5,13]**
Total expenditure (% GDP)	...	...	4.1	3.8	5.3	5.9	5.5	Dépenses totales (% du PIB)
General govt.(% expend. total)	...	...	3.2	4.8	6.5	6.5	6.5	Dépenses publiques (en % du totale)
Ireland								**Irlande**
Total expenditure (% GDP)	6.4	6.0	7.3	8.8	8.3	8.0	7.8	Dépenses totales (% du PIB)
General govt.(% expend. total)	11.4	14.5	16.5	9.3	13.4	13.4	13.4	Dépenses publiques (en % du totale)
Israel								**Israël**
Total expenditure (% GDP)	7.3	7.1	7.4	7.4	7.7	7.9	7.8	Dépenses totales (% du PIB)
General govt.(% expend. total)	9.8	9.3	9.6	11.2	11.8	11.8	11.6	Dépenses publiques (en % du totale)
Italy								**Italie**
Total expenditure (% GDP)	7.1	7.9	8.7	9.4	9.3	9.2	9.2	Dépenses totales (% du PIB)
General govt.(% expend. total)	9.7	12.5	14.1	14.6	13.8	13.7	13.7	Dépenses publiques (en % du totale)
Jamaica *								**Jamaïque ***
Total expenditure (% GDP)	4.2	5.8	4.1	5.3	5.7	5.9	5.4	Dépenses totales (% du PIB)
General govt.(% expend. total)	5.6	7.4	3.5	8.5	7.1	9.7	8.1	Dépenses publiques (en % du totale)
Japan								**Japon**
Total expenditure (% GDP)	6.6	7.5	8.2	9.6	10.2	10.2	10.2	Dépenses totales (% du PIB)
General govt.(% expend. total)	15.0	15.4	18.3	19.4	20.0	20.1	20.3	Dépenses publiques (en % du totale)
Jordan * [14]								**Jordanie * [14]**
Total expenditure (% GDP)	8.5	9.7	8.9	8.4	8.0	7.2	7.5	Dépenses totales (% du PIB)
General govt.(% expend. total)	14.8	13.7	12.3	19.5	17.5	13.7	13.7	Dépenses publiques (en % du totale)
Kazakhstan								**Kazakhstan**
Total expenditure (% GDP)	4.6	4.2	4.1	4.4	4.3	4.3	4.4	Dépenses totales (% du PIB)
General govt.(% expend. total)	11.5	9.2	9.3	11.4	10.9	10.9	10.9	Dépenses publiques (en % du totale)

Region, country or area	1995	2000	2005	2010	2012	2013	2014	Région, pays ou zone
Kenya [15]								**Kenya** [15]
Total expenditure (% GDP)	4.3	4.7	4.4	4.0	5.5	5.6	5.7	Dépenses totales (% du PIB)
General govt.(% expend. total)	6.4	10.6	7.6	5.9	13.4	13.1	12.8	Dépenses publiques (en % du totale)
Kiribati								**Kiribati**
Total expenditure (% GDP)	9.4	8.1	10.1	10.5	10.4	10.2	10.2	Dépenses totales (% du PIB)
General govt.(% expend. total)	10.9	10.4	9.2	10.2	8.4	8.0	5.8	Dépenses publiques (en % du totale)
Kuwait								**Koweït**
Total expenditure (% GDP)	3.7	2.5	2.4	2.8	2.6	2.6	3.0	Dépenses totales (% du PIB)
General govt.(% expend. total)	5.6	5.2	6.8	5.2	5.8	5.8	5.8	Dépenses publiques (en % du totale)
Kyrgyzstan								**Kirghizistan**
Total expenditure (% GDP)	6.0	4.7	5.8	6.7	7.0	6.7	6.5	Dépenses totales (% du PIB)
General govt.(% expend. total)	10.7	12.0	11.9	11.9	12.2	13.2	11.9	Dépenses publiques (en % du totale)
Lao People's Dem. Rep.								**Rép. dém. pop. lao**
Total expenditure (% GDP)	3.8	3.4	4.3	2.7	2.1	2.0	1.9	Dépenses totales (% du PIB)
General govt.(% expend. total)	7.6	5.2	4.1	5.2	3.4	3.4	3.4	Dépenses publiques (en % du totale)
Latvia								**Lettonie**
Total expenditure (% GDP)	5.8	6.0	6.4	6.6	5.9	5.7	5.9	Dépenses totales (% du PIB)
General govt.(% expend. total)	9.9	8.7	10.2	9.1	9.8	9.8	9.8	Dépenses publiques (en % du totale)
Lebanon								**Liban**
Total expenditure (% GDP)	12.6	10.9	8.4	7.2	7.0	6.6	6.4	Dépenses totales (% du PIB)
General govt.(% expend. total)	13.8	7.6	11.8	9.2	10.7	10.7	10.7	Dépenses publiques (en % du totale)
Lesotho								**Lesotho**
Total expenditure (% GDP)	7.5	6.9	6.3	10.9	11.1	11.1	10.6	Dépenses totales (% du PIB)
General govt.(% expend. total)	6.4	6.3	6.7	13.2	13.1	12.8	13.1	Dépenses publiques (en % du totale)
Liberia								**Libéria**
Total expenditure (% GDP)	...	5.9	8.0	11.9	10.2	9.3	10.0	Dépenses totales (% du PIB)
General govt.(% expend. total)	...	6.7	13.3	12.8	11.6	9.9	11.9	Dépenses publiques (en % du totale)
Libya [5]								**Libye** [5]
Total expenditure (% GDP)	3.3	3.4	2.7	3.1	4.3	4.3	5.0	Dépenses totales (% du PIB)
General govt.(% expend. total)	5.5	6.0	5.8	4.3	7.9	4.9	4.9	Dépenses publiques (en % du totale)
Lithuania								**Lituanie**
Total expenditure (% GDP)	5.4	6.5	5.8	7.1	6.7	6.6	6.6	Dépenses totales (% du PIB)
General govt.(% expend. total)	11.6	11.3	11.6	11.9	12.1	12.5	13.4	Dépenses publiques (en % du totale)
Luxembourg								**Luxembourg**
Total expenditure (% GDP)	5.6	7.5	7.9	7.7	7.2	7.1	6.9	Dépenses totales (% du PIB)
General govt.(% expend. total)	13.0	16.9	16.2	15.2	13.6	13.6	13.6	Dépenses publiques (en % du totale)
Madagascar *								**Madagascar** *
Total expenditure (% GDP)	4.1	5.1	5.0	4.9	3.5	4.1	3.0	Dépenses totales (% du PIB)
General govt.(% expend. total)	8.1	14.0	12.3	18.6	13.5	17.2	10.2	Dépenses publiques (en % du totale)
Malawi								**Malawi**
Total expenditure (% GDP)	5.0	6.1	8.2	10.0	12.1	11.0	11.4	Dépenses totales (% du PIB)
General govt.(% expend. total)	5.1	9.0	20.0	20.1	22.1	16.8	16.8	Dépenses publiques (en % du totale)
Malaysia								**Malaisie**
Total expenditure (% GDP)	3.0	3.0	3.3	4.0	4.0	4.0	4.2	Dépenses totales (% du PIB)
General govt.(% expend. total)	5.2	5.3	5.3	6.7	5.7	5.9	6.4	Dépenses publiques (en % du totale)
Maldives								**Maldives**
Total expenditure (% GDP)	6.0	8.0	9.5	7.9	9.2	11.2	13.7	Dépenses totales (% du PIB)
General govt.(% expend. total)	10.0	13.8	17.8	14.4	19.3	22.9	26.6	Dépenses publiques (en % du totale)
Mali								**Mali**
Total expenditure (% GDP)	5.0	6.3	6.3	6.4	6.5	6.6	6.9	Dépenses totales (% du PIB)
General govt.(% expend. total)	12.1	8.9	12.3	12.2	12.4	4.7	5.6	Dépenses publiques (en % du totale)
Malta								**Malte**
Total expenditure (% GDP)	5.7	6.8	8.8	8.3	10.0	9.9	9.7	Dépenses totales (% du PIB)
General govt.(% expend. total)	9.8	11.8	14.2	12.7	15.6	15.6	15.6	Dépenses publiques (en % du totale)
Marshall Islands								**Îles Marshall**
Total expenditure (% GDP)	30.8	21.9	17.4	17.3	16.7	17.2	17.1	Dépenses totales (% du PIB)
General govt.(% expend. total)	19.4	21.1	18.9	22.7	24.4	24.0	23.8	Dépenses publiques (en % du totale)
Mauritania								**Mauritanie**
Total expenditure (% GDP)	4.6	5.3	4.2	3.3	3.4	3.6	3.8	Dépenses totales (% du PIB)
General govt.(% expend. total)	9.2	10.8	7.1	6.8	4.6	6.0	6.0	Dépenses publiques (en % du totale)
Mauritius								**Maurice**
Total expenditure (% GDP)	3.6	3.8	4.5	5.3	4.8	4.8	4.8	Dépenses totales (% du PIB)
General govt.(% expend. total)	9.0	8.7	9.4	10.3	9.8	9.5	10.0	Dépenses publiques (en % du totale)
Mexico								**Mexique**
Total expenditure (% GDP)	5.1	5.0	6.0	6.4	6.2	6.3	6.3	Dépenses totales (% du PIB)
General govt.(% expend. total)	8.9	10.5	12.0	11.5	11.5	11.6	11.6	Dépenses publiques (en % du totale)

Expenditure on health *(continued)*
Percentage of GDP and of government expenditure

Dépenses de santé *(suite)*
Pourcentage du PIB et des dépenses du gouvernement

Region, country or area	1995	2000	2005	2010	2012	2013	2014	Région, pays ou zone
Micronesia (Fed. States of)								**Micronésie (États féd. de)**
Total expenditure (% GDP)	9.2	7.9	12.1	13.8	12.8	13.4	13.7	Dépenses totales (% du PIB)
General govt.(% expend. total)	11.3	10.9	19.1	18.6	17.8	20.4	21.2	Dépenses publiques (en % du totale)
Monaco								**Monaco**
Total expenditure (% GDP)	3.1	3.3	4.0	4.4	4.3	4.0	4.3	Dépenses totales (% du PIB)
General govt.(% expend. total)	14.3	14.2	16.3	18.8	18.8	18.8	18.8	Dépenses publiques (en % du totale)
Mongolia								**Mongolie**
Total expenditure (% GDP)	3.2	4.9	5.1	4.7	4.2	4.2	4.7	Dépenses totales (% du PIB)
General govt.(% expend. total)	9.8	11.1	9.9	8.4	5.8	5.5	6.7	Dépenses publiques (en % du totale)
Montenegro								**Monténégro**
Total expenditure (% GDP)	7.4	7.3	8.5	6.9	7.2	6.4	6.4	Dépenses totales (% du PIB)
General govt.(% expend. total)	16.9	16.9	14.3	9.5	9.8	9.8	9.8	Dépenses publiques (en % du totale)
Morocco								**Maroc**
Total expenditure (% GDP)	3.6	4.2	5.1	5.9	6.1	5.9	5.9	Dépenses totales (% du PIB)
General govt.(% expend. total)	4.9	4.8	4.4	6.5	6.0	5.8	6.0	Dépenses publiques (en % du totale)
Mozambique *								**Mozambique ***
Total expenditure (% GDP)	5.3	6.2	6.9	5.4	5.6	5.9	7.0	Dépenses totales (% du PIB)
General govt.(% expend. total)	14.0	17.0	18.2	10.4	8.8	8.8	8.8	Dépenses publiques (en % du totale)
Myanmar								**Myanmar**
Total expenditure (% GDP)	2.1	1.8	1.8	1.9	2.2	2.2	2.3	Dépenses totales (% du PIB)
General govt.(% expend. total)	1.6	1.3	1.1	1.8	3.3	3.3	3.6	Dépenses publiques (en % du totale)
Namibia								**Namibie**
Total expenditure (% GDP)	6.2	6.1	7.3	7.9	8.2	8.5	8.9	Dépenses totales (% du PIB)
General govt.(% expend. total)	13.2	13.9	12.8	13.9	13.9	13.9	13.9	Dépenses publiques (en % du totale)
Nauru								**Nauru**
Total expenditure (% GDP)	13.2	13.5	12.6	9.1	7.0	4.7	3.3	Dépenses totales (% du PIB)
General govt.(% expend. total)	5.9	11.2	18.5	10.3	11.5	11.5	5.2	Dépenses publiques (en % du totale)
Nepal								**Népal**
Total expenditure (% GDP)	5.3	5.4	5.7	6.4	5.9	5.7	5.8	Dépenses totales (% du PIB)
General govt.(% expend. total)	7.9	7.7	10.3	14.3	11.8	11.6	11.2	Dépenses publiques (en % du totale)
Netherlands								**Pays-Bas**
Total expenditure (% GDP)	7.4	7.4	9.6	10.5	11.0	11.0	10.9	Dépenses totales (% du PIB)
General govt.(% expend. total)	10.5	11.4	15.8	19.0	20.3	20.9	20.9	Dépenses publiques (en % du totale)
New Zealand								**Nouvelle-Zélande**
Total expenditure (% GDP)	6.9	7.5	8.2	11.2	11.5	11.2	11.0	Dépenses totales (% du PIB)
General govt.(% expend. total)	13.2	15.7	17.7	19.8	23.1	23.4	23.4	Dépenses publiques (en % du totale)
Nicaragua								**Nicaragua**
Total expenditure (% GDP)	6.5	5.4	6.1	6.6	8.0	8.4	9.0	Dépenses totales (% du PIB)
General govt.(% expend. total)	21.5	13.1	18.6	19.8	18.7	19.8	24.0	Dépenses publiques (en % du totale)
Niger								**Niger**
Total expenditure (% GDP)	6.7	6.0	7.1	6.4	6.1	5.1	5.8	Dépenses totales (% du PIB)
General govt.(% expend. total)	9.0	8.4	14.8	10.7	7.6	7.6	7.6	Dépenses publiques (en % du totale)
Nigeria [16]								**Nigéria** [16]
Total expenditure (% GDP)	2.8	2.8	4.1	3.5	3.3	3.7	3.7	Dépenses totales (% du PIB)
General govt.(% expend. total)	6.1	5.9	7.3	5.7	7.4	6.5	8.2	Dépenses publiques (en % du totale)
Niue								**Nioué**
Total expenditure (% GDP)	8.0	7.9	8.2	8.3	7.0	7.2	7.4	Dépenses totales (% du PIB)
General govt.(% expend. total)	8.1	6.6	7.2	6.5	4.8	5.6	5.9	Dépenses publiques (en % du totale)
Norway								**Norvège**
Total expenditure (% GDP)	7.7	8.3	8.9	9.3	9.2	9.4	9.7	Dépenses totales (% du PIB)
General govt.(% expend. total)	12.9	16.2	17.7	17.4	18.1	18.2	18.2	Dépenses publiques (en % du totale)
Oman								**Oman**
Total expenditure (% GDP)	3.6	3.1	2.6	2.7	2.5	2.8	3.6	Dépenses totales (% du PIB)
General govt.(% expend. total)	6.9	7.0	6.1	6.7	4.8	5.4	6.8	Dépenses publiques (en % du totale)
Pakistan * [17]								**Pakistan *** [17]
Total expenditure (% GDP)	2.5	2.8	2.9	3.0	2.8	2.7	2.6	Dépenses totales (% du PIB)
General govt.(% expend. total)	3.8	3.5	4.3	4.7	4.7	4.7	4.7	Dépenses publiques (en % du totale)
Palau								**Palaos**
Total expenditure (% GDP)	11.7	9.5	8.2	10.6	9.4	9.3	9.0	Dépenses totales (% du PIB)
General govt.(% expend. total)	13.9	12.0	16.0	16.9	16.5	17.8	18.1	Dépenses publiques (en % du totale)
Panama								**Panama**
Total expenditure (% GDP)	7.7	7.8	7.5	8.0	7.2	8.1	8.0	Dépenses totales (% du PIB)
General govt.(% expend. total)	16.1	21.3	12.7	15.1	12.7	14.5	14.6	Dépenses publiques (en % du totale)
Papua New Guinea								**Papouasie-Nvl-Guinée**
Total expenditure (% GDP)	2.9	4.0	6.4	4.2	4.6	4.8	4.3	Dépenses totales (% du PIB)
General govt.(% expend. total)	8.6	9.9	15.5	10.4	11.7	11.2	9.5	Dépenses publiques (en % du totale)

Expenditure on health *(continued)*
Percentage of GDP and of government expenditure

Dépenses de santé *(suite)*
Pourcentage du PIB et des dépenses du gouvernement

Region, country or area	1995	2000	2005	2010	2012	2013	2014	Région, pays ou zone
Paraguay [18]								**Paraguay [18]**
Total expenditure (% GDP)	5.9	8.1	6.1	9.1	10.3	10.5	9.8	Dépenses totales (% du PIB)
General govt.(% expend. total)	11.8	17.7	13.2	11.2	11.5	12.8	11.9	Dépenses publiques (en % du totale)
Peru *								**Pérou ***
Total expenditure (% GDP)	4.5	4.8	4.7	5.0	5.2	5.2	5.5	Dépenses totales (% du PIB)
General govt.(% expend. total)	13.1	14.1	14.9	13.3	13.9	14.4	15.0	Dépenses publiques (en % du totale)
Philippines								**Philippines**
Total expenditure (% GDP)	3.4	3.2	3.9	4.4	4.5	4.6	4.7	Dépenses totales (% du PIB)
General govt.(% expend. total)	7.4	8.4	8.9	9.3	8.2	8.9	10.0	Dépenses publiques (en % du totale)
Poland								**Pologne**
Total expenditure (% GDP)	5.4	5.5	6.2	6.9	6.6	6.4	6.4	Dépenses totales (% du PIB)
General govt.(% expend. total)	8.2	9.2	9.7	10.7	10.8	10.7	10.7	Dépenses publiques (en % du totale)
Portugal								**Portugal**
Total expenditure (% GDP)	7.4	9.1	10.0	10.4	9.7	9.6	9.5	Dépenses totales (% du PIB)
General govt.(% expend. total)	10.9	14.5	15.0	13.8	12.9	12.5	11.9	Dépenses publiques (en % du totale)
Qatar *								**Qatar ***
Total expenditure (% GDP)	3.7	2.2	3.0	2.1	2.2	2.1	2.2	Dépenses totales (% du PIB)
General govt.(% expend. total)	5.2	5.0	8.1	5.1	5.8	5.8	5.8	Dépenses publiques (en % du totale)
Republic of Korea								**République de Corée**
Total expenditure (% GDP)	3.7	4.2	5.3	6.8	7.0	7.2	7.4	Dépenses totales (% du PIB)
General govt.(% expend. total)	6.3	8.4	9.6	12.4	11.8	12.3	12.3	Dépenses publiques (en % du totale)
Republic of Moldova								**République de Moldova**
Total expenditure (% GDP)	9.1	6.7	9.2	12.1	11.9	10.1	10.3	Dépenses totales (% du PIB)
General govt.(% expend. total)	15.9	8.9	11.3	13.8	13.8	13.1	13.3	Dépenses publiques (en % du totale)
Romania								**Roumanie**
Total expenditure (% GDP)	3.2	4.3	5.5	5.8	5.5	5.6	5.6	Dépenses totales (% du PIB)
General govt.(% expend. total)	7.0	9.1	13.2	11.8	12.1	12.8	12.8	Dépenses publiques (en % du totale)
Russian Federation								**Fédération de Russie**
Total expenditure (% GDP)	5.4	5.4	5.2	6.8	6.9	7.1	7.1	Dépenses totales (% du PIB)
General govt.(% expend. total)	9.1	12.7	11.7	9.7	10.2	9.8	9.5	Dépenses publiques (en % du totale)
Rwanda								**Rwanda**
Total expenditure (% GDP)	4.3	4.2	6.8	7.9	7.7	7.7	7.5	Dépenses totales (% du PIB)
General govt.(% expend. total)	8.0	8.5	13.7	11.3	10.4	10.1	9.9	Dépenses publiques (en % du totale)
Saint Kitts and Nevis *								**Saint-Kitts-et-Nevis ***
Total expenditure (% GDP)	5.9	5.3	4.8	5.5	5.8	6.1	5.1	Dépenses totales (% du PIB)
General govt.(% expend. total)	10.0	9.6	4.5	6.8	7.1	7.2	6.9	Dépenses publiques (en % du totale)
Saint Lucia								**Sainte-Lucie**
Total expenditure (% GDP)	6.1	5.5	6.2	8.1	8.3	7.9	6.7	Dépenses totales (% du PIB)
General govt.(% expend. total)	11.3	11.7	9.5	13.1	10.2	11.3	11.5	Dépenses publiques (en % du totale)
Saint Vincent-Grenadines								**Saint-Vincent-Grenadines**
Total expenditure (% GDP)	6.7	3.7	3.7	4.7	8.5	5.2	8.6	Dépenses totales (% du PIB)
General govt.(% expend. total)	13.1	10.8	8.4	11.7	15.0	14.6	14.8	Dépenses publiques (en % du totale)
Samoa								**Samoa**
Total expenditure (% GDP)	4.6	5.3	4.5	5.7	5.0	6.9	7.2	Dépenses totales (% du PIB)
General govt.(% expend. total)	7.5	13.7	11.2	12.6	11.6	16.5	15.1	Dépenses publiques (en % du totale)
San Marino								**Saint-Marin**
Total expenditure (% GDP)	5.3	4.8	4.1	5.0	6.3	6.0	6.1	Dépenses totales (% du PIB)
General govt.(% expend. total)	17.7	20.4	12.4	14.6	13.2	13.2	13.2	Dépenses publiques (en % du totale)
Sao Tome and Principe								**Sao Tomé-et-Principe**
Total expenditure (% GDP)	7.6	8.9	10.0	5.2	8.6	9.8	8.4	Dépenses totales (% du PIB)
General govt.(% expend. total)	7.9	9.0	13.2	5.6	5.9	12.4	12.4	Dépenses publiques (en % du totale)
Saudi Arabia *								**Arabie saoudite ***
Total expenditure (% GDP)	2.9	4.2	3.4	3.5	3.9	4.2	4.7	Dépenses totales (% du PIB)
General govt.(% expend. total)	4.4	8.6	8.1	6.2	7.7	8.2	8.2	Dépenses publiques (en % du totale)
Senegal								**Sénégal**
Total expenditure (% GDP)	3.9	4.6	5.4	4.6	4.3	4.5	4.7	Dépenses totales (% du PIB)
General govt.(% expend. total)	7.1	10.1	12.4	8.8	7.5	7.9	8.0	Dépenses publiques (en % du totale)
Serbia								**Serbie**
Total expenditure (% GDP)	6.5	6.5	8.7	10.1	9.9	10.1	10.4	Dépenses totales (% du PIB)
General govt.(% expend. total)	13.6	13.6	14.3	14.3	13.1	13.9	13.9	Dépenses publiques (en % du totale)
Seychelles								**Seychelles**
Total expenditure (% GDP)	5.0	4.6	3.9	3.6	4.1	3.6	3.4	Dépenses totales (% du PIB)
General govt.(% expend. total)	8.0	7.9	10.1	8.9	10.4	9.7	9.7	Dépenses publiques (en % du totale)
Sierra Leone *								**Sierra Leone ***
Total expenditure (% GDP)	11.4	13.6	12.2	10.3	11.2	11.6	11.1	Dépenses totales (% du PIB)
General govt.(% expend. total)	14.6	14.6	14.6	11.9	9.7	10.8	10.8	Dépenses publiques (en % du totale)

Expenditure on health *(continued)*
Percentage of GDP and of government expenditure

Dépenses de santé *(suite)*
Pourcentage du PIB et des dépenses du gouvernement

Region, country or area	1995	2000	2005	2010	2012	2013	2014	Région, pays ou zone
Singapore								**Singapour**
Total expenditure (% GDP)	2.9	2.7	3.7	4.0	4.2	4.5	4.9	Dépenses totales (% du PIB)
General govt.(% expend. total)	9.3	7.1	7.9	9.8	11.1	12.7	14.1	Dépenses publiques (en % du totale)
Slovakia								**Slovaquie**
Total expenditure (% GDP)	6.1	5.5	7.0	8.5	8.1	8.0	8.1	Dépenses totales (% du PIB)
General govt.(% expend. total)	11.0	9.4	13.8	14.6	14.9	15.0	15.0	Dépenses publiques (en % du totale)
Slovenia								**Slovénie**
Total expenditure (% GDP)	7.5	8.3	8.5	9.1	9.4	9.3	9.2	Dépenses totales (% du PIB)
General govt.(% expend. total)	11.1	13.1	13.8	13.6	14.1	11.2	12.8	Dépenses publiques (en % du totale)
Solomon Islands								**Îles Salomon**
Total expenditure (% GDP)	3.2	4.6	7.8	7.5	5.5	5.4	5.1	Dépenses totales (% du PIB)
General govt.(% expend. total)	14.8	20.7	28.5	20.3	13.3	13.1	12.5	Dépenses publiques (en % du totale)
South Africa								**Afrique du Sud**
Total expenditure (% GDP)	8.3	8.1	7.8	8.5	8.8	8.8	8.8	Dépenses totales (% du PIB)
General govt.(% expend. total)	13.0	13.6	13.0	14.1	14.4	14.2	14.2	Dépenses publiques (en % du totale)
South Sudan [5]								**Soudan du sud [5]**
Total expenditure (% GDP)	...	...	...	...	2.8	2.6	2.7	Dépenses totales (% du PIB)
General govt.(% expend. total)	...	...	...	...	4.0	4.0	4.0	Dépenses publiques (en % du totale)
Spain								**Espagne**
Total expenditure (% GDP)	7.4	7.2	8.1	9.6	9.4	9.1	9.0	Dépenses totales (% du PIB)
General govt.(% expend. total)	12.1	13.2	15.3	15.5	14.1	14.5	14.5	Dépenses publiques (en % du totale)
Sri Lanka								**Sri Lanka**
Total expenditure (% GDP)	3.4	3.8	4.1	3.4	3.2	3.7	3.5	Dépenses totales (% du PIB)
General govt.(% expend. total)	5.3	6.9	7.8	6.8	6.2	11.5	11.2	Dépenses publiques (en % du totale)
Sudan * [5]								**Soudan * [5]**
Total expenditure (% GDP)	4.0	3.2	3.2	8.0	8.2	8.4	8.4	Dépenses totales (% du PIB)
General govt.(% expend. total)	11.7	8.5	5.5	11.6	11.6	11.6	11.6	Dépenses publiques (en % du totale)
Suriname								**Suriname**
Total expenditure (% GDP)	7.0	9.7	6.8	5.8	6.1	6.0	5.7	Dépenses totales (% du PIB)
General govt.(% expend. total)	18.3	14.0	11.6	11.9	12.1	11.8	11.8	Dépenses publiques (en % du totale)
Swaziland								**Swaziland**
Total expenditure (% GDP)	5.0	5.3	6.8	8.5	8.8	9.7	9.3	Dépenses totales (% du PIB)
General govt.(% expend. total)	12.4	10.5	13.0	15.4	18.1	19.7	16.6	Dépenses publiques (en % du totale)
Sweden								**Suède**
Total expenditure (% GDP)	8.0	8.2	9.1	9.5	11.8	12.0	11.9	Dépenses totales (% du PIB)
General govt.(% expend. total)	10.6	12.6	13.7	14.7	19.2	19.0	19.0	Dépenses publiques (en % du totale)
Switzerland								**Suisse**
Total expenditure (% GDP)	9.3	9.9	10.9	11.1	11.6	11.7	11.7	Dépenses totales (% du PIB)
General govt.(% expend. total)	14.4	15.4	18.3	21.0	22.0	22.6	22.7	Dépenses publiques (en % du totale)
Syrian Arab Republic [19]								**Rép. arabe syrienne [19]**
Total expenditure (% GDP)	5.5	4.9	4.1	3.3	3.3	3.3	3.3	Dépenses totales (% du PIB)
General govt.(% expend. total)	7.4	6.9	7.4	5.3	4.8	4.8	4.8	Dépenses publiques (en % du totale)
Tajikistan								**Tadjikistan**
Total expenditure (% GDP)	3.1	4.6	5.9	6.0	6.4	6.8	6.9	Dépenses totales (% du PIB)
General govt.(% expend. total)	7.4	6.5	5.9	5.9	7.5	7.3	6.8	Dépenses publiques (en % du totale)
Thailand								**Thaïlande**
Total expenditure (% GDP)	3.8	3.8	4.6	5.4	6.2	6.2	6.5	Dépenses totales (% du PIB)
General govt.(% expend. total)	11.7	12.1	16.4	19.5	22.0	21.7	23.2	Dépenses publiques (en % du totale)
TFYR of Macedonia								**ex-R.Y. de Macédoine**
Total expenditure (% GDP)	8.4	8.5	8.0	6.8	6.8	6.1	6.5	Dépenses totales (% du PIB)
General govt.(% expend. total)	13.3	15.0	14.7	13.2	13.3	13.2	12.9	Dépenses publiques (en % du totale)
Timor-Leste								**Timor-Leste**
Total expenditure (% GDP)	...	3.3	1.0	0.9	1.0	1.3	1.5	Dépenses totales (% du PIB)
General govt.(% expend. total)	...	11.6	10.5	4.5	2.1	2.9	2.4	Dépenses publiques (en % du totale)
Togo								**Togo**
Total expenditure (% GDP)	4.0	4.3	5.2	5.4	5.1	5.1	5.2	Dépenses totales (% du PIB)
General govt.(% expend. total)	7.7	7.7	9.8	8.6	7.4	8.4	7.8	Dépenses publiques (en % du totale)
Tonga								**Tonga**
Total expenditure (% GDP)	4.1	4.8	6.5	4.6	4.5	5.0	5.2	Dépenses totales (% du PIB)
General govt.(% expend. total)	9.9	15.4	26.6	13.0	14.1	13.7	13.5	Dépenses publiques (en % du totale)
Trinidad and Tobago *								**Trinité-et-Tobago ***
Total expenditure (% GDP)	5.1	4.2	5.3	5.3	5.8	6.0	5.9	Dépenses totales (% du PIB)
General govt.(% expend. total)	9.4	6.8	9.6	8.3	7.8	8.1	8.2	Dépenses publiques (en % du totale)
Tunisia *								**Tunisie ***
Total expenditure (% GDP)	5.9	5.4	5.6	6.5	7.2	7.3	7.0	Dépenses totales (% du PIB)
General govt.(% expend. total)	10.7	12.0	11.9	15.7	14.2	14.2	14.2	Dépenses publiques (en % du totale)

Region, country or area	1995	2000	2005	2010	2012	2013	2014	Région, pays ou zone
Turkey								**Turquie**
Total expenditure (% GDP)	2.5	4.9	5.4	5.6	5.2	5.4	5.4	Dépenses totales (% du PIB)
General govt.(% expend. total)	10.7	9.8	11.3	11.0	10.8	10.5	10.5	Dépenses publiques (en % du totale)
Turkmenistan								**Turkménistan**
Total expenditure (% GDP)	3.1	3.9	3.5	2.0	2.0	2.1	2.1	Dépenses totales (% du PIB)
General govt.(% expend. total)	9.2	13.7	10.3	8.7	8.7	8.7	8.7	Dépenses publiques (en % du totale)
Tuvalu								**Tuvalu**
Total expenditure (% GDP)	8.2	15.7	18.4	16.8	13.8	16.6	16.5	Dépenses totales (% du PIB)
General govt.(% expend. total)	6.3	8.5	22.9	16.1	17.3	20.2	16.9	Dépenses publiques (en % du totale)
Uganda [20]								**Ouganda** [20]
Total expenditure (% GDP)	5.7	6.8	9.4	11.0	7.6	7.5	7.2	Dépenses totales (% du PIB)
General govt.(% expend. total)	9.9	8.8	14.1	19.5	14.3	13.1	11.0	Dépenses publiques (en % du totale)
Ukraine								**Ukraine**
Total expenditure (% GDP)	7.0	5.6	6.4	7.8	7.5	7.7	7.1	Dépenses totales (% du PIB)
General govt.(% expend. total)	11.4	10.2	11.9	12.7	11.8	12.0	10.8	Dépenses publiques (en % du totale)
United Arab Emirates								**Émirats arabes unis**
Total expenditure (% GDP)	2.6	2.4	2.3	3.9	3.4	3.5	3.6	Dépenses totales (% du PIB)
General govt.(% expend. total)	8.3	7.8	8.7	8.7	8.7	8.7	8.7	Dépenses publiques (en % du totale)
United Kingdom								**Royaume-Uni**
Total expenditure (% GDP)	6.7	6.9	8.2	9.5	9.4	9.3	9.1	Dépenses totales (% du PIB)
General govt.(% expend. total)	13.0	15.2	15.3	15.9	16.2	16.5	16.5	Dépenses publiques (en % du totale)
United Rep. of Tanzania								**Rép.-Unie de Tanzanie**
Total expenditure (% GDP)	3.0	2.6	4.7	5.3	5.7	5.6	5.6	Dépenses totales (% du PIB)
General govt.(% expend. total)	9.2	9.2	19.0	10.9	15.9	13.4	12.3	Dépenses publiques (en % du totale)
United States								**États-Unis**
Total expenditure (% GDP)	13.1	13.1	15.2	17.0	17.0	16.9	17.1	Dépenses totales (% du PIB)
General govt.(% expend. total)	15.9	16.8	18.5	19.0	20.1	20.8	21.3	Dépenses publiques (en % du totale)
Uruguay *								**Uruguay** *
Total expenditure (% GDP)	12.6	7.8	11.1	8.6	8.7	8.7	8.6	Dépenses totales (% du PIB)
General govt.(% expend. total)	12.6	9.0	29.2	19.5	21.3	20.9	20.8	Dépenses publiques (en % du totale)
Uzbekistan								**Ouzbékistan**
Total expenditure (% GDP)	6.7	5.3	5.1	5.3	6.5	6.3	5.8	Dépenses totales (% du PIB)
General govt.(% expend. total)	9.5	8.7	7.3	8.7	9.6	9.7	10.7	Dépenses publiques (en % du totale)
Vanuatu								**Vanuatu**
Total expenditure (% GDP)	2.7	3.3	3.9	4.7	3.7	3.9	5.0	Dépenses totales (% du PIB)
General govt.(% expend. total)	7.8	10.5	17.9	15.9	13.5	14.1	17.9	Dépenses publiques (en % du totale)
Venezuela (Boliv. Rep. of)								**Venezuela (Rép. boliv. du)**
Total expenditure (% GDP)	5.4	4.9	4.7	5.0	4.8	4.9	5.3	Dépenses totales (% du PIB)
General govt.(% expend. total)	7.8	7.3	8.3	9.1	5.9	6.4	5.8	Dépenses publiques (en % du totale)
Viet Nam								**Viet Nam**
Total expenditure (% GDP)	5.2	4.9	5.4	6.4	7.0	7.2	7.1	Dépenses totales (% du PIB)
General govt.(% expend. total)	7.9	7.2	5.4	9.9	12.7	13.2	14.2	Dépenses publiques (en % du totale)
Yemen *								**Yémen** *
Total expenditure (% GDP)	4.2	4.1	4.6	5.2	5.7	5.8	5.6	Dépenses totales (% du PIB)
General govt.(% expend. total)	6.4	8.0	4.8	4.3	3.9	3.9	3.9	Dépenses publiques (en % du totale)
Zambia								**Zambie**
Total expenditure (% GDP)	5.0	7.2	7.6	4.4	4.9	5.0	5.0	Dépenses totales (% du PIB)
General govt.(% expend. total)	9.5	13.3	17.6	12.6	11.4	11.0	11.3	Dépenses publiques (en % du totale)
Zimbabwe [5]								**Zimbabwe** [5]
Total expenditure (% GDP)	...	7.0	6.4	5.4	6.7	6.9	6.4	Dépenses totales (% du PIB)
General govt.(% expend. total)	...	...	11.0	7.5	9.7	9.6	8.5	Dépenses publiques (en % du totale)

Source:

World Health Organization (WHO), Geneva, WHO Global Health Expenditure database, last accessed June 2016.

Source:

Organisation mondiale de la Santé (OMS), Genève, base de données de l'Observatoire sur les dépenses de santé mondiales, dernier accès juin 2016.

1   Non-profit institutions (such as NGOs) serving households are accounted for in "external assistance" and recorded under government expenditure. GDP includes both licit and illicit GDPs (for example, opium).

1   Les institutions sans but lucratif (telles que les ONG) desservant les ménages sont prises en compte à la rubrique « aide extérieure » et enregistrées comme dépenses publiques. Le PIB comprend à la fois le PIB licite et illicite (opium, par ex.).

2	Government expenditures include external assistance (external budget).	2	Les dépenses publiques comprennent l'aide extérieure (budget externe).
3	Adjustments for currency change (from old to new manat) were made for the entire Azerbaijan series starting from World Health Statistics 2008.	3	Des ajustements ont été apportés à l'ensemble de la série concernant l'Azerbaïdjan à partir des Statistiques sanitaires mondiales de 2008, compte tenu du changement de monnaie (passage du vieux manat au nouveau manat).
4	Funds previously included in social security were reclassified.	4	Les montants précédemment inclus dans la sécurité sociale ont été reclassés.
5	Estimates should be viewed with caution as these are derived from scarce data.	5	Les montants estimatifs sont à prendre avec prudence, car calculés à partir de données peu nombreuses.
6	National Accounts: The authorities have revised national accounts estimates, moving from the 1968 to the 1993 SNA. IMF country report 11/302.	6	« Comptabilité nationale : Les autorités ont révisé les estimations des comptes nationaux, passant du SCN de 1968 au SCN de 1993». Rapport du FMI 11/302 (Tchad).
7	Data is converted from SHA 2011.	7	Les données ont été converties du SCS de 2011.
8	Estimates are based on different national accounting bases (2000, 2005 and 2010).	8	Les montants estimatifs reposent sur des bases de comptabilité nationale différentes (2000, 2005 et 2010).
9	As a result of recent health-care reforms in Georgia, public compulsory insurance has since 2008 been implemented by private insurance companies. The voucher cost of this insurance is treated as general government health expenditure.	9	Du fait de réformes récentes des soins de santé en Géorgie, l'assurance publique obligatoire est réalisée depuis 2008 par des sociétés d'assurance privées. Le coût des coupons correspondant à cette assurance est traité comme dépense publique de santé.
10	In 2010, Ghana revalued its economy from 1993 base to 2006, increasing its overall GDP by more than 60 per cent. Estimates have been updated taking into account 2010 GDP rebasing data series.	10	En 2010, le Ghana a changé la base de calcul de son économie, passant de la base 1993 à la base 2006, ce qui accru son PIB global de plus de 60 pour cent. Les montants estimatifs ont été actualisés compte tenu du passage des séries de données à la base 2010.
11	Government expenditures show fluctuations due to variations in capital investment.	11	Les dépenses publiques manifestent des fluctuations du fait de variations des investissements.
12	Exchange rate changed in 2002 from multiple to a managed floating exchange rate. Inter-bank market rate used prior to 2002.	12	Les taux de change, qui étaient précédemment des taux multiples, sont depuis 2002 des taux flottants dirigés. Pour les données antérieures à 2002, les taux utilisés étaient les taux du marché interbancaire.
13	The estimates do not include expenditures for Northern Iraq.	13	Les montants estimatifs ne comprennent pas les dépenses pour le nord de l'Irak.
14	The public expenditure on health includes contributions from the United Nations Relief and Works Agency for Palestine Refugees in the Near East (UNRWA) made to Palestinian refugees residing in Jordanian territories.	14	Les dépenses publiques de santé comprennent les contributions de l'Office de secours et de travaux pour les réfugiés de Palestine dans le Proche-Orient (UNRWA) versées aux réfugiés de Palestine résidant en Jordanie.
15	In 2014, Kenya revalued its base year from 2002 to 2009, which increased its GDP by 25 per cent in 2013.	15	En 2014, le Kenya a changé son année de base, qui est passée de 2002 à 2009, de sorte que le PIB a augmenté de 25 pour cent pour 2013.
16	In 2014, a revised GDP series was published following a statistical rebasing exercise (the base year changed from 1990 to 2010). Nigeria has emerged as Africa's largest economy, with a GDP increase of 89% for year 2013.	16	En 2014, une série révisée a été publiée pour le PIB à la suite d'une modification de l'année de base utilisée pour les statistiques (de 1990 à 2010). Le Nigéria est de ce fait devenu l'économie la plus importante d'Afrique, avec une augmentation de 89 pour cent du PIB pour 2013.
17	Total level of government expenditure on health increased due to the inclusion of local government expenditure, as well as a more-comprehensive estimation of regional expenditure on health.	17	Le montant total des dépenses publiques de santé a augmenté du fait qu'on y a ajouté les dépenses des administrations publiques locales et rendu plus complètes les estimations des dépenses régionales de santé.
18	A study of Health Accounts is in the process (2012-2013)	18	Une étude des comptes de la santé est en cours (2012-2013)
19	The exchange rate used for the Syrian Arab Republic is the rate for non-commercial transactions from the Central Bank of Syria.	19	Le taux de change utilisé pour la République arabe syrienne est le taux de la Banque centrale syrienne servant pour les transactions non commerciales.
20	Unlike other countries, in Uganda fiscal year 2010/2011 corresponds to calendar year 2010.	20	À la différence d'autres pays, en Ouganda l'exercice budgétaire 2010/2011 correspond à l'année civile 2010.

9

Intentional homicides and other crimes
By type of crime per 100 000 population and homicide victims by sex

Homicides intentionnels et autres crimes
Par type de crime pour 100 000 habitants et victimes d'homicides par sexe

Country or area Pays ou zone	Year Année	Intentional homicides Homicides volontaires Victims (percentage) Victimes (pourcentage) Per 100,000 Pour 100 000	Male Hommes	Female Femmes	Other crimes (per 100 000 pop.) Autres infractions (pour 100 000 hab.) Assault Agression	Kidnapping Enlèvement	Theft Vol simple	Robbery Vol qualifié	Sexual Violence Violences Sexuelles
Afghanistan	2010	3.5	...	...	...	...	...	...	...
Afghanistan	2012	6.6	...	...	...	...	...	...	...
Albania	2005	5.0	86.4	13.6	5.4	...	96.9	6.5	2.9
Albanie	2010	4.4	85.0	15.0	6.1	0.3	137.1	8.1	2.7
	2012	5.5	83.4	16.6	5.4	0.1	185.5	11.3	1.5
	2014	4.0	...	...	4.6	0.1	275.5	10.3	3.4
Algeria	2005	0.6	...	...	# 99.0	0.3	...	# 75.6	...
Algérie	2010	0.7	...	...	# 114.8	0.5	# 141.7	# 51.3	# 10.5
	2014	1.5	...	...	# 138.3	0.9	# 119.9	# 54.0	# 7.7
Andorra	2004	1.3	...	...	...	...	...	...	...
Andorre	2010	1.2	100.0	0.0	116.1	0.0	1 323.2	9.5	11.8
	2012	0.0	0.0	0.0	205.5	0.0	1 269.6	7.6	22.7
	2014	0.0	...	...	169.0	1.4	1 286.0	9.6	22.0
Angola Angola	2012	9.8	...	...	...	...	...	...	...
Anguilla	2005	7.9	...	...	...	...	...	...	...
Anguilla	2008	7.5	...	...	...	...	...	...	...
Antigua and Barbuda	2005	3.6	...	...	...	...	...	...	...
Antigua-et-Barbuda	2010	6.9	...	...	...	...	...	...	...
	2012	11.2	...	...	...	...	...	...	...
Argentina	2005	...	...	...	363.0	...	775.7	907.7	...
Argentine	2008	...	...	...	359.7	...	703.8	957.9	26.3
	2014	7.6	...	...	...	...	...	...	...
Armenia	2005	1.9	81.0	19.0	6.0	0.7	81.7	8.9	1.7
Arménie	2010	1.9	77.3	22.7	5.2	1.3	123.6	12.0	2.8
	2012	2.2	64.8	35.2	5.3	1.3	134.9	10.4	3.5
	2013	2.0	...	...	5.9	...	155.7	...	3.9
Aruba	2005	12.0	...	...	...	...	...	...	...
Aruba	2010	3.9	...	...	...	...	...	...	...
Australia	2005	1.3	62.5	37.5	...	...	# 2 556.6	...	...
Australie	2010	1.1	61.3	38.7	324.5	# 2.8	# 2 153.4	# 66.0	# 85.1
	2011	1.1	64.8	35.2	308.2	# 3.0	# 2 223.8	# 60.6	# 82.6
	2012	1.1	67.3	32.7	...	# 2.8	# 2 236.2	# 57.4	# 84.9
	2014	1.0	...	...	...	# 2.3	# 2 054.1	# 41.9	# 87.5
Austria	2005	0.7	55.8	44.2	43.9	0.1	2 149.7	57.9	20.3
Autriche	2010	0.7	60.6	39.4	43.0	0.1	1 796.1	51.4	32.1
	2012	1.0	59.8	40.2	47.7	0.1	1 732.8	48.4	37.7
	2014	0.5	...	...	42.7	0.0	1 691.6	40.9	32.8
Azerbaijan	2005	2.2	...	...	1.9	0.3	25.1	2.3	...
Azerbaïdjan	2010	2.1	69.9	30.1	1.9	0.0	42.5	3.2	2.2
	2014	2.5	...	...	1.7	0.0	60.9	2.5	2.2
Bahamas	2005	15.8	84.6	15.4	...	7.3	440.4	...	133.3
Bahamas	2010	26.1	85.1	14.9	848.9	5.0	545.7	93.1	85.6
	2011	34.6	87.4	12.6	891.2	7.1	536.9	100.6	64.9
	2012	29.8	...	...	841.1	7.5	533.0	98.6	80.3
Bahrain	2005	0.5	...	...	406.9	1.4	1 114.1	48.4	...
Bahreïn	2007	0.4	...	...	256.6	0.0	706.8	28.7	14.1
	2008	0.5	...	...	327.5	...	681.0	27.4	15.7
	2010	0.9	...	...	...	...	...	...	...
	2011	0.5	...	...	...	...	...	...	...

9

Intentional homicides and other crimes *(continued)*
By type of crime per 100 000 population and homicide victims by sex

Homicides intentionnels et autres crimes *(suite)*
Par type de crime pour 100 000 habitants et victimes d'homicides par sexe

Country or area Pays ou zone	Year Année	Intentional homicides Homicides volontaires Victims (percentage) Victimes (pourcentage) Per 100,000 Pour 100 000	Male Hommes	Female Femmes	Other crimes (per 100 000 pop.) Autres infractions (pour 100 000 hab.) Assault Agression	Kidnapping Enlèvement	Theft Vol simple	Robbery Vol qualifié	Sexual Violence Violences Sexuelles
Bangladesh	2005	2.5	...	...	0.4	0.9	8.8	0.6	...
Bangladesh	2006	2.9	...	...	0.4	0.8	9.0	0.6	...
	2010	2.6	...	...	...	...	...	...	...
	2014	2.8	...	...	...	...	...	...	...
Barbados	2005	10.6	...	...	631.7	8.4	412.4	120.4	70.1
Barbade	2010	11.1	67.7	32.3	540.5	4.3	739.4	174.2	60.5
	2014	8.8			517.0	1.1	690.9	100.6	57.9
Belarus	2005	8.6	...	...	19.9	0.4	# 1 070.4	119.9	8.5
Bélarus	2009	5.0	59.3	40.7	13.9	0.2	# 866.3	60.2	5.4
	2010	5.1	...	...	12.2	0.2	# 775.3	49.9	3.9
	2014	3.6	...	...	8.7	0.3	# 426.6	25.9	4.7
Belgium	2005	2.1	...	...	638.3	8.7	2 058.2	# 1 814.0	57.3
Belgique	2010	1.7	...	...	711.3	10.3	2 155.6	# 1 695.3	64.4
	2014	1.8	...	...	616.1	10.0	1 909.7	# 1 529.3	57.4
Belize	2005	28.6	...	...		3.2	592.3	235.5	...
Belize	2009	61.8	90.7	9.3	165.3	0.6	375.6	173.6	64.3
	2010	40.1	87.5	12.5	...	...	259.9	148.3	40.4
	2011	37.7	90.3	9.7	...	0.9	283.1	135.8	36.8
	2014	34.4	...	...	...	...	...	...	...
Benin			...	...	...	...	...	...	...
Bénin	2012	6.3	...	...	...	...	...	...	...
Bermuda	2004	1.5	...	...	668.6	# 3.1	1 541.7	118.4	...
Bermudes	2005	3.1	...	...	...	...	...	...	...
	2010	10.9	...	...	964.8	# 1.6	950.7	154.8	143.9
	2013	4.8	...	...	841.1	# 6.4	774.2	51.0	111.5
Bhutan	2010	1.9	...	...	9.3	0.1	67.9	6.5	9.9
Bhoutan	2013	2.5	...	...	1.7	0.1	76.5	5.4	8.5
	2014	2.7	...	...	...	0.8	69.5	...	10.6
Bolivia (Plur. State of)	2005	7.2	...	...	66.0	1.1	41.6	93.9	18.9
Bolivie (État plur. de)	2010	10.6	...	...	85.2	0.9	50.9	125.7	39.8
	2012	12.4	...	...	72.5	1.0	50.8	140.7	47.1
Bosnia and Herzegovina	2002	...	72.2	27.8	...	...	...	...	...
Bosnie-Herzégovine	2005	...	...	...	# 35.7	0.2	# 480.4	19.7	...
	2008	...	...	...	# 38.4	0.2	# 168.1	19.4	4.4
	2010	1.5	69.6	30.4	# 13.2	0.0	# 296.3	...	3.8
	2011	1.3	68.6	31.4	# 17.0	0.0	# 175.8	25.4	4.4
	2014	1.3	...	...	# 15.5	0.1	# 155.4	26.9	3.5
Botswana	2005	15.6	...	...	...	...	...	...	...
Botswana	2010	14.8	...	...	852.5	0.1	1 451.1	117.3	...
	2014	...	...	...	758.9	0.1	1 335.9	76.5	123.9
Brazil	2010	20.9	...	...	360.3	0.2	696.2	544.3	23.4
Brésil	2013	23.6	...	...	323.9	0.2	873.8	495.7	27.5
	2014	24.6	...	...	...	...	...	...	...
British Virgin Islands	2004	17.7	...	...	...	...	...	...	...
Îles Vierges britanniques	2006	8.4	...	...	...	...	...	...	...
Brunei Darussalam	2005	0.6	...	...	146.5	0.0	332.1	2.5	...
Brunéi Darussalam	2006	0.8	...	...	122.5	0.0	308.6	0.5	...
	2010	0.3	...	...	...	...	...	...	...
	2013	0.5	...	...	...	...	...	...	...
Bulgaria	2005	2.6	73.9	26.1	47.6	2.5	561.3	48.9	13.5
Bulgarie	2010	2.0	81.6	18.4	41.0	1.6	681.5	50.5	9.4
	2012	1.9	82.3	17.7	33.0	1.0	608.8	40.5	9.5
	2014	1.6	...	...	35.2	1.3	572.8	32.3	8.5
Burkina Faso	2005	0.6	...	...	...	...	...	...	...
Burkina Faso	2010	0.6	...	...	...	...	...	...	...
	2012	0.7	...	...	...	...	...	...	...
Burundi	2010	3.7	...	...	3.2	0.3	...	32.1	7.4
Burundi	2014	4.0	...	...	5.1	0.7	7.5	41.6	11.7

Intentional homicides and other crimes *(continued)*
By type of crime per 100 000 population and homicide victims by sex

Homicides intentionnels et autres crimes *(suite)*
Par type de crime pour 100 000 habitants et victimes d'homicides par sexe

| Country or area
Pays ou zone | Year
Année | Intentional homicides
Homicides volontaires | | | Other crimes (per 100 000 pop.)
Autres infractions (pour 100 000 hab.) | | | | |
| | | Victims (percentage)
Victimes
(pourcentage) | | | | | | | |
		Per 100,000 Pour 100 000	Male Hommes	Female Femmes	Assault Agression	Kidnapping Enlèvement	Theft Vol simple	Robbery Vol qualifié	Sexual Violence Violences Sexuelles
Cabo Verde	2010	8.0	...	...	834.0	1.2	...	...	24.5
Cabo Verde	2013	10.6	...	...	791.1	5.5	746.6	763.7	41.0
Cambodia	2005	3.4	...	...	...	...	...	...	...
Cambodge	2010	2.3	...	...	...	...	...	...	...
	2011	1.8	...	...	...	...	...	...	...
Cameroon	#2005	...	...	...	17.3	...	66.3	...	...
Cameroun	2010	3.5	...	...	# 9.0	...	# 105.8	# 84.3	# 10.0
	2012	2.7	...	...	# 26.8	20.8	# 112.6	# 73.9	# 9.2
	#2014	...	...	...	12.7	...	64.2	33.4	3.1
Canada	2005	1.8	72.6	27.4	165.7	# 12.1	2 033.4	100.6	81.7
Canada	2010	1.4	71.4	28.6	162.4	# 12.6	1 594.8	89.3	78.8
	2011	1.5	69.8	30.2	156.4	# 11.0	1 494.5	86.3	77.2
	2013	1.4	...	...	139.8	# 9.2	1 381.1	66.0	75.0
	2014	...	...	...	135.0	# 9.2	1 374.5	58.8	73.6
Cayman Islands	2004	12.3	...	...	...	...	...	...	...
Îles Caïmanes	2009	14.7	...	...	...	...	...	...	...
Central African Rep. Rép. centrafricaine	2012	13.2	...	...	...	...	...	...	...
Chad Tchad	2012	9.2	...	...	...	...	...	...	...
Chile	2005	3.6	85.6	14.4	...	# 0.9	964.6	403.0	...
Chili	2010	3.2	82.3	17.7	130.8	# 1.5	1 098.9	480.0	76.5
	2011	3.7	81.9	18.1	133.5	# 1.9	1 210.0	534.7	91.2
	2014	3.6	...	...	98.2	# 1.5	1 082.1	598.7	70.8
China	2005	1.6	...	...	...	...	...	...	...
Chine	2010	1.0	...	...	...	...	...	...	...
	2012	0.8	...	...	...	...	...	...	...
China, Hong Kong SAR	2005	0.5	61.8	38.2	118.2	0.0	514.6	24.5	21.3
Chine, Hong Kong RAS	2010	0.5	45.7	54.3	108.7	0.0	491.1	11.1	26.3
	2011	0.2	47.1	52.9	104.5	0.0	497.2	10.4	25.6
	2013	0.9	...	...	92.4	0.0	441.1	7.0	24.9
China, Macao SAR	2005	1.5	...	...	...	...	...	...	...
Chine, Macao RAS	2010	0.7	...	...	# 343.4	# 0.4	# 526.0	# 33.3	# 8.4
	2014	1.0	...	...	# 308.5	# 0.0	# 467.7	# 20.8	# 9.5
Colombia	2005	39.5	91.8	8.2	70.3	1.8	158.8	93.9	10.6
Colombie	2010	32.7	92.0	8.0	115.7	0.6	200.0	133.8	14.8
	2011	34.1	91.6	8.4	134.0	0.7	217.9	146.2	17.6
	2013	32.6	...	...	175.2	0.6	289.7	197.4	24.5
	2014	27.9	...	...	172.1	...	286.4	197.5	26.5
Comoros Comores	2012	7.8	...	...	...	...	...	...	...
Congo Congo	2012	10.5	...	...	...	...	...	...	...
Cook Islands Îles Cook	2012	3.1	...	...	...	...	...	...	...
Costa Rica	2005	7.9	...	...	...	0.3	196.8	# 501.1	...
Costa Rica	2010	11.6	88.4	11.6	175.0	0.2	445.5	# 950.1	143.2
	2012	8.7	87.7	12.3	139.2	0.1	309.1	...	134.3
	2013	8.7	...	...	174.7	0.1	696.6	# 1 018.9	154.7
	2014	10.0	...	...	...	0.1	...	# 1 095.6	...
Côte d'Ivoire	2008	...	...	...	52.6	...	...	3.1	3.7
Côte d'Ivoire	2012	11.4	...	...	...	...	...	...	...
Croatia	2005	1.5	60.3	39.7	26.3	0.5	# 694.3	35.6	14.2
Croatie	2010	1.4	48.4	51.6	22.8	0.3	# 315.2	28.8	13.1
	2012	1.2	64.7	35.3	21.1	0.1	# 402.5	37.0	11.9
	2014	0.8	...	...	18.0	0.0	# 283.0	29.8	19.4

9

Intentional homicides and other crimes *(continued)*
By type of crime per 100 000 population and homicide victims by sex

Homicides intentionnels et autres crimes *(suite)*
Par type de crime pour 100 000 habitants et victimes d'homicides par sexe

Country or area Pays ou zone	Year Année	Intentional homicides Homicides volontaires Victims (percentage) Victims (pourcentage) Per 100,000 Pour 100 000	Male Hommes	Female Femmes	Other crimes (per 100 000 pop.) Autres infractions (pour 100 000 hab.) Assault Agression	Kidnapping Enlèvement	Theft Vol simple	Robbery Vol qualifié	Sexual Violence Violences Sexuelles
Cuba	2005	6.1	...	...	...	...	...	...	...
Cuba	2010	4.5	...	...	...	...	...	...	...
	2011	4.7	...	...	...	...	...	...	...
Cyprus	2005	1.9	90.0	10.0	12.8	1.6	116.3	8.1	10.8
Chypre	2010	0.7	87.5	12.5	15.7	2.4	151.3	14.1	7.2
	2012	1.9	77.3	22.7	13.1	2.7	137.3	15.3	6.1
	2014	0.1	...	...	11.6	1.8	83.7	9.0	4.4
Czech Republic	2005	1.1	58.3	41.7	211.9	0.1	1 506.5	54.2	18.1
République tchèque	2010	0.1	48.5	51.5	172.0	0.1	1 202.2	38.3	17.2
	2012	0.1	54.3	45.7	174.1	0.1	1 132.0	32.4	18.8
	2014	0.7	...	...	160.8	0.1	983.7	24.2	20.9
Dem. P. R. Korea R. p. dém. de Corée	2012	4.7	...	...	...	...	...	...	...
Dem. Rep. of the Congo Rép. dém. du Congo	2012	12.5	...	...	...	...	...	...	...
Denmark	2005	0.1	...	...	205.2	...	3 103.2	53.9	...
Danemark	2010	0.8	47.6	52.4	190.9	...	3 442.8	60.7	...
	2012	0.7	66.0	34.0	170.6	...	3 426.8	53.3	...
	2014	1.0	...	...	164.6	...	3 026.3	41.0	...
Djibouti Djibouti	2012	7.0	...	...	...	...	...	...	...
Dominica	2005	11.3	...	...	...	...	...	...	...
Dominique	2010	21.1	...	...	...	...	...	...	...
	2011	8.4	...	...	...	...	...	...	...
Dominican Republic	2005	25.9	...	...	...	0.2	...	...	...
Rép. dominicaine	2006	22.9	...	...	...	0.1	...	...	...
	2010	25.0	91.0	9.0	...	...	...	...	...
	2012	22.3	91.1	8.9	58.5	0.2	98.9	20.6	0.5
	2014	17.4	...	...	48.3	0.2	235.3	144.2	2.8
Ecuador	2005	15.4	...	...	56.5	0.3	# 43.4	346.7	...
Équateur	2010	17.7	...	...	30.2	0.2	# 33.6	362.3	24.4
	2014	8.2	...	...	46.7	0.2	# 152.8	570.6	19.0
Egypt	2005	0.7	87.9	12.1	...	0.0	46.7	0.6	...
Égypte	2010	2.2	89.3	10.7	0.2	0.1	94.0	0.8	0.2
	2011	3.2	87.8	12.2	0.4	0.3	104.0	3.2	0.1
El Salvador	2005	63.5	...	...	# 77.8	0.1	# 180.3	# 81.7	...
El Salvador	2010	66.0	85.9	14.1	# 64.4	0.5	# 160.2	# 89.7	# 36.9
	2012	42.7	89.0	11.0	# 69.6	0.2	# 168.9	# 90.9	# 57.1
	2014	64.2	...	...	# 64.2	0.3	# 123.3	# 77.5	# 41.9
Equatorial Guinea Guinée équatoriale	2012	3.4	...	...	...	...	...	...	...
Eritrea Érythrée	2012	9.7	...	...	...	...	...	...	...
Estonia	2005	8.3	74.8	25.2	9.7	# 0.1	2 246.3	97.8	29.7
Estonie	2010	5.3	76.6	23.4	7.7	# 0.0	1 895.7	45.0	20.6
	2014	3.1	...	...	5.8	# 0.0	1 195.7	27.4	25.8
Ethiopia Éthiopie	2012	8.0	...	...	...	...	...	...	...
Fiji	2005	...	60.9	39.1	...	...	...	...	...
Fidji	2008	3.6	41.2	58.8	...	...	...	...	...
	2010	2.8	...	...	...	...	...	...	...
	2012	3.0	...	...	...	...	...	...	...
Finland	2005	2.3	70.1	29.9	# 580.7	0.0	2 384.1	34.6	# 36.6
Finlande	2010	2.2	77.6	22.4	# 616.0	0.0	2 268.1	28.1	# 45.0
	2012	1.6	53.9	46.1	# 704.4	0.0	2 088.5	29.8	# 64.7
	2014	1.6	...	...	# 600.7	0.0	2 142.0	30.8	# 54.8

9 Intentional homicides and other crimes *(continued)*
By type of crime per 100 000 population and homicide victims by sex

Homicides intentionnels et autres crimes *(suite)*
Par type de crime pour 100 000 habitants et victimes d'homicides par sexe

Country or area / Pays ou zone	Year / Année	Intentional homicides / Homicides volontaires Per 100,000 / Pour 100 000	Victims (percentage) / Victimes (pourcentage) Male / Hommes	Female / Femmes	Other crimes (per 100 000 pop.) / Autres infractions (pour 100 000 hab.) Assault / Agression	Kidnapping / Enlèvement	Theft / Vol simple	Robbery / Vol qualifié	Sexual Violence / Violences Sexuelles
France	2005	1.6	...	...	# 242.7	3.3	# 1 363.5	# 203.5	# 39.0
France	2010	1.3	...	...	# 350.0	3.3	# 1 848.5	# 192.2	# 36.5
	2014	1.2	...	...	# 354.5	4.6	# 1 970.4	# 177.9	# 48.3
French Guiana	2005	22.1	...	...	...	...	...	...	...
Guyane française	2009	13.2	...	...	...	...	...	...	...
French Polynesia									
Polynésie française	2009	0.4	...	...	...	...	...	...	...
Gabon									
Gabon	2012	9.4	...	...	...	...	...	...	...
Gambia									
Gambie	2012	9.4	...	...	...	...	...	...	...
Georgia	2005	9.0	82.1	17.9	8.2	2.0	363.2	46.6	3.7
Géorgie	2010	4.4	85.0	15.0	3.0	0.1	267.5	15.0	3.9
	2011	...	75.7	24.3	...	...	...	...	...
	2014	2.7	...	...	5.7	...	236.9	11.7	3.2
Germany	2005	1.1	55.0	45.0	# 608.4	# 2.1	# 2 750.2	67.5	67.9
Allemagne	2010	0.1	48.0	52.0	# 177.7	# 6.1	# 1 516.7	59.9	58.3
	2011	0.9	52.7	47.3	# 172.9	# 6.1	# 1 586.7	59.7	58.5
	2014	0.9	...	...	# 155.9	# 6.2	# 1 624.0	56.4	58.3
Ghana	2005	1.8	...	...	...	...	...	...	...
Ghana	2010	1.7	...	...	...	...	...	...	...
	2011	1.7	...	...	...	...	...	...	...
Greece	2005	1.2	96.2	3.8	70.3	0.2	556.9	18.8	5.7
Grèce	2009	1.4	93.4	6.6	70.4	0.1	888.9	42.2	8.3
	2010	1.6	...	...	68.3	1.5	1 060.3	54.4	8.1
	2014	0.1	...	...	50.7	0.5	851.5	34.5	5.4
Greenland	2005	17.6	...	...	...	...	...	...	...
Groenland	2010	10.6	...	...	...	...	...	...	...
	2011	1.8	...	...	...	...	...	...	...
Grenada	2005	10.7	72.7	27.3	1 839.7	0.0	1 761.0	43.7	146.7
Grenade	2010	9.6	80.0	20.0	1 100.5	0.0	918.1	24.8	106.0
	2012	13.3	64.3	35.7	1 244.8	0.0	1 456.3	22.8	200.0
	2014	7.5	...	...	1 342.7	...	1 267.5	22.6	76.2
Guadeloupe	2005	5.1	...	...	...	...	...	...	...
Guadeloupe	2009	7.9	...	...	...	...	...	...	...
Guam	2005	4.4	...	...	...	...	...	...	...
Guam	2010	1.9	...	...	...	...	...	...	...
	2011	2.5	...	...	...	...	...	...	...
Guatemala	2004	35.0	89.0	11.0	50.0	0.4	40.2	# 93.8	2.8
Guatemala	2005	...	90.3	9.7	46.2	0.4	32.6	# 83.3	2.4
	2009	45.1	88.9	11.1	52.7	1.1	41.5	# 26.6	2.8
	2010	40.5	...	...	50.6	0.9	44.2	# 22.0	3.4
	2012	33.5	...	...	39.6	0.5	59.3	# 24.9	4.3
	2014	31.2	...	...	37.3	0.3	...	# 19.4	...
Guinea	2005	...	...	...	3.0	0.2	...	...	...
Guinée	2007	...	...	...	2.9	0.2	13.3	1.5	0.3
	2008	...	...	...	...	0.2	...	...	...
	2012	8.7	...	...	...	...	...	...	...
Guinea-Bissau									
Guinée-Bissau	2012	9.9	...	...	...	...	...	...	...
Guyana	2005	19.1	80.3	19.7	...	...	...	...	32.2
Guyana	2010	18.6	77.9	22.1	# 1 484.3	0.3	518.5	145.3	19.5
	2011	17.2	60.0	40.0	...	0.1	252.6	157.6	39.3
	2013	20.4	...	...	# 854.2	0.0	380.8	201.2	40.3
Haiti	2010	6.8	...	...	...	...	...	...	...
Haïti	2012	10.0	...	...	...	...	...	...	...
Holy See	2010	...	...	...	0.0	0.0	0.0	0.0	0.0
Saint-Siège	2014	...	...	...	0.0	0.0	0.0	0.0	0.0

Intentional homicides and other crimes *(continued)*
By type of crime per 100 000 population and homicide victims by sex

Homicides intentionnels et autres crimes *(suite)*
Par type de crime pour 100 000 habitants et victimes d'homicides par sexe

| Country or area
Pays ou zone | Year
Année | Intentional homicides
Homicides volontaires | | | Other crimes (per 100 000 pop.)
Autres infractions (pour 100 000 hab.) | | | | |
| | | Per 100,000
Pour 100 000 | Victims (percentage)
Victimes
(pourcentage) | | | | | | |
			Male Hommes	Female Femmes	Assault Agression	Kidnapping Enlèvement	Theft Vol simple	Robbery Vol qualifié	Sexual Violence Violences Sexuelles
Honduras	2005	46.7	...	...	...	...	...	...	...
Honduras	2009	71.5	93.1	6.9	...	...	66.4	...	...
	2010	83.1	...	...	...	...	59.7	...	...
	2011	93.2	93.2	6.8	47.8	0.5	53.4	325.5	47.0
	2013	86.1	...	...	22.1	0.7	57.7	237.9	33.2
	2014	74.6	...	...	...	...	...	...	...
Hungary	2005	1.6	53.4	46.6	122.2	0.1	1 250.5	29.5	# 8.5
Hongrie	2010	1.4	57.9	42.1	145.5	0.2	1 340.0	33.9	# 12.9
	2012	1.2	58.3	41.7	140.3	0.1	1 308.4	30.5	# 13.6
	2014	1.5	...	...	136.3	0.0	1 081.3	19.8	# 7.8
Iceland	2005	1.0	66.7	33.3	...	...	1 072.0	16.5	97.1
Islande	2008	0.0	0.0	0.0	24.8	...	1 397.3	13.9	118.7
	2010	0.6	100.0	0.0	17.9	...	1 547.0	13.2	...
	2012	0.3	0.0	100.0	22.9	...	1 282.3	15.5	84.4
	2013	0.3	...	...	28.3	...	1 260.6	15.1	138.9
India	2005	3.6	64.2	35.8	23.7	2.0	23.9	1.5	5.5
Inde	2010	3.4	59.5	40.5	23.5	3.1	26.8	1.9	5.9
	2012	3.4	59.2	40.8	26.3	3.8	26.7	2.5	6.3
	2013	3.3	...	...	26.2	5.1	29.1	2.9	9.1
	2014	3.2	...	...	...	...	...	...	...
Indonesia	2004	0.6	...	...	...	...	...	...	...
Indonésie	2010	0.4	...	...	13.2	0.2	10.1	4.1	...
	2014	0.5	...	...	14.5	0.1	9.6	4.6	2.2
Iran (Islamic Rep. of)	2004	...	...	...	...	...	158.4	...	...
Iran (Rép. islamique d')	2012	4.8	...	...	...	...	...	...	...
Iraq	2005	34.1	...	...	...	...	...	...	...
Iraq	2010	8.3	...	...	...	...	...	...	...
	2011	7.9	...	...	...	...	...	...	...
Ireland	2005	1.2	86.2	13.8	296.9	1.8	1 400.6	57.7	42.8
Irlande	2010	1.1	86.7	13.3	325.7	2.9	1 416.7	69.2	51.2
	2012	1.2	86.7	13.3	290.6	2.2	1 455.8	60.3	45.3
	2014	1.1	...	...	282.5	2.7	1 496.6	56.6	44.2
Israel	2005	2.5	...	...	774.0	6.3	1 897.3	65.6	69.6
Israël	2010	2.0	...	...	620.0	3.6	974.9	39.2	66.8
	2011	2.0	...	...	569.4	3.5	936.2	36.3	60.5
	2012	1.7	...	...	...	...	...	...	...
Italy	2005	1.0	77.7	22.3	96.5	0.6	# 1 870.8	112.0	6.9
Italie	2010	0.9	70.8	29.2	108.9	0.6	# 1 584.3	80.5	8.1
	2011	0.9	69.9	30.1	114.8	0.6	# 1 740.8	97.5	7.7
	2014	0.8	...	...	110.7	0.5	# 1 876.2	97.6	7.1
Jamaica	2004	55.2	...	...	...	...	...	...	...
Jamaïque	2005	...	88.9	11.1	...	1.2	...	...	40.8
	2008	59.6	89.8	10.2	...	1.6	...	97.9	54.2
	2010	52.8	90.0	10.0	# 51.3	...	...	104.1	54.2
	2011	41.2	89.5	10.5	...	...	...	112.4	73.0
	2013	43.3	...	...	# 178.0	0.8	75.4	96.3	95.8
	2014	36.1	...	...	...	...	...	81.3	81.6
Japan	2005	0.5	50.4	49.6	47.5	0.2	727.1	4.7	8.5
Japon	2010	0.4	49.0	51.0	20.9	0.1	485.5	3.2	6.6
	2011	0.3	47.1	52.9	20.4	0.1	458.0	2.9	6.4
	2014	0.3	...	...	21.0	0.2	356.2	2.4	6.8
Jordan	2005	# 1.3	...	...	275.2	0.5	132.5	11.6	...
Jordanie	2006	# 1.8	...	...	283.8	0.6	161.7	14.5	...
	2010	# 1.8	...	...	...	...	187.2	...	...
	2012	# 2.3	...	...	...	...	193.1	...	...
Kazakhstan	2005	# 11.7	...	...	...	0.6	408.6	84.5	...
Kazakhstan	2010	# 9.5	...	...	9.7	0.8	367.5	65.8	2.2
	2013	# 7.4	...	...	7.9	0.6	1 229.1	110.1	3.0

9 Intentional homicides and other crimes *(continued)*
By type of crime per 100 000 population and homicide victims by sex

Homicides intentionnels et autres crimes *(suite)*
Par type de crime pour 100 000 habitants et victimes d'homicides par sexe

Country or area Pays ou zone	Year Année	Intentional homicides Homicides volontaires			Other crimes (per 100 000 pop.) Autres infractions (pour 100 000 hab.)				
		Per 100,000 Pour 100 000	Victims (percentage) Victimes (pourcentage)		Assault Agression	Kidnapping Enlèvement	Theft Vol simple	Robbery Vol qualifié	Sexual Violence Violences Sexuelles
			Male Hommes	Female Femmes					
Kenya	2005	3.6	...	...	36.0	0.4	34.5	19.6	7.3
Kenya	2010	5.6	...	...	34.9	0.2	29.7	7.0	11.9
	2014	5.9	...	...	31.1	0.1	22.4	6.7	11.6
Kiribati	2010	3.9	...	...	...	...	...	...	...
Kiribati	2012	7.5	...	...	...	...	...	...	...
Kosovo	2010	5.8	...	...	205.9	2.4	834.3	31.0	8.2
Kosovo	2013	2.2	...	...	35.1	1.8	912.4	29.7	11.2
	2014	...	...	...	48.7	1.2	834.2	22.9	2.9
Kuwait	2005	...	...	...	26.3	12.4	432.5	19.0	16.4
Koweït	2009	1.7	...	...	23.8	12.4	282.1	22.8	17.7
	2010	2.0	...	...	...	...	...	...	...
	2012	1.8	...	...	...	...	...	...	...
Kyrgyzstan	2005	8.2	78.4	21.6	5.2	# 0.5	# 241.1	53.1	# 0.3
Kirghizistan	2009	7.8	68.7	31.3	6.7	# 0.1	# 203.3	43.6	# 0.6
	2010	19.6	...	...	9.8	# 1.5	# 180.9	51.2	...
	2013	3.6	...	...	5.3	# 0.7	# 173.6	28.1	# 1.3
	2014	3.7	...	...	6.5	# 0.9	# 154.5	...	# 1.3
Lao People's Dem. Rep. Rép. dém. pop. lao	2012	7.3	...	...	...	...	...	...	...
Latvia	2002	9.2	66.7	33.3	...	...	...	...	...
Lettonie	2005	5.7	...	...	# 49.4	0.3	# 1 090.3	97.1	# 32.4
	2010	3.3	...	...	# 58.6	0.4	# 1 227.4	51.3	# 11.1
	2012	4.8	49.0	51.0	# 65.0	0.0	# 1 047.6	46.3	# 16.0
	2014	3.9	...	...	# 28.6	0.9	# 1 034.4	40.7	# 27.8
Lebanon	2010	3.8	...	...	218.9	12.9	662.1	0.5	4.8
Liban	2014	4.3	...	...	151.7	18.4	506.2	0.4	3.6
Lesotho	2009	36.3	...	...	368.4	3.1	225.9	63.8	...
Lesotho	2010	38.0	...	...	...	...	...	...	...
Liberia	2010	3.3	...	...	...	...	...	...	...
Libéria	2012	3.2	...	...	...	...	...	...	...
Libya Libye	2012	2.5	...	...	...	...	...	...	...
Liechtenstein	2005	0.0	...	...	203.7	0.0	502.1	2.9	51.6
Liechtenstein	2010	2.8	100.0	0.0	212.3	0.0	479.7	5.5	16.5
	2014	2.7	...	...	321.8	0.0	573.9	2.7	21.5
Lithuania	2005	11.1	73.1	26.9	12.2	2.1	826.1	155.7	16.7
Lituanie	2010	7.0	65.8	34.2	7.8	2.2	849.5	87.3	18.9
	2012	6.7	73.8	26.2	5.9	2.2	778.3	63.7	16.8
	2014	5.5	...	...	7.0	1.5	738.7	57.9	15.0
Luxembourg	2005	...	50.0	50.0	...	...	...	...	...
Luxembourg	2010	2.0	50.0	50.0	487.3	5.3	1 421.2	74.2	# 41.5
	2011	0.8	100.0	0.0	527.7	9.0	1 687.6	76.5	# 41.7
	2014	0.7	...	...	572.6	9.3	1 843.2	110.5	# 51.4
Madagascar	2005	1.8	...	...	...	...	...	...	...
Madagascar	2010	0.6	...	...	...	...	...	...	...
Malawi	2005	1.6	49.5	50.5	...	...	...	...	...
Malawi	2010	3.5	69.2	30.8	...	...	...	...	...
	2012	1.8	87.5	12.5	...	...	...	...	...
Malaysia	2005	2.3	...	...	16.5	...	133.0	59.7	...
Malaisie	2006	2.2	...	...	21.8	...	141.4	81.6	...
	2010	1.9	...	...	...	...	...	...	...
Maldives	2004	...	...	...	...	0.3	...	155.7	...
Maldives	2010	1.8	...	...	444.4	7.8	1 203.0	164.5	157.3
	2013	0.9	...	...	367.1	2.3	1 903.4	207.6	163.2
Mali Mali	2012	10.2	...	...	...	...	...	...	...

Intentional homicides and other crimes *(continued)*
By type of crime per 100 000 population and homicide victims by sex

Homicides intentionnels et autres crimes *(suite)*
Par type de crime pour 100 000 habitants et victimes d'homicides par sexe

Country or area Pays ou zone	Year Année	Intentional homicides Homicides volontaires			Other crimes (per 100 000 pop.) Autres infractions (pour 100 000 hab.)				
		Per 100,000 Pour 100 000	Victims (percentage) Victimes (pourcentage)						Sexual Violence Violences Sexuelles
			Male Hommes	Female Femmes	Assault Agression	Kidnapping Enlèvement	Theft Vol simple	Robbery Vol qualifié	
Malta	2005	1.0	50.0	50.0	...	...	2 420.6	# 64.5	...
Malte	2010	0.1	75.0	25.0	# 43.7	# 0.0	1 885.4	# 47.6	# 22.1
	2012	2.9	75.0	25.0	# 41.9	# 0.0	2 091.0	# 61.8	# 19.7
	2014	1.4	...	...	# 38.3	# 0.0	1 962.5	# 45.5	# 18.4
Marshall Islands									
Îles Marshall	2012	4.7	...	...	...	...	...	...	...
Martinique	2005	4.8	...	...	...	...	...	...	...
Martinique	2009	2.8	...	...	...	...	...	...	...
Mauritania									
Mauritanie	2012	11.4	...	...	...	...	...	...	...
Mauritius	2005	...	...	...	10.0	...	# 1 172.3	110.5	25.7
Maurice	2010	2.6	75.8	24.2	19.1	2.8	# 864.0	86.9	34.6
	2011	2.7	76.5	23.5	18.4	4.0	# 729.9	65.0	37.2
Mayotte									
Mayotte	2009	5.9	...	...	...	...	...	...	...
Mexico	2005	9.0	...	...	# 217.9	# 0.3	# 77.3	# 470.1	# 26.1
Mexique	2010	21.7	...	...	# 194.5	# 1.1	# 111.3	# 622.3	# 28.8
	2013	18.6	...	...	# 171.1	# 1.5	# 72.9	# 588.9	# 31.6
	2014	15.7	...	...	...	...	...	...	...
Micronesia (Fed. States of)									
Micronésie (États féd. de)	2012	4.8	...	...	...	...	...	...	...
Monaco	2005	3.0	...	...	458.5	0.0	958.4	20.7	...
Monaco	2006	2.9	...	...	407.2	5.8	1 006.3	11.6	...
	2008	0.0	...	...	...	...	...	...	...
Mongolia	2005	15.8	...	...	118.4	0.0	264.5	23.1	13.4
Mongolie	2010	8.8	72.0	28.0	214.6	0.0	213.0	19.2	13.3
	2011	9.6	77.4	22.6	211.6	0.0	191.2	19.4	14.1
	2014	7.5	...	...	279.5	0.0	289.5	23.6	12.1
Montenegro	2005	...	...	...	# 25.1	...	249.8	# 14.0	5.7
Monténégro	2010	2.4	53.3	46.7	# 31.0	0.3	85.2	# 0.5	5.8
	2012	2.7	82.4	17.6	# 26.9	0.0	137.4	# 1.1	3.8
	2014	3.2	...	...	# 20.6	0.0	142.7	# 1.3	5.0
Montserrat	2005	20.9	...	...	...	...	...	...	...
Montserrat	2008	20.4	...	...	...	...	...	...	...
Morocco	2005	1.5	...	...	107.4	0.1	204.4	49.4	3.1
Maroc	2009	1.4	87.8	12.2	98.1	0.1	247.2	81.8	10.3
	2010	1.4	...	...	...	...	...	...	...
	2013	1.3	...	...	202.6	2.9	322.3	44.4	12.9
	2014	1.0	...	...	...	...	...	...	...
Mozambique	2005	5.2	...	...	6.0	...	26.3	32.3	3.7
Mozambique	2009	3.5	...	...	2.2	...	20.0	22.3	2.6
	2010	3.6	...	...	...	...	...	...	...
Myanmar	2005	1.5	...	...	...	...	...	...	...
Myanmar	2010	1.7	...	...	5.2	...	3.3	0.0	0.5
	2012	2.5	...	...	4.4	0.1	4.4	0.0	0.5
	2013	...	...	...	4.7	0.0	4.8	0.0	0.7
Namibia	2004	17.6	...	...	...	...	...	...	...
Namibie	2010	14.3	...	...	...	...	...	...	...
	2012	16.9	...	...	...	...	...	...	...
Nauru									
Nauru	2012	1.3	...	...	...	...	...	...	...
Nepal	2005	3.6	...	...	4.1	0.5	2.0	0.9	...
Népal	2006	2.5	...	...	4.1	0.1	2.1	0.6	...
	2010	3.0	...	...	...	...	...	...	...
	2011	2.9	...	...	...	...	...	...	...

Intentional homicides and other crimes *(continued)*
By type of crime per 100 000 population and homicide victims by sex

Homicides intentionnels et autres crimes *(suite)*
Par type de crime pour 100 000 habitants et victimes d'homicides par sexe

Country or area Pays ou zone	Year Année	Intentional homicides Homicides volontaires			Other crimes (per 100 000 pop.) Autres infractions (pour 100 000 hab.)				
		Per 100,000 Pour 100 000	Victims (percentage) Victimes (pourcentage)		Assault Agression	Kidnapping Enlèvement	Theft Vol simple	Robbery Vol qualifié	Sexual Violence Violences Sexuelles
			Male Hommes	Female Femmes					
Netherlands	2005	1.1	...	...	# 410.6	# 5.6	# 4 641.6	# 100.7	# 44.6
Pays-Bas	2010	0.9	...	...	# 362.4	# 3.9	# 3 981.0	# 97.0	# 57.8
	2013	0.7	...	...	# 314.4	# 3.0	# 3 835.4	# 78.0	# 49.5
	#2014	...	...	...	298.4	2.4	3 480.2	61.2	46.7
New Caledonia Nouvelle-Calédonie	2009	3.3	...	...	...	...	...	...	...
New Zealand	2005	1.5	...	...	234.2	# 7.0	3 088.4	56.2	59.6
Nouvelle-Zélande	2010	0.1	47.6	52.4	257.7	# 5.2	2 676.6	57.0	66.7
	2012	0.9	48.8	51.2	229.2	# 5.3	2 292.0	47.0	76.4
	2013	1.0	...	...	200.5	# 4.4	2 280.1	45.5	83.7
	2014	0.9	...	...	220.2	# 5.2	...	44.8	83.2
Nicaragua	2005	13.6	...	...	359.4	0.2	269.8	# 397.4	...
Nicaragua	2010	13.7	90.1	9.9	319.7	0.1	182.2	# 495.5	# 63.2
	2012	11.5	...	...	...	...	...	...	...
Niger Niger	2012	4.5	...	...	...	...	...	...	...
Nigeria	2010	...	...	...	11.3	0.5	12.8	1.4	1.1
Nigéria	2012	10.1	...	...	9.1	0.4	13.4	1.8	0.8
	2013	...	...	...	9.5	0.3	13.9	1.1	1.0
Niue Nioué	2012	3.1	...	...	...	...	...	...	...
Norway	2001	0.8	47.2	52.8	...	...	...	...	...
Norvège	2005	0.7	...	...	# 64.0	...	2 996.7	# 31.3	42.4
	2010	0.6	48.3	51.7	# 59.8	...	2 590.0	# 34.5	50.0
	2011	2.2	53.2	46.8	# 57.3	...	2 542.2	# 32.8	53.7
	2014	0.6	...	...	# 46.2	...	2 193.7	# 20.6	51.6
Oman	2002	0.6	...	...	...	...	...	...	...
Oman	2008	0.7	...	...	78.5	0.3	207.9	9.6	6.9
	2011	1.1	...	...	...	...	...	...	...
Other non-specified areas	2005	4.0	...	...	...	...	...	...	...
Autres zones non-spécifiées	2010	3.2	...	...	...	...	...	...	...
	2011	3.0	...	...	...	...	...	...	...
Pakistan	2005	6.3	...	...	...	...	...	...	...
Pakistan	2010	7.8	...	...	...	...	...	...	...
	2012	7.8	...	...	...	...	...	...	...
Palau Palaos	2012	3.1	...	...	...	...	...	...	...
Panama	2005	11.0	90.2	9.8	...	# 0.5	# 421.1	# 158.1	# 23.6
Panama	2010	21.0	92.5	7.5	167.4	# 1.1	# 504.4	# 262.6	# 27.8
	2012	17.5	94.6	5.4	145.9	# 0.5	# 542.3	# 268.1	# 65.2
	2013	17.4	...	...	111.3	# 5.7	# 541.0	...	# 61.0
Papua New Guinea	2005	9.9	...	...	...	...	...	...	...
Papouasie-Nvl-Guinée	2010	10.4	...	...	...	...	...	...	...
Paraguay	2005	18.6	...	...	38.6	0.0	...	...	...
Paraguay	2010	11.9	...	...	31.4	0.0	502.1	220.3	54.7
	2014	8.8	...	...	11.8	0.0	581.3	307.1	64.0
Peru	2005	11.1	...	...	53.2	# 1.8	189.9	165.1	...
Pérou	2010	9.2	...	...	188.0	# 1.6	190.1	193.4	18.0
	2014	6.7	...	...	211.1	# 2.3	289.3	250.5	18.1
Philippines	2005	# 7.5	...	...	...	0.1	# 13.0	# 9.0	...
Philippines	2010	# 9.6	86.0	14.0	...	0.0	# 78.2	# 38.6	2.3
	2011	# 9.2	88.0	12.0	...	0.0	# 59.0	# 32.6	1.8
	2012	# 8.8	...	...	...	0.0	# 45.4	# 28.1	1.9
	#2013	# 9.4	...	...	...	...	127.3	50.5	...
	#2014	# 9.9	...	...	...	...	...	...	...
Poland	2005	1.4	...	...	...	0.1	725.0	109.6	...
Pologne	2010	1.1	...	...	28.5	0.1	528.6	50.2	8.0
	2014	0.7	...	...	18.5	0.8	433.4	24.4	11.6

Intentional homicides and other crimes *(continued)*
By type of crime per 100 000 population and homicide victims by sex

Homicides intentionnels et autres crimes *(suite)*
Par type de crime pour 100 000 habitants et victimes d'homicides par sexe

Country or area Pays ou zone	Year Année	Intentional homicides Homicides volontaires Victims (percentage) Victimes (pourcentage) Per 100,000 Pour 100 000	Male Hommes	Female Femmes	Other crimes (per 100 000 pop.) Autres infractions (pour 100 000 hab.) Assault Agression	Kidnapping Enlèvement	Theft Vol simple	Robbery Vol qualifié	Sexual Violence Violences Sexuelles
Portugal	2005	1.3	...	...	374.2	4.2	876.1	192.9	15.4
Portugal	2010	1.2	...	...	291.1	4.7	898.1	193.1	20.8
	2014	0.9	...	...	239.1	3.6	875.2	149.9	23.8
Puerto Rico	2005	20.5	...	...	...	...	...	...	...
Porto Rico	2010	26.5	...	...	73.2	...	721.7	175.6	...
	2014	18.5	...	...	67.1	...	785.2	140.3	...
Qatar	2004	...	...	...	40.7	0.8	100.3	2.9	...
Qatar	2005	...	...	...	...	...	107.2	...	...
	2006	...	...	...	...	...	105.5	...	...
	2012	7.2	...	...	...	...	...	...	...
Republic of Korea	2005	...	...	...	...	...	396.5	10.9	...
République de Corée	2010	...	...	...	# 595.5	0.5	548.8	9.0	37.1
	2011	0.9	47.5	52.5	# 631.9	0.5	570.1	8.1	39.5
	2014	0.7	...	...	# 100.4	0.4	531.7	3.2	42.0
Republic of Moldova	2005	7.1	66.2	33.8	# 9.5	0.4	276.7	4.5	9.5
République de Moldova	2010	6.5	79.8	20.2	# 10.2	0.9	334.1	4.5	13.6
	2012	5.6	72.5	27.5	# 8.0	2.3	350.8	4.1	15.1
	2014	3.2	...	...	# 7.9	2.2	410.8	3.1	15.9
Réunion	2005	3.2	...	...	...	...	...	...	...
Réunion	2009	1.8	...	...	...	...	...	...	...
Romania	2005	2.1	63.6	36.4	# 41.6	# 1.3	# 201.8	# 15.5	...
Roumanie	2008	2.3	60.6	39.4	# 45.6	# 1.3	# 190.5	# 11.9	# 7.5
	2010	2.0	62.5	37.5	# 60.6	...	# 240.5	# 12.2	# 7.3
	2014	1.5	...	...	# 11.6	...	# 836.6	# 32.7	# 7.7
Russian Federation	2005	...	...	...	40.3	0.8	1 095.2	239.8	12.9
Fédération de Russie	2010	10.2	...	...	27.8	0.4	774.2	114.9	11.0
	2014	9.5	...	...	22.9	0.3	633.7	54.2	10.2
Rwanda	2010	4.9	...	...	19.5	0.1	...	20.1	...
Rwanda	2013	...	...	...	29.8	0.2	...	25.0	15.7
Saint Kitts and Nevis	2005	16.3	...	...	...	...	...	...	...
Saint-Kitts-et-Nevis	2010	40.1	90.5	9.5	296.1	3.8	1 018.1	166.2	118.4
	2011	64.2	...	...	258.5	5.7	926.5	124.5	120.8
	2012	33.6	...	...	...	...	...	...	...
Saint Lucia	2005	24.8	...	...	...	...	...	...	...
Sainte-Lucie	2010	24.8	...	...	...	...	...	...	...
	2012	21.6	...	...	...	...	...	...	...
Saint Pierre and Miquelon Saint-Pierre-et-Miquelon	2009	15.9	...	...	...	...	...	...	...
Saint Vincent-Grenadines	2005	21.1	73.1	26.9	1 151.3	1.8	1 777.5	67.1	104.8
Saint-Vincent-Grenadines	2010	22.9	88.0	12.0	1 197.4	1.8	1 739.0	107.0	153.7
	2012	25.6	...	...	1 017.1	3.7	1 923.5	146.3	160.1
	2013	...	...	...	...	...	...	136.3	209.5
Samoa	2010	8.6	...	...	...	...	...	...	...
Samoa	2013	3.2	...	...	...	...	...	...	...
San Marino	2005	0.0	...	...	...	...	...	...	...
Saint-Marin	2010	0.0	...	...	...	...	...	...	...
	2011	0.0	...	...	...	...	...	...	...
Sao Tome and Principe	2004	...	...	...	0.7	...	...	...	...
Sao Tomé-et-Principe	2010	3.5	...	...	0.0	0.0	9.4	0.6	0.0
	2011	3.4	...	...	2.3	0.0	11.5	0.6	0.0
Saudi Arabia Arabie saoudite	2012	6.2	...	...	...	...	...	...	...
Senegal	2005	...	...	...	2.5	...	...	24.4	...
Sénégal	2010	...	...	...	2.4	...	19.3	19.3	...
	2012	7.9	...	...	...	...	...	...	...

9

Intentional homicides and other crimes *(continued)*
By type of crime per 100 000 population and homicide victims by sex

Homicides intentionnels et autres crimes *(suite)*
Par type de crime pour 100 000 habitants et victimes d'homicides par sexe

Country or area Pays ou zone	Year Année	Intentional homicides Homicides volontaires Victims (percentage) Victimes (pourcentage) Per 100,000 Pour 100 000	Male Hommes	Female Femmes	Other crimes (per 100 000 pop.) Autres infractions (pour 100 000 hab.) Assault Agression	Kidnapping Enlèvement	Theft Vol simple	Robbery Vol qualifié	Sexual Violence Violences Sexuelles
Serbia	2005	1.6	69.4	30.6	17.6	0.3	# 182.5	32.5	5.8
Serbie	2010	1.4	66.7	33.3	15.4	0.2	# 175.6	40.0	5.4
	2012	1.2	64.9	35.1	14.7	0.1	# 199.3	42.2	4.3
	2014	1.3	...	...	12.9	0.1	# 300.6	36.3	3.9
Seychelles	2005	3.4	...	...	...	...	...	...	...
Seychelles	2010	2.1	...	...	...	...	...	...	...
Sierra Leone	2005	2.0	...	...	...	...	...	...	...
Sierra Leone	2008	3.3	...	...	359.9	...	190.2	3.3	11.3
	2010	2.8	...	...	...	...	...	...	...
	2012	1.9	...	...	...	...	...	...	...
Singapore	2005	0.5	...	...	13.5	0.0	470.9	24.3	...
Singapour	2010	0.4	84.2	15.8	8.4	0.0	359.8	10.2	31.1
	2011	0.3	62.5	37.5	7.8	0.0	332.3	7.6	29.8
	2014	0.3	...	...	9.3	...	288.2	4.2	27.9
Slovakia	2005	2.0	65.6	34.4	72.3	0.3	350.9	35.6	...
Slovaquie	2009	1.6	59.0	41.0	48.5	0.1	514.1	25.1	11.6
	2010	1.6	...	...	44.9	0.1	452.5	22.0	10.4
	2014	1.1	...	...	36.1	0.2	342.4	12.5	12.5
Slovenia	2005	1.0	50.0	50.0	115.4	0.1	1 477.0	21.5	15.6
Slovénie	2010	0.7	46.7	53.3	106.3	0.2	1 408.6	22.6	18.3
	2012	0.7	57.1	42.9	98.5	0.2	1 643.6	19.0	12.8
	2014	0.7	...	...	80.2	0.2	1 457.0	14.4	9.8
Solomon Islands	2005	5.5	...	...	258.5	...	219.3	5.1	43.5
Îles Salomon	2008	3.8	...	...	217.9	...	174.4	10.3	24.2
Somalia									
Somalie	2012	5.6	...	...	...	...	...	...	...
South Africa	2005	38.3	...	...	...	...	...	...	...
Afrique du Sud	2010	30.9	82.7	17.3	...	...	...	...	...
	2011	29.9	84.6	15.4	...	...	...	...	...
	2014	33.0	...	...	...	...	...	...	...
South Sudan									
Soudan du sud	2012	13.9	82.5	17.5	...	...	...	...	...
Spain	2005	1.2	...	...	# 394.1	0.5	...	...	...
Espagne	2010	0.8	61.0	39.0	# 53.8	0.3	# 310.3	181.1	21.4
	2012	0.8	65.7	34.3	# 37.2	0.3	# 354.4	207.1	19.3
	2014	0.7	...	...	# 36.3	0.2	# 335.6	153.2	20.5
Sri Lanka	2004	# 7.1	...	...	109.3	4.5	135.8	41.0	...
Sri Lanka	#2005	6.3	...	...	...	...	...	...	...
	2010	# 3.8	...	...	319.9	1.0	62.8	31.9	4.4
	2013	# 2.9	...	...	316.4	0.9	55.8	23.1	4.8
State of Palestine	2005	4.1	...	...	# 183.6	# 5.6	# 122.3	# 5.7	...
État de Palestine	2010	0.7	...	...	# 7.0	# 0.5	# 67.4	# 69.9	4.3
	2012	0.6	...	...	# 8.1	# 0.3	# 45.3	# 46.4	2.8
	2013	...	...	...	...	...	...	# 84.5	3.0
Sudan	2008	...	...	...	...	1.8	220.6	8.9	...
Soudan	2012	6.5	...	...	...	...	...	...	...
Suriname	2004	...	...	...	...	...	2 778.8	...	...
Suriname	2012	9.5	...	...	...	...	...	...	...
Swaziland	2004	13.3	...	...	1 332.4	8.8	1 904.6	309.8	...
Swaziland	2005	13.8	...	...	...	...	...	...	...
	2010	17.4	...	...	...	...	...	...	...
Sweden	2005	0.9	...	...	804.5	...	4 926.0	104.1	129.7
Suède	2010	0.1	68.1	31.9	936.4	...	3 921.9	98.3	187.1
	2014	0.9	...	...	858.7	...	3 971.8	86.2	219.2
Switzerland	2005	1.0	...	...	# 109.3	# 3.3	# 1 922.4	# 53.7	...
Suisse	2010	0.7	50.0	50.0	# 6.2	# 3.9	# 1 527.2	# 56.3	# 80.7
	2011	0.6	50.0	50.0	# 6.1	# 3.9	# 1 840.5	# 66.1	# 77.2
	2014	0.5	...	...	# 7.4	# 4.3	# 1 478.0	# 48.8	# 79.0

Intentional homicides and other crimes *(continued)*
By type of crime per 100 000 population and homicide victims by sex

Homicides intentionnels et autres crimes *(suite)*
Par type de crime pour 100 000 habitants et victimes d'homicides par sexe

Country or area Pays ou zone	Year Année	Intentional homicides Homicides volontaires Victims (percentage) Victimes (pourcentage) Per 100,000 Pour 100 000	Male Hommes	Female Femmes	Other crimes (per 100 000 pop.) Autres infractions (pour 100 000 hab.) Assault Agression	Kidnapping Enlèvement	Theft Vol simple	Robbery Vol qualifié	Sexual Violence Violences Sexuelles
Syrian Arab Republic	2004	2.4	...	...	36.0	0.8	65.3	3.1	...
Rép. arabe syrienne	2005	2.4	...	...	34.0	0.2	...	4.2	...
	2008	2.6	...	...	2.8	0.1	32.4	4.3	0.4
	2010	2.2	...	...	...	...	...	...	...
Tajikistan	2005	...	...	...	...	...	52.3	...	...
Tadjikistan	2010	2.0	87.3	12.7	72.4	1.9	46.9	3.2	2.6
	2011	1.9	86.5	13.5	48.2	2.2	47.3	3.7	2.6
	2013	1.4	...	...	...	...	...	...	...
Thailand	2005	7.3	...	...	# 42.9	0.0	92.4	...	...
Thaïlande	2010	5.5	...	...	# 13.7	0.0	81.5	# 2.8	2.9
	2014	3.9	...	...	# 18.7	0.0	58.9	# 2.1	1.8
TFYR of Macedonia	2005	2.2	...	...	...	1.2	...	34.7	...
ex-R.Y. de Macédoine	2010	2.1	65.1	34.9	12.0	0.8	237.4	29.3	6.0
	2011	1.5	86.7	13.3	11.6	0.7	245.2	24.8	6.1
	2014	1.6	...	...	10.4	0.4	240.1	14.7	6.5
Timor-Leste	2005	4.6	...	...	...	...	...	...	...
Timor-Leste	2010	3.7	...	...	...	...	...	...	...
Togo									
Togo	2012	9.2	...	...	...	...	...	...	...
Tonga	2005	4.0	100.0	0.0	...	...	...	...	...
Tonga	2010	0.1	100.0	0.0	...	...	...	...	...
	2012	0.1	0.0	100.0	...	...	...	...	...
Trinidad and Tobago	2005	29.8	89.9	10.1	61.8	4.5	212.2	375.3	61.9
Trinité-et-Tobago	2010	35.6	88.7	11.3	46.4	0.5	307.9	385.7	57.8
	2011	26.4	91.7	8.3	40.1	0.2	233.6	278.5	53.1
	2014	25.9	...	...	41.2	0.2	174.5	197.3	67.1
Tunisia	2005	2.6	...	...	...	...	...	...	...
Tunisie	2010	2.7	...	...	...	...	...	...	...
	2012	3.1	...	...	...	...	...	...	...
Turkey	2005	4.9	81.6	18.4	155.8	12.9	152.9	21.2	4.5
Turquie	2010	4.2	77.7	22.3	305.5	18.8	248.4	11.9	5.8
	2011	4.2	79.5	20.5	324.3	18.5	248.1	13.1	6.4
	2012	4.3	...	...	350.5	17.9	279.1	13.8	7.4
Turkmenistan	2004	...	...	...	...	2.7	...	3.8	...
Turkménistan	2005	...	...	...	1.9	...	35.0	3.4	...
	2006	...	...	...	1.7	...	29.8	2.9	...
	2012	4.3	...	...	...	...	...	...	...
Turks and Caicos Islands	2005	0.0	...	...	...	...	...	...	...
Îles Turques-et-Caïques	2009	6.6	...	...	...	...	...	...	...
Tuvalu	2005	0.0	...	...	...	...	...	...	...
Tuvalu	2010	10.2	...	...	...	...	...	...	...
	2012	20.3	...	...	...	...	...	...	...
Uganda	2005	8.9	...	...	125.1	...	196.3	26.1	50.3
Ouganda	2008	9.0	86.2	13.8	68.3	# 0.2	106.3	13.7	33.4
	2010	9.5	...	...	69.3	# 0.5	70.7	18.2	26.1
	2014	11.8	...	...	38.4	# 0.6	45.1	9.6	36.2
Ukraine	2005	6.5	72.8	27.2	14.3	0.4	398.9	100.5	4.1
Ukraine	2010	4.4	68.6	31.4	8.8	0.6	558.1	51.0	3.0
United Arab Emirates	2004	...	...	...	...	12.7	# 277.6	11.0	...
Émirats arabes unis	2005	...	...	...	16.0	19.2	...	9.6	...
	2006	...	...	...	14.7	18.8	...	10.8	...
	2010	0.8	...	...	...	...	...	...	...
	2014	0.7	...	...	3.5	0.8	# 23.1	7.9	4.2

Intentional homicides and other crimes *(continued)*
By type of crime per 100 000 population and homicide victims by sex

Homicides intentionnels et autres crimes *(suite)*
Par type de crime pour 100 000 habitants et victimes d'homicides par sexe

Country or area Pays ou zone	Year Année	Intentional homicides Homicides volontaires Per 100,000 Pour 100 000	Victims (percentage) Victimes (pourcentage) Male Hommes	Female Femmes	Other crimes (per 100 000 pop.) Autres infractions (pour 100 000 hab.) Assault Agression	Kidnapping Enlèvement	Theft Vol simple	Robbery Vol qualifié	Sexual Violence Violences Sexuelles
United Kingdom	2005	1.5	72.2	27.8	1 010.6[1]	5.2[1]	3 368.8[1]	183.3[1]	88.0[1]
Royaume-Uni	2010	1.2	69.6	30.4	658.8[1]	3.1[1]	2 654.9[1]	136.8[1]	81.5[1]
	2011	1.0	70.3	29.7	599.7[1]	2.7[1]	2 637.8[1]	133.0[1]	79.4[1]
	2013	0.9	...	...	564.6[1]	3.0[1]	2 329.6[1]	101.5[1]	99.3[1]
	2014[1]	...	...	...	649.1	3.8	2 209.0	87.5	136.9
United Rep. of Tanzania									
Rép.-Unie de Tanzanie	2012	7.9	...	...	...	...	...	...	...
United States	2004	5.5	78.0	22.0	288.7	...	2 363.3	136.8	...
États-Unis	2005	...	78.7	21.3	291.2	...	2 290.6	141.0	...
	2010	4.8	77.5	22.5	252.3	...	2 002.3	119.1	...
	2012	4.7	77.8	22.2	242.1	...	1 959.6	112.8	...
	2013	3.9	...	...	229.2	...	1 898.1	108.8	...
	2014	...	...	...	232.1	...	1 833.9	102.0	...
United States Virgin Is.	2005	32.5	...	...	...	...	...	...	...
Îles Vierges américaines	2010	52.6	...	...	...	...	...	...	...
Uruguay	2003	5.9	...	...	...	0.1	2 904.9	291.7	...
Uruguay	2004	5.8	...	...	...	...	3 179.5	277.5	...
	2005	5.7	...	...	...	...	...	291.7	...
	2010	6.1	...	...	15.4	...	2 823.3	409.8	...
	2013	7.6	...	...	12.5	0.6	2 859.9	490.6	40.3
	2014	7.8	...	...	11.9	...	3 096.1	542.8	44.9
Uzbekistan									
Ouzbékistan	2012	3.2	...	...	...	...	...	...	...
Vanuatu									
Vanuatu	2012	2.9	...	...	...	...	...	...	...
Venezuela (Boliv. Rep. of)	2005	37.2	...	...	...	...	...	...	...
Venezuela (Rép. boliv. du)	2010	45.1	...	...	...	...	...	...	...
	2014	62.0	...	...	...	...	...	...	...
Viet Nam	2005	1.3	...	...	...	...	...	...	...
Viet Nam	2010	1.5	...	...	...	...	...	...	...
	2011	1.5	...	...	...	...	...	...	...
Yemen	2005	4.6	...	...	...	0.5	...	...	0.0
Yémen	2009	4.3	...	...	0.1	0.2	...	1.9	0.3
	2010	4.7	...	...	...	...	...	...	...
	2013	6.7	...	...	...	...	...	...	...
Zambia									
Zambie	2010	5.8	77.8	22.2	...	...	...	...	...
Zimbabwe	2005	10.4	...	...	712.5	0.3	851.1	85.8	...
Zimbabwe	2008	...	...	...	379.7	1.6	677.7	65.2	...
	2010	5.1	...	...	...	...	...	...	...
	2012	6.7	...	...	...	...	...	...	...

Source:

United Nations Office on Drugs and Crime, Vienna, UNODC Statistics database, last accessed May 2016.

1   England and Wales only.

Source:

Office des Nations Unies contre la drogue et le crime, Vienna, la base de données de l'UNODC, dernier accès mai 2016.

1   Se rapporte seulement à l'Angleterre et au pays de Galles.

Part Two

Economic activity

Deuxième partie

Activité économique

Gross domestic product and gross domestic product per capita
Current and constant 2005 prices (millions of US dollars); per capita (US dollars); real annual rates of growth

Produit intérieur brut et produit intérieur brut par habitant
Prix courants et constants de 2005 (millions de dollars É.-U.); par habitant (dollars É.U.); taux de croissance réels

Region, country or area	1985	1995	2005	2010	2012	2013	2014	Région, pays ou zone
Total, countries or areas								**Total, tous pays ou zones**
GDP at current prices	13 502 356	30 860 249	47 264 846	65 644 956	74 221 881	76 176 342	78 037 088	PIB aux prix courants
GDP per capita	2 784	5 384	7 251	9 475	10 460	10 609	10 743	PIB par habitant
GDP at constant prices	26 013 322	34 592 400	47 264 846	52 895 088	55 617 816	56 813 235	58 254 247	PIB aux prix constants
Growth rates (annual)	3.7	2.9	3.6	4.1	2.2	2.1	2.5	Taux de croissance
Africa								**Afrique**
GDP at current prices	508 188	571 199	1 113 310	1 935 156	2 319 245	2 385 974	2 474 987	PIB aux prix courants
GDP per capita	925	794	1 212	1 856	2 113	2 119	2 143	PIB par habitant
GDP at constant prices	590 960	717 677	1 113 310	1 426 742	1 522 542	1 553 809	1 610 862	PIB aux prix constants
Growth rates (annual)	3.5	2.9	5.9	5.2	5.6	2.1	3.7	Taux de croissance
Northern America								**Amérique septentrionale**
GDP at current prices	4 711 499	8 269 809	14 264 417	16 586 595	17 995 911	18 510 183	19 141 500	PIB aux prix courants
GDP per capita	17 669	27 968	43 420	48 200	51 448	52 513	53 896	PIB par habitant
GDP at constant prices	7 660 653	10 207 106	14 264 417	14 846 071	15 432 040	15 668 365	16 048 712	PIB aux prix constants
Growth rates (annual)	4.3	2.8	3.3	2.6	2.2	1.5	2.4	Taux de croissance
Latin Amer. and the Carib.								**Amérique lat. et Caraïbes**
GDP at current prices	788 307	1 950 474	2 868 084	5 365 131	6 117 269	6 248 000	6 310 627	PIB aux prix courants
GDP per capita	1 946	4 011	5 097	8 963	9 985	10 085	10 075	PIB par habitant
GDP at constant prices	1 674 445	2 143 415	2 868 084	3 426 271	3 791 744	3 998 904	4 049 455	PIB aux prix constants
Growth rates (annual)	3.1	0.7	4.4	5.7	5.8	5.5	1.3	Taux de croissance
Asia								**Asie**
GDP at current prices	2 865 104	9 077 360	12 126 408	20 511 770	25 206 888	25 617 634	26 501 407	PIB aux prix courants
GDP per capita	1 011	2 612	3 074	4 919	5 917	5 951	6 093	PIB par habitant
GDP at constant prices	5 208 544	8 401 298	12 126 408	15 316 595	16 719 588	17 454 490	18 145 997	PIB aux prix constants
Growth rates (annual)	5.3	4.5	5.2	7.4	4.2	4.4	4.0	Taux de croissance
Europe								**Europe**
GDP at current prices	4 416 453	10 519 406	15 993 441	19 776 262	20 792 263	21 656 732	21 896 876	PIB aux prix courants
GDP per capita	5 735	14 471	21 887	26 826	28 141	29 289	29 594	PIB par habitant
GDP at constant prices	10 408 825	12 490 284	15 993 441	16 862 706	17 172 094	17 239 006	17 469 743	PIB aux prix constants
Growth rates (annual)	2.5	2.2	2.4	2.2	-0.1	0.4	1.3	Taux de croissance
Oceania								**Océanie**
GDP at current prices	212 805	472 002	899 186	1 470 043	1 790 305	1 757 819	1 711 690	PIB aux prix courants
GDP per capita	8 626	16 400	27 190	40 693	47 975	46 387	44 500	PIB par habitant
GDP at constant prices	469 894	632 621	899 186	1 016 704	1 078 850	1 106 213	1 137 334	PIB aux prix constants
Growth rates (annual)	3.7	4.0	3.0	2.3	2.5	2.5	2.8	Taux de croissance
Afghanistan								**Afghanistan**
GDP at current prices	3 322	3 236	6 622	16 078	21 331	21 610	21 122	PIB aux prix courants
GDP per capita	286	193	271	575	718	704	668	PIB par habitant
GDP at constant prices	7 501	4 532	6 622	10 393	12 529	13 341	13 627	PIB aux prix constants
Growth rates (annual)	0.3	49.9	9.9	3.2	10.9	6.5	2.1	Taux de croissance
Albania								**Albanie**
GDP at current prices	2 389	2 459	8 094	11 927	12 345	12 916	13 413	PIB aux prix courants
GDP per capita	805	792	2 626	4 110	4 285	4 480	4 642	PIB par habitant
GDP at constant prices	5 147	4 744	8 094	10 417	10 856	11 007	11 238	PIB aux prix constants
Growth rates (annual)	1.8	13.3	5.8	3.7	1.6	1.4	2.1	Taux de croissance
Algeria								**Algérie**
GDP at current prices	57 866	41 971	103 198	161 207	209 047	209 704	213 518	PIB aux prix courants
GDP per capita	2 564	1 452	3 102	4 473	5 584	5 492	5 484	PIB par habitant
GDP at constant prices	64 280	67 626	103 198	116 968	124 452	127 937	132 799	PIB aux prix constants
Growth rates (annual)	3.7	3.8	5.9	3.6	3.4	2.8	3.8	Taux de croissance
Andorra								**Andorre**
GDP at current prices	429	1 457	3 248	3 346	3 146	3 249	3 278	PIB aux prix courants
GDP per capita	9 614	22 825	39 990	39 639	39 666	42 807	45 033	PIB par habitant
GDP at constant prices	1 321	1 774	3 248	2 674	2 541	2 499	2 533	PIB aux prix constants
Growth rates (annual)	2.3	2.8	7.5	-7.4	-0.7	-1.6	1.4	Taux de croissance
Angola								**Angola**
GDP at current prices	9 125	6 642	36 971	83 369	125 430	142 738	146 676	PIB aux prix courants
GDP per capita	936	509	2 064	3 929	5 529	6 087	6 054	PIB par habitant
GDP at constant prices	18 016	17 679	36 971	55 286	60 612	63 165	66 194	PIB aux prix constants
Growth rates (annual)	3.5	15.0	15.0	3.6	7.6	4.2	4.8	Taux de croissance

10

Gross domestic product and gross domestic product per capita *(continued)*
Current and constant 2005 prices (millions of US dollars); per capita (US dollars); real annual rates of growth
Produit intérieur brut et produit intérieur brut par habitant *(suite)*
Prix courants et constants de 2005 (millions de dollars É.-U.); par habitant (dollars É.U.); taux de croissance réels

Region, country or area	1985	1995	2005	2010	2012	2013	2014	Région, pays ou zone
Anguilla								**Anguilla**
GDP at current prices	27	104	229	268	285	286	311	PIB aux prix courants
GDP per capita	4 040	10 578	18 130	19 469	20 195	20 000	21 493	PIB par habitant
GDP at constant prices	55	130	229	240	237	235	250	PIB aux prix constants
Growth rates (annual)	12.9	-2.3	13.1	-4.4	-6.3	-0.7	6.2	Taux de croissance
Antigua and Barbuda								**Antigua-et-Barbuda**
GDP at current prices	245	577	997	1 136	1 205	1 185	1 248	PIB aux prix courants
GDP per capita	3 723	8 446	12 080	13 017	13 526	13 169	13 731	PIB par habitant
GDP at constant prices	449	685	997	1 012	1 034	1 032	1 076	PIB aux prix constants
Growth rates (annual)	7.7	-4.4	6.1	-7.2	4.0	-0.1	4.3	Taux de croissance
Argentina								**Argentine**
GDP at current prices	105 516	308 941	222 911	464 757	609 569	623 932	543 490	PIB aux prix courants
GDP per capita	3 472	8 828	5 694	11 274	14 481	14 668	12 645	PIB par habitant
GDP at constant prices	135 114	177 795	222 911	294 401	321 649	330 930	332 482	PIB aux prix constants
Growth rates (annual)	-7.0	-2.8	9.2	9.5	0.8	2.9	0.5	Taux de croissance
Armenia								**Arménie**
GDP at current prices	...	1 287	4 900	9 260	9 958	10 439	10 889	PIB aux prix courants
GDP per capita	...	399	1 625	3 125	3 343	3 489	3 622	PIB par habitant
GDP at constant prices	...	2 110	4 900	5 922	6 641	6 872	7 108	PIB aux prix constants
Growth rates (annual)	...	6.9	13.9	2.2	7.2	3.5	3.4	Taux de croissance
Aruba								**Aruba**
GDP at current prices	385	1 321	2 331	2 391	2 533	2 586	2 664	PIB aux prix courants
GDP per capita	6 107	16 441	23 303	23 529	24 738	25 126	25 751	PIB par habitant
GDP at constant prices	760	1 811	2 331	2 062	2 106	2 204	2 230	PIB aux prix constants
Growth rates (annual)	9.1	2.5	1.2	-3.4	-1.4	4.7	1.2	Taux de croissance
Australia								**Australie**
GDP at current prices	181 741	392 176	762 114	1 291 430	1 574 801	1 528 761	1 471 439	PIB aux prix courants
GDP per capita	11 509	21 638	37 590	58 270	68 734	65 695	62 290	PIB par habitant
GDP at constant prices	390 490	532 753	762 114	870 367	925 491	949 051	975 012	PIB aux prix constants
Growth rates (annual)	4.1	3.9	3.0	2.3	2.5	2.5	2.7	Taux de croissance
Austria								**Autriche**
GDP at current prices	69 221	240 474	314 641	390 212	407 373	428 699	436 888	PIB aux prix courants
GDP per capita	9 107	30 162	38 208	46 498	48 179	50 513	51 296	PIB par habitant
GDP at constant prices	192 112	248 960	314 641	335 524	347 558	348 674	349 907	PIB aux prix constants
Growth rates (annual)	2.5	2.7	2.1	1.9	0.8	0.3	0.4	Taux de croissance
Azerbaijan								**Azerbaïdjan**
GDP at current prices	...	3 081	13 245	52 906	69 680	74 161	75 193	PIB aux prix courants
GDP per capita	...	396	1 547	5 814	7 443	7 808	7 808	PIB par habitant
GDP at constant prices	...	5 085	13 245	28 277	28 405	30 071	30 858	PIB aux prix constants
Growth rates (annual)	...	-11.8	28.0	4.6	2.1	5.9	2.6	Taux de croissance
Bahamas								**Bahamas**
GDP at current prices	2 256	4 009	7 706	7 910	8 234	8 432	8 510	PIB aux prix courants
GDP per capita	9 614	14 309	23 406	21 921	22 113	22 316	22 217	PIB par habitant
GDP at constant prices	4 942	5 540	7 706	7 617	7 833	7 835	7 915	PIB aux prix constants
Growth rates (annual)	4.8	4.4	3.4	1.5	2.2	<0.0	1.0	Taux de croissance
Bahrain								**Bahreïn**
GDP at current prices	4 475	6 787	15 969	25 713	30 756	32 898	33 850	PIB aux prix courants
GDP per capita	10 669	12 039	18 418	20 386	23 063	24 379	24 854	PIB par habitant
GDP at constant prices	6 462	9 882	15 969	20 928	22 134	23 331	24 378	PIB aux prix constants
Growth rates (annual)	-16.1	1.9	6.8	4.3	3.6	5.4	4.5	Taux de croissance
Bangladesh								**Bangladesh**
GDP at current prices	19 169	37 866	57 628	114 508	128 899	153 505	173 062	PIB aux prix courants
GDP per capita	206	320	403	755	830	977	1 088	PIB par habitant
GDP at constant prices	22 804	34 316	57 628	77 343	87 712	92 987	98 623	PIB aux prix constants
Growth rates (annual)	3.0	4.9	6.0	5.6	6.5	6.0	6.1	Taux de croissance
Barbados								**Barbade**
GDP at current prices	1 425	2 275	3 897	4 447	4 332	4 371	4 353	PIB aux prix courants
GDP per capita	5 569	8 585	14 223	15 906	15 385	15 473	15 360	PIB par habitant
GDP at constant prices	2 975	3 185	3 897	4 054	4 099	4 095	4 103	PIB aux prix constants
Growth rates (annual)	1.1	2.0	4.0	0.3	0.3	-0.1	0.2	Taux de croissance

10

Gross domestic product and gross domestic product per capita *(continued)*
Current and constant 2005 prices (millions of US dollars); per capita (US dollars); real annual rates of growth

Produit intérieur brut et produit intérieur brut par habitant *(suite)*
Prix courants et constants de 2005 (millions de dollars É.-U.); par habitant (dollars É.U.); taux de croissance réels

Region, country or area	1985	1995	2005	2010	2012	2013	2014	Région, pays ou zone
Belarus								**Bélarus**
GDP at current prices	...	13 856	30 210	55 221	63 615	73 098	76 139	PIB aux prix courants
GDP per capita	...	1 364	3 134	5 818	6 703	7 697	8 014	PIB par habitant
GDP at constant prices	...	15 516	30 210	42 921	46 091	46 556	47 301	PIB aux prix constants
Growth rates (annual)	...	-10.4	9.4	7.7	1.7	1.0	1.6	Taux de croissance
Belgium								**Belgique**
GDP at current prices	86 728	289 571	387 356	483 549	497 780	521 402	531 547	PIB aux prix courants
GDP per capita	8 766	28 496	36 676	44 241	44 928	46 749	47 348	PIB par habitant
GDP at constant prices	243 699	307 036	387 356	415 035	423 132	423 202	428 908	PIB aux prix constants
Growth rates (annual)	1.7	2.4	2.1	2.7	0.2	<0.0	1.3	Taux de croissance
Belize								**Belize**
GDP at current prices	209	587	1 114	1 397	1 574	1 624	1 699	PIB aux prix courants
GDP per capita	1 269	2 834	3 933	4 344	4 674	4 719	4 831	PIB par habitant
GDP at constant prices	323	639	1 114	1 261	1 337	1 357	1 405	PIB aux prix constants
Growth rates (annual)	1.0	4.2	3.0	3.3	3.8	1.5	3.6	Taux de croissance
Benin								**Bénin**
GDP at current prices	1 130	2 345	4 804	6 970	8 117	9 111	9 575	PIB aux prix courants
GDP per capita	263	392	587	733	808	883	903	PIB par habitant
GDP at constant prices	2 167	3 101	4 804	5 800	6 249	6 679	7 116	PIB aux prix constants
Growth rates (annual)	7.5	10.1	1.7	2.1	4.6	6.9	6.5	Taux de croissance
Bermuda								**Bermudes**
GDP at current prices	1 428	2 557	4 868	5 855	5 584	5 640	5 601	PIB aux prix courants
GDP per capita	24 239	40 782	74 752	91 555	88 389	89 843	89 795	PIB par habitant
GDP at constant prices	3 107	3 563	4 868	4 883	4 483	4 375	4 356	PIB aux prix constants
Growth rates (annual)	4.2	5.4	1.7	-2.5	-3.9	-2.4	-0.4	Taux de croissance
Bhutan								**Bhoutan**
GDP at current prices	172	289	819	1 585	1 824	1 781	1 965	PIB aux prix courants
GDP per capita	367	568	1 258	2 201	2 452	2 360	2 569	PIB par habitant
GDP at constant prices	174	399	819	1 288	1 460	1 490	1 586	PIB aux prix constants
Growth rates (annual)	5.7	6.8	7.1	11.7	5.1	2.0	6.4	Taux de croissance
Bolivia (Plurin. State of)								**Bolivie (État plurin. de)**
GDP at current prices	4 122	6 715	9 549	19 650	27 084	30 659	32 996	PIB aux prix courants
GDP per capita	664	887	1 046	1 981	2 645	2 948	3 124	PIB par habitant
GDP at constant prices	5 064	6 925	9 549	11 954	13 220	14 119	14 890	PIB aux prix constants
Growth rates (annual)	-1.0	4.7	4.4	4.1	5.1	6.8	5.5	Taux de croissance
Bosnia and Herzegovina								**Bosnie-Herzégovine**
GDP at current prices	...	2 034	11 225	17 164	17 207	18 155	18 491	PIB aux prix courants
GDP per capita	...	524	2 928	4 475	4 495	4 748	4 844	PIB par habitant
GDP at constant prices	...	3 196	11 225	12 950	12 946	13 256	13 421	PIB aux prix constants
Growth rates (annual)	...	20.8	3.9	0.8	-0.9	2.4	1.2	Taux de croissance
Botswana								**Botswana**
GDP at current prices	838	4 731	9 931	12 787	14 792	14 979	15 813	PIB aux prix courants
GDP per capita	708	3 001	5 328	6 244	6 936	6 882	7 123	PIB par habitant
GDP at constant prices	2 773	6 454	9 931	12 412	13 799	15 085	15 752	PIB aux prix constants
Growth rates (annual)	9.2	7.0	4.6	8.6	4.8	9.3	4.4	Taux de croissance
Brazil								**Brésil**
GDP at current prices	187 460	778 053	892 107	2 209 400	2 413 175	2 392 094	2 346 523	PIB aux prix courants
GDP per capita	1 370	4 781	4 733	11 124	11 923	11 711	11 387	PIB par habitant
GDP at constant prices	552 303	704 382	892 107	1 108 460	1 172 175	1 204 318	1 206 081	PIB aux prix constants
Growth rates (annual)	7.8	4.2	3.1	7.6	1.8	2.7	0.1	Taux de croissance
British Virgin Islands								**Îles Vierges britanniques**
GDP at current prices	90	397	870	894	909	916	902	PIB aux prix courants
GDP per capita	6 754	21 547	37 550	32 840	31 895	31 509	30 501	PIB par habitant
GDP at constant prices	185	530	870	929	872	869	867	PIB aux prix constants
Growth rates (annual)	1.4	24.2	14.3	1.3	-4.6	-0.3	-0.3	Taux de croissance
Brunei Darussalam								**Brunéi Darussalam**
GDP at current prices	4 425	5 245	10 561	13 707	19 048	18 094	17 104	PIB aux prix courants
GDP per capita	19 841	17 780	29 184	34 851	46 974	43 971	40 979	PIB par habitant
GDP at constant prices	8 356	8 922	10 561	10 914	11 426	11 183	10 921	PIB aux prix constants
Growth rates (annual)	-1.5	4.5	0.4	2.6	0.9	-2.1	-2.3	Taux de croissance

10
Gross domestic product and gross domestic product per capita *(continued)*
Current and constant 2005 prices (millions of US dollars); per capita (US dollars); real annual rates of growth
Produit intérieur brut et produit intérieur brut par habitant *(suite)*
Prix courants et constants de 2005 (millions de dollars É.-U.); par habitant (dollars É.U.); taux de croissance réels

Region, country or area	1985	1995	2005	2010	2012	2013	2014	Région, pays ou zone
Bulgaria								**Bulgarie**
GDP at current prices	17 400	14 434	29 821	49 939	53 575	55 628	56 718	PIB aux prix courants
GDP per capita	1 942	1 727	3 882	6 742	7 335	7 670	7 876	PIB par habitant
GDP at constant prices	26 007	24 556	29 821	34 705	35 338	35 791	36 345	PIB aux prix constants
Growth rates (annual)	2.7	2.9	7.2	0.1	0.2	1.3	1.5	Taux de croissance
Burkina Faso								**Burkina Faso**
GDP at current prices	1 569	2 404	5 463	8 980	11 166	12 126	12 756	PIB aux prix courants
GDP per capita	203	238	407	574	673	710	725	PIB par habitant
GDP at constant prices	2 088	2 926	5 463	7 139	8 103	8 604	9 034	PIB aux prix constants
Growth rates (annual)	8.5	5.7	8.7	8.4	6.5	6.2	5.0	Taux de croissance
Burundi								**Burundi**
GDP at current prices	1 168	1 000	1 117	2 032	2 327	2 520	2 869	PIB aux prix courants
GDP per capita	245	160	141	215	230	241	265	PIB par habitant
GDP at constant prices	1 018	1 089	1 117	2 122	2 722	2 845	2 978	PIB aux prix constants
Growth rates (annual)	11.7	-7.0	-0.9	15.7	13.2	4.5	4.7	Taux de croissance
Cabo Verde								**Cabo Verde**
GDP at current prices	157	558	1 105	1 664	1 752	1 838	1 855	PIB aux prix courants
GDP per capita	499	1 434	2 331	3 394	3 498	3 623	3 609	PIB par habitant
GDP at constant prices	365	569	1 105	1 413	1 485	1 501	1 528	PIB aux prix constants
Growth rates (annual)	8.6	7.5	6.5	1.5	1.1	1.0	1.8	Taux de croissance
Cambodia								**Cambodge**
GDP at current prices	1 059	3 309	6 293	11 242	14 038	15 450	16 778	PIB aux prix courants
GDP per capita	137	309	472	783	946	1 025	1 095	PIB par habitant
GDP at constant prices	1 410	2 851	6 293	8 693	9 984	10 730	11 489	PIB aux prix constants
Growth rates (annual)	4.7	5.9	13.2	6.0	7.3	7.5	7.1	Taux de croissance
Cameroon								**Cameroun**
GDP at current prices	8 436	8 913	16 588	23 622	26 472	29 568	32 051	PIB aux prix courants
GDP per capita	813	640	915	1 147	1 222	1 331	1 407	PIB par habitant
GDP at constant prices	14 129	10 993	16 588	19 147	20 855	22 015	23 320	PIB aux prix constants
Growth rates (annual)	2.4	4.1	2.3	3.3	4.6	5.6	5.9	Taux de croissance
Canada								**Canada**
GDP at current prices	362 965	602 003	1 164 179	1 614 072	1 832 716	1 838 964	1 785 387	PIB aux prix courants
GDP per capita	14 042	20 547	36 092	47 297	52 561	52 198	50 169	PIB par habitant
GDP at constant prices	678 512	842 770	1 164 179	1 240 064	1 301 322	1 327 394	1 359 773	PIB aux prix constants
Growth rates (annual)	4.7	2.7	3.2	3.4	1.9	2.0	2.4	Taux de croissance
Cayman Islands								**Îles Caïmanes**
GDP at current prices	416	1 296	3 042	3 267	3 417	3 482	3 480	PIB aux prix courants
GDP per capita	21 542	40 930	62 558	58 857	59 402	59 657	58 808	PIB par habitant
GDP at constant prices	899	1 801	3 042	2 953	3 024	3 061	3 023	PIB aux prix constants
Growth rates (annual)	3.5	4.8	6.5	-2.7	1.2	1.2	-1.3	Taux de croissance
Central African Rep.								**Rép. centrafricaine**
GDP at current prices	905	1 167	1 413	2 034	2 237	1 592	1 838	PIB aux prix courants
GDP per capita	344	350	348	458	484	338	383	PIB par habitant
GDP at constant prices	1 169	1 304	1 413	1 667	1 751	1 120	1 131	PIB aux prix constants
Growth rates (annual)	3.3	5.2	2.4	3.6	2.9	-36.0	1.0	Taux de croissance
Chad								**Tchad**
GDP at current prices	987	1 643	6 681	9 791	11 363	11 895	12 791	PIB aux prix courants
GDP per capita	194	235	664	823	894	905	941	PIB par habitant
GDP at constant prices	2 110	2 932	6 681	9 700	10 424	11 195	12 359	PIB aux prix constants
Growth rates (annual)	7.9	1.4	7.9	13.4	10.1	7.4	10.4	Taux de croissance
Chile								**Chili**
GDP at current prices	18 747	74 160	123 056	217 538	265 232	276 674	258 062	PIB aux prix courants
GDP per capita	1 548	5 225	7 645	12 785	15 253	15 742	14 528	PIB par habitant
GDP at constant prices	38 321	80 487	123 056	147 859	165 034	172 008	175 266	PIB aux prix constants
Growth rates (annual)	2.0	10.6	6.2	5.8	5.5	4.2	1.9	Taux de croissance
China [1]								**Chine** [1]
GDP at current prices	310 608	734 353	2 291 432	6 005 388	8 471 426	9 518 402	10 430 590	PIB aux prix courants
GDP per capita	295	598	1 755	4 478	6 250	6 986	7 617	PIB par habitant
GDP at constant prices	364 135	953 450	2 291 432	3 903 780	4 603 786	4 958 278	5 320 232	PIB aux prix constants
Growth rates (annual)	13.5	11.0	11.3	10.6	7.7	7.7	7.3	Taux de croissance

10

Gross domestic product and gross domestic product per capita *(continued)*
Current and constant 2005 prices (millions of US dollars); per capita (US dollars); real annual rates of growth
Produit intérieur brut et produit intérieur brut par habitant *(suite)*
Prix courants et constants de 2005 (millions de dollars É.-U.); par habitant (dollars É.U.); taux de croissance réels

Region, country or area	1985	1995	2005	2010	2012	2013	2014	Région, pays ou zone
China, Hong Kong SAR								**Chine, Hong Kong RAS**
GDP at current prices	35 700	144 652	181 569	228 639	262 629	275 743	290 896	PIB aux prix courants
GDP per capita	6 593	23 542	26 536	32 693	36 980	38 490	40 252	PIB par habitant
GDP at constant prices	69 050	129 722	181 569	220 057	234 574	241 780	247 828	PIB aux prix constants
Growth rates (annual)	0.8	2.4	7.4	6.8	1.7	3.1	2.5	Taux de croissance
China, Macao SAR								**Chine, Macao RAS**
GDP at current prices	1 297	6 955	11 793	28 360	42 992	51 314	55 502	PIB aux prix courants
GDP per capita	4 379	17 455	25 190	53 046	77 079	90 332	96 038	PIB par habitant
GDP at constant prices	3 497	6 936	11 793	20 684	27 388	30 307	30 200	PIB aux prix constants
Growth rates (annual)	0.7	3.3	8.6	27.5	9.2	10.7	-0.4	Taux de croissance
Colombia								**Colombie**
GDP at current prices	49 322	110 292	146 566	287 018	369 660	380 063	377 740	PIB aux prix courants
GDP per capita	1 590	2 946	3 386	6 251	7 885	8 028	7 904	PIB par habitant
GDP at constant prices	75 485	117 224	146 566	182 951	202 893	212 908	222 601	PIB aux prix constants
Growth rates (annual)	3.1	5.2	4.7	4.0	4.0	4.9	4.6	Taux de croissance
Comoros								**Comores**
GDP at current prices	114	232	386	530	571	619	648	PIB aux prix courants
GDP per capita	320	484	624	759	778	823	841	PIB par habitant
GDP at constant prices	284	319	386	452	477	494	509	PIB aux prix constants
Growth rates (annual)	2.3	3.6	2.8	2.2	3.0	3.5	3.0	Taux de croissance
Congo								**Congo**
GDP at current prices	2 161	2 116	6 087	12 281	13 656	14 022	14 077	PIB aux prix courants
GDP per capita	1 037	778	1 738	3 020	3 186	3 191	3 125	PIB par habitant
GDP at constant prices	4 437	4 428	6 087	7 878	8 457	8 734	9 332	PIB aux prix constants
Growth rates (annual)	-1.2	4.0	7.6	8.7	3.8	3.3	6.8	Taux de croissance
Cook Islands								**Îles Cook**
GDP at current prices	29	106	183	255	307	295	311	PIB aux prix courants
GDP per capita	1 614	5 786	9 411	12 579	14 981	14 317	15 003	PIB par habitant
GDP at constant prices	99	143	183	181	191	182	193	PIB aux prix constants
Growth rates (annual)	8.8	-4.4	-1.1	-3.0	4.4	-4.8	6.2	Taux de croissance
Costa Rica								**Costa Rica**
GDP at current prices	4 984	11 716	19 965	36 298	45 301	49 237	49 553	PIB aux prix courants
GDP per capita	1 826	3 337	4 700	7 986	9 733	10 462	10 415	PIB par habitant
GDP at constant prices	7 868	12 846	19 965	25 018	27 499	28 445	29 441	PIB aux prix constants
Growth rates (annual)	0.7	3.9	5.9	5.0	5.2	3.4	3.5	Taux de croissance
Côte d'Ivoire								**Côte d'Ivoire**
GDP at current prices	6 978	11 105	17 085	24 884	27 087	31 325	34 254	PIB aux prix courants
GDP per capita	687	771	942	1 236	1 284	1 449	1 546	PIB par habitant
GDP at constant prices	11 660	14 519	17 085	19 604	20 750	22 663	24 597	PIB aux prix constants
Growth rates (annual)	4.5	7.1	1.7	2.0	10.7	9.2	8.5	Taux de croissance
Croatia								**Croatie**
GDP at current prices	...	22 388	45 416	59 665	56 485	57 770	57 137	PIB aux prix courants
GDP per capita	...	4 849	10 374	13 823	13 176	13 525	13 425	PIB par habitant
GDP at constant prices	...	30 834	45 416	46 492	45 347	44 865	44 703	PIB aux prix constants
Growth rates (annual)	...	6.8	4.2	-1.7	-2.2	-1.1	-0.4	Taux de croissance
Cuba								**Cuba**
GDP at current prices	22 921	30 429	42 644	64 328	73 139	77 150	82 775	PIB aux prix courants
GDP per capita	2 273	2 790	3 787	5 689	6 448	6 790	7 274	PIB par habitant
GDP at constant prices	38 933	26 733	42 644	55 437	58 709	60 323	61 130	PIB aux prix constants
Growth rates (annual)	1.6	2.5	11.2	2.4	3.0	2.7	1.3	Taux de croissance
Curaçao								**Curaçao**
GDP at current prices	...	...	2 345	2 951	3 131	3 148	3 159	PIB aux prix courants
GDP per capita	...	...	18 120	19 995	20 516	20 403	20 283	PIB par habitant
GDP at constant prices	...	...	2 345	2 484	2 496	2 477	2 448	PIB aux prix constants
Growth rates (annual)	...	...	...	0.1	-0.1	-0.8	-1.1	Taux de croissance
Cyprus [2]								**Chypre** [2]
GDP at current prices	2 679	9 814	18 850	25 320	25 015	23 985	23 077	PIB aux prix courants
GDP per capita	4 951	15 082	25 523	30 526	28 954	27 462	26 147	PIB par habitant
GDP at constant prices	7 238	13 040	18 850	21 281	20 844	19 606	19 117	PIB aux prix constants
Growth rates (annual)	4.7	9.9	3.9	1.4	-2.4	-5.9	-2.5	Taux de croissance

10

Gross domestic product and gross domestic product per capita *(continued)*
Current and constant 2005 prices (millions of US dollars); per capita (US dollars); real annual rates of growth

Produit intérieur brut et produit intérieur brut par habitant *(suite)*
Prix courants et constants de 2005 (millions de dollars É.-U.); par habitant (dollars É.U.); taux de croissance réels

Region, country or area	1985	1995	2005	2010	2012	2013	2014	Région, pays ou zone
Czech Republic								**République tchèque**
GDP at current prices	...	59 536	135 990	207 016	206 442	208 328	205 270	PIB aux prix courants
GDP per capita	...	5 760	13 292	19 703	19 577	19 756	19 470	PIB par habitant
GDP at constant prices	...	102 694	135 990	153 349	154 957	154 138	157 188	PIB aux prix constants
Growth rates (annual)	...	6.2	6.4	2.3	-0.9	-0.5	2.0	Taux de croissance
Dem. P. R. Korea								**R. p. dém. de Corée**
GDP at current prices	12 075	4 849	13 031	13 945	15 907	16 565	17 396	PIB aux prix courants
GDP per capita	722	222	548	570	643	666	696	PIB par habitant
GDP at constant prices	15 015	12 063	13 031	12 958	13 233	13 375	13 514	PIB aux prix constants
Growth rates (annual)	3.7	-4.4	3.8	-0.5	1.3	1.1	1.0	Taux de croissance
Dem. Rep. of the Congo								**Rép. dém. du Congo**
GDP at current prices	7 524	8 947	11 965	21 672	29 306	32 672	35 909	PIB aux prix courants
GDP per capita	251	212	213	329	417	450	480	PIB par habitant
GDP at constant prices	17 720	12 119	11 965	15 669	17 934	19 455	21 297	PIB aux prix constants
Growth rates (annual)	0.5	0.7	6.1	7.1	7.1	8.5	9.5	Taux de croissance
Denmark								**Danemark**
GDP at current prices	62 221	185 008	264 559	319 812	325 012	338 927	346 119	PIB aux prix courants
GDP per capita	12 168	35 357	48 832	57 614	58 028	60 261	61 294	PIB par habitant
GDP at constant prices	176 929	213 309	264 559	265 134	267 992	267 339	270 712	PIB aux prix constants
Growth rates (annual)	4.0	3.1	2.4	1.6	-0.1	-0.2	1.3	Taux de croissance
Djibouti								**Djibouti**
GDP at current prices	369	510	709	1 067	1 255	1 393	1 589	PIB aux prix courants
GDP per capita	872	771	910	1 284	1 471	1 611	1 814	PIB par habitant
GDP at constant prices	520	602	709	961	1 052	1 105	1 171	PIB aux prix constants
Growth rates (annual)	-0.2	5.6	3.2	3.5	4.8	5.0	6.0	Taux de croissance
Dominica								**Dominique**
GDP at current prices	117	260	370	493	518	523	533	PIB aux prix courants
GDP per capita	1 587	3 639	5 251	6 927	7 220	7 265	7 361	PIB par habitant
GDP at constant prices	218	309	370	443	438	440	445	PIB aux prix constants
Growth rates (annual)	1.3	2.0	-0.3	1.2	-1.1	0.4	1.1	Taux de croissance
Dominican Republic								**Rép. dominicaine**
GDP at current prices	5 906	16 555	35 662	53 043	60 441	61 198	63 969	PIB aux prix courants
GDP per capita	910	2 098	3 861	5 359	5 952	5 952	6 147	PIB par habitant
GDP at constant prices	14 432	21 480	35 662	48 272	50 939	53 373	57 291	PIB aux prix constants
Growth rates (annual)	-2.1	5.5	9.3	8.3	2.6	4.8	7.3	Taux de croissance
Ecuador								**Équateur**
GDP at current prices	18 897	22 720	41 507	69 555	87 925	94 776	100 917	PIB aux prix courants
GDP per capita	2 089	1 986	3 022	4 657	5 702	6 052	6 346	PIB par habitant
GDP at constant prices	23 377	30 581	41 507	49 036	55 878	58 423	60 570	PIB aux prix constants
Growth rates (annual)	4.3	1.7	5.3	3.5	5.6	4.6	3.7	Taux de croissance
Egypt								**Égypte**
GDP at current prices	23 831	65 758	94 456	214 630	260 153	255 199	282 242	PIB aux prix courants
GDP per capita	483	1 053	1 260	2 616	3 037	2 913	3 151	PIB par habitant
GDP at constant prices	37 449	61 042	94 456	127 482	132 810	135 600	138 534	PIB aux prix constants
Growth rates (annual)	6.8	4.6	4.6	5.1	2.2	2.1	2.2	Taux de croissance
El Salvador								**El Salvador**
GDP at current prices	1 886	9 501	17 094	21 418	23 814	24 351	25 164	PIB aux prix courants
GDP per capita	383	1 700	2 874	3 547	3 922	3 999	4 120	PIB par habitant
GDP at constant prices	8 842	13 093	17 094	18 341	19 101	19 453	19 833	PIB aux prix constants
Growth rates (annual)	2.0	6.4	3.6	1.4	1.9	1.8	2.0	Taux de croissance
Equatorial Guinea								**Guinée équatoriale**
GDP at current prices	80	166	7 206	13 392	19 213	17 084	16 731	PIB aux prix courants
GDP per capita	255	371	11 513	18 378	24 832	21 433	20 382	PIB par habitant
GDP at constant prices	319	501	7 206	13 503	15 924	13 998	14 319	PIB aux prix constants
Growth rates (annual)	2.0	11.7	8.9	-0.8	9.5	-12.1	2.3	Taux de croissance
Eritrea								**Érythrée**
GDP at current prices	...	640	1 098	2 117	3 092	3 444	3 858	PIB aux prix courants
GDP per capita	...	202	262	451	632	689	755	PIB par habitant
GDP at constant prices	...	920	1 098	1 057	1 229	1 245	1 266	PIB aux prix constants
Growth rates (annual)	...	20.9	1.5	2.2	7.0	1.3	1.7	Taux de croissance

10

Gross domestic product and gross domestic product per capita *(continued)*
Current and constant 2005 prices (millions of US dollars); per capita (US dollars); real annual rates of growth
Produit intérieur brut et produit intérieur brut par habitant *(suite)*
Prix courants et constants de 2005 (millions de dollars É.-U.); par habitant (dollars É.U.); taux de croissance réels

Region, country or area	1985	1995	2005	2010	2012	2013	2014	Région, pays ou zone
Estonia								**Estonie**
GDP at current prices	...	4 423	14 003	19 505	23 135	25 247	26 485	PIB aux prix courants
GDP per capita	...	3 086	10 330	14 643	17 473	19 126	20 122	PIB par habitant
GDP at constant prices	...	7 398	14 003	13 750	15 559	15 803	16 263	PIB aux prix constants
Growth rates (annual)	...	4.3	9.4	2.5	5.2	1.6	2.9	Taux de croissance
Ethiopia								**Éthiopie**
GDP at current prices	...	7 587	12 164	26 311	42 210	46 628	53 638	PIB aux prix courants
GDP per capita	...	133	159	300	458	493	553	PIB par habitant
GDP at constant prices	...	7 125	12 164	20 386	25 062	27 691	30 443	PIB aux prix constants
Growth rates (annual)	...	6.1	11.8	12.6	8.6	10.5	9.9	Taux de croissance
Fiji								**Fidji**
GDP at current prices	1 193	2 081	3 112	3 251	3 978	4 196	4 532	PIB aux prix courants
GDP per capita	1 676	2 684	3 787	3 780	4 550	4 766	5 112	PIB par habitant
GDP at constant prices	1 798	2 486	3 112	3 226	3 375	3 531	3 665	PIB aux prix constants
Growth rates (annual)	-3.9	2.5	0.7	3.1	1.8	4.6	3.8	Taux de croissance
Finland								**Finlande**
GDP at current prices	55 914	134 196	204 431	247 800	256 706	269 190	272 217	PIB aux prix courants
GDP per capita	11 406	26 271	38 966	46 165	47 322	49 365	49 678	PIB par habitant
GDP at constant prices	121 263	140 425	204 431	212 913	215 272	212 859	211 997	PIB aux prix constants
Growth rates (annual)	3.5	4.2	2.8	3.0	-1.4	-1.1	-0.4	Taux de croissance
France [3]								**France** [3]
GDP at current prices	555 201	1 609 794	2 203 624	2 646 837	2 681 416	2 810 249	2 829 192	PIB aux prix courants
GDP per capita	9 775	26 859	34 833	40 667	40 795	42 557	42 651	PIB par habitant
GDP at constant prices	1 397 266	1 758 875	2 203 624	2 289 830	2 341 712	2 357 085	2 361 317	PIB aux prix constants
Growth rates (annual)	1.6	2.1	1.6	2.0	0.2	0.7	0.2	Taux de croissance
French Polynesia								**Polynésie française**
GDP at current prices	1 716	4 421	5 703	6 081	5 672	5 756	5 623	PIB aux prix courants
GDP per capita	9 795	20 542	22 374	22 684	20 719	20 796	20 099	PIB par habitant
GDP at constant prices	3 372	4 435	5 703	5 466	5 220	5 103	4 965	PIB aux prix constants
Growth rates (annual)	5.4	0.5	1.4	-2.5	-1.6	-2.3	-2.7	Taux de croissance
Gabon								**Gabon**
GDP at current prices	4 639	5 519	9 579	12 882	15 968	17 027	17 412	PIB aux prix courants
GDP per capita	5 589	5 080	6 952	8 354	9 897	10 317	10 317	PIB par habitant
GDP at constant prices	7 428	9 170	9 579	10 123	11 417	12 056	12 659	PIB aux prix constants
Growth rates (annual)	-2.3	5.0	1.1	6.8	5.3	5.6	5.0	Taux de croissance
Gambia								**Gambie**
GDP at current prices	654	786	624	952	913	904	851	PIB aux prix courants
GDP per capita	894	738	433	563	505	484	441	PIB par habitant
GDP at constant prices	324	431	624	784	794	832	846	PIB aux prix constants
Growth rates (annual)	3.4	0.5	-0.9	6.5	5.9	4.8	1.6	Taux de croissance
Georgia								**Géorgie**
GDP at current prices	...	2 703	6 411	11 638	15 847	16 141	16 530	PIB aux prix courants
GDP per capita	...	534	1 433	2 738	3 829	3 953	4 097	PIB par habitant
GDP at constant prices	...	3 388	6 411	8 281	9 445	9 758	10 223	PIB aux prix constants
Growth rates (annual)	...	2.6	9.6	6.2	6.4	3.3	4.8	Taux de croissance
Germany								**Allemagne**
GDP at current prices	729 751	2 591 447	2 861 339	3 417 095	3 539 615	3 745 317	3 868 291	PIB aux prix courants
GDP per capita	9 408	31 753	35 218	42 483	43 982	46 488	47 966	PIB par habitant
GDP at constant prices	1 944 168	2 529 418	2 861 339	3 042 360	3 166 488	3 175 919	3 226 726	PIB aux prix constants
Growth rates (annual)	2.3	1.7	0.7	4.1	0.4	0.3	1.6	Taux de croissance
Ghana								**Ghana**
GDP at current prices	6 605	10 361	17 199	32 174	41 939	47 806	37 177	PIB aux prix courants
GDP per capita	519	618	804	1 323	1 642	1 827	1 388	PIB par habitant
GDP at constant prices	6 894	10 740	17 199	23 168	28 878	30 990	32 242	PIB aux prix constants
Growth rates (annual)	5.1	4.0	6.2	7.9	9.3	7.3	4.0	Taux de croissance
Greece								**Grèce**
GDP at current prices	47 816	136 886	247 777	299 362	245 671	239 510	235 574	PIB aux prix courants
GDP per capita	4 826	12 864	22 383	26 782	22 113	21 665	21 414	PIB par habitant
GDP at constant prices	151 314	171 214	247 777	243 731	205 304	198 738	200 037	PIB aux prix constants
Growth rates (annual)	2.5	2.1	0.6	-5.5	-7.3	-3.2	0.7	Taux de croissance

10 Gross domestic product and gross domestic product per capita *(continued)*
Current and constant 2005 prices (millions of US dollars); per capita (US dollars); real annual rates of growth
Produit intérieur brut et produit intérieur brut par habitant *(suite)*
Prix courants et constants de 2005 (millions de dollars É.-U.); par habitant (dollars É.U.); taux de croissance réels

Region, country or area	1985	1995	2005	2010	2012	2013	2014	Région, pays ou zone
Greenland								**Groenland**
GDP at current prices	406	1 189	1 650	2 287	2 356	2 419	2 441	PIB aux prix courants
GDP per capita	7 632	21 298	28 969	40 450	41 815	42 994	43 430	PIB par habitant
GDP at constant prices	1 140	1 267	1 650	1 868	1 900	1 874	1 844	PIB aux prix constants
Growth rates (annual)	3.7	3.7	3.7	2.5	-0.5	-1.3	-1.6	Taux de croissance
Grenada								**Grenade**
GDP at current prices	137	280	701	777	809	848	884	PIB aux prix courants
GDP per capita	1 368	2 793	6 804	7 418	7 673	8 011	8 313	PIB par habitant
GDP at constant prices	282	408	701	669	666	680	697	PIB aux prix constants
Growth rates (annual)	8.1	2.1	13.3	-0.5	-1.3	2.2	2.4	Taux de croissance
Guatemala								**Guatemala**
GDP at current prices	9 967	13 066	27 211	41 338	50 388	53 851	58 827	PIB aux prix courants
GDP per capita	1 228	1 261	2 064	2 806	3 279	3 432	3 673	PIB par habitant
GDP at constant prices	13 566	19 310	27 211	32 557	34 919	36 210	37 749	PIB aux prix constants
Growth rates (annual)	-0.6	4.9	3.3	2.9	3.0	3.7	4.2	Taux de croissance
Guinea								**Guinée**
GDP at current prices	1 978	3 654	2 823	4 730	5 738	6 483	6 579	PIB aux prix courants
GDP per capita	390	465	292	429	493	543	536	PIB par habitant
GDP at constant prices	1 360	2 002	2 823	3 139	3 388	3 466	3 511	PIB aux prix constants
Growth rates (annual)	5.0	4.7	3.0	1.9	3.9	2.3	1.3	Taux de croissance
Guinea-Bissau								**Guinée-Bissau**
GDP at current prices	415	800	587	847	1 072	1 158	1 209	PIB aux prix courants
GDP per capita	440	677	401	519	625	659	672	PIB par habitant
GDP at constant prices	374	525	587	690	735	742	763	PIB aux prix constants
Growth rates (annual)	4.3	4.0	4.3	4.4	-2.2	0.9	2.9	Taux de croissance
Guyana								**Guyana**
GDP at current prices	737	991	1 315	2 259	2 851	2 990	3 086	PIB aux prix courants
GDP per capita	963	1 364	1 772	2 999	3 759	3 929	4 040	PIB par habitant
GDP at constant prices	888	1 110	1 315	1 603	1 768	1 835	1 892	PIB aux prix constants
Growth rates (annual)	2.4	5.0	-2.0	5.4	5.2	3.8	3.1	Taux de croissance
Haiti								**Haïti**
GDP at current prices	2 665	2 696	4 154	6 708	7 820	8 387	8 599	PIB aux prix courants
GDP per capita	417	345	448	671	760	804	813	PIB par habitant
GDP at constant prices	4 275	3 771	4 154	4 313	4 682	4 881	5 015	PIB aux prix constants
Growth rates (annual)	0.6	9.9	1.8	-5.5	2.9	4.2	2.7	Taux de croissance
Honduras								**Honduras**
GDP at current prices	4 342	4 724	9 757	15 839	18 529	18 511	19 497	PIB aux prix courants
GDP per capita	1 025	845	1 418	2 111	2 395	2 358	2 449	PIB par habitant
GDP at constant prices	4 822	6 694	9 757	11 648	12 594	12 945	13 345	PIB aux prix constants
Growth rates (annual)	4.2	4.1	6.1	3.7	4.1	2.8	3.1	Taux de croissance
Hungary								**Hongrie**
GDP at current prices	23 597	46 212	112 531	130 091	127 176	134 402	138 347	PIB aux prix courants
GDP per capita	2 232	4 464	11 146	12 990	12 771	13 542	13 989	PIB par habitant
GDP at constant prices	86 319	78 603	112 531	111 359	111 402	113 507	117 675	PIB aux prix constants
Growth rates (annual)	-0.3	1.5	4.4	0.7	-1.7	1.9	3.7	Taux de croissance
Iceland								**Islande**
GDP at current prices	3 008	7 182	16 749	13 237	14 195	15 377	17 036	PIB aux prix courants
GDP per capita	12 462	26 852	56 443	41 620	43 891	47 256	52 048	PIB par habitant
GDP at constant prices	9 015	10 678	16 749	17 822	18 391	19 108	19 457	PIB aux prix constants
Growth rates (annual)	3.3	0.1	6.0	-3.6	1.2	3.9	1.8	Taux de croissance
India								**Inde**
GDP at current prices	221 994	361 957	820 980	1 668 768	1 869 210	1 936 088	2 054 941	PIB aux prix courants
GDP per capita	284	377	717	1 356	1 479	1 513	1 586	PIB par habitant
GDP at constant prices	253 058	442 358	820 980	1 223 944	1 371 516	1 466 140	1 572 967	PIB aux prix constants
Growth rates (annual)	5.5	7.6	9.3	10.3	5.1	6.9	7.3	Taux de croissance
Indonesia								**Indonésie**
GDP at current prices	102 171	236 456	304 372	755 094	917 870	910 479	888 538	PIB aux prix courants
GDP per capita	617	1 195	1 345	3 125	3 701	3 624	3 492	PIB par habitant
GDP at constant prices	114 371	232 971	304 372	402 359	452 943	478 214	502 242	PIB aux prix constants
Growth rates (annual)	2.5	8.2	5.7	6.2	6.0	5.6	5.0	Taux de croissance

10 Gross domestic product and gross domestic product per capita *(continued)*
Current and constant 2005 prices (millions of US dollars); per capita (US dollars); real annual rates of growth
Produit intérieur brut et produit intérieur brut par habitant *(suite)*
Prix courants et constants de 2005 (millions de dollars É.-U.); par habitant (dollars É.U.); taux de croissance réels

Region, country or area	1985	1995	2005	2010	2012	2013	2014	Région, pays ou zone
Iran (Islamic Rep. of)								**Iran (Rép. islamique d')**
GDP at current prices	76 257	114 364	219 846	467 790	587 209	511 621	425 326	PIB aux prix courants
GDP per capita	1 613	1 896	3 135	6 300	7 711	6 631	5 443	PIB par habitant
GDP at constant prices	120 226	141 577	219 846	279 059	270 389	265 221	276 740	PIB aux prix constants
Growth rates (annual)	1.9	2.4	4.2	6.6	-6.6	-1.9	4.3	Taux de croissance
Iraq								**Iraq**
GDP at current prices	12 074	3 477	36 268	117 138	185 918	205 186	225 422	PIB aux prix courants
GDP per capita	775	172	1 342	3 795	5 641	6 016	6 391	PIB par habitant
GDP at constant prices	20 184	12 059	36 268	48 218	59 848	63 192	60 729	PIB aux prix constants
Growth rates (annual)	-<0.0	-18.3	4.4	5.5	12.6	5.6	-3.9	Taux de croissance
Ireland								**Irlande**
GDP at current prices	21 294	69 219	211 383	220 063	224 651	238 260	250 814	PIB aux prix courants
GDP per capita	5 981	18 989	50 285	47 660	48 127	51 005	53 648	PIB par habitant
GDP at constant prices	66 411	101 240	211 383	219 837	225 868	229 107	241 019	PIB aux prix constants
Growth rates (annual)	1.9	9.6	6.3	0.4	0.2	1.4	5.2	Taux de croissance
Israel								**Israël**
GDP at current prices	27 587	99 969	142 838	234 323	259 612	292 412	305 673	PIB aux prix courants
GDP per capita	6 757	18 750	21 630	31 578	33 740	37 403	38 500	PIB par habitant
GDP at constant prices	57 396	98 662	142 838	176 636	190 864	197 070	202 104	PIB aux prix constants
Growth rates (annual)	4.0	6.8	4.4	5.5	2.9	3.3	2.6	Taux de croissance
Italy								**Italie**
GDP at current prices	450 870	1 171 250	1 853 466	2 126 621	2 074 632	2 133 539	2 141 161	PIB aux prix courants
GDP per capita	7 922	20 505	31 598	35 689	34 729	35 695	35 812	PIB par habitant
GDP at constant prices	1 287 765	1 602 161	1 853 466	1 825 020	1 783 999	1 752 826	1 745 045	PIB aux prix constants
Growth rates (annual)	2.8	2.9	0.9	1.7	-2.8	-1.7	-0.4	Taux de croissance
Jamaica								**Jamaïque**
GDP at current prices	2 291	6 544	11 244	13 218	14 786	14 262	13 927	PIB aux prix courants
GDP per capita	991	2 629	4 199	4 822	5 352	5 143	5 004	PIB par habitant
GDP at constant prices	7 101	10 430	11 244	10 971	11 092	11 153	11 230	PIB aux prix constants
Growth rates (annual)	-4.6	1.0	0.9	-1.5	-0.6	0.5	0.7	Taux de croissance
Japan								**Japon**
GDP at current prices	1 384 532	5 333 927	4 571 867	5 498 719	5 957 250	4 919 588	4 602 419	PIB aux prix courants
GDP per capita	11 539	42 849	36 005	43 188	46 856	38 741	36 298	PIB par habitant
GDP at constant prices	3 018 214	4 132 181	4 571 867	4 651 105	4 710 639	4 785 425	4 780 944	PIB aux prix constants
Growth rates (annual)	6.3	1.9	1.3	4.7	1.7	1.6	-0.1	Taux de croissance
Jordan								**Jordanie**
GDP at current prices	5 119	6 732	12 589	26 425	30 937	33 594	35 827	PIB aux prix courants
GDP per capita	1 840	1 558	2 361	4 054	4 423	4 656	4 831	PIB par habitant
GDP at constant prices	5 982	7 894	12 589	17 034	17 938	18 445	19 016	PIB aux prix constants
Growth rates (annual)	-2.7	6.2	8.1	2.3	2.7	2.8	3.1	Taux de croissance
Kazakhstan								**Kazakhstan**
GDP at current prices	...	20 563	57 124	148 047	203 517	231 876	216 036	PIB aux prix courants
GDP per capita	...	1 291	3 697	9 077	12 099	13 560	12 436	PIB par habitant
GDP at constant prices	...	30 835	57 124	77 291	87 047	92 248	96 232	PIB aux prix constants
Growth rates (annual)	...	-8.2	9.7	7.3	5.0	6.0	4.3	Taux de croissance
Kenya								**Kenya**
GDP at current prices	9 110	13 428	21 506	40 000	50 410	54 931	60 936	PIB aux prix courants
GDP per capita	463	491	608	992	1 185	1 257	1 358	PIB par habitant
GDP at constant prices	11 339	16 134	21 506	27 424	30 426	32 156	33 871	PIB aux prix constants
Growth rates (annual)	4.3	4.4	5.9	8.4	4.6	5.7	5.3	Taux de croissance
Kiribati								**Kiribati**
GDP at current prices	30	56	112	156	188	181	180	PIB aux prix courants
GDP per capita	468	720	1 215	1 515	1 762	1 663	1 632	PIB par habitant
GDP at constant prices	76	89	112	111	115	117	122	PIB aux prix constants
Growth rates (annual)	-4.6	-0.6	5.0	-0.9	3.4	2.4	3.7	Taux de croissance
Kosovo								**Kosovo**
GDP at current prices	...	5 304	3 680	5 830	6 500	7 072	7 387	PIB aux prix courants
GDP per capita	...	2 351	2 108	3 298	3 644	3 985	4 183	PIB par habitant
GDP at constant prices	...	3 417	3 680	4 729	5 075	5 249	5 313	PIB aux prix constants
Growth rates (annual)	...	8.1	3.9	3.3	2.8	3.4	1.2	Taux de croissance

10

Gross domestic product and gross domestic product per capita *(continued)*
Current and constant 2005 prices (millions of US dollars); per capita (US dollars); real annual rates of growth

Produit intérieur brut et produit intérieur brut par habitant *(suite)*
Prix courants et constants de 2005 (millions de dollars É.-U.); par habitant (dollars É.U.); taux de croissance réels

Region, country or area	1985	1995	2005	2010	2012	2013	2014	Région, pays ou zone
Kuwait								**Koweït**
GDP at current prices	21 446	26 554	80 798	115 416	174 048	174 168	163 637	PIB aux prix courants
GDP per capita	12 359	16 221	35 694	37 724	50 897	48 465	43 600	PIB par habitant
GDP at constant prices	31 211	48 626	80 798	85 603	100 064	101 214	99 572	PIB aux prix constants
Growth rates (annual)	-4.3	1.4	10.6	-2.4	6.6	1.1	-1.6	Taux de croissance
Kyrgyzstan								**Kirghizistan**
GDP at current prices	...	1 492	2 460	4 794	6 605	7 335	7 404	PIB aux prix courants
GDP per capita	...	325	481	877	1 169	1 277	1 267	PIB par habitant
GDP at constant prices	...	1 555	2 460	3 056	3 235	3 576	3 705	PIB aux prix constants
Growth rates (annual)	...	-5.4	-0.2	-0.5	-0.1	10.5	3.6	Taux de croissance
Lao People's Dem. Rep.								**Rép. dém. pop. lao**
GDP at current prices	601	1 708	2 717	6 744	9 397	10 760	11 749	PIB aux prix courants
GDP per capita	163	352	473	1 077	1 452	1 635	1 756	PIB par habitant
GDP at constant prices	889	1 482	2 717	3 988	4 649	5 020	5 399	PIB aux prix constants
Growth rates (annual)	5.1	7.0	6.8	8.1	7.9	8.0	7.6	Taux de croissance
Latvia								**Lettonie**
GDP at current prices	...	5 404	16 903	23 743	28 029	30 219	31 286	PIB aux prix courants
GDP per capita	...	2 172	7 588	11 358	13 759	15 021	15 726	PIB par habitant
GDP at constant prices	...	8 867	16 903	16 524	18 254	18 805	19 249	PIB aux prix constants
Growth rates (annual)	...	-0.8	10.7	-3.8	4.0	3.0	2.4	Taux de croissance
Lebanon								**Liban**
GDP at current prices	2 275	11 506	21 490	38 420	44 100	47 221	49 631	PIB aux prix courants
GDP per capita	850	3 793	5 390	8 858	8 956	8 931	8 844	PIB par habitant
GDP at constant prices	16 152	16 338	21 490	31 042	32 192	33 162	33 825	PIB aux prix constants
Growth rates (annual)	24.3	6.5	2.7	8.0	2.8	3.0	2.0	Taux de croissance
Lesotho								**Lesotho**
GDP at current prices	248	859	1 368	2 187	2 384	2 135	2 081	PIB aux prix courants
GDP per capita	169	490	711	1 088	1 159	1 025	986	PIB par habitant
GDP at constant prices	621	979	1 368	1 763	1 926	2 014	2 083	PIB aux prix constants
Growth rates (annual)	3.4	2.0	2.7	7.9	5.0	4.6	3.4	Taux de croissance
Liberia								**Libéria**
GDP at current prices	881	171	608	1 074	1 734	1 946	2 122	PIB aux prix courants
GDP per capita	401	82	186	271	414	453	483	PIB par habitant
GDP at constant prices	802	117	608	1 099	1 257	1 359	1 451	PIB aux prix constants
Growth rates (annual)	-2.0	-4.3	9.5	10.8	8.2	8.1	6.8	Taux de croissance
Libya								**Libye**
GDP at current prices	29 887	28 292	45 451	80 942	101 166	65 825	41 319	PIB aux prix courants
GDP per capita	7 782	5 800	7 834	12 918	16 100	10 505	6 602	PIB par habitant
GDP at constant prices	33 739	32 935	45 451	60 501	52 657	25 208	19 150	PIB aux prix constants
Growth rates (annual)	8.3	-2.2	10.3	4.3	124.7	-52.1	-24.0	Taux de croissance
Liechtenstein								**Liechtenstein**
GDP at current prices	529	2 429	3 658	5 082	5 456	5 721	5 855	PIB aux prix courants
GDP per capita	19 409	78 764	104 968	140 102	148 300	154 448	157 040	PIB par habitant
GDP at constant prices	1 564	2 496	3 658	4 074	3 951	4 103	4 152	PIB aux prix constants
Growth rates (annual)	7.0	5.9	4.8	7.4	1.1	3.9	1.2	Taux de croissance
Lithuania								**Lituanie**
GDP at current prices	...	6 702	26 141	37 130	42 846	46 411	48 392	PIB aux prix courants
GDP per capita	...	1 847	7 819	11 890	14 204	15 659	16 591	PIB par habitant
GDP at constant prices	...	14 422	26 141	27 715	30 518	31 600	32 558	PIB aux prix constants
Growth rates (annual)	...	3.3	7.7	1.6	3.8	3.5	3.0	Taux de croissance
Luxembourg								**Luxembourg**
GDP at current prices	4 786	21 804	36 976	52 349	55 987	61 795	64 874	PIB aux prix courants
GDP per capita	13 044	53 421	80 762	103 071	105 144	113 408	116 560	PIB par habitant
GDP at constant prices	13 639	23 741	36 976	41 772	42 481	44 327	46 131	PIB aux prix constants
Growth rates (annual)	2.8	1.4	3.2	5.7	-0.8	4.3	4.1	Taux de croissance
Madagascar								**Madagascar**
GDP at current prices	2 858	3 160	5 039	8 733	9 920	10 603	10 674	PIB aux prix courants
GDP per capita	286	235	275	414	445	463	453	PIB par habitant
GDP at constant prices	3 312	3 730	5 039	5 798	6 058	6 196	6 408	PIB aux prix constants
Growth rates (annual)	1.2	1.7	4.6	0.4	3.0	2.3	3.4	Taux de croissance

Gross domestic product and gross domestic product per capita *(continued)*
Current and constant 2005 prices (millions of US dollars); per capita (US dollars); real annual rates of growth
Produit intérieur brut et produit intérieur brut par habitant *(suite)*
Prix courants et constants de 2005 (millions de dollars É.-U.); par habitant (dollars É.U.); taux de croissance réels

Region, country or area	1985	1995	2005	2010	2012	2013	2014	Région, pays ou zone
Malawi								**Malawi**
GDP at current prices	2 028	2 474	3 656	6 960	5 721	5 217	5 720	PIB aux prix courants
GDP per capita	281	252	287	471	364	322	343	PIB par habitant
GDP at constant prices	2 173	2 621	3 656	5 236	5 530	5 530	5 889	PIB aux prix constants
Growth rates (annual)	4.7	9.0	4.7	4.9	6.3	0.0	6.5	Taux de croissance
Malaysia								**Malaisie**
GDP at current prices	33 709	95 977	143 534	247 534	304 956	313 158	326 933	PIB aux prix courants
GDP per capita	2 138	4 631	5 564	8 803	10 508	10 628	10 933	PIB par habitant
GDP at constant prices	41 539	90 111	143 534	178 674	198 552	207 973	220 496	PIB aux prix constants
Growth rates (annual)	-1.1	9.8	5.3	7.4	5.6	4.7	6.0	Taux de croissance
Maldives								**Maldives**
GDP at current prices	165	562	1 120	2 332	2 535	2 705	3 032	PIB aux prix courants
GDP per capita	869	2 211	3 672	7 013	7 350	7 705	8 484	PIB par habitant
GDP at constant prices	252	575	1 120	1 703	1 975	2 149	2 332	PIB aux prix constants
Growth rates (annual)	13.0	7.1	-8.6	7.1	3.0	8.8	8.5	Taux de croissance
Mali								**Mali**
GDP at current prices	1 158	2 618	5 486	9 400	10 341	10 943	11 979	PIB aux prix courants
GDP per capita	148	272	426	620	642	660	701	PIB par habitant
GDP at constant prices	2 158	3 365	5 486	6 989	7 181	7 304	7 829	PIB aux prix constants
Growth rates (annual)	8.5	6.2	6.1	5.8	<0.0	1.7	7.2	Taux de croissance
Malta								**Malte**
GDP at current prices	1 156	3 697	6 393	8 741	9 257	10 003	10 536	PIB aux prix courants
GDP per capita	3 417	9 929	16 100	21 212	22 275	24 002	25 222	PIB par habitant
GDP at constant prices	2 554	4 496	6 393	7 066	7 397	7 593	7 862	PIB aux prix constants
Growth rates (annual)	2.6	6.3	3.8	3.5	2.5	2.6	3.5	Taux de croissance
Marshall Islands								**Îles Marshall**
GDP at current prices	44	121	139	178	200	207	209	PIB aux prix courants
GDP per capita	1 148	2 366	2 677	3 397	3 804	3 930	3 947	PIB par habitant
GDP at constant prices	71	137	139	151	158	163	164	PIB aux prix constants
Growth rates (annual)	-6.3	8.2	2.0	6.1	4.7	3.0	0.5	Taux de croissance
Mauritania								**Mauritanie**
GDP at current prices	1 085	1 681	2 184	4 338	4 845	5 058	5 092	PIB aux prix courants
GDP per capita	614	720	693	1 208	1 283	1 306	1 283	PIB par habitant
GDP at constant prices	1 145	1 504	2 184	2 798	3 095	3 271	3 481	PIB aux prix constants
Growth rates (annual)	3.4	-2.4	9.0	4.8	6.0	5.7	6.4	Taux de croissance
Mauritius								**Maurice**
GDP at current prices	1 103	4 092	6 489	9 718	11 442	11 928	12 616	PIB aux prix courants
GDP per capita	1 086	3 625	5 310	7 787	9 093	9 440	9 945	PIB par habitant
GDP at constant prices	2 293	4 193	6 489	8 155	8 746	9 024	9 352	PIB aux prix constants
Growth rates (annual)	6.9	4.4	1.8	4.1	3.2	3.2	3.6	Taux de croissance
Mexico								**Mexique**
GDP at current prices	219 661	319 551	864 810	1 049 925	1 184 504	1 258 775	1 294 695	PIB aux prix courants
GDP per capita	2 841	3 384	7 880	8 851	9 703	10 173	10 326	PIB par habitant
GDP at constant prices	515 589	604 407	864 810	952 037	1 029 312	1 043 670	1 066 952	PIB aux prix constants
Growth rates (annual)	2.8	-6.2	3.1	5.2	4.0	1.4	2.2	Taux de croissance
Micronesia (Fed. States of)								**Micronésie (États féd. de)**
GDP at current prices	108	222	250	294	326	315	308	PIB aux prix courants
GDP per capita	1 262	2 063	2 352	2 834	3 148	3 033	2 960	PIB par habitant
GDP at constant prices	160	244	250	248	253	243	239	PIB aux prix constants
Growth rates (annual)	16.6	4.2	2.2	3.2	0.1	-4.0	-1.5	Taux de croissance
Monaco								**Monaco**
GDP at current prices	1 059	3 070	4 203	5 362	5 743	6 554	7 060	PIB aux prix courants
GDP per capita	36 722	100 010	124 316	145 538	153 542	174 636	187 650	PIB par habitant
GDP at constant prices	2 665	3 355	4 203	4 609	4 983	5 461	5 854	PIB aux prix constants
Growth rates (annual)	1.6	2.1	1.6	2.2	1.0	9.6	7.2	Taux de croissance
Mongolia								**Mongolie**
GDP at current prices	1 218	1 678	2 926	7 189	12 293	12 582	12 067	PIB aux prix courants
GDP per capita	634	730	1 158	2 650	4 377	4 401	4 147	PIB par habitant
GDP at constant prices	1 774	1 859	2 926	4 005	5 276	5 891	6 348	PIB aux prix constants
Growth rates (annual)	5.7	6.3	7.3	6.4	12.3	11.6	7.8	Taux de croissance

10

Gross domestic product and gross domestic product per capita *(continued)*
Current and constant 2005 prices (millions of US dollars); per capita (US dollars); real annual rates of growth
Produit intérieur brut et produit intérieur brut par habitant *(suite)*
Prix courants et constants de 2005 (millions de dollars É.-U.); par habitant (dollars É.U.); taux de croissance réels

Region, country or area	1985	1995	2005	2010	2012	2013	2014	Région, pays ou zone
Montenegro								**Monténégro**
GDP at current prices	...	1 215	2 257	4 139	4 088	4 465	4 588	PIB aux prix courants
GDP per capita	...	1 958	3 662	6 655	6 552	7 147	7 337	PIB par habitant
GDP at constant prices	...	1 316	2 257	2 803	2 814	2 914	2 966	PIB aux prix constants
Growth rates (annual)	...	14.3	4.2	2.5	-2.7	3.5	1.8	Taux de croissance
Montserrat								**Montserrat**
GDP at current prices	44	69	49	56	63	60	63	PIB aux prix courants
GDP per capita	3 850	6 729	10 231	11 208	12 573	11 896	12 384	PIB par habitant
GDP at constant prices	92	103	49	51	55	56	58	PIB aux prix constants
Growth rates (annual)	5.9	-8.8	3.1	-2.8	3.5	2.0	3.7	Taux de croissance
Morocco [4]								**Maroc** [4]
GDP at current prices	15 111	38 728	62 545	93 217	98 266	107 235	110 009	PIB aux prix courants
GDP per capita	669	1 426	2 058	2 903	2 979	3 206	3 243	PIB par habitant
GDP at constant prices	31 288	40 658	62 545	79 350	86 026	90 091	92 269	PIB aux prix constants
Growth rates (annual)	6.3	-6.6	3.0	3.8	3.0	4.7	2.4	Taux de croissance
Mozambique								**Mozambique**
GDP at current prices	5 771	2 572	7 724	10 154	15 265	16 019	17 081	PIB aux prix courants
GDP per capita	440	162	366	418	593	605	628	PIB par habitant
GDP at constant prices	2 083	2 926	7 724	11 053	12 692	13 599	14 580	PIB aux prix constants
Growth rates (annual)	-8.8	2.2	8.7	6.7	7.2	7.1	7.2	Taux de croissance
Myanmar								**Myanmar**
GDP at current prices	6 606	7 764	11 931	41 445	61 014	62 141	66 478	PIB aux prix courants
GDP per capita	172	174	239	801	1 161	1 173	1 244	PIB par habitant
GDP at constant prices	3 669	4 374	11 931	20 286	22 991	24 929	27 097	PIB aux prix constants
Growth rates (annual)	2.9	6.9	13.6	10.2	7.3	8.4	8.7	Taux de croissance
Namibia								**Namibie**
GDP at current prices	1 537	4 011	7 261	11 273	13 020	12 932	13 429	PIB aux prix courants
GDP per capita	1 338	2 425	3 582	5 139	5 682	5 511	5 589	PIB par habitant
GDP at constant prices	3 474	4 992	7 261	8 992	9 946	10 454	10 922	PIB aux prix constants
Growth rates (annual)	0.2	4.1	1.8	6.0	5.2	5.1	4.5	Taux de croissance
Nauru								**Nauru**
GDP at current prices	33	35	26	62	121	153	182	PIB aux prix courants
GDP per capita	4 042	3 496	2 599	6 234	11 966	15 100	17 857	PIB par habitant
GDP at constant prices	138	60	26	41	56	71	80	PIB aux prix constants
Growth rates (annual)	-2.8	-7.9	-12.1	20.1	20.2	26.4	12.6	Taux de croissance
Nepal								**Népal**
GDP at current prices	2 741	4 534	8 259	16 281	17 927	18 227	19 489	PIB aux prix courants
GDP per capita	164	212	324	606	652	655	692	PIB par habitant
GDP at constant prices	3 416	5 499	8 259	10 299	11 160	11 621	12 246	PIB aux prix constants
Growth rates (annual)	6.1	3.5	3.1	4.8	4.8	4.1	5.4	Taux de croissance
Netherlands								**Pays-Bas**
GDP at current prices	142 011	446 514	678 517	836 390	828 947	864 169	879 319	PIB aux prix courants
GDP per capita	9 813	28 899	41 546	50 289	49 491	51 411	52 129	PIB par habitant
GDP at constant prices	388 883	513 546	678 517	722 835	727 093	723 491	730 806	PIB aux prix constants
Growth rates (annual)	2.6	3.1	2.2	1.4	-1.1	-0.5	1.0	Taux de croissance
Netherlands Ant. [former]								**Antilles néerland. [anc.]**
GDP at current prices	1 190	2 571	3 049	3 844	4 009	...	...	PIB aux prix courants
GDP per capita	7 024	14 734	18 831	21 267	21 351	...	...	PIB par habitant
GDP at constant prices	2 407	3 009	3 049	3 287	3 301	...	...	PIB aux prix constants
Growth rates (annual)	-2.1	1.6	1.1	-0.3	0.2	...	...	Taux de croissance
New Caledonia								**Nouvelle-Calédonie**
GDP at current prices	855	3 628	6 236	9 355	9 406	9 858	10 234	PIB aux prix courants
GDP per capita	5 553	19 178	27 270	37 976	37 160	38 437	39 392	PIB par habitant
GDP at constant prices	2 670	4 982	6 236	7 564	7 889	8 036	8 283	PIB aux prix constants
Growth rates (annual)	4.5	5.9	3.6	6.9	1.6	1.9	3.1	Taux de croissance
New Zealand								**Nouvelle-Zélande**
GDP at current prices	24 108	63 151	114 722	146 583	176 617	189 494	198 652	PIB aux prix courants
GDP per capita	7 376	17 185	27 746	33 551	39 815	42 437	44 189	PIB par habitant
GDP at constant prices	67 364	81 673	114 722	120 837	126 204	129 408	133 574	PIB aux prix constants
Growth rates (annual)	1.6	4.6	3.4	1.4	2.2	2.5	3.2	Taux de croissance

10 Gross domestic product and gross domestic product per capita *(continued)*
Current and constant 2005 prices (millions of US dollars); per capita (US dollars); real annual rates of growth
Produit intérieur brut et produit intérieur brut par habitant *(suite)*
Prix courants et constants de 2005 (millions de dollars É.-U.); par habitant (dollars É.U.); taux de croissance réels

Region, country or area	1985	1995	2005	2010	2012	2013	2014	Région, pays ou zone
Nicaragua								**Nicaragua**
GDP at current prices	3 618	4 132	6 321	8 741	10 460	10 851	11 806	PIB aux prix courants
GDP per capita	975	896	1 175	1 523	1 780	1 825	1 963	PIB par habitant
GDP at constant prices	4 589	4 236	6 321	7 154	7 990	8 350	8 743	PIB aux prix constants
Growth rates (annual)	-4.1	5.9	4.3	3.2	5.1	4.5	4.7	Taux de croissance
Niger								**Niger**
GDP at current prices	1 531	1 786	3 369	5 719	6 942	7 683	8 169	PIB aux prix courants
GDP per capita	224	191	250	351	394	418	427	PIB par habitant
GDP at constant prices	2 118	2 307	3 369	4 334	4 958	5 185	5 543	PIB aux prix constants
Growth rates (annual)	7.7	3.3	7.4	8.4	11.8	4.6	6.9	Taux de croissance
Nigeria								**Nigéria**
GDP at current prices	178 821	49 030	180 502	369 062	460 954	514 965	568 499	PIB aux prix courants
GDP per capita	2 131	452	1 293	2 315	2 740	2 980	3 203	PIB par habitant
GDP at constant prices	67 600	94 219	180 502	249 671	273 080	287 811	305 971	PIB aux prix constants
Growth rates (annual)	11.3	2.1	6.5	7.8	4.3	5.4	6.3	Taux de croissance
Norway								**Norvège**
GDP at current prices	65 598	152 028	308 722	428 527	509 705	522 746	500 519	PIB aux prix courants
GDP per capita	15 813	34 871	66 760	87 611	101 568	102 833	97 226	PIB par habitant
GDP at constant prices	177 364	231 758	308 722	323 263	335 366	338 716	346 217	PIB aux prix constants
Growth rates (annual)	5.4	4.2	2.6	0.6	2.7	1.0	2.2	Taux de croissance
Oman								**Oman**
GDP at current prices	10 281	13 650	31 082	58 641	76 341	78 183	81 797	PIB aux prix courants
GDP per capita	6 861	6 228	12 399	19 921	21 534	20 011	19 310	PIB par habitant
GDP at constant prices	16 487	25 248	31 082	41 164	43 598	45 304	46 615	PIB aux prix constants
Growth rates (annual)	14.5	4.8	2.5	4.8	7.1	3.9	2.9	Taux de croissance
Pakistan								**Pakistan**
GDP at current prices	38 840	77 266	117 708	174 508	214 642	221 286	251 255	PIB aux prix courants
GDP per capita	421	630	768	1 026	1 210	1 221	1 358	PIB par habitant
GDP at constant prices	47 300	78 593	117 708	139 224	148 067	154 595	162 961	PIB aux prix constants
Growth rates (annual)	7.6	5.0	7.7	1.6	3.5	4.4	5.4	Taux de croissance
Palau								**Palaos**
GDP at current prices	44	116	180	185	204	219	234	PIB aux prix courants
GDP per capita	3 235	6 717	9 038	9 050	9 851	10 455	11 068	PIB par habitant
GDP at constant prices	108	155	180	162	176	179	179	PIB aux prix constants
Growth rates (annual)	5.4	10.9	-0.5	-7.2	4.3	1.9	-0.1	Taux de croissance
Panama								**Panama**
GDP at current prices	5 725	9 042	15 465	28 917	39 955	44 856	49 166	PIB aux prix courants
GDP per capita	2 580	3 302	4 659	7 987	10 672	11 787	12 712	PIB par habitant
GDP at constant prices	7 198	9 986	15 465	21 961	26 820	28 597	30 328	PIB aux prix constants
Growth rates (annual)	5.1	1.8	7.2	5.8	9.2	6.6	6.1	Taux de croissance
Papua New Guinea								**Papouasie-Nvl-Guinée**
GDP at current prices	2 450	4 840	4 866	9 707	15 422	15 291	16 576	PIB aux prix courants
GDP per capita	666	1 026	799	1 418	2 155	2 092	2 221	PIB par habitant
GDP at constant prices	2 612	4 213	4 866	6 492	7 777	8 161	8 848	PIB aux prix constants
Growth rates (annual)	3.6	-3.3	3.9	7.6	7.7	4.9	8.4	Taux de croissance
Paraguay								**Paraguay**
GDP at current prices	4 017[5]	8 066[5]	8 735	20 048	24 595	28 897	30 985	PIB aux prix courants
GDP per capita	1 094[5]	1 694[5]	1 507	3 228	3 856	4 469	4 729	PIB par habitant
GDP at constant prices	4 740[5]	6 776[5]	8 735	11 148	11 488	13 122	13 693	PIB aux prix constants
Growth rates (annual)	4.0[5]	5.5[5]	2.1	13.1	-1.2	14.2	4.4	Taux de croissance
Peru								**Pérou**
GDP at current prices	14 101	51 796	76 080	147 070	192 806	200 643	201 809	PIB aux prix courants
GDP per capita	721	2 155	2 755	5 007	6 393	6 564	6 516	PIB par habitant
GDP at constant prices	46 436	54 299	76 080	106 185	119 762	126 658	129 657	PIB aux prix constants
Growth rates (annual)	2.1	7.4	6.3	8.5	6.0	5.8	2.4	Taux de croissance
Philippines								**Philippines**
GDP at current prices	34 052	82 121	103 072	199 591	250 240	272 067	284 582	PIB aux prix courants
GDP per capita	627	1 176	1 197	2 145	2 606	2 788	2 871	PIB par habitant
GDP at constant prices	49 280	69 129	103 072	131 138	145 183	155 609	165 095	PIB aux prix constants
Growth rates (annual)	-7.3	4.7	4.8	7.6	6.8	7.2	6.1	Taux de croissance

10
Gross domestic product and gross domestic product per capita *(continued)*
Current and constant 2005 prices (millions of US dollars); per capita (US dollars); real annual rates of growth
Produit intérieur brut et produit intérieur brut par habitant *(suite)*
Prix courants et constants de 2005 (millions de dollars É.-U.); par habitant (dollars É.U.); taux de croissance réels

Region, country or area	1985	1995	2005	2010	2012	2013	2014	Région, pays ou zone
Poland								**Pologne**
GDP at current prices	73 333	142 138	304 412	479 243	500 221	524 056	544 959	PIB aux prix courants
GDP per capita	1 956	3 683	7 914	12 424	12 956	13 570	14 111	PIB par habitant
GDP at constant prices	185 229	202 219	304 412	383 287	408 770	413 940	427 741	PIB aux prix constants
Growth rates (annual)	3.6	7.0	3.5	3.7	1.6	1.3	3.3	Taux de croissance
Portugal								**Portugal**
GDP at current prices	27 118	118 132	197 300	238 303	216 368	226 073	230 117	PIB aux prix courants
GDP per capita	2 731	11 721	18 826	22 514	20 577	21 614	22 122	PIB par habitant
GDP at constant prices	107 932	154 749	197 300	203 429	191 667	189 501	191 218	PIB aux prix constants
Growth rates (annual)	2.8	4.3	0.8	1.9	-4.0	-1.1	0.9	Taux de croissance
Puerto Rico								**Porto Rico**
GDP at current prices	20 574	43 246	83 915	98 381	101 565	102 526	103 676	PIB aux prix courants
GDP per capita	6 106	11 721	22 311	26 520	27 482	27 780	28 123	PIB par habitant
GDP at constant prices	39 130	62 809	83 915	78 370	78 112	78 083	78 034	PIB aux prix constants
Growth rates (annual)	2.1	4.5	-2.0	-0.4	<0.0	-<0.0	-0.1	Taux de croissance
Qatar								**Qatar**
GDP at current prices	6 153	8 138	44 530	125 122	190 290	203 235	211 817	PIB aux prix courants
GDP per capita	16 583	16 243	53 207	70 870	94 407	96 719	97 519	PIB par habitant
GDP at constant prices	12 865	17 694	44 530	101 936	122 168	129 891	137 884	PIB aux prix constants
Growth rates (annual)	-2.1	2.4	7.5	16.7	6.0	6.3	6.2	Taux de croissance
Republic of Korea								**République de Corée**
GDP at current prices	100 273	556 129	898 137	1 094 499	1 222 807	1 305 605	1 410 383	PIB aux prix courants
GDP per capita	2 476	12 454	18 866	22 296	24 649	26 192	28 166	PIB par habitant
GDP at constant prices	221 415	545 690	898 137	1 098 694	1 165 258	1 199 006	1 238 695	PIB aux prix constants
Growth rates (annual)	7.7	9.6	3.9	6.5	2.3	2.9	3.3	Taux de croissance
Republic of Moldova								**République de Moldova**
GDP at current prices	...	1 767	2 988	5 812	7 285	7 985	7 944	PIB aux prix courants
GDP per capita	...	407	719	1 423	1 788	1 960	1 951	PIB par habitant
GDP at constant prices	...	2 405	2 988	3 502	3 713	4 062	4 240	PIB aux prix constants
Growth rates (annual)	...	-1.4	7.5	7.1	-0.7	9.4	4.4	Taux de croissance
Romania								**Roumanie**
GDP at current prices	50 742	37 657	99 699	167 998	171 665	191 548	199 045	PIB aux prix courants
GDP per capita	2 196	1 640	4 657	8 276	8 607	9 677	10 129	PIB par habitant
GDP at constant prices	93 084	76 334	99 699	115 113	117 074	121 209	124 575	PIB aux prix constants
Growth rates (annual)	-0.1	7.1	4.2	-0.8	0.6	3.5	2.8	Taux de croissance
Russian Federation								**Fédération de Russie**
GDP at current prices	...	399 472	764 016	1 524 917	2 016 110	2 079 022	1 849 940	PIB aux prix courants
GDP per capita	...	2 694	5 320	10 652	14 070	14 501	12 898	PIB par habitant
GDP at constant prices	...	524 113	764 016	909 266	980 324	993 469	999 647	PIB aux prix constants
Growth rates (annual)	...	-4.1	6.4	4.5	3.4	1.3	0.6	Taux de croissance
Rwanda								**Rwanda**
GDP at current prices	1 813	1 230	2 581	5 699	7 220	7 522	7 903	PIB aux prix courants
GDP per capita	296	208	287	554	667	679	697	PIB par habitant
GDP at constant prices	1 612	1 057	2 581	3 847	4 514	4 725	5 054	PIB aux prix constants
Growth rates (annual)	4.4	33.5	9.4	7.3	8.8	4.7	7.0	Taux de croissance
Saint Kitts and Nevis								**Saint-Kitts-et-Nevis**
GDP at current prices	106	300	543	692	732	787	852	PIB aux prix courants
GDP per capita	2 524	6 988	11 054	13 227	13 642	14 499	15 510	PIB par habitant
GDP at constant prices	215	374	543	563	565	598	639	PIB aux prix constants
Growth rates (annual)	8.6	5.8	9.3	-3.2	-1.2	5.8	6.9	Taux de croissance
Saint Lucia								**Sainte-Lucie**
GDP at current prices	249	617	935	1 244	1 303	1 336	1 406	PIB aux prix courants
GDP per capita	1 970	4 199	5 656	7 014	7 201	7 327	7 655	PIB par habitant
GDP at constant prices	447	754	935	1 039	1 035	1 030	1 035	PIB aux prix constants
Growth rates (annual)	7.5	1.1	-1.7	-1.0	-1.6	-0.4	0.5	Taux de croissance
Saint Vincent-Grenadines								**Saint-Vincent-Grenadines**
GDP at current prices	133	312	551	681	693	721	729	PIB aux prix courants
GDP per capita	1 277	2 887	5 064	6 232	6 338	6 592	6 669	PIB par habitant
GDP at constant prices	255	389	551	589	595	605	608	PIB aux prix constants
Growth rates (annual)	6.2	7.8	2.5	-3.4	1.4	1.7	0.6	Taux de croissance

10

Gross domestic product and gross domestic product per capita *(continued)*
Current and constant 2005 prices (millions of US dollars); per capita (US dollars); real annual rates of growth
Produit intérieur brut et produit intérieur brut par habitant *(suite)*
Prix courants et constants de 2005 (millions de dollars É.-U.); par habitant (dollars É.U.); taux de croissance réels

Region, country or area	1985	1995	2005	2010	2012	2013	2014	Région, pays ou zone
Samoa								**Samoa**
GDP at current prices	85	196	434	680	800	805	824	PIB aux prix courants
GDP per capita	532	1 151	2 414	3 656	4 236	4 227	4 294	PIB par habitant
GDP at constant prices	251	273	434	460	465	468	475	PIB aux prix constants
Growth rates (annual)	3.9	6.0	5.1	4.4	-2.3	0.5	1.6	Taux de croissance
San Marino								**Saint-Marin**
GDP at current prices	318	1 020	2 027	2 139	1 801	1 865	1 845	PIB aux prix courants
GDP per capita	14 017	39 410	69 333	69 708	57 764	59 412	58 393	PIB par habitant
GDP at constant prices	859	1 305	2 027	1 842	1 542	1 496	1 481	PIB aux prix constants
Growth rates (annual)	2.8	9.3	2.3	-4.6	-7.5	-3.0	-1.0	Taux de croissance
Sao Tome and Principe								**Sao Tomé-et-Principe**
GDP at current prices	77	97	126	195	266	306	337	PIB aux prix courants
GDP per capita	748	772	824	1 142	1 488	1 676	1 811	PIB par habitant
GDP at constant prices	92	93	126	167	183	191	199	PIB aux prix constants
Growth rates (annual)	9.3	2.0	7.1	4.5	4.6	4.2	4.5	Taux de croissance
Saudi Arabia								**Arabie saoudite**
GDP at current prices	103 894	143 152	328 461	526 811	733 956	744 336	752 460	PIB aux prix courants
GDP per capita	7 776	7 593	13 274	18 754	24 883	24 646	24 362	PIB par habitant
GDP at constant prices	147 670	248 623	328 461	374 862	434 389	445 987	462 015	PIB aux prix constants
Growth rates (annual)	-9.8	0.2	5.6	4.8	5.4	2.7	3.6	Taux de croissance
Senegal								**Sénégal**
GDP at current prices	2 807	4 873	8 708	12 926	14 228	14 952	15 658	PIB aux prix courants
GDP per capita	436	559	773	998	1 033	1 051	1 067	PIB par habitant
GDP at constant prices	4 381	5 498	8 708	10 358	11 001	11 399	11 896	PIB aux prix constants
Growth rates (annual)	3.8	4.7	5.6	4.2	4.4	3.6	4.4	Taux de croissance
Serbia								**Serbie**
GDP at current prices	...	21 823[6]	26 252[7]	39 460[7]	40 742[7]	45 520[7]	43 866[7]	PIB aux prix courants
GDP per capita	...	2 861[6]	3 528[7]	5 412[7]	5 659[7]	6 355[7]	6 155[7]	PIB par habitant
GDP at constant prices	...	18 246[6]	26 252[7]	29 943[7]	30 054[7]	30 827[7]	30 268[7]	PIB aux prix constants
Growth rates (annual)	...	5.7[6]	5.5[7]	0.6[7]	-1.0[7]	2.6[7]	-1.8[7]	Taux de croissance
Seychelles								**Seychelles**
GDP at current prices	204	614	919	970	1 134	1 426	1 511	PIB aux prix courants
GDP per capita	2 922	8 005	10 357	10 421	11 999	14 973	15 759	PIB par habitant
GDP at constant prices	460	699	919	1 138	1 312	1 404	1 527	PIB aux prix constants
Growth rates (annual)	10.3	-0.6	9.0	5.9	6.8	7.0	8.7	Taux de croissance
Sierra Leone								**Sierra Leone**
GDP at current prices	1 376	1 179	1 651	2 578	3 787	4 929	4 893	PIB aux prix courants
GDP per capita	397	307	325	446	627	798	775	PIB par habitant
GDP at constant prices	1 724	1 548	1 651	2 130	2 600	3 123	3 341	PIB aux prix constants
Growth rates (annual)	2.3	-10.0	4.5	5.3	15.2	20.1	7.0	Taux de croissance
Singapore								**Singapour**
GDP at current prices	18 555	87 891	127 418	236 420	289 942	302 246	307 872	PIB aux prix courants
GDP per capita	6 850	25 237	28 343	46 549	54 711	55 920	55 910	PIB par habitant
GDP at constant prices	33 360	76 309	127 418	176 458	193 810	202 421	208 329	PIB aux prix constants
Growth rates (annual)	-0.7	7.0	7.5	15.2	3.4	4.4	2.9	Taux de croissance
Sint Maarten (Dutch part)								**Saint-Martin (partie néer.)**
GDP at current prices	...	...	705	892	986	1 023	1 059	PIB aux prix courants
GDP per capita	...	...	21 662	26 937	28 055	28 069	28 084	PIB par habitant
GDP at constant prices	...	...	705	785	798	808	822	PIB aux prix constants
Growth rates (annual)	...	...	...	1.1	1.4	1.3	1.7	Taux de croissance
Slovakia								**Slovaquie**
GDP at current prices	...	19 959	48 805	89 249	93 050	98 034	100 249	PIB aux prix courants
GDP per capita	...	3 722	9 063	16 507	17 182	18 090	18 486	PIB par habitant
GDP at constant prices	...	32 182	48 805	61 573	64 288	65 206	66 850	PIB aux prix constants
Growth rates (annual)	...	5.8	6.4	5.1	1.5	1.4	2.5	Taux de croissance
Slovenia								**Slovénie**
GDP at current prices	...	21 274	36 345	48 014	46 240	47 676	49 491	PIB aux prix courants
GDP per capita	...	10 683	18 204	23 393	22 415	23 090	23 954	PIB par habitant
GDP at constant prices	...	24 719	36 345	39 598	38 772	38 362	39 531	PIB aux prix constants
Growth rates (annual)	...	4.1	4.0	1.2	-2.7	-1.1	3.0	Taux de croissance

10

Gross domestic product and gross domestic product per capita *(continued)*
Current and constant 2005 prices (millions of US dollars); per capita (US dollars); real annual rates of growth
Produit intérieur brut et produit intérieur brut par habitant *(suite)*
Prix courants et constants de 2005 (millions de dollars É.-U.); par habitant (dollars É.U.); taux de croissance réels

Region, country or area	1985	1995	2005	2010	2012	2013	2014	Région, pays ou zone
Solomon Islands								**Îles Salomon**
GDP at current prices	160	365	429	720	976	1 009	1 103	PIB aux prix courants
GDP per capita	591	1 015	915	1 368	1 777	1 799	1 927	PIB par habitant
GDP at constant prices	252	425	429	599	654	674	684	PIB aux prix constants
Growth rates (annual)	2.8	10.0	12.8	10.6	2.6	3.0	1.5	Taux de croissance
Somalia								**Somalie**
GDP at current prices	810	1 122	2 316	1 071	1 306	1 399	1 375	PIB aux prix courants
GDP per capita	133	177	273	112	130	136	131	PIB par habitant
GDP at constant prices	2 480	1 796	2 316	2 628	2 766	2 838	2 912	PIB aux prix constants
Growth rates (annual)	9.5	0.0	3.0	2.6	2.6	2.6	2.6	Taux de croissance
South Africa								**Afrique du Sud**
GDP at current prices	59 083	155 461	257 772	375 348	397 388	366 060	349 819	PIB aux prix courants
GDP per capita	1 791	3 753	5 331	7 271	7 521	6 853	6 482	PIB par habitant
GDP at constant prices	164 193	186 154	257 772	300 266	316 792	323 800	328 738	PIB aux prix constants
Growth rates (annual)	-1.2	3.1	5.3	3.0	2.2	2.2	1.5	Taux de croissance
South Sudan								**Soudan du sud**
GDP at current prices	...	...	...	15 720	10 369	11 804	11 007	PIB aux prix courants
GDP per capita	...	...	...	1 563	944	1 031	924	PIB par habitant
GDP at constant prices	...	...	...	10 919	5 614	6 351	6 537	PIB aux prix constants
Growth rates (annual)	...	...	...	5.5	-46.1	13.1	2.9	Taux de croissance
Spain								**Espagne**
GDP at current prices	180 305	612 943	1 157 248	1 431 588	1 339 947	1 369 262	1 381 342	PIB aux prix courants
GDP per capita	4 655	15 414	26 388	30 720	28 731	29 475	29 861	PIB par habitant
GDP at constant prices	597 137	801 810	1 157 248	1 219 911	1 176 065	1 156 402	1 172 137	PIB aux prix constants
Growth rates (annual)	2.3	2.8	3.7	<0.0	-2.6	-1.7	1.4	Taux de croissance
Sri Lanka								**Sri Lanka**
GDP at current prices	6 005	13 363	24 406	49 566	59 391	67 206	74 941	PIB aux prix courants
GDP per capita	371	732	1 250	2 454	2 908	3 275	3 635	PIB par habitant
GDP at constant prices	10 188	15 678	24 406	33 254	38 278	41 053	44 078	PIB aux prix constants
Growth rates (annual)	5.0	5.5	6.2	8.0	6.3	7.2	7.4	Taux de croissance
State of Palestine								**État de Palestine**
GDP at current prices	1 005	3 283	4 832	8 913	11 279	12 476	12 766	PIB aux prix courants
GDP per capita	571	1 254	1 350	2 191	2 624	2 824	2 811	PIB par habitant
GDP at constant prices	1 463	3 324	4 832	6 167	7 368	7 532	7 504	PIB aux prix constants
Growth rates (annual)	-0.6	7.1	10.8	8.1	6.3	2.2	-0.4	Taux de croissance
Sudan								**Soudan**
GDP at current prices	...	...	...	53 945	60 785	72 067	81 894	PIB aux prix courants
GDP per capita	...	...	...	1 494	1 612	1 871	2 081	PIB par habitant
GDP at constant prices	...	...	...	35 905	35 009	36 165	37 286	PIB aux prix constants
Growth rates (annual)	...	...	...	6.9	-2.2	3.3	3.1	Taux de croissance
Sudan [former]								**Soudan [anc.]**
GDP at current prices	6 624	12 847	35 183	69 665	...	...	...	PIB aux prix courants
GDP per capita	294	426	878	1 509	...	...	...	PIB par habitant
GDP at constant prices	11 109	17 707	35 183	46 910	...	...	...	PIB aux prix constants
Growth rates (annual)	5.4	6.0	9.0	6.6	...	...	...	Taux de croissance
Suriname								**Suriname**
GDP at current prices	1 196	844	2 193	4 368	4 980	5 131	5 210	PIB aux prix courants
GDP per capita	3 221	1 894	4 457	8 431	9 422	9 618	9 680	PIB par habitant
GDP at constant prices	1 578	1 561	2 193	2 750	2 984	3 069	3 125	PIB aux prix constants
Growth rates (annual)	14.0	<0.0	3.9	5.1	3.1	2.8	1.8	Taux de croissance
Swaziland								**Swaziland**
GDP at current prices	504	1 932	3 200	4 548	4 981	4 629	4 482	PIB aux prix courants
GDP per capita	714	2 005	2 897	3 812	4 044	3 701	3 532	PIB par habitant
GDP at constant prices	1 262	2 410	3 200	3 635	3 843	3 959	4 061	PIB aux prix constants
Growth rates (annual)	2.2	3.6	4.4	1.4	4.2	3.0	2.6	Taux de croissance
Sweden								**Suède**
GDP at current prices	112 514	264 053	389 043	488 378	543 881	578 742	571 090	PIB aux prix courants
GDP per capita	13 473	29 915	43 083	52 053	56 990	60 134	58 856	PIB par habitant
GDP at constant prices	245 761	286 677	389 043	420 871	430 848	436 196	446 367	PIB aux prix constants
Growth rates (annual)	2.2	4.0	2.8	6.0	-0.3	1.2	2.3	Taux de croissance

10 Gross domestic product and gross domestic product per capita *(continued)*
Current and constant 2005 prices (millions of US dollars); per capita (US dollars); real annual rates of growth
Produit intérieur brut et produit intérieur brut par habitant *(suite)*
Prix courants et constants de 2005 (millions de dollars É.-U.); par habitant (dollars É.U.); taux de croissance réels

Region, country or area	1985	1995	2005	2010	2012	2013	2014	Région, pays ou zone
Switzerland								**Suisse**
GDP at current prices	107 495	341 768	407 543	581 209	665 408	684 919	701 037	PIB aux prix courants
GDP per capita	16 651	48 705	55 009	74 223	82 941	84 363	85 374	PIB par habitant
GDP at constant prices	290 505	337 990	407 543	454 938	468 356	476 641	485 647	PIB aux prix constants
Growth rates (annual)	3.7	0.5	3.0	3.0	1.1	1.8	1.9	Taux de croissance
Syrian Arab Republic								**Rép. arabe syrienne**
GDP at current prices	10 050	13 547	28 397	60 465	40 057	31 667	34 184	PIB aux prix courants
GDP per capita	942	945	1 566	2 918	2 005	1 639	1 821	PIB par habitant
GDP at constant prices	13 198	18 570	28 397	36 081	26 235	19 755	19 834	PIB aux prix constants
Growth rates (annual)	6.1	7.0	6.2	3.4	-22.4	-24.7	0.4	Taux de croissance
Tajikistan								**Tadjikistan**
GDP at current prices	...	1 218	2 312	5 642	7 633	8 506	9 242	PIB aux prix courants
GDP per capita	...	211	340	744	962	1 049	1 114	PIB par habitant
GDP at constant prices	...	1 454	2 312	3 162	3 483	3 742	3 992	PIB aux prix constants
Growth rates (annual)	...	-12.4	6.7	6.5	7.5	7.4	6.7	Taux de croissance
Thailand								**Thaïlande**
GDP at current prices	40 240	168 998	189 318	340 923	397 472	420 167	404 824	PIB aux prix courants
GDP per capita	773	2 852	2 874	5 112	5 918	6 229	5 977	PIB par habitant
GDP at constant prices	56 940	139 809	189 318	227 448	246 139	253 054	255 245	PIB aux prix constants
Growth rates (annual)	4.6	8.1	4.2	7.5	7.3	2.8	0.9	Taux de croissance
TFYR of Macedonia								**ex-R.Y. de Macédoine**
GDP at current prices	...	4 707	6 259	9 407	9 745	10 818	11 319	PIB aux prix courants
GDP per capita	...	2 409	3 064	4 561	4 710	5 220	5 453	PIB par habitant
GDP at constant prices	...	4 905	6 259	7 610	7 753	7 959	8 259	PIB aux prix constants
Growth rates (annual)	...	-1.1	4.7	3.4	-0.5	2.7	3.8	Taux de croissance
Timor-Leste								**Timor-Leste**
GDP at current prices	...	451	1 850	4 069	6 807	5 596	4 970	PIB aux prix courants
GDP per capita	...	527	1 869	3 849	6 177	4 955	4 294	PIB par habitant
GDP at constant prices	...	624	1 850	2 907	3 446	2 969	3 102	PIB aux prix constants
Growth rates (annual)	...	9.5	52.7	-3.3	5.2	-13.9	4.5	Taux de croissance
Togo								**Togo**
GDP at current prices	740	1 446	2 110	3 173	3 897	4 180	4 576	PIB aux prix courants
GDP per capita	228	338	378	496	578	603	643	PIB par habitant
GDP at constant prices	1 484	1 790	2 110	2 462	2 733	2 873	3 048	PIB aux prix constants
Growth rates (annual)	3.7	6.8	1.2	4.0	5.8	5.1	6.1	Taux de croissance
Tonga								**Tonga**
GDP at current prices	73	203	262	374	466	439	435	PIB aux prix courants
GDP per capita	779	2 120	2 593	3 597	4 445	4 178	4 122	PIB par habitant
GDP at constant prices	185	216	262	272	282	273	278	PIB aux prix constants
Growth rates (annual)	6.6	4.0	1.6	3.6	0.9	-3.1	2.0	Taux de croissance
Trinidad and Tobago								**Trinité-et-Tobago**
GDP at current prices	7 376	5 329	15 982	21 038	24 581	27 257	28 069	PIB aux prix courants
GDP per capita	6 299	4 246	12 323	15 840	18 322	20 217	20 723	PIB par habitant
GDP at constant prices	7 868	7 527	15 982	18 718	18 975	19 305	19 666	PIB aux prix constants
Growth rates (annual)	-4.1	4.0	6.2	-0.1	1.4	1.7	1.9	Taux de croissance
Tunisia								**Tunisie**
GDP at current prices	9 234	19 795	32 272	44 051	45 055	46 154	47 423	PIB aux prix courants
GDP per capita	1 261	2 172	3 194	4 140	4 141	4 194	4 261	PIB par habitant
GDP at constant prices	14 194	19 858	32 272	40 182	40 946	41 944	42 909	PIB aux prix constants
Growth rates (annual)	5.7	2.3	4.0	3.0	3.9	2.4	2.3	Taux de croissance
Turkey								**Turquie**
GDP at current prices	90 379	227 607	482 986	731 144	788 863	823 256	798 414	PIB aux prix courants
GDP per capita	1 838	3 889	7 117	10 111	10 539	10 801	10 299	PIB par habitant
GDP at constant prices	205 784	315 856	482 986	565 099	627 750	654 069	673 129	PIB aux prix constants
Growth rates (annual)	4.2	7.2	8.4	9.2	2.1	4.2	2.9	Taux de croissance
Turkmenistan								**Turkménistan**
GDP at current prices	...	2 190	14 182	22 148	35 164	41 013	47 932	PIB aux prix courants
GDP per capita	...	523	2 987	4 393	6 798	7 827	9 032	PIB par habitant
GDP at constant prices	...	8 891	14 182	23 227	29 598	32 617	35 977	PIB aux prix constants
Growth rates (annual)	...	-7.2	13.0	9.2	11.1	10.2	10.3	Taux de croissance

10

Gross domestic product and gross domestic product per capita *(continued)*
Current and constant 2005 prices (millions of US dollars); per capita (US dollars); real annual rates of growth
Produit intérieur brut et produit intérieur brut par habitant *(suite)*
Prix courants et constants de 2005 (millions de dollars É.-U.); par habitant (dollars É.U.); taux de croissance réels

Region, country or area	1985	1995	2005	2010	2012	2013	2014	Région, pays ou zone
Turks and Caicos Islands								**Îles Turques-et-Caïques**
GDP at current prices	58	191	579	687	716	736	797	PIB aux prix courants
GDP per capita	6 144	12 424	21 877	22 159	22 070	22 245	23 615	PIB par habitant
GDP at constant prices	94	258	579	634	646	655	685	PIB aux prix constants
Growth rates (annual)	10.7	7.3	14.4	1.0	-2.5	1.3	4.6	Taux de croissance
Tuvalu								**Tuvalu**
GDP at current prices	4	12	22	32	40	38	38	PIB aux prix courants
GDP per capita	407	1 275	2 259	3 238	4 044	3 880	3 796	PIB par habitant
GDP at constant prices	8	17	22	24	26	26	27	PIB aux prix constants
Growth rates (annual)	-1.8	-5.0	-3.9	-2.7	0.2	1.3	2.2	Taux de croissance
Uganda								**Ouganda**
GDP at current prices	4 669	7 037	10 984	19 687	24 412	25 496	27 465	PIB aux prix courants
GDP per capita	319	345	392	594	690	697	727	PIB par habitant
GDP at constant prices	2 949	5 585	10 984	16 326	17 898	18 683	19 565	PIB aux prix constants
Growth rates (annual)	-0.3	9.4	10.0	7.7	2.6	4.4	4.7	Taux de croissance
Ukraine								**Ukraine**
GDP at current prices	...	50 379	89 239	141 209	182 592	190 499	131 806[8]	PIB aux prix courants
GDP per capita	...	991	1 907	3 093	4 029	4 218	2 929[8]	PIB par habitant
GDP at constant prices	...	68 183	89 239	93 824	99 084	99 129	92 358[8]	PIB aux prix constants
Growth rates (annual)	...	-12.2	3.1	4.1	0.2	<0.0	-6.8[8]	Taux de croissance
United Arab Emirates								**Émirats arabes unis**
GDP at current prices	40 604	65 744	180 617	286 049	372 314	387 192	399 451	PIB aux prix courants
GDP per capita	30 067	27 974	40 299	34 342	41 588	42 831	43 963	PIB par habitant
GDP at constant prices	72 917	101 625	180 617	203 435	228 784	238 670	249 578	PIB aux prix constants
Growth rates (annual)	-2.4	8.2	4.9	1.6	7.2	4.3	4.6	Taux de croissance
United Kingdom								**Royaume-Uni**
GDP at current prices	489 256	1 237 624	2 418 949	2 403 581	2 630 473	2 712 296	2 988 893	PIB aux prix courants
GDP per capita	8 672	21 374	40 175	38 324	41 377	42 409	46 461	PIB par habitant
GDP at constant prices	1 392 626	1 797 821	2 418 949	2 466 786	2 545 100	2 600 071	2 676 520	PIB aux prix constants
Growth rates (annual)	4.1	2.5	3.0	1.5	1.2	2.2	2.9	Taux de croissance
United Rep. of Tanzania [9]								**Rép.-Unie de Tanzanie** [9]
GDP at current prices	11 654	7 575	18 072	31 105	38 809	44 333	48 030	PIB aux prix courants
GDP per capita	534	260	476	701	820	908	952	PIB par habitant
GDP at constant prices	7 120	10 467	18 072	24 275	27 540	29 540	31 598	PIB aux prix constants
Growth rates (annual)	4.6	3.6	7.4	6.4	5.1	7.3	7.0	Taux de croissance
United States								**États-Unis**
GDP at current prices	4 346 700	7 664 060	13 093 720	14 964 380	16 155 255	16 663 160	17 348 072	PIB aux prix courants
GDP per capita	18 059	28 782	44 215	48 291	51 319	52 543	54 306	PIB par habitant
GDP at constant prices	6 977 895	9 359 506	13 093 720	13 599 256	14 124 336	14 334 721	14 682 739	PIB aux prix constants
Growth rates (annual)	4.2	2.8	3.3	2.5	2.2	1.5	2.4	Taux de croissance
Uruguay								**Uruguay**
GDP at current prices	5 226	21 312	17 363	40 285	51 384	57 525	57 471	PIB aux prix courants
GDP per capita	1 735	6 609	5 221	11 938	15 127	16 879	16 807	PIB par habitant
GDP at constant prices	10 537	15 453	17 363	23 193	25 201	26 487	27 413	PIB aux prix constants
Growth rates (annual)	1.5	-1.5	7.5	7.8	3.3	5.1	3.5	Taux de croissance
Uzbekistan								**Ouzbékistan**
GDP at current prices	...	13 474	14 396	39 526	52 127	57 157	63 030	PIB aux prix courants
GDP per capita	...	594	555	1 425	1 823	1 969	2 139	PIB par habitant
GDP at constant prices	...	9 061	14 396	21 707	25 437	27 472	29 697	PIB aux prix constants
Growth rates (annual)	...	-0.9	7.0	8.5	8.2	8.0	8.1	Taux de croissance
Vanuatu								**Vanuatu**
GDP at current prices	133	273	395	701	782	802	812	PIB aux prix courants
GDP per capita	1 020	1 621	1 886	2 966	3 158	3 167	3 138	PIB par habitant
GDP at constant prices	240	321	395	504	519	529	548	PIB aux prix constants
Growth rates (annual)	1.1	4.7	5.3	1.6	1.8	2.0	3.6	Taux de croissance
Venezuela (Boliv. Rep. of)								**Venezuela (Rép. boliv. du)**
GDP at current prices	59 963	74 889	145 513	393 808	381 286	371 339	509 968	PIB aux prix courants
GDP per capita	3 425	3 375	5 436	13 582	12 772	12 265	16 615	PIB par habitant
GDP at constant prices	91 789	123 574	145 513	174 551	192 071	194 651	186 865	PIB aux prix constants
Growth rates (annual)	0.2	4.0	10.3	-1.5	5.6	1.3	-4.0	Taux de croissance

10

Gross domestic product and gross domestic product per capita *(continued)*
Current and constant 2005 prices (millions of US dollars); per capita (US dollars); real annual rates of growth

Produit intérieur brut et produit intérieur brut par habitant *(suite)*
Prix courants et constants de 2005 (millions de dollars É.-U.); par habitant (dollars É.U.); taux de croissance réels

Region, country or area	1985	1995	2005	2010	2012	2013	2014	Région, pays ou zone
Viet Nam								**Viet Nam**
GDP at current prices	4 797	20 736	57 633	115 932	155 820	171 222	186 205	PIB aux prix courants
GDP per capita	79	276	684	1 312	1 725	1 874	2 015	PIB par habitant
GDP at constant prices	15 942	28 984	57 633	78 282	87 531	92 277	97 799	PIB aux prix constants
Growth rates (annual)	5.6	9.5	7.5	6.4	5.2	5.4	6.0	Taux de croissance
Yemen								**Yémen**
GDP at current prices	...	5 936	19 041	30 907	32 075	34 714	37 131	PIB aux prix courants
GDP per capita	...	389	929	1 310	1 289	1 360	1 418	PIB par habitant
GDP at constant prices	...	9 843	19 041	23 604	21 006	21 687	21 646	PIB aux prix constants
Growth rates (annual)	...	16.7	5.1	5.7	2.0	3.2	-0.2	Taux de croissance
Zambia								**Zambie**
GDP at current prices	2 772	3 807	8 332	20 265	24 940	26 821	26 963	PIB aux prix courants
GDP per capita	395	411	692	1 456	1 687	1 759	1 715	PIB par habitant
GDP at constant prices	4 880	5 164	8 332	12 647	14 354	15 318	16 176	PIB aux prix constants
Growth rates (annual)	-<0.0	2.9	7.2	10.3	6.7	6.7	5.6	Taux de croissance
Zanzibar								**Zanzibar**
GDP at current prices	...	186	438	746	989	1 156	1 289	PIB aux prix courants
GDP per capita	...	235	408	588	759	835	935	PIB par habitant
GDP at constant prices	...	225	438	570	654	701	750	PIB aux prix constants
Growth rates (annual)	...	3.5	4.9	4.3	4.9	7.2	7.0	Taux de croissance
Zimbabwe								**Zimbabwe**
GDP at current prices	7 548	9 542	6 223	9 422	12 393	13 490	14 719	PIB aux prix courants
GDP per capita	852	817	479	674	851	905	965	PIB par habitant
GDP at constant prices	5 897	7 826	6 223	9 573	11 845	12 376	12 766	PIB aux prix constants
Growth rates (annual)	6.9	0.2	-4.1	11.4	10.6	4.5	3.1	Taux de croissance

Source:
United Nations Statistics Division, New York, national accounts estimates main aggregates database, last accessed March 2016.

Source:
Organisation des Nations Unies, Division de statistique, New York, la base de données d'estimation des comptes nationaux, dernier accès mars 2016.

1	For statistical purposes, the data for China do not include those for the Hong Kong Special Administrative Region (Hong Kong SAR), Macao Special Administrative Region (Macao SAR) and Taiwan Province of China.
2	Excluding northern Cyprus.
3	Including Guadeloupe, Martinique, Réunion and French Guiana.
4	Including Western Sahara.
5	Does not incorporate value added generated by binational hydroelectric plants.
6	Including Kosovo and Metohija.
7	Excluding Kosovo and Metohija.
8	Data for GDP and its components excludes the temporarily occupied territory of the Autonomous Republic of Crimea and Sevastopol.
9	Tanzania mainland only.

1 Pour la présentation des statistiques, les données pour la Chine ne comprennent pas la Région Administrative Spéciale de Hong Kong (Hong Kong RAS), la Région Administrative Spéciale de Macao (Macao RAS) et la province de Taiwan.
2 Chypre du nord non compris.
3 Y compris Guadeloupe, Martinique, Réunion et Guyane française.
4 Y compris les données de Sahara occidental.
5 Ne comprend pas la valeur ajoutée produite par les centrales hydroélectriques binationales.
6 Y compris Kosovo et Metohija.
7 Non compris Kosovo et Metohija.
8 Les données concernant le PIB et ses composantes excluent le territoire temporairement occupé de la République autonome de Crimée et de Sébastopol.
9 Tanzanie continentale seulement.

Gross value added by kind of economic activity
Percentage distribution, current prices
Valeur ajoutée brute par type d'activité économique
Répartition en pourcentage, aux prix courants

Country or area	1985	1995	2005	2010	2012	2013	2014	Pays ou zone
Afghanistan								**Afghanistan**
Agriculture	51.2	65.7	36.6	29.6	26.9	26.2	25.7	Agriculture
Industry	24.9	10.5	26.9	21.9	21.8	20.8	22.1	Industrie
Services	23.9	23.8	36.5	48.5	51.3	52.9	52.2	Services
Albania								**Albanie**
Agriculture	30.8[1]	50.8[1]	20.1	20.7	21.8	22.6	22.7	Agriculture
Industry	54.7[1]	31.3[1]	33.5	28.9	26.6	27.1	26.5	Industrie
Services	14.5[1]	17.9[1]	46.4	50.4	51.6	50.3	50.8	Services
Algeria								**Algérie**
Agriculture	8.4	10.4	8.0	8.6	9.0	10.2	10.7	Agriculture
Industry	49.8	47.9	59.7	51.4	49.3	45.9	43.9	Industrie
Services	41.9	41.7	32.3	40.0	41.7	43.8	45.4	Services
Andorra								**Andorre**
Agriculture[2]	0.5	0.5	0.4	0.5	0.6	0.6	0.6	Agriculture[2]
Industry[3]	18.2	18.6	17.7	15.0	12.5	11.5	12.6	Industrie[3]
Services[4,5]	81.3	80.9	81.9	84.5	86.9	87.9	86.8	Services[4,5]
Angola								**Angola**
Agriculture	13.8	7.4	5.0	6.0	5.1	5.3	4.9	Agriculture
Industry	43.5	67.4	59.8	57.2	60.2	53.6	56.1	Industrie
Services	42.8	25.2	35.2	36.8	34.7	41.1	38.9	Services
Anguilla								**Anguilla**
Agriculture	5.6	3.3	2.7	2.0	2.5	2.3	2.2	Agriculture
Industry	16.6	14.5	19.3	15.8	14.1	14.5	16.5	Industrie
Services	77.8	82.2	78.0	82.2	83.5	83.2	81.3	Services
Antigua and Barbuda								**Antigua-et-Barbuda**
Agriculture	2.2	1.9	2.0	1.9	2.2	2.3	2.2	Agriculture
Industry	11.4	14.4	16.8	18.5	16.1	17.6	18.1	Industrie
Services	86.4	83.7	81.2	79.6	81.7	80.2	79.8	Services
Argentina								**Argentine**
Agriculture	6.5	4.8	8.4	8.2	6.7	7.4	8.2	Agriculture
Industry	37.4	26.9	34.7	30.9	29.8	28.8	28.8	Industrie
Services	56.1	68.3	56.9	60.9	63.5	63.8	63.0	Services
Armenia								**Arménie**
Agriculture	...	41.9	20.6	18.8[2]	20.9[2]	21.6[2]	21.7[2]	Agriculture
Industry	...	29.3	44.7	36.3[3]	32.2[3]	31.1[3]	29.7[3]	Industrie
Services	...	28.8	34.6	45.0[4,5]	46.8[4,5]	47.4[4,5]	48.6[4,5]	Services
Aruba								**Aruba**
Agriculture[6]	0.5	0.5	0.4	0.5	0.5	0.5	0.5	Agriculture[6]
Industry[7]	15.9	15.6	19.6	15.4	15.4	15.4	15.4	Industrie[7]
Services	83.5	83.9	80.0	84.2	84.1	84.2	84.2	Services
Australia								**Australie**
Agriculture[2]	4.5	3.7	3.0	2.5	2.4	2.5	2.4	Agriculture[2]
Industry[3]	35.5	28.5	27.9	28.5	27.1	27.1	27.4	Industrie[3]
Services[4,5]	60.0	67.8	69.1	69.0	70.5	70.4	70.1	Services[4,5]
Austria								**Autriche**
Agriculture[2]	3.6	2.4	1.4	1.4	1.5	1.4	1.4	Agriculture[2]
Industry[3]	34.6	32.2	30.4	28.7	28.8	28.3	28.0	Industrie[3]
Services[4,5]	61.8	65.4	68.2	69.9	69.7	70.3	70.6	Services[4,5]
Azerbaijan								**Azerbaïdjan**
Agriculture[2]	...	26.9	9.8	5.9	5.5	5.7	5.7	Agriculture[2]
Industry[3]	...	32.9	63.2	64.0	63.2	60.8	57.9	Industrie[3]
Services[4,5]	...	40.3	27.0	30.1	31.3	33.5	36.4	Services[4,5]
Bahamas								**Bahamas**
Agriculture	2.2	3.6	2.1	2.2	2.2	1.8	1.6	Agriculture
Industry	13.9	12.7	14.6	15.2	16.6	17.8	19.9	Industrie
Services	84.0	83.7	83.4	82.6	81.2	80.4	78.5	Services
Bahrain[1]								**Bahreïn**[1]
Agriculture	0.8	0.7	0.3	0.3	0.3	0.3	0.3	Agriculture
Industry	38.9	36.1	42.8	45.5	48.0	48.9	47.4	Industrie
Services	60.2	63.2	56.9	54.2	51.7	50.8	52.3	Services

Gross value added by kind of economic activity *(continued)*
Percentage distribution, current prices

Valeur ajoutée brute par type d'activité économique *(suite)*
Répartition en pourcentage, aux prix courants

Country or area	1985	1995	2005	2010	2012	2013	2014	Pays ou zone
Bangladesh								**Bangladesh**
Agriculture	35.7	26.4	20.1	17.8	17.1	16.3	16.1	Agriculture
Industry	21.5	24.6	27.2	26.1	26.7	27.6	27.6	Industrie
Services	42.8	49.1	52.6	56.0	56.2	56.1	56.3	Services
Barbados								**Barbade**
Agriculture	4.5	3.5	1.8	1.5	1.6	1.8	1.7	Agriculture
Industry	23.0	16.6	16.9	14.4	12.9	11.9	11.2	Industrie
Services	72.5	79.9	81.3	84.0	85.4	86.2	87.1	Services
Belarus								**Bélarus**
Agriculture	...	16.8	9.8	10.2	9.5	7.9	8.6	Agriculture
Industry	...	37.7	43.4	40.8	41.8	41.0	41.3	Industrie
Services	...	45.5	46.8	49.0	48.7	51.1	50.1	Services
Belgium								**Belgique**
Agriculture [2]	2.6	1.4	0.9	0.9	0.9	0.8	0.7	Agriculture [2]
Industry [3]	31.4	29.0	25.1	23.2	22.5	22.2	22.1	Industrie [3]
Services [4,5]	66.0	69.6	74.0	76.0	76.7	77.0	77.2	Services [4,5]
Belize								**Belize**
Agriculture	19.8	19.8	14.7	12.6	14.4	14.7	13.9	Agriculture
Industry	24.3	23.5	16.5	20.7	19.3	18.3	19.7	Industrie
Services	55.9	56.8	68.8	66.7	66.3	67.0	66.4	Services
Benin								**Bénin**
Agriculture	32.4	23.7	27.1	25.4	24.9	23.6	23.0	Agriculture
Industry	15.5	32.5	30.3	24.7	22.6	22.5	22.7	Industrie
Services	52.0	43.8	42.5	49.9	52.6	53.9	54.3	Services
Bermuda								**Bermudes**
Agriculture	0.8	0.8	0.8	0.7	0.7	0.7	0.6	Agriculture
Industry	10.4	10.4	9.7	7.1	5.4	5.3	5.3	Industrie
Services	88.9	88.8	89.5	92.2	93.9	94.0	94.1	Services
Bhutan								**Bhoutan**
Agriculture	42.3	33.2	23.2	17.5	17.0	17.1	17.1	Agriculture
Industry	21.3	33.5	37.3	44.6	44.3	44.6	44.0	Industrie
Services	36.4	33.3	39.5	37.9	38.8	38.3	39.0	Services
Bolivia (Plur. State of)								**Bolivie (État plur. de)**
Agriculture	28.7	16.4	13.9	12.4	12.3	12.7	12.4	Agriculture
Industry	28.5	32.1	30.9	35.8	36.9	36.2	35.0	Industrie
Services	42.8	51.5	55.2	51.8	50.8	51.1	52.6	Services
Bosnia and Herzegovina								**Bosnie-Herzégovine**
Agriculture [2]	...	22.0	9.8	8.0	7.3	8.0	7.1	Agriculture [2]
Industry [3]	...	31.8	25.3	26.4	25.5	26.1	26.0	Industrie [3]
Services [4,5]	...	46.2	64.9	65.6	67.2	65.9	66.9	Services [4,5]
Botswana								**Botswana**
Agriculture	7.4	4.9	2.0	2.8	2.9	2.5	2.4	Agriculture
Industry	54.3	46.5	47.6	35.7	34.7	37.3	37.9	Industrie
Services	38.3	48.6	50.3	61.6	62.4	60.1	59.7	Services
Brazil								**Brésil**
Agriculture	11.0	5.5[2]	5.5[2]	4.9[2]	5.3[2]	5.6[2]	5.6[2]	Agriculture
Industry	42.2	26.0[3]	28.6[3]	27.4[3]	25.4[3]	24.4[3]	23.4[3]	Industrie
Services	46.7	68.5[4,5]	65.9[4,5]	67.8[4,5]	69.4[4,5]	70.0[4,5]	71.0[4,5]	Services
British Virgin Islands								**Îles Vierges britanniques**
Agriculture	4.3	1.8	1.1	1.0	1.0	1.0	1.0	Agriculture
Industry	13.1	13.3	11.9	10.0	11.1	11.2	11.1	Industrie
Services	82.6	84.9	87.1	89.0	87.9	87.7	87.9	Services
Brunei Darussalam								**Brunéi Darussalam**
Agriculture [2]	0.4	1.1	0.9	0.7	0.6	0.7	0.8	Agriculture [2]
Industry [3]	78.5	53.4	72.1	67.4	71.5	68.7	66.8	Industrie [3]
Services [4,5]	21.1	45.5	27.0	31.9	27.8	30.6	32.4	Services [4,5]
Bulgaria								**Bulgarie**
Agriculture [2]	13.0	13.4	8.7	5.1	5.4	5.5	5.3	Agriculture [2]
Industry [3]	53.7	25.6	28.8	27.8	30.1	27.9	28.4	Industrie [3]
Services [4,5]	33.3	61.0	62.5	67.1	64.5	66.6	66.4	Services [4,5]

Gross value added by kind of economic activity *(continued)*
Percentage distribution, current prices

Valeur ajoutée brute par type d'activité économique *(suite)*
Répartition en pourcentage, aux prix courants

Country or area	1985	1995	2005	2010	2012	2013	2014	Pays ou zone
Burkina Faso								**Burkina Faso**
Agriculture	34.9	34.3	38.9	35.6	34.7	37.9	35.4	Agriculture
Industry	19.8	24.8	17.9	23.0	24.7	21.0	24.2	Industrie
Services	45.4	40.9	43.1	41.4	40.6	41.1	40.3	Services
Burundi								**Burundi**
Agriculture	56.1	48.2	43.0	40.7	38.0	38.9	41.1	Agriculture
Industry	17.0	19.2	17.8	16.3	16.9	16.3	17.9	Industrie
Services	26.9	32.6	39.2	43.0	45.1	44.8	40.9	Services
Cabo Verde								**Cabo Verde**
Agriculture	16.5	15.7	11.7	9.2	9.8	9.3	9.4	Agriculture
Industry	24.3	30.1	22.8	20.8	19.7	20.4	20.2	Industrie
Services	59.3	54.1	65.4	70.1	70.5	70.3	70.4	Services
Cambodia								**Cambodge**
Agriculture	47.0	51.4	32.4	36.0	35.6	33.5	30.5	Agriculture
Industry	13.6	12.9	26.4	23.3	24.4	25.6	27.1	Industrie
Services	39.4	35.7	41.2	40.7	40.1	40.8	42.4	Services
Cameroon								**Cameroun**
Agriculture	19.8	23.7	20.4	23.3	23.0	22.7	22.0	Agriculture
Industry	32.4	29.8	31.8	29.7	30.1	29.7	29.9	Industrie
Services	47.8	46.5	47.8	47.0	46.9	47.6	48.1	Services
Canada								**Canada**
Agriculture	3.5	2.9	1.8	1.5^2	1.8^2	1.7^2	1.7^2	Agriculture
Industry	35.1	30.7	32.4	27.7^3	28.2^3	28.2^3	28.4^3	Industrie
Services	61.4	66.4	65.8	$70.8^{4,5}$	$70.0^{4,5}$	$70.1^{4,5}$	$69.9^{4,5}$	Services
Cayman Islands								**Îles Caïmanes**
Agriculture[2]	0.3	0.2	0.2	0.3	0.3	0.3	0.3	Agriculture[2]
Industry[3]	9.8	9.2	9.1	7.7	7.5	7.5	7.5	Industrie[3]
Services[4,5]	89.9	90.5	90.6	92.0	92.2	92.2	92.2	Services[4,5]
Central African Rep.								**Rép. centrafricaine**
Agriculture	35.7	36.1	45.0	41.2	41.6	41.7	41.9	Agriculture
Industry	22.2	26.9	18.6	24.0	23.7	23.9	23.9	Industrie
Services	42.1	37.0	36.5	34.8	34.7	34.4	34.3	Services
Chad								**Tchad**
Agriculture	32.0	27.9	26.1	35.9	24.3	23.9	24.4	Agriculture
Industry	14.3	14.4	34.9	36.7	45.2	42.6	40.0	Industrie
Services	53.7	57.6	39.0	27.4	30.5	33.5	35.5	Services
Chile[1]								**Chili**[1]
Agriculture	5.4	5.4	4.1	3.5	3.2	3.2	3.3	Agriculture
Industry	37.0	41.4	40.3	39.6	36.6	35.1	35.1	Industrie
Services	57.6	53.2	55.6	57.0	60.1	61.7	61.5	Services
China[1,8]								**Chine**[1,8]
Agriculture	28.4	19.9	12.1	9.9	9.8	9.7	9.5	Agriculture
Industry	42.7	46.8	47.0	46.4	45.2	43.9	42.9	Industrie
Services	29.0	33.3	40.9	43.7	45.0	46.4	47.7	Services
China, Hong Kong SAR								**Chine, Hong Kong RAS**
Agriculture[9]	0.5	0.1	0.1	0.1	0.1	0.1	0.1	Agriculture[9]
Industry	26.9	14.1	8.7	7.0	6.9	7.1	7.3	Industrie
Services[10]	72.6	85.7	91.3	93.0	93.0	92.9	92.7	Services[10]
China, Macao SAR								**Chine, Macao RAS**
Industry	24.7	15.1	15.2	7.3	6.3	5.7	6.1	Industrie
Services	75.3	84.9	84.8	92.7	93.7	94.3	93.9	Services
Colombia								**Colombie**
Agriculture	12.3	9.3	8.4	7.1	6.3	6.0	6.4	Agriculture
Industry	31.2	30.0	32.8	35.0	37.8	37.2	36.5	Industrie
Services	56.5	60.7	58.8	57.9	55.9	56.8	57.1	Services
Comoros								**Comores**
Agriculture	36.1	39.3	41.0	38.5	35.9	34.9	33.6	Agriculture
Industry	14.1	15.5	11.7	11.5	11.0	10.8	10.6	Industrie
Services	49.8	45.3	47.2	50.0	53.1	54.3	55.8	Services

Gross value added by kind of economic activity *(continued)*
Percentage distribution, current prices

Valeur ajoutée brute par type d'activité économique *(suite)*
Répartition en pourcentage, aux prix courants

Country or area	1985	1995	2005	2010	2012	2013	2014	Pays ou zone
Congo								**Congo**
Agriculture	7.6	10.9	4.6	3.7	4.0	4.5	5.1	Agriculture
Industry	54.9	46.9	73.4	78.1	76.2	73.4	70.9	Industrie
Services	37.5	42.2	22.0	18.2	19.8	22.1	24.0	Services
Cook Islands[1]								**Îles Cook**[1]
Agriculture	6.9	7.4	6.9	4.9	4.5	6.8	9.0	Agriculture
Industry	8.8	7.8	9.6	8.5	9.5	6.5	7.5	Industrie
Services	84.2	84.8	83.5	86.6	86.0	86.7	83.5	Services
Costa Rica								**Costa Rica**
Agriculture	14.6	13.3	8.6	6.8	5.8	5.3	5.2	Agriculture
Industry	33.3	28.7	27.8	24.9	24.0	23.8	23.2	Industrie
Services	52.1	58.0	63.6	68.2	70.2	70.9	71.6	Services
Côte d'Ivoire								**Côte d'Ivoire**
Agriculture	28.1	26.7	25.2	27.0	24.8	24.5	25.4	Agriculture
Industry	21.8	22.4	25.5	24.7	25.9	26.1	24.0	Industrie
Services	50.1	50.9	49.3	48.3	49.3	49.4	50.6	Services
Croatia								**Croatie**
Agriculture[2]	...	7.2	5.0	4.9	4.5	4.4	4.3	Agriculture[2]
Industry[3]	...	32.3	29.0	27.1	27.1	26.6	26.3	Industrie[3]
Services[4,5]	...	60.5	66.0	68.1	68.5	69.0	69.4	Services[4,5]
Cuba								**Cuba**
Agriculture	12.5	8.8	5.6	5.0	5.0	5.0	5.0	Agriculture
Industry	19.2	22.9	19.4	20.5	20.5	20.5	20.5	Industrie
Services	68.2	68.4	75.0	74.5	74.5	74.5	74.5	Services
Curaçao								**Curaçao**
Agriculture[6]	...	...	0.6	0.5	0.4	0.4	0.4	Agriculture[6]
Industry	...	...	16.4	16.1	18.9	19.2	19.3	Industrie
Services	...	...	83.0	83.5	80.7	80.5	80.3	Services
Cyprus[11]								**Chypre**[11]
Agriculture[2]	7.1	4.8	2.7	2.3	2.5	2.5	2.3	Agriculture[2]
Industry[3]	27.1	22.1	20.1	16.6	13.6	11.7	10.8	Industrie[3]
Services[4,5]	65.9	73.1	77.3	81.1	83.9	85.8	86.9	Services[4,5]
Czech Republic								**République tchèque**
Agriculture[2]	...	4.4	2.4	1.7	2.6	2.7	2.7	Agriculture[2]
Industry[3]	...	39.0	37.7	36.8	37.0	36.9	38.0	Industrie[3]
Services[4,5]	...	56.7	59.8	61.5	60.4	60.4	59.3	Services[4,5]
Dem. P. R. Korea								**R. p. dém. de Corée**
Agriculture	28.3	27.6	25.0	20.8	23.4	22.4	21.8	Agriculture
Industry	48.5	42.0	42.8	48.2	47.2	47.6	46.9	Industrie
Services	23.2	30.3	32.2	31.0	29.4	30.0	31.3	Services
Dem. Rep. of the Congo								**Rép. dém. du Congo**
Agriculture	31.8	57.0	22.3	22.4	21.8	20.8	21.5	Agriculture
Industry	31.0	17.0	32.9	40.5	42.9	44.4	43.5	Industrie
Services	37.1	26.0	44.9	37.0	35.3	34.8	35.0	Services
Denmark								**Danemark**
Agriculture[2]	4.5	3.3	1.3	1.4	1.9	1.5	1.6	Agriculture[2]
Industry[3]	25.5	25.5	26.2	22.8	23.7	23.5	22.8	Industrie[3]
Services[4,5]	70.0	71.2	72.4	75.8	74.3	75.0	75.6	Services[4,5]
Djibouti								**Djibouti**
Agriculture	3.1[12]	3.2[12]	3.6	3.6	3.5	3.6	3.2	Agriculture
Industry	21.2[13]	15.4[13]	16.2	19.0	18.3	19.2	22.7	Industrie
Services	75.7	81.3	80.2	77.4	78.2	77.3	74.0	Services
Dominica								**Dominique**
Agriculture	22.7	13.7	13.2	13.8	16.4	16.8	16.7	Agriculture
Industry	13.7	16.0	15.0	14.4	14.2	13.7	13.8	Industrie
Services	63.6	70.3	71.9	71.8	69.4	69.5	69.4	Services
Dominican Republic								**Rép. dominicaine**
Agriculture[2]	13.7	10.9	8.1	6.5	6.3	6.3	6.2	Agriculture[2]
Industry[3]	43.7	37.8	32.7	30.1	29.0	28.8	28.7	Industrie[3]
Services[4,5]	42.5	51.3	59.2	63.4	64.8	64.9	65.1	Services[4,5]

11

Gross value added by kind of economic activity *(continued)*
Percentage distribution, current prices
Valeur ajoutée brute par type d'activité économique *(suite)*
Répartition en pourcentage, aux prix courants

Country or area	1985	1995	2005	2010	2012	2013	2014	Pays ou zone
Ecuador								**Équateur**
Agriculture	29.9	25.1	10.0	10.2	9.0	9.1	9.1	Agriculture
Industry	22.8	21.2	33.4	36.3	39.2	39.2	39.1	Industrie
Services	47.3	53.6	56.6	53.5	51.8	51.8	51.8	Services
Egypt [14]								**Égypte** [14]
Agriculture	18.3	16.8	14.4	14.0	14.5	14.5	14.5	Agriculture
Industry	31.7	32.7	36.9	37.5	39.2	39.2	39.9	Industrie
Services	50.0	50.9	48.8	48.5	46.3	46.3	45.6	Services
El Salvador [1]								**El Salvador** [1]
Agriculture	18.2	14.0	10.2	12.1	11.5	10.6	11.0	Agriculture
Industry	21.9	28.7	28.7	25.9	26.1	26.3	25.8	Industrie
Services	59.9	57.3	61.1	62.1	62.5	63.1	63.3	Services
Equatorial Guinea								**Guinée équatoriale**
Agriculture	60.7	55.3	2.0	1.6	1.3	1.6	1.7	Agriculture
Industry	8.1	28.2	95.2	94.9	95.5	94.7	94.1	Industrie
Services	31.2	16.5	2.8	3.4	3.2	3.7	4.2	Services
Eritrea								**Érythrée**
Agriculture	...	20.9	24.2	19.1	16.9	17.6	17.2	Agriculture
Industry	...	16.8	21.9	23.1	23.2	23.5	23.6	Industrie
Services	...	62.3	53.9	57.8	59.9	58.9	59.2	Services
Estonia								**Estonie**
Agriculture [2]	...	5.7	3.5	3.2	4.0	3.5	3.4	Agriculture [2]
Industry [3]	...	31.9	29.8	28.0	28.8	28.9	28.1	Industrie [3]
Services [4,5]	...	62.4	66.7	68.8	67.2	67.6	68.4	Services [4,5]
Ethiopia								**Éthiopie**
Agriculture	...	57.3	45.2	45.3	47.7	44.6	41.9	Agriculture
Industry	...	10.2	13.1	10.4	10.2	11.8	15.3	Industrie
Services	...	32.4	41.7	44.3	42.1	43.6	42.8	Services
Fiji								**Fidji**
Agriculture	16.7	16.4	12.8	10.2	11.4	12.1	11.5	Agriculture
Industry	16.3	17.8	17.9	19.9	19.8	19.6	18.7	Industrie
Services	67.0	65.8	69.2	69.9	68.8	68.3	69.9	Services
Finland								**Finlande**
Agriculture [2]	7.8	4.3	2.6	2.7	2.7	3.0	2.8	Agriculture [2]
Industry [3]	35.1	33.7	33.5	30.0	27.0	26.9	26.5	Industrie [3]
Services [4,5]	57.1	62.0	63.8	67.3	70.2	70.1	70.6	Services [4,5]
France [15]								**France** [15]
Agriculture [2]	3.8	2.7	1.9	1.8	1.8	1.6	1.7	Agriculture [2]
Industry [3]	28.4	24.5	21.5	19.6	19.7	19.7	19.4	Industrie [3]
Services [4,5]	67.9	72.7	76.6	78.6	78.5	78.7	78.9	Services [4,5]
French Polynesia								**Polynésie française**
Agriculture	4.7	5.2	3.4	2.5	2.6	2.6	2.7	Agriculture
Industry	21.0	12.5	12.7	12.3	12.4	12.1	12.1	Industrie
Services	74.3	82.3	83.9	85.2	85.0	85.3	85.3	Services
Gabon								**Gabon**
Agriculture	5.0	7.3	5.2	4.6	3.7	3.6	3.9	Agriculture
Industry	58.5	54.5	62.6	55.6	57.5	55.2	50.9	Industrie
Services	36.5	38.2	32.2	39.8	38.8	41.2	45.2	Services
Gambia								**Gambie**
Agriculture	23.3	20.6	28.6	30.7	23.5	22.6	19.7	Agriculture
Industry	10.7	14.3	14.9	13.1	15.2	15.0	15.1	Industrie
Services	66.0	65.1	56.5	56.3	61.2	62.4	65.1	Services
Georgia								**Géorgie**
Agriculture	...	44.4	16.5	8.3	8.5	9.3	9.1	Agriculture
Industry	...	14.3	26.5	22.0	24.2	23.7	24.0	Industrie
Services	...	41.3	57.0	69.8	67.3	67.0	66.9	Services
Germany								**Allemagne**
Agriculture [2]	1.6	1.0	0.8	0.7	0.8	0.8	0.7	Agriculture [2]
Industry [3]	39.4	32.9	29.4	30.2	30.8	30.3	30.3	Industrie [3]
Services [4,5]	59.0	66.0	69.8	69.1	68.4	68.9	69.0	Services [4,5]

Gross value added by kind of economic activity *(continued)*
Percentage distribution, current prices

Valeur ajoutée brute par type d'activité économique *(suite)*
Répartition en pourcentage, aux prix courants

Country or area	1985	1995	2005	2010	2012	2013	2014	Pays ou zone
Ghana								**Ghana**
Agriculture	37.8	34.0	31.8	29.8[2]	22.9[2]	22.4[2]	22.0[2]	Agriculture
Industry	15.8	20.9	20.3	19.1[3]	28.0[3]	27.8[3]	28.4[3]	Industrie
Services	46.4	45.1	47.9	51.1[4,5]	49.2[4,5]	49.8[4,5]	49.6[4,5]	Services
Greece								**Grèce**
Agriculture[2]	10.2	8.1	4.8	3.3	3.7	3.7	3.8	Agriculture[2]
Industry[3]	27.0	21.6	19.8	15.7	16.2	16.4	15.8	Industrie[3]
Services[4,5]	62.7	70.3	75.4	81.1	80.1	79.9	80.4	Services[4,5]
Greenland								**Groenland**
Agriculture	10.2	10.2	10.2	7.5	9.6	9.8	10.6	Agriculture
Industry	14.8	14.8	15.3	17.6	17.7	15.9	13.9	Industrie
Services	75.0	75.0	74.5	74.9	72.7	74.3	75.4	Services
Grenada								**Grenade**
Agriculture	16.1	9.5	3.4	5.2	5.5	5.5	6.5	Agriculture
Industry	16.3	18.6	26.1	16.8	14.1	15.0	14.0	Industrie
Services	67.7	71.9	70.5	78.1	80.4	79.5	79.5	Services
Guatemala								**Guatemala**
Agriculture	16.8	15.8	13.1	11.4	10.9	10.9	11.2	Agriculture
Industry	29.7	29.0	28.6	28.0	28.6	28.1	28.1	Industrie
Services	53.5	55.3	58.3	60.6	60.5	60.9	60.8	Services
Guinea								**Guinée**
Agriculture	18.3	25.7	24.3	22.1	21.0	21.2	18.0	Agriculture
Industry	31.9	27.2	34.8	44.7	41.7	40.4	41.4	Industrie
Services	49.8	47.1	40.9	33.2	37.3	38.5	40.6	Services
Guinea-Bissau								**Guinée-Bissau**
Agriculture	46.5	55.1	45.4	46.4	47.3	44.9	46.0	Agriculture
Industry	15.6	12.2	14.7	13.4	14.3	15.4	14.3	Industrie
Services	37.9	32.7	39.9	40.3	38.4	39.7	39.7	Services
Guyana								**Guyana**
Agriculture	20.0	37.1	25.7[16]	17.6	17.1	18.2	18.2	Agriculture
Industry	31.6	33.3	28.7	34.5	34.8	34.4	32.8	Industrie
Services	48.4	29.6	45.6	47.9	48.1	47.4	49.1	Services
Haiti								**Haïti**
Agriculture	33.3	24.3	22.4	21.0	18.7	18.6	17.6	Agriculture
Industry	24.7	31.4	32.9	33.7	36.1	36.8	37.7	Industrie
Services	42.0	44.3	44.8	45.4	45.2	44.7	44.8	Services
Honduras								**Honduras**
Agriculture	20.4	20.4	13.1	11.9	13.8	12.4	13.0	Agriculture
Industry	23.8	29.9	27.6	26.2	26.6	25.8	24.8	Industrie
Services	55.8	49.7	59.3	62.0	59.6	61.9	62.2	Services
Hungary								**Hongrie**
Agriculture[2]	16.0	8.4	4.3	3.6	4.6	4.6	4.5	Agriculture[2]
Industry[3]	42.9	30.6	31.4	30.2	30.3	30.1	31.2	Industrie[3]
Services[4,5]	41.1	61.0	64.3	66.3	65.1	65.3	64.4	Services[4,5]
Iceland								**Islande**
Agriculture[2]	11.2	10.9	5.8	7.4	7.7	6.8	6.0	Agriculture[2]
Industry[3]	34.9	29.1	25.1	24.8	23.5	23.5	23.6	Industrie[3]
Services[4,5]	53.8	59.9	69.2	67.8	68.8	69.6	70.2	Services[4,5]
India								**Inde**
Agriculture[2]	32.4	27.3	19.3	18.7	18.0	18.0	17.0	Agriculture[2]
Industry[3]	31.7	33.6	34.2	33.1	31.9	30.7	30.0	Industrie[3]
Services[4,5]	35.9	39.1	46.4	48.2	50.0	51.3	53.0	Services[4,5]
Indonesia								**Indonésie**
Agriculture[2]	19.8	14.2	12.1	14.3	13.7	13.7	13.7	Agriculture[2]
Industry[3]	33.1	38.3	43.1	43.9	44.6	43.6	42.9	Industrie[3]
Services[4,5]	47.1	47.6	44.8	41.8	41.8	42.6	43.3	Services[4,5]
Iran (Islamic Rep. of)								**Iran (Rép. islamique d')**
Agriculture	13.1	12.4	6.4	6.7	7.7	8.9	7.4	Agriculture
Industry	31.4	38.7	45.7	40.3	40.2	39.4	41.6	Industrie
Services	55.4	48.9	47.9	53.1	52.1	51.7	50.9	Services

Gross value added by kind of economic activity *(continued)*
Percentage distribution, current prices
Valeur ajoutée brute par type d'activité économique *(suite)*
Répartition en pourcentage, aux prix courants

Country or area	1985	1995	2005	2010	2012	2013	2014	Pays ou zone
Iraq								**Iraq**
Agriculture	13.9[17]	12.3	6.9	5.1	4.1	4.8	4.1	Agriculture
Industry	42.2[17,18]	66.0	63.3	55.4	60.2	59.0	59.0	Industrie
Services	43.9[18,19]	21.7	29.9	39.4	35.7	36.3	36.9	Services
Ireland								**Irlande**
Agriculture[2]	9.4	6.4	1.2	1.1	1.2	1.4	1.6	Agriculture[2]
Industry[3]	30.6	32.1	34.5	26.7	26.9	26.1	25.6	Industrie[3]
Services[4,5]	59.9	61.5	64.4	72.3	71.8	72.5	72.8	Services[4,5]
Israel								**Israël**
Agriculture[2]	4.2	2.0	1.7	1.7	1.3	1.3	1.3	Agriculture[2]
Industry[3]	27.1	26.3	22.8	22.6	21.9	22.4	22.1	Industrie[3]
Services[4,5]	68.7	71.7	75.4	75.7	76.8	76.2	76.6	Services[4,5]
Italy								**Italie**
Agriculture[2]	4.6	3.3	2.2	2.0	2.2	2.3	2.2	Agriculture[2]
Industry[3]	33.0	29.1	25.8	24.4	23.8	23.6	23.5	Industrie[3]
Services[4,5]	62.4	67.6	71.9	73.7	74.0	74.1	74.3	Services[4,5]
Jamaica								**Jamaïque**
Agriculture	6.0	8.9	5.7	5.9	6.5	6.8	6.7	Agriculture
Industry	30.3	29.3	23.9	20.1	20.2	20.4	20.7	Industrie
Services	63.7	61.7[20]	70.4	74.1	73.3	72.8	72.6	Services
Japan								**Japon**
Agriculture[2]	3.0	1.8	1.2	1.2	1.2	1.2	1.2	Agriculture[2]
Industry[3]	37.7	33.1	28.1	27.5	26.0	26.2	26.7	Industrie[3]
Services[4,5]	59.4	65.2	70.6	71.3	72.8	72.6	72.0	Services[4,5]
Jordan								**Jordanie**
Agriculture	5.0	4.3	3.0	3.2	3.0	3.2	3.6	Agriculture
Industry	26.5	27.5	26.9	29.1	28.5	28.1	28.2	Industrie
Services	68.5	68.3	70.1	67.6	68.6	68.7	68.2	Services
Kazakhstan								**Kazakhstan**
Agriculture	...	12.9	6.6	4.6[2]	4.7[2]	4.9[2]	4.5[2]	Agriculture
Industry	...	30.2	39.2	41.8[3]	39.5[3]	36.9[3]	36.1[3]	Industrie
Services	...	56.9	54.2	53.6[4,5]	55.8[4,5]	58.2[4,5]	59.4[4,5]	Services
Kenya								**Kenya**
Agriculture[2]	28.7	27.6	23.2	27.1	28.3	28.6	29.5	Agriculture[2]
Industry[3]	25.2	20.9	22.3	20.3	20.1	19.5	18.8	Industrie[3]
Services[4,5]	46.1	51.5	54.5	52.6	51.6	51.9	51.7	Services[4,5]
Kiribati								**Kiribati**
Agriculture	44.0	26.9	21.8	24.3	25.3	24.6	25.2	Agriculture
Industry	9.4	9.1	9.3	10.2	11.2	11.0	10.9	Industrie
Services	46.6	64.0	68.9	65.6	63.5	64.4	63.9	Services
Kosovo								**Kosovo**
Agriculture	...	7.2	17.2[2]	16.2[2]	14.8[2]	14.4[2]	14.3[2]	Agriculture
Industry	...	32.6	26.6[3]	28.4[3]	28.1[3]	28.2[3]	26.7[3]	Industrie
Services	...	60.2	56.2[4,5]	55.4[4,5]	57.1[4,5]	57.4[4,5]	59.0[4,5]	Services
Kuwait								**Koweït**
Agriculture	0.6	0.4	0.3	0.4	0.3	0.3	0.4	Agriculture
Industry	56.4	52.8	60.2	58.2	66.8	65.5	61.9	Industrie
Services	43.0	46.7	39.5	41.4	32.9	34.2	37.7	Services
Kyrgyzstan								**Kirghizistan**
Agriculture[2]	...	43.1	31.3	18.8	18.5	16.4	16.6	Agriculture[2]
Industry[3]	...	21.8	22.1	28.2	24.7	27.9	25.8	Industrie[3]
Services[4,5]	...	35.1	46.6	53.0	56.7	55.7	57.6	Services[4,5]
Lao People's Dem. Rep.								**Rép. dém. pop. lao**
Agriculture	37.5	41.8	36.3	29.7	25.0	24.1	23.9	Agriculture
Industry	15.6	15.3	23.2	28.9	36.0	34.1	33.5	Industrie
Services	46.9	42.9	40.5	41.4	39.0	41.7	42.5	Services
Latvia								**Lettonie**
Agriculture[2]	...	8.9	4.3	4.4	3.7	3.4	3.3	Agriculture[2]
Industry[3]	...	30.3	22.9	23.8	24.0	23.7	23.4	Industrie[3]
Services[4,5]	...	60.8	72.8	71.7	72.3	72.9	73.4	Services[4,5]

Gross value added by kind of economic activity *(continued)*
Percentage distribution, current prices

Valeur ajoutée brute par type d'activité économique *(suite)*
Répartition en pourcentage, aux prix courants

Country or area	1985	1995	2005	2010	2012	2013	2014	Pays ou zone
Lebanon								**Liban**
Agriculture	3.7^2	5.7^2	4.0^2	4.3^2	4.0^2	4.2^2	3.2	Agriculture
Industry[3]	22.4	31.7	16.7	15.7	17.2	18.8	23.6	Industrie[3]
Services[4,5]	73.8	62.6	79.3	80.1	78.8	77.0	73.2	Services[4,5]
Lesotho								**Lesotho**
Agriculture	22.4	13.2	8.9	8.3	7.4	8.5	8.0	Agriculture
Industry	12.3	26.5	32.8	31.2	31.1	29.5	31.2	Industrie
Services	65.3	60.3	58.2	60.5	61.5	62.0	60.8	Services
Liberia								**Libéria**
Agriculture	35.9	80.5	68.8	70.0	70.1	70.1	70.1	Agriculture
Industry	24.7	5.2	9.8	11.3	11.3	11.3	11.3	Industrie
Services	39.4	14.3	21.5	18.7	18.6	18.6	18.6	Services
Libya								**Libye**
Agriculture	3.5	6.7	2.2	2.5	2.3	2.2	2.0	Agriculture
Industry	58.8	40.3	75.7	74.0	65.1	63.2	59.6	Industrie
Services	37.7	53.0	22.2	23.5	32.6	34.7	38.4	Services
Liechtenstein								**Liechtenstein**
Agriculture[2]	2.5	1.5	0.9	0.8	0.8	0.8	0.8	Agriculture[2]
Industry[3]	32.2	29.7	38.6	39.0	38.6	39.5	38.1	Industrie[3]
Services[4,5]	65.4	68.8	60.5	60.2	60.6	59.7	61.1	Services[4,5]
Lithuania								**Lituanie**
Agriculture[2]	...	11.1	4.8	3.3	4.4	4.0	3.4	Agriculture[2]
Industry[3]	...	31.5	32.7	29.1	30.7	30.1	30.5	Industrie[3]
Services[4,5]	...	57.4	62.5	67.6	64.8	66.0	66.0	Services[4,5]
Luxembourg								**Luxembourg**
Agriculture[2]	2.0	1.0	0.4	0.3	0.4	0.3	0.3	Agriculture[2]
Industry[3]	28.5	21.7	16.6	12.9	12.2	11.7	11.9	Industrie[3]
Services[4,5]	69.5	77.2	83.0	86.8	87.4	88.0	87.8	Services[4,5]
Madagascar								**Madagascar**
Agriculture	34.5	32.3	28.1	27.6	27.6	25.7	25.7	Agriculture
Industry	13.1	14.8	18.6	19.5	19.2	18.8	18.6	Industrie
Services	52.4	52.8	53.3	52.9	53.2	55.5	55.6	Services
Malawi								**Malawi**
Agriculture[2]	50.0	31.9	37.1	31.9	28.3	27.5	29.0	Agriculture[2]
Industry[3]	28.1	21.9	16.8	16.4	14.0	16.3	15.7	Industrie[3]
Services[4,5]	21.8	46.1	46.1	51.7	57.7	56.2	55.3	Services[4,5]
Malaysia								**Malaisie**
Agriculture	19.9	12.6	8.4	10.5	10.2	9.4	9.2	Agriculture
Industry	38.5	37.5	46.9	41.6	41.2	41.0	41.0	Industrie
Services	41.6	49.9	44.7	48.0	48.6	49.6	49.8	Services
Maldives[1]								**Maldives**[1]
Agriculture[6]	9.7	6.6	7.5	4.1	3.6	3.9	3.5	Agriculture[6]
Industry	9.1	10.6	14.8	14.9	18.6	16.7	18.1	Industrie
Services	81.3	82.7	77.7	81.0	77.8	79.5	78.4	Services
Mali								**Mali**
Agriculture	40.3	37.7	37.5	40.5	41.8	38.2	39.5	Agriculture
Industry	14.7^{21}	19.0	24.0	20.1	22.3	22.4	22.3	Industrie
Services	45.0	43.2	38.5	39.4	35.8	39.4	38.2	Services
Malta								**Malte**
Agriculture[2]	4.3	2.8	2.2	1.7	1.5	1.5	1.3	Agriculture[2]
Industry[3]	34.2	28.7	23.3	19.1	17.9	16.3	15.3	Industrie[3]
Services[4,5]	61.5	68.5	74.5	79.3	80.6	82.2	83.4	Services[4,5]
Marshall Islands								**Îles Marshall**
Agriculture	14.0^{14}	14.9^{14}	9.1^{14}	15.4	21.9	22.0	19.9	Agriculture
Industry	13.1^{14}	15.0^{14}	9.2^{14}	11.6	10.9	10.3	11.3	Industrie
Services	72.8^{14}	70.0^{14}	81.6^{14}	73.0	67.2	67.7	68.7	Services
Mauritania								**Mauritanie**
Agriculture	36.1	43.6	29.8	21.3	20.7	20.8	20.6	Agriculture
Industry	27.0	24.7	32.4	40.9	42.9	43.6	44.5	Industrie
Services	36.9	31.6	37.9	37.8	36.3	35.7	35.0	Services

Gross value added by kind of economic activity *(continued)*
Percentage distribution, current prices
Valeur ajoutée brute par type d'activité économique *(suite)*
Répartition en pourcentage, aux prix courants

Country or area	1985	1995	2005	2010	2012	2013	2014	Pays ou zone
Mauritius								**Maurice**
Agriculture	14.9	9.8	5.7	3.6^2	3.5^2	3.2^2	3.0^2	Agriculture
Industry	32.5	31.3	26.6	26.6^3	25.1^3	24.6^3	23.5^3	Industrie
Services	52.5	58.9	67.8	$69.8^{4,5}$	$71.5^{4,5}$	$72.2^{4,5}$	$73.5^{4,5}$	Services
Mexico								**Mexique**
Agriculture [2]	7.8	4.8	3.2	3.3	3.3	3.3	3.3	Agriculture [2]
Industry [3]	45.3	37.3	38.8	38.3	39.6	37.8	37.7	Industrie [3]
Services [4,5]	46.8	57.9	58.1	58.3	57.0	58.9	59.0	Services [4,5]
Micronesia (Fed. States of)								**Micronésie (États féd. de)**
Agriculture	24.5	24.9	24.1	26.2	29.9	28.1	27.5	Agriculture
Industry	7.4	7.2	5.7	7.8	8.9	7.9	6.1	Industrie
Services	68.1	68.0	70.2	65.9	61.2	63.9	66.4	Services
Monaco [1]								**Monaco** [1]
Industry [3]	12.8	12.8	12.2	12.9	12.9	14.0	12.8	Industrie [3]
Services [4,5]	87.2	87.2	87.8	87.1	87.1	86.0	87.2	Services [4,5]
Mongolia								**Mongolie**
Agriculture [2]	10.1^1	28.9^1	17.8	13.1	12.7	15.1	15.6	Agriculture [2]
Industry [3]	33.1^1	34.3^1	37.4	37.0	34.6	34.4	36.2	Industrie [3]
Services [4,5]	56.7^1	36.7^1	44.8	50.0	52.7	50.4	48.1	Services [4,5]
Montenegro								**Monténégro**
Agriculture	...	12.2	10.4	9.2^2	8.9^2	9.8^2	10.0^2	Agriculture
Industry	...	23.9	20.7	20.5^3	17.8^3	18.8^3	17.7^3	Industrie
Services	...	63.9	68.9	$70.3^{4,5}$	$73.4^{4,5}$	$71.4^{4,5}$	$72.3^{4,5}$	Services
Montserrat								**Montserrat**
Agriculture	4.0	4.9	0.9	1.1	1.4	1.3	1.5	Agriculture
Industry	17.7	12.3	16.7	13.3	13.0	15.0	13.9	Industrie
Services	78.3	82.8	82.5	85.7	85.6	83.6	84.6	Services
Morocco [22]								**Maroc** [22]
Agriculture	17.4	15.1	13.1	14.4	13.4	14.7	13.0	Agriculture
Industry	32.8	29.6	28.9	28.6	28.6	28.7	29.3	Industrie
Services	49.8	55.3	58.0	56.9	58.0	56.6	57.7	Services
Mozambique								**Mozambique**
Agriculture	47.5	33.5^2	25.4^2	28.9^2	27.0^2	25.9^2	24.6^2	Agriculture
Industry	13.2	14.4^3	20.7^3	18.6^3	18.6^3	18.2^3	20.6^3	Industrie
Services	39.3	$52.1^{4,5}$	$53.8^{4,5}$	$52.5^{4,5}$	$54.4^{4,5}$	$55.9^{4,5}$	$54.7^{4,5}$	Services
Myanmar [1]								**Myanmar** [1]
Agriculture	48.2	60.0	46.7	36.9	30.6	29.5	27.9	Agriculture
Industry	13.1	9.9	17.5	26.5	32.4	32.4	34.4	Industrie
Services [23]	38.7	30.1	35.8	36.7	37.0	38.1	37.7	Services [23]
Namibia								**Namibie**
Agriculture	10.0	12.7	11.2	9.2	8.6	6.2	6.0	Agriculture
Industry	43.7	27.4	28.8	29.8	32.0	32.9	34.4	Industrie
Services	46.3	59.9	60.0	61.0	59.4	60.9	59.6	Services
Nauru [1]								**Nauru** [1]
Agriculture	6.0	6.4	7.8	4.2	2.6	3.4	3.1	Agriculture
Industry	32.9	27.7	-6.5^{24}	47.4	66.0	55.5	58.2	Industrie
Services	61.1	65.8	98.7	48.4	31.5	41.1	38.6	Services
Nepal								**Népal**
Agriculture	49.0	38.9	35.2	35.4	35.2	33.8	32.5	Agriculture
Industry	11.5	17.7	17.1	15.1	15.0	15.2	15.1	Industrie
Services	39.5	43.4	47.7	49.5	49.8	51.0	52.4	Services
Netherlands								**Pays-Bas**
Agriculture [2]	4.1	3.4	2.0	1.9	1.8	1.9	1.8	Agriculture [2]
Industry [3]	33.1	27.0	24.0	22.1	22.1	21.9	21.2	Industrie [3]
Services [4,5]	62.9	69.6	74.0	76.0	76.1	76.2	77.0	Services [4,5]
New Caledonia								**Nouvelle-Calédonie**
Agriculture	1.8	1.8	1.7	1.4	1.4	1.4	1.4	Agriculture
Industry	25.9	22.0	26.6	25.5	24.6	25.0	24.9	Industrie
Services	72.2	76.2	71.7	73.1	74.0	73.6	73.7	Services

Gross value added by kind of economic activity *(continued)*
Percentage distribution, current prices
Valeur ajoutée brute par type d'activité économique *(suite)*
Répartition en pourcentage, aux prix courants

Country or area	1985	1995	2005	2010	2012	2013	2014	Pays ou zone
New Zealand								**Nouvelle-Zélande**
Agriculture	7.4	7.0	4.9	7.1	6.1	6.7	6.6	Agriculture
Industry	34.1	27.8	25.8	23.0	23.0	23.0	23.0	Industrie
Services	58.5	65.3	69.3	69.9	70.9	70.3	70.5	Services
Nicaragua								**Nicaragua**
Agriculture	14.8	22.3	17.7	18.8	19.9	19.1	20.5	Agriculture
Industry	33.9	22.2	23.0	24.3	26.5	25.9	25.7	Industrie
Services	51.3	55.5	59.3	56.9	53.6	55.0	53.8	Services
Niger								**Niger**
Agriculture	37.0	35.9	45.5[2]	43.8[2]	40.1[2]	38.0[2]	39.2[2]	Agriculture
Industry	20.2	14.3	11.8[3]	16.7[3]	22.2[3]	23.3[3]	20.9[3]	Industrie
Services	42.8	49.7	42.7[4,5]	39.4[4,5]	37.7[4,5]	38.6[4,5]	39.9[4,5]	Services
Nigeria								**Nigéria**
Agriculture[2]	23.2	27.0	25.6	23.9	22.1	21.0	20.2	Agriculture[2]
Industry[3]	18.2	25.1	23.7	25.3	27.3	26.0	24.9	Industrie[3]
Services[4,5]	58.6	47.9	50.7	50.8	50.6	53.0	54.8	Services[4,5]
Norway								**Norvège**
Agriculture[2]	3.2	3.0	1.6	1.8	1.3	1.5	1.7	Agriculture[2]
Industry[3]	40.5	33.2	42.6	39.1	41.1	39.8	38.4	Industrie[3]
Services[4,5]	56.3	63.8	55.8	59.2	57.6	58.7	59.9	Services[4,5]
Oman								**Oman**
Agriculture	2.3	2.9	1.6	1.4	1.1	1.1	1.2	Agriculture
Industry	63.3	49.7	62.1	62.8	65.8	63.9	60.8	Industrie
Services	34.4	47.4	36.3	35.9	33.1	35.0	37.9	Services
Pakistan								**Pakistan**
Agriculture[2]	30.7	28.2	24.3	24.3	24.5	25.1	25.1	Agriculture[2]
Industry[3]	16.9	18.0	21.1	20.6	22.1	21.1	21.3	Industrie[3]
Services[4,5]	52.4	53.8	54.6	55.1	53.4	53.8	53.6	Services[4,5]
Palau								**Palaos**
Agriculture[2]	14.1	7.0	4.2	4.6	4.8	4.7	4.3	Agriculture[2]
Industry[3]	19.0	11.3	17.6	9.9	9.7	8.4	8.6	Industrie[3]
Services[4,5]	66.8	81.7	78.2	85.5	85.5	87.0	87.1	Services[4,5]
Panama								**Panama**
Agriculture	7.3	7.4	6.8	3.9	3.3	3.1	3.3	Agriculture
Industry	22.1	19.8	16.3	20.3	22.1	25.3	22.7	Industrie
Services	70.6	72.8	76.9	75.8	74.6	71.7	74.1	Services
Papua New Guinea								**Papouasie-Nvl-Guinée**
Agriculture	33.7	35.1	34.0	31.5	28.5	27.4	25.5	Agriculture
Industry	27.0	33.3	44.3	45.1	44.4	45.2	49.1	Industrie
Services	39.3	31.7	21.7	23.4	27.1	27.4	25.4	Services
Paraguay								**Paraguay**
Agriculture	26.5	22.8	19.6	22.5	18.1	21.6	20.9	Agriculture
Industry	24.2	25.7	34.8	30.1	30.4	28.4	28.8	Industrie
Services	49.3	51.5	45.7	47.4	51.5	50.0	50.3	Services
Peru								**Pérou**
Agriculture[2]	10.8	9.0	7.5	7.5	7.5	7.3	7.3	Agriculture[2]
Industry[3]	35.2	33.4	37.7	39.2	39.4	38.0	35.7	Industrie[3]
Services[4,5]	54.0	57.6	54.7	53.2	53.1	54.7	57.0	Services[4,5]
Philippines								**Philippines**
Agriculture[2]	21.3	18.9	12.7	12.3	11.9	11.2	11.3	Agriculture[2]
Industry[3]	38.1	35.0	33.8	32.6	31.1	31.1	31.2	Industrie[3]
Services[4,5]	40.5	46.1	53.5	55.1	57.0	57.6	57.5	Services[4,5]
Poland								**Pologne**
Agriculture[2]	14.6	5.6	3.3	2.9	3.0	3.2	2.9	Agriculture[2]
Industry[3]	51.4	37.4	32.1	33.3	33.6	32.3	32.5	Industrie[3]
Services[4,5]	34.0	57.0	64.6	63.8	63.4	64.5	64.6	Services[4,5]
Portugal								**Portugal**
Agriculture[2]	13.3	5.4	2.6	2.2	2.2	2.4	2.3	Agriculture[2]
Industry[3]	26.8	28.2	24.6	22.6	21.8	21.5	21.5	Industrie[3]
Services[4,5]	59.9	66.4	72.7	75.2	76.0	76.2	76.1	Services[4,5]

Country or area	1985	1995	2005	2010	2012	2013	2014	Pays ou zone
Puerto Rico[1]								**Porto Rico**[1]
Agriculture	2.5	1.0	0.6	0.8	0.8	0.8	0.8	Agriculture
Industry	45.3	47.6	47.3	50.7	49.9	49.8	50.8	Industrie
Services	52.2	51.4	52.1	48.4	49.3	49.4	48.3	Services
Qatar[1]								**Qatar**[1]
Agriculture	0.9	1.0	0.1	0.1	0.1	0.1	0.1	Agriculture
Industry	56.6	52.1	73.8	66.8	70.8	68.2	65.1	Industrie
Services	42.5	46.9	26.0	33.1	29.1	31.7	34.8	Services
Republic of Korea								**République de Corée**
Agriculture[2]	13.4	5.9	3.1	2.5	2.5	2.3	2.3	Agriculture[2]
Industry[3]	37.1	39.5	37.5	38.3	38.1	38.4	38.2	Industrie[3]
Services[4,5]	49.5	54.6	59.4	59.3	59.5	59.3	59.4	Services[4,5]
Republic of Moldova								**République de Moldova**
Agriculture	...	32.2	19.1	14.1	13.1	14.5	14.9	Agriculture
Industry	...	31.4	22.2	19.5	20.4	20.7	20.5	Industrie
Services	...	36.4	58.7	66.4	66.5	64.8	64.7	Services
Romania								**Roumanie**
Agriculture[2]	15.5	19.2	9.5	6.3	5.3	6.2	5.4	Agriculture[2]
Industry[3]	54.0	38.4	36.0	41.3	37.1	34.4	34.4	Industrie[3]
Services[4,5]	30.4	42.5	54.5	52.4	57.6	59.4	60.3	Services[4,5]
Russian Federation								**Fédération de Russie**
Agriculture	...	7.2	5.0	3.9	3.9	4.0	4.1	Agriculture
Industry	...	39.3	38.1	34.7	37.0	35.9	36.8	Industrie
Services	...	53.5	57.0	61.4	59.1	60.0	59.1	Services
Rwanda								**Rwanda**
Agriculture[2]	53.8	44.5	41.3	34.7	35.3	35.1	35.0	Agriculture[2]
Industry[3]	15.7	12.4	12.5	13.8	15.2	15.7	15.2	Industrie[3]
Services[4,5]	30.5	43.1	46.2	51.5	49.5	49.2	49.8	Services[4,5]
Saint Kitts and Nevis								**Saint-Kitts-et-Nevis**
Agriculture	5.7	3.3	1.9	1.6	1.6	1.6	1.5	Agriculture
Industry	19.1	24.0	25.4	26.4	26.0	25.9	25.1	Industrie
Services	75.2	72.6	72.7	72.0	72.4	72.5	73.4	Services
Saint Lucia								**Sainte-Lucie**
Agriculture	12.0	7.5	3.5	2.9	2.9	3.0	2.7	Agriculture
Industry	16.0	16.2	18.7	15.5	14.8	14.1	13.3	Industrie
Services	72.1	76.3	77.8	81.6	82.3	82.9	84.0	Services
Saint Vincent-Grenadines								**Saint-Vincent-Grenadines**
Agriculture	15.1	11.2	6.2	7.1	7.1	7.5	7.7	Agriculture
Industry	19.6	20.9	18.6	19.2	17.6	17.6	16.9	Industrie
Services	65.3	67.9	75.1	73.7	75.3	74.9	75.4	Services
Samoa[1]								**Samoa**[1]
Agriculture	22.4	20.1	12.3	9.1	9.4	9.6	9.2	Agriculture
Industry	27.3	27.1	30.6	25.9	26.0	25.8	24.9	Industrie
Services	50.3	52.7	57.2	65.0	64.7	64.6	65.8	Services
San Marino								**Saint-Marin**
Agriculture[2]	0.1	0.1	0.1	0.1	0.1	0.1	<0.0	Agriculture[2]
Industry[3]	40.9	40.8	38.2	35.6	34.0	34.6	34.6	Industrie[3]
Services[4,5]	59.0	59.1	61.7	64.4	66.0	65.4	65.4	Services[4,5]
Sao Tome and Principe								**Sao Tomé-et-Principe**
Agriculture	27.6	26.4	18.3	22.6	21.5	22.0	22.4	Agriculture
Industry	17.9	19.6	14.9	13.6	11.7	10.8	9.9	Industrie
Services	54.5	53.9	66.8	63.8	66.8	67.2	67.7	Services
Saudi Arabia								**Arabie saoudite**
Agriculture	3.6	5.9	3.2	2.4	1.8	1.9	1.9	Agriculture
Industry	41.2	48.6	61.8	58.4	62.9	60.0	57.5	Industrie
Services	55.2	45.5	35.0	39.2	35.3	38.1	40.6	Services
Senegal								**Sénégal**
Agriculture	18.1	19.4	16.8	17.5	15.6	15.6	15.3	Agriculture
Industry	22.2	24.7	23.6	23.4	24.4	23.8	24.3	Industrie
Services	59.6	55.9	59.6	59.2	60.0	60.6	60.4	Services

Gross value added by kind of economic activity *(continued)*
Percentage distribution, current prices

Valeur ajoutée brute par type d'activité économique *(suite)*
Répartition en pourcentage, aux prix courants

Country or area	1985	1995	2005	2010	2012	2013	2014	Pays ou zone
Serbia[25]								**Serbie**[25]
Agriculture[2]	...	20.9	12.0	10.2	9.0	9.4	9.7	Agriculture[2]
Industry[3]	...	35.5	29.3	28.4	30.3	31.7	29.8	Industrie[3]
Services[4,5]	...	43.6	58.7	61.4	60.7	59.0	60.5	Services[4,5]
Seychelles								**Seychelles**
Agriculture	6.7	5.9	3.8[2]	2.7[2]	2.3[2]	2.9[2]	2.4[2]	Agriculture
Industry	9.2	13.5	19.4[3]	16.5[3]	15.6[3]	14.8[3]	15.9[3]	Industrie
Services	84.1	80.6	76.8[4,5]	80.8[4,5]	82.1[4,5]	82.3[4,5]	81.6[4,5]	Services
Sierra Leone								**Sierra Leone**
Agriculture	41.0	48.1	51.0	55.2	52.0	49.0	52.3	Agriculture
Industry	20.4	9.1	11.6	8.1	14.9	22.7	15.2	Industrie
Services	38.7	42.8	37.4	36.7	33.1	28.4	32.4	Services
Singapore								**Singapour**
Agriculture[2,6]	1.0	0.2	0.1	<0.0	<0.0	<0.0	<0.0	Agriculture[2,6]
Industry[3,7]	33.6	33.9	32.4	27.6	26.4	24.8	24.9	Industrie[3,7]
Services[4,5]	65.4	66.0	67.6	72.3	73.6	75.1	75.0	Services[4,5]
Sint Maarten (Dutch part)								**Saint-Martin (partie néerl.)**
Agriculture	...	...	0.5	0.2	0.1	0.1	0.1	Agriculture
Industry	...	...	16.0	13.0	12.7	12.2	12.0	Industrie
Services	...	...	83.5	86.8	87.2	87.7	87.9	Services
Slovakia								**Slovaquie**
Agriculture[2]	...	5.6	3.6	2.8	3.5	4.0	4.4	Agriculture[2]
Industry[3]	...	36.8	36.1	35.2	35.3	33.0	33.6	Industrie[3]
Services[4,5]	...	57.5	60.3	62.0	61.2	63.1	62.0	Services[4,5]
Slovenia								**Slovénie**
Agriculture[2]	...	4.3	2.6	2.0	2.1	2.1	2.2	Agriculture[2]
Industry[3]	...	34.7	34.1	30.6	31.7	32.3	33.2	Industrie[3]
Services[4,5]	...	61.0	63.3	67.4	66.2	65.6	64.6	Services[4,5]
Solomon Islands								**Îles Salomon**
Agriculture	50.4	44.7	30.4	28.7	27.9	28.4	28.2	Agriculture
Industry	8.3	14.4	7.5	13.3	15.7	15.2	15.8	Industrie
Services	41.3	41.0	62.1	57.9	56.5	56.4	55.9	Services
Somalia								**Somalie**
Agriculture	66.1	60.1	60.1	60.2	60.2	60.2	60.2	Agriculture
Industry	7.6	7.3	7.4	7.4	7.4	7.4	7.4	Industrie
Services	26.3	32.6	32.6	32.5	32.5	32.5	32.5	Services
South Africa								**Afrique du Sud**
Agriculture	5.2	3.9	2.7	2.6	2.4	2.3	2.5	Agriculture
Industry	43.4	34.9	30.3	30.2	29.7	29.9	29.5	Industrie
Services	51.5	61.3	67.1	67.2	67.9	67.8	68.0	Services
South Sudan								**Soudan du sud**
Agriculture	...	...	...	5.1	4.1	4.8	4.1	Agriculture
Industry	...	...	...	55.4	60.2	59.0	59.0	Industrie
Services	...	...	...	39.4	35.7	36.3	36.9	Services
Spain								**Espagne**
Agriculture[2]	5.7	4.2	3.0	2.6	2.4	2.8	2.5	Agriculture[2]
Industry[3]	34.9	30.7	30.4	26.0	23.6	23.3	22.4	Industrie[3]
Services[4,5]	59.4	65.1	66.5	71.4	74.0	73.9	75.1	Services[4,5]
Sri Lanka								**Sri Lanka**
Agriculture	27.1[14]	20.6[14]	11.8[1]	12.8[1]	11.0[1]	10.8[1]	9.9[1]	Agriculture
Industry	28.5[14]	29.7[14]	30.2[1]	29.4[1]	31.5[1]	32.5[1]	33.8[1]	Industrie
Services	44.4[14]	49.7[14]	58.0[1]	57.8[1]	57.5[1]	56.8[1]	56.3[1]	Services
State of Palestine								**État de Palestine**
Agriculture[2]	13.2	12.6	5.8	6.4	5.2	4.7	5.5	Agriculture[2]
Industry[3]	30.7	31.6	25.8	23.3	24.3	23.0	23.4	Industrie[3]
Services[4,5]	56.1	55.8	68.3	70.2	70.5	72.2	71.1	Services[4,5]
Sudan								**Soudan**
Agriculture	...	...	...	42.9	34.1	34.9	32.2	Agriculture
Industry	...	...	...	13.7	18.6	19.8	21.9	Industrie
Services	...	...	...	43.4	47.4	45.2	45.9	Services

Gross value added by kind of economic activity *(continued)*
Percentage distribution, current prices
Valeur ajoutée brute par type d'activité économique *(suite)*
Répartition en pourcentage, aux prix courants

Country or area	1985	1995	2005	2010	2012	2013	2014	Pays ou zone
Suriname								**Suriname**
Agriculture	8.6[14]	29.9	11.3	10.2	9.2	8.8	9.9	Agriculture
Industry	28.5[14]	27.5	37.4	37.9	38.6	35.3	32.6	Industrie
Services	62.9[14]	42.6	51.3	51.9	52.2	55.8	57.5	Services
Swaziland								**Swaziland**
Agriculture	12.6	10.0	6.8	6.5	6.2	6.0	6.0	Agriculture
Industry	26.8	42.4	43.7	43.2	44.2	44.2	44.3	Industrie
Services	60.6	47.7	49.5	50.3	49.6	49.7	49.7	Services
Sweden								**Suède**
Agriculture[2]	4.5	2.8	1.1	1.6	1.5	1.4	1.4	Agriculture[2]
Industry[3]	33.7	31.3	29.7	28.9	26.9	26.1	26.0	Industrie[3]
Services[4,5]	61.9	65.9	69.2	69.4	71.6	72.5	72.6	Services[4,5]
Switzerland								**Suisse**
Agriculture[2]	2.5	1.5	0.9	0.7	0.7	0.7	0.8	Agriculture[2]
Industry[3]	32.2	29.7	26.8	26.3	26.2	26.1	26.3	Industrie[3]
Services[4,5]	65.4	68.8	72.3	73.0	73.2	73.2	73.0	Services[4,5]
Syrian Arab Republic[1]								**Rép. arabe syrienne**[1]
Agriculture	21.0	28.2	20.3	19.7	21.0	20.4	20.6	Agriculture
Industry	21.9	18.1	31.2	30.7	29.6	30.3	30.2	Industrie
Services	57.1	53.7	48.5	49.6	49.4	49.3	49.2	Services
Tajikistan								**Tadjikistan**
Agriculture	...	31.6	23.8	21.8	26.2	23.1	27.2	Agriculture
Industry	...	48.4	30.7	27.9	24.8	25.9	25.5	Industrie
Services	...	20.0	45.6	50.3	49.0	51.0	47.3	Services
Thailand[1]								**Thaïlande**[1]
Agriculture	15.8	9.1	9.2	10.5	11.6	11.3	10.5	Agriculture
Industry	31.8	37.6	38.6	40.0	37.5	37.0	36.8	Industrie
Services	52.3	53.3	52.2	49.4	51.0	51.7	52.7	Services
TFYR of Macedonia								**ex-R.Y. de Macédoine**
Agriculture[2]	...	12.5	11.3	11.7	10.5	11.5	10.2	Agriculture[2]
Industry[3]	...	24.7	23.7	24.4	24.4	25.4	24.8	Industrie[3]
Services[4,5]	...	62.8	64.9	63.9	65.1	63.0	65.0	Services[4,5]
Timor-Leste								**Timor-Leste**
Agriculture	...	20.0	7.3[2]	4.7[2]	3.7[2]	4.6[2]	3.8[2]	Agriculture
Industry	...	49.3	76.7[3]	81.5[3]	85.3[3]	81.1[3]	84.1[3]	Industrie
Services	...	30.7	16.0[4,5]	13.8[4,5]	11.0[4,5]	14.3[4,5]	12.1[4,5]	Services
Togo								**Togo**
Agriculture	37.9	41.9	43.3	46.1	47.7	44.6	46.7	Agriculture
Industry[26]	24.4	24.6	19.0	18.2	20.6	20.5	19.4	Industrie[26]
Services[27]	37.8	33.4	37.6	35.7	31.7	34.9	33.9	Services[27]
Tonga								**Tonga**
Agriculture	37.8	22.1	20.2	18.3	18.8	19.9	19.4	Agriculture
Industry	14.9	22.0	19.1	20.0	21.2	18.2	18.2	Industrie
Services	47.3	55.9	60.6	61.7	60.0	61.9	62.4	Services
Trinidad and Tobago								**Trinité-et-Tobago**
Agriculture	2.9	1.9	0.5	0.5	0.4	0.4	0.4	Agriculture
Industry	42.5	41.7	56.7	53.3	51.4	52.4	52.0	Industrie
Services	54.6	56.3	42.8	46.2	48.2	47.2	47.6	Services
Tunisia[14]								**Tunisie**[14]
Agriculture	14.4	10.5	10.0	8.1	9.4	9.2	9.3	Agriculture
Industry	34.7	29.6	28.8	31.1	30.5	29.9	29.1	Industrie
Services	50.8	59.9	61.3	60.8	60.1	60.8	61.6	Services
Turkey								**Turquie**
Agriculture[2]	15.3	11.9	10.6	9.5	8.8	8.3	8.0	Agriculture[2]
Industry[3]	33.7	38.2	28.0	26.4	26.7	26.6	27.1	Industrie[3]
Services[4,5]	51.0	50.0	61.4	64.2	64.5	65.1	64.9	Services[4,5]
Turkmenistan								**Turkménistan**
Agriculture	...	16.9	18.8	14.5	14.5	14.5	14.5	Agriculture
Industry	...	65.3	37.6	48.4	48.4	48.4	48.4	Industrie
Services	...	17.9	43.6	37.0	37.0	37.0	37.0	Services

Gross value added by kind of economic activity *(continued)*
Percentage distribution, current prices

Valeur ajoutée brute par type d'activité économique *(suite)*
Répartition en pourcentage, aux prix courants

Country or area	1985	1995	2005	2010	2012	2013	2014	Pays ou zone
Turks and Caicos Islands								**Îles Turques-et-Caïques**
Agriculture	1.3	1.3	1.2	0.6	0.5	0.6	0.6	Agriculture
Industry	16.3	16.4	19.7	12.4	10.5	11.0	11.0	Industrie
Services	82.4	82.3	79.1	86.9	89.0	88.5	88.4	Services
Tuvalu								**Tuvalu**
Agriculture	10.8	24.0	21.6	27.6	24.5	25.5	25.9	Agriculture
Industry	13.5	14.0	8.5	5.7	5.6	9.2	8.3	Industrie
Services	75.6	62.0	69.9	66.7	70.0	65.3	65.8	Services
Uganda								**Ouganda**
Agriculture [2]	50.4	40.6	29.1	26.2	27.9	27.2	27.0	Agriculture [2]
Industry [3]	8.7	15.1	19.9	20.4	22.1	22.2	21.9	Industrie [3]
Services [4,5]	40.9	44.3	51.0	53.4	49.9	50.5	51.1	Services [4,5]
Ukraine								**Ukraine**
Agriculture [2]	...	14.5	10.0	8.4	9.0	10.0	9.5[28]	Agriculture [2]
Industry [3]	...	41.8	34.1	29.0	28.1	25.9	27.6[28]	Industrie [3]
Services [4,5]	...	43.7	55.8	62.7	63.0	64.1	63.0[28]	Services [4,5]
United Arab Emirates [1]								**Émirats arabes unis** [1]
Agriculture	0.9	1.7	1.4	0.8	0.7	0.7	0.6	Agriculture
Industry	60.4	47.3	53.8	52.4	57.1	54.8	52.2	Industrie
Services	38.7	51.0	44.8	46.8	42.3	44.5	47.2	Services
United Kingdom								**Royaume-Uni**
Agriculture [2]	1.4	1.5	0.7	0.7	0.7	0.7	0.7	Agriculture [2]
Industry [3]	36.0	30.0	23.3	20.8	20.8	21.3	21.0	Industrie [3]
Services [4,5]	62.6	68.5	76.0	78.5	78.6	78.0	78.4	Services [4,5]
United Rep. of Tanzania [29]								**Rép.-Unie de Tanzanie** [29]
Agriculture	27.6	34.2[2]	30.1[2]	31.7[2]	32.8[2]	32.9[2]	31.1[2]	Agriculture
Industry	13.0	16.7[3]	20.8[3]	21.5[3]	23.0[3]	23.9[3]	24.7[3]	Industrie
Services	59.4	49.2[4,5]	49.1[4,5]	46.8[4,5]	44.2[4,5]	43.2[4,5]	44.2[4,5]	Services [5]
United States								**États-Unis**
Agriculture [2]	1.6	1.2	1.0	1.1	1.2	1.4	1.2	Agriculture [2]
Industry [3]	28.4	24.3	21.5	20.2	20.3	20.4	20.4	Industrie [3]
Services [4,5]	70.0	74.5	77.5	78.8	78.5	78.3	78.4	Services [4,5]
Uruguay								**Uruguay**
Agriculture	12.5	8.1	9.8	8.0	9.0	8.8	7.9	Agriculture
Industry	30.8	25.9	26.6	27.3	25.3	26.1	27.1	Industrie
Services	56.7	66.0	63.6	64.7	65.7	65.1	65.0	Services
Uzbekistan								**Ouzbékistan**
Agriculture	...	31.4	29.5	19.8	19.2	19.1	19.3	Agriculture
Industry	...	30.9	29.1	33.4	33.0	32.8	32.9	Industrie
Services	...	37.7	41.4	46.8	47.8	48.0	47.8	Services
Vanuatu								**Vanuatu**
Agriculture [2]	32.6	30.5	24.1	21.9	26.7	26.7	25.9	Agriculture [2]
Industry [3]	5.9	9.1	8.5	13.0	8.0	8.4	8.9	Industrie [3]
Services [4,5]	61.5	60.4	67.4	65.0	65.3	64.9	65.2	Services [4,5]
Venezuela (Boliv. Rep. of)								**Venezuela (Rép. boliv. du)**
Agriculture	6.4	5.9	4.0	5.7	5.3	5.2	5.2	Agriculture
Industry	51.9	47.1	56.9	51.0	47.5	47.0	48.6	Industrie
Services	41.7	47.0	39.2	43.4	47.2	47.9	46.2	Services
Viet Nam [1]								**Viet Nam** [1]
Agriculture	42.9	27.2	21.0[2]	18.9[2]	19.7[2]	18.4[2]	18.1[2]	Agriculture
Industry	23.4	28.8	41.5[3]	38.2[3]	38.6[3]	38.3[3]	38.5[3]	Industrie
Services	33.7	44.1	37.5[4,5]	42.9[4,5]	41.7[4,5]	43.3[4,5]	43.4[4,5]	Services
Yemen								**Yémen**
Agriculture	...	17.3	9.6	12.1	14.9	15.0	14.7	Agriculture
Industry	...	27.3	43.8	38.6	36.0	36.0	37.3	Industrie
Services	...	55.4	46.6	49.2	49.0	49.0	48.0	Services
Zambia								**Zambie**
Agriculture	13.3	15.1	15.5	10.2	10.1	9.4	9.3	Agriculture
Industry	50.8	36.8	28.6	34.5	33.4	32.8	31.1	Industrie
Services	35.9	48.1	55.9	55.3	56.5	57.8	59.6	Services

Country or area	1985	1995	2005	2010	2012	2013	2014	Pays ou zone
Zanzibar								**Zanzibar**
Agriculture [2]	...	21.3	23.0	32.4	33.1	33.4	30.8	Agriculture [2]
Industry [3]	...	22.5	17.5	19.3	20.4	19.8	18.5	Industrie [3]
Services [4,5]	...	56.2	59.5	48.4	46.5	46.8	50.7	Services [4,5]
Zimbabwe								**Zimbabwe**
Agriculture	20.8	14.0	12.3	14.5	13.1	11.9	13.2	Agriculture
Industry	27.8	28.8	41.9	30.7	31.4	30.9	31.1	Industrie
Services	51.4	57.2	45.9	54.8	55.5	57.2	55.7	Services

Source:

United Nations Statistics Division, New York, national accounts analysis of main aggregates (AMA) database, last accessed March 2016.

Source:

Organisation des Nations Unies, Division de statistique, New York, base de données des estimations des principaux agrégats des comptes nationaux, dernier accès mars 2016.

1	At producers' prices.	1	Aux prix à la production.
2	Excludes irrigation canals and landscaping care.	2	Exclut les canaux d'irrigation et l'aménagement paysager.
3	Excludes publishing activities, includes irrigation canals.	3	Ne comprend pas les activités d'édition, comprend les canaux d'irrigation.
4	Excludes repair of personal and household goods.	4	Exclut la réparation de biens personnels et articles ménagers.
5	Includes publishing activities and landscaping care.	5	Y compris les activités d'édition et l'aménagement paysager.
6	Includes mining and quarrying.	6	Y compris les industries extractives.
7	Excludes mining and quarrying.	7	Non compris les industries extractives.
8	For statistical purposes, the data for China do not include those for the Hong Kong Special Administrative Region (Hong Kong SAR), Macao Special Administrative Region (Macao SAR) and Taiwan Province of China.	8	Pour la présentation des statistiques, les données pour la Chine ne comprennent pas la Région Administrative Spéciale de Hong Kong (Hong Kong RAS), la Région Administrative Spéciale de Macao (Macao RAS) et la province de Taiwan.
9	Excludes hunting and forestry.	9	Non compris la chasse et la sylviculture.
10	Excluding waste management.	10	Gestion des déchets non-compris.
11	Excludes northern Cyprus.	11	Exclut Chypre du nord.
12	Excludes hunting.	12	Non compris la chasse.
13	Excludes gas.	13	Non compris le gaz.
14	At factor cost.	14	Au coût des facteurs.
15	Including Guadeloupe, Martinique, Réunion and French Guiana.	15	Y compris Guadeloupe, Martinique, Réunion et Guyane française.
16	Includes processing of sugar and rice.	16	Traitement du sucre et du riz inclus.
17	Agricultural services and related activities such as cotton ginning and pressing are included in Industry.	17	Les services agricoles et les activités connexes (par exemple l'égrenage du coton, tressage) figurent sous « Industrie».
18	Gas distribution is included in "Services".	18	la distribution du gaz est inclue dans la catégorie "Services".
19	Distribution of petroleum products is included in "Services".	19	La distribution de produits pétroliers est inclue dans la catégorie "Services".
20	Excludes repair of motor vehicles and motorcycles, personal and household goods.	20	Non compris les réparations de véhicules à moteur et de motocycles, et d'articles personnels et ménagers.
21	Includes handicrafts.	21	Y compris l'artisanat.
22	Including Western Sahara.	22	Y compris les données de Sahara occidental.
23	Includes gas and water.	23	Y compris le gaz et l'eau.
24	Negative value is due to negative value added in electricity and utilities and a sharp decline in mining.	24	Valeur négative est due à la valeur ajoutée négative de l'électricité et des services publics et une forte baisse du secteur minier.
25	Excluding Kosovo and Metohija.	25	Non compris Kosovo et Metohija.
26	Construction refers to buildings and public works.	26	Construction se rapporte aux bâtiments et travaux publics.
27	Refers to trade only.	27	Se rapporte seulement au commerce
28	Data for GDP and its components excludes the temporarily occupied territory of the Autonomous Republic of Crimea and Sevastopol.	28	Les données concernant le PIB et ses composantes excluent le territoire temporairement occupé de la République autonome de Crimée et de Sébastopol.
29	Tanzania mainland only.	29	Tanzanie continentale seulement.

12

Balance of payments summary
Millions of US dollars

Résumé de la balance des paiements
Millions de dollars É.-U.

Country or area	1985	1995	2005	2010	2012	2013	2014	Pays ou zone
Afghanistan								**Afghanistan**
Current account	-243	...	...	-3 711	-8 309	-8 113	-6 855	Compte des transac. courantes
Capital account, n.i.e.	...	...	...	2 668	2 382	2 791	2 053	Compte de capital, n.i.a.
Financial account, n.i.e.	-101	...	...	272	340	97	-88	Compte financier, n.i.a.
Reserves and related items	26	...	...	-152	-1 918	-2 457	-1 814	Réserves et postes apparentés
Albania								**Albanie**
Current account	-36	-12	-571	-1 353	-1 258	#-1 395	-1 703	Compte des transac. courantes
Capital account, n.i.e.	...	389	123	112	104	#64	115	Compte de capital, n.i.a.
Financial account, n.i.e.	-13	411	-393	-723	-908	#-1 224	-1 306	Compte financier, n.i.a.
Reserves and related items	-18	21	148	-76	100	#151	68	Réserves et postes apparentés
Algeria								**Algérie**
Current account	1 015	...	21 180	12 220	12 212	1 039	-9 682	Compte des transac. courantes
Capital account, n.i.e.	...	...	-3	4	-9	<0	-3	Compte de capital, n.i.a.
Financial account, n.i.e.	121	...	4 084	-4 286	-2 346	112	-3 781	Compte financier, n.i.a.
Reserves and related items	1 020	...	16 904	15 207	12 040	107	-5 941	Réserves et postes apparentés
Angola								**Angola**
Current account	195	-295	5 138	7 506	13 853	8 348	-3 722	Compte des transac. courantes
Capital account, n.i.e.	...	...	...	1	<0	...	...	Compte de capital, n.i.a.
Financial account, n.i.e.	-454	925	3 126	1 661	9 016	7 948	462	Compte financier, n.i.a.
Reserves and related items	-6	-1 239	1 438	5 200	4 365	593	-3 897	Réserves et postes apparentés
Anguilla								**Anguilla**
Current account	...	-10	-52	-51	-55	-48	...	Compte des transac. courantes
Capital account, n.i.e.	...	1	2	4	5	6	...	Compte de capital, n.i.a.
Financial account, n.i.e.	...	1	-47	-37	-57	-36	...	Compte financier, n.i.a.
Reserves and related items	...	4	5	2	2	1	...	Réserves et postes apparentés
Antigua and Barbuda								**Antigua-et-Barbuda**
Current account	-23	-1	-171	-167	-167	-204	...	Compte des transac. courantes
Capital account, n.i.e.	...	4	211	17	2	12	...	Compte de capital, n.i.a.
Financial account, n.i.e.	-20	11	25	-138	-144	-201	...	Compte financier, n.i.a.
Reserves and related items	-3	14	7	-20	-14	12	...	Réserves et postes apparentés
Argentina								**Argentine**
Current account	-952	-5 118	5 274	1 360	-1 170	-4 568	-5 877	Compte des transac. courantes
Capital account, n.i.e.	...	14	89	89	48	33	58	Compte de capital, n.i.a.
Financial account, n.i.e.	-638	-5 466	-1 955	-9 831	-671	5 559	-2 605	Compte financier, n.i.a.
Reserves and related items	-846	-2 311	7 654	10 727	-3 630	-13 736	-850	Réserves et postes apparentés
Armenia								**Arménie**
Current account	...	-221	-124	-1 261	-1 058	-845	-849	Compte des transac. courantes
Capital account, n.i.e.	...	8	84	99	108	84	70	Compte de capital, n.i.a.
Financial account, n.i.e.	...	-227	-442	-1 027	-763	-1 724	-299	Compte financier, n.i.a.
Reserves and related items	...	30	239	-291	4	662	-545	Réserves et postes apparentés
Aruba								**Aruba**
Current account	...	-<0	105	-460	91	-331	-142	Compte des transac. courantes
Capital account, n.i.e.	...	...	16	4	...	...	...	Compte de capital, n.i.a.
Financial account, n.i.e.	...	-42	170	-435	37	-275	-133	Compte financier, n.i.a.
Reserves and related items	...	43	-22	-11	66	-43	9	Réserves et postes apparentés
Australia								**Australie**
Current account	...	#-18 671	-43 342	-44 714	-66 346	-51 156	-44 138	Compte des transac. courantes
Capital account, n.i.e.	...	#-300	-121	-287	-409	-461	-358	Compte de capital, n.i.a.
Financial account, n.i.e.	...	#-19 262	-49 428	-44 535	-75 380	-63 899	-45 380	Compte financier, n.i.a.
Reserves and related items	...	#396	7 254	430	2 536	5 507	3 644	Réserves et postes apparentés
Austria								**Autriche**
Current account	...	...	6 245	11 480	6 143	8 375	8 434	Compte des transac. courantes
Capital account, n.i.e.	...	...	-84	-363	-533	-649	-609	Compte de capital, n.i.a.
Financial account, n.i.e.	...	...	917	3 406	5 383	13 902	1 791	Compte financier, n.i.a.
Reserves and related items	...	...	-750	1 445	1 244	526	2 890	Réserves et postes apparentés
Azerbaijan								**Azerbaïdjan**
Current account	...	-401	167	15 040	14 976	#12 232	10 209	Compte des transac. courantes
Capital account, n.i.e.	...	-2	<0	...	-7	#-13	-7	Compte de capital, n.i.a.
Financial account, n.i.e.	...	-400	-78	12 695	11 944	#8 106	7 182	Compte financier, n.i.a.
Reserves and related items	...	58	161	1 370	1 106	#2 233	214	Réserves et postes apparentés

12

Summary of balance of payments *(continued)*
Millions of US dollars

Résumé de la balance des paiements *(suite)*
Millions de dollars É.-U.

Country or area	1985	1995	2005	2010	2012	2013	2014	Pays ou zone
Bahamas								**Bahamas**
Current account	-3	-146	-701	-814	-1 505	-1 494	-1 897	Compte des transac. courantes
Financial account, n.i.e.	12	-105	-822	-1 145	-1 314	-981	-1 509	Compte financier, n.i.a.
Reserves and related items	19	-3	-88	48	-75	-69	52	Réserves et postes apparentés
Bahrain								**Bahreïn**
Current account	39	237	1 474	770	2 222	2 560	1 124	Compte des transac. courantes
Capital account, n.i.e.	...	157	50	50	100	100	100	Compte de capital, n.i.a.
Financial account, n.i.e.	476	1 727	1 380	-352	1 767	2 363	-3 707	Compte financier, n.i.a.
Reserves and related items	357	169	294	1 280	673	173	556	Réserves et postes apparentés
Bangladesh								**Bangladesh**
Current account	-458	-824	#508	2 109	2 576	2 058	-1 677	Compte des transac. courantes
Capital account, n.i.e.	...	...	#263	603	500	725	361	Compte de capital, n.i.a.
Financial account, n.i.e.	-442	-179	#206	-2 765	-2 699	-3 628	-4 461	Compte financier, n.i.a.
Reserves and related items	-84	-512	#-104	1 020	3 543	5 277	4 571	Réserves et postes apparentés
Barbados								**Barbade**
Current account	51	9	-467	-236	-411	-248	...	Compte des transac. courantes
Capital account, n.i.e.	...	...	...	...	-7	-7	...	Compte de capital, n.i.a.
Financial account, n.i.e.	<0	26	-418	-257	-479	-46	...	Compte financier, n.i.a.
Reserves and related items	14	42	22	-45	28	-156	...	Réserves et postes apparentés
Belarus								**Bélarus**
Current account	...	-458	459	-8 280	-1 862	-7 567	-5 197	Compte des transac. courantes
Capital account, n.i.e.	...	7	...	...	4	10	8	Compte de capital, n.i.a.
Financial account, n.i.e.	...	-204	37	-6 101	-1 068	-8 402	-3 524	Compte financier, n.i.a.
Reserves and related items	...	-78	548	-1 477	107	-94	-2 146	Réserves et postes apparentés
Belgium								**Belgique**
Current account	...	...	7 703	7 973	-230	-1 100	-1 187	Compte des transac. courantes
Capital account, n.i.e.	...	...	-922	-1 142	3 018	-552	-1 351	Compte de capital, n.i.a.
Financial account, n.i.e.	...	...	11 397	5 747	1 651	589	-1 633	Compte financier, n.i.a.
Reserves and related items	...	...	-2 176	820	668	-472	-1 356	Réserves et postes apparentés
Belize								**Belize**
Current account	9	-17	-151	-46	-33	-73	-136	Compte des transac. courantes
Capital account, n.i.e.	...	...	3	5	22	38	34	Compte de capital, n.i.a.
Financial account, n.i.e.	-9	1	-144	-27	-67	-136	-164	Compte financier, n.i.a.
Reserves and related items	2	4	-11	5	55	117	85	Réserves et postes apparentés
Benin								**Bénin**
Current account	-39	-213	-270	-618	-577	-673	-885	Compte des transac. courantes
Capital account, n.i.e.	...	84	79	115	157	186	253	Compte de capital, n.i.a.
Financial account, n.i.e.	7	125	-111	-176	1	-415	-732	Compte financier, n.i.a.
Reserves and related items	-29	-254	-52	-294	-395	-53	106	Réserves et postes apparentés
Bermuda								**Bermudes**
Current account	...	...	...	696	927	841	827	Compte des transac. courantes
Financial account, n.i.e.	...	...	...	711	834	927	890	Compte financier, n.i.a.
Reserves and related items	...	...	...	3	-5	8	-9	Réserves et postes apparentés
Bhutan								**Bhoutan**
Current account	...	...	...	-323	-378	-472	-484	Compte des transac. courantes
Capital account, n.i.e.	...	...	...	150	95	263	276	Compte de capital, n.i.a.
Financial account, n.i.e.	...	...	...	-195	-297	-357	-228	Compte financier, n.i.a.
Reserves and related items	...	...	...	90	-159	168	71	Réserves et postes apparentés
Bolivia (Plurinational State of)								**Bolivie (État plurinational de)**
Current account	-286	-303	622	874	1 970	1 054	-16	Compte des transac. courantes
Financial account, n.i.e.	295	-505	-181	-860	-536	-2 022	924	Compte financier, n.i.a.
Reserves and related items	-387	92	437	924	1 712	1 124	972	Réserves et postes apparentés
Bosnia and Herzegovina								**Bosnie-Herzégovine**
Current account	...	...	-1 844	-1 031	-1 520	-999	-1 451	Compte des transac. courantes
Capital account, n.i.e.	...	...	276	264	220	229	347	Compte de capital, n.i.a.
Financial account, n.i.e.	...	...	-1 828	-607	-1 100	-1 130	-1 327	Compte financier, n.i.a.
Reserves and related items	...	...	491	-60	-82	524	377	Réserves et postes apparentés
Botswana								**Botswana**
Current account	82	300	1 634	-356	46	1 322	2 496	Compte des transac. courantes
Capital account, n.i.e.	...	12	...	...	...	...	...	Compte de capital, n.i.a.
Financial account, n.i.e.	-123	34	384	363	-568	482	-296	Compte financier, n.i.a.
Reserves and related items	254	207	2 033	-1 033	-132	986	1 260	Réserves et postes apparentés

12 Summary of balance of payments *(continued)*
Millions of US dollars

Résumé de la balance des paiements *(suite)*
Millions de dollars É.-U.

Country or area	1985	1995	2005	2010	2012	2013	2014	Pays ou zone
Brazil								**Brésil**
Current account	-280	-18 136	13 984	#-75 760	-74 059	-74 769	-104 181	Compte des transac. courantes
Capital account, n.i.e.	...	...	52	#242	208	322	231	Compte de capital, n.i.a.
Financial account, n.i.e.	8 676	-29 306	-13 144	#-125 000	-92 827	-66 410	-111 431	Compte financier, n.i.a.
Reserves and related items	-9 479	12 969	27 566	#49 080	18 899	-5 924	10 833	Réserves et postes apparentés
Brunei Darussalam								**Brunéi Darussalam**
Current account	...	...	4 033	#5 016	5 684	3 778	4 750	Compte des transac. courantes
Capital account, n.i.e.	...	...	-8	...	...	...	...	Compte de capital, n.i.a.
Financial account, n.i.e.	...	...	82	#4 842	2 761	2 698	4 372	Compte financier, n.i.a.
Reserves and related items	...	...	-63	#76	703	178	246	Réserves et postes apparentés
Bulgaria								**Bulgarie**
Current account	-136	-26	-3 347	#504	-210	983	693	Compte des transac. courantes
Capital account, n.i.e.	...	...	290	#398	699	624	1 254	Compte de capital, n.i.a.
Financial account, n.i.e.	165	-327	-4 985	#2 594	-1 471	1 916	-2 079	Compte financier, n.i.a.
Reserves and related items	298	445	709	#524	2 679	-755	2 325	Réserves et postes apparentés
Burkina Faso								**Burkina Faso**
Current account	...	...	-819	-549	-162	-1 345	-998	Compte des transac. courantes
Capital account, n.i.e.	...	...	209	200	270	483	405	Compte de capital, n.i.a.
Financial account, n.i.e.	...	...	-149	382	136	-444	-325	Compte financier, n.i.a.
Reserves and related items	...	...	-466	-733	-19	-424	-283	Réserves et postes apparentés
Burundi								**Burundi**
Current account	-41	10	#-222	-301	-255	-253	...	Compte des transac. courantes
Capital account, n.i.e.	...	...	#24	78	154	104	...	Compte de capital, n.i.a.
Financial account, n.i.e.	-66	-21	#11	-217	-123	-167	...	Compte financier, n.i.a.
Reserves and related items	16	37	#-293	-9	28	15	...	Réserves et postes apparentés
Cabo Verde								**Cabo Verde**
Current account	-9	-62	-41	-223	-245	-107	-168	Compte des transac. courantes
Capital account, n.i.e.	...	21	21	38	12	6	7	Compte de capital, n.i.a.
Financial account, n.i.e.	-18	-45	-83	-298	-308	-173	-222	Compte financier, n.i.a.
Reserves and related items	14	-32	65	28	50	70	91	Réserves et postes apparentés
Cambodia								**Cambodge**
Current account	...	-186	#-321	-410	-1 038	-1 607	-1 657	Compte des transac. courantes
Capital account, n.i.e.	...	78	#83	331	277	342	278	Compte de capital, n.i.a.
Financial account, n.i.e.	...	-122	#-311	-256	-1 180	-1 666	-2 165	Compte financier, n.i.a.
Reserves and related items	...	26	#74	150	375	352	754	Réserves et postes apparentés
Cameroon								**Cameroun**
Current account	-562	90	-495	-856	-956	-1 128	...	Compte des transac. courantes
Capital account, n.i.e.	...	20	204	147	117	97	...	Compte de capital, n.i.a.
Financial account, n.i.e.	-512	-168	78	-527	-1 092	-1 089	...	Compte financier, n.i.a.
Reserves and related items	59	152	-397	7	96	-84	...	Réserves et postes apparentés
Canada								**Canada**
Current account	-5 839	-5 061	21 910	-56 626	-59 942	-54 665	-37 475	Compte des transac. courantes
Capital account, n.i.e.	-1	-433	-191	-121	-140	-52	-104	Compte de capital, n.i.a.
Financial account, n.i.e.	-9 924	-4 332	19 070	-58 803	-61 003	-58 721	-38 939	Compte financier, n.i.a.
Reserves and related items	-63	2 711	1 335	3 815	1 729	4 751	5 245	Réserves et postes apparentés
Chile								**Chili**
Current account	-1 413	-1 350	1 449	3 581	-9 624	-10 125	-2 995	Compte des transac. courantes
Capital account, n.i.e.	...	...	41	6 240	12	11	10	Compte de capital, n.i.a.
Financial account, n.i.e.	1 394	-2 357	-1 550	5 943	-8 954	-11 563	-3 783	Compte financier, n.i.a.
Reserves and related items	-2 877	1 139	1 711	3 023	-366	312	1 057	Réserves et postes apparentés
China								**Chine**
Current account	-11 417	1 618	#132 378	237 810	215 392	148 204	219 678	Compte des transac. courantes
Capital account, n.i.e.	...	...	#4 102	4 630	4 272	3 052	-33	Compte de capital, n.i.a.
Financial account, n.i.e.	-8 971	-38 674	#-91 247	-282 234	36 038	-343 048	-38 272	Compte financier, n.i.a.
Reserves and related items	-2 440	22 469	#250 975	471 659	96 555	431 382	117 784	Réserves et postes apparentés
China, Hong Kong SAR								**Chine, Hong Kong RAS**
Current account	...	...	21 575	16 012	4 147	4 153	3 788	Compte des transac. courantes
Capital account, n.i.e.	...	...	-74	-571	-185	-208	-96	Compte de capital, n.i.a.
Financial account, n.i.e.	...	...	21 143	3 251	-15 817	3 460	-8 516	Compte financier, n.i.a.
Reserves and related items	...	...	-1 773	7 616	24 360	7 464	17 939	Réserves et postes apparentés

12

Summary of balance of payments *(continued)*
Millions of US dollars
Résumé de la balance des paiements *(suite)*
Millions de dollars É.-U.

Country or area	1985	1995	2005	2010	2012	2013	2014	Pays ou zone
China, Macao SAR								**Chine, Macao RAS**
Current account	...	...	2 965	12 092	17 956	21 938	21 082	Compte des transac. courantes
Capital account, n.i.e.	...	...	515	20	-<0	-1	-2	Compte de capital, n.i.a.
Financial account, n.i.e.	...	...	320	1 629	15 070	18 023	11 944	Compte financier, n.i.a.
Reserves and related items	...	...	1 126	5 158	3 771	-572	77	Réserves et postes apparentés
Colombia								**Colombie**
Current account	-1 809	-4 516	-1 892	-8 663	-11 306	-12 367	-19 567	Compte des transac. courantes
Financial account, n.i.e.	-2 236	-4 560	-3 231	-12 405	-17 155	-18 786	-24 338	Compte financier, n.i.a.
Reserves and related items	154	-4	1 724	3 117	5 321	6 941	4 436	Réserves et postes apparentés
Comoros								**Comores**
Current account	-14	-19	-29	-91	-44	...	...	Compte des transac. courantes
Capital account, n.i.e.	...	...	15	30	50	...	...	Compte de capital, n.i.a.
Financial account, n.i.e.	-19	-11	-1	32	-18	...	...	Compte financier, n.i.a.
Reserves and related items	6	-10	-6	-90	31	...	...	Réserves et postes apparentés
Congo								**Congo**
Current account	-161	-625	696		...	...	...	Compte des transac. courantes
Capital account, n.i.e.	...	19	11	...	...	...	...	Compte de capital, n.i.a.
Financial account, n.i.e.	-39	-81	227	...	...	...	...	Compte financier, n.i.a.
Reserves and related items	-81	-581	510	...	...	...	...	Réserves et postes apparentés
Costa Rica								**Costa Rica**
Current account	-291	-358	-981	-1 179	-2 117	-2 314	-2 008	Compte des transac. courantes
Capital account, n.i.e.	...	...	16	54	46	9	8	Compte de capital, n.i.a.
Financial account, n.i.e.	287	-517	-878	-1 875	-4 453	-3 277	-2 271	Compte financier, n.i.a.
Reserves and related items	-435	216	57	561	2 110	461	-113	Réserves et postes apparentés
Côte d'Ivoire								**Côte d'Ivoire**
Current account	...	...	40	465	-321	-633	...	Compte des transac. courantes
Capital account, n.i.e.	...	...	185	1 178	8 111	192	...	Compte de capital, n.i.a.
Financial account, n.i.e.	...	...	483	162	8 383	-124	...	Compte financier, n.i.a.
Reserves and related items	...	...	-315	1 427	-626	-17	...	Réserves et postes apparentés
Croatia								**Croatie**
Current account	...	-1 442	-2 479	-894	-209	550	412	Compte des transac. courantes
Capital account, n.i.e.	...	...	64	77	63	63	69	Compte de capital, n.i.a.
Financial account, n.i.e.	...	-1 143	-4 768	-1 885	-698	-3 004	488	Compte financier, n.i.a.
Reserves and related items	...	40	1 019	18	78	2 473	-765	Réserves et postes apparentés
Curaçao								**Curaçao**
Current account	...	...	...	...	-875	-662	-376	Compte des transac. courantes
Capital account, n.i.e.	...	...	...	...	29	23	8	Compte de capital, n.i.a.
Financial account, n.i.e.	...	...	...	...	-707	-596	-517	Compte financier, n.i.a.
Reserves and related items	...	...	...	...	-119	-35	187	Réserves et postes apparentés
Cyprus								**Chypre**
Current account	-180	-205	-971	-2 728	-1 441	-1 074	-1 053	Compte des transac. courantes
Capital account, n.i.e.	...	...	35	78	46	326	195	Compte de capital, n.i.a.
Financial account, n.i.e.	-112	213	-1 421	-2 096	-669	174	-926	Compte financier, n.i.a.
Reserves and related items	-30	-363	703	-296	-60	-386	-225	Réserves et postes apparentés
Czech Republic								**République tchèque**
Current account	...	-1 374	-1 210	-7 351	-3 158	-1 106	1 343	Compte des transac. courantes
Capital account, n.i.e.	...	6	186	1 953	2 719	4 216	1 565	Compte de capital, n.i.a.
Financial account, n.i.e.	...	-8 225	-6 588	-8 521	-3 516	-6 075	-1 103	Compte financier, n.i.a.
Reserves and related items	...	7 453	3 879	2 076	4 186	9 614	3 540	Réserves et postes apparentés
Dem. Rep. of the Congo								**Rép. dém. du Congo**
Current account	...	...	-389	-2 174	-1 260	-3 109	-3 040	Compte des transac. courantes
Capital account, n.i.e.	...	...	-93	-161	486	193	515	Compte de capital, n.i.a.
Financial account, n.i.e.	...	...	-334	9 464	-1 345	-2 996	-2 699	Compte financier, n.i.a.
Reserves and related items	...	...	-43	-10 650	601	59	76	Réserves et postes apparentés
Denmark								**Danemark**
Current account	-2 767	1 855	11 104	18 183	18 750	#24 248	21 421	Compte des transac. courantes
Capital account, n.i.e.	...	...	518	83	91	#10	-56	Compte de capital, n.i.a.
Financial account, n.i.e.	-4 603	432	10 579	-5 327	17 346	#30 149	35 340	Compte financier, n.i.a.
Reserves and related items	1 532	2 498	-1 506	4 280	1 852	#-647	-7 999	Réserves et postes apparentés

Summary of balance of payments *(continued)*
Millions of US dollars

Résumé de la balance des paiements *(suite)*
Millions de dollars É.-U.

Country or area	1985	1995	2005	2010	2012	2013	2014	Pays ou zone
Djibouti								**Djibouti**
Current account	...	78	20	50	-148	-309	...	Compte des transac. courantes
Capital account, n.i.e.	...	5	27	55	52	50	...	Compte de capital, n.i.a.
Financial account, n.i.e.	...	12	33	-11	17	-241	...	Compte financier, n.i.a.
Reserves and related items	...	-17	-31	-7	-36	144	...	Réserves et postes apparentés
Dominica								**Dominique**
Current account	-6	-40	-76	-80	-92	-72	...	Compte des transac. courantes
Capital account, n.i.e.	...	22	15	30	9	5	...	Compte de capital, n.i.a.
Financial account, n.i.e.	-7	-42	-57	-55	-81	-43	...	Compte financier, n.i.a.
Reserves and related items	-1	8	11	1	3	-5	...	Réserves et postes apparentés
Dominican Republic								**Rép. dominicaine**
Current account	-108	-183	-473	#-4 006	-3 971	-2 537	-2 026	Compte des transac. courantes
Capital account, n.i.e.	...	...	...	#38	41	41	26	Compte de capital, n.i.a.
Financial account, n.i.e.	-44	-254	-1 641	#-5 147	-3 596	-4 148	-3 305	Compte financier, n.i.a.
Reserves and related items	92	146	748	#72	-441	1 340	617	Réserves et postes apparentés
Ecuador								**Équateur**
Current account	76	-1 000	474	-1 586	-168	-968	-590	Compte des transac. courantes
Capital account, n.i.e.	...	17	16	15	14	12	13	Compte de capital, n.i.a.
Financial account, n.i.e.	1 122	43	246	-393	636	-2 893	-318	Compte financier, n.i.a.
Reserves and related items	-878	-1 459	865	-1 282	-690	1 792	-474	Réserves et postes apparentés
Egypt								**Égypte**
Current account	-2 166	-254	2 103	-4 504	-6 972	-3 534	-5 972	Compte des transac. courantes
Capital account, n.i.e.	...	...	-40	-39	-119	-83	-141	Compte de capital, n.i.a.
Financial account, n.i.e.	-1 381	1 845	-5 591	-6 470	-3 458	-5 837	-1 186	Compte financier, n.i.a.
Reserves and related items	-200	-1 827	5 226	-218	-5 792	903	-3 259	Réserves et postes apparentés
El Salvador								**El Salvador**
Current account	-189	-262	-622	-533	-1 235	-1 574	-1 199	Compte des transac. courantes
Capital account, n.i.e.	...	...	94	232	201	101	64	Compte de capital, n.i.a.
Financial account, n.i.e.	-3	-438	-787	33	-2 031	-1 037	-520	Compte financier, n.i.a.
Reserves and related items	-163	148	-190	-297	652	-327	-33	Réserves et postes apparentés
Estonia								**Estonie**
Current account	...	-158	-1 386	344	-570	-26	259	Compte des transac. courantes
Capital account, n.i.e.	...	-1	103	681	784	697	287	Compte de capital, n.i.a.
Financial account, n.i.e.	...	-233	-1 520	2 210	566	609	222	Compte financier, n.i.a.
Reserves and related items	...	84	386	-1 112	91	12	164	Réserves et postes apparentés
Ethiopia								**Éthiopie**
Current account	106	39	-1 568	-425	-2 985	...	...	Compte des transac. courantes
Capital account, n.i.e.	...	3	...	...	...	...	...	Compte de capital, n.i.a.
Financial account, n.i.e.	-225	25	-759	-2 369	-667	...	...	Compte financier, n.i.a.
Reserves and related items	162	-105	-323	-987	330	...	...	Réserves et postes apparentés
Euro Area								**Zone euro**
Current account	...	...	19 191	12 240	171 381	305 395	#315 010	Compte des transac. courantes
Capital account, n.i.e.	...	...	14 237	7 293	7 268	27 907	#26 485	Compte de capital, n.i.a.
Financial account, n.i.e.	...	...	71 432	12 215	213 634	345 144	#384 550	Compte financier, n.i.a.
Reserves and related items	...	...	-22 911	122	-1 288	-10 076	#19 192	Réserves et postes apparentés
Faroe Islands								**Îles Féroé**
Current account	...	...	31	144	...	...	...	Compte des transac. courantes
Fiji								**Fidji**
Current account	19	-113	#-212	-142	-56	-561	...	Compte des transac. courantes
Capital account, n.i.e.	...	116	#3	3	4	5	...	Compte de capital, n.i.a.
Financial account, n.i.e.	2	-88	#-222	-291	-276	-334	...	Compte financier, n.i.a.
Reserves and related items	-4	93	#-105	138	101	74	...	Réserves et postes apparentés
Finland								**Finlande**
Current account	-806	5 231	#7 788	3 168	-5 019	-4 450	-2 686	Compte des transac. courantes
Capital account, n.i.e.	...	66	#324	238	265	308	247	Compte de capital, n.i.a.
Financial account, n.i.e.	-1 433	4 284	#3 966	6 482	-22 842	-10 452	-10 128	Compte financier, n.i.a.
Reserves and related items	583	-372	#-180	-2 118	688	1 052	-284	Réserves et postes apparentés
France								**France**
Current account	-35	10 840	-137	-22 034	-32 171	-22 501	-27 481	Compte des transac. courantes
Capital account, n.i.e.	...	442	1 161	1 637	709	2 552	2 942	Compte de capital, n.i.a.
Financial account, n.i.e.	-2 441	7 517	-2 114	-7 003	-58 002	-21 242	-18 164	Compte financier, n.i.a.
Reserves and related items	2 696	712	-9 028	7 784	5 492	-2 018	1 111	Réserves et postes apparentés

12
Summary of balance of payments *(continued)*
Millions of US dollars
Résumé de la balance des paiements *(suite)*
Millions de dollars É.-U.

Country or area	1985	1995	2005	2010	2012	2013	2014	Pays ou zone
French Polynesia								**Polynésie française**
Current account	...	...	9	-18	78	159	208	Compte des transac. courantes
Capital account, n.i.e.	...	...	-1	-1	-1	4	-1	Compte de capital, n.i.a.
Financial account, n.i.e.	...	...	32	-119	-109	175	271	Compte financier, n.i.a.
Reserves and related items	...	...	...	...	...	-<0	...	Réserves et postes apparentés
Gabon								**Gabon**
Current account	-162	515	1 983	...	...	...	...	Compte des transac. courantes
Capital account, n.i.e.	...	5	...	...	...	...	...	Compte de capital, n.i.a.
Financial account, n.i.e.	-164	516	1 342	...	...	...	...	Compte financier, n.i.a.
Reserves and related items	-61	-228	226	...	...	...	...	Réserves et postes apparentés
Gambia								**Gambie**
Current account	8	-8	-50	21	-15	...	...	Compte des transac. courantes
Financial account, n.i.e.	4	-25	-43	25	24	...	...	Compte financier, n.i.a.
Reserves and related items	-8	2	-42	-93	-65	...	...	Réserves et postes apparentés
Georgia								**Géorgie**
Current account	...	...	-757	-1 328	-1 913	-1 006	-1 820	Compte des transac. courantes
Capital account, n.i.e.	...	...	54	155	126	123	108	Compte de capital, n.i.a.
Financial account, n.i.e.	...	...	-725	-994	-1 280	-741	-1 612	Compte financier, n.i.a.
Reserves and related items	...	...	49	-212	-562	-170	-176	Réserves et postes apparentés
Germany								**Allemagne**
Current account	17 994	-32 186	132 197	193 330	240 862	239 296	280 339	Compte des transac. courantes
Capital account, n.i.e.	-741	-3 118	-2 997	1 617	1 777	1 481	3 918	Compte de capital, n.i.a.
Financial account, n.i.e.	18 494	-44 430	123 096	121 608	200 558	275 331	326 627	Compte financier, n.i.a.
Reserves and related items	2 220	7 224	-2 600	2 134	1 700	1 156	-3 296	Réserves et postes apparentés
Ghana								**Ghana**
Current account	-134	-144	-1 105	-2 747	-4 912	-6 689	-3 698	Compte des transac. courantes
Capital account, n.i.e.	...	...	331	338	283	349	...	Compte de capital, n.i.a.
Financial account, n.i.e.	-85	-459	-834	-4 146	-3 368	-5 427	-3 753	Compte financier, n.i.a.
Reserves and related items	14	183	125	561	-1 235	-1 190	-85	Réserves et postes apparentés
Greece								**Grèce**
Current account	-3 276	-2 864	-18 233	-30 275	-6 172	-4 947	-4 872	Compte des transac. courantes
Capital account, n.i.e.	...	...	2 563	2 776	3 010	4 032	3 355	Compte de capital, n.i.a.
Financial account, n.i.e.	-2 924	-3 162	-15 633	-14 180	-1 616	9 794	-4 395	Compte financier, n.i.a.
Reserves and related items	-396	-23	-104	-13 936	-2 137	-6 456	5 472	Réserves et postes apparentés
Grenada								**Grenade**
Current account	2	-42	-193	-204	-193	-213	...	Compte des transac. courantes
Capital account, n.i.e.	...	9	24	36	36	36	...	Compte de capital, n.i.a.
Financial account, n.i.e.	-6	-7	-144	-125	-115	-155	...	Compte financier, n.i.a.
Reserves and related items	6	6	-27	-16	-2	32	...	Réserves et postes apparentés
Guatemala								**Guatemala**
Current account	-246	-572	-1 301	-563	-1 310	-1 351	-1 230	Compte des transac. courantes
Capital account, n.i.e.	...	62	...	3	...	...	...	Compte de capital, n.i.a.
Financial account, n.i.e.	124	-495	-612	-1 584	-2 262	-2 620	-1 797	Compte financier, n.i.a.
Reserves and related items	-327	-152	109	677	499	702	73	Réserves et postes apparentés
Guinea								**Guinée**
Current account	...	-216	-160	-329	-1 102	-1 240	...	Compte des transac. courantes
Capital account, n.i.e.	...	...	21	5	56	29	...	Compte de capital, n.i.a.
Financial account, n.i.e.	...	-109	-91	-313	-23	-666	...	Compte financier, n.i.a.
Reserves and related items	...	-72	71	40	-764	-184	...	Réserves et postes apparentés
Guinea-Bissau								**Guinée-Bissau**
Current account	...	...	-46	-99	-83	-53	...	Compte des transac. courantes
Capital account, n.i.e.	...	...	36	64	31	32	...	Compte de capital, n.i.a.
Financial account, n.i.e.	...	...	9	908	-2	-27	...	Compte financier, n.i.a.
Reserves and related items	...	...	-23	-946	-58	14	...	Réserves et postes apparentés
Guyana								**Guyana**
Current account	-97	-135	-96	-246	-367	-456	-385	Compte des transac. courantes
Capital account, n.i.e.	...	11	52	27	29	7	4	Compte de capital, n.i.a.
Financial account, n.i.e.	38	-71	-127	-160	-362	-290	-382	Compte financier, n.i.a.
Reserves and related items	-139	-43	14	-195	-111	49	22	Réserves et postes apparentés

12
Summary of balance of payments *(continued)*
Millions of US dollars

Résumé de la balance des paiements *(suite)*
Millions de dollars É.-U.

Country or area	1985	1995	2005	2010	2012	2013	2014	Pays ou zone
Haiti								**Haïti**
Current account	-95	-87	-356	-1 942	-1 419	-1 287	-1 365	Compte des transac. courantes
Capital account, n.i.e.	...	...	...	658	76	20	26	Compte de capital, n.i.a.
Financial account, n.i.e.	-46	-99	-15	752	-606	-1 495	-226	Compte financier, n.i.a.
Reserves and related items	-2	137	-331	-1 670	-730	-739	-1 046	Réserves et postes apparentés
Honduras								**Honduras**
Current account	-309	-201	-304	-682	-1 581	-1 763	-1 444	Compte des transac. courantes
Capital account, n.i.e.	...	...	581	48	35	28	32	Compte de capital, n.i.a.
Financial account, n.i.e.	-175	-115	-689	-1 341	-1 435	-2 501	-1 658	Compte financier, n.i.a.
Reserves and related items	-172	-41	788	567	-294	483	456	Réserves et postes apparentés
Hungary								**Hongrie**
Current account	-455	#-1 577	-7 883	346	2 189	5 294	3 044	Compte des transac. courantes
Capital account, n.i.e.	...	#60	740	2 365	3 250	4 895	5 079	Compte de capital, n.i.a.
Financial account, n.i.e.	-1 066	#-4 979	-14 977	-2 783	5 318	125	5 328	Compte financier, n.i.a.
Reserves and related items	536	#5 381	4 904	4 162	621	8 342	1 202	Réserves et postes apparentés
Iceland								**Islande**
Current account	-115	#-47	-2 339	-308	-944	883	551	Compte des transac. courantes
Capital account, n.i.e.	...	#-4	-6	-11	-10	-11	-14	Compte de capital, n.i.a.
Financial account, n.i.e.	-233	#-71	-6 286	24 632	-4 895	1 965	-5 806	Compte financier, n.i.a.
Reserves and related items	64	...	71	-21 859	886	226	9 735	Réserves et postes apparentés
India								**Inde**
Current account	-4 177	-5 563	-10 284	-54 516	-91 471	-49 226	-27 452	Compte des transac. courantes
Capital account, n.i.e.	...	...	...	50	-597	962	-195	Compte de capital, n.i.a.
Financial account, n.i.e.	-3 281	-3 861	-25 284	-69 597	-85 646	-59 178	-68 338	Compte financier, n.i.a.
Reserves and related items	-397	-733	14 554	14 127	-4 023	10 928	37 583	Réserves et postes apparentés
Indonesia								**Indonésie**
Current account	-1 923	-6 431	278	#5 144	-24 418	-29 109	-27 516	Compte des transac. courantes
Capital account, n.i.e.	...	...	334	#50	51	45	27	Compte de capital, n.i.a.
Financial account, n.i.e.	-1 782	-10 259	2 587	#-26 476	-24 858	-21 926	-45 340	Compte financier, n.i.a.
Reserves and related items	510	1 573	-2 111	#30 342	215	-7 325	15 248	Réserves et postes apparentés
Iraq								**Iraq**
Current account	...	...	-7 513	6 488	29 541	...	...	Compte des transac. courantes
Capital account, n.i.e.	...	...	3 889	25	7	...	...	Compte de capital, n.i.a.
Financial account, n.i.e.	...	...	1 350	-7 703	16 177	...	...	Compte financier, n.i.a.
Reserves and related items	...	...	-4 523	5 066	9 255	...	...	Réserves et postes apparentés
Ireland								**Irlande**
Current account	...	...	-7 150	2 319	9 245	14 438	#8 914	Compte des transac. courantes
Capital account, n.i.e.	...	...	418	-827	-2 557	100	#180	Compte de capital, n.i.a.
Financial account, n.i.e.	...	...	2 501	-8 378	7 334	17 726	#-2 526	Compte financier, n.i.a.
Reserves and related items	...	...	-1 776	-42	-8 470	-4 421	#11 309	Réserves et postes apparentés
Israel								**Israël**
Current account	988	-4 790	4 043	7 855	4 414	7 954	12 927	Compte des transac. courantes
Capital account, n.i.e.	151	285	253	143	161	250	217	Compte de capital, n.i.a.
Financial account, n.i.e.	-377	-4 179	8 282	17	6 614	5 161	8 573	Compte financier, n.i.a.
Reserves and related items	399	458	1 623	12 076	-135	4 672	7 159	Réserves et postes apparentés
Italy								**Italie**
Current account	-4 088	25 096	-29 744	-74 385	-9 229	19 233	39 908	Compte des transac. courantes
Capital account, n.i.e.	222	1 721	1 628	69	5 132	329	4 187	Compte de capital, n.i.a.
Financial account, n.i.e.	-98	3 040	-25 286	-114 150	-17 181	12 233	60 774	Compte financier, n.i.a.
Reserves and related items	-7 637	2 585	-1 030	1 335	1 880	1 999	-1 214	Réserves et postes apparentés
Jamaica								**Jamaïque**
Current account	-273	-99	-1 071	-934	-1 376	-1 281	-1 110	Compte des transac. courantes
Capital account, n.i.e.	...	21	<0	4	6	19	9	Compte de capital, n.i.a.
Financial account, n.i.e.	-217	-108	-1 310	-437	202	-986	-2 142	Compte financier, n.i.a.
Reserves and related items	-71	27	230	-348	-823	-179	800	Réserves et postes apparentés
Japan								**Japon**
Current account	...	...	170 123	220 888	60 117	41 132	24 021	Compte des transac. courantes
Capital account, n.i.e.	...	...	-4 878	-4 964	-1 017	-7 681	-1 894	Compte de capital, n.i.a.
Financial account, n.i.e.	...	...	126 662	203 004	91 341	-48 260	42 515	Compte financier, n.i.a.
Reserves and related items	...	...	22 325	43 854	-38 261	38 776	8 477	Réserves et postes apparentés

12

Summary of balance of payments *(continued)*
Millions of US dollars
Résumé de la balance des paiements *(suite)*
Millions de dollars É.-U.

Country or area	1985	1995	2005	2010	2012	2013	2014	Pays ou zone
Jordan								**Jordanie**
Current account	...	...	-2 271	-1 882	-4 711	-3 462	-2 419	Compte des transac. courantes
Capital account, n.i.e.	...	...	8	<0	3	2	4	Compte de capital, n.i.a.
Financial account, n.i.e.	...	...	-1 849	-2 539	-1 699	-6 952	-3 680	Compte financier, n.i.a.
Reserves and related items	...	...	261	1 460	-3 654	4 584	1 986	Réserves et postes apparentés
Kazakhstan								**Kazakhstan**
Current account	...	-213	-1 036	1 386	1 058	927	4 643	Compte des transac. courantes
Capital account, n.i.e.	...	...	5	7 898	15	-6	31	Compte de capital, n.i.a.
Financial account, n.i.e.	...	-1 163	2 294	10 632	4 319	-338	-7 362	Compte financier, n.i.a.
Reserves and related items	...	299	-1 944	4 706	-4 306	-2 380	3 920	Réserves et postes apparentés
Kenya								**Kenya**
Current account	-118	-1 578	-252	-2 369	-4 255	-4 872	-6 339	Compte des transac. courantes
Capital account, n.i.e.	...	124	103	240	235	98	24	Compte de capital, n.i.a.
Financial account, n.i.e.	-32	266	-511	-2 128	-4 713	-4 924	-6 833	Compte financier, n.i.a.
Reserves and related items	-52	-14	117	142	1 216	394	1 378	Réserves et postes apparentés
Kiribati								**Kiribati**
Current account	...	...	...	-3	-8	15	45	Compte des transac. courantes
Capital account, n.i.e.	...	...	...	13	22	23	32	Compte de capital, n.i.a.
Financial account, n.i.e.	...	...	...	-7	24	15	97	Compte financier, n.i.a.
Reserves and related items	...	...	...	10	-17	12	2	Réserves et postes apparentés
Kosovo								**Kosovo**
Current account	...	...	-308	-768	-484	#-451	-564	Compte des transac. courantes
Capital account, n.i.e.	...	...	20	27	19	#39	27	Compte de capital, n.i.a.
Financial account, n.i.e.	...	...	-27	-442	-399	#-157	-131	Compte financier, n.i.a.
Reserves and related items	...	...	-40	46	216	#-30	-58	Réserves et postes apparentés
Kuwait								**Koweït**
Current account	4 798	5 017	30 071	36 989	79 122	69 493	53 968	Compte des transac. courantes
Capital account, n.i.e.	...	-194	710	2 096	4 243	4 461	3 991	Compte de capital, n.i.a.
Financial account, n.i.e.	2 334	-157	32 762	45 577	80 717	68 676	57 005	Compte financier, n.i.a.
Reserves and related items	545	-139	619	611	3 358	3 356	1 253	Réserves et postes apparentés
Kyrgyzstan								**Kirghizistan**
Current account	...	-235	-62	-317	-1 675	-1 684	-1 788	Compte des transac. courantes
Capital account, n.i.e.	...	2	43	109	184	297	82	Compte de capital, n.i.a.
Financial account, n.i.e.	...	-260	-85	-496	-869	-938	-792	Compte financier, n.i.a.
Reserves and related items	...	-81	68	103	192	266	-127	Réserves et postes apparentés
Lao People's Dem. Rep.								**Rép. dém. pop. lao**
Current account	-164	-346	-174	29	-413	-376	...	Compte des transac. courantes
Financial account, n.i.e.	11	-90	-162	-477	-715	-777	...	Compte financier, n.i.a.
Reserves and related items	-136	-151	-6	103	-107	-76	...	Réserves et postes apparentés
Latvia								**Lettonie**
Current account	...	-16	-1 988	563	-924	-720	-629	Compte des transac. courantes
Capital account, n.i.e.	...	...	212	470	825	763	1 006	Compte de capital, n.i.a.
Financial account, n.i.e.	...	-636	-2 593	900	-2 328	-227	1 235	Compte financier, n.i.a.
Reserves and related items	...	-33	525	579	2 539	520	-155	Réserves et postes apparentés
Lebanon								**Liban**
Current account	...	...	-2 748	-7 552	-10 033	-11 716	-12 207	Compte des transac. courantes
Capital account, n.i.e.	...	...	27	39	73	1 539	1 428	Compte de capital, n.i.a.
Financial account, n.i.e.	...	...	-3 786	-3 516	-6 412	-5 360	-8 166	Compte financier, n.i.a.
Reserves and related items	...	...	458	3 059	617	2 055	3 308	Réserves et postes apparentés
Lesotho								**Lesotho**
Current account	-12	-323	166	-158	-381	-165	-228	Compte des transac. courantes
Capital account, n.i.e.	...	44	21	156	224	133	61	Compte de capital, n.i.a.
Financial account, n.i.e.	-16	-349	116	179	-253	-130	-233	Compte financier, n.i.a.
Reserves and related items	6	98	44	-209	109	212	125	Réserves et postes apparentés
Liberia								**Libéria**
Current account	55	...	-208	-737	-480	-536	-1 217	Compte des transac. courantes
Capital account, n.i.e.	...	...	...	1 594	37	33	117	Compte de capital, n.i.a.
Financial account, n.i.e.	150	...	-61	-447	-796	-794	-451	Compte financier, n.i.a.
Reserves and related items	-203	...	-186	2 003	31	19	-100	Réserves et postes apparentés
Libya								**Libye**
Current account	1 906	1 672	14 945	16 801	23 836	-108	...	Compte des transac. courantes
Financial account, n.i.e.	-784	207	-392	10 339	7 890	4 109	...	Compte financier, n.i.a.
Reserves and related items	2 362	1 701	13 840	4 170	13 409	-6 912	...	Réserves et postes apparentés

Summary of balance of payments *(continued)*
Millions of US dollars

Résumé de la balance des paiements *(suite)*
Millions de dollars É.-U.

Country or area	1985	1995	2005	2010	2012	2013	2014	Pays ou zone
Lithuania								**Lituanie**
Current account	...	-614	-1 878	-119	-510	726	1 685	Compte des transac. courantes
Capital account, n.i.e.	...	-39	261	1 422	1 240	1 455	1 287	Compte de capital, n.i.a.
Financial account, n.i.e.	...	-534	-1 951	448	463	2 217	-802	Compte financier, n.i.a.
Reserves and related items	...	168	712	491	120	-575	1 615	Réserves et postes apparentés
Luxembourg								**Luxembourg**
Current account	...	...	4 107	3 585	3 401	3 502	3 616	Compte des transac. courantes
Capital account, n.i.e.	...	...	1 278	-263	-503	-1 012	-1 300	Compte de capital, n.i.a.
Financial account, n.i.e.	...	...	5 453	3 264	2 922	2 420	2 449	Compte financier, n.i.a.
Reserves and related items	...	...	-69	34	-22	74	-130	Réserves et postes apparentés
Madagascar								**Madagascar**
Current account	-184	-276	-767	-896	-759	-622	...	Compte des transac. courantes
Capital account, n.i.e.	...	40	286	75	120	134	...	Compte de capital, n.i.a.
Financial account, n.i.e.	-6	198	-15	-943	-678	-274	...	Compte financier, n.i.a.
Reserves and related items	-167	-330	-378	-41	-161	-397	...	Réserves et postes apparentés
Malawi								**Malawi**
Current account	-127	-78	-507	-969	-745	-1 264	-1 078	Compte des transac. courantes
Capital account, n.i.e.	...	...	363	710	360	611	574	Compte de capital, n.i.a.
Financial account, n.i.e.	4	-88	-150	-121	-68	-1 105	-1 010	Compte financier, n.i.a.
Reserves and related items	-26	-75	41	114	-10	177	188	Réserves et postes apparentés
Malaysia								**Malaisie**
Current account	-600	-8 644	19 980	#25 644	16 316	11 205	14 472	Compte des transac. courantes
Capital account, n.i.e.	...	...	...	#-34	79	-5	81	Compte de capital, n.i.a.
Financial account, n.i.e.	-1 929	-7 643	9 806	#5 961	7 501	6 175	24 703	Compte financier, n.i.a.
Reserves and related items	1 148	-1 763	3 620	#-37	1 343	4 399	-11 080	Réserves et postes apparentés
Maldives								**Maldives**
Current account	-6	-18	-273	-196	-184	-127	-125	Compte des transac. courantes
Capital account, n.i.e.	...	...	...	9	17	8	7	Compte de capital, n.i.a.
Financial account, n.i.e.	4	-68	-223	-153	-188	-67	-507	Compte financier, n.i.a.
Reserves and related items	2	17	-23	82	-30	69	253	Réserves et postes apparentés
Mali								**Mali**
Current account	...	...	#-438	-1 190	-273	-375	...	Compte des transac. courantes
Capital account, n.i.e.	...	...	#149	230	92	210	...	Compte de capital, n.i.a.
Financial account, n.i.e.	...	...	#-333	-752	-110	14	...	Compte financier, n.i.a.
Reserves and related items	...	...	#16	-177	-81	-213	...	Réserves et postes apparentés
Malta								**Malte**
Current account	-26	-380	-418	-420	116	364	422	Compte des transac. courantes
Capital account, n.i.e.	...	13	197	171	173	177	189	Compte de capital, n.i.a.
Financial account, n.i.e.	25	-39	-356	-91	768	-289	628	Compte financier, n.i.a.
Reserves and related items	-67	-307	219	27	164	-47	30	Réserves et postes apparentés
Marshall Islands								**Îles Marshall**
Current account	...	...	#-3	-14	-20	-29	-5	Compte des transac. courantes
Capital account, n.i.e.	...	...	#6	19	9	13	9	Compte de capital, n.i.a.
Financial account, n.i.e.	...	...	#-4	19	-29	-34	-14	Compte financier, n.i.a.
Reserves and related items	...	...	...	0	<0	<0	0	Réserves et postes apparentés
Mauritania								**Mauritanie**
Current account	-116	22	...	...	-1 498	-1 349	-1 535	Compte des transac. courantes
Capital account, n.i.e.	...	...	...	...	41	5	16	Compte de capital, n.i.a.
Financial account, n.i.e.	-91	10	...	...	-1 807	-1 588	-1 192	Compte financier, n.i.a.
Reserves and related items	-31	-6	...	...	210	-63	-385	Réserves et postes apparentés
Mauritius								**Maurice**
Current account	-30	-22	-324	-1 006	-828	#-750	-714	Compte des transac. courantes
Financial account, n.i.e.	24	-25	-142	-1 065	-1 395	#-1 061	-1 367	Compte financier, n.i.a.
Reserves and related items	-3	109	-165	209	192	#541	756	Réserves et postes apparentés
Mexico								**Mexique**
Current account	800	-1 576	-9 046	-4 867	-15 923	-29 680	-24 036	Compte des transac. courantes
Financial account, n.i.e.	612	10 487	-14 385	-47 777	-53 065	-66 422	-58 362	Compte financier, n.i.a.
Reserves and related items	-2 729	-16 312	9 996	20 698	17 517	17 778	16 722	Réserves et postes apparentés
Micronesia (Fed. States of)								**Micronésie (États féd. de)**
Current account	...	...	...	-25	-24	-2	22	Compte des transac. courantes
Capital account, n.i.e.	...	...	...	64	68	42	21	Compte de capital, n.i.a.
Financial account, n.i.e.	...	...	...	24	34	15	28	Compte financier, n.i.a.
Reserves and related items	...	...	...	5	5	1	11	Réserves et postes apparentés

12 Summary of balance of payments *(continued)*
Millions of US dollars
Résumé de la balance des paiements *(suite)*
Millions de dollars É.-U.

Country or area	1985	1995	2005	2010	2012	2013	2014	Pays ou zone
Mongolia								**Mongolie**
Current account	-814	39	-5	-887	-3 362	-3 192	-1 405	Compte des transac. courantes
Capital account, n.i.e.	...	...	...	142	120	126	100	Compte de capital, n.i.a.
Financial account, n.i.e.	-755	16	-46	-1 592	-4 809	-1 324	-963	Compte financier, n.i.a.
Reserves and related items	25	32	-34	875	1 704	-1 867	-471	Réserves et postes apparentés
Montenegro								**Monténégro**
Current account	...	...	...	-952	-769	#-649	-699	Compte des transac. courantes
Capital account, n.i.e.	...	...	...	-1	10	#3	...	Compte de capital, n.i.a.
Financial account, n.i.e.	...	...	...	-710	-496	#-420	-320	Compte financier, n.i.a.
Reserves and related items	...	...	...	13	53	#105	161	Réserves et postes apparentés
Montserrat								**Montserrat**
Current account	...	-2	-16	-19	-14	-27	...	Compte des transac. courantes
Capital account, n.i.e.	...	7	7	13	20	30	...	Compte de capital, n.i.a.
Financial account, n.i.e.	...	-2	-10	-9	-<0	-8	...	Compte financier, n.i.a.
Reserves and related items	...	1	-<0	3	6	8	...	Réserves et postes apparentés
Morocco								**Maroc**
Current account	-891	-1 296	949	-4 209	-9 843	-8 692	...	Compte des transac. courantes
Financial account, n.i.e.	-815	984	88	-1 364	-1 884	-3 056	...	Compte financier, n.i.a.
Reserves and related items	-32	-1 895	449	-3 012	-8 368	-5 258	...	Réserves et postes apparentés
Mozambique								**Mozambique**
Current account	...	...	-761	-1 679	-6 790	-6 253	-5 797	Compte des transac. courantes
Capital account, n.i.e.	...	...	188	355	490	423	375	Compte de capital, n.i.a.
Financial account, n.i.e.	...	...	-427	-1 471	-6 646	-6 204	-5 283	Compte financier, n.i.a.
Reserves and related items	...	...	-121	201	377	396	-107	Réserves et postes apparentés
Myanmar								**Myanmar**
Current account	-205	-258	582	1 574	-1 260	#-389	-1 641	Compte des transac. courantes
Capital account, n.i.e.	...	...	...	...	-3	#6 467	-1	Compte de capital, n.i.a.
Financial account, n.i.e.	-149	-243	-165	-1 117	-2 315	#948	-1 564	Compte financier, n.i.a.
Reserves and related items	-15	-32	142	559	6 916	#2 617	1 111	Réserves et postes apparentés
Namibia								**Namibie**
Current account	...	176	267	-717	-909	-654	-1 505	Compte des transac. courantes
Capital account, n.i.e.	...	40	80	113	149	129	138	Compte de capital, n.i.a.
Financial account, n.i.e.	...	205	875	64	-179	1 881	12	Compte financier, n.i.a.
Reserves and related items	...	24	-401	-1 023	-1 081	-2 559	-1 437	Réserves et postes apparentés
Nepal								**Népal**
Current account	-122	-356	153	-128	577	1 160	488	Compte des transac. courantes
Capital account, n.i.e.	...	...	40	185	202	167	139	Compte de capital, n.i.a.
Financial account, n.i.e.	-26	-369	162	-285	118	-67	-11	Compte financier, n.i.a.
Reserves and related items	-93	15	171	162	717	1 457	833	Réserves et postes apparentés
Netherlands								**Pays-Bas**
Current account	4 248	25 773	41 599	61 820	89 546	94 907	93 401	Compte des transac. courantes
Capital account, n.i.e.	-39	-497	84	-4 121	-11 874	665	-1 112	Compte de capital, n.i.a.
Financial account, n.i.e.	2 373	18 839	34 149	50 863	57 282	88 620	73 320	Compte financier, n.i.a.
Reserves and related items	771	-1 911	-1 790	492	2 767	-129	-1 596	Réserves et postes apparentés
Netherlands Antilles [former]								**Antilles néerlandaises [anc.]**
Current account	403	128	-106	...	...	...	...	Compte des transac. courantes
Capital account, n.i.e.	...	63	96	...	...	...	...	Compte de capital, n.i.a.
Financial account, n.i.e.	324	142	-27	...	...	...	...	Compte financier, n.i.a.
Reserves and related items	72	60	48	...	...	...	...	Réserves et postes apparentés
New Caledonia								**Nouvelle-Calédonie**
Current account	...	...	-112	-1 360	-1 934	-1 861	-1 469	Compte des transac. courantes
Capital account, n.i.e.	...	...	9	2	7	8	11	Compte de capital, n.i.a.
Financial account, n.i.e.	...	...	-19	-1 279	-2 109	-2 060	-1 519	Compte financier, n.i.a.
New Zealand								**Nouvelle-Zélande**
Current account	...	...	-8 025	-3 430	-6 869	-5 782	-6 137	Compte des transac. courantes
Capital account, n.i.e.	...	...	...	4 354	-6	2	33	Compte de capital, n.i.a.
Financial account, n.i.e.	...	...	-11 367	-99	-6 392	1 280	-3 062	Compte financier, n.i.a.
Reserves and related items	...	...	2 417	847	515	-842	-140	Réserves et postes apparentés
Nicaragua								**Nicaragua**
Current account	-771	-722	#-784	-780	-1 113	-1 200	-838	Compte des transac. courantes
Capital account, n.i.e.	...	227	#479	264	238	231	276	Compte de capital, n.i.a.
Financial account, n.i.e.	-110	612	#-403	-548	-1 872	-1 408	-1 326	Compte financier, n.i.a.
Reserves and related items	9	-964	#34	202	-3	112	313	Réserves et postes apparentés

Summary of balance of payments *(continued)*
Millions of US dollars
Résumé de la balance des paiements *(suite)*
Millions de dollars É.-U.

Country or area	1985	1995	2005	2010	2012	2013	2014	Pays ou zone
Niger								**Niger**
Current account	...	...	-312	-1 136	-1 022	-1 150	...	Compte des transac. courantes
Capital account, n.i.e.	...	...	49	196	273	571	...	Compte de capital, n.i.a.
Financial account, n.i.e.	...	...	-174	-1 031	-1 082	-689	...	Compte financier, n.i.a.
Reserves and related items	...	...	33	106	310	98	...	Réserves et postes apparentés
Nigeria								**Nigéria**
Current account	2 604	-2 578	36 529	13 111	17 374	19 049	1 268	Compte des transac. courantes
Capital account, n.i.e.	...	-46	7 336	...	...	...	...	Compte de capital, n.i.a.
Financial account, n.i.e.	3 678	46	15 184	7 890	1 278	-6 706	-4 639	Compte financier, n.i.a.
Reserves and related items	-1 209	-2 774	11 336	-9 693	11 098	-980	-8 384	Réserves et postes apparentés
Norway								**Norvège**
Current account	3 030	5 233	49 967	50 258	#63 501	52 364	47 163	Compte des transac. courantes
Capital account, n.i.e.	...	-170	-279	-164	#-225	-243	-185	Compte de capital, n.i.a.
Financial account, n.i.e.	-1 506	542	46 895	36 853	#44 171	41 013	55 408	Compte financier, n.i.a.
Reserves and related items	3 460	575	4 511	3 543	#1 296	3 138	6 930	Réserves et postes apparentés
Oman								**Oman**
Current account	...	...	5 178	4 884	7 740	5 246	4 056	Compte des transac. courantes
Capital account, n.i.e.	...	...	-16	-65	-86	-112	-130	Compte de capital, n.i.a.
Financial account, n.i.e.	...	...	1 501	4 216	6 185	-6 034	1 922	Compte financier, n.i.a.
Reserves and related items	...	...	2 809	1 510	1 036	12 300	1 117	Réserves et postes apparentés
Pakistan								**Pakistan**
Current account	-1 083	-3 349	#-3 606	-1 354	-2 342	-4 416	-3 544	Compte des transac. courantes
Capital account, n.i.e.	...	...	#202	109	191	329	1 963	Compte de capital, n.i.a.
Financial account, n.i.e.	-634	-2 449	#-3 811	-3 127	-537	-2 049	-8 650	Compte financier, n.i.a.
Reserves and related items	-419	-1 204	#260	694	-2 017	-2 556	7 099	Réserves et postes apparentés
Palau								**Palaos**
Current account	...	...	#-40	-19	-34	-29	-48	Compte des transac. courantes
Capital account, n.i.e.	...	...	#51	30	26	24	21	Compte de capital, n.i.a.
Financial account, n.i.e.	...	...	#7	6	3	-2	-15	Compte financier, n.i.a.
Reserves and related items	...	...	...	0	<0	<0	0	Réserves et postes apparentés
Panama								**Panama**
Current account	75	-471	-1 064	-3 113	-4 177	-4 401	-4 794	Compte des transac. courantes
Capital account, n.i.e.	...	9	16	43	17	28	24	Compte de capital, n.i.a.
Financial account, n.i.e.	83	-116	-1 785	-1 528	-2 736	-4 368	-5 699	Compte financier, n.i.a.
Reserves and related items	-128	-331	497	-1 276	-958	-110	397	Réserves et postes apparentés
Papua New Guinea								**Papouasie-Nvl-Guinée**
Current account	-122	492	539	-914	-2 833	-3 973	1 932	Compte des transac. courantes
Capital account, n.i.e.	...	...	33	37	25	24	10	Compte de capital, n.i.a.
Financial account, n.i.e.	-124	445	606	-1 037	-1 882	-2 742	2 762	Compte financier, n.i.a.
Reserves and related items	-1	-39	-30	85	-126	180	-105	Réserves et postes apparentés
Paraguay								**Paraguay**
Current account	-252	-217	-68	-57	-501	477	-117	Compte des transac. courantes
Capital account, n.i.e.	...	11	20	40	51	61	141	Compte de capital, n.i.a.
Financial account, n.i.e.	-39	-1 012	-565	-155	-873	47	-1 700	Compte financier, n.i.a.
Reserves and related items	-140	48	163	319	-24	1 036	1 138	Réserves et postes apparentés
Peru								**Pérou**
Current account	102	-4 625	1 148	-3 782	-5 237	-8 474	-8 031	Compte des transac. courantes
Capital account, n.i.e.	32	65	6	7	9	12	12	Compte de capital, n.i.a.
Financial account, n.i.e.	200	-3 718	-24	-13 391	-20 196	-10 168	-5 648	Compte financier, n.i.a.
Reserves and related items	-1 594	-590	1 411	10 970	15 149	1 650	-3 307	Réserves et postes apparentés
Philippines								**Philippines**
Current account	-36	-1 980	#1 990	7 179	6 949	11 384	10 917	Compte des transac. courantes
Capital account, n.i.e.	...	...	#79	88	95	134	101	Compte de capital, n.i.a.
Financial account, n.i.e.	-328	-5 309	#2 583	-13 775	-6 747	2 230	10 519	Compte financier, n.i.a.
Reserves and related items	838	1 235	#1 662	17 528	9 236	5 085	-2 858	Réserves et postes apparentés
Poland								**Pologne**
Current account	-982	854	-7 981	-25 875	-18 605	-6 749	-11 124	Compte des transac. courantes
Capital account, n.i.e.	...	285	996	8 612	10 958	11 962	13 305	Compte de capital, n.i.a.
Financial account, n.i.e.	1 476	-9 260	-15 171	-46 129	-22 662	-6 963	-4 915	Compte financier, n.i.a.
Reserves and related items	-2 340	9 835	8 143	15 109	11 188	946	371	Réserves et postes apparentés

12 Summary of balance of payments *(continued)*
Millions of US dollars
Résumé de la balance des paiements *(suite)*
Millions de dollars É.-U.

Country or area	1985	1995	2005	2010	2012	2013	2014	Pays ou zone
Portugal								**Portugal**
Current account	380	-132	-19 537	-24 199	-4 407	3 189	1 209	Compte des transac. courantes
Capital account, n.i.e.	...	...	2 782	3 317	4 505	3 716	3 393	Compte de capital, n.i.a.
Financial account, n.i.e.	-580	-3 025	-15 564	-21 653	10 387	10 482	4 761	Compte financier, n.i.a.
Reserves and related items	707	-300	-1 743	1 270	-10 373	-3 986	-186	Réserves et postes apparentés
Qatar								**Qatar**
Current account	...	...	...	...	62 000	60 461	49 662	Compte des transac. courantes
Capital account, n.i.e.	...	...	...	...	-3 507	-1 961	-2 722	Compte de capital, n.i.a.
Financial account, n.i.e.	...	...	...	...	38 289	47 366	43 647	Compte financier, n.i.a.
Reserves and related items	...	...	...	...	16 079	9 064	1 293	Réserves et postes apparentés
Republic of Korea								**République de Corée**
Current account	-2 079	-9 752	12 655	28 850	50 835	81 148	84 373	Compte des transac. courantes
Capital account, n.i.e.	...	...	-1	-63	-42	-27	-9	Compte de capital, n.i.a.
Financial account, n.i.e.	-3 861	-18 562	-1 035	-3 781	38 398	63 809	71 448	Compte financier, n.i.a.
Reserves and related items	201	7 039	19 864	26 899	13 460	15 982	17 412	Réserves et postes apparentés
Republic of Moldova								**République de Moldova**
Current account	...	-88	-248	-545	-700	-561	-676	Compte des transac. courantes
Capital account, n.i.e.	...	...	...	1	21	58	95	Compte de capital, n.i.a.
Financial account, n.i.e.	...	69	-200	-399	-810	-656	-20	Compte financier, n.i.a.
Reserves and related items	...	-175	114	-111	259	239	-655	Réserves et postes apparentés
Romania								**Roumanie**
Current account	1 381	-1 780	#-8 541	-8 479	-8 199	-2 077	-951	Compte des transac. courantes
Capital account, n.i.e.	...	32	#714	258	2 423	4 048	5 206	Compte de capital, n.i.a.
Financial account, n.i.e.	1 580	-812	#-14 162	-6 392	-4 478	-6 754	3	Compte financier, n.i.a.
Reserves and related items	-317	-480	#6 804	-1 156	230	8 825	3 960	Réserves et postes apparentés
Russian Federation								**Fédération de Russie**
Current account	...	6 963	84 389	67 452	71 282	34 801	58 432	Compte des transac. courantes
Capital account, n.i.e.	...	786	-12 387	-41	-5 218	-395	-42 005	Compte de capital, n.i.a.
Financial account, n.i.e.	...	6 290	2 048	21 529	25 675	46 213	130 185	Compte financier, n.i.a.
Reserves and related items	...	-8 325	64 968	36 749	30 020	-22 078	-107 546	Réserves et postes apparentés
Rwanda								**Rwanda**
Current account	-64	57	-98	-412	-821	#-815	-1 047	Compte des transac. courantes
Capital account, n.i.e.	...	...	93	286	171	#235	337	Compte de capital, n.i.a.
Financial account, n.i.e.	-69	11	-82	-214	-411	#-751	-591	Compte financier, n.i.a.
Reserves and related items	2	53	88	72	-212	#228	-90	Réserves et postes apparentés
Saint Kitts and Nevis								**Saint-Kitts-et-Nevis**
Current account	-7	-45	-65	-139	-85	-63	...	Compte des transac. courantes
Capital account, n.i.e.	...	6	12	56	164	128	...	Compte de capital, n.i.a.
Financial account, n.i.e.	-9	-25	-29	-142	110	62	...	Compte financier, n.i.a.
Reserves and related items	2	2	-7	33	-14	34	...	Réserves et postes apparentés
Saint Lucia								**Sainte-Lucie**
Current account	-13	-36	-129	-203	-183	-100	...	Compte des transac. courantes
Capital account, n.i.e.	...	12	4	42	31	27	...	Compte de capital, n.i.a.
Financial account, n.i.e.	-12	-31	-123	-157	-160	-57	...	Compte financier, n.i.a.
Reserves and related items	1	6	-17	21	16	-40	...	Réserves et postes apparentés
Saint Vincent-Grenadines								**Saint-Vincent-Grenadines**
Current account	4	-40	-102	-208	-193	-210	...	Compte des transac. courantes
Capital account, n.i.e.	...	5	12	52	33	11	...	Compte de capital, n.i.a.
Financial account, n.i.e.	-1	-33	-108	-179	-187	-210	...	Compte financier, n.i.a.
Reserves and related items	6	-1	-3	31	21	24	...	Réserves et postes apparentés
Samoa								**Samoa**
Current account	2	9	#-48	-44	1	-37	-49	Compte des transac. courantes
Capital account, n.i.e.	...	...	#35	31	15	37	39	Compte de capital, n.i.a.
Financial account, n.i.e.	<0	6	#-8	-46	10	-29	-24	Compte financier, n.i.a.
Reserves and related items	5	2	#-1	24	-<0	-21	-25	Réserves et postes apparentés
Sao Tome and Principe								**Sao Tomé-et-Principe**
Current account	-18	...	-36	-88	-99	-83	-104	Compte des transac. courantes
Capital account, n.i.e.	...	...	66	42	39	28	29	Compte de capital, n.i.a.
Financial account, n.i.e.	-1	...	4	-75	-65	-9	17	Compte financier, n.i.a.
Reserves and related items	-11	...	31	13	-11	6	-2	Réserves et postes apparentés

12

Summary of balance of payments *(continued)*
Millions of US dollars

Résumé de la balance des paiements *(suite)*
Millions de dollars É.-U.

Country or area	1985	1995	2005	2010	2012	2013	2014	Pays ou zone
Saudi Arabia								**Arabie saoudite**
Current account	-12 932	-5 318	#90 060	66 751	164 764	135 442	73 758	Compte des transac. courantes
Capital account, n.i.e.	...	...	...	...	...	-271	-335	-329 Compte de capital, n.i.a.
Financial account, n.i.e.	-12 222	-6 533	#-8 361	-2 657	6 369	57 382	57 357	Compte financier, n.i.a.
Reserves and related items	-709	1 215	#63 969	35 255	115 773	69 127	7 477	Réserves et postes apparentés
Senegal								**Sénégal**
Current account	...	...	#-676	-589	...	...	...	Compte des transac. courantes
Capital account, n.i.e.	...	...	#200	302	...	...	...	Compte de capital, n.i.a.
Financial account, n.i.e.	...	...	#-118	224	...	...	...	Compte financier, n.i.a.
Reserves and related items	...	...	#-362	-529	...	...	...	Réserves et postes apparentés
Serbia								**Serbie**
Current account	...	...	...	-2 692	-4 730	-2 795	-2 635	Compte des transac. courantes
Capital account, n.i.e.	...	...	...	-1	-11	20	9	Compte de capital, n.i.a.
Financial account, n.i.e.	...	...	...	-402	-3 131	-4 007	-669	Compte financier, n.i.a.
Reserves and related items	...	...	...	-1 676	-1 183	1 838	-1 597	Réserves et postes apparentés
Seychelles								**Seychelles**
Current account	-19	-3	-188	-214	-161	-159	-310	Compte des transac. courantes
Capital account, n.i.e.	...	1	30	275	64	71	39	Compte de capital, n.i.a.
Financial account, n.i.e.	-16	1	-129	-164	-228	-134	-236	Compte financier, n.i.a.
Reserves and related items	-<0	-32	-29	302	22	95	41	Réserves et postes apparentés
Sierra Leone								**Sierra Leone**
Current account	3	-118	-171	-746	-1 286	-775	-1 317	Compte des transac. courantes
Capital account, n.i.e.	...	...	37	79	100	93	163	Compte de capital, n.i.a.
Financial account, n.i.e.	-4	-98	-63	-325	-1 042	-753	-624	Compte financier, n.i.a.
Reserves and related items	-74	1	-130	-319	-134	-69	-505	Réserves et postes apparentés
Singapore								**Singapour**
Current account	-4	#14 417	27 868	55 943	49 774	54 084	58 772	Compte des transac. courantes
Financial account, n.i.e.	-698	#5 383	15 878	18 277	22 779	36 071	49 615	Compte financier, n.i.a.
Reserves and related items	1 337	#8 594	12 283	42 353	26 222	18 097	6 819	Réserves et postes apparentés
Sint Maarten (Dutch part)								**Saint-Martin (partie néer.)**
Current account	...	...	...	...	93	4	-47	Compte des transac. courantes
Capital account, n.i.e.	...	...	...	...	12	7	4	Compte de capital, n.i.a.
Financial account, n.i.e.	...	...	...	...	243	87	-76	Compte financier, n.i.a.
Reserves and related items	...	...	...	...	-86	-21	64	Réserves et postes apparentés
Slovakia								**Slovaquie**
Current account	...	390	-5 125	-4 211	890	1 887	176	Compte des transac. courantes
Capital account, n.i.e.	...	46	-13	1 392	1 815	1 422	937	Compte de capital, n.i.a.
Financial account, n.i.e.	...	-1 211	-6 035	-3 177	429	-1 712	-2 706	Compte financier, n.i.a.
Reserves and related items	...	1 791	1 294	37	-21	96	552	Réserves et postes apparentés
Slovenia								**Slovénie**
Current account	...	-75	-681	-55	1 191	2 684	3 451	Compte des transac. courantes
Capital account, n.i.e.	...	-6	-137	72	52	96	-224	Compte de capital, n.i.a.
Financial account, n.i.e.	...	-516	-844	-1 879	-154	1 781	2 981	Compte financier, n.i.a.
Reserves and related items	...	240	206	-31	-42	5	124	Réserves et postes apparentés
Solomon Islands								**Îles Salomon**
Current account	-28	8	-90	-144	26	-39	-50	Compte des transac. courantes
Capital account, n.i.e.	...	1	28	50	98	87	71	Compte de capital, n.i.a.
Financial account, n.i.e.	-14	8	9	-209	10	-9	8	Compte financier, n.i.a.
Reserves and related items	-15	-1	-18	103	79	59	1	Réserves et postes apparentés
South Africa								**Afrique du Sud**
Current account	2 261	-2 493	-8 015	-5 492	-19 678	-21 194	-19 086	Compte des transac. courantes
Capital account, n.i.e.	...	...	30	31	29	25	22	Compte de capital, n.i.a.
Financial account, n.i.e.	2 092	-4 295	-12 612	-11 304	-21 937	-13 970	-15 412	Compte financier, n.i.a.
Reserves and related items	-742	911	5 766	3 796	1 198	499	1 398	Réserves et postes apparentés
Spain								**Espagne**
Current account	2 785	-1 967	-87 006	-56 363	-3 420	20 713	12 808	Compte des transac. courantes
Capital account, n.i.e.	...	5 861	9 048	6 501	6 642	9 017	5 983	Compte de capital, n.i.a.
Financial account, n.i.e.	3 217	6 675	-73 517	-58 118	-814	46 251	23 520	Compte financier, n.i.a.
Reserves and related items	-2 275	-6 414	-1 920	1 051	2 927	689	4 879	Réserves et postes apparentés

12
Summary of balance of payments *(continued)*
Millions of US dollars
Résumé de la balance des paiements *(suite)*
Millions de dollars É.-U.

Country or area	1985	1995	2005	2010	2012	2013	2014	Pays ou zone
Sri Lanka								**Sri Lanka**
Current account	-418	-770	-743	-1 127	#-4 009	-2 541	-2 018	Compte des transac. courantes
Capital account, n.i.e.	...	116	242	150	#130	71	58	Compte de capital, n.i.a.
Financial account, n.i.e.	-373	-730	-67	952	#-4 244	-4 628	-4 204	Compte financier, n.i.a.
Reserves and related items	-88	239	-498	-2 782	#-23	1 566	2 268	Réserves et postes apparentés
State of Palestine								**État de Palestine**
Current account	...	-984	-1 365	-1 307	-1 821	-2 384	-1 387	Compte des transac. courantes
Capital account, n.i.e.	...	262	386	828	560	546	322	Compte de capital, n.i.a.
Financial account, n.i.e.	...	-557	-892	-238	-1 100	-1 614	-1 051	Compte financier, n.i.a.
Reserves and related items	...	...	7	36	138	16	-20	Réserves et postes apparentés
Sudan								**Soudan**
Current account	149	-500	-2 473	-1 725	-6 259	-5 426	#-4 852	Compte des transac. courantes
Capital account, n.i.e.	...	...	165	314	629	314	#213	Compte de capital, n.i.a.
Financial account, n.i.e.	444	-474	-1 991	-2 363	-2 928	-2 227	#-1 366	Compte financier, n.i.a.
Reserves and related items	-444	-3	542	-25	-332	-1 499	#-1 592	Réserves et postes apparentés
Suriname								**Suriname**
Current account	...	...	-144	651	162	-196	-415	Compte des transac. courantes
Capital account, n.i.e.	...	...	15	54	-7	<0	-<0	Compte de capital, n.i.a.
Financial account, n.i.e.	...	...	21	502	-487	-429	-688	Compte financier, n.i.a.
Reserves and related items	...	...	20	35	180	-148	-150	Réserves et postes apparentés
Swaziland								**Swaziland**
Current account	-38	-30	-103	-389	98	150	58	Compte des transac. courantes
Capital account, n.i.e.	...	-<0	-3	14	113	26	78	Compte de capital, n.i.a.
Financial account, n.i.e.	-22	25	-148	-88	74	189	339	Compte financier, n.i.a.
Reserves and related items	-5	24	1	-231	108	90	-128	Réserves et postes apparentés
Sweden								**Suède**
Current account	-1 010	4 940	26 423	29 402	36 048	38 805	32 640	Compte des transac. courantes
Capital account, n.i.e.	...	...	392	-662	-910	-1 439	-605	Compte de capital, n.i.a.
Financial account, n.i.e.	2 893	5 052	27 017	37 447	9 986	4 986	13 032	Compte financier, n.i.a.
Reserves and related items	-4 651	-1 664	412	-1 176	548	14 910	486	Réserves et postes apparentés
Switzerland								**Suisse**
Current account	6 039	20 703	57 530	86 601	68 589	76 145	61 539	Compte des transac. courantes
Capital account, n.i.e.	...	-462	-2 297	-4 460	-2 359	763	-11 286	Compte de capital, n.i.a.
Financial account, n.i.e.	7 272	11 231	97 549	-16 225	-91 484	99 737	17 981	Compte financier, n.i.a.
Reserves and related items	1 228	29	-17 673	125 382	184 517	13 994	35 900	Réserves et postes apparentés
Syrian Arab Republic								**Rép. arabe syrienne**
Current account	...	...	295	-367	...	...	...	Compte des transac. courantes
Capital account, n.i.e.	...	...	18	50	...	...	...	Compte de capital, n.i.a.
Financial account, n.i.e.	...	...	162	-1 252	...	...	...	Compte financier, n.i.a.
Reserves and related items	...	...	14	2 076	...	...	...	Réserves et postes apparentés
Tajikistan								**Tadjikistan**
Current account	...	...	-19	-370	-248	-203	#-946	Compte des transac. courantes
Capital account, n.i.e.	...	...	...	69	71	27	#129	Compte de capital, n.i.a.
Financial account, n.i.e.	...	...	-101	-248	-156	-44	#-358	Compte financier, n.i.a.
Reserves and related items	...	...	6	106	75	9	#-86	Réserves et postes apparentés
Thailand								**Thaïlande**
Current account	-1 537	-13 582	#-7 647	9 945	-1 458	-5 068	15 413	Compte des transac. courantes
Capital account, n.i.e.	...	...	...	245	232	285	100	Compte de capital, n.i.a.
Financial account, n.i.e.	-1 538	-21 937	#-7 864	-23 570	-12 728	2 692	16 500	Compte financier, n.i.a.
Reserves and related items	105	7 159	#5 417	31 246	5 236	-5 130	-1 216	Réserves et postes apparentés
TFYR of Macedonia								**ex-R.Y. de Macédoine**
Current account	...	...	-159	-198	-319	-177	-104	Compte des transac. courantes
Capital account, n.i.e.	...	...	<0	4	12	20	4	Compte de capital, n.i.a.
Financial account, n.i.e.	...	...	-576	-272	-470	-90	-717	Compte financier, n.i.a.
Reserves and related items	...	...	411	78	185	-48	660	Réserves et postes apparentés
Timor-Leste								**Timor-Leste**
Current account	...	...	...	1 671	2 736	2 390	1 106	Compte des transac. courantes
Capital account, n.i.e.	...	...	...	31	23	20	-3	Compte de capital, n.i.a.
Financial account, n.i.e.	...	...	...	1 547	2 249	2 569	1 402	Compte financier, n.i.a.
Reserves and related items	...	...	...	156	422	-202	-390	Réserves et postes apparentés

Summary of balance of payments *(continued)*
Millions of US dollars

Résumé de la balance des paiements *(suite)*
Millions de dollars É.-U.

Country or area	1985	1995	2005	2010	2012	2013	2014	Pays ou zone
Togo								**Togo**
Current account	...	...	-204	-200	-294	-568	...	Compte des transac. courantes
Capital account, n.i.e.	...	...	51	1 388	286	315	...	Compte de capital, n.i.a.
Financial account, n.i.e.	...	...	-31	1 177	345	-307	...	Compte financier, n.i.a.
Reserves and related items	...	...	-110	14	-337	49	...	Réserves et postes apparentés
Tonga								**Tonga**
Current account	2	...	-21	-80	-39	-33	...	Compte des transac. courantes
Capital account, n.i.e.	...	...	13	34	44	33	...	Compte de capital, n.i.a.
Financial account, n.i.e.	1	...	-3	-43	-<0	-11	...	Compte financier, n.i.a.
Reserves and related items	1	...	-17	-14	15	8	...	Réserves et postes apparentés
Trinidad and Tobago								**Trinité-et-Tobago**
Current account	-48	294	3 881	4 172	...	...	...	Compte des transac. courantes
Financial account, n.i.e.	-20	215	1 510	3 763	...	...	...	Compte financier, n.i.a.
Reserves and related items	-301	84	1 387	436	...	...	...	Réserves et postes apparentés
Tunisia								**Tunisie**
Current account	-581	-774	-299	-2 104	-3 721	-3 879	-4 302	Compte des transac. courantes
Capital account, n.i.e.	...	47	127	82	449	115	287	Compte de capital, n.i.a.
Financial account, n.i.e.	-381	-958	-1 136	-1 757	-4 564	-2 815	-3 865	Compte financier, n.i.a.
Reserves and related items	-226	97	936	-170	1 366	-848	-132	Réserves et postes apparentés
Turkey								**Turquie**
Current account	-1 013	-2 338	-21 449	-45 312	-48 535	-64 658	-46 526	Compte des transac. courantes
Capital account, n.i.e.	...	...	...	-51	-58	-96	-70	Compte de capital, n.i.a.
Financial account, n.i.e.	-1 065	-4 643	-42 685	-60 099	-71 068	-72 721	-42 762	Compte financier, n.i.a.
Reserves and related items	-784	4 660	23 176	14 971	22 820	10 774	-484	Réserves et postes apparentés
Tuvalu								**Tuvalu**
Current account	...	...	-4	-14	6	7	...	Compte des transac. courantes
Capital account, n.i.e.	...	...	2	9	7	4	...	Compte de capital, n.i.a.
Financial account, n.i.e.	...	...	2	3	10	-2	...	Compte financier, n.i.a.
Reserves and related items	...	...	-1	-4	2	10	...	Réserves et postes apparentés
Uganda								**Ouganda**
Current account	5	-339	-13	-1 659	-1 711	-1 845	-2 674	Compte des transac. courantes
Capital account, n.i.e.	...	48	...	...	22	80	95	Compte de capital, n.i.a.
Financial account, n.i.e.	-81	-211	-565	-1 070	-1 622	-1 367	-2 116	Compte financier, n.i.a.
Reserves and related items	33	-51	102	-91	488	181	24	Réserves et postes apparentés
Ukraine								**Ukraine**
Current account	...	-1 152	#2 534	-3 016	-14 335	-16 518	-4 596	Compte des transac. courantes
Capital account, n.i.e.	...	6	#-43	188	40	-60	400	Compte de capital, n.i.a.
Financial account, n.i.e.	...	526	#-8 126	-6 508	-8 727	-19 241	9 644	Compte financier, n.i.a.
Reserves and related items	...	-1 624	#10 725	5 045	-4 174	2 005	-13 308	Réserves et postes apparentés
United Kingdom								**Royaume-Uni**
Current account	3 314	-13 436	-30 054	-67 601	-86 448	-122 223	-151 882	Compte des transac. courantes
Capital account, n.i.e.	...	486	-1 441	47	-546	-739	-2 066	Compte de capital, n.i.a.
Financial account, n.i.e.	1 788	-3 669	-22 280	-57 253	-82 925	-116 667	-177 591	Compte financier, n.i.a.
Reserves and related items	2 570	-853	1 732	10 010	11 627	6 956	10 140	Réserves et postes apparentés
United Rep. of Tanzania								**Rép.-Unie de Tanzanie**
Current account	-375	-646	-1 570	#-2 211	-3 769	-5 015	-5 021	Compte des transac. courantes
Capital account, n.i.e.	...	191	393	#538	777	713	536	Compte de capital, n.i.a.
Financial account, n.i.e.	72	-67	-1 422	#-3 061	-3 880	-5 021	-4 172	Compte financier, n.i.a.
Reserves and related items	-487	-359	-595	#349	295	510	-238	Réserves et postes apparentés
United States								**États-Unis**
Current account	-124 455	-113 561	-745 445	-441 963	-449 669	-376 763	-389 525	Compte des transac. courantes
Capital account, n.i.e.	...	-222	13 115	-158	6 904	-413	-46	Compte de capital, n.i.a.
Financial account, n.i.e.	-108 512	-92 562	-686 624	-438 809	-445 708	-392 733	-236 062	Compte financier, n.i.a.
Reserves and related items	3 835	9 747	-14 096	1 825	4 464	-3 088	-3 583	Réserves et postes apparentés
Uruguay								**Uruguay**
Current account	-98	-213	24	-756	-2 617	-2 842	-2 516	Compte des transac. courantes
Capital account, n.i.e.	...	...	4	...	40	201	12	Compte de capital, n.i.a.
Financial account, n.i.e.	75	-422	-924	-1 057	-6 246	-4 520	-4 908	Compte financier, n.i.a.
Reserves and related items	66	228	778	-387	3 263	2 901	1 337	Réserves et postes apparentés

12

Summary of balance of payments *(continued)*
Millions of US dollars

Résumé de la balance des paiements *(suite)*
Millions de dollars É.-U.

Country or area	1985	1995	2005	2010	2012	2013	2014	Pays ou zone
Vanuatu								**Vanuatu**
Current account	-10	-18	-53	#-64	-72	-5	19	Compte des transac. courantes
Capital account, n.i.e.	10	31	20	#21	23	21	32	Compte de capital, n.i.a.
Financial account, n.i.e.	-5	-25	-38	#-155	-69	-131	-3	Compte financier, n.i.a.
Reserves and related items	-<0	5	-11	#-15	2	10	12	Réserves et postes apparentés
Venezuela (Boliv. Rep. of)								**Venezuela (Rép. boliv. du)**
Current account	3 327	2 014	#25 447	5 585	2 586	4 604	3 598	Compte des transac. courantes
Capital account, n.i.e.	...	...	...	-211	...	...	...	Compte de capital, n.i.a.
Financial account, n.i.e.	629	2 964	#16 430	9 583	1 310	5 596	641	Compte financier, n.i.a.
Reserves and related items	1 699	-1 444	#5 425	-7 939	-846	-4 410	-609	Réserves et postes apparentés
Viet Nam								**Viet Nam**
Current account	...	...	-560	-4 276	9 429	#7 745	9 359	Compte des transac. courantes
Financial account, n.i.e.	...	...	-3 087	-6 201	-7 970	#280	-5 571	Compte financier, n.i.a.
Reserves and related items	...	...	2 130	-1 765	11 860	#557	8 375	Réserves et postes apparentés
Yemen								**Yémen**
Current account	...	...	624	-1 398	-335	-1 530	-842	Compte des transac. courantes
Capital account, n.i.e.	...	...	202	88	<0	...	...	Compte de capital, n.i.a.
Financial account, n.i.e.	...	...	606	317	-602	227	698	Compte financier, n.i.a.
Reserves and related items	...	...	434	-1 561	1 148	-1 040	-1 326	Réserves et postes apparentés
Zambia								**Zambie**
Current account	-395	...	#-363	1 377	1 248	-218	-401	Compte des transac. courantes
Capital account, n.i.e.	...	...	#287	150	223	278	51	Compte de capital, n.i.a.
Financial account, n.i.e.	-363	...	#1 888	1 564	1 268	277	-699	Compte financier, n.i.a.
Reserves and related items	-181	...	#-2 086	-66	166	-263	301	Réserves et postes apparentés

Source:
International Monetary Fund (IMF), Washington, D.C., the database on International Financial Statistics, last accessed March 2016.

Source:
Fonds monétaire international (FMI), Washington, D.C., la base de données de Statistiques Financières Internationales, dernier accès mars 2016.

Labour force and unemployment
Labour force and unemployment rate by sex (percent)
Population active et chômage
Force de travail et taux de chômage par sexe (pourcentage)

Region, country or area [&] Région, pays ou zone [&]	Year Année	Male and Female Hommes et femmes		Male Hommes		Female Femmes	
		Labor force Force de travail	Unemploy-ment Chômage	Labor force Force de travail	Unemploy-ment Chômage	Labor force Force de travail	Unemploy-ment Chômage
Total, all countries or areas *	**2005**	**64.7**	**6.2**	**77.6**	**5.8**	**51.9**	**6.7**
Total, tous pays ou zones *	**2010**	**63.2**	**6.1**	**76.5**	**5.8**	**50.0**	**6.5**
	2014	**62.9**	**5.8**	**76.1**	**5.5**	**49.6**	**6.3**
Africa *	2005	64.5	8.7	75.4	7.5	53.8	10.3
Afrique *	2010	64.8	8.1	75.4	6.9	54.5	9.5
	2014	65.2	8.1	75.5	7.0	55.1	9.5
Northern Africa *	2005	47.7	12.8	74.3	10.2	21.4	21.7
Afrique septentrionale *	2010	47.8	10.4	73.8	7.9	22.0	18.5
	2014	48.0	12.5	73.8	10.0	22.5	20.5
Sub-Saharan Africa *	2005	69.6	7.9	75.8	6.7	63.7	9.1
Afrique subsaharienne *	2010	69.9	7.6	75.8	6.7	64.1	8.6
	2014	70.1	7.3	75.9	6.2	64.3	8.4
Americas *	2005	65.4	7.0	77.1	6.0	54.2	8.2
Amériques *	2010	64.9	8.1	75.9	7.6	54.4	8.9
	2014	64.2	6.4	74.7	5.7	54.2	7.2
Northern America *	2005	65.2	5.3	72.2	5.4	58.5	5.3
Amérique septentrionale *	2010	64.0	9.5	70.2	10.4	58.0	8.4
	2014	62.6	6.3	68.7	6.4	56.6	6.2
Latin America and the Caribbean *	2005	65.5	8.0	80.3	6.4	51.4	10.4
Amérique latine et Caraïbes *	2010	65.6	7.3	79.6	6.0	52.1	9.1
	2014	65.2	6.4	78.4	5.3	52.6	7.9
Central Asia *	2005	63.5	9.7	73.8	9.5	53.9	9.9
Asie centrale *	2010	64.4	9.0	75.3	9.0	54.1	9.1
	2014	65.0	8.6	76.4	8.5	54.3	8.6
Eastern Asia *	2005	71.8	4.2	78.9	4.6	64.4	3.6
Asie orientale *	2010	69.5	4.3	77.0	4.8	61.9	3.7
	2014	69.5	4.5	77.0	4.9	61.9	4.0
South-eastern Asia *	2005	70.4	6.6	82.7	6.1	58.5	7.2
Asie du Sud-est *	2010	70.5	4.9	82.1	4.7	59.1	5.1
	2014	70.2	4.2	81.9	4.2	58.8	4.2
Southern Asia *	2005	59.9	5.0	82.8	4.6	35.8	6.0
Asie méridionale *	2010	55.6	4.2	80.5	3.8	29.5	5.3
	2014	54.4	4.2	79.4	3.9	28.2	5.0
Western Asia *[1]	2005	49.0	10.6	74.5	8.6	18.7	19.9
Asie occidentale *[1]	2010	50.9	9.5	75.6	6.8	20.0	22.2
	2014	51.9	10.1	76.5	7.5	20.9	21.6
Caucasus *[2]	2005	50.5	10.4	69.9	10.2	32.3	10.7
Caucase *[2]	2010	52.6	10.3	70.6	10.1	35.7	10.5
	2014	54.0	9.4	71.3	8.9	37.8	10.4
Eastern Europe *	2005	58.8	8.6	65.8	8.8	52.7	8.5
Europe orientale *	2010	59.6	8.0	67.3	8.5	52.9	7.4
	2014	60.2	6.8	68.3	7.2	53.2	6.3
Northern Europe *	2005	62.1	5.5	69.0	5.7	55.6	5.2
Europe septentrionale *	2010	62.2	8.6	68.3	9.5	56.4	7.5
	2014	62.5	6.8	68.2	7.1	57.1	6.4
Southern Europe *	2005	53.2	10.1	64.4	8.4	42.7	12.6
Europe méridionale *	2010	53.4	14.3	62.9	13.4	44.4	15.5
	2014	53.0	18.8	61.3	17.8	45.2	20.0
Western Europe *	2005	58.2	9.0	65.8	8.8	51.0	9.3
Europe occidentale *	2010	58.8	7.2	65.3	7.3	52.7	7.1
	2014	59.1	7.0	64.8	7.2	53.6	6.7
Oceania *	2005	65.8	5.1	72.9	4.9	58.9	5.4
Océanie *	2010	66.3	5.6	72.6	5.4	60.1	5.9
	2014	65.6	6.1	71.2	5.9	60.1	6.4
Afghanistan *	2005	51.7	10.3	84.3	9.5	16.4	14.7
Afghanistan *	2010	51.3	10.1	84.0	9.3	16.5	14.1
	2014	52.4	9.1	83.7	8.4	18.9	12.5

13 Labour force and unemployment *(continued)*
Labour force and unemployment rate by sex (percent)
Population active et chômage *(suite)*
Force de travail et taux de chômage par sexe (pourcentage)

Region, country or area [&] Région, pays ou zone [&]	Year Année	Male and Female Hommes et femmes		Male Hommes		Female Femmes	
		Labor force Force de travail	Unemploy- ment Chômage	Labor force Force de travail	Unemploy- ment Chômage	Labor force Force de travail	Unemploy- ment Chômage
Albania * Albanie *	2005	58.0	13.0	68.5	13.2	47.6	12.7
	2010	53.9	14.2	63.6	12.3	44.1	17.0
	2014	50.4	17.5	60.6	19.0	40.4	15.3
Algeria * Algérie *	2005	42.5	15.3	71.8	13.2	12.8	27.1
	2010	42.3	10.0	70.0	8.1	14.4	19.3
	2014	43.6	10.6	70.3	9.0	16.7	17.2
American Samoa [3] Samoa américaines [3]	2000	52.9	5.1	58.8	4.9	41.2	6.0
	#2005	59.9	...	...	...	...	...
	2010	52.8	#9.2	...	...	...	...
Angola * Angola *	2005	69.0	7.4	76.9	6.8	61.4	8.2
	2010	68.2	7.7	77.4	7.0	59.3	8.5
	2014	68.4	7.6	77.2	6.9	59.9	8.4
Anguilla Anguilla	2002	72.3	7.8	77.2	6.3	67.2	9.5
Antigua and Barbuda Antigua-et-Barbuda	2001	71.7	8.4	78.4	8.0	65.9	8.8
Argentina * Argentine *	2005	62.2	10.6	76.6	9.2	48.8	12.6
	2010	60.5	7.4	74.9	6.5	47.0	8.6
	2014	60.9	7.3	74.4	6.4	48.3	8.5
Armenia * Arménie *	2005	57.6	17.2	67.7	15.3	48.5	19.7
	2010	61.6	19.0	72.3	17.3	50.9	21.4
	2014	63.0	16.2	73.0	14.4	54.6	18.3
Aruba Aruba	2001	...	6.9	...	6.5	...	7.4
	#2010[4]	63.9[5,6]	10.6	68.9[5,6]	10.8	59.5[5,6]	10.4
	#2011	63.8	...	69.6	...	58.8	...
Australia * Australie *	2005	64.5	5.0	72.2	4.9	57.0	5.2
	2010	65.5	5.2	72.5	5.1	58.7	5.3
	2014	64.8	6.1	71.0	6.0	58.7	6.2
Austria * Autriche *	2005	59.0	5.6	67.3	5.4	51.2	5.8
	2010	60.5	4.8	67.6	5.0	53.9	4.6
	2014	60.2	5.6	66.1	5.8	54.7	5.4
Azerbaijan * Azerbaïdjan *	2005	63.3	7.3	66.3	6.2	60.5	8.4
	2010	64.5	5.6	67.4	4.4	61.8	6.8
	2014	64.7	4.9	68.0	4.0	61.6	5.8
Bahamas * Bahamas *	2005	72.6	10.2	77.8	9.2	67.7	11.3
	2010	74.2	14.7	79.2	13.4	69.5	16.2
	2014	74.1	15.1	79.1	14.8	69.4	15.4
Bahrain * Bahreïn *	2005	66.8	1.3	84.5	0.5	36.2	4.5
	2010	71.0	1.1	87.0	0.4	41.3	3.8
	2014	69.2	1.2	85.2	0.5	39.4	4.1
Bangladesh * Bangladesh *	2005	66.4	4.3	84.1	3.0	48.1	6.6
	2010	62.3	4.5	82.3	4.1	42.0	5.4
	2014	62.1	4.4	80.9	4.2	43.0	4.9
Barbados * Barbade *	2005	69.5	9.1	75.4	7.5	64.2	10.8
	2010	68.0	10.8	73.1	11.1	63.5	10.5
	2014	66.6	12.3	71.1	12.2	62.6	12.4
Belarus * Bélarus *	2005	59.6	6.1	66.6	7.6	53.8	4.6
	2010	60.5	6.0	67.7	7.5	54.5	4.5
	2014	60.7	5.9	68.1	7.3	54.6	4.4
Belgium * Belgique *	2005	53.4	8.4	61.4	7.6	45.8	9.5
	2010	54.1	8.3	60.8	8.1	47.8	8.5
	2014	53.6	8.5	59.3	9.0	48.2	7.9
Belize * Belize *	2005	63.8	11.0	81.8	7.5	45.7	17.4
	2010	67.2	13.3	82.6	9.4	51.9	19.4
	2014	69.7	11.6	83.5	6.6	56.1	18.9

Region, country or area & Région, pays ou zone &	Year Année	Male and Female Hommes et femmes		Male Hommes		Female Femmes	
		Labor force Force de travail	Unemploy- ment Chômage	Labor force Force de travail	Unemploy- ment Chômage	Labor force Force de travail	Unemploy- ment Chômage
Benin * Bénin *	2005 2010 2014	72.1 71.5 71.6	1.1 1.0 1.1	77.0 73.7 73.4	0.9 0.9 0.9	67.4 69.3 70.0	1.3 1.2 1.2
Bermuda Bermudes	2000 #2009[3] 2010[3] 2012[3,5] 2013[3,5]	73.8 82.0 84.0 76.1 ...	2.7[3] 4.5 ... 8.4 6.7	79.2 86.0 87.0 80.0 ...	3.1[3] 6.0 ... 8.7 ...	67.4 79.0 81.0 72.6 ...	2.3[3] 3.0 ... 8.2 ...
Bhutan * Bhoutan *	2005 2010 2014	72.0 69.7 66.0	3.1 3.3 2.6	77.7 74.5 72.3	2.9 2.6 2.1	65.2 63.8 58.3	3.4 4.3 3.4
Bolivia (Plurinational State of) * Bolivie (État plurinational de) *	2005 2010 2014	71.0 72.3 73.1	5.4 2.6 3.2	81.4 82.3 82.5	4.6 2.0 2.6	60.7 62.5 63.8	6.5 3.3 4.1
Bosnia and Herzegovina * Bosnie-Herzégovine *	2005 2010 2014	44.9 46.9 46.2	25.9 27.2 31.2	58.3 59.6 58.2	24.4 25.5 29.9	31.8 34.6 34.5	28.4 30.0 33.3
Botswana * Botswana *	2005 2010 2014	76.1 76.8 77.2	20.5 17.9 17.9	80.9 81.0 81.1	18.2 14.4 15.2	71.4 72.7 73.3	23.0 21.6 20.9
Brazil * Brésil *	2005 2010 2014	70.5 68.8 67.3	9.3 7.9 6.8	82.4 80.7 78.7	7.1 5.8 5.2	59.1 57.6 56.4	12.3 10.6 9.0
Brunei Darussalam * Brunéi Darussalam *	2005 2010 2014	66.9 65.9 63.7	2.4 1.7 1.6	78.1 76.9 75.5	2.2 1.6 1.5	55.2 54.1 51.1	2.6 1.8 1.7
Bulgaria * Bulgarie *	2005 2010 2014	50.2 53.3 54.1	10.1 10.2 11.4	56.3 59.5 60.0	10.4 10.8 12.3	44.6 47.5 48.6	9.8 9.5 10.4
Burkina Faso * Burkina Faso *	2005 2010 2014	83.5 83.6 83.6	2.7 3.2 3.0	91.0 91.0 90.7	3.4 4.1 3.8	76.5 76.6 76.7	1.9 2.3 2.1
Burundi * Burundi *	2005 2010 2014	82.9 82.8 83.6	1.7 1.6 1.6	81.8 81.6 82.6	1.4 1.3 1.3	84.0 84.0 84.6	2.0 1.9 2.0
Cabo Verde * Cabo Verde *	2005 2010 2014	64.4 66.4 68.0	10.8 10.7 10.7	82.2 82.8 83.9	9.4 9.3 9.3	48.0 50.9 52.9	13.2 12.8 12.7
Cambodia * Cambodge *	2005 2010 2014	80.9 84.9 80.8	1.1 0.3 0.4	86.6 88.6 86.6	1.3 0.4 0.5	75.9 81.6 75.5	0.8 0.3 0.3
Cameroon * Cameroun *	2005 2010 2014	73.2 75.8 75.9	4.4 4.1 4.4	79.5 81.2 81.0	3.7 3.4 3.7	67.1 70.5 70.9	5.2 4.8 5.2
Canada * Canada *	2005 2010 2014	66.6 66.6 65.7	6.8 8.1 6.9	72.6 71.5 70.5	7.1 8.9 7.4	60.9 61.8 61.1	6.5 7.3 6.4
Cayman Islands Îles Caïmanes	2005 #2009 2010 2013	... 84.2[6] 81.8 83.1	3.5 6.0 6.7 6.3	... 88.0[6] ... 85.6	... 7.0 ... 6.7	... 80.6[6] ... 80.6	... 5.0 ... 5.8
Central African Rep. * Rép. centrafricaine *	2005 2010 2014	77.6 77.3 78.0	7.7 7.7 7.6	85.7 85.2 84.6	7.1 7.1 7.3	69.9 69.8 71.6	8.4 8.4 8.0
Chad * Tchad *	2005 2010 2014	71.8 71.6 71.6	5.6 5.5 5.6	79.5 79.3 79.3	4.6 4.5 4.6	64.2 64.1 64.0	6.8 6.7 6.8

Region, country or area & Région, pays ou zone &	Year Année	Male and Female Hommes et femmes		Male Hommes		Female Femmes	
		Labor force Force de travail	Unemploy- ment Chômage	Labor force Force de travail	Unemploy- ment Chômage	Labor force Force de travail	Unemploy- ment Chômage
Channel Islands * Îles Anglo-Normandes *	2005	59.7	4.2	68.8	3.1	51.1	5.5
	2010	59.5	4.3	67.0	3.2	52.2	5.6
	2014	58.3	4.6	66.2	3.6	50.6	5.8
Chile * Chili *	2005	55.3	8.0	73.1	7.0	38.2	9.8
	2010	60.4	8.1	74.7	7.1	46.8	9.6
	2014	62.3	6.4	74.5	6.0	50.6	6.9
China * Chine *	2005	73.5	4.1	79.8	4.6	66.9	3.5
	2010	70.9	4.2	77.8	4.7	63.7	3.6
	2014	71.0	4.6	77.9	5.0	63.7	4.0
China, Hong Kong SAR * Chine, Hong Kong RAS *	2005	60.9	5.6	71.3	6.6	51.7	4.4
	2010	59.5	4.3	68.5	5.0	51.8	3.5
	2014	60.7	3.3	68.8	3.6	53.7	3.0
China, Macao SAR * Chine, Macao RAS *	2005	66.2	4.2	74.3	4.4	59.0	4.0
	2010	71.0	2.8	77.1	3.5	65.5	2.1
	2014	71.1	1.7	76.4	2.0	66.3	1.4
Colombia * Colombie *	2005	66.7	12.0	81.2	9.2	53.0	16.1
	2010	67.6	12.0	80.3	9.2	55.5	15.9
	2014	68.5	9.1	79.8	6.9	57.9	11.9
Comoros * Comores *	2005	55.5	19.6	78.7	17.9	32.2	23.9
	2010	56.6	19.4	79.0	17.7	34.1	23.5
	2014	57.3	19.4	79.3	17.5	35.1	23.5
Congo * Congo *	2005	69.5	7.3	71.5	6.6	67.4	8.0
	2010	69.9	7.2	72.3	6.5	67.5	7.9
	2014	69.8	7.2	72.6	6.5	67.1	7.9
Cook Islands Îles Cook	2001	74.5[7]	13.1	75.5[7]	11.7	63.0[7]	14.8
	#2006[6]	70.2	6.9	76.1	6.7	64.2	7.3
	2011[6]	71.0	8.2	76.6	8.2	65.4	8.1
Costa Rica * Costa Rica *	2005	61.8	6.6	80.2	4.9	43.6	9.6
	2010	60.5	7.3	77.1	6.0	44.1	9.6
	2014	61.6	9.6	76.6	8.0	46.7	12.2
Côte d'Ivoire * Côte d'Ivoire *	2005	67.0	9.4	82.0	8.5	50.5	10.9
	2010	67.1	9.4	81.4	8.5	51.8	10.8
	2014	67.0	9.4	80.9	8.6	52.3	10.9
Croatia * Croatie *	2005	53.3	12.6	61.1	11.5	46.2	13.9
	2010	52.2	11.6	59.2	11.1	45.8	12.2
	2014	52.4	17.3	58.9	16.4	46.5	18.3
Cuba * Cuba *	2005	52.8	1.9	67.3	1.7	38.2	2.3
	2010	55.7	2.5	68.9	2.4	42.4	2.7
	2014	55.8	2.7	68.8	2.4	42.7	3.1
Curaçao Curaçao	2005	59.4	18.2	65.4	17.1	54.8	19.2
	2008	59.0	10.3	66.5	8.0	53.2	12.5
	#2009[5]	...	9.6	...	7.8	...	11.2
	2013[5]	...	13.0	...	10.5	...	15.4
Cyprus * Chypre *	2005	63.4	5.3	73.1	4.5	53.6	6.4
	2010	64.4	6.3	71.3	6.2	57.1	6.5
	2014	63.7	16.1	69.9	16.9	57.2	15.1
Czech Republic * République tchèque *	2005	59.5	7.9	68.9	6.4	50.8	9.8
	2010	58.4	7.3	68.0	6.4	49.3	8.5
	2014	59.4	6.1	68.3	5.1	51.0	7.4
Dem. P. R. Korea * R. p. dém. de Corée *	2005	81.3	6.8	88.6	6.7	74.5	6.9
	2010	80.2	6.7	86.9	6.6	73.9	6.7
	2014	79.5	6.7	85.9	6.6	73.6	6.7
Dem. Rep. of the Congo * Rép. dém. du Congo *	2005	72.0	3.7	73.0	3.0	70.9	4.4
	2010	71.5	3.7	72.2	3.0	70.7	4.4
	2014	71.1	3.8	71.7	3.1	70.5	4.5

13

Labour force and unemployment *(continued)*
Labour force and unemployment rate by sex (percent)
Population active et chômage *(suite)*
Force de travail et taux de chômage par sexe (pourcentage)

Region, country or area [&] Région, pays ou zone [&]	Year Année	Male and Female Hommes et femmes		Male Hommes		Female Femmes	
		Labor force Force de travail	Unemploy- ment Chômage	Labor force Force de travail	Unemploy- ment Chômage	Labor force Force de travail	Unemploy- ment Chômage
Denmark * Danemark *	2005 2010 2014	65.9 64.4 62.1	4.8 7.5 6.6	71.4 69.1 66.3	4.4 8.4 6.4	60.5 59.8 58.1	5.2 6.5 6.8
Djibouti * Djibouti *	2001 2005 2010 2014	48.9 49.6 51.2 52.2	43.3 59.4 59.4 55.7	66.4 66.2 66.8 67.9	38.9	31.3 33.0 35.6 36.5	52.7 ... ...
Dominica Dominique	2001	57.7	11.0	70.2	12.0	45.0	9.5
Dominican Republic * Rép. dominicaine *	2005 2010 2014	64.6 63.9 65.3	18.0 12.4 14.5	80.7 78.7 78.7	11.4 8.4 9.1	48.8 49.3 52.2	28.7 18.6 22.4
Ecuador * Équateur *	2005 2010 2014	69.8 65.1 64.1	4.8 5.0 3.8	84.6 80.6 79.6	4.1 4.0 3.1	55.3 49.9 48.9	5.8 6.6 5.0
Egypt * Égypte *	2005 2010 2014	47.8 49.1 49.3	11.2 9.0 13.0	75.3 75.6 75.9	7.0 5.3 9.1	20.5 22.7 22.6	26.6 21.4 26.0
El Salvador * El Salvador *	2005 2010 2014	60.6 61.3 62.7	7.2 7.0 6.2	78.5 78.9 79.1	9.3 8.4 7.5	45.2 46.4 49.0	4.2 5.0 4.6
Equatorial Guinea * Guinée équatoriale *	2005 2010 2014	81.4 81.8 82.0	7.4 7.8 8.6	92.6 92.3 92.1	7.0 7.3 8.2	69.3 70.4 71.2	8.0 8.4 9.2
Eritrea * Érythrée *	2005 2010 2014	82.7 83.6 83.9	8.5 8.2 8.3	89.2 89.8 90.3	7.8 7.5 7.5	76.5 77.6 77.6	9.2 8.9 9.1
Estonia * Estonie *	2005 2010 2014	59.0 61.4 61.8	8.0 16.7 7.4	65.6 67.7 69.3	9.1 19.3 8.0	53.4 56.0 55.4	6.8 14.0 6.8
Ethiopia * Éthiopie *	2005 2010 2014	84.5 83.6 82.9	5.4 5.2 5.3	90.8 90.0 89.1	2.7 3.0 3.0	78.4 77.5 76.9	8.4 7.7 7.9
Falkland Is. (Malvinas) [8] Îles Falkland (Malvinas) [8]	2013	#81.9	1.2	#86.0	1.1	#77.2	1.4
Faroe Islands Îles Féroé	2005[3] 2010[5,9] 2013[5,9]	... 81.7 82.9	3.2 #6.4 4.0	... 85.3 86.0	2.6 #5.1 3.4	... 77.5 79.4	3.9 #8.0 4.9
Fiji * Fidji *	2005 2010 2014	56.2 54.9 54.4	4.6 8.9 8.3	73.8 71.8 71.3	3.7 7.1 6.6	38.1 37.4 37.0	6.4 12.5 11.7
Finland * Finlande *	2005 2010 2014	61.0 60.3 58.8	8.4 8.4 8.7	65.4 64.6 62.4	8.3 9.2 9.4	56.8 56.2 55.4	8.5 7.6 8.0
France * France *	2005 2010 2014	55.8 55.9 55.5	8.5 8.9 10.3	62.0 61.3 60.4	7.8 8.7 10.5	50.1 50.9 50.9	9.3 9.1 10.0
French Guiana * Guyane française *	2005 2010 2014	53.4 54.6 54.8	24.8 21.0 22.7	61.5 61.4 60.8	22.9 17.8 18.8	45.7 48.0 49.0	27.2 25.0 27.4
French Polynesia * Polynésie française *	2005 2010 2014	56.7 55.7 55.2	15.7 11.6 18.0	65.8 64.1 63.3	14.3 10.9 16.9	47.2 47.0 46.8	17.6 12.6 19.6
Gabon * Gabon *	2005 2010 2014	59.1 47.8 48.4	21.7 20.4 20.4	64.1 56.3 57.2	15.3 14.7 14.6	54.1 39.0 39.5	29.2 28.8 29.1

13

Labour force and unemployment *(continued)*
Labour force and unemployment rate by sex (percent)
Population active et chômage *(suite)*
Force de travail et taux de chômage par sexe (pourcentage)

Region, country or area [&] Région, pays ou zone [&]	Year Année	Male and Female Hommes et femmes		Male Hommes		Female Femmes	
		Labor force Force de travail	Unemploy-ment Chômage	Labor force Force de travail	Unemploy-ment Chômage	Labor force Force de travail	Unemploy-ment Chômage
Gambia * Gambie *	2005	77.5	30.3	83.5	22.3	71.7	39.2
	2010	77.6	29.8	83.2	21.7	72.2	38.6
	2014	77.3	30.2	82.8	22.2	72.2	39.0
Georgia * Géorgie *	2005	63.6	13.8	73.3	14.8	55.4	12.7
	2010	65.1	16.3	75.4	17.5	56.2	15.0
	2014	66.8	12.4	78.1	13.9	57.0	10.6
Germany * Allemagne *	2005	58.5	11.2	66.9	11.4	50.6	10.9
	2010	59.4	7.1	66.5	7.5	52.8	6.6
	2014	60.4	5.0	66.5	5.3	54.5	4.7
Ghana * Ghana *	2005	69.6	4.0	71.7	3.8	67.6	4.2
	2010	73.3	4.2	75.2	3.9	71.6	4.5
	2014	76.8	5.9	78.4	5.5	75.3	6.2
Gibraltar Gibraltar	2001	68.4	...	78.8	...	57.5	...
Greece * Grèce *	2005	52.9	10.2	64.2	6.4	42.1	15.8
	2010	53.7	12.5	63.7	9.9	44.2	16.0
	2014	51.8	26.5	60.1	23.7	44.0	30.1
Greenland Groenland	2005	...	9.3	...	10.4	...	8.0
	2006	...	8.4	...	10.0	...	6.7
	#2013[10,11]	...	9.7	...	9.6	...	9.8
Grenada Grenade	2001	...	10.2	...	9.6	...	10.9
Guadeloupe * Guadeloupe *	2005	52.4	25.9	56.5	22.4	48.9	29.4
	2010	54.5	23.8	57.7	21.2	51.8	26.2
	2014	54.7	26.2	59.1	24.0	51.2	28.3
Guam * Guam *	2005	63.0	7.0	69.7	6.4	56.1	7.8
	2010	62.5	8.2	70.1	7.4	54.8	9.3
	2014	62.2	10.8	68.8	10.2	55.4	11.6
Guatemala * Guatemala *	2005	65.3	2.4	87.6	2.0	45.1	3.0
	2010	62.7	3.7	85.7	3.4	41.8	4.3
	2014	61.3	2.8	83.4	2.6	41.0	3.2
Guinea * Guinée *	2005	71.7	1.7	80.9	2.4	62.7	0.8
	2010	73.5	1.6	80.9	2.3	66.1	0.8
	2014	82.3	1.7	85.1	2.5	79.5	0.9
Guinea-Bissau * Guinée-Bissau *	2005	72.3	7.6	78.8	7.0	65.9	8.4
	2010	72.4	7.5	78.4	6.9	66.6	8.3
	2014	72.6	7.6	78.2	7.0	67.1	8.3
Guyana * Guyana *	2005	60.1	10.2	82.1	8.6	38.7	13.5
	2010	60.0	10.9	79.7	9.0	40.4	14.9
	2014	59.6	11.2	77.4	9.0	41.8	15.3
Haiti * Haïti *	2005	63.7	7.4	69.4	6.4	58.4	8.5
	2010	65.2	7.7	70.5	6.8	60.2	8.7
	2014	66.1	6.9	71.2	6.0	61.3	7.9
Honduras * Honduras *	2005	59.4	4.2	83.1	3.2	36.3	6.4
	2010	64.2	4.8	84.6	3.9	44.2	6.4
	2014	65.5	3.9	84.4	3.3	47.0	5.0
Hungary * Hongrie *	2005	50.1	7.2	58.3	7.0	42.9	7.4
	2010	50.5	11.2	58.3	11.6	43.7	10.7
	2014	53.9	7.7	62.5	7.6	46.3	7.9
Iceland * Islande *	2005	75.5	2.5	80.2	2.5	70.8	2.5
	2010	74.5	7.6	78.6	8.3	70.4	6.8
	2014	74.3	4.9	77.7	5.0	70.9	4.7
India * Inde *	2005	60.6	4.4	83.1	4.1	36.8	5.0
	2010	55.3	3.5	80.5	3.2	28.6	4.3
	2014	53.7	3.5	79.1	3.4	26.7	3.9

13

Labour force and unemployment *(continued)*
Labour force and unemployment rate by sex (percent)
Population active et chômage *(suite)*
Force de travail et taux de chômage par sexe (pourcentage)

Region, country or area [&] Région, pays ou zone [&]	Year Année	Male and Female Hommes et femmes		Male Hommes		Female Femmes	
		Labor force Force de travail	Unemploy- ment Chômage	Labor force Force de travail	Unemploy- ment Chômage	Labor force Force de travail	Unemploy- ment Chômage
Indonesia * Indonésie *	2005	67.6	11.2	85.3	9.5	50.0	14.1
	2010	68.1	7.1	84.3	6.2	51.9	8.5
	2014	67.4	5.9	83.9	5.5	50.8	6.5
Iran (Islamic Rep. of) * Iran (Rép. islamique d') *	2005	47.4	12.1	74.2	10.5	19.4	18.4
	2010	43.1	13.5	69.9	11.8	16.0	20.8
	2014	44.3	10.6	72.4	8.6	16.0	19.3
Iraq * Iraq *	2005	41.2	18.0	68.9	16.1	13.7	27.6
	2010	41.9	15.2	69.3	13.1	14.5	25.1
	2014	42.3	16.0	69.6	13.9	15.0	25.5
Ireland * Irlande *	2005	62.6	4.3	72.9	4.6	52.4	3.9
	2010	61.2	13.9	69.2	17.1	53.3	9.9
	2014	60.4	11.3	68.2	12.8	52.7	9.4
Isle of Man Île de Man	2001	63.3	1.6	71.4	1.7	55.7	1.5
	2006	62.8	2.4	69.9	2.8	56.3	2.0
	#2011[3]	63.3	3.3[10]	69.3	3.8[10]	57.5	2.8[10]
	#2013[12]	...	2.6	...	3.4	...	1.7
Israel * Israël *	2005	62.2	8.0	69.3	7.5	55.4	8.6
	2010	63.2	6.0	69.7	6.0	57.1	6.0
	2014	64.2	5.9	69.5	5.9	59.0	5.9
Italy * Italie *	2005	49.0	7.7	61.0	6.2	37.8	10.0
	2010	48.0	8.4	59.0	7.5	37.8	9.7
	2014	48.5	12.7	58.3	11.8	39.4	13.9
Jamaica * Jamaïque *	2005	66.0	10.9	75.1	7.1	57.3	15.6
	2010	62.9	12.4	71.0	9.4	55.1	16.1
	2014	64.8	13.7	72.1	10.3	57.6	17.9
Japan * Japon *	2005	60.5	4.4	73.3	4.5	48.4	4.2
	2010	60.2	4.9	71.6	5.3	49.4	4.3
	2014	59.5	3.5	70.5	3.6	49.3	3.3
Jordan * Jordanie *	2005	41.1	14.8	67.6	12.7	12.3	27.0
	2010	42.4	12.5	67.6	10.3	15.4	22.9
	2014	39.9	11.9	64.3	10.0	14.1	21.2
Kazakhstan * Kazakhstan *	2005	69.5	8.1	75.2	6.9	64.4	9.3
	2010	70.4	5.8	75.8	5.0	65.5	6.6
	2014	70.8	5.1	76.5	4.4	65.8	5.8
Kenya * Kenya *	2005	64.8	9.5	69.8	8.3	59.9	10.8
	2010	66.2	9.2	71.4	8.0	61.0	10.6
	2014	66.9	9.2	72.0	7.9	61.9	10.7
Kiribati Kiribati	2005	#26.4[13]	14.7[14]	#32.8[13]	12.3[14]	#20.4[13]	18.2[14]
	2010	59.3[13]	30.6[14]	66.8[13]	27.6[14]	52.3[13]	34.1[14]
Kosovo [12] Kosovo [12]	2005	48.7	41.4	68.3	32.9	29.7	60.5
	2009	47.7	#45.4	66.9	#56.4	28.7	#40.7
	#2012	37.1	30.9	55.5	28.1	17.9	40.0
	#2013	40.6	...	60.2	...	21.1	...
Kuwait * Koweït *	2005	67.7	2.0	81.9	2.0	45.0	2.1
	2010	68.3	1.8	83.8	1.8	46.9	1.7
	2014	68.9	3.5	84.1	3.7	48.3	2.9
Kyrgyzstan * Kirghizistan *	2005	64.9	8.1	76.1	7.4	54.1	9.0
	2010	64.1	8.6	76.6	7.7	52.2	9.9
	2014	62.7	8.2	76.8	7.3	49.2	9.4
Lao People's Dem. Rep. * Rép. dém. pop. lao *	2005	78.8	1.4	78.8	1.6	78.9	1.2
	2010	77.5	1.4	77.0	1.7	77.9	1.2
	2014	77.2	1.5	76.7	1.8	77.6	1.3
Latvia * Lettonie *	2005	57.8	10.0	66.3	10.1	50.9	9.9
	2010	59.0	19.5	65.4	22.8	53.8	16.3
	2014	60.2	10.8	67.5	11.8	54.2	9.8

13

Labour force and unemployment *(continued)*
Labour force and unemployment rate by sex (percent)
Population active et chômage *(suite)*
Force de travail et taux de chômage par sexe (pourcentage)

Region, country or area [&] Région, pays ou zone [&]	Year Année	Male and Female Hommes et femmes		Male Hommes		Female Femmes	
		Labor force Force de travail	Unemploy-ment Chômage	Labor force Force de travail	Unemploy-ment Chômage	Labor force Force de travail	Unemploy-ment Chômage
Lebanon * Liban *	2005 2010 2014	46.1 46.7 46.9	8.2 6.2 6.7	70.6 70.0 70.2	7.6 4.4 5.3	20.3 22.0 23.2	10.5 12.2 11.0
Lesotho * Lesotho *	2005 2010 2014	68.4 65.6 66.2	36.1 25.6 26.4	75.5 73.2 73.6	28.6 21.9 23.5	61.8 58.6 59.1	44.3 30.0 29.9
Liberia * Libéria *	2005 2010 2014	60.7 61.2 60.9	4.9 3.7 4.0	62.9 64.3 63.8	5.7 3.4 3.9	58.5 58.1 58.0	4.2 4.1 4.0
Libya * Libye *	2005 2010 2014	53.2 54.3 53.4	20.5 19.7 20.6	75.9 77.7 78.9	17.1 16.4 17.2	28.6 29.8 27.9	30.1 28.7 30.2
Liechtenstein Liechtenstein	2005 2010 2013	63.0 61.6 61.9	... 2.6[12] 2.6[12]	73.7 70.9 70.6	... 2.3[12] 2.3[12]	52.8 52.6 53.5	... 3.1[12] 3.1[12]
Lithuania * Lituanie *	2005 2010 2014	56.1 56.9 59.2	8.3 17.8 10.7	62.7 62.1 65.3	8.1 21.2 12.2	50.5 52.6 54.1	8.5 14.5 9.3
Luxembourg * Luxembourg *	2005 2010 2014	54.8 57.0 59.0	4.5 4.4 5.9	64.7 65.4 66.0	3.5 3.9 6.0	45.4 48.8 52.0	5.8 5.1 5.8
Madagascar * Madagascar *	2005 2010 2014	86.6 89.0 86.5	2.6 3.8 1.8	89.1 91.0 89.1	1.8 2.8 1.4	84.1 87.2 83.9	3.5 4.9 2.3
Malawi * Malawi *	2005 2010 2014	82.2 82.1 80.9	7.7 6.7 6.6	85.8 80.6 80.7	7.4 6.3 6.2	78.7 83.6 81.2	8.0 7.0 7.0
Malaysia * Malaisie *	2005 2010 2014	61.2 59.8 63.2	3.5 3.4 2.9	77.8 75.9 77.5	3.3 3.3 2.6	44.2 43.8 49.2	3.8 3.6 3.2
Maldives * Maldives *	2005 2010 2014	62.4 65.7 67.6	13.4 11.7 11.6	74.5 76.3 78.4	8.8 10.9 10.0	50.4 55.2 57.0	20.0 12.8 13.9
Mali * Mali *	2005 2010 2014	54.0 65.8 66.0	9.6 7.3 8.2	70.1 81.4 82.2	7.4 5.6 6.2	38.3 50.3 49.8	13.6 10.1 11.5
Malta * Malte *	2005 2010 2014	48.5 49.3 52.3	6.9 6.8 5.9	68.8 66.0 66.1	6.3 6.7 6.2	29.1 32.6 38.5	8.2 6.9 5.4
Marshall Islands Îles Marshall	2006[3] 2011	44.6 41.3	... #4.7	... 53.3	... #4.9	... 29.0	... #4.5
Martinique * Martinique *	2005 2010 2014	49.2 52.3 52.1	18.7 21.0 23.0	53.0 55.8 52.3	17.0 19.5 22.4	46.0 49.5 52.0	20.3 22.4 23.6
Mauritania * Mauritanie *	2005 2010 2014	46.8 47.1 47.1	32.1 30.9 30.9	65.9 65.5 65.2	30.4 28.8 28.9	27.8 28.8 29.0	36.1 35.7 35.3
Mauritius * Maurice *	2005 2010 2014	58.7 58.6 60.6	9.6 7.7 7.7	76.8 74.4 75.1	5.8 4.6 5.4	41.1 43.2 46.7	16.5 12.8 11.2
Mexico * Mexique *	2005 2010 2014	60.6 62.0 62.1	3.5 5.2 4.9	80.9 80.7 79.5	3.4 5.2 4.9	41.0 43.8 45.3	3.6 5.2 4.9
Monaco [15] Monaco [15]	2000	45.7	3.6	57.3	2.5	35.1	5.2

13 Labour force and unemployment *(continued)*
Labour force and unemployment rate by sex (percent)
Population active et chômage *(suite)*
Force de travail et taux de chômage par sexe (pourcentage)

Region, country or area [&] Région, pays ou zone [&]	Year Année	Male and Female Hommes et femmes		Male Hommes		Female Femmes	
		Labor force Force de travail	Unemploy-ment Chômage	Labor force Force de travail	Unemploy-ment Chômage	Labor force Force de travail	Unemploy-ment Chômage
Mongolia *	2005	60.3	6.7	65.2	7.0	55.4	6.3
Mongolie *	2010	61.3	6.5	67.6	7.0	55.1	5.9
	2014	62.2	7.3	68.4	7.1	56.2	7.5
Montenegro *	2005	50.6	19.3	59.0	18.5	42.6	20.5
Monténégro *	2010	50.7	19.7	58.9	18.9	42.9	20.7
	2014	49.0	18.0	56.2	17.2	42.0	19.0
Montserrat	#2001	85.2	9.5	89.2	...	80.6	...
Montserrat	2011	...	5.6	...	7.7	...	3.3
Morocco *	2005	51.7	11.0	77.2	10.8	27.6	11.5
Maroc *	2010	49.7	9.1	75.2	8.9	25.7	9.7
	2014	49.1	9.9	74.2	9.8	25.2	10.3
Mozambique *	2005	85.0	21.6	82.8	19.1	87.0	23.7
Mozambique *	2010	81.6	22.1	78.7	20.0	84.3	24.0
	2014	79.1	22.3	75.3	20.4	82.6	24.0
Myanmar *	2005	78.0	5.4	80.8	5.0	75.4	5.9
Myanmar *	2010	78.5	4.2	81.5	3.8	75.8	4.5
	2014	78.1	4.6	81.2	4.2	75.3	5.0
Namibia *	2005	61.4	23.9	67.7	21.1	55.5	27.0
Namibie *	2010	61.2	22.1	66.3	20.5	56.5	23.8
	2014	58.9	26.9	62.9	24.2	55.4	29.8
Nauru	2002[3]	78.0	22.7	86.8	17.1	69.6	29.6
Nauru	2011	64.0	#23.0	78.9	#21.4	49.3	#25.5
Nepal *	2005	84.6	2.8	89.0	3.0	80.4	2.5
Népal *	2010	83.4	2.6	87.5	2.9	79.8	2.4
	2014	83.0	3.1	86.8	3.5	79.7	2.8
Netherlands *	2005	64.5	4.7	72.5	4.4	56.7	5.1
Pays-Bas *	2010	64.6	4.5	71.2	4.4	58.1	4.6
	2014	64.0	6.8	70.6	7.0	57.7	6.6
Netherlands Antilles [former] [5,13,16]	#2009	56.3	...	61.1	...	52.6	...
Antilles néerland. [anc.] [5,13,16]	2011	57.9	...	62.2	...	54.5	...
New Caledonia *	2005	58.8	16.9	69.9	14.6	47.8	20.2
Nouvelle-Calédonie *	2010	56.7	15.1	67.3	13.0	45.8	18.2
	2014	56.3	15.6	66.9	13.4	45.6	18.9
New Zealand *	2005	67.3	3.8	74.8	3.5	60.3	4.2
Nouvelle-Zélande *	2010	67.7	6.5	74.2	6.2	61.5	6.8
	2014	67.8	5.8	73.4	5.1	62.6	6.6
Nicaragua *	2005	61.5	5.6	80.4	5.7	43.7	5.5
Nicaragua *	2010	63.1	8.0	80.5	7.6	46.8	8.7
	2014	64.0	5.6	80.2	5.5	48.9	5.8
Niger *	2005	64.6	3.1	90.7	3.2	39.2	2.8
Niger *	2010	64.7	2.4	90.1	2.7	39.8	1.7
	2014	64.7	2.7	89.5	2.9	40.2	2.1
Nigeria *	2005	54.9	7.3	62.1	6.6	47.6	8.3
Nigéria *	2010	55.6	7.3	63.2	6.6	47.9	8.3
	2014	56.2	4.8	63.8	4.3	48.3	5.4
Niue	2001	75.5	2.2	76.7	2.3	74.8	2.1
Nioué	#2002	...	9.7	...	...	...	...
Northern Mariana Islands [3]	2003	81.8	4.6	82.5	5.0	81.3	4.3
Îles Mariannes du Nord [3]	2005	...	6.5	...	7.3	...	5.8
	#2010	72.3	11.2	77.6	9.7	66.6	13.0
Norway *	2005	65.3	4.4	70.5	4.6	60.3	4.2
Norvège *	2010	65.7	3.5	70.0	4.0	61.4	2.9
	2014	64.9	3.5	68.6	3.7	61.2	3.3
Oman *	2005	55.6	7.0	77.1	5.1	25.4	14.8
Oman *	2010	60.9	6.6	80.7	5.2	27.6	13.4
	2014	67.7	6.3	84.4	5.1	29.5	13.9

Region, country or area [&] Région, pays ou zone [&]	Year Année	Male and Female Hommes et femmes		Male Hommes		Female Femmes	
		Labor force Force de travail	Unemploy-ment Chômage	Labor force Force de travail	Unemploy-ment Chômage	Labor force Force de travail	Unemploy-ment Chômage
Other non-specified areas * Autres zones non-spécifiées *	2005	58.0	4.1	67.6	4.4	48.1	3.7
	2010	58.2	5.2	66.5	5.8	49.9	4.4
	2014	57.4	4.0	65.0	4.3	50.0	3.6
Pakistan * Pakistan *	2005	52.6	7.1	84.1	6.0	19.3	12.3
	2010	54.0	5.3	82.8	4.2	23.9	9.4
	2014	53.7	5.6	82.1	4.6	24.1	9.0
Palau [3] Palaos [3]	2005	67.5	4.2	75.4	3.7	58.1	4.9
Panama * Panama *	2005	64.5	9.8	81.1	7.7	47.9	13.3
	2010	65.4	6.5	82.0	5.3	49.0	8.6
	2014	65.5	4.8	80.5	4.0	50.5	6.1
Papua New Guinea * Papouasie-Nvl-Guinée *	2005	72.9	3.4	74.4	3.0	71.3	3.8
	2010	71.7	3.6	72.9	3.2	70.5	4.1
	2014	70.3	3.3	71.0	2.9	69.6	3.7
Paraguay * Paraguay *	2005	70.2	5.8	85.7	4.9	54.3	7.2
	2010	69.8	5.7	84.6	4.7	54.6	7.3
	2014	71.4	6.0	84.5	4.7	58.0	8.0
Peru * Pérou *	2005	68.1	5.2	78.9	4.8	57.5	5.7
	2010	74.9	4.0	84.3	3.6	65.6	4.5
	2014	73.9	3.3	82.5	3.0	65.5	3.7
Philippines * Philippines *	2005	64.9	7.7	80.0	7.7	49.8	7.7
	2010	64.9	7.4	79.6	7.6	50.2	7.0
	2014	64.6	6.6	78.8	6.9	50.4	6.1
Poland * Pologne *	2005	54.7	17.7	62.5	16.5	47.5	19.1
	2010	56.0	9.6	64.3	9.3	48.3	10.0
	2014	56.9	9.0	65.3	8.5	49.2	9.6
Portugal * Portugal *	2005	62.0	7.6	69.3	6.7	55.4	8.7
	2010	61.0	10.8	66.8	9.8	55.8	11.9
	2014	58.7	13.9	64.4	13.5	53.7	14.3
Puerto Rico * Porto Rico *	2005	48.9	11.3	61.0	12.1	38.1	10.2
	2010	43.9	16.1	53.7	18.7	35.1	12.6
	2014	42.3	13.9	51.4	15.2	34.2	12.1
Qatar * Qatar *	2005	79.6	1.0	94.1	0.3	43.5	5.0
	2010	86.6	0.4	95.7	0.2	51.5	2.3
	2014	84.6	0.2	94.0	0.1	53.3	0.9
Republic of Korea * République de Corée *	2005	60.9	3.7	72.6	4.0	49.5	3.3
	2010	60.3	3.7	71.7	4.0	49.3	3.3
	2014	60.8	3.5	71.8	3.6	50.0	3.4
Republic of Moldova * République de Moldova *	2005	49.0	7.3	50.4	8.7	47.9	6.0
	2010	41.8	7.4	45.3	9.0	38.6	5.7
	2014	41.6	3.9	45.0	4.7	38.5	3.1
Réunion * Réunion *	2005	56.9	30.1	64.5	27.4	49.8	33.3
	2010	55.3	28.9	62.2	27.7	49.1	30.2
	2014	54.1	29.5	60.4	28.4	48.4	30.8
Romania * Roumanie *	2005	54.5	7.2	62.1	7.8	47.4	6.4
	2010	55.1	7.0	63.6	7.6	47.3	6.3
	2014	56.1	6.8	65.1	7.4	47.7	6.1
Russian Federation * Fédération de Russie *	2005	61.9	7.1	68.6	7.3	56.2	6.9
	2010	63.0	7.3	70.7	7.8	56.5	6.8
	2014	63.4	5.2	71.6	5.5	56.7	4.9
Rwanda * Rwanda *	2005	84.1	2.5	82.5	2.2	85.6	2.7
	2010	85.3	2.2	84.0	2.0	86.4	2.3
	2014	84.9	2.7	83.1	2.5	86.5	2.8
Saint Helena Sainte-Hélène	2005	...	5.2	...	...	...	...
	2010	...	2.0	...	...	...	...
Saint Kitts and Nevis Saint-Kitts-et-Nevis	2001	#68.8	5.1	#62.8	4.3	#74.9	5.9

13

Labour force and unemployment *(continued)*
Labour force and unemployment rate by sex (percent)
Population active et chômage *(suite)*
Force de travail et taux de chômage par sexe (pourcentage)

Region, country or area [&] Région, pays ou zone [&]	Year Année	Male and Female Hommes et femmes		Male Hommes		Female Femmes	
		Labor force Force de travail	Unemploy- ment Chômage	Labor force Force de travail	Unemploy- ment Chômage	Labor force Force de travail	Unemploy- ment Chômage
Saint Lucia *	2005	70.1	18.7	77.6	14.9	63.1	23.1
Sainte-Lucie *	2010	68.5	20.6	75.5	16.1	61.9	25.8
	2014	69.5	20.8	76.4	17.2	63.0	25.0
Saint Vincent-Grenadines *	2005	65.8	19.4	78.0	20.0	53.3	18.7
Saint-Vincent-Grenadines *	2010	66.7	19.1	77.9	19.0	55.4	19.2
	2014	66.9	19.7	77.4	20.2	56.3	19.1
Samoa *	2005	49.4	4.7	69.2	4.3	28.3	5.8
Samoa *	2010	42.8	2.1	60.5	1.9	24.1	2.6
	2014	41.0	6.6	58.0	6.2	23.1	7.9
San Marino	2003	66.2	...	79.2	...	53.8	...
Saint-Marin	2010	...	4.4	...	2.1	...	7.3
	#2014[4]	...	6.6	...	4.3	...	9.5
Sao Tome and Principe *	2005	57.9	14.8	72.9	13.1	43.4	17.6
Sao Tomé-et-Principe *	2010	59.7	13.8	75.2	12.2	44.6	16.4
	2014	60.4	13.9	76.1	12.2	45.2	16.5
Saudi Arabia *	2005	51.3	5.8	75.4	3.9	17.5	17.4
Arabie saoudite *	2010	52.1	5.6	75.8	3.6	18.3	17.1
	2014	54.6	5.9	78.8	2.9	20.0	22.8
Senegal *	2005	52.6	8.5	71.2	6.8	35.6	11.7
Sénégal *	2010	55.4	9.3	69.6	7.3	42.4	12.4
	2014	56.9	9.7	70.0	7.4	44.9	12.8
Serbia *	2005	54.1	20.8	63.9	18.9	44.9	23.4
Serbie *	2010	50.6	19.2	59.1	17.4	42.6	21.6
	2014	51.5	18.9	60.2	18.2	43.4	19.7
Seychelles	2005	72.1	5.5	...	6.1	...	4.9
Seychelles	#2011[5,17]	65.0	4.1	68.3	3.8	61.9	4.5
Sierra Leone *	2005	65.9	3.4	66.7	4.6	65.1	2.3
Sierra Leone *	2010	66.5	3.4	67.9	4.5	65.1	2.3
	2014	66.7	3.4	68.5	4.5	65.0	2.3
Singapore *	2005	64.9	4.1	76.7	3.9	53.5	4.4
Singapour *	2010	67.2	3.1	77.5	2.9	57.3	3.3
	2014	67.4	2.8	76.7	2.6	58.5	3.0
Slovakia *	2005	59.5	16.3	68.5	15.5	51.2	17.3
Slovaquie *	2010	59.0	14.4	67.8	14.2	50.8	14.6
	2014	59.6	13.2	68.4	12.8	51.4	13.7
Slovenia *	2005	59.2	6.5	66.0	6.1	52.8	7.0
Slovénie *	2010	59.3	7.2	65.5	7.4	53.2	7.0
	2014	57.7	9.7	63.3	8.9	52.4	10.6
Solomon Islands *	2005	67.9	39.7	74.3	39.2	61.3	40.3
Îles Salomon *	2010	67.7	35.6	74.2	35.0	61.2	36.2
	2014	67.4	33.9	73.6	33.4	61.2	34.4
Somalia *	2005	54.2	7.5	76.8	7.0	32.4	8.5
Somalie *	2010	54.0	7.5	76.3	7.0	32.4	8.5
	2014	54.3	7.5	75.9	7.1	33.2	8.6
South Africa *	2005	53.7	23.8	61.5	20.0	46.7	28.4
Afrique du Sud *	2010	51.3	24.7	59.4	22.9	43.8	26.9
	2014	52.7	24.9	59.8	23.2	46.1	27.0
South Sudan	*2005	73.8	...	77.1	...	70.5	...
Soudan du sud	2008	*73.6	13.7[18]	*76.6	12.9[18]	*70.6	14.6[18]
	*2010	73.4	...	76.1	...	70.8	...
	*2014	73.2	...	75.2	...	71.1	...
Spain *	2005	56.9	9.1	68.3	7.1	45.9	11.9
Espagne *	2010	59.3	19.9	67.5	19.6	51.4	20.3
	2014	58.7	24.4	65.1	23.6	52.5	25.4
Sri Lanka *	2005	54.8	7.7	76.2	5.7	34.4	12.0
Sri Lanka *	2010	54.8	4.9	76.6	3.5	34.6	7.7
	2014	51.9	4.3	76.0	3.0	30.1	7.2

13

Labour force and unemployment *(continued)*
Labour force and unemployment rate by sex (percent)
Population active et chômage *(suite)*
Force de travail et taux de chômage par sexe (pourcentage)

Region, country or area [&] Région, pays ou zone [&]	Year Année	Male and Female Hommes et femmes		Male Hommes		Female Femmes	
		Labor force Force de travail	Unemploy- ment Chômage	Labor force Force de travail	Unemploy- ment Chômage	Labor force Force de travail	Unemploy- ment Chômage
State of Palestine * État de Palestine *	2005	40.6	23.6	66.8	24.7	14.1	18.4
	2010	40.7	23.7	66.2	24.1	14.8	22.0
	2014	43.4	26.9	68.8	24.1	17.6	38.4
Sudan * Soudan *	2005	49.1	14.0	73.9	11.8	24.5	20.4
	2010	47.9	13.0	72.9	10.8	23.2	19.7
	2014	48.1	13.3	72.1	11.3	24.2	19.4
Suriname * Suriname *	2005	52.4	8.5	67.2	6.5	37.6	12.1
	2010	54.1	7.6	68.4	5.4	39.9	11.2
	2014	54.5	6.9	68.6	3.8	40.5	12.2
Swaziland * Swaziland *	2005	49.0	25.5	64.3	24.0	35.1	28.1
	2010	49.9	28.0	62.7	26.8	37.8	29.8
	2014	51.4	26.1	63.6	24.6	39.7	28.3
Sweden * Suède *	2005	63.7	7.5	68.1	7.6	59.3	7.4
	2010	63.5	8.6	68.1	8.7	59.0	8.5
	2014	64.6	8.0	68.3	8.2	60.9	7.8
Switzerland * Suisse *	2005	67.0	4.4	75.0	3.9	59.5	5.0
	2010	67.9	4.5	75.3	4.1	60.8	4.9
	2014	68.7	4.5	74.9	4.4	62.7	4.7
Syrian Arab Republic * Rép. arabe syrienne *	2005	46.8	9.2	76.1	6.5	16.3	22.6
	2010	43.2	8.4	72.7	5.8	13.3	22.6
	2014	42.1	13.2	71.6	9.9	12.3	32.5
Tajikistan * Tadjikistan *	2005	66.2	11.8	74.7	13.0	57.7	10.2
	2010	67.3	11.6	76.3	12.7	58.3	10.2
	2014	68.3	10.9	77.3	11.7	59.3	9.9
Thailand * Thaïlande *	2005	73.6	1.3	81.4	1.4	66.0	1.2
	2010	72.5	1.0	81.0	1.0	64.4	1.0
	2014	71.5	0.8	80.3	0.9	63.1	0.8
TFYR of Macedonia * ex-R.Y. de Macédoine *	2005	53.5	37.3	64.6	36.6	42.6	38.4
	2010	56.1	32.0	69.3	31.9	43.1	32.2
	2014	55.9	28.0	68.2	27.6	43.9	28.6
Timor-Leste * Timor-Leste *	2005	49.3	6.7	66.2	5.7	32.0	8.6
	2010	40.8	3.9	55.3	3.1	26.1	5.6
	2014	41.3	4.4	55.6	3.7	26.7	6.0
Togo * Togo *	2005	80.1	8.0	80.4	7.3	79.8	8.6
	2010	80.7	7.8	80.3	7.1	81.0	8.5
	2014	80.8	7.7	80.5	7.0	81.1	8.4
Tonga * Tonga *	2005	63.8	5.0	75.0	3.4	53.1	7.3
	2010	63.9	5.2	75.2	3.5	53.1	7.4
	2014	63.4	5.3	74.3	3.7	52.8	7.5
Trinidad and Tobago * Trinité-et-Tobago *	2005	64.7	8.0	76.2	5.8	53.5	11.0
	2010	63.8	5.9	75.4	5.1	52.5	7.0
	2014	62.9	3.3	73.7	2.8	52.6	3.9
Tunisia * Tunisie *	2005	45.7	14.2	68.0	13.1	24.0	17.3
	2010	46.7	13.0	69.7	11.5	24.5	17.1
	2014	47.6	15.3	71.2	13.3	25.0	20.7
Turkey * Turquie *	2005	46.1	10.6	70.3	10.5	23.4	10.9
	2010	48.5	10.7	70.7	10.5	27.6	11.3
	2014	50.3	9.9	71.5	9.1	30.3	11.7
Turkmenistan * Turkménistan *	2005	60.5	10.4	74.8	10.4	47.0	10.3
	2010	60.6	10.4	75.7	10.4	46.4	10.4
	2014	61.8	10.1	77.3	10.1	47.1	10.0
Turks and Caicos Islands Îles Turques-et-Caïques	2005	#62.0	8.0	...	...	...	...
	#2008	64.0	8.3	...	...	...	...
Tuvalu Tuvalu	2005	58.2	6.5	69.6	4.9	47.9	8.6

13 Labour force and unemployment *(continued)*
Labour force and unemployment rate by sex (percent)
Population active et chômage *(suite)*
Force de travail et taux de chômage par sexe (pourcentage)

Region, country or area [&] Région, pays ou zone [&]	Year Année	Male and Female Hommes et femmes		Male Hommes		Female Femmes	
		Labor force Force de travail	Unemploy-ment Chômage	Labor force Force de travail	Unemploy-ment Chômage	Labor force Force de travail	Unemploy-ment Chômage
Uganda * Ouganda *	2005	78.4	2.0	79.7	1.8	77.1	2.2
	2010	82.9	4.2	85.4	3.5	80.6	4.9
	2014	85.0	3.8	87.6	3.3	82.3	4.3
Ukraine * Ukraine *	2005	58.3	7.2	65.6	7.5	52.2	6.8
	2010	58.5	8.1	65.9	9.3	52.4	6.8
	2014	58.9	9.3	67.2	10.9	52.0	7.5
United Arab Emirates * Émirats arabes unis *	2005	77.8	3.1	91.2	2.6	37.1	7.0
	2010	80.7	4.2	91.3	3.4	42.4	10.5
	2014	80.1	3.8	91.4	3.0	41.9	9.4
United Kingdom * Royaume-Uni *	2005	61.9	4.8	69.3	5.2	54.9	4.3
	2010	62.1	7.8	68.7	8.6	55.8	6.9
	2014	62.7	6.1	68.8	6.4	56.9	5.8
United Rep. of Tanzania * Rép.-Unie de Tanzanie *	2005	89.6	3.2	90.6	2.2	88.7	4.1
	2010	84.8	2.9	86.0	2.0	83.6	3.8
	2014	78.6	3.1	83.4	2.2	74.0	4.1
United States * États-Unis *	2005	65.1	5.2	72.2	5.2	58.3	5.2
	2010	63.7	9.7	70.0	10.6	57.6	8.6
	2014	62.2	6.3	68.5	6.3	56.1	6.2
United States Virgin Is. * Îles Vierges américaines *	2005	64.9	9.0	76.1	7.7	54.9	10.6
	2010	64.0	8.9	73.9	7.7	55.2	10.4
	2014	62.1	9.1	71.8	7.9	53.6	10.5
Uruguay * Uruguay *	2005	62.8	12.2	74.1	9.5	52.7	15.6
	2010	65.4	7.2	76.7	5.3	55.1	9.6
	2014	65.3	6.6	76.3	5.1	55.4	8.3
Uzbekistan * Ouzbékistan *	2005	59.2	10.6	72.1	10.6	47.2	10.6
	2010	60.6	10.5	74.4	10.5	47.6	10.4
	2014	61.6	10.2	75.9	10.1	48.2	10.2
Vanuatu * Vanuatu *	2005	73.6	5.1	81.7	4.6	65.3	5.7
	2010	71.0	4.3	80.7	3.8	61.5	5.0
	2014	71.0	4.5	80.6	3.9	61.6	5.1
Venezuela (Boliv. Rep. of) * Venezuela (Rép. boliv. du) *	2005	66.5	11.4	81.5	9.9	51.7	13.7
	2010	64.8	8.6	79.3	8.2	50.6	9.2
	2014	64.7	7.0	78.4	6.5	51.4	7.8
Viet Nam * Viet Nam *	2005	77.0	2.1	81.8	2.0	72.5	2.3
	2010	76.9	2.6	81.6	2.6	72.5	2.7
	2014	78.2	1.9	83.0	1.8	73.6	2.0
Western Sahara * Sahara occidental *	2005	55.9	6.9	83.1	5.5	24.1	12.7
	2010	57.6	6.8	83.5	5.2	27.7	12.2
	2014	57.7	6.7	83.7	5.1	28.2	12.1
Yemen * Yémen *	2005	47.5	16.1	71.3	13.2	23.6	24.8
	2010	48.2	17.8	71.4	9.7	24.8	41.2
	2014	49.3	16.4	72.8	11.9	25.6	29.2
Zambia * Zambie *	2005	79.5	15.9	85.6	15.2	73.5	16.7
	2010	75.0	14.0	79.8	13.1	70.4	14.9
	2014	75.3	9.6	80.9	9.4	69.8	9.8
Zimbabwe * Zimbabwe *	2005	86.3	10.5	90.2	12.2	82.5	8.7
	2010	83.2	7.4	88.7	9.1	78.0	5.5
	2014	82.3	11.3	87.2	7.1	77.6	15.8

Source:
International Labour Office (ILO), Geneva, Key Indicators of the Labour Market (KILM 9th edition), last accessed March 2016.

& Ages 15 years and over unless indicated otherwise.

Source :
Bureau international du travail (BIT), Genève, Indicateurs clés du marché du travail (ICMT 9e édition), dernier accès mars 2016.

& Sauf indication contraire, Age 15 ans et plus.

13 Labour force and unemployment *(continued)*
Labour force and unemployment rate by sex (percent)
Population active et chômage *(suite)*
Force de travail et taux de chômage par sexe (pourcentage)

1	Data excludes Armenia, Azerbaijan, Cyprus, Georgia, Israel and Turkey.	1	Les données excluent l'Arménie, l'Azerbaïdjan, Chypre, la Géorgie, l'Israël et la Turquie.
2	Refers to Armenia, Azerbaijan, Cyprus, Georgia, Israel and Turkey.	2	Les données se rapportent à l'Arménie, l'Azerbaïdjan, Chypre, la Géorgie, l'Israël et la Turquie.
3	Persons aged 16 years and over.	3	Personnes âgées de 16 ans et plus.
4	Persons aged 14 years and over.	4	Personnes âgées de 14 ans et plus.
5	Excluding the institutional population.	5	Non compris la population dans les institutions.
6	Resident population (de jure).	6	Population résidente (de droit).
7	Persons aged 15 to 69 years.	7	Personnes âgées de 15 à 69 ans.
8	Persons aged 16 to 65 years.	8	Personnes âgées de 16 à 65 ans.
9	Persons aged 15 to 74 years.	9	Personnes âgées de 15 à 74 ans.
10	Nationals, residents.	10	Ressortissants, résidents.
11	Persons aged 18 to 64 years.	11	Personnes âgées de 18 à 64 ans.
12	Persons aged 15 to 64 years.	12	Personnes âgées de 15 à 64 ans.
13	Persons present (de facto).	13	Personnes présentes (de facto)
14	De facto population.	14	Population de fait.
15	Persons aged 17 years and over.	15	Personnes âgées de 17 ans et plus.
16	Main city or metropolitan area.	16	Ville principale ou zone métropolitaine.
17	Excluding some areas.	17	Certaines régions sont exclues.
18	Persons aged 10 years and over.	18	Personnes âgées de 10 ans et plus.

14

Employment by economic activity
Percentage of persons employed by sex and ISIC 4 categories; agriculture (agr.), industry (ind.) and services (ser.)

Emploi par activité économique
Personnes employées par sexe et branches de la CITI rév. 4 ; agriculture (agr.), industrie (ind.) et services (ser.), pourcentage

Region, country or area [+] / Région, pays ou zone [+]	Year / Année	Male and Female Hommes et femmes			Male Hommes			Female Femmes		
		Agr.	Ind.	Ser.	Agr.	Ind.	Ser.	Agr.	Ind.	Ser.
Total, all countries or areas *	2005	36.7	20.7	42.6	35.3	23.7	41.0	38.8	16.2	45.0
Total, tous pays ou zones *	2010	32.7	21.6	45.7	32.2	25.0	42.9	33.5	16.4	50.1
	2014	29.8	22.0	48.2	29.4	26.0	44.6	30.3	15.7	53.9
Africa *	2005	53.9	14.9	31.2	46.6	20.4	32.9	64.2	7.0	28.8
Afrique *	2010	51.2	15.8	32.9	44.5	21.8	33.8	60.7	7.5	31.8
	2014	49.7	16.0	34.3	43.3	22.0	34.7	58.5	7.7	33.7
Northern Africa *	2005	25.1	24.2	50.7	24.4	25.2	50.4	27.8	20.2	52.0
Afrique septentrionale *	2010	25.1	26.3	48.6	24.6	28.4	47.0	26.9	18.6	54.5
	2014	26.8	25.1	48.1	25.1	27.1	47.7	33.0	17.8	49.2
Sub-Saharan Africa *	2005	59.6	13.0	27.4	53.1	19.0	27.8	67.4	5.8	26.8
Afrique subsaharienne *	2010	56.4	13.7	29.9	50.1	19.9	29.9	63.7	6.5	29.8
	2014	53.9	14.4	31.7	48.1	20.7	31.2	60.7	6.9	32.4
Americas *	2005	10.9	21.1	68.0	14.4	27.7	57.9	6.1	11.8	82.0
Amériques *	2010	11.0	19.9	69.0	13.9	25.3	60.7	7.2	12.6	80.2
	2014	10.1	20.3	69.6	12.6	25.8	61.6	6.7	13.0	80.3
Northern America *	2005	1.7	20.4	77.9	2.4	29.6	68.0	0.9	9.5	89.7
Amérique septentrionale *	2010	1.7	18.0	80.3	2.4	26.9	70.7	0.8	7.9	91.3
	2014	1.6	18.0	80.4	2.2	26.7	71.0	0.8	7.8	91.4
Latin America and the Caribbean *	2005	17.2	21.5	61.2	21.6	26.5	51.8	10.3	13.7	76.0
Amérique latine et Caraïbes *	2010	16.7	21.1	62.1	20.1	24.5	55.4	11.7	16.0	72.2
	2014	15.1	21.7	63.1	18.2	25.2	56.5	10.6	16.5	72.8
Central Asia *	2005	36.9	20.2	42.9	37.5	25.0	37.5	36.3	14.0	49.7
Asie centrale *	2010	32.1	22.1	45.8	32.3	27.4	40.3	31.9	15.2	52.9
	2014	29.6	21.9	48.5	28.8	28.0	43.2	30.6	13.9	55.5
Eastern Asia *	2005	40.7	22.1	37.2	40.8	21.8	37.4	40.6	22.4	37.0
Asie orientale *	2010	32.9	24.0	43.2	34.7	24.5	40.7	30.6	23.2	46.2
	2014	27.5	24.5	48.0	31.0	26.0	43.0	23.1	22.6	54.3
South-eastern Asia *	2005	44.8	18.0	37.2	42.3	20.1	37.6	48.3	15.1	36.6
Asie du Sud-est *	2010	40.0	18.8	41.2	37.1	21.3	41.6	43.9	15.3	40.8
	2014	35.6	19.7	44.7	33.0	23.9	43.1	39.2	14.0	46.9
Southern Asia *	2005	52.8	18.9	28.3	46.1	20.6	33.2	69.4	14.5	16.1
Asie méridionale *	2010	49.0	21.6	29.5	43.4	23.3	33.3	65.2	16.5	18.3
	2014	45.4	22.6	32.0	39.1	24.9	36.0	64.1	15.6	20.3
Western Asia * [1]	2005	14.7	22.9	62.4	12.4	26.1	61.5	27.1	5.5	67.4
Asie occidentale * [1]	2010	11.1	24.1	64.9	10.1	27.2	62.7	16.5	6.4	77.1
	2014	11.1	24.9	64.0	10.0	28.3	61.7	17.5	6.3	76.2
Caucasus * [2]	2005	29.7	21.6	48.7	23.8	26.4	49.8	41.6	12.0	46.4
Caucase * [2]	2010	25.1	22.7	52.2	19.4	28.5	52.1	35.8	12.0	52.3
	2014	22.8	23.9	53.2	17.9	30.1	51.9	31.7	12.7	55.6
Eastern Europe *	2005	14.0	30.1	55.9	15.5	37.2	47.3	12.4	22.4	65.3
Europe orientale *	2010	11.7	28.8	59.5	14.6	38.4	47.1	8.6	18.4	73.0
	2014	10.6	28.3	61.1	13.1	38.5	48.4	7.9	17.2	74.9
Northern Europe *	2005	2.7	23.1	74.0	3.8	33.9	62.1	1.6	10.5	87.8
Europe septentrionale *	2010	2.1	20.0	77.7	3.0	30.4	66.4	1.1	8.3	90.4
	2014	2.2	19.5	78.2	3.0	29.3	67.5	1.2	8.5	90.2
Southern Europe *	2005	8.5	29.4	62.2	8.6	38.7	52.7	8.2	15.5	76.3
Europe méridionale *	2010	7.6	25.9	66.5	8.1	35.7	56.2	6.8	12.6	80.6
	2014	7.1	23.5	69.4	8.2	32.5	59.4	5.8	11.8	82.4
Western Europe *	2005	3.0	26.3	70.7	3.8	36.8	59.4	2.2	13.4	84.4
Europe occidentale *	2010	2.4	24.7	72.9	3.0	35.7	61.3	1.7	11.8	86.5
	2014	2.2	23.9	74.0	2.8	34.7	62.6	1.5	11.5	87.0
Oceania *	2005	16.6	17.6	63.2	14.1	25.3	57.7	19.6	8.2	69.9
Océanie *	2010	16.3	17.2	63.9	13.6	25.4	58.1	19.5	7.5	70.7
	2014	15.7	18.2	63.5	12.6	27.4	57.2	19.4	7.4	71.0
Albania [3]	2005	58.5	13.5	27.8	...	...	...	...	...	...
Albanie [3]	#2009[4]	44.1	19.9	36.0	34.4	27.1	38.4	56.9	10.3	32.8
Algeria	2004[3]	20.7	26.0	53.1	20.4	25.6	53.8	22.3	28.2	49.5
Algérie	#2011	10.8	30.9	58.4	12.3	32.2	55.5	2.9	23.8	73.2
American Samoa [3,5]	2000	3.1	41.7	51.9	...	...	...	...	...	...
Samoa américaines [3,5]	#2010	3.0	23.2	73.8	...	...	...	...	...	...
Anguilla # [3]										
Anguilla # [3]	2001	2.9	18.9	76.7	4.8	32.1	61.7	0.7	3.7	93.9

Employment by economic activity *(continued)*
Percentage of persons employed by sex and ISIC 4 categories; agriculture (agr.), industry (ind.) and services (ser.)

Emploi par activité économique *(suite)*
Personnes employées par sexe et branches de la CITI rév. 4 ; agriculture (agr.), industrie (ind.) et services (ser.), pourcentage

Region, country or area [+] Région, pays ou zone [+]	Year Année	Male and Female Hommes et femmes			Male Hommes			Female Femmes		
		Agr.	Ind.	Ser.	Agr.	Ind.	Ser.	Agr.	Ind.	Ser.
Antigua and Barbuda [3]	2005	2.8	15.6	81.6	4.4	26.1	69.5	1.2	5.0	93.8
Antigua-et-Barbuda [3]	2008	2.8	15.6	81.6	4.4	26.1	69.5	1.2	5.0	93.8
Argentina	2005[3,6,7]	1.1	23.5	75.1	1.5	32.7	65.6	0.6	11.0	88.1
Argentine	#2010[4,6,8]	3.0	24.1	72.5	4.6	34.2	60.6	0.5	9.2	90.0
	2014[9]	0.5	24.0	74.7	0.8	35.1	63.3	0.2	8.5	90.7
Armenia	2005[3]	46.2	15.9	37.8	46.1	21.1	32.8	46.4	9.8	43.8
Arménie	#2010[10]	38.6	17.4	44.0	31.4	27.0	41.7	47.2	5.8	47.0
	#2013[4,10]	36.3	17.0	46.7	28.7	26.5	44.8	44.5	6.6	48.8
Aruba # [3]	2000	0.5	16.4	82.3	0.8	26.7	71.6	0.2	4.5	94.7
Aruba # [3]	2010[11]	0.6	14.5	84.4	1.0	24.4	74.1	0.3	4.5	94.8
	2011	0.6	14.0	85.1	0.8	24.4	74.4	0.4	3.3	96.1
Australia	2005	3.6	21.5	68.1	4.5	31.3	57.5	2.5	9.5	81.1
Australie	2010	3.2	21.4	68.4	4.0	32.0	57.3	2.3	8.7	81.7
	2013	2.6	20.8	69.5	3.4	31.4	58.2	1.7	8.3	82.8
Austria	#2005[3]	4.9	27.8	67.3	4.8	40.1	55.1	5.0	13.0	82.0
Autriche	2010	4.7	25.2	70.1	4.7	37.1	58.1	4.6	11.6	83.8
	2014	4.3	26.1	69.7	4.4	37.9	57.7	4.1	12.8	83.1
Azerbaijan	2005[3]	39.3	12.1	48.6	41.1	15.2	43.7	37.4	8.7	54.0
Azerbaïdjan	#2010[4]	38.2	13.7	48.1	32.3	20.9	46.8	44.5	6.1	49.4
	2013[4]	37.1	14.4	48.5	31.1	22.2	46.7	43.5	6.0	50.4
	2014[4]	36.8	14.3	48.9	...	...	...	...	...	...
Bahamas [3]	2005	3.5	17.8	78.4	6.4	29.6	63.5	0.3	5.1	94.3
Bahamas [3]	#2009	2.9	16.0	80.8	5.1	27.9	66.5	0.5	3.7	95.5
	#2011	3.7	12.9	83.0	5.6	21.7	72.3	1.8	4.1	93.7
Bahrain	#2004[3]	0.8	15.0	84.2	1.0	18.2	80.8	0.1	3.0	96.9
Bahreïn	2010	1.1	35.3	62.4	1.3	42.1	55.3	0.0	9.0	89.7
Bangladesh	2005[3]	48.1	14.5	37.4	41.8	15.1	43.0	68.1	12.5	19.4
Bangladesh	#2010[12]	47.5	17.7	35.3	40.1	19.6	41.1	64.8	13.3	21.9
Barbados	2004[13]	3.3	17.3	69.7	4.2	26.2	61.8	2.5	7.8	78.2
Barbade	#2010	2.8	19.6	77.5	3.8	29.8	66.3	1.9	8.9	89.2
	#2013[4]	2.7	18.9	78.4	3.7	27.3	69.0	1.6	10.2	88.2
Belarus	2009[3]	10.5	33.7	49.9	13.1	43.3	37.4	7.8	24.0	62.5
Bélarus	#2010	10.5	34.1	55.3	...	...	...	...	...	...
	#2013[3,5]	9.6	33.2	57.2	...	...	...	...	...	...
Belgium	#2005[3]	2.0	24.8	73.2	2.4	35.2	62.4	1.4	11.4	87.2
Belgique	2010	1.3	23.5	75.2	1.7	34.6	63.8	0.9	10.1	89.0
	2014	1.1	21.5	77.4	1.4	33.0	65.6	0.8	8.3	90.9
Belize # [3,11]										
Belize # [3,11]	2005	19.5	17.9	62.5	28.0	21.8	50.1	3.3	10.3	86.2
Benin	2003[13]	42.7	9.5	46.2	53.1	9.8	34.5	32.7	9.2	57.5
Bénin	#2010[14]	45.1	10.4	44.0	53.2	14.0	32.1	37.8	7.2	54.7
Bermuda	2004[13]	1.7	12.1	86.2	3.0	20.5	76.6	0.3	3.1	96.6
Bermudes	#2010[3,5]	1.4	12.9	85.7	2.4	22.6	75.0	0.4	3.2	96.4
	#2012[3,15]	1.7	9.6	88.7	3.0	16.5	80.5	0.3	2.6	97.1
	#2013[3,4,5]	1.6	10.3	87.6	...	...	...	...	...	...
Bhutan [3]	#2005[16]	43.6	17.2	39.2	32.6	23.7	43.7	63.0	5.8	31.6
Bhoutan [3]	2010	59.5	6.7	33.8	54.0	6.8	39.3	65.3	6.7	28.1
	#2013[17]	56.3	11.0	32.7	48.5	11.6	39.9	65.7	10.2	24.2
Bolivia (Plurinational State of) [6]	2005[3]	38.6	19.4	42.0	38.2	26.3	35.5	39.1	10.8	50.0
Bolivie (État plurinational de) [6]	2008[13]	34.3	19.1	46.5	34.1	26.3	39.7	34.7	10.4	54.8
	#2009[3]	32.1	20.0	47.9	...	...	...	...	...	...
Botswana	#2003[3,18,19]	21.2	22.6	56.1	28.6	28.0	43.4	12.9	16.5	70.6
Botswana	2010	26.4	17.5	56.1	30.7	23.6	45.6	21.3	10.3	68.5
Brazil [3,6]	2005	20.5	21.4	57.9	23.7	27.3	48.7	16.0	13.3	70.6
Brésil [3,6]	#2009	17.0	22.1	60.7	20.5	28.8	50.4	12.2	13.2	74.6
	#2010[9]	...	24.1	75.3	...	...	...	...	...	...
	2013[4]	# 14.5	22.6	62.9	# 17.7	# 30.4	# 51.8	# 10.1	# 12.0	# 77.9
	#2014[9]	...	22.9	76.6	...	...	...	...	...	...
British Virgin Islands [3]										
Îles Vierges britanniques [3]	2010	0.5	11.1	87.4	...	...	...	...	...	...
Brunei Darussalam	2001[3]	1.4	21.4	77.2	2.1	28.6	69.3	0.3	11.2	88.4
Brunéi Darussalam	#2014	0.6	18.7	80.8	0.6	24.3	74.9	0.4	11.0	88.5

Employment by economic activity *(continued)*
Percentage of persons employed by sex and ISIC 4 categories; agriculture (agr.), industry (ind.) and services (ser.)

Emploi par activité économique *(suite)*
Personnes employées par sexe et branches de la CITI rév. 4 ; agriculture (agr.), industrie (ind.) et services (ser.), pourcentage

Region, country or area [+] Région, pays ou zone [+]	Year Année	Male and Female Hommes et femmes			Male Hommes			Female Femmes		
		Agr.	Ind.	Ser.	Agr.	Ind.	Ser.	Agr.	Ind.	Ser.
Bulgaria	2005[3]	8.4	34.5	57.0	10.1	39.3	50.4	6.4	29.1	64.3
Bulgarie	2010	6.7	33.4	59.9	8.1	41.1	50.7	5.1	24.9	69.9
	2014	6.9	30.3	62.8	9.0	36.5	54.5	4.5	23.4	72.1
Burkina Faso [3]	#2005	84.8	3.1	12.2	82.3	3.9	13.7	87.2	2.1	10.2
Burkina Faso [3]	2007[14]	67.4	10.1	22.3	68.2	9.7	22.2	67.3	6.2	26.4
Cambodia	2004[13]	53.7	13.2	33.1	51.6	12.4	36.0	55.8	14.1	30.1
Cambodge	#2010[14]	54.1	16.2	29.6	52.9	17.0	30.1	55.4	15.5	29.1
Cameroon # [13]										
Cameroun # [13]	2001	61.3	9.1	22.6	57.7	10.9	23.5	64.7	7.4	21.8
Canada [3]	2005	2.7	22.0	75.2	3.7	32.1	64.1	1.6	10.6	87.8
Canada [3]	2010	2.2	19.7	78.1	3.1	30.0	66.9	1.3	8.4	90.3
	2014	2.1	19.8	78.2	2.9	30.0	67.1	1.2	8.4	90.4
Cayman Islands	2005[3]	1.7	22.2	75.5	2.6	39.3	57.3	0.6	3.3	95.7
Îles Caïmanes	#2010	0.6	14.9	84.2	1.1	26.2	72.5	0.1	3.2	96.4
	#2013	0.8	15.5	83.6	1.5	29.0	69.5	0.2	2.0	97.7
Chile	2003[13]	13.6	23.4	63.0	18.1	29.4	52.6	4.9	11.8	83.4
Chili	#2005[3]	14.8	41.8	43.5	...	...	...	...	...	...
	#2010[3]	11.1	23.6	65.3	14.5	32.1	53.4	5.9	10.5	83.5
	#2013[3]	9.2	23.7	67.1	12.4	32.7	54.9	4.6	10.6	84.8
China [5]	2005[3,8]	3.9	43.4	49.5	4.0	45.4	47.6	3.8	40.1	52.7
Chine [5]	2007[3,8]	3.5	44.5	48.6	3.6	47.0	46.2	3.5	40.5	52.5
	#2010	2.9	44.3	48.8	...	...	...	...	...	...
	2011	2.5	46.9	47.0	...	...	...	...	...	...
China, Hong Kong SAR	2005[13,20]	0.3	15.1	84.7	0.3	22.2	77.5	0.2	6.5	93.3
Chine, Hong Kong RAS	2007[13,20]	0.2	14.1	85.7	0.2	21.3	78.5	0.1	5.9	94.0
	2008	0.2	12.8	87.0	...	19.2	79.8	...	4.6	95.2
	#2010[4]	...	11.4	80.1	...	18.1	80.2	...	4.0	79.9
	2014[4]	...	11.7	79.9	...	18.8	79.5	...	4.0	79.3
China, Macao SAR [3]	2005[11]	0.1	25.0	74.7	0.2	26.8	73.0	0.2	23.1	76.7
Chine, Macao RAS [3]	#2010[5,21]	0.2	13.7	86.1	0.2	19.8	80.0	0.2	7.6	92.2
	#2014[5]	0.0	15.7	84.3	0.0	25.7	74.3	0.0	4.7	95.3
Colombia [3]	2005[6]	21.8	19.3	58.9	30.4	21.2	48.3	8.1	16.3	75.6
Colombie [3]	#2010[4,19,22]	18.5	19.8	61.7	26.5	22.6	50.8	6.6	15.7	77.7
	#2014	16.3	19.6	64.1	23.2	23.5	53.3	6.6	14.2	79.2
Congo [13]										
Congo [13]	2005	35.4	20.6	42.2	31.3	20.0	45.9	39.3	21.2	38.7
Cook Islands [23]										
Îles Cook [23]	2011	4.3	11.7	84.0	6.4	17.7	75.9	1.8	5.0	93.3
Costa Rica	2005[3,19]	15.2	21.5	62.9	20.8	26.4	52.3	4.8	12.6	82.5
Costa Rica	2010	15.0	19.5	64.7	21.0	24.1	53.8	4.8	11.6	83.1
	#2013	12.7	19.0	68.2	18.4	23.7	57.7	3.4	11.4	85.1
Croatia	#2005[3]	14.3	29.7	56.0	13.2	38.5	48.2	15.5	18.9	65.6
Croatie	2010	12.5	28.0	59.3	11.8	38.7	49.3	13.4	15.3	71.2
	2014	8.7	27.2	63.9	10.0	37.1	52.6	7.1	15.5	77.2
Cuba [13,24]	2005	20.2	19.1	60.6	26.5	22.5	51.0	9.4	13.2	77.4
Cuba [13,24]	#2010	18.5	17.0	64.5	24.7	20.4	54.9	8.4	11.5	80.1
	2013	18.6	17.2	64.2	25.1	20.8	54.1	7.7	11.3	81.0
Curaçao [3]	2000	1.1	18.0	80.9	2.0	28.6	69.4	0.2	5.6	94.2
Curaçao [3]	2005	0.9	15.3	83.8	...	...	...	...	...	...
	2008	1.1	17.6	81.3	...	...	...	...	...	...
Cyprus	2005[3]	3.7	24.3	72.0	4.3	35.1	60.7	2.9	10.4	86.6
Chypre	2010	2.8	20.7	76.5	3.7	30.7	65.2	1.9	9.3	88.7
	2014	3.9	16.5	79.6	5.9	25.7	68.2	1.8	7.1	91.1
Czech Republic	#2005[3]	3.9	39.7	56.3	4.8	49.7	45.4	2.8	26.6	70.6
République tchèque	2010	3.1	38.3	58.7	3.9	49.4	46.7	1.9	23.4	74.7
	2014	2.7	38.3	58.9	3.6	49.5	46.9	1.6	23.8	74.6
Denmark	#2005[3]	3.0	23.9	72.9	4.2	34.0	61.5	1.6	12.4	85.9
Danemark	2010	2.3	19.7	77.9	3.7	29.3	66.9	0.8	9.3	89.8
	2014	2.3	19.3	78.0	3.6	28.5	67.6	0.9	9.3	89.5
Dominica # [13]										
Dominique # [13]	2001	21.0	20.0	58.8	29.4	26.8	43.9	8.3	9.7	81.5
Dominican Republic [6,13]	2005	14.6	22.3	58.8	20.6	26.0	47.7	2.8	15.1	80.7
Rép. dominicaine [6,13]	2010	12.0	14.9	52.4	19.4	20.9	46.8	1.8	6.7	60.1
	#2014	14.5	17.4	41.9	21.2	21.8	42.7	2.8	9.5	40.4

14

Employment by economic activity *(continued)*
Percentage of persons employed by sex and ISIC 4 categories; agriculture (agr.), industry (ind.) and services (ser.)

Emploi par activité économique *(suite)*
Personnes employées par sexe et branches de la CITI rév. 4 ; agriculture (agr.), industrie (ind.) et services (ser.), pourcentage

Region, country or area [+] Région, pays ou zone [+]	Year Année	Male and Female Hommes et femmes			Male Hommes			Female Femmes		
		Agr.	Ind.	Ser.	Agr.	Ind.	Ser.	Agr.	Ind.	Ser.
Ecuador	2005[6,13]	31.5	17.2	51.3	34.6	21.9	43.5	26.7	9.9	63.3
Équateur	2010[6,13]	28.2	18.6	53.1	32.6	23.3	44.0	21.3	11.2	67.5
	#2013	25.3	20.4	54.3	28.5	25.6	45.9	20.3	12.3	67.5
Egypt	2005[3,14]	30.9	21.5	47.5	27.1	25.4	47.2	46.5	5.0	48.4
Égypte	#2010[25]	28.2	25.3	46.3	24.7	30.1	45.1	42.8	5.9	51.2
	#2013	28.0	24.1	47.9	24.1	29.0	46.9	42.9	5.0	52.1
El Salvador	2005[6,13]	20.0	22.2	54.8	30.9	23.7	40.8	4.8	20.1	74.5
El Salvador	#2010[5,13]	20.8	21.4	57.8	32.0	23.5	44.5	5.4	18.4	76.2
	2013	19.6	20.4	60.1	30.7	22.5	46.8	4.5	17.5	78.0
Estonia	2005[26]	5.2	33.8	61.1	7.1	43.9	49.1	3.3	23.9	72.9
Estonie	#2010	4.0	30.9	64.9	5.6	43.7	50.5	2.6	18.5	78.8
	2014	3.9	30.5	65.5	5.3	42.3	52.1	2.3	18.0	79.7
Ethiopia	#2005[3,6]	79.3	6.6	13.0	83.2	5.1	10.4	74.8	8.3	16.0
Éthiopie	#2006[3,6,8]	8.6	22.1	69.2	11.7	26.8	61.4	5.5	17.4	77.1
	2013	72.7	7.4	19.9	79.5	6.9	13.6	64.7	8.0	27.3
Faroe Islands [5,13] Îles Féroé [5,13]	2005	11.1	22.2	66.7	20.0	33.3	53.3	8.3	8.3	83.3
Fiji [13,27]	2005	1.5	30.4	68.1	1.4	32.8	65.8	0.4	26.6	73.0
Fidji [13,27]	2007	1.0	25.7	73.2	1.3	27.7	71.0	0.4	22.0	77.6
Finland	#2005[3]	4.6	25.9	69.3	6.3	38.5	55.0	2.9	12.3	84.5
Finlande	2010	4.1	23.3	72.0	5.5	36.2	57.8	2.7	9.9	86.9
	2014	3.9	22.0	73.7	5.5	34.9	59.2	2.2	8.6	88.7
France	#2005[3]	3.6	23.8	72.3	4.7	33.9	61.0	2.2	12.1	85.3
France	2010	2.9	22.2	74.4	3.9	33.2	62.6	1.8	10.2	87.6
	2014	2.8	20.5	75.8	3.8	30.6	64.5	1.6	9.7	87.8
French Guiana [4,14]	2010	...	14.1	51.5	...	...	...	...	...	...
Guyane française [4,14]	2012	...	14.9	58.3	...	...	...	...	...	...
French Polynesia [28]	2005	3.5	17.2	79.2	...	...	...	...	...	...
Polynésie française [28]	2010	2.7	15.9	81.4	...	...	...	...	...	...
	2011	2.8	15.8	81.4	...	...	...	...	...	...
Gabon # [13] Gabon # [13]	2005	24.2	11.8	64.0	17.3	18.5	64.1	33.7	2.5	63.8
Gambia Gambie	2014	31.5	13.9	54.6	26.1	22.1	51.8	38.2	3.8	58.0
Georgia [3]	2005	54.3	9.3	36.4	51.7	13.7	34.5	57.2	4.4	38.4
Géorgie [3]	2007	53.4	10.4	36.2	50.5	16.5	33.0	56.6	3.7	39.6
Germany	#2005[3]	2.3	30.0	67.8	2.8	41.4	55.8	1.7	16.1	82.3
Allemagne	2010	1.5	28.5	69.9	1.9	40.7	57.4	1.1	14.4	84.5
	2014	1.3	28.3	70.4	1.7	40.7	57.7	0.9	14.3	84.8
Ghana	2010	41.5	15.4	43.1	45.5	17.4	37.1	37.7	13.5	48.8
Ghana	#2013	44.7	14.4	40.9	48.2	17.8	34.0	41.4	11.3	47.3
Greece	#2005[3]	11.5	22.6	65.9	10.6	30.6	58.8	13.0	10.0	77.0
Grèce	2010	11.7	19.8	68.5	11.5	28.0	60.4	12.1	7.9	80.0
	2014	13.0	15.1	71.8	13.5	20.5	65.9	12.4	7.6	79.9
Greenland [14,29] Groenland [14,29]	2011	4.6	12.6	82.5	8.1	20.3	71.3	0.3	3.3	96.2
Guadeloupe [4,14]	2010	...	13.8	64.4	...	...	...	...	...	...
Guadeloupe [4,14]	2012	3.3	13.5	65.5	...	...	...	...	...	...
Guam [5,30]	2000	0.5	11.7	87.8	...	...	...	...	...	...
Guam [5,30]	2010	0.3	14.0	85.7	...	...	...	...	...	...
Guatemala	2004[6,13]	38.3	19.5	42.2	50.6	19.0	30.4	15.3	20.3	64.4
Guatemala	#2006[3,6]	33.2	22.8	44.0	43.8	24.1	32.1	16.0	20.6	63.3
	2013	32.7	17.1	50.2	44.6	18.5	36.9	10.9	14.5	74.6
Guernsey Guernesey	2013	1.5	13.6	84.8	1.0	12.2	40.9	0.5	1.4	44.3
Guinea	2010	74.3	5.8	19.9	67.4	10.3	22.4	81.7	1.0	17.3
Guinée	2012	74.8	5.6	19.3	68.0	9.9	21.7	82.0	0.9	17.1
Guyana # [3] Guyana # [3]	2002	21.4	24.5	50.5	27.3	29.5	39.2	7.1	12.3	77.7
Honduras [6,13]	2005	39.2	20.9	39.7	51.3	19.7	28.7	13.1	23.3	63.4
Honduras [6,13]	#2010	37.5	18.5	43.3	51.4	18.2	29.6	12.5	19.0	67.9
	2013	35.8	18.8	45.4	50.2	18.4	31.3	10.3	19.4	70.2

Employment by economic activity *(continued)*
Percentage of persons employed by sex and ISIC 4 categories; agriculture (agr.), industry (ind.) and services (ser.)

Emploi par activité économique *(suite)*
Personnes employées par sexe et branches de la CITI rév. 4 ; agriculture (agr.), industrie (ind.) et services (ser.), pourcentage

Region, country or area [+] Région, pays ou zone [+]	Year Année	Male and Female Hommes et femmes			Male Hommes			Female Femmes		
		Agr.	Ind.	Ser.	Agr.	Ind.	Ser.	Agr.	Ind.	Ser.
Hungary	#2005[3]	4.8	32.6	62.5	6.7	42.2	51.0	2.6	21.3	76.0
Hongrie	2010	4.5	30.8	64.7	6.5	40.4	53.1	2.3	19.9	77.7
	2014	4.6	30.5	64.5	6.3	40.1	53.1	2.6	19.2	77.7
Iceland	2005[3]	6.3	21.6	71.3	9.4	31.6	58.6	2.9	10.3	85.9
Islande	2010	5.4	18.2	75.9	8.4	27.9	63.3	2.2	7.8	89.8
	2014	4.2	18.2	77.1	6.3	27.7	64.8	1.9	8.0	89.2
India	2005[3,31]	55.8	19.0	25.2	49.3	21.0	29.8	70.9	14.4	14.7
Inde	2010[3,31]	51.1	22.4	26.6	46.1	24.0	29.9	65.3	17.8	17.0
	2012[31]	47.1	24.8	28.1	42.9	26.1	31.0	59.7	20.9	19.4
	#2013	49.7	21.5	28.7	...	...	...	...	...	...
Indonesia	2005[3]	44.0	18.7	37.2	43.8	20.3	35.9	44.3	15.9	39.7
Indonésie	#2010[3]	38.3	19.3	42.3	38.8	21.6	39.5	37.6	15.5	46.9
	2013[3]	# 34.8	# 20.4	# 44.8	35.6	23.6	40.8	33.4	15.0	51.6
	2014[13]	34.3	21.0	44.8	...	...	...	...	...	...
Iran (Islamic Rep. of) [6]	#2005[3]	24.7	30.3	44.8	22.6	30.8	46.5	33.6	28.3	38.0
Iran (Rép. islamique d') [6]	2010[3,4]	19.2	32.2	48.6	17.5	33.8	48.8	28.0	24.4	47.6
	2014[4]	17.9	33.8	48.3	17.2	35.4	47.4	21.8	24.3	53.8
Iraq [3]	2004	17.0	17.8	65.2	14.0	19.8	66.2	32.6	7.4	60.1
Iraq [3]	#2008	23.4	18.2	58.3	17.1	21.6	61.3	50.7	3.7	45.6
Ireland [13]	2005	6.0	27.6	65.8	9.4	39.1	50.7	1.4	11.9	86.3
Irlande [13]	2010	4.8	20.2	74.4	7.9	30.0	61.3	1.2	8.9	89.4
	2014	6.1	18.4	75.2	9.9	26.5	63.2	1.6	8.7	89.4
Isle of Man [3]	2001	1.4	16.1	82.5	2.2	25.0	72.9	0.5	5.5	94.0
Île de Man [3]	2006	1.9	14.8	83.3	2.9	24.0	73.0	0.7	3.9	95.4
Israel	2005[3]	2.0	21.4	75.7	3.1	30.8	64.9	0.8	10.5	88.0
Israël	#2010[3]	1.6	20.2	77.2	2.4	29.7	66.8	0.8	9.6	88.8
	2014	1.1	17.6	79.7	0.8	13.7	37.4	0.3	3.9	42.5
Italy	2005[3]	4.0	30.9	65.1	4.5	39.7	55.8	3.2	17.5	79.3
Italie	2010	3.6	28.8	67.6	4.3	38.9	56.9	2.7	14.2	83.2
	2014	3.5	27.1	69.5	4.3	37.0	58.7	2.3	13.5	84.2
Jamaica [11]	2005[13]	18.1	17.7	64.0	25.1	26.9	47.8	8.5	5.3	86.0
Jamaïque [11]	#2010[3,16]	20.2	15.9	63.8	28.5	22.7	48.8	9.4	6.9	83.6
	#2013[3]	18.2	15.2	66.5	25.8	22.2	52.0	8.2	5.9	85.9
Japan	2005	4.4	28.4	66.0	4.4	36.1	58.3	4.5	17.5	76.8
Japon	#2010	4.0	25.4	69.5	4.2	33.2	61.8	3.9	14.8	80.2
	2013	3.7	25.8	69.1	4.0	34.3	60.4	3.3	14.5	80.6
Jersey										
Jersey	2013	4.1	11.7	70.2	...	...	...	...	...	...
Jordan	#2005[3]	3.4	20.7	76.0	3.6	22.2	74.4	2.0	11.3	86.8
Jordanie	2010	2.0	17.5	80.5	2.3	19.5	78.2	1.0	7.6	91.4
	2014	1.8	18.5	79.6	2.1	20.6	77.4	0.5	7.3	92.2
Kazakhstan	2005[3]	32.4	18.0	49.6	33.4	24.6	42.0	31.3	10.9	57.8
Kazakhstan	#2010	28.3	18.7	53.0	29.1	25.6	45.3	27.4	11.5	61.1
	#2013[4]	24.2	19.8	56.0	25.1	27.1	47.8	23.2	12.2	64.5
Kenya [13]										
Kenya [13]	2005	61.1	6.7	32.2	54.5	10.8	34.6	68.0	2.3	29.7
Kiribati	2005[13]	7.1	8.4	81.1	8.7	10.9	77.3	4.7	4.5	87.8
Kiribati	2010	22.1	16.1	61.8	32.1	9.1	58.8	9.0	25.3	65.7
Kosovo [14,25]										
Kosovo [14,25]	2012	4.6	28.4	67.1	4.3	33.3	62.3	5.4	10.1	84.7
Kuwait	#2005[3]	2.7	20.6	76.1	3.6	26.7	69.1	0.0	2.2	97.0
Koweït	2008	2.5	25.1	72.3	...	...	...	...	...	...
	#2014	1.2	40.2	58.6	0.8	28.2	71.0	0.0	2.4	97.6
Kyrgyzstan	2005[3]	38.5	17.6	43.9	39.3	23.0	37.7	37.4	10.3	52.3
Kirghizistan	2006[3]	36.3	19.4	44.3	36.9	25.7	37.4	35.4	10.7	53.9
	2008[3]	34.0	20.6	45.3	...	...	...	...	...	...
	#2013	31.7	20.2	48.1	30.8	26.3	42.9	32.9	11.0	56.0
Lao People's Dem. Rep.										
Rép. dém. pop. lao	2010	71.3	8.3	20.2	69.2	9.9	21.0	73.5	6.7	19.5
Latvia	#2005[3]	11.6	26.7	61.4	15.7	35.7	48.3	7.4	17.5	75.1
Lettonie	2010	8.3	23.4	68.3	11.5	33.8	54.4	5.4	13.9	80.2
	2014	7.3	24.1	68.6	10.4	35.1	54.4	4.2	13.1	82.6
Lesotho # [3,6]										
Lesotho # [3,6]	2008	12.1	41.7	45.5	18.0	47.6	33.8	4.6	34.2	60.4

Employment by economic activity *(continued)*
Percentage of persons employed by sex and ISIC 4 categories; agriculture (agr.), industry (ind.) and services (ser.)

Emploi par activité économique *(suite)*
Personnes employées par sexe et branches de la CITI rév. 4 ; agriculture (agr.), industrie (ind.) et services (ser.), pourcentage

Region, country or area [+] Région, pays ou zone [+]	Year Année	Male and Female Hommes et femmes			Male Hommes			Female Femmes		
		Agr.	Ind.	Ser.	Agr.	Ind.	Ser.	Agr.	Ind.	Ser.
Liberia Libéria	2010	46.5	10.4	41.2	46.5	15.7	36.7	46.4	5.3	45.7
Lithuania	#2005[3]	14.2	29.1	56.1	16.9	37.0	45.4	11.4	20.8	67.2
Lituanie	2010	8.7	24.5	66.4	11.2	33.3	55.1	6.4	16.7	76.6
	2014	9.0	24.8	65.8	11.5	33.5	54.3	6.6	16.3	76.9
Luxembourg	#2005[3]	1.8	17.2	80.7	2.2	25.5	72.0	1.1	5.7	93.0
Luxembourg	2010	1.0	12.7	81.2	1.3	18.5	74.4	0.6	4.4	90.2
	2014	1.3	10.8	85.7	1.8	16.0	79.9	0.7	4.3	93.0
Madagascar	#2005[3,32]	82.0	3.4	14.6	81.5	5.1	13.4	82.5	1.6	15.9
Madagascar	2012[6]	75.3	7.9	16.9	77.2	7.7	15.2	73.2	8.1	18.7
Malawi Malawi	2013	64.1	7.4	28.5	58.5	9.6	31.8	69.9	5.0	25.1
Malaysia [14]	2005[3]	14.6	29.7	55.6	17.1	32.7	50.2	10.2	24.4	65.4
Malaisie [14]	#2010	13.3	27.6	59.2	16.0	31.3	52.8	8.5	21.0	70.5
	#2013	12.7	27.9	59.3	15.4	32.8	51.8	8.3	19.8	71.9
	#2014	12.2	27.4	60.3	...	...	...	...	...	...
Maldives # [3]	2003	17.3	23.4	56.7	21.8	17.8	57.2	9.0	33.8	55.7
Maldives # [3]	2010[33,34]	14.6	15.7	67.0	19.6	15.1	62.6	6.6	16.6	74.2
Mali [3]	2004	41.5	16.5	41.9	49.8	17.8	32.4	29.9	14.7	55.4
Mali [3]	#2006	66.0	5.6	28.3	67.8	8.0	24.1	63.9	2.7	33.3
Malta	2005[3]	2.0	30.0	67.8	2.7	35.9	61.3	0.4	17.1	82.7
Malte	2010	1.3	25.7	72.6	1.9	32.3	65.6	0.2	13.3	85.4
	2014	1.2	21.6	77.1	1.7	28.6	69.6	0.4	10.6	88.7
Marshall Islands # [3] Îles Marshall # [3]	2010	11.0	9.4	79.6	...	...	...	...	...	...
Martinique [4,14]	2010	4.1	11.9	65.3	...	...	...	...	...	...
Martinique [4,14]	2012	3.9	11.8	69.0	...	...	...	...	...	...
Mauritius	2005[3]	10.0	32.4	57.5	10.5	34.2	55.1	8.9	28.8	62.3
Maurice	#2007[3,5]	9.1	32.3	58.0	9.9	35.5	54.0	7.6	25.8	66.1
	2010	...	27.5	49.5	...	...	...	...	...	...
	2014[5,29]	8.0	# 29.8	# 63.5	7.7	32.0	59.9	8.5	26.1	69.5
Mexico [13]	2005[19]	14.9	25.5	59.0	20.6	29.4	49.3	4.9	18.8	75.8
Mexique [13]	#2010[11]	13.1	25.5	60.6	18.6	30.1	50.5	4.0	17.8	77.5
	#2013	13.4	23.6	62.4	19.4	28.4	51.5	3.7	15.9	79.8
Mongolia	2005[3,5]	39.9	16.8	43.3	43.0	18.9	38.1	36.8	14.8	48.4
Mongolie	#2010	33.5	16.2	50.2	34.7	20.3	45.0	32.2	11.7	56.1
	2012	35.0	18.2	46.8	35.8	24.2	40.0	34.2	11.6	54.3
Montenegro	#2005[3,14]	8.6	19.2	72.1	8.5	26.1	65.3	8.9	9.2	82.0
Monténégro	2010[3]	6.2	20.0	73.8	7.1	27.6	65.4	5.0	9.9	85.0
	#2014	5.7	17.5	73.0	6.5	25.6	64.9	4.6	7.6	83.1
Morocco [3]	2005	45.4	19.5	35.0	39.6	20.8	39.5	61.4	15.9	22.6
Maroc [3]	#2008	40.9	21.7	37.2	34.2	24.0	41.6	59.2	15.4	25.2
	#2010	40.2	22.1	37.6	...	...	...	...	...	...
	#2012	39.2	21.4	39.4	31.9	24.7	43.4	59.9	12.1	27.9
Mozambique [13] Mozambique [13]	2003	80.5	3.4	16.1	69.2	7.0	23.8	89.9	0.4	9.7
Namibia	2004[3,35]	29.9	14.8	55.1	33.7	19.1	47.1	25.2	9.3	65.4
Namibie	#2010[3]	31.2	12.5	54.7	32.3	18.9	47.0	29.9	5.4	63.0
	2013[4]	31.4	14.4	54.2	31.9	23.6	44.6	30.9	5.3	63.8
Nepal	#2001[3,6]	65.7	13.4	20.7	60.2	13.1	26.4	72.8	13.7	13.3
Népal	#2008[3,4]	73.9	10.8	15.3	62.1	15.5	22.4	84.3	6.8	9.0
	2013	66.5	11.2	22.4	...	...	...	...	...	...
Netherlands	2005[3]	3.0	19.7	72.6	3.9	29.1	62.4	2.0	8.1	85.0
Pays-Bas	2010	2.6	16.1	71.9	3.4	24.5	61.1	1.6	6.3	84.4
	2014	2.0	15.1	75.3	2.6	23.4	66.5	1.2	5.6	85.4
Netherlands Antilles [former]	2009	0.8	16.0	83.2	1.6	28.2	70.1	0.2	4.4	95.4
Antilles néerlandaises [anc.]	2011	0.7	15.3	84.0	1.4	27.7	71.0	0.2	4.1	95.7
	2013	0.2	15.9	83.2	...	...	...	...	...	...
New Caledonia [3,11]	2005	3.1	21.3	42.7	...	...	...	...	...	...
Nouvelle-Calédonie [3,11]	2008	2.7	22.4	43.0	...	...	...	...	...	...
New Zealand	2005[3]	7.1	22.2	70.6	8.9	32.2	58.6	5.0	10.6	84.2
Nouvelle-Zélande	#2010	6.8	20.9	71.9	9.0	30.6	59.9	4.4	9.8	85.5
	2013	6.4	20.2	73.0	8.3	29.8	61.4	4.2	9.3	86.1

Employment by economic activity *(continued)*
Percentage of persons employed by sex and ISIC 4 categories; agriculture (agr.), industry (ind.) and services (ser.)

Emploi par activité économique *(suite)*
Personnes employées par sexe et branches de la CITI rév. 4 ; agriculture (agr.), industrie (ind.) et services (ser.), pourcentage

Region, country or area [+] Région, pays ou zone [+]	Year Année	Male and Female Hommes et femmes			Male Hommes			Female Femmes		
		Agr.	Ind.	Ser.	Agr.	Ind.	Ser.	Agr.	Ind.	Ser.
Nicaragua	2005[6,13]	28.9	19.7	51.4	41.3	20.1	38.6	8.3	19.1	72.6
Nicaragua	2008[6,13]	28.2	19.6	52.2	41.2	20.1	38.7	6.2	18.8	75.0
	#2010[3,11]	32.2	16.5	51.3	...	...	...	...	...	...
Niger [13]										
Niger [13]	2005	56.9	11.1	31.1	64.1	8.3	26.5	37.8	18.4	43.0
Nigeria	#2004[13]	44.6	11.5	41.7	49.1	11.8	37.0	38.7	11.2	47.9
Nigéria	2007[3]	48.6	8.5	42.9	53.9	9.0	37.0	41.5	7.7	50.8
Niue [3]	2001	9.0	20.4	70.1	11.4	28.7	59.4	5.8	8.7	85.1
Nioué [3]	#2002	4.8	9.3	85.9	...	...	...	...	...	...
Northern Mariana Islands # [5,13]										
Îles Mariannes du Nord # [5,13]	2000	1.5	47.2	45.8	...	...	...	...	...	...
Norway	#2005[3]	3.2	21.0	75.8	4.7	32.6	62.7	1.6	8.1	90.3
Norvège	2010	2.4	19.8	77.7	3.7	31.5	64.8	1.0	7.1	91.9
	2014	2.1	20.5	77.0	3.2	32.5	63.9	1.0	7.4	91.4
Oman [3]	2000[19,36]	6.4	11.2	82.2	6.6	10.7	82.4	5.4	14.0	80.3
Oman [3]	#2010	5.2	36.9	57.9	6.1	43.1	50.7	0.5	6.3	93.2
Other non-specified areas	2005	5.9	36.4	57.7	7.3	42.9	49.7	4.0	27.4	68.6
Autres zones non-spécifiées	2010	5.2	35.9	58.8	6.5	43.6	49.8	3.6	26.1	70.3
	2014	4.9	36.1	58.9	6.4	44.5	49.1	3.2	25.6	71.2
Pakistan	2005[6,13]	43.0	20.3	36.6	38.1	21.4	40.5	67.3	15.0	17.6
Pakistan	#2010[3]	45.0	20.9	34.2	28.6	18.4	31.2	16.3	2.5	3.0
	2014	43.5	22.5	34.0	34.2	25.5	40.4	74.0	14.4	11.7
Palau [5]	2000[13]	7.1	13.8	79.1	9.6	20.7	69.7	3.1	2.6	94.4
Palaos [5]	2008[3]	2.4	11.8	85.9	...	...	...	...	...	...
Panama	2005[3]	15.7	17.2	67.1	22.2	21.8	56.0	4.3	9.1	86.5
Panama	#2010[3]	17.4	18.7	63.9	23.3	24.1	52.6	7.7	9.7	82.6
	#2012[4]	16.7	18.2	65.0	22.0	24.3	53.7	8.5	8.8	82.7
Papua New Guinea [3,6]										
Papouasie-Nvl-Guinée [3,6]	2000	72.3	3.6	22.7	...	...	...	...	...	...
Paraguay [6]	2005[13]	32.4	15.7	51.9	38.8	20.0	41.2	22.3	9.0	68.7
Paraguay [6]	#2010[3,4,37]	26.8	18.8	54.3	31.5	25.2	43.3	19.1	8.5	72.3
	2014[3,4,37]	22.8	18.9	58.2	26.9	25.6	47.4	16.6	9.0	74.3
Peru [3,11]	2005[8]	10.8	18.6	70.6	13.1	24.6	62.3	8.0	11.1	80.9
Pérou [3,11]	2008[8]	8.1	20.8	71.1	10.3	27.6	62.0	5.5	12.6	81.9
	#2010[38]	...	23.9	74.7	...	...	...	...	...	...
	2014[38]	...	22.9	75.9	...	...	...	...	...	...
Philippines	2005[3,39]	36.0	15.6	48.5	43.6	17.8	38.6	23.7	12.0	64.3
Philippines	#2010[3,40]	33.2	15.0	51.8	40.5	18.2	41.2	21.8	9.9	68.3
	2012	32.2	15.3	52.6	39.4	18.6	42.0	21.0	10.1	68.9
	2014	30.4	15.9	53.6	...	...	...	...	...	...
Poland	2005[3]	16.5	29.7	53.8	17.1	39.7	43.2	15.7	17.4	66.9
Pologne	2010	12.6	30.6	56.7	12.9	42.2	44.8	12.2	16.2	71.5
	2014	11.2	30.8	57.9	12.2	42.2	45.4	10.0	16.6	73.2
Portugal	#2005[3]	8.0	32.1	60.0	6.9	43.1	49.9	9.1	19.4	71.4
Portugal	2010	7.1	28.8	64.0	7.4	39.7	52.8	6.8	16.8	76.3
	2014	5.5	24.9	69.5	7.0	34.0	59.0	4.1	15.5	80.4
Puerto Rico [5,13]	2005	2.1	19.0	79.0	3.3	25.4	71.6	0.4	11.0	88.6
Porto Rico [5,13]	#2010	1.6	13.7	84.7	2.8	18.2	78.8	0.2	8.5	91.3
	2011	1.6	13.5	84.9	2.8	18.1	79.1	0.2	8.3	91.3
Qatar	#2004[3]	2.7	41.0	56.2	3.2	47.6	49.2	0.1	3.1	96.9
Qatar	2009[3,27]	1.6	58.4	40.0	1.7	64.1	34.1	0.0	6.1	93.9
	#2013[41,42]	1.4	51.6	46.8	1.6	58.4	39.9	0.0	4.3	95.1
Republic of Korea	2005[3]	7.9	26.8	65.2	7.2	34.1	58.6	8.9	16.6	74.4
République de Corée	#2010[17]	6.6	24.9	68.5	6.4	32.5	61.2	6.9	14.4	78.7
	#2013	6.1	24.4	69.5	6.0	32.2	61.8	6.2	13.6	80.2
Republic of Moldova [3]	2005	40.6	16.0	43.3	41.1	20.9	37.9	40.3	11.5	48.2
République de Moldova [3]	#2010[4,37,41]	27.5	18.7	53.8	30.5	24.9	44.5	24.5	12.3	63.2
	2013[4,37]	28.8	17.7	53.5	32.1	23.0	44.9	25.4	12.3	62.3
Réunion [4]	2010[14]	4.2	14.1	67.1	...	...	...	...	...	...
Réunion [4]	#2012	4.3	12.7	81.7	6.3	20.0	72.0	2.1	4.1	93.0
Romania	#2005[3]	28.8	32.1	39.1	28.7	36.7	34.6	29.0	26.3	44.7
Roumanie	2010	27.7	29.7	42.6	27.2	36.3	36.4	28.4	21.1	50.5
	2014	25.4	30.1	44.5	25.2	36.6	38.2	25.6	21.7	52.6

14

Employment by economic activity *(continued)*
Percentage of persons employed by sex and ISIC 4 categories; agriculture (agr.), industry (ind.) and services (ser.)

Emploi par activité économique *(suite)*
Personnes employées par sexe et branches de la CITI rév. 4 ; agriculture (agr.), industrie (ind.) et services (ser.), pourcentage

Region, country or area [+] Région, pays ou zone [+]	Year Année	Male and Female Hommes et femmes			Male Hommes			Female Femmes		
		Agr.	Ind.	Ser.	Agr.	Ind.	Ser.	Agr.	Ind.	Ser.
Russian Federation [3,43]	2005	10.2	29.8	60.0	12.3	38.1	49.6	8.0	21.2	70.7
Fédération de Russie [3,43]	#2010[4,37]	7.9	27.7	64.4	10.0	37.3	52.7	5.7	17.8	76.5
	2014	6.7	27.5	65.8	8.2	37.9	53.9	5.2	16.6	78.2
Rwanda	2005[3]	78.8	3.8	16.4	...	...	...	...	...	...
Rwanda	2012	# 75.3	# 6.7	# 16.2	65.5	11.4	21.3	84.4	2.4	11.5
Saint Helena [3,35]										
Sainte-Hélène [3,35]	2008	7.3	20.0	72.7	11.8	33.9	54.3	1.7	2.9	95.4
Saint Kitts and Nevis # [13]										
Saint-Kitts-et-Nevis # [13]	2001	0.2	48.8	42.1	0.4	51.6	34.7	0.1	45.5	50.6
Saint Lucia [3]	2005	13.1	19.3	62.5	16.9	26.3	51.4	8.2	10.3	76.6
Sainte-Lucie [3]	2006	12.9	19.6	59.6	16.0	26.9	48.3	8.5	9.2	75.6
Saint Vincent-Grenadines [13]	2001	22.2	14.2	60.5	...	...	...	...	...	...
Saint-Vincent-Grenadines [13]	#2008	23.8	16.2	55.2	17.3	29.5	48.9	8.4	3.5	81.0
Samoa	2001[3]	39.9	19.7	38.7	50.3	14.8	33.3	15.8	30.9	51.2
Samoa	#2006[3]	35.4	21.8	38.1	48.1	15.3	32.3	8.8	35.4	50.3
	#2012	5.4	14.8	79.9	6.5	19.0	74.5	3.5	7.6	89.0
San Marino	2005[3]	0.5	39.3	60.2	0.6	50.4	49.1	0.3	23.4	76.3
Saint-Marin	#2010	0.3	34.3	65.4	0.4	45.6	54.0	0.2	18.7	81.1
Sao Tome and Principe	2000[3,6]	27.9	19.2	52.4	30.6	26.3	42.6	22.8	5.9	70.7
Sao Tomé-et-Principe	#2012	26.1	21.4	46.9	26.8	23.4	44.6	23.7	14.1	55.9
Saudi Arabia	2002[3]	4.7	21.0	74.4	5.3	24.0	70.7	0.6	1.2	98.1
Arabie saoudite	2009[3]	4.1	20.4	75.5	4.7	23.3	72.0	0.2	1.5	98.4
	2013	4.9	24.2	70.9	5.6	27.5	66.9	0.2	1.7	98.0
Senegal	2001[13]	45.6	12.4	36.1	46.9	15.7	31.7	43.7	7.5	42.5
Sénégal	2006[3]	33.7	14.8	36.1	34.1	20.2	32.7	33.0	4.9	42.3
	#2011	46.1	18.1	22.4	43.8	23.7	21.8	49.1	10.7	23.3
Serbia	2005[3]	23.3	27.6	49.1	23.3	34.4	42.3	23.3	17.5	59.2
Serbie	#2010	22.2	26.0	51.8	23.2	33.4	43.4	20.9	16.1	63.0
	2013[4,37]	21.3	25.9	52.9	22.4	32.3	45.3	19.8	17.1	63.1
Seychelles [4,37]										
Seychelles [4,37]	2011	3.6	17.9	78.2	6.4	25.3	68.1	0.7	10.4	88.5
Sierra Leone [3,6]										
Sierra Leone [3,6]	2004	68.5	6.5	25.0	66.0	10.3	23.7	71.1	2.5	26.4
Singapore	2004[3,44,45]	0.8	24.0	75.2	1.1	27.2	71.6	0.4	19.2	80.3
Singapour	2005[3,44,45]	1.1	21.7	77.3	...	...	...	...	...	...
	2006[46]	1.3	22.1	76.7	1.6	26.0	72.5	0.8	16.8	82.4
	#2008	...	23.8	76.2	...	28.6	71.4	...	17.4	82.6
	2010	...	30.4	68.9	...	...	...	...	...	...
	2013	...	18.6	80.1	...	22.2	76.1	...	14.1	85.1
	2014	...	28.3	70.6	...	...	...	...	...	...
Slovakia	#2005[3]	4.7	38.9	56.2	6.4	49.7	43.7	2.6	25.3	71.9
Slovaquie	2010	3.2	37.2	59.5	4.4	50.1	45.5	1.8	21.1	77.0
	2014	3.5	35.5	60.9	4.9	47.9	47.1	1.7	20.0	78.3
Slovenia	#2005[3]	7.3	37.7	54.5	7.4	47.7	44.3	7.2	25.9	66.6
Slovénie	2010	7.0	33.1	59.5	7.2	43.6	48.8	6.7	20.9	72.1
	2014	7.7	31.5	60.2	7.7	42.6	49.1	7.6	18.2	73.5
South Africa [14]	2005[13]	5.8	25.3	59.1	6.8	34.1	55.0	4.5	14.2	64.3
Afrique du Sud [14]	#2010[3]	4.9	24.5	70.6	5.7	33.0	61.2	3.8	13.3	82.8
	#2014[3]	4.6	23.5	71.9	5.7	32.4	61.9	3.2	12.0	84.7
Spain	2005[3]	5.2	29.7	65.1	6.3	41.3	52.4	3.5	12.5	84.1
Espagne	2010	4.1	23.1	72.8	5.5	33.8	60.7	2.4	9.6	88.0
	2014	4.2	19.5	76.3	5.9	28.8	65.4	2.2	8.5	89.3
Sri Lanka	#2005[3,6]	30.7	25.6	38.4	28.9	24.7	26.0	34.6	27.6	25.8
Sri Lanka	2010[3,6,47]	32.7	24.2	40.4	30.2	23.9	28.3	37.8	24.8	27.2
	#2014	30.4	25.5	43.4	28.6	26.2	44.5	33.9	24.1	41.3
State of Palestine	2005	14.6	26.3	59.1	11.0	29.8	59.2	32.5	8.4	59.1
État de Palestine	2010[13,48]	11.8	24.6	63.6	9.9	28.0	62.1	21.4	7.8	70.8
	#2013	10.5	28.4	61.1	8.5	32.0	59.6	20.9	10.4	68.7
Sudan										
Soudan	2011	44.6	15.3	40.1	39.8	18.7	41.3	60.7	3.8	35.6
Suriname	2004[3]	8.0	23.0	64.3	9.9	30.8	54.6	4.5	8.4	82.2
Suriname	#2013	3.2	22.4	72.9	4.2	33.0	61.2	1.7	5.8	91.0

Employment by economic activity *(continued)*
Percentage of persons employed by sex and ISIC 4 categories; agriculture (agr.), industry (ind.) and services (ser.)

Emploi par activité économique *(suite)*
Personnes employées par sexe et branches de la CITI rév. 4 ; agriculture (agr.), industrie (ind.) et services (ser.), pourcentage

Region, country or area [+] Région, pays ou zone [+]	Year Année	Male and Female Hommes et femmes			Male Hommes			Female Femmes		
		Agr.	Ind.	Ser.	Agr.	Ind.	Ser.	Agr.	Ind.	Ser.
Sweden	#2005[3]	2.0	22.0	75.6	3.0	33.7	63.0	0.9	9.3	89.4
Suède	2010	1.8	20.0	77.9	2.7	31.2	65.8	0.9	7.6	91.2
	2014	1.7	18.7	79.0	2.5	29.4	67.6	0.8	7.2	91.3
Switzerland	2005[3]	3.5	22.5	73.4	4.3	32.1	63.0	2.6	11.0	85.9
Suisse	2010	3.0	21.3	71.2	3.6	30.9	61.6	2.3	10.1	82.5
	2014	3.2	20.1	73.9	3.9	28.8	64.9	2.4	10.2	84.1
Syrian Arab Republic [3]	2003	27.0	25.6	47.4	22.6	29.1	48.3	49.1	8.1	42.8
Rép. arabe syrienne [3]	#2010	15.2	32.9	53.0	14.6	36.8	50.7	22.2	9.2	68.6
	2011	13.2	31.4	55.3	13.4	34.5	52.1	12.3	9.5	78.2
Tajikistan	2004[3]	55.5	17.9	26.2	41.8	27.1	30.8	75.1	4.8	19.7
Tadjikistan	#2009[10]	52.9	15.6	31.1	41.1	23.4	35.0	68.5	5.4	25.9
Thailand	2005[3]	42.6	20.2	37.1	44.2	21.6	34.1	40.7	18.7	40.5
Thaïlande	#2010[3]	38.2	20.6	41.0	40.1	22.7	37.1	36.0	18.2	45.7
	#2013	41.9	20.3	37.5	44.1	22.6	33.1	39.3	17.7	42.8
TFYR of Macedonia	2005[3]	19.5	32.3	48.2	19.8	33.8	46.5	19.2	29.9	50.9
ex-R.Y. de Macédoine	#2008[3]	18.9	31.6	49.5	18.7	33.7	47.5	19.1	28.1	52.8
	2014	18.0	30.6	51.4	17.9	33.4	48.7	18.1	26.2	55.7
Timor-Leste	2001[3,14]	81.6	3.9	11.2	82.3	4.3	9.6	79.9	3.1	15.1
Timor-Leste	#2010	50.6	9.2	39.8	50.8	10.1	38.6	50.2	7.3	42.4
Togo [13]										
Togo [13]	2006	54.1	6.8	37.5	60.5	9.5	28.6	48.2	4.4	45.7
Tonga [3]										
Tonga [3]	2003	31.8	30.6	37.5	50.6	13.5	35.8	4.6	55.3	40.0
Trinidad and Tobago [13]	2005	4.3	31.0	64.3	6.1	41.1	52.3	1.7	15.8	82.0
Trinité-et-Tobago [13]	2008	3.8	32.2	63.8	5.2	43.8	50.8	1.8	15.3	82.0
Tunisia [3,49]	2010	17.6	32.7	48.8	...	...	...	...	...	...
Tunisie [3,49]	#2014	14.8	33.5	51.5	...	...	...	...	...	...
Turkey	#2005[3]	29.5	24.8	45.8	21.7	28.1	50.2	51.6	15.1	33.3
Turquie	2010	22.4	26.8	50.8	16.1	31.0	52.9	38.3	16.2	45.5
	2014	19.7	28.4	51.9	14.6	33.1	52.4	31.7	17.4	50.9
Turks and Caicos Islands [3]	2005	1.4	16.8	70.9	...	...	...	...	...	...
Îles Turques-et-Caïques [3]	#2008	1.2	23.1	74.0	...	...	...	...	...	...
Uganda	#2003[3,6]	68.7	7.8	23.5	61.7	10.3	28.0	75.7	5.3	19.0
Ouganda	#2005[3,50]	71.6	4.5	23.2	...	...	...	...	...	...
	2009[51]	73.8	7.1	19.1	69.8	9.7	20.5	77.5	4.6	17.8
	2013	71.9	4.4	20.2	...	...	...	...	...	...
Ukraine	2005[3,52]	19.4	24.2	56.4	...	...	...	...	...	...
Ukraine	#2009[3]	20.0	25.7	53.5	20.8	35.8	42.8	19.2	15.3	64.5
	2014	14.8	26.1	59.1	16.5	36.9	46.6	13.0	14.5	72.5
United Arab Emirates # [3]	2005	4.9	39.8	54.5	5.6	44.8	48.9	0.1	6.2	92.4
Émirats arabes unis # [3]	2009[53]	3.8	23.1	73.0	4.6	27.1	68.3	0.1	6.4	93.5
United Kingdom	#2005[3]	1.3	22.2	76.2	1.8	33.3	64.6	0.7	9.4	89.7
Royaume-Uni	2010	1.1	19.2	79.0	1.5	29.6	68.2	0.5	7.4	91.4
	2014	1.1	18.9	79.1	1.5	28.7	68.9	0.6	8.0	90.6
United Rep. of Tanzania	#2002[6,13]	81.0	3.8	15.1	76.2	6.1	17.8	86.2	1.5	12.4
Rép.-Unie de Tanzanie	2006[3]	74.6	5.0	20.3	71.2	7.3	21.5	78.0	2.8	19.2
	#2007[3,54]	69.0	4.2	26.5	...	...	...	...	...	...
	2014	66.9	6.4	26.6	64.0	9.6	26.4	70.0	3.2	26.8
United States [5]	2005[3,16]	1.6	20.6	77.8	2.2	30.2	67.6	0.8	9.6	89.6
États-Unis [5]	#2010	1.6	17.2	81.2	2.3	25.9	71.7	0.8	7.4	91.9
Uruguay	2005[3,8,11,18]	4.6[21]	21.9	73.5	7.1[21]	29.1	63.8	1.5	12.8	85.7
Uruguay	#2010[3]	11.6	21.4	67.0	16.7	29.0	54.2	5.2	11.9	82.8
	2013	9.3	21.6	69.1	13.3	30.6	56.2	4.4	10.3	85.3
Vanuatu										
Vanuatu	2009	60.5	7.0	31.1	59.2	10.5	29.4	62.3	2.5	33.3
Venezuela (Boliv. Rep. of) [13]	2005	9.7	20.8	68.7	14.4	26.8	57.9	2.0	11.2	86.0
Venezuela (Rép. boliv. du) [13]	#2010	8.7	22.1	68.9	13.1	29.0	57.5	1.8	11.0	86.9
	2011	7.9	21.9	70.0	11.9	29.1	58.7	1.6	10.6	87.6
	#2013[30]	7.4	21.3	71.1	...	...	...	...	...	...
Viet Nam	2004[3]	57.9	17.4	24.8	55.9	20.8	23.3	60.0	13.7	26.3
Viet Nam	#2006[13]	51.7	20.2	28.2	49.6	24.4	26.0	53.8	15.9	30.3
	#2013[17]	46.8	21.2	32.0	44.9	25.2	29.9	48.8	17.0	34.2

Employment by economic activity *(continued)*
Percentage of persons employed by sex and ISIC 4 categories; agriculture (agr.), industry (ind.) and services (ser.)

Emploi par activité économique *(suite)*
Personnes employées par sexe et branches de la CITI rév. 4 ; agriculture (agr.), industrie (ind.) et services (ser.), pourcentage

Region, country or area [+] Région, pays ou zone [+]	Year Année	Male and Female Hommes et femmes			Male Hommes			Female Femmes		
		Agr.	Ind.	Ser.	Agr.	Ind.	Ser.	Agr.	Ind.	Ser.
Yemen	#2004[3]	31.0	16.3	52.7	30.0	16.5	53.1	41.8	9.3	48.9
Yémen	2010[55]	24.7	18.8	56.2	24.5	19.1	56.1	28.0	14.6	57.4
Zambia	#2005[13]	72.2	7.1	20.6	65.9	10.9	23.7	78.9	3.1	17.3
Zambie	2008	71.3	7.5	21.0	...	...	...	...	...	...
	#2012	52.2	9.5	38.3	51.0	15.7	33.3	53.4	3.4	43.1
Zimbabwe	#2004[3]	64.8	9.3	15.3	58.8	14.0	17.3	71.1	4.4	13.2
Zimbabwe	2011	65.8	9.1	25.0	59.9	15.2	24.8	71.6	3.0	25.2

Source:

International Labour Organization (ILO), Geneva, Key Indicators of the Labour Market (KILM 9th edition), last accessed March 2016.

Source:

Bureau international du travail (BIT), Genève, Indicateurs clés du marché du travail (ICMT 9e édition), dernier accès mars 2016.

+ Ages 15 years and over unless indicated otherwise. Table is in ISIC Rev. 4 unless otherwise indicated.

+ Sauf indication contraire, Age 15 ans et plus. Sauf indication contraire, la classification utilisée dans ce tableau est la CITI Rév. 4.

1 Data excludes Armenia, Azerbaijan, Cyprus, Georgia, Israel and Turkey.
2 Refers to Armenia, Azerbaijan, Cyprus, Georgia, Israel and Turkey.
3 Data classified according to ISIC Rev. 3.
4 Excluding the institutional population.
5 Persons aged 16 years and over.
6 Persons aged 10 years and over.
7 28 urban agglomerations.
8 Urban areas.
9 Main cities or metropolitan areas.
10 Persons aged 15 to 75 years.
11 Persons aged 14 years and over.
12 Economic activity refers to jobs. As a result, the sum of the categories may not add up to the total.
13 Data classified according to ISIC Rev. 2.
14 Persons aged 15 to 64 years.
15 Persons aged 13 years and over.
16 Services include activities not adequately defined.
17 Excluding overseas territories.
18 Excluding conscripts.
19 Persons aged 12 years and over.
20 Excluding marine and institutional populations.

21 Agriculture includes mining and quarrying.
22 Age 10+ in rural areas.
23 Resident population (de jure).
24 Labour force data for males refers to ages 17-59 and data for females to ages 17-54.
25 Civilians only.
26 Persons aged 15 to 74 years.
27 Paid employment only.
28 Employees only.
29 Nationals, residents.
30 Excluding armed forces.
31 Excluding Leh and Kargil of Jammu and Kashmir districts, some villages in Nagaland, Andaman and Nicobar Islands.

32 Persons aged 6 years and over.
33 Nonstandard reference group coverage.
34 Nonstandard geographical coverage.
35 Persons aged 15 to 69 years.
36 Omani nationals only.

1 Les données excluent l'Arménie, l'Azerbaïdjan, Chypre, la Géorgie, l'Israël et la Turquie.
2 Les données se rapportent à l'Arménie, l'Azerbaïdjan, Chypre, la Géorgie, l'Israël et la Turquie.
3 Données classifiées selon la CITI, Rév. 3.
4 Non compris la population dans les institutions.
5 Personnes âgées de 16 ans et plus.
6 Personnes âgées de 10 ans et plus.
7 28 agglomérations urbaines.
8 Zones urbaines.
9 Villes principales ou zones métropolitaines.
10 Personnes âgées de 15 à 75 ans.
11 Population âgée de plus de 14 ans.
12 L'activité économique se réfère aux emplois. Par conséquent, si l'on ajoute les catégories, il se peut que cela ne corresponde pas au total.
13 Données classifiées selon la CITI, Rév. 2.
14 Personnes âgées de 15 à 64 ans.
15 Personnes âgées de 13 ans et plus.
16 Les services comprennent les activités mal définies.
17 Non compris les départements d'outre-mer.
18 Non compris les conscrits.
19 Personnes âgées de 12 ans et plus.
20 A l'exclusion des populations marines et de la population institutionnelle.

21 L'agriculture englobe les industries extractives.
22 Personnes âgées de 10 ans et plus dans les zones rurales.
23 Population résidente (de droit).
24 Les données relatives à la main d'œuvre portent sur les hommes âgés de 17 à 59 ans et sur les femmes âgées de 17 à 54 ans.
25 Civils uniquement.
26 Personnes âgées de 15 à 74 ans.
27 Emploi rémunéré seulement.
28 Salariés uniquement.
29 Ressortissants, résidents.
30 Non compris les militaires.
31 Les districts de Leh et Kargil dans l'état de Jammu-et-Cachemire, quelques villages dans l'etat de Nagaland ou dans les îles Andaman-et-Nicobar ne sont pas inclus.

32 Personnes âgées de 6 ans et plus.
33 Couverture du groupe de référence non standard.
34 Couverture géographique hors normes.
35 Personnes âgées de 15 à 69 ans.
36 Ressortissants omanais seulement.

14

Employment by economic activity *(continued)*
Percentage of persons employed by sex and ISIC 4 categories; agriculture (agr.), industry (ind.) and services (ser.)

Emploi par activité économique *(suite)*
Personnes employées par sexe et branches de la CITI rév. 4 ; agriculture (agr.), industrie (ind.) et services (ser.), pourcentage

37	Excluding some areas.
38	Main city or metropolitan area.
39	Excluding regular military living in barracks.
40	Nonstandard population coverage.
41	Persons aged 15 to 76 years.
42	Excluding persons temporarily absent from work.
43	Persons aged 15 to 72 years.
44	Data refer to permanent residents.
45	Agriculture includes mining and quarrying, electricity, gas and water supply and not classifiable by economic activity.
46	Agriculture includes Mining and quarrying, Electricity, gas, steam and air conditioning supply, and Water supply; sewerage, waste management and remediation activities.
47	Excluding the Northern province.
48	Services includes electricity, gas and water.
49	Industry includes repairs.
50	Persons aged 14 to 64 years.
51	Persons aged 10 to 76 years.
52	Persons aged 15 to 70 years.
53	Excluding labour camps.
54	Excluding Zanzibar.
55	Persons aged 15 to 65 years.

37	Certaines régions sont exclues.
38	Ville principale ou zone métropolitaine.
39	Ne comprend pas les militaires qui vivent en casernes.
40	Couverture de la population hors normes.
41	Personnes âgées de 15 à 76 ans.
42	Non compris les personnes temporairement absentes de leur travail.
43	Personnes âgées de 15 à 72 ans.
44	Les données concernent les résidents permanents.
45	L'agriculture englobe les industries extractives, la production et distribution d'électricité, de gaz et d'eau, qui ne peuvent être classées parmi les activités économiques.
46	L'agriculture englobe les industries extractives, la production et distribution d'électricité, de gaz, de vapeur et d'air conditionné, ainsi que l'alimentation en eau, les réseaux d'assainissement, la gestion des déchets et les activités de dépollution.
47	Non compris la province du nord.
48	Les services englobent l'électricité, le gaz et l'eau.
49	L'industrie englobe les réparations.
50	Personnes âgées de 14 à 64 ans.
51	Personnes âgées de 10 à 76 ans.
52	Personnes âgées de 15 à 70 ans.
53	Non compris les camps de travail.
54	Zanzibar non compris.
55	Personnes âgées de 15 à 65 ans.

15

Consumer price indices
General and food (Index base: 2000 = 100)

Indices des prix à la consommation
Généraux et alimentation (Indices base: 2000 = 100)

Country or area	1985	1995	2005	2010	2012	2013	2014	Pays ou zone
Albania								**Albanie**
General	...	55.0	116.7	134.6	142.1	144.8	147.2	Généraux
Food	...	228.5[1,2]	#114.3[3]	136.7	146.5	152.7	156.1	Alimentation
Algeria								**Algérie**
General	...	72.7	116.7	145.9	169.3	176.2	183.0	Généraux
Food	...	74.3	116.6	157.1	188.0	193.6	202.4	Alimentation
American Samoa								**Samoa américaines**
General[4]	63.6	90.1	122.1	155.9	...	...	...	Généraux [4]
Food	72.7	97.5	130.4	192.1	...	...	...	Alimentation
Andorra [5]								**Andorre** [5]
General	...	...	112.9	126.2	131.4	132.1	132.0	Généraux
Food[6]	...	...	112.9	123.3	128.7	131.7	133.2	Alimentation [6]
Angola [7]								**Angola** [7]
General	...	<0.0	1 846.0	3 437.6[8]	4 302.1	4 679.6	...	Généraux
Food	...	<0.0	1 957.0	...	5 565.0	...	...	Alimentation
Anguilla [5]								**Anguilla** [5]
General	...	...	113.5	138.3	148.5	148.7	148.3	Généraux
Food	...	...	105.0	144.8	157.7	160.5	163.2	Alimentation
Antigua and Barbuda								**Antigua-et-Barbuda**
General	...	91.9[9]	110.2	123.3	131.5	132.9	...	Généraux
Food	...	91.6[9]	109.5	140.9	152.2	155.8	...	Alimentation
Argentina [10,11]								**Argentine** [10,11]
General	<0.0	100.5	161.7	248.8	300.6	332.5	...	Généraux
Food	<0.0	106.0	183.3	287.0	343.9	369.9	...	Alimentation
Armenia								**Arménie**
General	...	68.1	117.0	154.2	169.5	179.3	184.6	Généraux
Food	...	85.2	126.8	169.7	192.8	204.0	207.5	Alimentation
Aruba								**Aruba**
General	57.6	86.8[8]	116.8	138.8	145.7	142.3	142.9	Généraux
Food	50.9	87.9[8]	118.7	158.0	169.1	169.4	173.1	Alimentation
Australia								**Australie**
General	54.9	91.0	116.1	134.4	141.4[8]	144.9	148.5	Généraux
Food	54.8	87.2	120.0	145.8	150.4[8]	151.6	155.4	Alimentation
Austria								**Autriche**
General	71.5	93.3	110.6	121.1	128.2	130.7	132.8	Généraux
Food	77.7[12]	94.9[12]	111.0[12]	125.4	134.9	139.6	142.3	Alimentation
Azerbaijan								**Azerbaïdjan**
General	...	87.1	124.8	204.4	...	...	...	Généraux
Food[2]	...	95.1	134.1	236.0	...	...	...	Alimentation [2]
Bahamas								**Bahamas**
General	60.8[13]	94.1[8,13]	#110.4	124.9	131.4	131.9	133.5	Généraux
Food	...	92.1[8,13]	111.2[13]	#120.8[14]	...	127.4[14]	129.6[14]	Alimentation
Bahrain								**Bahreïn**
General	#101.3[16,17]	106.9[17]	#104.9	119.9	122.8	126.9	130.2	Généraux
Food[15]	#105.0[16,17]	107.1[17]	#101.3	137.0	145.3	148.8	153.0	Alimentation [15]
Bangladesh [18]								**Bangladesh** [18]
General	...	78.9	126.7	183.2	220.5	246.5[8]	263.7	Généraux
Food[6]	...	77.6	127.8	195.9	237.4	251.9[8]	271.8	Alimentation [6]
Barbados								**Barbade**
General	62.8	88.3[8]	112.5	149.0	170.4	173.5	176.7	Généraux
Food	59.1	84.4[8]	123.1	178.6	206.7	212.5	215.2	Alimentation
Belarus								**Bélarus**
General	...	2.2[19]	384.3[19]	622.9[8]	...	...	...	Généraux
Food	...	2.0[12,19]	358.2[12,19]	#29 099.4[20]	...	...	...	Alimentation
Belgium								**Belgique**
General	73.5	92.1	111.0	122.8	130.8	132.2	132.7[8]	Généraux
Food	76.2	86.1	112.5	129.4	136.5	141.4	140.9[8]	Alimentation
Belize								**Belize**
General	73.0	94.4	113.3	128.3	132.1	132.8	134.1	Généraux
Food	73.0	94.3[2]	111.8[2]	137.0[2]	...	...	...	Alimentation
Benin [21]								**Bénin** [21]
General	...	84.1	114.9	133.7[8]	146.6	148.1	146.4	Généraux
Food	...	78.4	114.4	141.6[8]	156.6	160.3	159.4	Alimentation
Bermuda								**Bermudes**
General	60.6[8]	89.1	116.0	135.6	142.6	145.1	148.0	Généraux
Food	62.9[8]	88.4	111.4	134.7	142.6	146.9	151.4	Alimentation

Country or area	1985	1995	2005	2010	2012	2013	2014	Pays ou zone
Bhutan								**Bhoutan**
General	26.9	70.3	116.8	156.1	188.4	204.9[8]	221.8	Généraux
Food	30.9	74.1	108.8	163.9	205.7	223.7[8]	246.6	Alimentation
Bolivia (Plurin. State of)[22]								**Bolivie (État plurin. de)**[22]
General	...	73.8	116.6	159.7	183.4	193.9	205.1	Généraux
Food	...	79.8	115.7	127.8	151.3	164.6	178.0	Alimentation
Bosnia and Herzegovina[14]								**Bosnie-Herzégovine**[14]
General	...	...	100.0	117.7	124.5	124.4	123.3	Généraux
Food	...	...	100.0	122.8	132.5	132.5	128.9	Alimentation
Botswana								**Botswana**
General	22.7	67.0	146.1	227.2	265.0	280.6	293.0	Généraux
Food	23.0	67.4	137.9	246.1	283.9	299.2	308.3	Alimentation
Brazil								**Brésil**
General	<0.0[17]	#69.7	151.4	190.4	214.0	227.2	241.6	Généraux
Food	<0.0[17]	#83.9[12]	151.0[12]	204.6[12]	240.8[12]	267.7[12]	287.8[12]	Alimentation
British Virgin Islands								**Îles Vierges britanniques**
General	56.1	82.1	110.4	...	...	...	...	Généraux
Food	61.7	85.3	112.1	...	...	...	...	Alimentation
Brunei Darussalam								**Brunéi Darussalam**
General	92.7[17,18]	#95.7	100.5	105.1[8]	107.6	105.5[8]	105.3	Généraux
Food	96.1[17,18]	#93.5	102.2	111.6[8]	118.7	112.0[8]	111.8	Alimentation
Bulgaria								**Bulgarie**
General	...	2.9[8]	129.6	178.2	191.3	193.0	190.2	Généraux
Food	...	100.0[8,20]	#117.0	161.7	178.1	183.1	181.3	Alimentation
Burkina Faso[23]								**Burkina Faso**[23]
General	66.8	88.5	116.0	131.4[8]	140.2	140.9	140.6	Généraux
Food	78.9	85.7	120.2	152.7[8]	170.7	170.1	164.7	Alimentation
Burundi[24]								**Burundi**[24]
General	18.2	41.3	144.8	236.0	305.6	331.7	344.5[8]	Généraux
Food	18.8	40.5	139.4	237.2	296.5	326.3	330.7[8]	Alimentation
Cabo Verde								**Cabo Verde**
General	71.8[4,17,25]	#81.8	104.9	127.1	136.2	138.3	137.9	Généraux
Food	66.9[17,25]	#139.8[17]	#96.4[26]	122.6[26]	132.4[26]	133.3[26]	131.8[26]	Alimentation
Cambodia[27]								**Cambodge**[27]
General	...	73.0	114.1	165.3	179.5	184.8	...	Généraux
Food	...	73.5[2]	116.6[2]	#104.0[28]	114.2[28]	118.7[28]	...	Alimentation
Cameroon								**Cameroun**
General	...	86.3	110.5	128.9	135.8[8]	138.5	141.2	Généraux
Food	...	84.9	110.3	139.4	151.2[8]	156.2	157.2	Alimentation
Canada								**Canada**
General	66.1	91.8	112.2	122.1	127.6	128.8	131.3	Généraux
Food	70.3	93.2	114.1	132.0	140.2	141.9	145.3	Alimentation
Cayman Islands								**Îles Caïmanes**
General	54.1	84.0[8]	116.8	124.5	127.7	130.5	132.1	Généraux
Food	67.8	89.7[8]	117.0	148.6	160.3	165.5	169.3	Alimentation
Central African Rep.[29]								**Rép. centrafricaine**[29]
General[4]	74.7	95.3	111.7	138.1	148.0	150.2	...	Généraux [4]
Food	78.3	95.9	110.9	#270.7[30]	293.8[30]	298.3[30]	...	Alimentation
Chad[31]								**Tchad**[31]
General	...	85.4	125.7	146.8	161.1	161.5	...	Généraux
Food	...	82.8	137.7	168.7	192.1	188.6	...	Alimentation
Chile								**Chili**
General	16.8[32]	77.8[32]	113.6[32]	#101.4[30]	108.0[30]	109.9[30]	114.7[8,30]	Généraux
Food	16.1[12,32]	83.4[12,32]	107.4[12,32]	#102.2[30]	117.4[30]	122.6[30]	131.2[8,30]	Alimentation
China[33]								**Chine**[33]
General	...	...	106.9	123.5	133.4	136.9	139.6	Généraux
Food	...	...	116.3	165.2	193.6	202.6	208.9	Alimentation
China, Hong Kong SAR								**Chine, Hong Kong RAS**
General	69.6[17]	157.3[17]	#93.6	104.4	114.3	119.3	...	Généraux
Food	69.2[17]	149.5[17]	#98.4	119.3	135.0	140.9	...	Alimentation
China, Macao SAR								**Chine, Macao RAS**
General	...	145.3[4,17]	#99.0	124.1	139.3	147.0	155.9[8]	Généraux
Food	...	97.5	101.3	147.2	172.8	184.2	195.5[8]	Alimentation
Colombia[34]								**Colombie**[34]
General	32.5[17]	#47.7	139.6	177.4	190.7	194.3	199.6	Généraux
Food	...	52.3	143.1	191.7	209.2	211.0	216.7	Alimentation
Congo[35,36]								**Congo**[35,36]
General	48.9	76.8	109.6	139.2[8]	145.2	153.9	...	Généraux
Food	54.3	75.6	95.6	#143.3[14]	161.4[14]	173.3[14]	...	Alimentation

Consumer price indices *(continued)*
General and food (Index base: 2000 = 100)
Indices des prix à la consommation *(suite)*
Généraux et alimentation (Indices base: 2000 = 100)

Country or area	1985	1995	2005	2010	2012	2013	2014	Pays ou zone
Cook Islands [37]								**Îles Cook** [37]
General	75.8[38]	#95.9	118.5	144.0	151.5	154.4	157.6	Généraux
Food	75.5[38]	#95.8[39]	122.3[39]	151.5[39]	159.9[39]	164.1[39]	169.5[39]	Alimentation
Costa Rica [40]								**Costa Rica** [40]
General	...	55.1	169.9	267.7	293.4	308.7	322.7	Généraux
Food	...	53.3[2]	176.5[2]	#140.2[41]	153.3[41]	159.2[41]	166.0[41]	Alimentation
Côte d'Ivoire [35,42]								**Côte d'Ivoire** [35,42]
General	78.0[34,43]	...	#117.1	132.9[8]	141.3	144.9	...	Généraux
Food	68.3[34,43]	...	#114.3[15]	#109.1[28]	...	122.5[28]	...	Alimentation
Croatia								**Croatie**
General	...	79.4	114.0	132.9	140.5	143.6[8]	143.3	Généraux
Food	...	86.5	110.4	128.7	138.1	...	140.2	Alimentation
Cuba								**Cuba**
General	...	...	108.9	...	...	...	...	Généraux
Food [12,44]	...	...	110.3	...	...	...	...	Alimentation [12,44]
Curaçao								**Curaçao**
General	...	87.0	109.6	130.1	137.4	139.2	141.3	Généraux
Food	...	84.6	123.3	197.4	225.5	229.2	237.0	Alimentation
Cyprus								**Chypre**
General	58.9	86.7	114.6	129.3	136.8	136.2	134.4	Généraux
Food	82.6[15,17]	127.9[15,17]	#120.9	149.5	156.9	157.2	154.9	Alimentation
Czech Republic								**République tchèque**
General	...	72.1	111.7	128.6	135.1	137.3	137.8	Généraux
Food	...	231.8[17]	#110.3[45]	115.7[46]	129.1[46]	135.8[46]	138.5[46]	Alimentation
Denmark								**Danemark**
General	66.8	89.2	110.2	122.4	128.8	129.9	130.6	Généraux
Food	75.0	90.3	107.3	124.2	134.6	135.2	134.0	Alimentation
Dominica								**Dominique**
General	66.3	93.1	#105.8[5]	122.5[5,8]	125.7[5]	125.5[5]	...	Généraux
Food	68.7	94.8	#107.4[5]	145.2[5,8]	155.0[5]	157.5[5]	...	Alimentation
Dominican Republic								**Rép. dominicaine**
General	9.8	72.9	230.4	314.0	353.2	370.2	381.3	Généraux
Food	9.0[15]	78.9[15]	233.2[15]	320.4[15]	356.7	373.3	387.8	Alimentation
Ecuador								**Équateur**
General	0.4	15.1[8]	175.4[8]	218.8	240.3	246.8	255.7	Généraux
Food	0.4	14.1[8,15]	#100.0[14]	141.5[14]	159.8[14]	162.7[14]	170.0[14]	Alimentation
Egypt								**Égypte**
General	17.1	79.6	133.7	231.6[8]	273.1	298.9	328.9	Généraux
Food	17.2[2]	78.6[2]	#105.0[47]	225.3[8,47]	284.1[47]	319.4[47]	358.7[47]	Alimentation
El Salvador [22]								**El Salvador** [22]
General	15.7	82.7	118.0	139.7[8]	149.4	150.6	152.3	Généraux
Food	29.3[2,17]	#83.6[3]	122.6[3]	#100.0[3,48]	#107.3[3,48]	109.8[3,48]	113.3[3,48]	Alimentation
Equatorial Guinea [49]								**Guinée équatoriale** [49]
General [43]	...	149.8	...	...	...	...	...	Généraux [43]
Food [17]	...	141.6	...	...	...	...	...	Alimentation [17]
Estonia								**Estonie**
General	...	4 251.5[43]	#119.1	150.6	164.3	168.9	168.7	Généraux
Food	...	2 837.5[43]	#118.3	153.4	174.6	181.7	181.7	Alimentation
Ethiopia								**Éthiopie**
General	93.6[4,17,50]	183.9[4,17,50]	#138.0[5]	313.0[5]	517.6[5,8]	559.4[5]	600.8[5]	Généraux
Food	103.6[17,50]	198.6[17,50]	#153.3[5]	359.7[5]	641.2[5,8]	678.1[5]	714.9[5]	Alimentation
Falkland Is. (Malvinas) [51]								**Îles Falkland (Malvinas)** [51]
General	54.4[4]	84.5	...	...	...	...	...	Généraux
Food [17]	82.3	123.8	...	...	...	...	...	Alimentation [17]
Faroe Islands								**Îles Féroé**
General	61.1	83.0	109.3	121.3	126.7	125.9	124.7	Généraux
Food	48.9	76.9	111.1	127.7	132.8	133.4	133.4	Alimentation
Fiji								**Fidji**
General	51.5	86.1	115.2	145.9	165.4	169.4	163.8[8]	Généraux
Food	54.7	88.0	117.1	161.8	184.0	190.1	191.0[8]	Alimentation
Finland								**Finlande**
General	92.5	92.6	106.2[8]	116.5	123.8	125.7	127.0	Généraux
Food	85.9[17]	#98.0[46]	109.2[8,46]	120.7[46]	134.9[46]	142.1[46]	142.4[46]	Alimentation
France								**France**
General	72.4	94.1	109.9	118.5	123.4	124.5	125.1	Généraux
Food	75.8	93.1	111.0	121.4	127.4	129.1	128.4	Alimentation
French Guiana								**Guyane française**
General	85.3[17,52]	#96.2	108.1	119.2	123.4	125.2	125.7	Généraux
Food	85.8[17,52]	#93.7	110.6	123.7	129.3	133.6	135.6	Alimentation

15

Consumer price indices *(continued)*
General and food (Index base: 2000 = 100)
Indices des prix à la consommation *(suite)*
Généraux et alimentation (Indices base: 2000 = 100)

Country or area	1985	1995	2005	2010	2012	2013	2014	Pays ou zone
French Polynesia								**Polynésie française**
General	78.9	94.6	105.8	116.1	119.6	121.3	121.7	Généraux
Food	83.7	94.7	113.2	#104.8[28]	110.8[28]	114.5[28]	115.5[28]	Alimentation
Gabon [35,53]								**Gabon** [35,53]
General	90.4[43]	#90.7	104.9	122.6	127.4	128.1	...	Généraux
Food	91.3[43]	#90.5[6]	105.5[6]	#116.7[41]	...	127.7[41]	...	Alimentation
Gambia [54]								**Gambie** [54]
General	25.2	90.9	156.5	193.1	210.9	222.9	...	Généraux
Food	24.3	90.3	169.2	221.1	247.2	263.8	...	Alimentation
Georgia [55]								**Géorgie** [55]
General	...	52.2	132.5	189.3	...	...	...	Généraux
Food [2]	...	57.6	149.6	...	...	...	...	Alimentation [2]
Germany								**Allemagne**
General	...	93.5	108.3	117.2	122.2	123.8[8]	125.0	Généraux
Food	...	#98.8	105.3	118.5	125.8	130.7[8]	132.1	Alimentation
Ghana								**Ghana**
General	2.5	31.9	250.7	510.6	607.2[8]	678.0	783.0	Généraux
Food	3.4	41.4	244.7	424.9	462.3[8]	495.6	529.9	Alimentation
Gibraltar								**Gibraltar**
General	60.8	93.5	110.9	128.5	136.9	139.7	142.2	Généraux
Food	63.7	90.7	117.4	149.5	164.1	167.4	169.8	Alimentation
Greece								**Grèce**
General	44.9[17]	#79.0	118.2[8]	138.6	145.3	144.0	142.1	Généraux
Food	46.3[17]	#82.3	117.4[8]	135.1	141.4	141.4	139.0	Alimentation
Greenland								**Groenland**
General	67.3	94.8	113.3	131.7	140.3	142.1	144.0	Généraux
Food	83.2[2,17]	113.4[2,17]	#114.5	141.1	151.6	154.0	157.5	Alimentation
Grenada								**Grenade**
General	71.4	92.2	112.4	135.7[8]	142.7	142.0	...	Généraux
Food	65.6[2]	93.2[2]	110.4	148.9[8]	159.5	161.0	...	Alimentation
Guadeloupe								**Guadeloupe**
General	88.0[17,22]	#95.6	112.1	122.0	127.5	128.6	129.1	Généraux
Food	89.1[17,22]	#96.6	116.0	128.2	133.7	136.5	137.5	Alimentation
Guam								**Guam**
General	39.1	90.6	116.3	154.0	164.1	164.0	165.3	Généraux
Food	58.3[17]	225.7[17]	#140.8[12]	177.1[12]	198.1[12]	203.8[12]	213.5[12]	Alimentation
Guatemala								**Guatemala**
General	...	69.3[56]	143.8	192.8	212.6	221.8	229.4	Généraux
Food	...	75.3[56]	160.5[57]	229.5[57]	272.2[57]	295.0[57]	317.0[57]	Alimentation
Guinea [58]								**Guinée** [58]
General	...	81.0	185.3	438.7	613.4	686.3	753.0	Généraux
Food	...	82.1	206.9	589.0	867.5	...	...	Alimentation
Guinea-Bissau [59]								**Guinée-Bissau** [59]
General [26]	...	...	104.3	122.2[8]	...	...	...	Généraux [26]
Food	...	39.2[15]	#104.7[15,26]	#99.4[28]	109.9[28]	110.8[28]	109.4[28]	Alimentation
Guyana [60]								**Guyana** [60]
General	51.5[38]	#75.6	128.3	177.3[8]	...	...	...	Généraux
Food	48.2[38]	#78.5[2]	121.7[2]	178.6[2,8]	...	...	...	Alimentation
Haiti								**Haïti**
General	80.8[11,17]	227.6[11,17]	#251.9[8]	374.9	431.9	457.2	478.1	Généraux
Food	84.8[11,17]	216.4[11,17]	#263.2[8,15]	397.5[15]	467.3[15]	497.5[15]	516.5[15]	Alimentation
Honduras								**Honduras**
General	12.3	47.6	149.5	207.8	233.3	245.4	260.4	Généraux
Food [3]	11.6	52.1	137.5	198.6	215.9	226.3	242.4	Alimentation [3]
Hungary								**Hongrie**
General	...	49.5	133.1	173.1	190.1	192.9	192.5	Généraux
Food	...	56.4	134.5	191.7	216.4	221.0	220.2	Alimentation
Iceland [61]								**Islande** [61]
General	...	87.0	122.3	182.3	199.5	207.2	211.4	Généraux
Food	...	85.2	106.7	161.9	178.3	187.4	188.4	Alimentation
India [62]								**Inde** [62]
General	28.0	69.4	121.5	182.7	217.4	241.1	256.4	Généraux
Food	27.8	73.2	115.0	193.8	227.9	259.6	276.8	Alimentation
Indonesia								**Indonésie**
General	20.1	44.1	156.0	227.2	249.6	267.0	282.5[8]	Généraux
Food	14.9	36.7	140.3	247.0	283.8	317.7	338.9[8]	Alimentation
Iran (Islamic Rep. of)								**Iran (Rép. islamique d')**
General	4.4	40.1	192.9	386.3	593.5	778.5[8]	912.7	Généraux
Food	4.8[2]	42.3[2]	186.3[2]	#164.8[41]	283.1[41]	403.1[8,41]	450.9[41]	Alimentation

Consumer price indices *(continued)*
General and food (Index base: 2000 = 100)

Indices des prix à la consommation *(suite)*
Généraux et alimentation (Indices base: 2000 = 100)

Country or area	1985	1995	2005	2010	2012	2013	2014	Pays ou zone
Iraq								**Iraq**
General	54.8[17]	...	#322.6	660.2[8]	739.4	753.3	770.1	Généraux
Food	46.8[17]	...	...	#330.1[8]	362.3	363.7	373.5	Alimentation
Ireland								**Irlande**
General	66.2	88.2	118.8	127.5	133.0[8]	133.7	133.9	Généraux
Food	66.6	86.3	111.2	113.8	115.8[8]	117.0	114.4	Alimentation
Isle of Man								**Île de Man**
General	55.9	88.1	117.5	139.8	152.4	156.6	160.8	Généraux
Food	46.9	82.5	131.0	179.0	205.0	216.2	228.4	Alimentation
Israel								**Israël**
General	13.8	73.5	108.5	123.6	130.1	132.1[8]	132.7	Généraux
Food	16.1	72.2	109.9	138.6	144.4	151.6[8]	150.6	Alimentation
Italy								**Italie**
General	59.7[2]	100.5[44]	112.4[44]	123.3[44]	131.2	132.8	133.1	Généraux
Food	57.4[12]	93.0[12]	113.7[46]	127.9[46]	134.3[46]	137.5[46]	137.7[46]	Alimentation
Jamaica								**Jamaïque**
General	5.9	57.9	165.6	296.4	340.6	372.5	403.3	Généraux
Food	5.6	63.8	161.4	316.8	378.1	425.3	462.0	Alimentation
Japan								**Japon**
General	86.1[8]	98.5[8]	97.8[8]	97.4	97.1	97.4	100.1	Généraux
Food	88.1[8]	99.4[8]	97.8[8]	101.0	100.7	100.6	104.4	Alimentation
Jersey [63]								**Jersey** [63]
General	...	82.9[64]	122.6[64]	142.4[64]	153.6	155.9	158.5	Généraux
Food	...	88.4[64]	114.4[64]	148.3[64]	162.0	165.4	167.1	Alimentation
Jordan								**Jordanie**
General	45.6	87.3	112.7	149.0	163.0	171.9	176.8	Généraux
Food	44.1	85.6	113.4[2]	167.7[2]	182.7[2]	189.5[2]	193.0[2]	Alimentation
Kazakhstan								**Kazakhstan**
General	...	46.6	140.3	...	...	...	...	Généraux
Food [15]	...	53.6	148.2	...	...	...	...	Alimentation [15]
Kenya								**Kenya**
General	68.8[17,34,65]	#72.4[34,65]	149.1[34,65]	#133.7[8,41]	166.7[41]	176.2[41]	188.4[41]	Généraux
Food	73.2[17,34,65]	#73.5[34,65]	164.8[34,65]	#150.4[8,41]	199.3[41]	213.8[41]	...	Alimentation
Kiribati [66]								**Kiribati** [66]
General	58.8	93.9	110.0	...	...	...	...	Généraux
Food	58.3	89.7	112.7	...	...	...	...	Alimentation
Kosovo [26]								**Kosovo** [26]
General	...	...	97.6	113.1	124.4	126.6	127.2	Généraux
Food	...	...	96.5	117.0	132.5	135.2	135.0	Alimentation
Kuwait								**Koweït**
General	68.1	91.3	108.8	141.5	152.5	152.8[8]	157.2	Généraux
Food [3]	72.4	88.9	119.4	162.1	187.6	185.0[8]	190.3	Alimentation [3]
Kyrgyzstan								**Kirghizistan**
General	...	34.4	122.2		...	...	...	Généraux
Food	...	30.7	118.3	...	...	...	...	Alimentation
Lao People's Dem. Rep.								**Rép. dém. pop. lao**
General	...	168.8[17,67]	#163.0[8]	207.7	233.0	247.8	258.0	Généraux
Food	...	149.9[17,67]	#160.2	231.2[8]	268.9	...	...	Alimentation
Latvia								**Lettonie**
General	...	71.3[8]	121.9	169.0	180.4	180.3	181.4	Généraux
Food	...	86.0[8]	#130.5	188.7	209.4	212.1	211.6	Alimentation
Lebanon								**Liban**
General	...	...	105.0[68]	#105.2[28]	117.7[28]	124.3[28]	125.7[8,28]	Généraux
Food [28]	...	...	...	107.0	120.6	123.7	128.1[8]	Alimentation [28]
Lesotho								**Lesotho**
General	19.3	68.0	140.1	197.8[8]	220.4	231.3	243.7	Généraux
Food	53.2[17]	#68.0[3,20]	#152.0[3]	#181.0[69]	212.7[69]	224.2[69]	236.1[69]	Alimentation
Lithuania								**Lituanie**
General	...	69.0	104.3[8]	134.3	144.2	145.7[8]	145.8	Généraux
Food	...	78.8	105.3[8]	145.8	162.6	165.4[8]	166.6	Alimentation
Luxembourg								**Luxembourg**
General	91.8[4,8,17]	#92.4[44]	112.0[8]	124.9	132.6	134.9	135.7	Généraux
Food	90.9[8,17]	#92.9[44]	114.8[8,46]	131.0[46]	137.8[46]	142.8[46]	143.6[46]	Alimentation
Madagascar								**Madagascar**
General	10.6[4,70]	61.2[4,70]	#152.6	263.5	306.8	324.7	344.4	Généraux
Food	10.7[70]	59.4[70]	#170.1[6]	257.2[6]	306.6[6]	321.7[6]	339.1[6]	Alimentation
Malawi								**Malawi**
General	...	27.3	198.5	309.1	403.4	513.4[8]	635.6[8]	Généraux
Food	...	29.1	181.0	271.2	332.5	409.9[8]	496.8[8]	Alimentation

Country or area	1985	1995	2005	2010	2012	2013	2014	Pays ou zone
Malaysia								**Malaisie**
General	64.1	85.7	109.1[8]	124.4	130.5	133.2	137.4	Généraux
Food	55.4	78.2	108.8[8]	134.5	144.8	149.9	155.0	Alimentation
Maldives[71]								**Maldives**[71]
General	...	87.2	107.1[8]	146.1	183.9[8]	191.3	196.0	Généraux
Food	...	81.0[15]	#100.0[14]	155.4[14]	219.4[8,14]	235.8[14]	238.2[14]	Alimentation
Mali[72]								**Mali**[72]
General	...	92.7	112.3	130.9[8]	141.9	141.1	142.3	Généraux
Food	...	134.6[17]	#115.1[15]	#107.3[28]	121.8[28]	117.8[28]	117.5[28]	Alimentation
Malta								**Malte**
General	70.3	88.7	112.8	126.8[8]	133.4	135.2	135.6	Généraux
Food	74.4	91.0[46]	112.0[46]	138.3[8,46]	150.4[46]	157.6[46]	158.4[46]	Alimentation
Marshall Islands[73]								**Îles Marshall**[73]
General	...	82.5[8]	106.9	...	...	...	...	Généraux
Food	...	81.6	106.3	...	...	...	...	Alimentation
Martinique								**Martinique**
General	70.1	95.0	111.2	121.4	126.3	127.9	128.9	Généraux
Food	73.1	94.3[2]	118.2	132.5	137.3	141.9	143.7	Alimentation
Mauritania								**Mauritanie**
General	...	78.7	139.3	185.1	205.2	213.6	...	Généraux
Food	...	76.2	149.3	210.5	233.5	243.8	...	Alimentation
Mauritius								**Maurice**
General	36.8	74.0	128.1	175.8	194.4	201.3[8]	207.8	Généraux
Food	37.1	77.5	129.5	205.8	222.6	230.2[8]	240.8	Alimentation
Mexico								**Mexique**
General	1.3	41.7	127.2	157.9	170.0	176.5	183.6	Généraux
Food[15]	1.5	41.5	129.4	174.0	196.9	207.1	217.5	Alimentation[15]
Mongolia								**Mongolie**
General	...	...	137.6[74]	#166.2[75]	...	...	...	Généraux
Food	...	...	139.7[15,74]	#180.7[75]	...	...	...	Alimentation
Montenegro[14]								**Monténégro**[14]
General	...	...	100.0	121.7	...	...	...	Généraux
Food	...	...	100.0	124.1	...	...	...	Alimentation
Morocco								**Maroc**
General	79.5[17]	#91.2	107.2	#108.4[75]	110.8[75]	112.9[75]	113.4[75]	Généraux
Food	82.6[17]	#142.5[2]	156.9[2]	#114.7[75]	119.1[75]	121.7[75]	120.2[75]	Alimentation
Mozambique								**Mozambique**
General	...	53.3[76]	#173.3	287.0	327.5	341.5	350.2	Généraux
Food	...	63.9[76]	#173.9	329.4[8]	...	...	...	Alimentation
Myanmar								**Myanmar**
General	42.3[17,77]	330.3[17,77]	#297.1	#155.7[75]	165.9[75]	...	...	Généraux
Food	40.2[17,77]	353.9[17,77]	#303.2	#153.7[75]	157.3[75]	...	...	Alimentation
Namibia								**Namibie**
General	...	67.5[78]	#114.1[69]	160.4[69]	179.5[69]	189.4[8,69]	199.6[69]	Généraux
Food	...	75.1[78]	#112.0[69]	179.0[69]	204.6[69]	217.9[8,69]	235.9[69]	Alimentation
Nepal								**Népal**
General	58.6[17]	169.6[17]	#123.3	192.8[8]	230.2	251.0	271.4	Généraux
Food	56.9[17]	171.1[17]	#120.3	211.3[8]	256.1	283.0	313.7	Alimentation
Netherlands								**Pays-Bas**
General	75.5	89.8	113.1	122.1	128.0	131.2	132.5	Généraux
Food	90.6[20]	100.0[2,20]	#92.9	101.8	106.1	108.7	108.6	Alimentation
New Caledonia[79]								**Nouvelle-Calédonie**[79]
General	73.5	94.0	107.6	119.3	124.4	126.0	126.2	Généraux
Food	74.8	91.1	109.7	124.8	133.0	135.7	136.3	Alimentation
New Zealand								**Nouvelle-Zélande**
General	53.8	93.1	113.0	129.9	136.6	138.1	139.8	Généraux
Food	56.6	91.2	112.2	138.1	144.8	145.4	146.1	Alimentation
Nicaragua								**Nicaragua**
General	...	58.5[11,80]	#147.4[81]	234.3[8,81]	271.4[81]	290.8[81]	308.3[81]	Généraux
Food	...	65.0[11,80]	#149.0[81]	263.8[8,81]	312.5[81]	341.9[81]	368.8[81]	Alimentation
Niger[35,82]								**Niger**[35,82]
General	71.2	87.3	113.5	128.2[8]	132.6	135.6	134.4	Généraux
Food	73.3	80.7	120.7[2]	#132.5[8,75]	141.5[75]	147.7[75]	145.9[75]	Alimentation
Nigeria[83]								**Nigéria**[83]
General	2.8	56.1	207.4	337.9[8]	420.2	455.9	492.6	Généraux
Food[3]	3.2	64.2	216.3	355.8[8]	436.7	479.1	524.5	Alimentation[3]
Niue								**Nioué**
General	57.2	92.1	118.7	...	...	...	...	Généraux
Food	62.2	93.6	122.0	...	...	...	...	Alimentation

Country or area	1985	1995	2005	2010	2012	2013	2014	Pays ou zone
Norfolk Island								**Île Norfolk**
General	...	85.6	125.2	156.4	...	...	...	Généraux
Food	...	88.7	129.9	173.7	...	...	...	Alimentation
Northern Mariana Is.[84]								**Îles Mariannes du Nord**[84]
General	...	93.2	99.8	...	...	...	...	Généraux
Food	...	100.3	93.9	...	...	...	...	Alimentation
Norway								**Norvège**
General	58.6	89.2	109.1	122.0	124.5	127.2	129.7	Généraux
Food	75.4[17]	105.0[17]	#103.1	116.8	118.1	119.4	123.4	Alimentation
Oman								**Oman**
General	...	101.2[8,85]	#101.7	133.7	143.2	145.0	146.2[8]	Généraux
Food	...	98.6[2,8,85]	#105.7[2]	154.4[2]	164.9[2]	168.3[2]	#104.5[86]	Alimentation
Other non-specified areas								**Autres zones non-spécifiées**
General	...	93.2	103.4	109.7	113.5	114.4[8]	115.7	Généraux
Food	...	...	110.7	123.0	131.0	132.7[8]	137.7	Alimentation
Pakistan								**Pakistan**
General	29.9	70.5	129.5	234.0	273.8	295.1	316.3	Généraux
Food	28.4	72.0	132.1	268.0	333.6	361.8	385.4	Alimentation
Panama								**Panama**
General	98.3[17,87]	#94.3[87]	#103.3[22,26]	127.1[22,26]	142.2[22,26]	148.0[22,26]	...	Généraux
Food	97.8[17,87]	#97.6[87]	#105.6[22,26]	142.8[22,26]	163.5[22,26]	172.9[22,26]	...	Alimentation
Papua New Guinea								**Papouasie-Nvl-Guinée**
General	31.5	57.1	145.7	189.0	209.5	217.0	...	Généraux
Food	32.2	54.8	151.2	211.2	...	...	...	Alimentation
Paraguay[88]								**Paraguay**[88]
General	29.1[17]	#65.6	149.5	211.5	237.3	243.7	255.9	Généraux
Food	25.8[17]	#72.8	156.2	268.6	298.7	308.3	328.2	Alimentation
Peru[11,89]								**Pérou**[11,89]
General	<0.0	71.7	110.1	126.4[8]	135.4	139.2	143.7	Généraux
Food	0.0[17]	#76.5	107.6	131.8[8]	146.0	150.8	156.3	Alimentation
Philippines								**Philippines**
General	68.6[17]	#70.9	129.8	166.1	179.5	184.7	192.4	Généraux
Food	68.4[17]	#74.9[15]	123.8[15]	166.1[15]	#139.9[75]	143.8[75]	153.4[75]	Alimentation
Poland								**Pologne**
General	...	556.7[17]	#114.6	131.2	142.0	143.3	143.3	Généraux
Food[12]	...	461.6[17]	#110.6	131.1	142.8	145.4	144.7	Alimentation[12]
Portugal								**Portugal**
General[4]	36.3	87.8	116.7	126.9	135.2	135.5[8]	134.9	Généraux[4]
Food	40.4	89.9	111.3	116.7	122.9	125.2[8]	123.6	Alimentation
Puerto Rico								**Porto Rico**
General	55.6	75.9	110.6	131.1	136.8	138.2	139.0	Généraux
Food	34.8	60.5	109.5[12]	134.2[12]	140.2[12]	142.3[12]	145.3[12]	Alimentation
Qatar[69]								**Qatar**[69]
General	...	...	119.0	161.7	167.8	173.1	178.3	Généraux
Food[2]	...	...	106.5	151.5	164.1	167.9	168.5	Alimentation[2]
Republic of Korea								**République de Corée**
General	46.8[8]	82.3[8]	#100.0[14]	116.1[14]	123.4[14]	124.8[14]	126.6[14]	Généraux
Food	42.9[8]	82.2[8]	#100.0[14]	123.8[14]	139.2[14]	139.9[14]	140.8[14]	Alimentation
Republic of Moldova								**République de Moldova**
General	...	36.8	#99.5[14]	152.5[14]	171.7[14]	179.7[14]	188.8[14]	Généraux
Food	...	41.5	#100.0[14]	140.1[14]	158.3[14]	168.9[14]	179.8[14]	Alimentation
Réunion								**Réunion**
General	70.4[22]	93.9	110.4	120.6	124.7	126.4	126.7	Généraux
Food	78.3[22]	94.3	108.8	125.2	132.0	136.4	137.6	Alimentation
Romania[19]								**Roumanie**[19]
General	...	8.4	231.7	312.7	341.8	355.4	359.2	Généraux
Food	...	10.7	213.8	266.2	287.5	296.0	291.2	Alimentation
Russian Federation								**Fédération de Russie**
General	...	20.6	199.7	325.2	370.7	395.7	...	Généraux
Food	...	22.2	189.7	327.8	377.7[12]	...	...	Alimentation
Rwanda								**Rwanda**
General	23.6[90]	77.2[90,91]	138.3[90]	#100.0[48]	113.7[48]	120.4[48]	123.3[8,48]	Généraux
Food	25.8[90]	82.6[90,91]	156.1[90]	#100.0[48]	118.8[48]	129.0[48]	132.2[8,48]	Alimentation
Saint Helena								**Sainte-Hélène**
General	62.0	89.0	115.8	154.6[8]	173.0	175.8	179.5	Généraux
Food	68.0	95.5	112.7	139.5[8]	151.1	153.4	156.4	Alimentation
Saint Kitts and Nevis								**Saint-Kitts-et-Nevis**
General	65.0[92]	82.4[92]	#111.0[5]	135.9[5]	147.1[5]	147.9[5]	149.4[5]	Généraux
Food	63.8[92]	82.0[92]	#109.3[5]	141.5[5]	159.9[5]	164.6[5]	165.7[5]	Alimentation

15 Consumer price indices *(continued)*
General and food (Index base: 2000 = 100)
Indices des prix à la consommation *(suite)*
Généraux et alimentation (Indices base: 2000 = 100)

Country or area	1985	1995	2005	2010	2012	2013	2014	Pays ou zone
Saint Lucia								**Sainte-Lucie**
General	61.0	89.8	111.9	128.1	137.2	139.2	144.1	Généraux
Food	59.7	94.8	112.1	128.6	140.8	147.9	151.0	Alimentation
Saint Pierre and Miquelon								**Saint-Pierre-et-Miquelon**
General	87.8[17]	...	#114.0	...	...	...	...	Généraux
Food	89.0[17]	...	#109.7	...	...	...	...	Alimentation
Saint Vincent-Grenadines [93]								**Saint-Vincent-Grenadines** [93]
General	...	92.2	109.0	134.8	154.0	155.3	155.6	Généraux
Food	...	97.1	112.1	148.1	158.0	161.1	163.7	Alimentation
Samoa								**Samoa**
General[4]	49.1	93.9	133.0	174.3	187.8	188.9	...	Généraux [4]
Food	55.4	104.5	146.7	...	223.2	224.7	...	Alimentation
San Marino								**Saint-Marin**
General	49.3	86.4	#103.1[26]	118.0[26]	124.9[26]	126.9[26]	128.3[26]	Généraux
Food	53.9	89.0	#108.9[26]	138.0[26]	151.8[26]	160.2[26]	164.7[26]	Alimentation
Sao Tome and Principe [94]								**Sao Tomé-et-Principe** [94]
General	...	...	561.0	1 432.1	1 811.4	1 958.2	2 095.0	Généraux
Food[6]	...	...	484.0	1 442.9	1 888.5	2 059.0	2 203.7	Alimentation [6]
Saudi Arabia [95]								**Arabie saoudite** [95]
General	92.3	101.7	100.2	129.7	142.3	135.1[8]	138.7	Généraux
Food	87.6	98.9[2]	107.5[2]	149.8[2]	164.5[2]	170.3[2,8]	175.8[2]	Alimentation
Senegal								**Sénégal**
General	66.6[96]	93.3[96]	107.7[96]	#99.2[28]	104.3[28]	105.3[28]	103.8[28]	Généraux
Food	67.1[96]	95.0[96]	#100.0[14,96]	#99.5[28]	106.8[28]	107.6[28]	108.5[28]	Alimentation
Serbia								**Serbie**
General	...	...	330.0	#137.4[75]	...	...	...	Généraux
Food	...	...	280.9	#142.7[75]	...	...	...	Alimentation
Serbia and Montenegro [fmr.] [19]								**Serbie-et-Monténégro [anc.]** [19]
General	...	12.3	...	...	...		...	Généraux
Food	...	11.7	...	...	...		...	Alimentation
Seychelles								**Seychelles**
General	90.5[17]	#86.6	115.0	215.1	236.3	246.6	250.0	Généraux
Food	88.7[17]	#97.8	110.3	236.4	261.6	280.0	281.9	Alimentation
Sierra Leone								**Sierra Leone**
General	...	39.2[97]	#131.8[26]	238.8[8,26]	312.8[26]	345.4[26]	370.2[26]	Généraux
Food	...	43.4[97]	#137.6[26]	243.1[8,26]	343.7[26]	380.9[26]	408.0[26]	Alimentation
Singapore								**Singapour**
General	78.9	95.5	103.2	117.4	129.2	132.3	...	Généraux
Food	84.2	94.3	104.6	122.0	128.7	131.4	...	Alimentation
Slovakia								**Slovaquie**
General	...	67.5	132.8	152.9	164.7	167.0	166.9	Généraux
Food	...	78.8	114.7	128.1	141.6	146.5	145.4	Alimentation
Slovenia								**Slovénie**
General	...	67.4	130.6	150.4	157.1	159.9	160.2	Généraux
Food	...	70.9	124.3	153.1	166.5	172.4	171.9	Alimentation
Solomon Islands [98]								**Îles Salomon** [98]
General	20.5[8]	63.7	149.3	225.9	257.0	270.9	...	Généraux
Food	19.6[8]	59.1	145.1	227.5	248.9	255.4	...	Alimentation
South Africa								**Afrique du Sud**
General	20.8[8]	72.4	128.0	141.5	157.0	166.0[8]	176.1	Généraux
Food	16.5[3,8]	71.8[3]	137.9[3]	#111.0[28]	127.4[28]	134.7[8,28]	145.0[28]	Alimentation
Spain								**Espagne**
General	49.9	87.9	#113.6[5]	127.7[5]	135.1[5]	137.0[5]	136.8[5]	Généraux
Food	72.4[2,17]	#93.2	#116.7[5]	130.9[5]	136.6[5]	140.4[5]	140.0[5]	Alimentation
Sri Lanka [99]								**Sri Lanka** [99]
General	22.1	64.8	159.7	#219.1[69]	252.0[69]	269.4[69]	278.3[69]	Généraux
Food	21.3	62.8	163.0	#234.2[69]	268.0[69]	289.1[69]	300.0[69]	Alimentation
State of Palestine								**État de Palestine**
General	...	...	119.1	148.4	157.0	159.7	162.4[8]	Généraux
Food	...	...	113.3	157.8	165.0	166.4	167.0[8]	Alimentation
Suriname [5,100]								**Suriname** [5,100]
General	0.1	18.0	171.4	248.6	307.2	313.1	323.7	Généraux
Food	0.1	24.2	174.2	273.1	333.2	340.7	355.1	Alimentation
Swaziland								**Swaziland**
General	20.7[4,34,101]	#69.1	139.9	#126.9[8,41]	146.6[41]	154.9[41]	163.7[41]	Généraux
Food	17.8[34,101]	#69.2	169.2	...	...	...	...	Alimentation
Sweden								**Suède**
General	74.1[17]	#97.7	107.5	116.0	120.5	120.4	120.2	Généraux
Food	75.6[17]	#104.3	105.3	120.9	124.2	127.0	127.4	Alimentation

Country or area	1985	1995	2005	2010	2012	2013	2014	Pays ou zone
Switzerland								**Suisse**
General	72.9	96.4	104.3	109.0	108.5	108.2	108.2	Généraux
Food	81.1	97.6	105.5	107.9	103.3	104.6	105.6	Alimentation
Syrian Arab Republic								**Rép. arabe syrienne**
General	14.8[102]	#92.9	121.9	173.2	249.4	472.8	...	Généraux
Food	14.8[102]	#97.5	122.5	190.7[3]	283.9[3]	587.3[3]	...	Alimentation
Thailand								**Thaïlande**
General	53.1	81.3	111.8	129.2	138.2[8]	141.2	143.9	Généraux
Food	47.9	80.0	114.8	153.4	173.7[8]	179.6	186.6	Alimentation
TFYR of Macedonia								**ex-R.Y. de Macédoine**
General	...	90.8[19]	108.8[19]	125.3[8]		...	...	Généraux
Food	...	98.2[19]	102.7[19]	125.0[8]		...	...	Alimentation
Togo [103]								**Togo [103]**
General	51.2	88.2	113.7	134.2[8]	142.6	145.1	145.4	Généraux
Food	100.8[17]	#100.0[20]	#114.0	152.0[8]	159.1	160.0	153.0	Alimentation
Tonga								**Tonga**
General [4]	#42.0	82.8	160.4	209.7	225.4	227.1	232.8	Généraux [4]
Food	#45.2	80.4	165.5	218.7	236.5	240.9	249.7	Alimentation
Trinidad and Tobago								**Trinité-et-Tobago**
General	37.0	82.6	126.5	195.9	225.0	236.7	250.1	Généraux
Food	15.2	61.0	198.4	497.1	654.0	711.2	782.5	Alimentation
Tunisia								**Tunisie**
General	46.6	85.3	113.8	139.1[8]	152.0	161.2	170.0	Généraux
Food [12]	45.2	84.2	115.2	147.2[8]	164.1	177.2	188.9	Alimentation [12]
Turkey								**Turquie**
General	11.0[17]	1 872.3[17]	#380.5[8]	577.9	670.0	720.2	783.9	Généraux
Food	11.5[17]	1 938.2[2,17]	#112.1[26]	186.2[26]	214.5[26]	234.0[26]	263.5[26]	Alimentation
Tuvalu [104]								**Tuvalu [104]**
General	61.1	90.3	117.0	...	...	...	...	Généraux
Food	66.3	94.3	127.4	...	...	...	...	Alimentation
Uganda								**Ouganda**
General	...	78.5	124.1	186.3	252.0	265.8	277.2	Généraux
Food	...	75.1	126.1	218.4	313.7	324.7	339.6	Alimentation
Ukraine								**Ukraine**
General	...	27.5	146.9	286.8[8]	311.7	310.9	...	Généraux
Food	...	29.4[15]	157.5[15]	122.7[8]	127.7	125.0	...	Alimentation
United Kingdom								**Royaume-Uni**
General	55.5	87.6	112.7	131.3	142.5	146.9	150.3	Généraux
Food	66.2	95.5	107.3	136.0	148.8	154.3	154.3	Alimentation
United Rep. of Tanzania [105]								**Rép.-Unie de Tanzanie [105]**
General	4.1	55.2	128.3	192.0[8]	251.0	270.8	287.4	Généraux
Food	3.7	53.1	138.6	#100.0[48]	139.8[48]	151.7[48]	163.0[48]	Alimentation
United States [106]								**États-Unis [106]**
General	62.5[8]	88.5	113.4	126.6	133.3	135.3	137.5	Généraux
Food	93.1[12,107]	#88.4	113.6	130.9	139.3	141.3	144.7	Alimentation
Uruguay [108]								**Uruguay [108]**
General	20.0	#3 872.8	162.3	230.0	269.9	293.4	319.9	Généraux
Food	0.3	56.5	165.7	261.2	313.8	347.1	382.2	Alimentation
Vanuatu								**Vanuatu**
General	67.4[17,34]	#89.2	111.7	133.1	136.1	138.1	139.2	Généraux
Food	63.5[17,34]	#94.0	108.1	140.5	144.6	146.7	149.2	Alimentation
Venezuela (Boliv. Rep. of)								**Venezuela (Rép. boliv. du)**
General	0.6[8,11,109]	17.1[11,109]	255.0[11,109]	#156.2[28]	238.5[28]	335.4[28]	543.9[28]	Généraux
Food	0.4[8,11,109]	20.9[11,109]	332.5[11,109]	#170.1[28]	275.8[28]	425.2[28]	794.0[28]	Alimentation
Viet Nam								**Viet Nam**
General	...	82.9	125.5[8]	207.8[8]	269.1	286.8	298.5	Généraux
Food	...	84.6[6]	#77.3	151.8[8]	...	...	...	Alimentation
Yemen								**Yémen**
General	...	...	174.4	292.0[8]	383.5	425.6	...	Généraux
Food	...	...	57.9	110.4[8]	140.1	155.9	...	Alimentation
Zambia								**Zambie**
General	<0.0[34]	28.2	251.4	419.6	474.6[8]	507.7	547.3	Généraux
Food	0.1[34]	30.9	254.5	384.3	#119.2[30]	126.9[30]	136.1[30]	Alimentation
Zanzibar [17]								**Zanzibar [17]**
General	20.6	305.6	612.3	...	...	...	...	Généraux
Food	21.7	324.2	590.3	...	...	...	...	Alimentation
Zimbabwe								**Zimbabwe**
General	...	21.3	#0.1[110]	#103.1[30]	110.6[30]	112.4[8,30]	112.2[30]	Généraux
Food	...	19.2	#0.1[110]	#104.0[30]	113.1[30]	115.0[8,30]	111.4[30]	Alimentation

Source:
International Labour Organization (ILO), Geneva, the LABORSTA database, last accessed April 2016.

Source:
Bureau international du Travail (BIT), Genève, LABORSTA base de données du BIT, dernier accès avril 2016.

1	Index base: 1992=100.
2	Including tobacco.
3	Food only.
4	Excluding `Rent´.
5	Index base: 2001=100.
6	Including beverages and tobacco.
7	Luanda
8	Series linked to former series.
9	January-November.
10	Buenos Aires
11	Metropolitan areas.
12	Including alcoholic beverages.
13	New Providence
14	Index base: 2005=100.
15	Including alcoholic beverages and tobacco.
16	June-December.
17	Index base: 1990=100.
18	Government officials.
19	Annual average is weighted mean of monthly data.
20	Index base: 1995=100.
21	Cotonou
22	Urban areas.
23	Ouagadougou
24	Bujumbura
25	Praia
26	Index base: 2003=100.
27	Phnom Penh
28	Index base: 2008=100.
29	Bangui
30	Index base: 2009=100.
31	N'Djamena
32	Santiago
33	Index base period: the same month of 2000=100.
34	Low income group.
35	African population.
36	Brazzaville
37	Rarotonga
38	Index base: 1988=100.
39	Excluding beverages.
40	Central area.
41	Index base: 2007=100.
42	Abidjan
43	Index base: 1991=100.
44	Excluding tobacco.
45	Including tobacco, beverages and public catering.
46	Excluding alcoholic beverages and tobacco.
47	Index base: 2004=100.
48	Index base: 2010=100.
49	Malabo
50	Addis Ababa
51	Stanley
52	Cayenne
53	Libreville
54	Banjul, Kombo St. Mary
55	Five cities.

1	Indices base: 1992=100.
2	Y compris le tabac.
3	Alimentation seulement.
4	Non compris le groupe "Loyer".
5	Indices base: 2001=100.
6	Y compris les boissons et le tabac.
7	Luanda
8	Série enchaînée à la précédente.
9	Janvier-novembre.
10	Buenos Aires
11	Région métropolitaines.
12	Y compris les boissons alcoolisées.
13	Nouvelle Providence
14	Indices base: 2005=100.
15	Y compris les boissons alcoolisées et le tabac.
16	Juin-décembre.
17	Indices base: 1990=100.
18	Fonctionnaires.
19	La moyenne annuelle est la moyenne pondérée des données mensuelles.
20	Indices base: 1995=100.
21	Cotonou
22	Zones urbaines.
23	Ouagadougou
24	Bujumbura
25	Praia
26	Indices base: 2003=100.
27	Phnom Penh
28	Indices base: 2008=100.
29	Bangui
30	Indices base: 2009=100.
31	N'Djamena
32	Santiago
33	Indices base: le même mois de 2000=100.
34	Groupe à faible revenu.
35	Population Africaine.
36	Brazzaville
37	Rarotonga
38	Indices base: 1988=100.
39	Non compris les boissons.
40	Région centrale.
41	Indices base: 2007=100.
42	Abidjan
43	Indices base: 1991=100.
44	Non compris le tabac.
45	Y compris le tabac, les boissons et la restauration.
46	Non compris les boissons alcoolisées et le tabac.
47	Indices base: 2004=100.
48	Indices base: 2010=100.
49	Malabo
50	Addis Ababa
51	Stanley
52	Cayenne
53	Libreville
54	Banjul, Kombo St. Mary
55	Cinq villes.

56	Guatemala city.	56	Ville de Guatemala.
57	Including meals outside home.	57	Y compris les repas pris à l'extérieur.
58	Conakry	58	Conakry
59	Bissau	59	Bissau
60	Georgetown	60	Georgetown
61	Annual averages are based on the months February-December and January of the following year.	61	Les moyennes annuelles sont basées sur les mois de février-décembre et de janvier de l'année suivante.
62	Industrial workers.	62	Ouvriers industriels.
63	Index base: June 2000=100.	63	Indices base: 2000 juin=100.
64	June.	64	juin.
65	Nairobi	65	Nairobi
66	Tarawa	66	Tarawa
67	Vientiane	67	Vientiane
68	Beirut	68	Beyrouth
69	Index base: 2002=100.	69	Indices base: 2002=100.
70	Antananarivo	70	Antananarivo
71	Male	71	Male
72	Bamako	72	Bamako
73	Majuro	73	Majuro
74	Ulan Bator	74	Ulan Bator
75	Index base: 2006=100.	75	Indices base: 2006=100.
76	Maputo	76	Maputo
77	Yangon	77	Yangon
78	windhoek	78	windhoek
79	Nouméa	79	Nouméa
80	Managua	80	Managua
81	Index base: 1999=100.	81	Indices base: 1999=100.
82	Niamey	82	Niamey
83	Rural and urban areas.	83	Régions rurales et urbaines.
84	Saipan	84	Saipan
85	Muscat	85	Muscat
86	Index base: 2012=100.	86	Indices base: 2012=100.
87	Panamá	87	Panama
88	Asunción	88	Asunción
89	Lima	89	Lima
90	Kigali	90	Kigali
91	April-December.	91	Avril-décembre.
92	Saint Kitts	92	Saint Kitts
93	Saint Vincent	93	Saint Vincent
94	Index base: 1996=100.	94	Indices base: 1996=100.
95	All cities.	95	Ensemble des villes.
96	Dakar	96	Dakar
97	Freetown	97	Freetown
98	Honiara	98	Honiara
99	Colombo	99	Colombo
100	Paramaribo	100	Paramaribo
101	Mbabane-Manzini	101	Mbabane-Manzini
102	Damas	102	Damas
103	Lomé	103	Lomé
104	Funafuti	104	Funafuti
105	Tanzania mainland only.	105	Tanzanie continentale seulement.
106	All urban consumers.	106	Tous les consommateurs urbains.
107	Index base: 1987=100.	107	Indices base: 1987=100.
108	Montevideo	108	Montevideo
109	Caracas	109	Caracas
110	Due to lack of space, multiply each figure by 1,000. Annual average is calculated as geometric mean of monthly indices. Index base: 2005=100.	110	En raison du manque de place, multiplier chaque chiffre par 1000. La moyenne annuelle est calculée en tant que moyenne géométrique des indices mensuels. Indices base: 2005=100.

Agricultural production indices
Index base: 2004 - 2006 = 100

Indices de la production agricole
Indices base: 2004 - 2006 = 100

Région, pays ou zone	1975	1985	1995	2005	2010	2011	2012	2013	Région, pays ou zone
Total, all countries or areas									**Total, tous pays ou zones**
Agriculture (gross)	**51**	**66**	**78**	**100**	**113**	**117**	**118**	**122**	**Agriculture (brut)**
Food (gross)	**51**	**65**	**78**	**100**	**113**	**117**	**118**	**122**	**Prod. ailmentaires (brut)**
Africa									**Afrique**
Agriculture (gross)	43	51	69	100	117	118	123	127	Agriculture (brut)
Food (gross)	42	50	68	100	118	119	124	128	Produits ailmentaires (brut)
Americas									**Amériques**
Agriculture (gross)	52	67	78	100	111	112	112	118	Agriculture (brut)
Food (gross)	51	66	78	100	112	113	112	120	Produits ailmentaires (brut)
Asia									**Asie**
Agriculture (gross)	34	49	73	100	119	124	127	129	Agriculture (brut)
Food (gross)	34	48	73	100	119	124	128	130	Produits ailmentaires (brut)
Europe									**Europe**
Agriculture (gross)	102	116	97	99	99	105	100	105	Agriculture (brut)
Food (gross)	100	114	97	99	99	105	100	106	Produits ailmentaires (brut)
Oceania									**Océanie**
Agriculture (gross)	60	71	85	105	102	108	115	114	Agriculture (brut)
Food (gross)	57	67	83	105	103	108	115	114	Produits ailmentaires (brut)
Afghanistan									**Afghanistan**
Agriculture (gross)	82	70	84	106	116	111	123	120	Agriculture (brut)
Food (gross)	79	69	82	106	116	111	123	120	Produits ailmentaires (brut)
Albania									**Albanie**
Agriculture (gross)	51	65	86	98	119	125	132	127	Agriculture (brut)
Food (gross)	48	61	85	98	119	125	132	128	Produits ailmentaires (brut)
Algeria									**Algérie**
Agriculture (gross)	40	50	65	99	125	135	144	158	Agriculture (brut)
Food (gross)	40	50	65	99	125	135	144	159	Produits ailmentaires (brut)
American Samoa									**Samoa américaines**
Agriculture (gross)	62	48	50	107	106	96	98	106	Agriculture (brut)
Food (gross)	62	48	50	107	106	96	98	106	Produits ailmentaires (brut)
Angola									**Angola**
Agriculture (gross)	48	37	46	102	172	189	157	212	Agriculture (brut)
Food (gross)	36	35	45	102	173	190	158	213	Produits ailmentaires (brut)
Antigua and Barbuda									**Antigua-et-Barbuda**
Agriculture (gross)	62	102	104	95	89	92	89	89	Agriculture (brut)
Food (gross)	61	101	104	95	89	92	89	89	Produits ailmentaires (brut)
Argentina									**Argentine**
Agriculture (gross)	51	63	74	103	115	117	106	120	Agriculture (brut)
Food (gross)	50	62	73	103	115	116	106	120	Produits ailmentaires (brut)
Armenia									**Arménie**
Agriculture (gross)	...	...	73	103	102	113	123	128	Agriculture (brut)
Food (gross)	...	...	73	103	102	113	122	128	Produits ailmentaires (brut)
Australia									**Australie**
Agriculture (gross)	61	69	87	108	100	108	117	116	Agriculture (brut)
Food (gross)	59	66	85	108	102	107	117	115	Produits ailmentaires (brut)
Austria									**Autriche**
Agriculture (gross)	86	94	93	100	103	108	101	101	Agriculture (brut)
Food (gross)	86	94	93	100	103	108	101	101	Produits ailmentaires (brut)
Azerbaijan									**Azerbaïdjan**
Agriculture (gross)	...	...	60	104	118	125	132	137	Agriculture (brut)
Food (gross)	...	...	56	103	122	129	137	142	Produits ailmentaires (brut)
Bahamas									**Bahamas**
Agriculture (gross)	71	67	80	99	126	129	132	134	Agriculture (brut)
Food (gross)	71	67	80	99	126	129	132	134	Produits ailmentaires (brut)
Bahrain									**Bahreïn**
Agriculture (gross)	96	115	119	92	115	166	169	207	Agriculture (brut)
Food (gross)	96	115	119	92	115	166	169	207	Produits ailmentaires (brut)
Bangladesh									**Bangladesh**
Agriculture (gross)	49	58	67	103	129	133	134	136	Agriculture (brut)
Food (gross)	48	56	67	103	130	132	133	136	Produits ailmentaires (brut)
Barbados									**Barbade**
Agriculture (gross)	99	106	97	106	92	93	97	97	Agriculture (brut)
Food (gross)	99	106	97	106	92	93	97	97	Produits ailmentaires (brut)

Région, pays ou zone	1975	1985	1995	2005	2010	2011	2012	2013	Région, pays ou zone
Belarus									**Bélarus**
Agriculture (gross)	...	...	91	98	117	114	121	116	Agriculture (brut)
Food (gross)	...	...	90	98	117	114	121	116	Produits ailmentaires (brut)
Belgium									**Belgique**
Agriculture (gross)	...	...	...	100	100	102	97	101	Agriculture (brut)
Food (gross)	...	...	...	100	100	102	97	101	Produits ailmentaires (brut)
Belize									**Belize**
Agriculture (gross)	28	39	65	97	93	87	103	105	Agriculture (brut)
Food (gross)	28	39	65	97	93	87	103	105	Produits ailmentaires (brut)
Benin									**Bénin**
Agriculture (gross)	25	40	67	102	115	126	132	145	Agriculture (brut)
Food (gross)	26	41	63	102	122	131	138	150	Produits ailmentaires (brut)
Bermuda									**Bermudes**
Agriculture (gross)	90	97	104	98	116	116	120	120	Agriculture (brut)
Food (gross)	90	97	104	98	116	116	120	120	Produits ailmentaires (brut)
Bhutan									**Bhoutan**
Agriculture (gross)	46	62	73	106	95	110	100	97	Agriculture (brut)
Food (gross)	46	62	73	106	94	110	100	97	Produits ailmentaires (brut)
Bolivia (Plurinational State of)									**Bolivie (État plurinational de)**
Agriculture (gross)	41	50	71	100	120	122	129	134	Agriculture (brut)
Food (gross)	40	49	71	100	120	122	130	134	Produits ailmentaires (brut)
Bosnia and Herzegovina									**Bosnie-Herzégovine**
Agriculture (gross)	...	...	64	98	107	109	98	113	Agriculture (brut)
Food (gross)	...	...	64	98	108	110	99	114	Produits ailmentaires (brut)
Botswana									**Botswana**
Agriculture (gross)	82	93	110	101	124	142	130	130	Agriculture (brut)
Food (gross)	82	92	110	101	124	143	130	130	Produits ailmentaires (brut)
Brazil									**Brésil**
Agriculture (gross)	31	49	67	99	122	128	126	135	Agriculture (brut)
Food (gross)	30	48	68	99	123	128	127	136	Produits ailmentaires (brut)
British Virgin Islands									**Îles Vierges britanniques**
Agriculture (gross)	76	94	91	100	102	103	104	104	Agriculture (brut)
Food (gross)	76	94	91	100	102	103	104	104	Produits ailmentaires (brut)
Brunei Darussalam									**Brunéi Darussalam**
Agriculture (gross)	32	53	46	75	139	148	153	167	Agriculture (brut)
Food (gross)	31	53	45	75	140	149	154	167	Produits ailmentaires (brut)
Bulgaria									**Bulgarie**
Agriculture (gross)	168	171	131	91	106	107	99	116	Agriculture (brut)
Food (gross)	162	168	134	91	107	108	101	118	Produits ailmentaires (brut)
Burkina Faso									**Burkina Faso**
Agriculture (gross)	27	39	59	103	116	105	122	123	Agriculture (brut)
Food (gross)	30	43	65	103	123	113	129	124	Produits ailmentaires (brut)
Burundi									**Burundi**
Agriculture (gross)	71	81	83	92	108	102	95	129	Agriculture (brut)
Food (gross)	82	91	95	100	114	112	99	145	Produits ailmentaires (brut)
Cabo Verde									**Cabo Verde**
Agriculture (gross)	22	35	68	98	113	89	91	90	Agriculture (brut)
Food (gross)	22	35	68	98	113	89	91	90	Produits ailmentaires (brut)
Cambodia									**Cambodge**
Agriculture (gross)	26	37	63	105	148	170	175	177	Agriculture (brut)
Food (gross)	25	36	61	105	148	170	175	177	Produits ailmentaires (brut)
Cameroon									**Cameroun**
Agriculture (gross)	46	53	69	103	139	143	150	154	Agriculture (brut)
Food (gross)	45	51	66	102	143	148	155	159	Produits ailmentaires (brut)
Canada									**Canada**
Agriculture (gross)	54	68	81	102	102	102	105	115	Agriculture (brut)
Food (gross)	53	68	81	102	103	102	106	115	Produits ailmentaires (brut)
Cayman Islands									**Îles Caïmanes**
Agriculture (gross)	236	237	139	94	105	110	113	113	Agriculture (brut)
Food (gross)	236	237	139	94	105	110	113	113	Produits ailmentaires (brut)
Central African Rep.									**Rép. centrafricaine**
Agriculture (gross)	50	56	76	99	114	119	121	123	Agriculture (brut)
Food (gross)	46	52	73	99	113	118	119	121	Produits ailmentaires (brut)
Chad									**Tchad**
Agriculture (gross)	46	50	69	104	104	97	120	115	Agriculture (brut)
Food (gross)	42	50	68	104	110	101	124	120	Produits ailmentaires (brut)

Région, pays ou zone	1975	1985	1995	2005	2010	2011	2012	2013	Région, pays ou zone
Chile									**Chili**
Agriculture (gross)	36	44	77	100	109	115	114	117	Agriculture (brut)
Food (gross)	35	43	77	100	109	115	114	117	Produits ailmentaires (brut)
China [1]									**Chine** [1]
Agriculture (gross)	26	40	67	100	120	124	129	131	Agriculture (brut)
Food (gross)	25	39	67	100	120	125	129	132	Produits ailmentaires (brut)
China, Hong Kong SAR									**Chine, Hong Kong RAS**
Agriculture (gross)	164	223	84	99	56	57	59	60	Agriculture (brut)
Food (gross)	164	223	84	99	56	57	59	60	Produits ailmentaires (brut)
China, Macao SAR									**Chine, Macao RAS**
Agriculture (gross)	59	66	68	100	95	81	86	88	Agriculture (brut)
Food (gross)	59	66	68	100	95	81	86	88	Produits ailmentaires (brut)
Colombia									**Colombie**
Agriculture (gross)	49	58	82	99	101	103	107	114	Agriculture (brut)
Food (gross)	45	55	79	99	103	105	110	115	Produits ailmentaires (brut)
Comoros									**Comores**
Agriculture (gross)	57	63	88	96	114	110	112	114	Agriculture (brut)
Food (gross)	57	63	88	96	114	110	112	113	Produits ailmentaires (brut)
Congo									**Congo**
Agriculture (gross)	48	61	72	100	123	125	132	136	Agriculture (brut)
Food (gross)	48	61	72	100	123	126	132	136	Produits ailmentaires (brut)
Cook Islands									**Îles Cook**
Agriculture (gross)	349	311	158	100	104	96	97	97	Agriculture (brut)
Food (gross)	348	310	157	100	104	96	97	97	Produits ailmentaires (brut)
Costa Rica									**Costa Rica**
Agriculture (gross)	39	47	77	98	113	117	121	122	Agriculture (brut)
Food (gross)	37	43	75	98	115	119	122	125	Produits ailmentaires (brut)
Côte d'Ivoire									**Côte d'Ivoire**
Agriculture (gross)	41	57	79	100	107	111	120	123	Agriculture (brut)
Food (gross)	39	54	80	98	109	114	122	124	Produits ailmentaires (brut)
Croatia									**Croatie**
Agriculture (gross)	...	...	97	98	100	100	88	96	Agriculture (brut)
Food (gross)	...	...	98	98	100	100	88	96	Produits ailmentaires (brut)
Cuba									**Cuba**
Agriculture (gross)	95	130	80	97	88	96	96	100	Agriculture (brut)
Food (gross)	94	129	80	97	88	96	97	101	Produits ailmentaires (brut)
Cyprus									**Chypre**
Agriculture (gross)	65	90	111	98	83	87	82	81	Agriculture (brut)
Food (gross)	65	90	111	98	84	87	82	81	Produits ailmentaires (brut)
Czech Republic									**République tchèque**
Agriculture (gross)	...	...	108	100	91	96	89	93	Agriculture (brut)
Food (gross)	...	...	108	100	91	97	89	93	Produits ailmentaires (brut)
Dem. P. R. Korea									**R. p. dém. de Corée**
Agriculture (gross)	66	87	75	101	98	99	101	103	Agriculture (brut)
Food (gross)	66	87	74	101	98	98	101	102	Produits ailmentaires (brut)
Dem. Rep. of the Congo									**Rép. dém. du Congo**
Agriculture (gross)	82	103	108	100	106	111	116	117	Agriculture (brut)
Food (gross)	79	100	107	100	107	111	116	117	Produits ailmentaires (brut)
Denmark									**Danemark**
Agriculture (gross)	68	85	94	101	101	103	103	100	Agriculture (brut)
Food (gross)	68	85	94	101	101	103	103	100	Produits ailmentaires (brut)
Djibouti									**Djibouti**
Agriculture (gross)	24	69	79	95	119	132	135	134	Agriculture (brut)
Food (gross)	24	69	79	95	119	132	135	134	Produits ailmentaires (brut)
Dominica									**Dominique**
Agriculture (gross)	101	118	116	95	111	110	114	113	Agriculture (brut)
Food (gross)	102	119	117	95	112	110	115	114	Produits ailmentaires (brut)
Dominican Republic									**Rép. dominicaine**
Agriculture (gross)	64	77	77	99	128	123	127	133	Agriculture (brut)
Food (gross)	61	73	76	99	130	124	128	136	Produits ailmentaires (brut)
Ecuador									**Équateur**
Agriculture (gross)	38	47	79	98	121	122	117	116	Agriculture (brut)
Food (gross)	37	45	78	98	123	124	118	117	Produits ailmentaires (brut)
Egypt									**Égypte**
Agriculture (gross)	31	41	67	99	109	114	119	118	Agriculture (brut)
Food (gross)	29	39	67	99	110	114	120	120	Produits ailmentaires (brut)

Région, pays ou zone	1975	1985	1995	2005	2010	2011	2012	2013	Région, pays ou zone
El Salvador									**El Salvador**
Agriculture (gross)	82	79	88	99	108	104	114	110	Agriculture (brut)
Food (gross)	60	65	81	99	106	104	115	118	Produits ailmentaires (brut)
Equatorial Guinea									**Guinée équatoriale**
Agriculture (gross)	67	70	88	100	111	112	116	117	Agriculture (brut)
Food (gross)	59	58	84	101	113	114	117	118	Produits ailmentaires (brut)
Eritrea									**Érythrée**
Agriculture (gross)	...	...	78	106	106	109	111	110	Agriculture (brut)
Food (gross)	...	...	78	106	106	109	111	110	Produits ailmentaires (brut)
Estonia									**Estonie**
Agriculture (gross)	...	...	110	103	110	118	123	127	Agriculture (brut)
Food (gross)	...	...	110	103	110	118	123	127	Produits ailmentaires (brut)
Ethiopia									**Éthiopie**
Agriculture (gross)	...	...	59	102	137	139	146	147	Agriculture (brut)
Food (gross)	...	...	57	103	137	138	148	148	Produits ailmentaires (brut)
Falkland Is. (Malvinas)									**Îles Falkland (Malvinas)**
Agriculture (gross)	99	103	112	103	95	95	95	95	Agriculture (brut)
Food (gross)	115	119	131	100	103	103	103	103	Produits ailmentaires (brut)
Faroe Islands									**Îles Féroé**
Agriculture (gross)	96	96	102	100	102	102	103	104	Agriculture (brut)
Food (gross)	96	96	102	100	102	102	103	104	Produits ailmentaires (brut)
Fiji									**Fidji**
Agriculture (gross)	63	88	107	99	82	93	85	86	Agriculture (brut)
Food (gross)	63	88	107	99	82	93	85	86	Produits ailmentaires (brut)
Finland									**Finlande**
Agriculture (gross)	99	106	95	102	94	98	94	99	Agriculture (brut)
Food (gross)	99	106	95	102	94	98	94	99	Produits ailmentaires (brut)
France									**France**
Agriculture (gross)	83	99	99	100	97	99	98	97	Agriculture (brut)
Food (gross)	83	99	99	100	97	99	98	97	Produits ailmentaires (brut)
French Guiana									**Guyane française**
Agriculture (gross)	22	52	106	96	90	90	94	89	Agriculture (brut)
Food (gross)	22	52	106	96	90	90	94	89	Produits ailmentaires (brut)
French Polynesia									**Polynésie française**
Agriculture (gross)	106	91	90	104	99	104	105	105	Agriculture (brut)
Food (gross)	106	91	90	104	99	104	105	105	Produits ailmentaires (brut)
Gabon									**Gabon**
Agriculture (gross)	51	70	85	100	117	118	121	122	Agriculture (brut)
Food (gross)	54	74	88	100	115	116	118	120	Produits ailmentaires (brut)
Gambia									**Gambie**
Agriculture (gross)	78	64	63	94	134	88	108	98	Agriculture (brut)
Food (gross)	78	63	63	94	134	88	108	98	Produits ailmentaires (brut)
Georgia									**Géorgie**
Agriculture (gross)	...	...	119	121	67	76	71	86	Agriculture (brut)
Food (gross)	...	...	117	121	68	77	72	88	Produits ailmentaires (brut)
Germany									**Allemagne**
Agriculture (gross)	95	110	92	100	102	104	105	105	Agriculture (brut)
Food (gross)	95	110	92	100	102	104	105	105	Produits ailmentaires (brut)
Ghana									**Ghana**
Agriculture (gross)	37	36	64	100	125	131	139	143	Agriculture (brut)
Food (gross)	37	36	64	100	125	131	139	143	Produits ailmentaires (brut)
Greece									**Grèce**
Agriculture (gross)	79	94	104	103	83	84	85	86	Agriculture (brut)
Food (gross)	82	96	102	103	87	87	88	88	Produits ailmentaires (brut)
Greenland									**Groenland**
Agriculture (gross)	61	103	100	100	99	99	99	99	Agriculture (brut)
Food (gross)	59	103	101	99	99	99	99	99	Produits ailmentaires (brut)
Grenada									**Grenade**
Agriculture (gross)	136	135	133	77	90	95	97	97	Agriculture (brut)
Food (gross)	136	135	133	77	90	95	97	97	Produits ailmentaires (brut)
Guadeloupe									**Guadeloupe**
Agriculture (gross)	137	117	81	103	89	96	96	100	Agriculture (brut)
Food (gross)	137	117	81	103	89	96	96	100	Produits ailmentaires (brut)
Guam									**Guam**
Agriculture (gross)	66	71	78	100	97	96	98	97	Agriculture (brut)
Food (gross)	66	71	78	100	97	96	98	97	Produits ailmentaires (brut)

Région, pays ou zone	1975	1985	1995	2005	2010	2011	2012	2013	Région, pays ou zone
Guatemala									**Guatemala**
Agriculture (gross)	41	48	72	98	128	131	141	147	Agriculture (brut)
Food (gross)	35	44	72	98	131	132	144	150	Produits ailmentaires (brut)
Guinea									**Guinée**
Agriculture (gross)	41	49	72	101	113	116	122	123	Agriculture (brut)
Food (gross)	42	50	72	101	113	116	122	123	Produits ailmentaires (brut)
Guinea-Bissau									**Guinée-Bissau**
Agriculture (gross)	38	53	69	99	122	128	137	140	Agriculture (brut)
Food (gross)	38	52	69	99	122	128	138	141	Produits ailmentaires (brut)
Guyana									**Guyana**
Agriculture (gross)	78	68	99	94	108	113	113	129	Agriculture (brut)
Food (gross)	78	68	99	94	108	113	113	129	Produits ailmentaires (brut)
Haiti									**Haïti**
Agriculture (gross)	91	104	83	102	111	112	110	118	Agriculture (brut)
Food (gross)	90	103	83	102	113	113	112	121	Produits ailmentaires (brut)
Honduras									**Honduras**
Agriculture (gross)	39	55	72	103	111	116	122	121	Agriculture (brut)
Food (gross)	41	57	72	104	109	112	115	118	Produits ailmentaires (brut)
Hungary									**Hongrie**
Agriculture (gross)	108	126	91	97	80	87	79	88	Agriculture (brut)
Food (gross)	107	126	91	96	80	87	79	88	Produits ailmentaires (brut)
Iceland									**Islande**
Agriculture (gross)	105	107	86	99	110	111	115	114	Agriculture (brut)
Food (gross)	104	105	85	99	110	111	115	114	Produits ailmentaires (brut)
India									**Inde**
Agriculture (gross)	44	60	81	100	125	132	135	140	Agriculture (brut)
Food (gross)	44	60	81	100	123	131	134	139	Produits ailmentaires (brut)
Indonesia									**Indonésie**
Agriculture (gross)	31	50	77	98	123	127	135	137	Agriculture (brut)
Food (gross)	31	50	77	98	124	128	136	138	Produits ailmentaires (brut)
Iran (Islamic Rep. of)									**Iran (Rép. islamique d')**
Agriculture (gross)	26	43	70	103	108	109	112	113	Agriculture (brut)
Food (gross)	25	42	69	103	108	109	113	114	Produits ailmentaires (brut)
Iraq									**Iraq**
Agriculture (gross)	62	103	98	104	104	113	114	129	Agriculture (brut)
Food (gross)	60	103	98	104	104	113	114	129	Produits ailmentaires (brut)
Ireland									**Irlande**
Agriculture (gross)	76	92	97	99	100	102	95	99	Agriculture (brut)
Food (gross)	76	92	97	99	100	101	95	99	Produits ailmentaires (brut)
Israel									**Israël**
Agriculture (gross)	54	69	80	100	104	106	111	110	Agriculture (brut)
Food (gross)	52	65	78	100	105	106	112	111	Produits ailmentaires (brut)
Italy									**Italie**
Agriculture (gross)	88	95	95	100	97	95	88	90	Agriculture (brut)
Food (gross)	88	95	95	100	97	96	89	91	Produits ailmentaires (brut)
Jamaica									**Jamaïque**
Agriculture (gross)	74	76	104	96	99	103	106	106	Agriculture (brut)
Food (gross)	75	77	106	96	99	103	107	106	Produits ailmentaires (brut)
Japan									**Japon**
Agriculture (gross)	103	117	110	101	97	96	98	98	Agriculture (brut)
Food (gross)	100	115	110	101	97	96	98	98	Produits ailmentaires (brut)
Jordan									**Jordanie**
Agriculture (gross)	20	45	80	98	129	134	132	137	Agriculture (brut)
Food (gross)	19	44	79	97	129	134	132	137	Produits ailmentaires (brut)
Kazakhstan									**Kazakhstan**
Agriculture (gross)	...	...	88	100	107	142	110	127	Agriculture (brut)
Food (gross)	...	...	88	100	108	144	111	126	Produits ailmentaires (brut)
Kenya									**Kenya**
Agriculture (gross)	36	53	71	103	123	118	121	123	Agriculture (brut)
Food (gross)	36	52	69	103	124	119	123	123	Produits ailmentaires (brut)
Kiribati									**Kiribati**
Agriculture (gross)	41	72	63	95	118	116	120	121	Agriculture (brut)
Food (gross)	41	72	63	95	118	116	120	121	Produits ailmentaires (brut)
Kuwait									**Koweït**
Agriculture (gross)	17	55	53	97	134	162	168	171	Agriculture (brut)
Food (gross)	17	54	54	97	135	162	168	172	Produits ailmentaires (brut)

Région, pays ou zone	1975	1985	1995	2005	2010	2011	2012	2013	Région, pays ou zone
Kyrgyzstan									**Kirghizistan**
Agriculture (gross)	...	...	66	98	105	107	107	112	Agriculture (brut)
Food (gross)	...	...	65	98	107	108	108	114	Produits ailmentaires (brut)
Lao People's Dem. Rep.									**Rép. dém. pop. lao**
Agriculture (gross)	28	42	53	100	130	138	155	156	Agriculture (brut)
Food (gross)	26	43	52	100	128	136	152	153	Produits ailmentaires (brut)
Latvia									**Lettonie**
Agriculture (gross)	...	...	113	105	109	110	125	117	Agriculture (brut)
Food (gross)	...	...	113	105	109	110	125	118	Produits ailmentaires (brut)
Lebanon									**Liban**
Agriculture (gross)	41	60	111	97	95	92	95	97	Agriculture (brut)
Food (gross)	41	60	112	97	94	92	95	97	Produits ailmentaires (brut)
Lesotho									**Lesotho**
Agriculture (gross)	82	90	88	102	110	109	99	111	Agriculture (brut)
Food (gross)	83	90	85	102	111	109	99	112	Produits ailmentaires (brut)
Liberia									**Libéria**
Agriculture (gross)	73	91	49	105	100	104	104	101	Agriculture (brut)
Food (gross)	71	89	64	104	120	125	126	121	Produits ailmentaires (brut)
Libya									**Libye**
Agriculture (gross)	49	66	89	101	110	110	112	111	Agriculture (brut)
Food (gross)	48	65	89	101	110	111	112	111	Produits ailmentaires (brut)
Liechtenstein									**Liechtenstein**
Agriculture (gross)	47	94	94	100	100	104	103	100	Agriculture (brut)
Food (gross)	47	94	94	100	100	104	103	100	Produits ailmentaires (brut)
Lithuania									**Lituanie**
Agriculture (gross)	...	...	97	106	99	107	122	118	Agriculture (brut)
Food (gross)	...	...	97	106	99	107	122	118	Produits ailmentaires (brut)
Luxembourg									**Luxembourg**
Agriculture (gross)	...	...	...	99	93	93	91	94	Agriculture (brut)
Food (gross)	...	...	...	99	93	93	91	94	Produits ailmentaires (brut)
Madagascar									**Madagascar**
Agriculture (gross)	70	78	87	103	124	125	130	119	Agriculture (brut)
Food (gross)	68	76	86	103	124	125	130	119	Produits ailmentaires (brut)
Malawi									**Malawi**
Agriculture (gross)	38	48	60	85	155	164	173	187	Agriculture (brut)
Food (gross)	38	45	52	85	157	167	182	193	Produits ailmentaires (brut)
Malaysia									**Malaisie**
Agriculture (gross)	31	47	70	100	111	120	119	121	Agriculture (brut)
Food (gross)	20	38	67	100	115	124	125	128	Produits ailmentaires (brut)
Maldives									**Maldives**
Agriculture (gross)	87	109	97	89	72	70	66	64	Agriculture (brut)
Food (gross)	87	109	97	89	72	70	66	64	Produits ailmentaires (brut)
Mali									**Mali**
Agriculture (gross)	37	47	66	103	142	136	146	141	Agriculture (brut)
Food (gross)	39	48	63	103	154	145	154	150	Produits ailmentaires (brut)
Malta									**Malte**
Agriculture (gross)	65	76	91	97	99	95	91	92	Agriculture (brut)
Food (gross)	65	76	91	97	99	95	91	92	Produits ailmentaires (brut)
Marshall Islands									**Îles Marshall**
Agriculture (gross)	...	...	173	97	175	97	98	103	Agriculture (brut)
Food (gross)	...	...	173	97	175	97	98	103	Produits ailmentaires (brut)
Martinique									**Martinique**
Agriculture (gross)	99	98	94	99	81	92	97	101	Agriculture (brut)
Food (gross)	99	98	94	99	81	92	97	101	Produits ailmentaires (brut)
Mauritania									**Mauritanie**
Agriculture (gross)	50	67	78	100	113	111	120	120	Agriculture (brut)
Food (gross)	50	67	78	100	113	111	120	120	Produits ailmentaires (brut)
Mauritius									**Maurice**
Agriculture (gross)	67	89	94	98	99	98	96	94	Agriculture (brut)
Food (gross)	66	85	92	98	99	98	96	94	Produits ailmentaires (brut)
Mexico									**Mexique**
Agriculture (gross)	45	65	78	98	108	108	113	115	Agriculture (brut)
Food (gross)	43	64	77	98	108	108	113	115	Produits ailmentaires (brut)
Micronesia (Fed. States of)									**Micronésie (États féd. de)**
Agriculture (gross)	...	...	85	101	98	93	99	102	Agriculture (brut)
Food (gross)	...	...	85	101	98	93	99	102	Produits ailmentaires (brut)

Région, pays ou zone	1975	1985	1995	2005	2010	2011	2012	2013	Région, pays ou zone
Mongolia									**Mongolie**
Agriculture (gross)	132	136	115	97	115	126	133	145	Agriculture (brut)
Food (gross)	132	136	113	97	115	127	135	147	Produits ailmentaires (brut)
Montenegro									**Monténégro**
Agriculture (gross)	...	...	...	...	91	104	93	98	Agriculture (brut)
Food (gross)	...	...	...	...	91	104	93	98	Produits ailmentaires (brut)
Montserrat									**Montserrat**
Agriculture (gross)	64	80	96	98	99	101	101	101	Agriculture (brut)
Food (gross)	64	80	96	98	99	101	101	101	Produits ailmentaires (brut)
Morocco									**Maroc**
Agriculture (gross)	38	51	57	93	126	131	124	134	Agriculture (brut)
Food (gross)	38	50	56	93	127	131	124	135	Produits ailmentaires (brut)
Mozambique									**Mozambique**
Agriculture (gross)	63	52	68	96	148	158	151	157	Agriculture (brut)
Food (gross)	65	57	74	95	153	162	155	161	Produits ailmentaires (brut)
Myanmar									**Myanmar**
Agriculture (gross)	28	45	53	99	135	133	130	132	Agriculture (brut)
Food (gross)	27	44	53	99	135	132	129	131	Produits ailmentaires (brut)
Namibia									**Namibie**
Agriculture (gross)	73	70	90	103	90	88	91	90	Agriculture (brut)
Food (gross)	72	70	90	104	91	89	92	90	Produits ailmentaires (brut)
Nauru									**Nauru**
Agriculture (gross)	61	67	87	101	104	103	103	106	Agriculture (brut)
Food (gross)	61	67	87	101	104	103	103	106	Produits ailmentaires (brut)
Nepal									**Népal**
Agriculture (gross)	39	51	76	100	114	123	141	132	Agriculture (brut)
Food (gross)	38	51	76	100	114	123	141	132	Produits ailmentaires (brut)
Netherlands									**Pays-Bas**
Agriculture (gross)	72	97	106	100	111	114	111	113	Agriculture (brut)
Food (gross)	72	97	106	100	112	114	111	113	Produits ailmentaires (brut)
Netherlands Antilles [former]									**Antilles néerlandaises [anc.]**
Agriculture (gross)	127	103	89	101	121	119	124	124	Agriculture (brut)
Food (gross)	127	103	89	101	121	119	124	124	Produits ailmentaires (brut)
New Caledonia									**Nouvelle-Calédonie**
Agriculture (gross)	80	92	89	100	98	97	99	103	Agriculture (brut)
Food (gross)	74	90	89	100	98	97	100	103	Produits ailmentaires (brut)
New Zealand									**Nouvelle-Zélande**
Agriculture (gross)	57	74	80	99	104	104	110	108	Agriculture (brut)
Food (gross)	54	70	78	99	105	106	111	110	Produits ailmentaires (brut)
Nicaragua									**Nicaragua**
Agriculture (gross)	76	58	59	104	118	128	130	138	Agriculture (brut)
Food (gross)	62	50	57	102	119	127	131	140	Produits ailmentaires (brut)
Niger									**Niger**
Agriculture (gross)	28	35	53	102	145	128	138	132	Agriculture (brut)
Food (gross)	28	35	53	102	145	128	138	132	Produits ailmentaires (brut)
Nigeria									**Nigéria**
Agriculture (gross)	29	33	67	100	104	96	106	110	Agriculture (brut)
Food (gross)	29	33	67	100	104	97	107	111	Produits ailmentaires (brut)
Niue									**Nioué**
Agriculture (gross)	92	98	87	101	93	87	91	92	Agriculture (brut)
Food (gross)	92	98	87	101	93	87	91	92	Produits ailmentaires (brut)
Norway									**Norvège**
Agriculture (gross)	89	104	103	99	102	98	101	102	Agriculture (brut)
Food (gross)	89	104	103	99	102	98	101	102	Produits ailmentaires (brut)
Oman									**Oman**
Agriculture (gross)	20	46	70	112	122	115	127	124	Agriculture (brut)
Food (gross)	20	45	70	112	122	115	127	125	Produits ailmentaires (brut)
Pakistan									**Pakistan**
Agriculture (gross)	33	49	78	100	110	116	116	95	Agriculture (brut)
Food (gross)	34	48	78	101	113	118	117	94	Produits ailmentaires (brut)
Panama									**Panama**
Agriculture (gross)	77	88	92	99	107	110	115	116	Agriculture (brut)
Food (gross)	78	89	92	99	107	111	116	117	Produits ailmentaires (brut)
Papua New Guinea									**Papouasie-Nvl-Guinée**
Agriculture (gross)	51	65	78	100	112	118	117	118	Agriculture (brut)
Food (gross)	51	64	77	99	112	117	118	118	Produits ailmentaires (brut)

Région, pays ou zone	1975	1985	1995	2005	2010	2011	2012	2013	Région, pays ou zone
Paraguay									**Paraguay**
Agriculture (gross)	30	56	72	97	137	144	112	157	Agriculture (brut)
Food (gross)	28	50	68	98	142	149	116	163	Produits ailmentaires (brut)
Peru									**Pérou**
Agriculture (gross)	43	47	63	99	128	133	139	144	Agriculture (brut)
Food (gross)	42	46	62	100	130	133	140	147	Produits ailmentaires (brut)
Philippines									**Philippines**
Agriculture (gross)	46	57	76	100	112	115	119	120	Agriculture (brut)
Food (gross)	45	56	75	100	113	115	119	120	Produits ailmentaires (brut)
Poland									**Pologne**
Agriculture (gross)	122	119	103	99	101	103	107	106	Agriculture (brut)
Food (gross)	121	118	103	99	101	103	107	106	Produits ailmentaires (brut)
Portugal									**Portugal**
Agriculture (gross)	80	82	98	97	104	100	100	104	Agriculture (brut)
Food (gross)	80	82	98	97	105	101	100	104	Produits ailmentaires (brut)
Puerto Rico									**Porto Rico**
Agriculture (gross)	140	125	118	97	105	109	110	111	Agriculture (brut)
Food (gross)	140	124	117	97	107	110	112	112	Produits ailmentaires (brut)
Qatar									**Qatar**
Agriculture (gross)	11	41	100	95	120	132	133	134	Agriculture (brut)
Food (gross)	11	41	100	95	120	132	133	134	Produits ailmentaires (brut)
Republic of Korea									**République de Corée**
Agriculture (gross)	49	71	92	100	102	99	101	104	Agriculture (brut)
Food (gross)	47	70	92	100	102	99	101	104	Produits ailmentaires (brut)
Republic of Moldova									**République de Moldova**
Agriculture (gross)	...	...	132	100	93	100	77	101	Agriculture (brut)
Food (gross)	...	...	130	100	93	100	77	101	Produits ailmentaires (brut)
Réunion									**Réunion**
Agriculture (gross)	55	66	85	99	104	105	106	107	Agriculture (brut)
Food (gross)	55	66	85	99	104	105	106	107	Produits ailmentaires (brut)
Romania									**Roumanie**
Agriculture (gross)	89	116	96	95	91	100	79	97	Agriculture (brut)
Food (gross)	88	115	96	95	91	100	79	97	Produits ailmentaires (brut)
Russian Federation									**Fédération de Russie**
Agriculture (gross)	...	...	100	100	94	116	108	117	Agriculture (brut)
Food (gross)	...	...	100	100	94	116	108	117	Produits ailmentaires (brut)
Rwanda									**Rwanda**
Agriculture (gross)	50	77	53	100	145	156	166	170	Agriculture (brut)
Food (gross)	50	75	52	101	146	157	168	172	Produits ailmentaires (brut)
Saint Kitts and Nevis									**Saint-Kitts-et-Nevis**
Agriculture (gross)	156	184	139	91	34	41	41	42	Agriculture (brut)
Food (gross)	155	184	139	91	34	41	41	42	Produits ailmentaires (brut)
Saint Lucia									**Sainte-Lucie**
Agriculture (gross)	190	256	302	89	91	96	99	94	Agriculture (brut)
Food (gross)	189	256	302	89	91	96	99	94	Produits ailmentaires (brut)
Saint Pierre and Miquelon									**Saint-Pierre-et-Miquelon**
Agriculture (gross)	2	35	75	92	109	108	109	110	Agriculture (brut)
Food (gross)	2	35	75	92	109	108	109	110	Produits ailmentaires (brut)
Saint Vincent-Grenadines									**Saint-Vincent-Grenadines**
Agriculture (gross)	63	123	102	104	120	114	118	119	Agriculture (brut)
Food (gross)	64	124	102	104	120	114	118	119	Produits ailmentaires (brut)
Samoa									**Samoa**
Agriculture (gross)	97	113	80	102	106	106	113	110	Agriculture (brut)
Food (gross)	97	113	80	102	106	106	114	110	Produits ailmentaires (brut)
Sao Tome and Principe									**Sao Tomé-et-Principe**
Agriculture (gross)	62	51	64	100	102	100	102	111	Agriculture (brut)
Food (gross)	62	51	64	100	102	100	102	111	Produits ailmentaires (brut)
Saudi Arabia									**Arabie saoudite**
Agriculture (gross)	21	52	74	100	108	108	110	107	Agriculture (brut)
Food (gross)	21	52	74	100	108	109	110	108	Produits ailmentaires (brut)
Senegal									**Sénégal**
Agriculture (gross)	93	74	89	110	151	108	128	126	Agriculture (brut)
Food (gross)	94	74	90	110	152	109	129	126	Produits ailmentaires (brut)
Serbia									**Serbie**
Agriculture (gross)	...	...	...	...	103	104	82	105	Agriculture (brut)
Food (gross)	...	...	...	...	103	104	82	105	Produits ailmentaires (brut)

Région, pays ou zone	1975	1985	1995	2005	2010	2011	2012	2013	Région, pays ou zone
Serbia and Montenegro [former]									**Serbie-et-Monténégro [anc.]**
Agriculture (gross)	...	...	98	96	...	...	...	...	Agriculture (brut)
Food (gross)	...	...	98	96	...	...	...	...	Produits ailmentaires (brut)
Seychelles									**Seychelles**
Agriculture (gross)	125	137	153	98	91	102	105	110	Agriculture (brut)
Food (gross)	130	140	155	98	94	105	109	114	Produits ailmentaires (brut)
Sierra Leone									**Sierra Leone**
Agriculture (gross)	49	53	54	93	149	156	161	169	Agriculture (brut)
Food (gross)	49	51	52	93	149	155	161	169	Produits ailmentaires (brut)
Singapore									**Singapour**
Agriculture (gross)	974	857	164	90	92	101	105	112	Agriculture (brut)
Food (gross)	968	857	164	90	92	101	105	112	Produits ailmentaires (brut)
Slovakia									**Slovaquie**
Agriculture (gross)	...	...	114	102	82	90	82	89	Agriculture (brut)
Food (gross)	...	...	113	102	82	90	82	89	Produits ailmentaires (brut)
Slovenia									**Slovénie**
Agriculture (gross)	...	...	96	99	92	91	85	81	Agriculture (brut)
Food (gross)	...	...	96	99	92	91	85	81	Produits ailmentaires (brut)
Solomon Islands									**Îles Salomon**
Agriculture (gross)	42	73	77	103	110	112	115	118	Agriculture (brut)
Food (gross)	42	73	77	103	110	112	115	118	Produits ailmentaires (brut)
Somalia									**Somalie**
Agriculture (gross)	68	88	87	100	104	107	111	117	Agriculture (brut)
Food (gross)	68	88	87	100	104	107	111	117	Produits ailmentaires (brut)
South Africa									**Afrique du Sud**
Agriculture (gross)	65	75	74	102	118	117	120	122	Agriculture (brut)
Food (gross)	63	73	74	102	118	117	121	123	Produits ailmentaires (brut)
Spain									**Espagne**
Agriculture (gross)	61	74	70	95	103	107	92	109	Agriculture (brut)
Food (gross)	61	74	70	94	104	107	92	110	Produits ailmentaires (brut)
Sri Lanka									**Sri Lanka**
Agriculture (gross)	69	90	96	102	124	119	122	136	Agriculture (brut)
Food (gross)	65	90	98	102	125	120	124	141	Produits ailmentaires (brut)
State of Palestine									**État de Palestine**
Agriculture (gross)	...	...	49	107	81	91	96	93	Agriculture (brut)
Food (gross)	...	...	49	107	81	91	96	93	Produits ailmentaires (brut)
Sudan [former]									**Soudan [anc.]**
Agriculture (gross)	37	47	62	101	99	107	102	113	Agriculture (brut)
Food (gross)	34	45	62	101	100	108	102	115	Produits ailmentaires (brut)
Suriname									**Suriname**
Agriculture (gross)	91	139	126	99	137	135	136	146	Agriculture (brut)
Food (gross)	90	139	126	99	137	135	136	146	Produits ailmentaires (brut)
Swaziland									**Swaziland**
Agriculture (gross)	63	89	83	103	104	106	111	112	Agriculture (brut)
Food (gross)	60	87	83	103	104	106	112	113	Produits ailmentaires (brut)
Sweden									**Suède**
Agriculture (gross)	99	113	100	100	94	96	96	96	Agriculture (brut)
Food (gross)	99	113	101	100	94	96	96	96	Produits ailmentaires (brut)
Switzerland									**Suisse**
Agriculture (gross)	97	108	103	99	103	109	104	101	Agriculture (brut)
Food (gross)	97	108	103	99	103	109	104	101	Produits ailmentaires (brut)
Syrian Arab Republic									**Rép. arabe syrienne**
Agriculture (gross)	31	49	65	100	89	98	89	79	Agriculture (brut)
Food (gross)	30	48	64	99	92	103	94	82	Produits ailmentaires (brut)
Tajikistan									**Tadjikistan**
Agriculture (gross)	...	...	70	99	123	130	145	154	Agriculture (brut)
Food (gross)	...	...	67	99	137	144	157	170	Produits ailmentaires (brut)
Thailand									**Thaïlande**
Agriculture (gross)	41	62	79	98	113	121	129	129	Agriculture (brut)
Food (gross)	45	66	80	98	115	122	131	129	Produits ailmentaires (brut)
TFYR of Macedonia									**ex-R.Y. de Macédoine**
Agriculture (gross)	...	...	88	100	116	113	109	115	Agriculture (brut)
Food (gross)	...	...	89	99	116	114	110	115	Produits ailmentaires (brut)
Timor-Leste									**Timor-Leste**
Agriculture (gross)	89	82	94	98	121	106	123	113	Agriculture (brut)
Food (gross)	98	86	98	98	126	113	132	119	Produits ailmentaires (brut)

Région, pays ou zone	1975	1985	1995	2005	2010	2011	2012	2013	Région, pays ou zone
Togo									**Togo**
Agriculture (gross)	46	55	79	96	123	136	133	121	Agriculture (brut)
Food (gross)	47	54	75	99	129	141	137	125	Produits ailmentaires (brut)
Tokelau									**Tokélaou**
Agriculture (gross)	92	74	89	102	120	122	125	125	Agriculture (brut)
Food (gross)	92	74	89	102	120	122	125	125	Produits ailmentaires (brut)
Tonga									**Tonga**
Agriculture (gross)	149	106	94	97	136	130	130	131	Agriculture (brut)
Food (gross)	149	106	94	97	136	130	130	131	Produits ailmentaires (brut)
Trinidad and Tobago									**Trinité-et-Tobago**
Agriculture (gross)	109	83	91	99	96	93	95	96	Agriculture (brut)
Food (gross)	106	82	91	99	96	94	95	96	Produits ailmentaires (brut)
Tunisia									**Tunisie**
Agriculture (gross)	53	60	59	101	107	107	119	118	Agriculture (brut)
Food (gross)	53	60	59	101	107	107	119	118	Produits ailmentaires (brut)
Turkey									**Turquie**
Agriculture (gross)	53	69	83	101	110	116	122	125	Agriculture (brut)
Food (gross)	53	69	82	101	111	117	124	127	Produits ailmentaires (brut)
Turkmenistan									**Turkménistan**
Agriculture (gross)	...	...	64	104	102	96	100	103	Agriculture (brut)
Food (gross)	...	...	53	102	106	101	106	109	Produits ailmentaires (brut)
Tuvalu									**Tuvalu**
Agriculture (gross)	48	81	78	101	111	104	106	108	Agriculture (brut)
Food (gross)	48	81	78	101	111	104	106	108	Produits ailmentaires (brut)
Uganda									**Ouganda**
Agriculture (gross)	69	56	77	100	112	114	111	112	Agriculture (brut)
Food (gross)	67	55	76	100	112	113	111	111	Produits ailmentaires (brut)
Ukraine									**Ukraine**
Agriculture (gross)	...	...	105	100	106	129	121	138	Agriculture (brut)
Food (gross)	...	...	104	100	106	129	121	138	Produits ailmentaires (brut)
United Arab Emirates									**Émirats arabes unis**
Agriculture (gross)	7	22	68	105	111	69	70	73	Agriculture (brut)
Food (gross)	7	22	67	105	111	69	70	73	Produits ailmentaires (brut)
United Kingdom									**Royaume-Uni**
Agriculture (gross)	86	105	107	100	102	104	98	100	Agriculture (brut)
Food (gross)	86	105	107	100	102	104	98	99	Produits ailmentaires (brut)
United Rep. of Tanzania									**Rép.-Unie de Tanzanie**
Agriculture (gross)	43	58	68	98	129	139	148	154	Agriculture (brut)
Food (gross)	41	58	68	96	131	140	149	155	Produits ailmentaires (brut)
United States									**États-Unis**
Agriculture (gross)	64	78	84	100	106	104	103	108	Agriculture (brut)
Food (gross)	65	79	84	100	107	105	104	110	Produits ailmentaires (brut)
United States Virgin Is.									**Îles Vierges américaines**
Agriculture (gross)	120	132	94	98	118	123	123	123	Agriculture (brut)
Food (gross)	120	132	94	98	118	123	123	123	Produits ailmentaires (brut)
Uruguay									**Uruguay**
Agriculture (gross)	57	59	74	101	120	122	129	131	Agriculture (brut)
Food (gross)	55	56	72	101	122	123	130	133	Produits ailmentaires (brut)
Uzbekistan									**Ouzbékistan**
Agriculture (gross)	...	...	81	100	127	136	143	152	Agriculture (brut)
Food (gross)	...	...	75	99	135	145	155	166	Produits ailmentaires (brut)
Vanuatu									**Vanuatu**
Agriculture (gross)	80	101	102	100	132	127	130	138	Agriculture (brut)
Food (gross)	80	101	102	100	132	127	130	138	Produits ailmentaires (brut)
Venezuela (Boliv. Rep. of)									**Venezuela (Rép. boliv. du)**
Agriculture (gross)	47	65	79	101	109	122	124	134	Agriculture (brut)
Food (gross)	46	64	78	101	109	122	125	135	Produits ailmentaires (brut)
Viet Nam									**Viet Nam**
Agriculture (gross)	22	37	58	100	120	125	134	136	Agriculture (brut)
Food (gross)	23	39	60	100	119	124	132	133	Produits ailmentaires (brut)
Wallis and Futuna Islands									**Îles Wallis-et-Futuna**
Agriculture (gross)	73	78	83	104	101	98	101	103	Agriculture (brut)
Food (gross)	73	78	83	104	101	98	101	103	Produits ailmentaires (brut)
Western Sahara									**Sahara occidental**
Agriculture (gross)	62	79	91	100	102	107	109	106	Agriculture (brut)
Food (gross)	62	79	91	100	102	107	109	106	Produits ailmentaires (brut)

Région, pays ou zone	1975	1985	1995	2005	2010	2011	2012	2013	Région, pays ou zone
Yemen									**Yémen**
Agriculture (gross)	41	45	65	98	136	133	141	138	Agriculture (brut)
Food (gross)	41	46	66	98	137	134	141	138	Produits ailmentaires (brut)
Zambia									**Zambie**
Agriculture (gross)	52	54	62	101	162	173	181	179	Agriculture (brut)
Food (gross)	59	60	69	98	170	181	183	187	Produits ailmentaires (brut)
Zimbabwe									**Zimbabwe**
Agriculture (gross)	80	104	85	93	101	105	106	107	Agriculture (brut)
Food (gross)	78	97	71	91	96	100	99	98	Produits ailmentaires (brut)

Source:

Food and Agriculture Organization of the United Nations (FAO), Rome, FAOSTAT database, last accessed July 2016.

Source:

Organisation des Nations Unies pour l'alimentation et l'agriculture (FAO), Rome, la base de données FAOSTAT, dernier accès juillet 2016.

1   For statistical purposes, the data for China do not include those for the Hong Kong Special Administrative Region (Hong Kong SAR) and Macao Special Administrative Region (Macao SAR).

1   Pour la présentation des statistiques, les données pour la Chine ne comprennent pas la Région Administrative Spéciale de Hong Kong (Hong Kong RAS) et la Région Administrative Spéciale de Macao (Macao RAS).

Total imports, exports and balance of trade
Imports CIF, exports FOB and balance: millions of US dollars

Total des importations, des exportations et balance commerciale
Importations CIF, exportations FOB et balance: en millions de dollars É.-U.

Region, country or area &	Sys.[t]	1985	1995	2005	2010	2012	2013	2014	2015	Région, pays ou zone &
Total, all countries or areas										**Total, tous pays ou zones**
Imports		2 037 010	5 158 729	10 602 415	15 157 638	18 125 887	18 423 181	18 597 330	16 449 902	Importations
Exports		1 985 237	5 128 232	10 356 386	15 106 207	18 081 309	18 468 451	18 642 803	16 462 567	Exportations
Balance		-51 773	-30 497	-246 029	-51 431	-44 578	45 270	45 473	12 666	Balance
Developed economies [1,2]										**Economies développées [1,2]**
Imports		1 443 773	3 537 095	6 895 244	8 629 159	9 855 374	9 872 786	10 018 920	8 873 357	Importations
Exports		1 357 287	3 562 735	6 154 577	7 996 035	9 071 029	9 199 392	9 327 971	8 227 042	Exportations
Balance		-86 486	25 640	-740 666	-633 124	-784 344	-673 394	-690 949	-646 315	Balance
Asia and Oceania [3]										**Asie et Océanie [3]**
Imports		159 672	407 463	660 631	917 455	1 173 988	1 105 373	1 082 073	884 976	Importations
Exports		205 529	509 761	722 686	1 014 394	1 092 678	1 008 667	972 098	847 424	Exportations
Balance		45 857	102 298	62 055	96 939	-81 309	-96 706	-109 975	-37 552	Balance
Europe [4]										**Europe [4]**
Imports		852 201	2 193 618	4 183 214	5 347 880	5 879 533	5 973 133	6 054 716	5 245 534	Importations
Exports		841 781	2 275 591	4 169 654	5 315 137	5 977 073	6 152 446	6 261 507	5 465 140	Exportations
Balance		-10 419	81 974	-13 560	-32 742	97 540	179 313	206 791	219 606	Balance
Northern America										**Amérique septentrionale**
Imports		431 901	936 015	2 051 399	2 363 824	2 801 853	2 794 280	2 882 132	2 742 846	Importations
Exports		309 977	777 383	1 262 237	1 666 503	2 001 278	2 038 279	2 094 366	1 914 478	Exportations
Balance		-121 924	-158 632	-789 162	-697 321	-800 575	-756 001	-787 766	-828 369	Balance
South-Eastern Europe [5]										**Europe centrale et de l'est [5]**
Imports		37 151	22 109	82 987	125 499	145 630	152 532	159 084	138 551	Importations
Exports		36 211	16 352	49 008	89 833	107 537	122 799	129 370	111 583	Exportations
Balance		-940	-5 757	-33 979	-35 666	-38 093	-29 733	-29 714	-26 968	Balance
CIS§										**CEI§**
Imports		83 140	79 286	188 606	385 468	530 798	536 057	479 860	335 858	Importations
Exports		87 281	110 047	336 611	575 087	777 787	762 333	720 560	499 926	Exportations
Balance		4 141	30 761	148 005	189 620	246 989	226 276	240 701	164 069	Balance
CIS Asia§										**CEI Asie§**
Imports		...	10 689	34 762	56 330	80 313	95 612	92 943	81 339	Importations
Exports		...	13 160	43 839	99 189	135 352	132 114	130 330	89 596	Exportations
Balance		...	2 471	9 077	42 859	55 039	36 503	37 387	8 257	Balance
CIS Europe§										**CEI Europe§**
Imports		83 140	68 597	153 844	329 137	450 485	440 445	386 917	254 519	Importations
Exports		87 281	96 887	292 772	475 898	642 436	630 218	590 230	410 331	Exportations
Balance		4 141	28 290	138 928	146 761	191 951	189 773	203 313	155 812	Balance
Northern Africa [6]										**Afrique septentrionale [6]**
Imports		26 197	44 604	80 241	161 273	208 466	211 607	209 267	184 225	Importations
Exports		30 993	33 951	110 314	164 438	199 678	177 149	147 739	102 065	Exportations
Balance		4 796	-10 653	30 074	3 165	-8 787	-34 458	-61 527	-82 160	Balance
Sub-Saharan Africa [7]										**Afrique subsaharienne [7]**
Imports		40 208	71 444	166 410	290 980	354 078	379 648	400 346	387 776	Importations
Exports		48 003	71 824	204 321	341 766	420 737	423 889	420 106	369 496	Exportations
Balance		7 796	380	37 911	50 786	66 659	44 241	19 760	-18 281	Balance
Latin America and the Caribbean										**Amérique latine et Caraïbes**
Imports		85 828	247 024	515 014	855 985	1 077 809	1 117 268	1 117 274	1 004 102	Importations
Exports		110 049	224 762	560 598	861 934	1 087 352	1 093 134	1 061 292	910 776	Exportations
Balance		24 222	-22 262	45 583	5 950	9 542	-24 134	-55 982	-93 326	Balance
Latin America										**Amérique latine**
Imports		64 659	225 801	477 423	805 297	1 021 295	1 063 528	1 063 755	950 308	Importations
Exports		94 775	215 718	543 078	839 739	1 057 353	1 061 585	1 028 659	877 232	Exportations
Balance		30 116	-10 082	65 655	34 442	36 058	-1 943	-35 096	-73 076	Balance
Caribbean										**Caraïbes**
Imports		21 168	21 223	37 591	50 688	56 514	53 740	53 519	53 795	Importations
Exports		15 274	9 044	17 520	22 196	29 999	31 548	32 633	33 545	Exportations
Balance		-5 894	-12 180	-20 071	-28 492	-26 516	-22 192	-20 886	-20 250	Balance
Eastern Asia [8]										**Asie orientale [8]**
Imports		126 636	569 111	1 416 437	2 521 545	3 137 294	3 285 271	3 333 544	2 893 496	Importations
Exports		121 436	563 296	1 537 821	2 717 456	3 352 356	3 546 624	3 717 587	3 568 653	Exportations
Balance		-5 200	-5 815	121 383	195 911	215 061	261 352	384 043	675 156	Balance
Southern Asia										**Asie méridionale**
Imports		39 566	73 400	235 858	505 555	655 725	625 801	632 523	573 584	Importations
Exports		28 294	64 470	187 109	373 614	453 797	461 490	470 526	437 134	Exportations
Balance		-11 272	-8 930	-48 750	-131 941	-201 927	-164 311	-161 996	-136 450	Balance

17

Total imports, exports and balance of trade *(continued)*
Imports CIF, exports FOB and balance: millions of US dollars

Total des importations, des exportations et balance commerciale *(suite)*
Importations CIF, exportations FOB et balance: en millions de dollars É.-U.

Region, country or area &	Sys.[1]	1985	1995	2005	2010	2012	2013	2014	2015	Région, pays ou zone &
South-eastern Asia										**Asie du Sud-est**
Imports		66 582	354 347	599 966	954 566	1 226 074	1 244 020	1 235 942	1 088 135	Importations
Exports		72 499	320 603	654 475	1 049 619	1 252 135	1 269 326	1 288 054	1 162 334	Exportations
Balance		5 917	-33 744	54 510	95 053	26 061	25 305	52 112	74 199	Balance
Western Asia [9]										**Asie occidentale** [9]
Imports		84 539	153 325	410 471	710 884	915 188	976 582	988 166	948 822	Importations
Exports		91 236	154 678	554 227	925 948	1 347 118	1 401 162	1 345 041	1 060 313	Exportations
Balance		6 697	1 354	143 756	215 064	431 929	424 581	356 874	111 492	Balance
Oceania [10]										**Océanie** [10]
Imports		3 391	6 983	11 181	16 723	19 451	21 609	22 403	21 995	Importations
Exports		1 948	5 511	7 326	10 476	11 783	11 154	14 555	13 245	Exportations
Balance		-1 443	-1 472	-3 855	-6 248	-7 668	-10 455	-7 848	-8 751	Balance
Afghanistan										**Afghanistan**
Imports	G	1 194	50	2 471	5 154	6 200	5 400	...	...	Importations
Exports	G	567	26	384	388	350	500	...	...	Exportations
Balance	G	-627	-24	-2 087	-4 766	-5 850	-4 900	...	...	Balance
Albania										**Albanie**
Imports	G	64	713	2 619	4 592	4 882	4 902	5 230	4 320	Importations
Exports	G	53	202	659	1 550	1 968	2 332	2 431	1 930	Exportations
Balance	G	-10	-511	-1 960	-3 042	-2 914	-2 571	-2 799	-2 391	Balance
Algeria										**Algérie**
Imports	S	9 841	10 788	20 383	40 228	50 352	54 965	58 367	51 763	Importations
Exports	S	12 841	10 448	46 693	57 786	72 857	65 555	61 413	35 278	Exportations
Balance	S	3 000	-340	26 310	17 558	22 505	10 590	3 046	-16 485	Balance
American Samoa [11]										**Samoa américaines** [11]
Imports	S	296	416	543	434	514	479	...	...	Importations
Exports	S	201	272	456	316	419	386	...	...	Exportations
Balance	S	-95	-144	-87	-118	-96	-92	...	...	Balance
Andorra										**Andorre**
Imports	S	...	793	1 796	1 518	1 396	1 455	1 556	1 295	Importations
Exports	S	...	37	142	54	68	99	98	89	Exportations
Balance	S	...	-756	-1 654	-1 464	-1 327	-1 356	-1 458	-1 206	Balance
Angola [12]										**Angola** [12]
Imports	S	1 401	1 468	8 353	16 667	23 717	26 344	28 587	20 095	Importations
Exports	S	2 260	3 592	24 109	50 595	71 093	68 247	59 170	33 165	Exportations
Balance	S	859	2 124	15 756	33 928	47 376	41 903	30 583	13 070	Balance
Anguilla										**Anguilla**
Imports	S	...	32	133	157	150	145	152	158	Importations
Exports	S	...	1	7	12	8	4	2	2	Exportations
Balance	S	...	-32	-126	-145	-142	-141	-150	-157	Balance
Antigua and Barbuda										**Antigua-et-Barbuda**
Imports	G	166	346	526	501	535	515	553	488	Importations
Exports	G	21	53	121	35	29	32	25	26	Exportations
Balance	G	-146	-293	-405	-466	-506	-483	-528	-462	Balance
Argentina										**Argentine**
Imports	S	3 814	20 026	28 693	48 048	68 505	74 002	65 323	59 789	Importations
Exports	S	8 396	20 963	40 351	64 722	75 219	83 026	71 936	59 706	Exportations
Balance	S	4 582	937	11 658	16 674	6 713	9 024	6 613	-83	Balance
Armenia										**Arménie**
Imports	S	...	674	1 768	3 783	4 267	4 386	4 402	3 254	Importations
Exports	S	...	271	950	1 011	1 428	1 479	1 519	1 487	Exportations
Balance	S	...	-403	-818	-2 771	-2 839	-2 907	-2 882	-1 767	Balance
Aruba										**Aruba**
Imports	S	...	567	1 028	1 069	1 258	1 303	1 265	...	Importations
Exports	S	...	15	102	125	173	167	116	...	Exportations
Balance	S	...	-552	-927	-945	-1 085	-1 136	-1 150	...	Balance
Australia [12]										**Australie** [12]
Imports	G	23 163	57 423	118 790	193 201	250 560	232 596	227 648	200 041	Importations
Exports	G	22 613	53 115	105 820	212 337	256 675	252 981	241 238	188 372	Exportations
Balance	G	-550	-4 308	-12 971	19 136	6 115	20 385	13 590	-11 668	Balance
Austria										**Autriche**
Imports	S	20 996	66 400	119 950	150 601	169 657	172 596	171 388	147 439	Importations
Exports	S	17 247	57 655	117 722	144 889	158 821	166 546	169 186	145 851	Exportations
Balance	S	-3 749	-8 745	-2 228	-5 712	-10 836	-6 050	-2 202	-1 589	Balance

17 Total imports, exports and balance of trade *(continued)*
Imports CIF, exports FOB and balance: millions of US dollars

Total des importations, des exportations et balance commerciale *(suite)*
Importations CIF, exportations FOB et balance: en millions de dollars É.-U.

Region, country or area &	Sys.¹	1985	1995	2005	2010	2012	2013	2014	2015	Région, pays ou zone &
Azerbaijan										**Azerbaïdjan**
Imports	G	...	668	4 211	6 601	9 653	10 713	9 188	9 221	Importations
Exports	G	...	637	4 347	21 360	23 908	23 975	21 829	11 425	Exportations
Balance	G	...	-30	136	14 760	14 255	13 263	12 641	2 203	Balance
Bahamas ¹³										**Bahamas ¹³**
Imports	G	3 078	1 243	2 230	2 887	3 658	3 366	3 791	...	Importations
Exports	G	2 728	176	562	621	829	812	689	...	Exportations
Balance	G	-349	-1 067	-1 668	-2 265	-2 829	-2 554	-3 101	...	Balance
Bahrain										**Bahreïn**
Imports	G	3 107	3 715	9 393	9 800	14 900	13 000	...	...	Importations
Exports	G	2 897	4 113	10 242	15 400	20 500	17 500	...	...	Exportations
Balance	G	-210	397	849	5 600	5 600	4 500	...	...	Balance
Bangladesh										**Bangladesh**
Imports	G	2 505	6 501	12 881	26 071	34 133	33 576	35 249	...	Importations
Exports	G	986	3 173	7 233	14 195	25 113	27 033	21 058	...	Exportations
Balance	G	-1 519	-3 328	-5 648	-11 877	-9 020	-6 543	-14 191	...	Balance
Barbados										**Barbade**
Imports	G	611	771	1 604	1 562	1 806	1 759	1 739	1 618	Importations
Exports	G	357	239	359	429	570	463	474	483	Exportations
Balance	G	-254	-532	-1 245	-1 133	-1 236	-1 296	-1 265	-1 135	Balance
Belarus										**Bélarus**
Imports	G	...	5 564	16 708	34 884	46 404	42 999	40 502	30 312	Importations
Exports	G	...	4 803	15 979	25 284	46 060	37 232	36 081	26 686	Exportations
Balance	G	...	-760	-729	-9 601	-345	-5 766	-4 422	-3 626	Balance
Belgium ¹⁴										**Belgique ¹⁴**
Imports	S	56 211	159 716	319 101	391 333	439 492	451 921	455 390	380 222	Importations
Exports	S	53 762	175 884	335 738	407 055	446 637	467 831	474 090	401 192	Exportations
Balance	S	-2 449	16 169	16 638	15 721	7 145	15 910	18 700	20 970	Balance
Belize										**Belize**
Imports	G	128	257	593	709	882	930	1 005	...	Importations
Exports	G	90	143	208	280	340	315	303	...	Exportations
Balance	G	-38	-114	-385	-430	-541	-616	-701	...	Balance
Benin										**Bénin**
Imports	S	332	746	1 018	1 494	2 202	2 148	3 567	2 369	Importations
Exports	S	151	417	574	437	1 402	1 154	968	624	Exportations
Balance	S	-181	-329	-445	-1 057	-800	-995	-2 599	-1 745	Balance
Bermuda										**Bermudes**
Imports	G	402	693	964	970	885	1 005	962	929	Importations
Exports	G	23	56	49	15	17	22	12	9	Exportations
Balance	G	-379	-637	-915	-955	-868	-983	-950	-920	Balance
Bhutan										**Bhoutan**
Imports	G	84	113	387	854	992	911	810	...	Importations
Exports	G	22	103	258	641	535	544	555	...	Exportations
Balance	G	-62	-9	-129	-213	-457	-367	-255	...	Balance
Bolivia (Plurinational State of)										**Bolivie (État plurinational de)**
Imports	G	691	1 424	2 341	5 590	8 578	9 338	10 421	9 480	Importations
Exports	G	623	1 101	2 791	6 179	10 312	11 189	12 266	8 261	Exportations
Balance	G	-68	-323	450	589	1 733	1 851	1 845	-1 219	Balance
Bosnia and Herzegovina										**Bosnie-Herzégovine**
Imports	S	...	...	7 072	9 204	10 018	10 303	10 988	8 983	Importations
Exports	S	...	...	2 400	4 802	5 160	5 688	5 893	5 096	Exportations
Balance	S	...	...	-4 672	-4 402	-4 858	-4 615	-5 095	-3 887	Balance
Botswana										**Botswana**
Imports	G	...	...	3 172	5 666	8 114	8 424	8 077	7 237	Importations
Exports	G	...	...	4 455	4 692	5 971	7 774	8 509	6 309	Exportations
Balance	G	...	...	1 283	-975	-2 143	-649	431	-928	Balance
Brazil										**Brésil**
Imports	G	14 333	54 137	77 628	191 537	228 377	244 677	239 156	178 832	Importations
Exports	G	25 639	46 506	118 529	201 915	242 580	242 179	225 101	191 134	Exportations
Balance	G	11 306	-7 631	40 901	10 378	14 203	-2 498	-14 055	12 302	Balance
Brunei Darussalam										**Brunéi Darussalam**
Imports	S	613	2 087	1 447	2 539	3 563	3 612	3 597	...	Importations
Exports	S	2 974	2 389	6 242	8 908	12 982	11 447	10 588	...	Exportations
Balance	S	2 361	302	4 794	6 369	9 418	7 835	6 990	...	Balance

17

Total imports, exports and balance of trade *(continued)*
Imports CIF, exports FOB and balance: millions of US dollars

Total des importations, des exportations et balance commerciale *(suite)*
Importations CIF, exportations FOB et balance: en millions de dollars É.-U.

Region, country or area &	Sys.[t]	1985	1995	2005	2010	2012	2013	2014	2015	Région, pays ou zone &
Bulgaria										**Bulgarie**
Imports	S	13 657	5 651	18 162	25 473	32 712	34 350	34 730	28 779	Importations
Exports	S	13 348	5 353	11 739	20 571	26 670	29 492	30 930	25 756	Exportations
Balance	S	-308	-298	-6 423	-4 902	-6 042	-4 858	-3 799	-3 024	Balance
Burkina Faso										**Burkina Faso**
Imports	G	333	455	1 255	2 157	3 420	4 163	3 351	...	Importations
Exports	G	70	276	467	1 319	2 183	2 356	2 487	...	Exportations
Balance	G	-263	-180	-788	-837	-1 237	-1 807	-864	...	Balance
Burundi										**Burundi**
Imports	S	186	234	267	509	751	811	769	724	Importations
Exports	S	112	106	95	100	132	99	124	113	Exportations
Balance	S	-74	-129	-172	-409	-619	-712	-645	-611	Balance
Cabo Verde										**Cabo Verde**
Imports	G	81	252	438	743	766	727	653	...	Importations
Exports	G	5	9	18	45	53	69	115	...	Exportations
Balance	G	-76	-243	-420	-698	-713	-658	-538	...	Balance
Cambodia										**Cambodge**
Imports	S	...	...	3 927	6 791	11 000	13 000	13 500	...	Importations
Exports	S	...	...	3 200	5 143	8 200	9 100	10 800	...	Exportations
Balance	S	...	...	-727	-1 648	-2 800	-3 900	-2 700	...	Balance
Cameroon										**Cameroun**
Imports	S	1 151	934	2 880	5 051	6 515	6 657	7 553	...	Importations
Exports	S	722	1 631	2 849	3 881	4 585	4 521	5 153	...	Exportations
Balance	S	-429	697	-31	-1 170	-1 930	-2 136	-2 400	...	Balance
Canada [12]										**Canada** [12]
Imports	G	78 699	163 952	314 566	392 119	462 423	461 925	465 958	423 778	Importations
Exports	G	90 953	192 204	360 673	387 481	454 833	458 397	469 981	409 003	Exportations
Balance	G	12 254	28 251	46 107	-4 638	-7 590	-3 528	4 023	-14 775	Balance
Cayman Islands										**Îles Caïmanes**
Imports	G	147	390	1 191	828	910	929	976	916	Importations
Exports	G	2	4	52	13	20	30	26	20	Exportations
Balance	G	-145	-386	-1 138	-815	-890	-899	-950	-895	Balance
Central African Rep.										**Rép. centrafricaine**
Imports	S	114	174	173	244	276	250	...	...	Importations
Exports	S	92	171	129	91	112	140	...	...	Exportations
Balance	S	-21	-3	-44	-153	-163	-111	...	...	Balance
Chad										**Tchad**
Imports	S	323	488	954	2 507	2 600	2 997	3 496	...	Importations
Exports	S	95	317	3 095	3 411	3 901	4 496	4 194	...	Exportations
Balance	S	-228	-171	2 141	903	1 301	1 498	698	...	Balance
Chile										**Chili**
Imports	S	3 271	15 898	32 735	57 928	79 080	80 443	72 433	62 797	Importations
Exports	S	3 804	16 024	41 267	68 996	79 712	77 877	74 547	64 087	Exportations
Balance	S	533	126	8 532	11 068	632	-2 566	2 113	1 290	Balance
China										**Chine**
Imports	S	42 252	132 079	660 206	1 396 200	1 818 170	1 949 300	1 963 110	1 680 790	Importations
Exports	S	27 350	148 780	761 953	1 578 270	2 048 940	2 210 250	2 343 190	2 284 480	Exportations
Balance	S	-14 902	16 701	101 747	182 070	230 770	260 950	380 080	603 690	Balance
China, Hong Kong SAR										**Chine, Hong Kong RAS**
Imports	G	29 703	192 751	299 533	433 111	504 405	523 558	544 112	521 984	Importations
Exports	G	30 183	173 750	289 337	390 143	442 799	458 959	473 659	465 077	Exportations
Balance	G	480	-19 001	-10 196	-42 968	-61 606	-64 599	-70 453	-56 907	Balance
China, Macao SAR										**Chine, Macao RAS**
Imports	G	773	2 042	3 913	5 513	8 877	10 141	11 262	10 603	Importations
Exports	G	901	1 997	2 476	870	1 021	1 138	1 241	1 339	Exportations
Balance	G	129	-44	-1 438	-4 643	-7 856	-9 002	-10 021	-9 264	Balance
Colombia										**Colombie**
Imports	G	4 141	13 853	21 204	40 683	58 633	59 397	64 060	54 058	Importations
Exports	G	3 552	10 126	21 146	39 710	59 573	58 657	54 788	35 606	Exportations
Balance	G	-589	-3 727	-59	-973	941	-740	-9 272	-18 451	Balance
Comoros										**Comores**
Imports	S	36	63	98	190	300	285	...	...	Importations
Exports	S	16	11	12	18	25	25	...	...	Exportations
Balance	S	-20	-51	-86	-172	-275	-260	...	...	Balance

Total imports, exports and balance of trade *(continued)*
Imports CIF, exports FOB and balance: millions of US dollars

Total des importations, des exportations et balance commerciale *(suite)*
Importations CIF, exportations FOB et balance: en millions de dollars É.-U.

Region, country or area &	Sys.¹	1985	1995	2005	2010	2012	2013	2014	2015	Région, pays ou zone &
Congo										**Congo**
Imports	S	598	670	1 343	2 987	5 200	5 500	6 200	...	Importations
Exports	S	1 087	1 173	4 745	8 200	11 000	9 800	8 614	...	Exportations
Balance	S	489	503	3 402	5 213	5 800	4 300	2 414	...	Balance
Cook Islands										**Îles Cook**
Imports	G	25	49	81	91	112	116	121	110	Importations
Exports	G	3	5	5	5	5	11	18	14	Exportations
Balance	G	-21	-44	-76	-85	-106	-105	-103	-96	Balance
Costa Rica										**Costa Rica**
Imports	S	1 098	4 090	9 812	13 557	17 513	17 923	17 229	15 425	Importations
Exports	S	976	3 476	7 026	9 343	11 151	11 542	11 217	9 525	Exportations
Balance	S	-122	-614	-2 786	-4 214	-6 362	-6 381	-6 012	-5 900	Balance
Côte d'Ivoire										**Côte d'Ivoire**
Imports	S	1 749	2 970	5 860	7 863	9 774	12 628	10 722	...	Importations
Exports	S	2 945	3 762	7 693	10 285	10 861	13 687	12 634	...	Exportations
Balance	S	1 196	792	1 834	2 423	1 087	1 060	1 911	...	Balance
Croatia										**Croatie**
Imports	G	...	7 352	18 560	20 051	20 762	20 961	22 523	20 581	Importations
Exports	G	...	4 517	8 773	11 806	12 347	11 928	13 686	12 844	Exportations
Balance	G	...	-2 834	-9 788	-8 244	-8 415	-9 033	-8 837	-7 737	Balance
Cuba										**Cuba**
Imports	S	9 536	2 805	8 130	...	...	...	...	...	Importations
Exports	S	7 086	1 625	2 159	...	...	...	...	...	Exportations
Balance	S	-2 450	-1 180	-5 972	...	...	...	...	...	Balance
Cyprus										**Chypre**
Imports	G	1 247	3 694	6 282	8 647	7 379	6 419	6 828	5 667	Importations
Exports	G	476	1 229	1 303	1 507	1 828	2 136	1 924	1 931	Exportations
Balance	G	-771	-2 465	-4 979	-7 139	-5 551	-4 283	-4 904	-3 736	Balance
Czech Republic										**République tchèque**
Imports	S	...	26 385	76 343	126 600	141 515	144 320	154 233	141 337	Importations
Exports	S	...	21 686	77 988	133 020	157 167	162 302	175 017	158 631	Exportations
Balance	S	...	-4 699	1 645	6 420	15 652	17 983	20 784	17 294	Balance
Dem. Rep. of the Congo										**Rép. dém. du Congo**
Imports	S	792	871	2 270	4 500	6 100	6 300	6 500	...	Importations
Exports	S	950	1 563	2 190	5 300	6 300	6 300	6 600	...	Exportations
Balance	S	158	692	-80	800	200	0	100	...	Balance
Denmark										**Danemark**
Imports	S	18 252	45 736	72 505	83 170	92 296	98 374	99 127	84 521	Importations
Exports	S	17 096	51 488	81 912	95 758	106 125	111 349	110 494	94 231	Exportations
Balance	S	-1 155	5 751	9 407	12 589	13 829	12 975	11 367	9 710	Balance
Djibouti										**Djibouti**
Imports	G	201	177	277	420	580	560	...	...	Importations
Exports	G	14	14	40	100	95	120	...	...	Exportations
Balance	G	-187	-163	-238	-320	-485	-440	...	...	Balance
Dominica										**Dominique**
Imports	S	55	113	165	224	208	203	230	218	Importations
Exports	S	28	45	41	37	34	35	36	30	Exportations
Balance	S	-27	-68	-124	-187	-174	-168	-194	-188	Balance
Dominican Republic [12,15]										**Rép. dominicaine** [12,15]
Imports	G	1 293	3 164	7 207	12 885	14 939	13 876	13 838	...	Importations
Exports	G	737	770	1 395	2 711	4 129	4 474	4 677	...	Exportations
Balance	G	-556	-2 395	-5 811	-10 174	-10 810	-9 401	-9 162	...	Balance
Ecuador										**Équateur**
Imports	G	1 767	4 193	10 287	20 591	25 304	27 021	27 726	21 518	Importations
Exports	G	2 905	4 307	10 100	17 415	23 765	24 751	25 724	18 331	Exportations
Balance	G	1 138	114	-187	-3 176	-1 539	-2 270	-2 002	-3 187	Balance
Egypt [16]										**Égypte** [16]
Imports	G	5 495	11 739	19 816	52 923	65 774	59 662	61 010	...	Importations
Exports	G	1 838	3 435	10 652	26 438	29 409	28 493	24 736	...	Exportations
Balance	G	-3 657	-8 304	-9 163	-26 485	-36 365	-31 169	-36 275	...	Balance
El Salvador										**El Salvador**
Imports	S	961	3 329	6 834	8 548	10 270	10 772	10 513	10 416	Importations
Exports	S	679	1 652	3 387	4 472	5 340	5 491	5 273	5 485	Exportations
Balance	S	-282	-1 677	-3 448	-4 077	-4 929	-5 281	-5 240	-4 931	Balance

17 Total imports, exports and balance of trade *(continued)*
Imports CIF, exports FOB and balance: millions of US dollars

Total des importations, des exportations et balance commerciale *(suite)*
Importations CIF, exportations FOB et balance: en millions de dollars É.-U.

Region, country or area &	Sys.[t]	1985	1995	2005	2010	2012	2013	2014	2015	Région, pays ou zone &
Equatorial Guinea										**Guinée équatoriale**
Imports	G	20	50	1 310	5 680	5 987	6 990	6 492	...	Importations
Exports	G	17	86	7 062	9 964	15 467	13 981	11 587	...	Exportations
Balance	G	-3	36	5 753	4 285	9 480	6 990	5 094	...	Balance
Estonia										**Estonie**
Imports	S	...	2 400	10 188	12 282	17 797	18 142	17 992	14 508	Importations
Exports	S	...	1 663	7 676	11 607	16 083	16 291	15 931	12 929	Exportations
Balance	S	...	-736	-2 513	-675	-1 714	-1 851	-2 061	-1 579	Balance
Ethiopia										**Éthiopie**
Imports	G	993	1 141	4 081	8 535	11 914	11 510	16 244	...	Importations
Exports	G	333	422	926	2 270	3 186	3 005	3 495	...	Exportations
Balance	G	-660	-719	-3 154	-6 265	-8 728	-8 505	-12 750	...	Balance
Faroe Islands										**Îles Féroé**
Imports	G	254	314	743	780	1 144	1 110	1 045	...	Importations
Exports	G	183	362	599	839	945	1 080	1 110	...	Exportations
Balance	G	-71	48	-144	59	-199	-29	65	...	Balance
Fiji										**Fidji**
Imports	G	442	892	1 607	1 817	2 254	2 827	2 655	...	Importations
Exports	G	236	544	705	841	1 224	1 108	1 220	...	Exportations
Balance	G	-205	-348	-903	-976	-1 030	-1 718	-1 435	...	Balance
Finland										**Finlande**
Imports	G	13 234	28 114	58 474	68 773	76 558	77 590	76 767	60 363	Importations
Exports	G	13 620	39 574	65 238	69 492	73 114	74 446	74 335	59 728	Exportations
Balance	G	386	11 460	6 764	719	-3 444	-3 144	-2 432	-636	Balance
France [17]										**France** [17]
Imports	S	108 379	281 497	490 611	608 657	667 251	673 425	669 184	563 310	Importations
Exports	S	101 709	284 914	443 619	516 955	558 597	568 507	568 051	493 831	Exportations
Balance	S	-6 669	3 417	-46 992	-91 703	-108 654	-104 918	-101 133	-69 479	Balance
French Guiana [17]										**Guyane française** [17]
Imports	S	257	752	...	...	...	...	...	...	Importations
Exports	S	38	131	...	...	...	...	...	...	Exportations
Balance	S	-220	-622	...	...	...	...	...	...	Balance
French Polynesia										**Polynésie française**
Imports	S	549	1 008	1 723	1 740	1 706	1 801	1 762	1 527	Importations
Exports	S	41	194	217	153	139	152	170	130	Exportations
Balance	S	-508	-814	-1 506	-1 587	-1 567	-1 649	-1 592	-1 397	Balance
Gabon										**Gabon**
Imports	S	863	884	1 472	2 984	3 630	3 886	3 105	...	Importations
Exports	S	1 980	2 718	5 068	8 691	7 704	9 514	8 949	...	Exportations
Balance	S	1 117	1 834	3 596	5 706	4 075	5 628	5 844	...	Balance
Gambia										**Gambie**
Imports	G	93	182	260	300	380	348	...	...	Importations
Exports	G	43	16	8	15	18	8	...	...	Exportations
Balance	G	-50	-166	-252	-285	-362	-340	...	...	Balance
Georgia										**Géorgie**
Imports	G	...	489	2 488	5 257	8 037	8 012	8 593	7 707	Importations
Exports	G	...	155	865	1 677	2 376	2 910	2 861	2 204	Exportations
Balance	G	...	-333	-1 622	-3 580	-5 661	-5 102	-5 732	-5 503	Balance
Germany [18,19]										**Allemagne** [18,19]
Imports	S	158 548	464 366	780 514	1 056 170	1 164 626	1 192 751	1 209 307	1 052 121	Importations
Exports	S	184 003	523 909	977 970	1 261 577	1 408 370	1 451 631	1 492 545	1 326 839	Exportations
Balance	S	25 455	59 544	197 456	205 408	243 744	258 880	283 238	274 718	Balance
Ghana										**Ghana**
Imports	G	866	1 897	5 344	11 032	13 626	12 793	...	...	Importations
Exports	G	623	1 755	2 801	7 960	11 976	13 691	...	...	Exportations
Balance	G	-243	-142	-2 543	-3 072	-1 649	898	...	...	Balance
Gibraltar										**Gibraltar**
Imports		91	408	502	627	605	748	705	...	Importations
Exports		14	116	199	259	253	279	267	...	Exportations
Balance		-77	-292	-302	-368	-353	-469	-438	...	Balance
Greece										**Grèce**
Imports	S	10 134	25 931	57 762	69 199	63 713	62 419	64 190	48 306	Importations
Exports	S	4 539	11 025	18 445	28 203	35 452	36 236	36 007	28 705	Exportations
Balance	S	-5 596	-14 906	-39 317	-40 996	-28 261	-26 183	-28 183	-19 601	Balance

Total imports, exports and balance of trade *(continued)*
Imports CIF, exports FOB and balance: millions of US dollars

Total des importations, des exportations et balance commerciale *(suite)*
Importations CIF, exportations FOB et balance: en millions de dollars É.-U.

Region, country or area &	Sys.[i]	1985	1995	2005	2010	2012	2013	2014	2015	Région, pays ou zone &
Greenland										**Groenland**
Imports	G	296	421	593	808	850	780	762	580	Importations
Exports	G	174	364	402	380	480	490	537	353	Exportations
Balance	G	-122	-57	-190	-428	-370	-290	-225	-227	Balance
Grenada										**Grenade**
Imports	S	69	124	334	317	336	368	336	353	Importations
Exports	S	22	22	28	24	35	33	40	34	Exportations
Balance	S	-47	-102	-306	-293	-301	-336	-296	-319	Balance
Guadeloupe [17]										**Guadeloupe [17]**
Imports	S	620	1 890	...	...	...	...	...	...	Importations
Exports	S	72	159	...	...	...	...	...	...	Exportations
Balance	S	-548	-1 731	...	...	...	...	...	...	Balance
Guam										**Guam**
Imports	G	350	...	...	698	693	687	707	...	Importations
Exports	G	58	...	52	46	46	45	41	40	Exportations
Balance	G	-292	...	...	-652	-647	-642	-666	...	Balance
Guatemala										**Guatemala**
Imports	S	1 175	3 293	8 810	12 051	14 873	14 368	14 921	14 998	Importations
Exports	S	1 057	1 991	3 477	5 907	7 139	6 975	7 366	7 176	Exportations
Balance	S	-118	-1 302	-5 333	-6 145	-7 734	-7 392	-7 555	-7 822	Balance
Guinea										**Guinée**
Imports	S	448	819	820	1 405	2 300	2 150	2 115	...	Importations
Exports	S	492	702	890	1 471	1 400	1 300	1 428	...	Exportations
Balance	S	44	-117	70	66	-900	-850	-687	...	Balance
Guinea-Bissau										**Guinée-Bissau**
Imports	G	...	133	120	197	250	240	...	...	Importations
Exports	G	12	44	89	120	130	210	...	...	Exportations
Balance	G	...	-89	-31	-77	-120	-30	...	...	Balance
Guyana										**Guyana**
Imports	S	226	528	788	1 397	1 997	1 750	1 780	...	Importations
Exports	S	206	468	553	880	1 415	1 380	1 160	...	Exportations
Balance	S	-19	-60	-235	-517	-581	-370	-620	...	Balance
Haiti										**Haïti**
Imports	G	442	654	1 449	3 147	3 170	3 400	3 734	...	Importations
Exports	G	168	112	470	579	814	885	950	...	Exportations
Balance	G	-273	-542	-979	-2 568	-2 356	-2 516	-2 785	...	Balance
Honduras										**Honduras**
Imports	S	888	1 643	4 853	7 079	9 464	9 169	9 311	9 424	Importations
Exports	S	780	1 220	1 892	2 712	4 427	3 923	4 063	3 911	Exportations
Balance	S	-108	-423	-2 960	-4 367	-5 037	-5 246	-5 247	-5 513	Balance
Hungary										**Hongrie**
Imports	S	8 224	15 377	65 783	87 612	94 282	99 091	103 942	91 361	Importations
Exports	S	8 538	12 802	62 179	94 759	103 047	108 426	112 438	100 324	Exportations
Balance	S	314	-2 575	-3 604	7 147	8 765	9 335	8 496	8 963	Balance
Iceland										**Islande**
Imports	G	906	1 756	4 554	3 920	4 772	4 787	5 240	5 307	Importations
Exports	G	815	1 804	2 944	4 604	5 064	4 990	4 980	4 740	Exportations
Balance	G	-90	49	-1 610	685	292	204	-260	-567	Balance
India [20]										**Inde [20]**
Imports	G	15 935	34 710	142 865	350 192	489 689	465 424	462 909	393 740	Importations
Exports	G	9 144	30 628	99 618	226 334	296 827	314 802	322 477	267 758	Exportations
Balance	G	-6 791	-4 082	-43 247	-123 858	-192 863	-150 622	-140 432	-125 981	Balance
Indonesia										**Indonésie**
Imports	S	10 262	40 655	75 725	135 323	190 992	186 351	178 182	142 691	Importations
Exports	S	18 590	45 417	86 995	158 074	188 516	182 659	176 341	150 358	Exportations
Balance	S	8 329	4 762	11 270	22 751	-2 476	-3 692	-1 841	7 667	Balance
Iran (Islamic Rep. of) [21,22]										**Iran (Rép. islamique d') [21,22]**
Imports	S	11 635	13 882	40 041	65 404	53 451	49 709	53 569	...	Importations
Exports	S	13 328	18 360	56 252	101 316	95 500	82 000	88 800	...	Exportations
Balance	S	1 693	4 478	16 211	35 912	42 049	32 291	35 231	...	Balance
Iraq										**Iraq**
Imports		7 619	...	...	43 915	57 000	61 000	59 000	...	Importations
Exports		...	...	...	52 483	94 400	89 550	88 968	...	Exportations
Balance		...	...	...	8 567	37 400	28 550	29 968	...	Balance

17

Total imports, exports and balance of trade *(continued)*
Imports CIF, exports FOB and balance: millions of US dollars

Total des importations, des exportations et balance commerciale *(suite)*
Importations CIF, exportations FOB et balance: en millions de dollars É.-U.

Region, country or area &	Sys.[1]	1985	1995	2005	2010	2012	2013	2014	2015	Région, pays ou zone &
Ireland										**Irlande**
Imports	G	10 019	41 987	71 497	60 692	63 230	65 996	70 769	71 469	Importations
Exports	G	10 362	56 677	107 923	118 951	117 771	115 333	118 637	122 098	Exportations
Balance	G	343	14 689	36 426	58 260	54 541	49 337	47 867	50 629	Balance
Israel [23]										**Israël** [23]
Imports	S	9 875	29 579	47 142	61 209	75 392	74 861	75 483	64 990	Importations
Exports	S	6 260	19 046	42 770	58 392	63 191	66 607	68 553	63 607	Exportations
Balance	S	-3 615	-10 533	-4 371	-2 817	-12 201	-8 254	-6 931	-1 382	Balance
Italy										**Italie**
Imports	S	87 720	206 059	384 837	486 968	489 096	477 292	470 392	407 924	Importations
Exports	S	76 742	234 020	372 962	446 852	501 534	517 628	528 041	458 478	Exportations
Balance	S	-10 978	27 960	-11 875	-40 116	12 438	40 336	57 648	50 554	Balance
Jamaica										**Jamaïque**
Imports	G	1 111	2 808	4 458	5 201	6 485	6 200	5 840	...	Importations
Exports	G	566	1 420	1 499	1 331	1 709	1 574	1 444	...	Exportations
Balance	G	-545	-1 388	-2 959	-3 870	-4 776	-4 626	-4 396	...	Balance
Japan										**Japon**
Imports	G	130 516	335 990	514 987	692 435	885 610	832 424	811 882	648 316	Importations
Exports	G	177 194	443 258	594 940	769 772	798 621	714 613	690 202	624 787	Exportations
Balance	G	46 678	107 268	79 953	77 337	-86 989	-117 811	-121 681	-23 529	Balance
Jordan										**Jordanie**
Imports	G	2 733	3 696	10 506	15 085	20 691	21 701	22 952	20 016	Importations
Exports	G	789	1 769	4 302	7 023	7 926	7 896	8 376	7 849	Exportations
Balance	G	-1 944	-1 928	-6 204	-8 062	-12 765	-13 804	-14 576	-12 166	Balance
Kazakhstan										**Kazakhstan**
Imports	G	...	3 807	17 979	24 024	35 307	45 966	41 202	30 179	Importations
Exports	G	...	5 250	28 301	57 244	88 575	81 912	79 117	45 722	Exportations
Balance	G	...	1 444	10 322	33 220	53 268	35 945	37 915	15 543	Balance
Kenya										**Kenya**
Imports	G	1 436	3 006	6 149	12 074	16 288	16 358	18 397	16 097	Importations
Exports	G	958	1 890	3 293	5 149	6 126	5 856	6 046	5 908	Exportations
Balance	G	-479	-1 116	-2 856	-6 925	-10 162	-10 503	-12 351	-10 189	Balance
Kiribati										**Kiribati**
Imports	G	31	35	74	73	100	112	...	...	Importations
Exports	G	4	7	4	4	6	8	...	...	Exportations
Balance	G	-27	-28	-70	-69	-94	-105	...	...	Balance
Kuwait										**Koweït**
Imports	S	6 007	7 792	15 801	22 691	27 259	29 299	31 020	31 903	Importations
Exports	S	10 600	12 785	44 869	66 619	114 513	114 115	100 810	55 151	Exportations
Balance	S	4 593	4 992	29 068	43 927	87 254	84 816	69 791	23 248	Balance
Kyrgyzstan										**Kirghizistan**
Imports	S	...	522	1 102	3 223	5 576	6 070	5 732	4 070	Importations
Exports	S	...	409	687	1 779	1 955	2 058	1 897	1 441	Exportations
Balance	S	...	-113	-415	-1 444	-3 622	-4 012	-3 836	-2 628	Balance
Lao People's Dem. Rep.										**Rép. dém. pop. lao**
Imports	S	193	589	882	2 060	3 055	3 020	3 300	...	Importations
Exports	S	54	311	553	1 746	2 271	2 264	2 650	...	Exportations
Balance	S	-139	-278	-329	-314	-784	-756	-650	...	Balance
Latvia										**Lettonie**
Imports	S	...	1 818	8 592	11 143	16 078	16 781	16 790	13 899	Importations
Exports	S	...	1 305	5 108	8 850	12 683	13 317	13 600	11 499	Exportations
Balance	S	...	-513	-3 483	-2 292	-3 395	-3 464	-3 190	-2 400	Balance
Lebanon										**Liban**
Imports	G	2 203	5 480	9 633	18 460	21 287	21 236	21 138	18 439	Importations
Exports	G	288	656	2 337	5 021	4 485	4 059	4 548	3 982	Exportations
Balance	G	-1 915	-4 825	-7 296	-13 439	-16 802	-17 176	-16 589	-14 458	Balance
Lesotho										**Lesotho**
Imports	G	...	...	1 410	2 206	1 598	2 284	2 207	...	Importations
Exports	G	...	...	650	801	676	934	924	...	Exportations
Balance	G	...	...	-760	-1 404	-922	-1 350	-1 283	...	Balance
Liberia										**Libéria**
Imports	S	284	...	310	710	1 076	1 210	1 046	...	Importations
Exports	S	436	...	131	222	459	540	583	...	Exportations
Balance	S	151	...	-179	-488	-617	-670	-463	...	Balance

17

Total imports, exports and balance of trade *(continued)*
Imports CIF, exports FOB and balance: millions of US dollars

Total des importations, des exportations et balance commerciale *(suite)*
Importations CIF, exportations FOB et balance: en millions de dollars É.-U.

Region, country or area &	Sys.ᵗ	1985	1995	2005	2010	2012	2013	2014	2015	Région, pays ou zone &
Libya										**Libye**
Imports	G	4 101	4 137	6 058	10 506	22 996	27 010	18 994	...	Importations
Exports	G	12 314	7 704	31 278	46 016	58 954	43 986	20 994	...	Exportations
Balance	G	8 213	3 567	25 220	35 510	35 959	16 975	2 000	...	Balance
Lithuania										**Lituanie**
Imports	G	...	3 013	15 510	23 385	31 988	34 814	35 243	28 151	Importations
Exports	G	...	2 039	11 782	20 726	29 625	32 604	32 399	25 481	Exportations
Balance	G	...	-974	-3 729	-2 658	-2 363	-2 210	-2 844	-2 670	Balance
Luxembourg [24]										**Luxembourg** [24]
Imports	S	...	...	17 908	21 738	24 180	23 912	23 545	19 308	Importations
Exports	S	...	...	12 672	14 293	13 989	14 086	15 069	13 099	Exportations
Balance	S	...	...	-5 236	-7 444	-10 190	-9 826	-8 476	-6 209	Balance
Madagascar										**Madagascar**
Imports	S	402	543	1 680	2 546	2 486	3 201	3 254	...	Importations
Exports	S	274	369	831	1 082	1 236	1 951	2 142	...	Exportations
Balance	S	-128	-174	-849	-1 464	-1 250	-1 250	-1 112	...	Balance
Malawi										**Malawi**
Imports	G	295	475	1 163	2 162	2 334	2 831	2 960	...	Importations
Exports	G	248	405	508	1 130	1 183	1 196	1 370	...	Exportations
Balance	G	-47	-69	-655	-1 032	-1 151	-1 636	-1 590	...	Balance
Malaysia										**Malaisie**
Imports	G	12 253	77 691	114 410	164 622	196 393	205 898	208 874	175 962	Importations
Exports	G	15 316	73 916	140 870	198 612	227 538	228 331	234 139	199 876	Exportations
Balance	G	3 063	-3 775	26 459	33 990	31 145	22 434	25 265	23 915	Balance
Maldives										**Maldives**
Imports	G	53	268	742	1 091	1 554	1 733	1 993	...	Importations
Exports	G	23	50	104	74	162	167	145	...	Exportations
Balance	G	-30	-218	-638	-1 017	-1 393	-1 567	-1 848	...	Balance
Mali										**Mali**
Imports	S	303	774	1 544	3 430	2 940	3 699	3 951	...	Importations
Exports	S	125	443	1 092	1 996	2 163	2 601	2 097	...	Exportations
Balance	S	-178	-331	-453	-1 434	-776	-1 098	-1 854	...	Balance
Malta										**Malte**
Imports	G	759	2 942	3 807	5 735	7 923	7 479	8 122	6 442	Importations
Exports	G	400	1 913	2 376	3 721	5 697	5 182	4 836	3 851	Exportations
Balance	G	-359	-1 029	-1 432	-2 014	-2 226	-2 297	-3 286	-2 591	Balance
Marshall Islands										**Îles Marshall**
Imports	G	...	75	68	...	...	...	...	...	Importations
Exports	G	...	23	...	...	...	...	...	...	Exportations
Balance	G	...	-52	...	...	...	...	...	...	Balance
Martinique [17]										**Martinique** [17]
Imports	S	683	1 963	...	...	...	...	...	...	Importations
Exports	S	162	224	...	...	...	...	...	...	Exportations
Balance	S	-520	-1 739	...	...	...	...	...	...	Balance
Mauritania										**Mauritanie**
Imports	S	233	...	1 344	1 708	2 971	3 975	3 622	3 657	Importations
Exports	S	374	...	556	1 799	2 624	2 685	2 293	1 630	Exportations
Balance	S	140	...	-787	91	-347	-1 290	-1 329	-2 027	Balance
Mauritius										**Maurice**
Imports	G	529	1 976	3 157	4 387	5 355	5 399	5 610	4 794	Importations
Exports	G	440	1 538	2 138	2 262	2 649	2 872	3 079	2 686	Exportations
Balance	G	-88	-438	-1 018	-2 125	-2 706	-2 527	-2 531	-2 108	Balance
Mexico [12,25]										**Mexique** [12,25]
Imports	G	18 359	72 453	221 414	301 482	370 746	381 202	399 977	395 232	Importations
Exports	G	26 757	79 542	213 891	298 138	370 889	380 107	397 658	380 763	Exportations
Balance	G	8 398	7 089	-7 523	-3 344	143	-1 095	-2 319	-14 469	Balance
Micronesia (Fed. States of) [12]										**Micronésie (États féd. de)** [12]
Imports	S	...	100	128	168	194	188	...	...	Importations
Exports	S	...	43	13	23	52	35	...	...	Exportations
Balance	S	...	-56	-115	-145	-142	-153	...	...	Balance
Mongolia										**Mongolie**
Imports	G	1 096	415	1 184	3 278	6 739	6 355	5 237	3 797	Importations
Exports	G	689	473	1 065	2 899	4 385	4 273	5 775	4 670	Exportations
Balance	G	-406	58	-119	-379	-2 354	-2 082	538	872	Balance

17

Total imports, exports and balance of trade *(continued)*
Imports CIF, exports FOB and balance: millions of US dollars

Total des importations, des exportations et balance commerciale *(suite)*
Importations CIF, exportations FOB et balance: en millions de dollars É.-U.

Region, country or area &	Sys.[t]	1985	1995	2005	2010	2012	2013	2014	2015	Région, pays ou zone &
Montenegro										**Monténégro**
Imports	S	...	...	...	2 186	2 309	2 354	2 369	2 039	Importations
Exports	S	...	...	...	437	471	498	447	352	Exportations
Balance	S	...	...	...	-1 749	-1 838	-1 856	-1 921	-1 687	Balance
Montserrat										**Montserrat**
Imports	S	18	...	30	29	37	42	41	39	Importations
Exports	S	3	...	1	1	2	6	3	3	Exportations
Balance	S	-15	...	-28	-28	-35	-36	-38	-36	Balance
Morocco										**Maroc**
Imports	S	3 850	10 024	20 790	35 385	44 885	45 641	46 057	...	Importations
Exports	S	2 165	6 882	11 190	17 765	21 444	22 049	23 836	...	Exportations
Balance	S	-1 685	-3 142	-9 601	-17 620	-23 441	-23 592	-22 221	...	Balance
Mozambique										**Mozambique**
Imports	S	424	704	2 408	3 864	8 688	10 099	8 717	7 908	Importations
Exports	S	77	168	1 745	2 333	3 856	4 024	4 725	3 198	Exportations
Balance	S	-347	-536	-663	-1 530	-4 832	-6 075	-3 991	-4 710	Balance
Myanmar										**Myanmar**
Imports	G	283	1 335	1 908	4 760	9 151	12 043	16 227	...	Importations
Exports	G	303	851	3 776	8 661	8 877	11 233	11 299	...	Exportations
Balance	G	20	-483	1 868	3 901	-274	-810	-4 928	...	Balance
Namibia										**Namibie**
Imports	G	...	...	2 739	6 510	7 321	7 568	...	...	Importations
Exports	G	...	...	2 779	5 290	5 481	5 740	...	...	Exportations
Balance	G	...	...	40	-1 219	-1 840	-1 828	...	...	Balance
Nepal										**Népal**
Imports	G	427	1 230	2 282	5 495	6 499	6 428	...	...	Importations
Exports	G	151	341	863	950	960	926	...	...	Exportations
Balance	G	-275	-890	-1 419	-4 545	-5 539	-5 502	...	...	Balance
Netherlands										**Pays-Bas**
Imports	S	73 151	176 874	310 600	440 024	500 643	513 108	508 207	419 117	Importations
Exports	S	77 894	196 276	349 844	492 742	554 707	567 658	574 233	471 087	Exportations
Balance	S	4 743	19 402	39 244	52 718	54 064	54 550	66 027	51 970	Balance
Netherlands Antilles [former] [26]										**Antilles néerlandaises [anc.]** [26]
Imports	S	...	1 841	1 950	2 687	...	...	...	...	Importations
Exports	S	...	1 522	608	811	...	...	...	...	Exportations
Balance	S	...	-319	-1 342	-1 876	...	...	...	...	Balance
New Caledonia										**Nouvelle-Calédonie**
Imports	S	342	951	1 774	3 312	3 245	3 240	3 323	2 717	Importations
Exports	S	268	471	1 090	1 493	1 321	1 196	1 565	1 288	Exportations
Balance	S	-74	-480	-684	-1 820	-1 923	-2 044	-1 758	-1 429	Balance
New Zealand										**Nouvelle-Zélande**
Imports	G	5 994	14 050	26 854	31 819	37 818	40 354	42 542	36 620	Importations
Exports	G	5 722	13 388	21 926	32 285	37 383	41 074	40 658	34 265	Exportations
Balance	G	-272	-662	-4 928	466	-435	720	-1 884	-2 354	Balance
Nicaragua										**Nicaragua**
Imports	G	964	975	2 595	4 229	5 847	5 647	5 874	5 899	Importations
Exports	G	302	466	858	1 845	2 644	2 408	2 626	2 423	Exportations
Balance	G	-663	-509	-1 737	-2 384	-3 204	-3 239	-3 248	-3 476	Balance
Niger										**Niger**
Imports	S	380	375	934	2 179	1 799	1 909	2 247	...	Importations
Exports	S	259	288	490	642	1 503	1 613	1 498	...	Exportations
Balance	S	-120	-87	-444	-1 537	-296	-295	-749	...	Balance
Nigeria										**Nigéria**
Imports	G	8 877	8 222	21 314	44 235	35 703	44 598	...	...	Importations
Exports	G	12 537	12 342	55 145	84 000	114 000	...	...	...	Exportations
Balance	G	3 660	4 121	33 831	39 765	78 297	...	...	...	Balance
Niue										**Nioué**
Imports	G	2	...	...	...	...	...	...	...	Importations
Exports	G	<0	...	<0	...	...	...	...	...	Exportations
Balance	G	-2	...	...	...	...	...	...	...	Balance
Northern Mariana Islands										**Îles Mariannes du Nord**
Imports	G	...	628	...	...	...	...	...	...	Importations
Exports	G	...	941	...	...	...	...	...	...	Exportations
Balance	G	...	313	...	...	...	...	...	...	Balance

Total imports, exports and balance of trade *(continued)*
Imports CIF, exports FOB and balance: millions of US dollars

Total des importations, des exportations et balance commerciale *(suite)*
Importations CIF, exportations FOB et balance: en millions de dollars É.-U.

Region, country or area &	Sys.ᴵ	1985	1995	2005	2010	2012	2013	2014	2015	Région, pays ou zone &
Norway										**Norvège**
Imports	G	15 560	32 972	55 473	77 326	87 316	89 988	88 053	75 677	Importations
Exports	G	19 991	41 997	103 737	130 669	161 026	153 188	142 301	103 413	Exportations
Balance	G	4 431	9 024	48 265	53 344	73 710	63 201	54 247	27 736	Balance
Oman										**Oman**
Imports	G	3 153	4 248	8 827	19 775	29 447	34 333	29 432	29 007	Importations
Exports	G	3 938	6 068	18 692	36 601	53 174	56 429	52 834	34 734	Exportations
Balance	G	785	1 821	9 865	16 827	23 727	22 096	23 402	5 727	Balance
Pakistan										**Pakistan**
Imports	G	5 890	11 461	25 356	37 783	44 105	44 647	47 434	43 795	Importations
Exports	G	2 740	7 992	16 050	21 410	24 567	25 121	24 706	22 089	Exportations
Balance	G	-3 150	-3 469	-9 306	-16 373	-19 537	-19 526	-22 729	-21 706	Balance
Palau										**Palaos**
Imports	S	...	60	...	103	136	145	162	...	Importations
Exports	S	...	14	...	...	...	...	...	...	Exportations
Balance	S	...	-47	...	...	...	...	...	...	Balance
Panama										**Panama**
Imports	S	1 392	2 511	4 180	9 145	12 494	13 024	13 705	12 136	Importations
Exports	S	336	625	1 018	832	822	844	818	696	Exportations
Balance	S	-1 056	-1 886	-3 162	-8 313	-11 672	-12 180	-12 887	-11 440	Balance
Papua New Guinea										**Papouasie-Nvl-Guinée**
Imports	G	1 007	1 451	1 728	3 950	5 500	...	...	...	Importations
Exports	G	912	2 653	3 276	5 742	6 328	5 951	8 852	...	Exportations
Balance	G	-95	1 202	1 548	1 792	828	...	...	...	Balance
Paraguay										**Paraguay**
Imports	S	502	3 144	3 790	10 040	11 502	12 142	12 169	10 215	Importations
Exports	S	304	919	3 153	6 517	7 283	9 456	9 636	8 357	Exportations
Balance	S	-198	-2 225	-637	-3 524	-4 219	-2 686	-2 533	-1 858	Balance
Peru [12]										**Pérou** [12]
Imports	S	1 529	7 750	12 084	28 818	41 089	42 199	40 766	...	Importations
Exports	S	2 979	5 491	17 368	35 565	45 600	41 484	37 870	33 970	Exportations
Balance	S	1 449	-2 258	5 284	6 747	4 510	-715	-2 897	...	Balance
Philippines										**Philippines**
Imports	G	5 456	28 328	46 963	58 533	65 845	65 645	68 700	70 086	Importations
Exports	G	4 612	17 492	41 255	51 541	52 072	56 647	62 148	58 653	Exportations
Balance	G	-844	-10 836	-5 708	-6 992	-13 772	-8 999	-6 552	-11 433	Balance
Poland										**Pologne**
Imports	S	11 136	29 064	100 759	178 149	196 198	205 174	219 859	194 134	Importations
Exports	S	11 423	22 890	89 214	159 829	183 523	202 107	216 666	198 232	Exportations
Balance	S	287	-6 173	-11 545	-18 320	-12 675	-3 067	-3 193	4 098	Balance
Portugal										**Portugal**
Imports	S	7 654	33 315	53 398	75 576	72 307	75 068	77 742	66 458	Importations
Exports	S	5 686	23 212	32 129	48 738	58 256	62 840	64 061	55 402	Exportations
Balance	S	-1 968	-10 103	-21 269	-26 838	-14 051	-12 227	-13 681	-11 056	Balance
Qatar										**Qatar**
Imports	S	1 139	3 398	10 061	23 240	25 223	27 038	30 471	32 609	Importations
Exports	S	3 419	3 651	25 763	74 800	132 985	136 855	131 261	77 893	Exportations
Balance	S	2 280	253	15 702	51 560	107 761	109 817	100 789	45 283	Balance
Republic of Korea										**République de Corée**
Imports	G	31 129	134 999	267 559	425 212	519 585	515 585	525 514	436 499	Importations
Exports	G	30 282	124 985	284 422	466 384	547 879	559 632	572 665	526 756	Exportations
Balance	G	-847	-10 014	16 863	41 172	28 294	44 047	47 151	90 257	Balance
Republic of Moldova										**République de Moldova**
Imports	G	...	841	2 292	3 855	5 213	5 493	5 317	3 986	Importations
Exports	G	...	739	1 091	1 542	2 162	2 399	2 340	1 968	Exportations
Balance	G	...	-102	-1 201	-2 314	-3 051	-3 094	-2 978	-2 019	Balance
Réunion [17]										**Réunion** [17]
Imports	S	841	2 625	...	...	...	...	...	...	Importations
Exports	S	97	207	...	...	...	...	...	...	Exportations
Balance	S	-744	-2 418	...	...	...	...	...	...	Balance
Romania										**Roumanie**
Imports	S	11 267	10 278	40 463	61 885	70 260	73 452	77 882	69 858	Importations
Exports	S	12 167	7 910	27 730	49 357	57 904	65 881	69 891	60 605	Exportations
Balance	S	900	-2 368	-12 733	-12 528	-12 355	-7 571	-7 991	-9 253	Balance

17

Total imports, exports and balance of trade *(continued)*
Imports CIF, exports FOB and balance: millions of US dollars

Total des importations, des exportations et balance commerciale *(suite)*
Importations CIF, exportations FOB et balance: en millions de dollars É.-U.

Region, country or area [&]	Sys.[i]	1985	1995	2005	2010	2012	2013	2014	2015	Région, pays ou zone [&]
Russian Federation										**Fédération de Russie**
Imports	G	...	46 709	98 708	229 655	314 150	314 967	286 669	182 719	Importations
Exports	G	...	78 217	241 473	397 668	525 383	527 266	497 909	343 543	Exportations
Balance	G	...	31 508	142 766	168 013	211 233	212 299	211 240	160 824	Balance
Rwanda										**Rwanda**
Imports	G	298	241	432	1 401	1 999	2 480	2 457	...	Importations
Exports	G	131	52	125	255	470	689	736	...	Exportations
Balance	G	-167	-189	-307	-1 146	-1 529	-1 792	-1 721	...	Balance
Saint Helena [27]										**Sainte-Hélène** [27]
Imports	G	...	...	12	...	...	...	...	...	Importations
Exports	G	...	...	1	...	...	...	...	...	Exportations
Balance	G	...	...	-12	...	...	...	...	...	Balance
Saint Kitts and Nevis										**Saint-Kitts-et-Nevis**
Imports	S	51	133	210	228	226	249	268	374	Importations
Exports	S	18	19	30	45	50	50	53	58	Exportations
Balance	S	-33	-114	-180	-183	-176	-199	-215	-315	Balance
Saint Lucia										**Sainte-Lucie**
Imports	S	125	307	479	601	683	598	556	570	Importations
Exports	S	62	124	89	228	156	171	168	180	Exportations
Balance	S	-63	-183	-390	-373	-527	-427	-388	-390	Balance
Saint Vincent-Grenadines										**Saint-Vincent-Grenadines**
Imports	S	79	136	241	345	357	378	362	334	Importations
Exports	S	63	43	40	44	44	48	48	46	Exportations
Balance	S	-16	-93	-201	-301	-314	-330	-314	-288	Balance
Samoa										**Samoa**
Imports	S	51	95	187	278	308	326	...	...	Importations
Exports	S	16	9	12	13	34	24	...	...	Exportations
Balance	S	-35	-86	-175	-264	-274	-302	...	...	Balance
Sao Tome and Principe										**Sao Tomé-et-Principe**
Imports	S	10	29	50	112	140	140	172	...	Importations
Exports	S	6	5	7	11	11	12	17	...	Exportations
Balance	S	-4	-24	-43	-101	-129	-128	-155	...	Balance
Saudi Arabia										**Arabie saoudite**
Imports	S	23 600	28 053	59 458	106 864	155 592	168 155	173 908	170 089	Importations
Exports	S	27 491	49 973	180 736	251 147	388 371	375 872	342 481	202 411	Exportations
Balance	S	3 891	21 920	121 278	144 283	232 779	207 718	168 573	32 321	Balance
Senegal										**Sénégal**
Imports	G	813	1 413	3 190	4 442	5 884	6 067	6 047	5 163	Importations
Exports	G	555	994	1 067	2 059	2 381	2 440	2 617	2 296	Exportations
Balance	G	-258	-419	-2 122	-2 383	-3 502	-3 627	-3 430	-2 868	Balance
Serbia										**Serbie**
Imports	S	...	...	...	16 686	18 927	20 551	20 608	18 172	Importations
Exports	S	...	...	...	9 766	11 348	14 609	14 844	13 355	Exportations
Balance	S	...	...	...	-6 920	-7 579	-5 942	-5 765	-4 817	Balance
Seychelles										**Seychelles**
Imports	G	99	233	675	989	1 074	1 098	1 144	...	Importations
Exports	G	28	53	340	400	497	578	539	...	Exportations
Balance	G	-71	-180	-335	-589	-577	-520	-605	...	Balance
Sierra Leone										**Sierra Leone**
Imports	S	155	134	341	776	1 603	1 617	1 568	...	Importations
Exports	S	127	42	159	319	1 122	1 910	1 552	...	Exportations
Balance	S	-28	-91	-182	-458	-482	292	-16	...	Balance
Singapore										**Singapour**
Imports	G	26 288	124 502	200 050	310 791	379 723	373 016	366 247	296 745	Importations
Exports	G	22 815	118 263	229 652	351 867	408 393	410 250	405 295	346 638	Exportations
Balance	G	-3 473	-6 239	29 602	41 076	28 670	37 234	39 048	49 893	Balance
Slovakia										**Slovaquie**
Imports	S	...	9 226	36 168	66 110	79 077	83 632	83 500	74 862	Importations
Exports	S	...	8 596	31 997	64 012	79 882	85 244	85 923	75 406	Exportations
Balance	S	...	-630	-4 171	-2 098	805	1 612	2 423	544	Balance
Slovenia										**Slovénie**
Imports	S	...	9 492	19 626	26 305	28 392	29 380	30 052	25 769	Importations
Exports	S	...	8 316	17 896	24 717	27 080	28 629	30 522	26 616	Exportations
Balance	S	...	-1 175	-1 730	-1 588	-1 312	-751	471	847	Balance

Total imports, exports and balance of trade *(continued)*
Imports CIF, exports FOB and balance: millions of US dollars

Total des importations, des exportations et balance commerciale *(suite)*
Importations CIF, exportations FOB et balance: en millions de dollars É.-U.

Region, country or area &	Sys.[1]	1985	1995	2005	2010	2012	2013	2014	2015	Région, pays ou zone &
Solomon Islands										**Îles Salomon**
Imports	S	83	154	185	398	497	537	499	467	Importations
Exports	S	70	168	100	217	488	487	458	401	Exportations
Balance	S	-13	14	-85	-181	-9	-50	-41	-66	Balance
Somalia										**Somalie**
Imports	G	111	...	...	...	...	...	...	...	Importations
Exports	G	89	...	...	...	...	...	...	...	Exportations
Balance	G	-22	...	...	...	...	...	...	...	Balance
South Africa [12,28]										**Afrique du Sud** [12,28]
Imports	G	10 311	27 017	54 848	80 132	101 415	101 262	99 924	85 716	Importations
Exports	G	16 340	27 856	51 640	81 826	87 372	83 543	91 193	81 648	Exportations
Balance	G	6 029	839	-3 208	1 695	-14 043	-17 719	-8 731	-4 068	Balance
Spain										**Espagne**
Imports	S	29 965	113 316	287 644	315 548	325 836	333 932	351 452	304 314	Importations
Exports	S	24 249	91 041	191 000	246 274	286 219	310 996	318 860	277 423	Exportations
Balance	S	-5 716	-22 275	-96 644	-69 274	-39 618	-22 936	-32 592	-26 892	Balance
Sri Lanka										**Sri Lanka**
Imports	G	1 843	5 185	8 833	13 512	19 102	17 973	19 652	...	Importations
Exports	G	1 333	3 798	6 347	8 307	9 784	10 397	11 199	...	Exportations
Balance	G	-510	-1 387	-2 487	-5 205	-9 318	-7 576	-8 452	...	Balance
State of Palestine										**État de Palestine**
Imports	S	...	...	2 668	3 959	4 697	4 580	5 055	4 942	Importations
Exports	S	...	...	335	576	782	839	865	912	Exportations
Balance	S	...	...	-2 332	-3 383	-3 915	-3 740	-4 190	-4 030	Balance
Sudan [former] [29]										**Soudan [anc.]** [29]
Imports	G	771	1 219	6 757	10 045	9 230	9 918	9 211	...	Importations
Exports	G	374	556	4 824	11 404	4 067	4 790	4 350	...	Exportations
Balance	G	-397	-663	-1 933	1 360	-5 164	-5 128	-4 861	...	Balance
Suriname										**Suriname**
Imports	G	3	1	829	1 380	1 963	2 141	1 982	1 952	Importations
Exports	G	3	<0	789	1 851	2 659	2 380	2 113	1 589	Exportations
Balance	G	<0	-<0	-40	471	696	239	131	-364	Balance
Swaziland										**Swaziland**
Imports	G	...	...	1 897	1 710	1 946	1 525	...	...	Importations
Exports	G	...	...	1 761	1 557	1 897	1 894	...	...	Exportations
Balance	G	...	...	-136	-153	-49	370	...	...	Balance
Sweden										**Suède**
Imports	G	28 553	64 753	111 580	148 474	164 113	159 665	159 535	136 296	Importations
Exports	G	30 467	79 813	130 885	158 090	172 725	167 620	162 588	139 459	Exportations
Balance	G	1 913	15 060	19 305	9 616	8 612	7 955	3 053	3 163	Balance
Switzerland										**Suisse**
Imports	S	30 711	77 006	119 784	166 924	188 618	191 705	195 148	172 869	Importations
Exports	S	27 446	78 061	126 099	185 790	213 982	217 079	227 605	210 884	Exportations
Balance	S	-3 266	1 055	6 314	18 866	25 364	25 374	32 457	38 015	Balance
Syrian Arab Republic										**Rép. arabe syrienne**
Imports	S	3 967	4 709	10 862	16 950	7 800	5 800	...	...	Importations
Exports	S	1 637	3 563	8 708	14 000	4 000	3 000	...	...	Exportations
Balance	S	-2 329	-1 146	-2 154	-2 950	-3 800	-2 800	...	...	Balance
Tajikistan										**Tadjikistan**
Imports	G	...	860	1 330	2 657	3 778	4 151	4 297	3 435	Importations
Exports	G	...	779	909	1 195	1 360	1 162	977	891	Exportations
Balance	G	...	-81	-421	-1 462	-2 418	-2 989	-3 320	-2 544	Balance
Thailand										**Thaïlande**
Imports	S	9 242	70 787	118 143	185 121	250 587	249 652	227 997	201 901	Importations
Exports	S	7 121	56 440	110 163	193 366	227 752	224 863	225 190	211 033	Exportations
Balance	S	-2 121	-14 347	-7 980	8 245	-22 835	-24 789	-2 807	9 132	Balance
TFYR of Macedonia										**ex-R.Y. de Macédoine**
Imports	S	...	1 719	3 228	5 474	6 522	6 620	7 277	6 400	Importations
Exports	S	...	1 204	2 041	3 351	4 015	4 299	4 934	4 490	Exportations
Balance	S	...	-515	-1 187	-2 123	-2 507	-2 321	-2 343	-1 910	Balance
Timor-Leste										**Timor-Leste**
Imports	S	...	...	102	246	664	523	547	578	Importations
Exports	S	...	...	43	42	77	53	39	45	Exportations
Balance	S	...	...	-58	-205	-587	-470	-508	-533	Balance

17 Total imports, exports and balance of trade *(continued)*
Imports CIF, exports FOB and balance: millions of US dollars

Total des importations, des exportations et balance commerciale *(suite)*
Importations CIF, exportations FOB et balance: en millions de dollars É.-U.

Region, country or area &	Sys.[1]	1985	1995	2005	2010	2012	2013	2014	2015	Région, pays ou zone &
Togo										**Togo**
Imports	S	289	594	1 054	996	1 793	2 108	...	...	Importations
Exports	S	191	378	659	641	997	1 048	...	...	Exportations
Balance	S	-99	-215	-396	-356	-796	-1 059	...	...	Balance
Tonga										**Tonga**
Imports	G	41	77	120	159	199	198	218	...	Importations
Exports	G	5	15	10	8	17	22	23	...	Exportations
Balance	G	-36	-63	-110	-151	-182	-176	-196	...	Balance
Trinidad and Tobago										**Trinité-et-Tobago**
Imports	S	1 534	1 714	5 694	6 483	9 400	8 799	8 750	...	Importations
Exports	S	2 139	2 454	9 941	10 188	13 100	12 700	11 600	...	Exportations
Balance	S	605	740	4 247	3 705	3 700	3 902	2 850	...	Balance
Tunisia										**Tunisie**
Imports	G	2 757	7 903	13 177	22 218	24 447	24 317	24 828	20 221	Importations
Exports	G	1 738	5 475	10 494	16 427	17 008	17 061	16 756	14 073	Exportations
Balance	G	-1 019	-2 428	-2 683	-5 791	-7 439	-7 256	-8 072	-6 148	Balance
Turkey										**Turquie**
Imports	S	11 343	35 710	116 774	185 544	236 545	251 661	242 177	207 191	Importations
Exports	S	7 957	21 599	73 476	113 883	152 462	151 803	157 614	144 047	Exportations
Balance	S	-3 386	-14 111	-43 298	-71 661	-84 083	-99 858	-84 563	-63 144	Balance
Turkmenistan										**Turkménistan**
Imports	G	...	777	...	...	...	...	...	...	Importations
Exports	G	...	1 939	...	...	...	...	...	...	Exportations
Balance	G	...	1 162	...	...	...	...	...	...	Balance
Turks and Caicos Islands										**Îles Turques-et-Caïques**
Imports	G	...	...	304	302	347	345	414	...	Importations
Exports	G	...	...	15	16	15	6	6	...	Exportations
Balance	G	...	...	-289	-286	-332	-339	-408	...	Balance
Tuvalu										**Tuvalu**
Imports	G	3	6	13	...	...	...	...	...	Importations
Exports	G	<0	<0	<0	...	...	...	...	...	Exportations
Balance	G	-3	-6	-13	...	...	...	...	...	Balance
Uganda										**Ouganda**
Imports	G	315	1 056	2 049	4 709	5 230	4 927	5 086	4 761	Importations
Exports	G	472	461	1 017	3 115	2 861	2 847	2 667	2 698	Exportations
Balance	G	157	-595	-1 033	-1 594	-2 369	-2 080	-2 420	-2 063	Balance
Ukraine										**Ukraine**
Imports	G	...	15 484	36 136	60 742	84 718	76 987	54 429	37 502	Importations
Exports	G	...	13 128	34 228	51 405	68 831	63 321	53 902	38 135	Exportations
Balance	G	...	-2 356	-1 908	-9 337	-15 887	-13 666	-527	633	Balance
United Arab Emirates										**Émirats arabes unis**
Imports	G	6 549	20 984	84 654	165 000	220 000	245 000	262 000	...	Importations
Exports	G	14 043	27 753	117 287	220 000	300 000	365 000	359 000	...	Exportations
Balance	G	7 494	6 769	32 633	55 000	80 000	120 000	97 000	...	Balance
United Kingdom										**Royaume-Uni**
Imports	G	109 612	265 237	508 644	562 493	648 671	645 516	663 718	...	Importations
Exports	G	101 434	241 999	382 887	410 006	476 284	476 991	477 934	...	Exportations
Balance	G	-8 178	-23 238	-125 757	-152 487	-172 387	-168 525	-185 785	...	Balance
United Rep. of Tanzania										**Rép.-Unie de Tanzanie**
Imports	G	860	1 679	3 292	7 708	11 266	12 235	...	...	Importations
Exports	G	246	685	1 676	3 522	5 075	5 043	...	...	Exportations
Balance	G	-614	-994	-1 616	-4 186	-6 191	-7 191	...	...	Balance
United States [30]										**États-Unis** [30]
Imports	G	352 463	770 852	1 735 060	1 969 180	2 336 520	2 329 060	2 412 550	2 315 300	Importations
Exports	G	218 815	584 743	901 082	1 278 490	1 545 710	1 579 050	1 623 410	1 504 580	Exportations
Balance	G	-133 648	-186 109	-833 978	-690 690	-790 810	-750 010	-789 140	-810 720	Balance
Uruguay										**Uruguay**
Imports	G	708	2 867	3 879	8 619	10 642	10 990	10 901	9 095	Importations
Exports	G	909	2 106	3 405	6 707	8 601	8 844	9 475	7 742	Exportations
Balance	G	201	-761	-474	-1 912	-2 041	-2 146	-1 425	-1 354	Balance
Uzbekistan										**Ouzbékistan**
Imports	G	...	2 893	3 666	8 386	...	13 799	...	...	Importations
Exports	G	...	3 720	4 749	11 587	...	15 087	...	...	Exportations
Balance	G	...	827	1 083	3 201	...	1 288	...	...	Balance

17 Total imports, exports and balance of trade *(continued)*
Imports CIF, exports FOB and balance: millions of US dollars

Total des importations, des exportations et balance commerciale *(suite)*
Importations CIF, exportations FOB et balance: en millions de dollars É.-U.

Region, country or area &	Sys.ᵗ	1985	1995	2005	2010	2012	2013	2014	2015	Région, pays ou zone &
Vanuatu										**Vanuatu**
Imports	G	70	95	149	285	296	313	313	367	Importations
Exports	G	31	28	38	49	55	39	63	39	Exportations
Balance	G	-39	-67	-111	-237	-241	-275	-250	-328	Balance
Venezuela (Boliv. Rep. of)										**Venezuela (Rép. boliv. du)**
Imports	G	8 106	12 650	24 027	33 815	43 501	46 363	44 478	40 146	Importations
Exports	G	14 438	18 457	51 859	65 745	97 877	88 753	74 714	37 236	Exportations
Balance	G	6 332	5 807	27 832	31 930	54 376	42 390	30 236	-2 910	Balance
Viet Nam										**Viet Nam**
Imports	G	1 857	8 155	36 408	83 779	115 101	131 260	148 770	162 825	Importations
Exports	G	699	5 449	31 726	71 658	115 458	132 478	149 565	162 061	Exportations
Balance	G	-1 159	-2 707	-4 682	-12 121	357	1 218	795	-764	Balance
Wallis and Futuna Islands										**Îles Wallis-et-Futuna**
Imports	S	...	14	...	...	...	...	...	...	Importations
Exports	S	...	<0	...	...	...	...	...	...	Exportations
Balance	S	...	-13	...	...	...	...	...	...	Balance
Yemen										**Yémen**
Imports	S	...	1 817	5 401	9 746	11 975	12 500	...	...	Importations
Exports	S	...	1 917	5 604	8 497	8 500	9 500	...	...	Exportations
Balance	S	...	101	204	-1 249	-3 475	-3 000	...	...	Balance
Zambia										**Zambie**
Imports	S	654	683	2 575	5 319	8 810	10 177	9 545	8 451	Importations
Exports	S	482	1 070	2 132	7 206	9 375	10 600	9 696	6 983	Exportations
Balance	S	-172	388	-443	1 888	565	423	151	-1 468	Balance
Zimbabwe										**Zimbabwe**
Imports	G	867	2 651	2 350	3 800	4 400	4 300	4 200	...	Importations
Exports	G	1 114	2 121	1 850	3 199	3 800	3 552	3 438	...	Exportations
Balance	G	248	-530	-500	-601	-600	-748	-762	...	Balance
ANCOM§ ³¹										**ANCOM§ ³¹**
Imports		8 470	27 220	45 917	95 681	133 604	137 955	142 974	121 140	Importations
Exports		10 058	21 025	51 405	98 869	139 249	136 081	130 648	96 168	Exportations
Balance		1 589	-6 195	5 488	3 188	5 645	-1 875	-12 326	-24 972	Balance
APEC§										**CEAP§**
Imports		804 180	2 402 261	5 079 942	7 352 092	9 105 185	9 203 730	9 284 723	8 239 215	Importations
Exports		740 385	2 347 079	4 682 109	7 233 306	8 699 870	8 865 744	9 055 196	8 288 364	Exportations
Balance		-63 794	-55 183	-397 833	-118 787	-405 314	-337 986	-229 527	49 148	Balance
ASEAN§										**ANASE§**
Imports		66 582	354 347	599 864	954 320	1 225 410	1 243 497	1 235 395	1 087 557	Importations
Exports		72 499	320 603	654 432	1 049 577	1 252 058	1 269 272	1 288 015	1 162 289	Exportations
Balance		5 917	-33 744	54 568	95 258	26 648	25 776	52 620	74 731	Balance
CACM§										**MCCA§**
Imports		5 087	13 330	32 904	45 464	57 967	57 879	57 847	56 162	Importations
Exports		3 794	8 805	16 640	24 279	30 701	30 340	30 545	28 520	Exportations
Balance		-1 293	-4 525	-16 264	-21 186	-27 266	-27 539	-27 302	-27 642	Balance
CARICOM§										**CARICOM§**
Imports		7 698	9 185	19 630	25 011	31 742	30 698	30 967	28 601	Importations
Exports		6 475	5 320	14 732	16 573	21 785	20 883	19 107	17 051	Exportations
Balance		-1 222	-3 865	-4 899	-8 438	-9 957	-9 815	-11 860	-11 550	Balance
COMESA§										**COMESA§**
Imports		17 471	30 902	62 254	127 911	169 244	170 341	169 739	157 107	Importations
Exports		20 100	21 847	64 023	117 014	132 277	118 867	92 093	72 873	Exportations
Balance		2 629	-9 055	1 769	-10 897	-36 967	-51 474	-77 646	-84 234	Balance
ECOWAS§										**CEDEAO§**
Imports		15 067	19 324	43 800	81 957	83 517	96 405	109 357	123 291	Importations
Exports		18 571	21 920	71 282	111 531	150 650	154 063	151 988	148 652	Exportations
Balance		3 504	2 596	27 481	29 574	67 133	57 658	42 631	25 361	Balance
EMCCA§										**CEMAC§**
Imports		3 069	3 199	8 131	19 453	24 208	26 281	27 066	28 621	Importations
Exports		3 994	6 095	22 948	34 237	42 770	42 451	38 666	35 901	Exportations
Balance		925	2 895	14 818	14 784	18 562	16 170	11 600	7 280	Balance
European Union (EU)										**Union européenne (UE)**
Imports		...	...	4 065 269	5 192 789	5 706 032	5 797 562	5 882 409	5 093 034	Importations
Exports		...	...	3 976 706	5 064 357	5 682 138	5 873 238	5 987 893	5 232 703	Exportations
Balance		...	...	-88 563	-128 432	-23 894	75 677	105 484	139 669	Balance

17

Total imports, exports and balance of trade *(continued)*
Imports CIF, exports FOB and balance: millions of US dollars

Total des importations, des exportations et balance commerciale *(suite)*
Importations CIF, exportations FOB et balance: en millions de dollars É.-U.

Region, country or area &	Sys.[t]	1985	1995	2005	2010	2012	2013	2014	2015	Région, pays ou zone &
Extra-EU-28 [32]										**Extra-UE-28** [32]
Imports		...	...	1 470 442	2 029 009	2 311 903	2 234 492	2 233 120	1 914 903	Importations
Exports		...	...	1 303 309	1 791 433	2 162 563	2 305 992	2 259 963	1 985 139	Exportations
Balance		...	...	-167 132	-237 576	-149 340	71 500	26 843	70 235	Balance
LAIA§										**ALAI§**
Imports		68 491	213 710	450 393	767 217	966 821	1 008 125	1 007 530	894 998	Importations
Exports		97 728	207 793	527 036	817 386	1 029 838	1 036 602	1 006 236	860 402	Exportations
Balance		29 237	-5 918	76 643	50 169	63 017	28 476	-1 295	-34 595	Balance
LLDCs [31]										**PDSL** [31]
Imports		17 382	30 895	84 872	165 892	220 674	238 675	252 236	251 434	Importations
Exports		11 498	21 884	80 274	155 440	203 008	211 932	200 886	184 932	Exportations
Balance		-5 884	-9 011	-4 598	-10 452	-17 666	-26 743	-51 350	-66 502	Balance
MERCOSUR§ [31]										**MERCOSUR§** [31]
Imports		28 154	94 247	140 358	297 650	371 105	397 511	382 447	307 557	Importations
Exports		50 309	90 052	220 088	351 785	441 870	443 446	403 128	312 435	Exportations
Balance		22 156	-4 195	79 729	54 135	70 766	45 935	20 680	4 878	Balance
NAFTA§										**ALENA§**
Imports		449 521	1 007 257	2 271 040	2 662 781	3 169 689	3 172 187	3 278 485	3 134 310	Importations
Exports		336 525	856 488	1 475 646	1 964 109	2 371 432	2 417 554	2 491 049	2 294 346	Exportations
Balance		-112 996	-150 769	-795 394	-698 672	-798 257	-754 633	-787 436	-839 964	Balance
OECD§ [31]										**OCDE§** [31]
Imports		1 463 774	3 807 884	7 529 584	9 594 770	11 053 916	11 089 896	11 244 896	9 964 274	Importations
Exports		1 377 380	3 813 206	6 780 943	8 955 040	10 222 810	10 370 097	10 532 036	9 350 041	Exportations
Balance		-86 394	5 322	-748 641	-639 730	-831 106	-719 800	-712 860	-614 233	Balance
OPEC§ [31]										**OPEP§** [31]
Imports		100 903	156 670	389 171	728 480	931 090	992 854	1 024 464	993 610	Importations
Exports		156 154	215 345	748 887	1 245 995	1 752 832	1 748 779	1 638 606	1 266 149	Exportations
Balance		55 251	58 674	359 715	517 515	821 742	755 925	614 142	272 538	Balance
World exc. intra-EU27										**Monde excl. intra-UE27**
Imports		...	...	8 007 587	11 993 859	14 731 758	14 860 111	14 948 041	13 271 771	Importations
Exports		...	...	7 682 990	11 833 284	14 561 734	14 901 204	14 914 873	13 215 003	Exportations
Balance		...	...	-324 597	-160 575	-170 024	41 093	-33 168	-56 767	Balance

Source:

United Nations Statistics Division, New York, Trade Statistics Branch, last accessed July 2016.

Source:

Organisation des Nations Unies, Division de statistique, New York, la base de données pour les statistiques du commerce extérieur, dernier accès juillet 2016.

& The regional totals for imports and exports have been adjusted to exclude the re-exports of countries or areas comprising each region. § For member states of this grouping, see Annex I – Other groupings. The totals have been calculated for all periods shown according to the current composition. t Systems of trade: Two systems of recording trade, the General trade system (G) and the Special trade system (S), are in common use. They differ mainly in the way warehoused and re-exported goods are recorded. See the Technical notes for an explanation of the trade systems.

& Les totaux régionaux pour importations et exportations ont été ajustés pour exclure les réexportations des pays ou zones qui comprennent la région. § Pour les Etats membres de ce groupements, voir annexe I – Autres groupements. Les totales ont été calculés pour toutes les périodes données suivant la composition présente. t Systèmes de commerce : Deux systèmes d'enregistrement du commerce sont couramment utilisés, le Commerce général (G) et le Commerce spécial (S). Ils ne diffèrent que par la façon dont sont enregistrées les marchandises entreposées et les marchandises réexportées. Voir les Notes techniques pour une explication des Systèmes de commerce.

1   This classification is intended for statistical convenience and does not, necessarily, express a judgement about the stage reached by a particular country in the development process.

2   Developed Economies of America, Europe, and the Asia-Pacific region.

3   Comprises of Australia, Japan and New Zealand.

4   Excludes CIS countries in Europe and South Eastern Europe.

5   Includes Albania, Bosnia and Herzegovina, Bulgaria, Montenegro, Romania, Serbia, and The Former Yugoslav Republic of Macedonia.

1   Cette classification est utilisée pour plus de commodité dans la présentation des statistiques et n'implique pas nécessairement un jugement quant au stade de développement auquel est parvenu un pays donné.

2   Économies développées de l'Amérique, de l'Europe, et de la région Asie-Pacifique.

3   Comprend l'Australie, le Japon et la Nouvelle-Zélande.

4   Exclut les pays membres de la CEI en Europe et Europe du Sud-est.

5   Comprend l'Albanie, la Bosnie-Herzégovine, la Bulgarie, le Monténégro, la Roumanie, la Serbie et l'ex-République yougoslave de Macédoine.

17

Total imports, exports and balance of trade *(continued)*
Imports CIF, exports FOB and balance: millions of US dollars

Total des importations, des exportations et balance commerciale *(suite)*
Importations CIF, exportations FOB et balance: en millions de dollars É.-U.

6	Excludes Sudan.	6	Exclut le Soudan.
7	Includes Sudan.	7	Comprend le Soudan.
8	Excludes Japan.	8	Exclut le Japon.
9	Excludes Armenia, Azerbaijan and Georgia.	9	Exclut l'Arménie, l'Azerbaïdjan et la Géorgie.
10	Excludes Australia and New Zealand.	10	Exclut l'Australie et la Nouvelle-Zélande.
11	Year ending 30 September.	11	Année finissant le 30 septembre.
12	Imports FOB.	12	Importations FOB.
13	Trade statistics exclude certain oil and chemical products.	13	Les statistiques commerciales font exclusion de certains produits pétroliers et chimiques.
14	Economic Union of Belgium and Luxembourg. Intertrade between the two countries is excluded. Beginning January 1997, data refer to Belgium only and include trade between Belgium and Luxembourg.	14	L'Union économique belgo-luxembourgeoise. Non compris le commerce entre ces pays. A partir de janvier 1997, les données se rapportent à Belgique seulement et recouvrent les échanges entre la Belgique et le Luxembourg
15	Export and import values exclude trade in the processing zone.	15	Les valeurs à l'exportation et à l'importation excluent le commerce de la zone de transformation.
16	Imports exclude petroleum imported without stated value. Exports cover domestic exports.	16	Non compris le pétrole brute dont la valeur des importations ne sont pas stipulée. Les exportations sont les exportations d'intérieur.
17	Beginning 1997, trade data for France include the import and export values of French Guiana, Guadeloupe, Martinique, and Réunion.	17	A partir de 1997, les valeurs de commerce pour la France comprennent les valeurs des importations et des exportations de la Guyane française, la Guadeloupe, la Martinique, et la Réunion.
18	Prior to 1991, data refer to the Federal Republic of Germany.	18	Avant 1991, les données se rapportent à la République Fédérale d'Allemagne.
19	Prior to January 1991, excludes trade conducted in accordance with the supplementary protocol to the treaty on the basis of relations between the Federal Republic of Germany and the former German Democratic Republic.	19	Avant janvier 1991, non compris le commerce effectué en accord avec le protocole additionnel au traité définissant la base des relations entre la République Fédérale d'Allemagne et l'ancienne République Démocratique Allemande.
20	Excluding military goods, fissionable materials, bunkers, ships and aircraft.	20	A l'exclusion des marchandises militaires, des matières fissibles, des soutes, des bateaux et de l'avion.
21	Year ending 20 March of the year stated.	21	Année finissant le 20 mars de l'année indiquée.
22	Data include oil and gas. The value of oil exports and total exports are rough estimates based on information published in various petroleum industry journals.	22	Les données comprennent le pétrole et le gaz. La valeur des exportations de pétrole et des exportations totales sont des évaluations grossières basées sur l'information publiée à divers journaux d'industrie de pétrole.
23	Imports and exports net of returned goods. The figures also exclude Judea and Samaria and the Gaza area.	23	Importations et exportations nets, ne comprenant pas les marchandises retournées. Sont également exclues les données de la Judée et de Samara et ainsi que la zone de Gaza.
24	Prior to 1997, included under Belgium. See also footnote for Belgium.	24	Avant 1997, inclus sous la Belgique. Voir également l'apostille pour la Belgique.
25	Trade data include maquiladoras and exclude goods from customs-bonded warehouses. Total exports include revaluation and exports of silver.	25	Les statistiques du commerce extérieur comprennent maquiladoras et ne comprennent pas les marchandises provenant des entrepôts en douane. Les exportations comprennent la réévaluation et les données sur les exportations d'argent.
26	The Netherlands Antilles was dissolved on October 10, 2010. Beginning 2011, data are reported separately for Curaçao, Sint Maarten (Dutch part), Bonaire, Saint Eustatius and Saba.	26	Les Antilles néerlandaises a été dissous le 10 Octobre 2010. Début 2011, les données sont présentées séparément pour Curaçao, Sint Maarten (partie néerlandaise), Bonaire, Saint-Eustache et Saba.
27	Year ending 31 March of the following year.	27	Année finissant le 31 Mars de l'année suivante.
28	Exports include gold.	28	Les exportations comprennent l'or.
29	Year ending June 30 through 1994. Year ending December 31 thereafter.	29	Année finissant juin 30 à 1994. Année finissant décembre 31 ensuite.
30	Including the trade of the U.S. Virgin Islands and Puerto Rico but excluding shipments of merchandise between the United States and its other possessions (Guam and American Samoa). Data include imports and exports of non-monetary gold.	30	Y compris le commerce des Iles Vierges américaines et de Porto Rico mais non compris les échanges de marchandises, entre les Etats-Unis et leurs autres possessions (Guam et Samoa américaines). Les données comprennent les importations et exportations d'or non-monétaire.
31	The figures for the country groupings aim to always reflect the membership of the grouping of the latest year published.	31	Les chiffres pour les groupes de pays visent à toujours refléter la composition du groupement de la dernière année de publication.
32	Excluding intra-EU trade.	32	Non compris le commerce de l'intra-UE.

Part Three

Energy, environment and infrastructure

Troisième partie

Énergie, environnement et infrastructures

Production, trade and supply of energy
Petajoules and gigajoules per capita
Production, commerce et fourniture d'énergie
pétajoules et gigajoules par habitant

Region, country or area	1990	1995	2000	2005	2010	2011	2012	2013	Région, pays ou zone
World									**Monde**
Primary Energy production	357 710	378 635	408 171	474 857	532 611	547 240	556 304	567 050	Production d'énergie primaire
Net imports	-6 747	-8 893	-10 690	-14 084	-14 795	-15 380	-14 871	-16 621	Importations nettes
Changes in stocks	3 911	128	-2 792	1 131	-454	2 202	630	228	Variations des stocks
Total supply	347 052	369 614	400 273	459 642	518 270	529 657	540 803	550 201	Approvisionnement total
Supply per capita	62	65	65	71	75	76	76	77	Approvisionnement par habitant
Africa									**Afrique**
Primary Energy production	27 016	32 075	37 222	45 034	47 702	44 906	47 033	45 472	Production d'énergie primaire
Net imports	-12 183	-13 034	-16 231	-20 559	-19 483	-15 604	-17 251	-15 179	Importations nettes
Changes in stocks	-116	307	-17	-3	-32	220	-40	135	Variations des stocks
Total supply	14 948	18 733	21 007	24 477	28 250	29 082	29 821	30 156	Approvisionnement total
Supply per capita	24	26	26	27	27	27	27	27	Approvisionnement par habitant
America, North									**Amérique du Nord**
Primary Energy production	89 399	93 074	96 017	97 592	100 278	103 349	104 444	108 024	Production d'énergie primaire
Net imports	7 822	9 185	15 648	19 970	12 674	9 844	6 391	3 243	Importations nettes
Changes in stocks	1 795	-931	-2 245	-173	-515	111	233	-1 712	Variations des stocks
Total supply	95 426	103 190	113 910	117 735	113 467	113 082	110 603	112 979	Approvisionnement total
Supply per capita	223	226	234	229	207	205	198	201	Approvisionnement par habitant
America, South									**Amérique du Sud**
Primary Energy production	16 214	20 331	24 088	26 662	29 526	30 512	30 757	31 052	Production d'énergie primaire
Net imports	-3 728	-5 787	-7 302	-8 054	-7 360	-7 897	-7 376	-7 191	Importations nettes
Changes in stocks	88	-8	-24	-323	-105	38	-196	-44	Variations des stocks
Total supply	12 399	14 551	16 810	18 930	22 271	22 578	23 578	23 905	Approvisionnement total
Supply per capita	42	45	48	51	56	56	58	58	Approvisionnement par habitant
Asia									**Asie**
Primary Energy production	104 990	131 214	145 551	190 838	238 944	253 226	257 518	264 336	Production d'énergie primaire
Net imports	-11 084	-8 857	-9 656	-9 595	-1 069	-2 683	3 665	5 424	Importations nettes
Changes in stocks	1 551	577	-538	1 081	1 534	2 013	876	1 745	Variations des stocks
Total supply	92 354	121 780	136 434	180 162	236 341	248 530	260 307	268 015	Approvisionnement total
Supply per capita	29	35	37	46	57	59	61 ,	62	Approvisionnement par habitant
Europe									**Europe**
Primary Energy production	112 686	93 201	94 583	102 385	102 277	101 823	102 517	102 943	Production d'énergie primaire
Net imports	15 176	13 388	12 215	10 673	8 133	8 377	7 441	5 733	Importations nettes
Changes in stocks	390	79	266	585	-1 195	263	-138	59	Variations des stocks
Total supply	127 473	106 510	106 533	112 472	111 605	109 937	110 095	108 618	Approvisionnement total
Supply per capita	117	146	147	154	152	149	149	147	Approvisionnement par habitant
Oceania									**Océanie**
Primary Energy production	7 406	8 741	10 710	12 346	13 883	13 425	14 035	15 222	Production d'énergie primaire
Net imports	-2 750	-3 787	-5 364	-6 516	-7 689	-7 419	-7 741	-8 651	Importations nettes
Changes in stocks	203	103	-233	-37	-141	-443	-105	44	Variations des stocks
Total supply	4 452	4 851	5 579	5 867	6 335	6 449	6 398	6 528	Approvisionnement total
Supply per capita	163	167	179	175	174	174	170	171	Approvisionnement par habitant
Afghanistan									**Afghanistan**
Primary Energy production	19	16	18	23	41	62	56	59	Production d'énergie primaire
Net imports	*28	*13	*8	*14	96	124	253	*263	Importations nettes
Changes in stocks *	0	0	0	...	...	...	...	...	Variations des stocks *
Total supply	*46	*29	25	36	137	185	308	*321	Approvisionnement total
Supply per capita	*4	*1	1	1	5	6	10	*10	Approvisionnement par habitant
Albania									**Albanie**
Primary Energy production	99	43	34	48	69	62	70	85	Production d'énergie primaire
Net imports	*13	4	31	43	*22	*33	*18	*27	Importations nettes
Changes in stocks	*23	0	...	0	0	0	0	10	Variations des stocks
Total supply	89	47	65	92	91	95	88	102	Approvisionnement total
Supply per capita	27	15	21	29	31	33	30	35	Approvisionnement par habitant
Algeria									**Algérie**
Primary Energy production	4 380	4 748	6 556	7 534	6 200	6 008	5 928	5 676	Production d'énergie primaire
Net imports	*-3 266	*-3 486	*-5 415	-5 953	-4 551	-4 292	-4 043	-3 714	Importations nettes
Changes in stocks	*41	-8	-27	4	9	1	-5	3	Variations des stocks
Total supply	1 073	1 269	1 169	1 578	1 641	1 715	1 890	1 959	Approvisionnement total
Supply per capita	42	45	38	48	46	47	50	51	Approvisionnement par habitant

18
Production, trade and supply of energy *(continue)*
Petajoules and gigajoules per capita
Production, commerce et fourniture d'énergie *(suite)*
pétajoules et gigajoules par habitant

Region, country or area	1990	1995	2000	2005	2010	2011	2012	2013	Région, pays ou zone
Andorra									**Andorre**
Primary Energy production	...	*0	*0	0	1	1	0	1	Production d'énergie primaire
Net imports	0	6	*9	*10	*9	*9	*9	*9	Importations nettes
Changes in stocks	...	...	*0	0	...	...	...	...	Variations des stocks
Total supply	0	7	9	10	10	9	*9	*9	Approvisionnement total
Supply per capita	0	107	138	130	114	111	*115	*121	Approvisionnement par habitant
Angola									**Angola**
Primary Energy production	1 198	1 586	1 811	2 951	4 109	3 827	4 030	4 067	Production d'énergie primaire
Net imports	*-960	-1 224	-1 475	-2 565	-3 547	-3 240	-3 374	-3 424	Importations nettes
Changes in stocks	...	17	21	5	1	1	3	1	Variations des stocks
Total supply	238	344	316	381	561	586	654	642	Approvisionnement total
Supply per capita	23	28	23	23	26	27	29	27	Approvisionnement par habitant
Anguilla *									**Anguilla ***
Primary Energy production	...	...	0	0	0	0	0	0	Production d'énergie primaire
Net imports	...	...	1	2	2	2	2	2	Importations nettes
Total supply	...	...	1	2	2	2	2	2	Approvisionnement total
Supply per capita	...	...	79	132	155	144	140	136	Approvisionnement par habitant
Antigua and Barbuda									**Antigua-et-Barbuda**
Primary Energy production *	3	3	...	...	...	...	...	...	Production d'énergie primaire *
Net imports *	5	5	5	7	8	8	8	8	Importations nettes *
Changes in stocks	0	...	...	...	...	...	...	...	Variations des stocks
Total supply *	7	7	5	6	8	7	8	8	Approvisionnement total *
Supply per capita *	113	107	65	71	86	84	85	84	Approvisionnement par habitant *
Argentina									**Argentine**
Primary Energy production	2 064	2 722	3 404	3 609	3 343	3 248	3 160	2 981	Production d'énergie primaire
Net imports	-144	-535	-866	-683	-21	137	205	402	Importations nettes
Changes in stocks	-8	-6	-17	8	16	29	-17	5	Variations des stocks
Total supply	1 928	2 194	2 556	2 919	3 307	3 356	3 382	3 378	Approvisionnement total
Supply per capita	59	63	69	75	80	81	80	79	Approvisionnement par habitant
Armenia									**Arménie**
Primary Energy production	...	10	27	36	52	55	52	52	Production d'énergie primaire
Net imports	...	59	57	69	*70	*79	*92	*90	Importations nettes
Changes in stocks	...	...	...	...	*3	0	0	0	Variations des stocks
Total supply	...	68	84	105	*119	135	145	143	Approvisionnement total
Supply per capita	...	21	27	34	*40	45	49	48	Approvisionnement par habitant
Aruba									**Aruba**
Primary Energy production	0	0	*5	*5	*5	*5	*4	*1	Production d'énergie primaire
Net imports *	23	23	20	23	22	23	13	12	Importations nettes *
Changes in stocks	0	...	...	...	...	...	...	...	Variations des stocks
Total supply	*24	*22	*25	27	*27	*27	*17	*12	Approvisionnement total
Supply per capita	*382	*275	*275	270	*266	*263	*162	*120	Approvisionnement par habitant
Australia [1]									**Australie [1]**
Primary Energy production	6 638	7 913	9 843	11 586	13 019	12 603	13 185	14 368	Production d'énergie primaire
Net imports	-2 694	-3 752	-5 412	-6 717	-7 912	-7 684	-7 984	-8 944	Importations nettes
Changes in stocks	203	104	-221	-28	-158	-449	-83	38	Variations des stocks
Total supply	3 741	4 057	4 651	4 898	5 264	5 367	5 284	5 386	Approvisionnement total
Supply per capita	219	224	243	240	238	238	231	231	Approvisionnement par habitant
Austria									**Autriche**
Primary Energy production	340	367	409	418	507	483	539	507	Production d'énergie primaire
Net imports	712	737	775	1 003	886	965	874	856	Importations nettes
Changes in stocks	14	-14	-10	0	-39	60	26	-27	Variations des stocks
Total supply	1 038	1 119	1 194	1 421	1 432	1 387	1 387	1 391	Approvisionnement total
Supply per capita	135	141	149	173	171	165	164	164	Approvisionnement par habitant
Azerbaijan									**Azerbaïdjan**
Primary Energy production	...	628	803	1 155	2 759	2 525	2 458	2 486	Production d'énergie primaire
Net imports	...	-91	-319	-547	-2 239	-2 009	-1 887	-1 907	Importations nettes
Changes in stocks	...	0	-2	35	34	-11	13	11	Variations des stocks
Total supply	...	536	485	573	486	527	558	567	Approvisionnement total
Supply per capita	...	69	60	67	53	57	60	60	Approvisionnement par habitant

18

Production, trade and supply of energy *(continue)*
Petajoules and gigajoules per capita
Production, commerce et fourniture d'énergie *(suite)*
pétajoules et gigajoules par habitant

Region, country or area	1990	1995	2000	2005	2010	2011	2012	2013	Région, pays ou zone
Bahamas									**Bahamas**
Primary Energy production	0	...	...	0	0	0	0	0	Production d'énergie primaire
Net imports *	35	21	23	26	35	28	27	44	Importations nettes *
Changes in stocks *	8	-1	0	...	...	...	...	...	Variations des stocks *
Total supply	*27	*23	*23	*25	36	28	28	45	Approvisionnement total
Supply per capita	*106	*83	*78	*79	99	76	75	118	Approvisionnement par habitant
Bahrain									**Bahreïn**
Primary Energy production	294	341	706	672	840	863	830	920	Production d'énergie primaire
Net imports	*-87	*-94	-388	-377	*-320	*-337	*-304	*-347	Importations nettes
Changes in stocks	*-6	-10	-6	-28	0	-1	-6	-2	Variations des stocks
Total supply	213	257	324	323	520	526	532	574	Approvisionnement total
Supply per capita	433	459	508	446	412	403	399	425	Approvisionnement par habitant
Bangladesh									**Bangladesh**
Primary Energy production	682	792	857	1 027	1 300	1 307	1 347	1 410	Production d'énergie primaire
Net imports	*97	*125	139	165	163	206	212	212	Importations nettes
Changes in stocks	-6	-4	-6	3	-22	-20	-36	-2	Variations des stocks
Total supply	785	921	1 001	1 189	1 486	1 533	1 596	1 625	Approvisionnement total
Supply per capita	7	8	8	8	10	10	10	10	Approvisionnement par habitant
Barbados									**Barbade**
Primary Energy production	5	5	6	5	4	3	3	3	Production d'énergie primaire
Net imports	9	8	10	*12	*16	*17	*17	*17	Importations nettes
Changes in stocks	-1	0	-1	0	0	0	0	0	Variations des stocks
Total supply	16	13	16	18	20	20	20	19	Approvisionnement total
Supply per capita	62	49	61	66	73	72	70	69	Approvisionnement par habitant
Belarus									**Bélarus**
Primary Energy production	...	138	147	159	174	180	173	167	Production d'énergie primaire
Net imports	...	912	881	962	980	1 031	1 116	982	Importations nettes
Changes in stocks	...	6	-1	-4	-8	-26	9	7	Variations des stocks
Total supply	...	1 043	1 029	1 125	1 163	1 238	1 280	1 142	Approvisionnement total
Supply per capita	...	102	102	115	122	130	135	120	Approvisionnement par habitant
Belgium									**Belgique**
Primary Energy production	544	495	570	577	645	670	590	619	Production d'énergie primaire
Net imports	1 457	1 771	1 835	1 877	1 893	1 733	1 639	1 740	Importations nettes
Changes in stocks	-4	29	-21	19	7	12	-24	21	Variations des stocks
Total supply	2 005	2 237	2 426	2 436	2 530	2 390	2 254	2 338	Approvisionnement total
Supply per capita	202	222	238	234	232	217	203	210	Approvisionnement par habitant
Belize									**Belize**
Primary Energy production	4	4	5	4	13	13	12	*10	Production d'énergie primaire
Net imports	4	6	*6	*7	-2	1	1	3	Importations nettes
Changes in stocks	...	...	...	...	0	0	0	0	Variations des stocks
Total supply	8	10	11	*11	12	13	13	13	Approvisionnement total
Supply per capita	44	44	45	*38	38	41	38	39	Approvisionnement par habitant
Benin									**Bénin**
Primary Energy production	74	79	62	70	86	89	91	94	Production d'énergie primaire
Net imports	*1	*10	20	37	66	69	72	76	Importations nettes
Changes in stocks	0	0	0	2	0	0	0	0	Variations des stocks
Total supply	74	89	83	105	153	157	163	170	Approvisionnement total
Supply per capita	16	16	13	14	16	16	16	16	Approvisionnement par habitant
Bermuda									**Bermudes**
Primary Energy production *	...	...	...	...	1	1	1	1	Production d'énergie primaire *
Net imports	8	*6	*6	*5	*8	*5	*6	*4	Importations nettes
Total supply	8	*6	*7	5	*8	6	7	6	Approvisionnement total
Supply per capita	140	*105	*106	86	*131	95	108	88	Approvisionnement par habitant
Bhutan									**Bhoutan**
Primary Energy production	41	43	46	53	73	73	72	75	Production d'énergie primaire
Net imports	-3	*-4	*-2	-5	-14	*-13	-11	*-12	Importations nettes
Total supply	37	39	44	48	59	59	61	63	Approvisionnement total
Supply per capita	67	76	77	73	81	81	82	83	Approvisionnement par habitant

18

Production, trade and supply of energy *(continue)*
Petajoules and gigajoules per capita
Production, commerce et fourniture d'énergie *(suite)*
pétajoules et gigajoules par habitant

Region, country or area	1990	1995	2000	2005	2010	2011	2012	2013	Région, pays ou zone
Bolivia (Plurinational State of)									**Bolivie (État plurinational de)**
Primary Energy production	175	233	231	586	659	706	804	912	Production d'énergie primaire
Net imports	-74	-89	-68	-366	-397	-425	-485	-577	Importations nettes
Changes in stocks	*2	*0	-14	0	-1	6	1	-1	Variations des stocks
Total supply	99	144	177	220	263	275	319	336	Approvisionnement total
Supply per capita	15	19	21	24	27	27	31	32	Approvisionnement par habitant
Bonaire, Sint Eustatius and Saba *									**Bonaire, Saint-Eustache et Saba ***
Net imports	...	...	...	...	...	...	6	5	Importations nettes
Total supply	...	...	...	...	...	...	5	5	Approvisionnement total
Supply per capita	...	...	...	...	...	...	215	213	Approvisionnement par habitant
Bosnia and Herzegovina									**Bosnie-Herzégovine**
Primary Energy production	...	28	126	152	182	193	188	192	Production d'énergie primaire
Net imports	...	29	52	54	84	99	93	79	Importations nettes
Changes in stocks	...	...	-1	-1	-3	-2	3	5	Variations des stocks
Total supply	...	56	179	206	269	294	278	266	Approvisionnement total
Supply per capita	...	17	48	54	70	77	73	70	Approvisionnement par habitant
Botswana									**Botswana**
Primary Energy production	24	27	28	29	30	25	41	42	Production d'énergie primaire
Net imports	15	17	29	36	47	51	53	54	Importations nettes
Changes in stocks	0	...	...	0	0	0	18	10	Variations des stocks
Total supply	39	44	57	65	77	76	76	86	Approvisionnement total
Supply per capita	28	28	32	35	38	36	36	39	Approvisionnement par habitant
Brazil									**Brésil**
Primary Energy production	4 490	5 038	6 308	8 344	10 050	10 220	10 305	10 360	Production d'énergie primaire
Net imports	1 555	1 918	1 713	900	859	1 009	1 235	1 760	Importations nettes
Changes in stocks	86	-106	1	-8	-8	57	-91	-30	Variations des stocks
Total supply	5 960	7 062	8 020	9 252	10 916	11 172	11 631	12 149	Approvisionnement total
Supply per capita	40	44	46	50	55	56	57	59	Approvisionnement par habitant
British Virgin Islands									**Îles Vierges britanniques**
Primary Energy production	0	0	0	0	0	0	0	0	Production d'énergie primaire
Net imports *	1	1	1	2	2	2	2	2	Importations nettes *
Total supply *	1	1	1	2	2	2	2	2	Approvisionnement total *
Supply per capita *	56	66	72	85	89	89	87	85	Approvisionnement par habitant *
Brunei Darussalam									**Brunéi Darussalam**
Primary Energy production	668	752	813	848	775	780	773	710	Production d'énergie primaire
Net imports	*-594	-683	-744	-772	-648	-617	-612	-582	Importations nettes
Changes in stocks	*-33	-1	-4	-1	-7	2	1	0	Variations des stocks
Total supply	107	71	73	76	136	162	160	127	Approvisionnement total
Supply per capita	424	246	224	210	345	406	396	310	Approvisionnement par habitant
Bulgaria									**Bulgarie**
Primary Energy production	407	428	410	444	442	517	491	443	Production d'énergie primaire
Net imports	730	546	358	384	291	288	275	265	Importations nettes
Changes in stocks	-44	7	-5	2	-9	1	-1	6	Variations des stocks
Total supply	1 181	966	773	826	741	804	767	703	Approvisionnement total
Supply per capita	134	116	97	107	100	109	105	97	Approvisionnement par habitant
Burkina Faso									**Burkina Faso**
Primary Energy production	85	96	69	98	118	120	121	123	Production d'énergie primaire
Net imports	*9	10	12	16	*25	*29	*34	*42	Importations nettes
Changes in stocks	*0	1	-1	0	0	0	0	0	Variations des stocks
Total supply	93	104	83	114	143	149	156	164	Approvisionnement total
Supply per capita	10	10	7	8	9	9	9	10	Approvisionnement par habitant
Burundi									**Burundi**
Primary Energy production	54	62	51	79	86	87	54	54	Production d'énergie primaire
Net imports	*4	*5	*5	3	4	4	4	4	Importations nettes
Changes in stocks	0	0	0	0	0	0	0	0	Variations des stocks
Total supply	58	66	55	81	89	91	58	58	Approvisionnement total
Supply per capita	10	11	9	11	9	9	6	6	Approvisionnement par habitant
Cabo Verde									**Cabo Verde**
Primary Energy production	1	1	1	2	1	1	1	1	Production d'énergie primaire
Net imports	*1	*2	*3	*6	6	8	6	6	Importations nettes
Changes in stocks	0	...	...	...	...	...	...	...	Variations des stocks
Total supply	*2	*3	*4	*6	8	8	8	7	Approvisionnement total
Supply per capita	*7	*7	*9	*14	16	17	16	15	Approvisionnement par habitant

Production, trade and supply of energy *(continue)*
Petajoules and gigajoules per capita
Production, commerce et fourniture d'énergie *(suite)*
pétajoules et gigajoules par habitant

Region, country or area	1990	1995	2000	2005	2010	2011	2012	2013	Région, pays ou zone
Cambodia									**Cambodge**
Primary Energy production	103	98	114	105	152	159	165	171	Production d'énergie primaire
Net imports	*22	22	28	39	70	73	78	79	Importations nettes
Total supply	124	119	142	144	222	232	242	250	Approvisionnement total
Supply per capita	13	11	11	11	15	16	16	17	Approvisionnement par habitant
Cameroon									**Cameroun**
Primary Energy production	501	457	532	442	351	341	352	376	Production d'énergie primaire
Net imports	*-308	*-202	-268	-152	-73	-59	-61	-69	Importations nettes
Changes in stocks	*0	*0	0	-5	-13	0	0	0	Variations des stocks
Total supply	193	255	264	295	291	281	291	307	Approvisionnement total
Supply per capita	16	18	17	17	14	13	13	14	Approvisionnement par habitant
Canada									**Canada**
Primary Energy production	11 346	14 458	15 441	16 674	16 265	16 771	17 147	17 868	Production d'énergie primaire
Net imports	-2 538	-5 047	-5 377	-5 654	-6 199	-6 276	-6 794	-7 600	Importations nettes
Changes in stocks	170	-146	-352	-171	-309	-120	-69	-163	Variations des stocks
Total supply	8 638	9 556	10 415	11 192	10 375	10 614	10 422	10 431	Approvisionnement total
Supply per capita	312	326	340	347	304	308	299	296	Approvisionnement par habitant
Cayman Islands									**Îles Caïmanes**
Net imports	4	*6	*7	*7	*8	*9	*7	8	Importations nettes
Total supply	4	*5	*7	7	8	8	8	8	Approvisionnement total
Supply per capita	138	*159	*166	130	144	149	134	131	Approvisionnement par habitant
Central African Rep.									**Rép. centrafricaine**
Primary Energy production	28	28	19	19	19	19	19	19	Production d'énergie primaire
Net imports *	3	3	4	3	4	4	5	5	Importations nettes *
Changes in stocks	0	*0	*0	...	...	...	...	...	Variations des stocks
Total supply	31	31	22	22	23	23	23	23	Approvisionnement total
Supply per capita	11	9	6	5	5	5	5	5	Approvisionnement par habitant
Chad									**Tchad**
Primary Energy production	43	48	55	433	324	309	288	275	Production d'énergie primaire
Net imports *	2	1	2	-367	-252	-235	-213	-198	Importations nettes *
Changes in stocks	0	0	...	...	...	...	...	...	Variations des stocks
Total supply	45	50	57	66	73	74	75	77	Approvisionnement total
Supply per capita	7	7	7	7	6	6	6	6	Approvisionnement par habitant
Chile									**Chili**
Primary Energy production	326	344	358	390	384	412	544	626	Production d'énergie primaire
Net imports	284	437	722	791	888	992	1 002	1 012	Importations nettes
Changes in stocks	24	10	7	3	-6	3	-5	26	Variations des stocks
Total supply	587	772	1 073	1 178	1 278	1 401	1 551	1 613	Approvisionnement total
Supply per capita	45	54	70	72	75	81	89	92	Approvisionnement par habitant
China [2]									**Chine** [2]
Primary Energy production	32 727	39 692	40 783	62 586	90 442	98 279	101 749	108 185	Production d'énergie primaire
Net imports	-1 316	*-397	*1 426	4 443	14 078	15 680	18 131	20 976	Importations nettes
Changes in stocks	959	132	-251	721	1 269	2 023	1 348	1 970	Variations des stocks
Total supply	30 452	39 163	42 461	66 308	103 251	111 935	118 532	127 191	Approvisionnement total
Supply per capita	27	32	33	51	77	83	87	93	Approvisionnement par habitant
China, Hong Kong SAR									**Chine, Hong Kong RAS**
Net imports	321	382	571	578	679	599	583	592	Importations nettes
Changes in stocks	5	*-16	1	-2	128	25	9	7	Variations des stocks
Total supply	316	398	570	579	551	574	573	585	Approvisionnement total
Supply per capita	54	65	84	85	79	81	81	82	Approvisionnement par habitant
China, Macao SAR									**Chine, Macao RAS**
Primary Energy production *	...	...	...	2	2	2	2	2	Production d'énergie primaire *
Net imports	15	18	23	26	39	45	46	46	Importations nettes
Changes in stocks	0	0	0	0	0	0	0	0	Variations des stocks
Total supply	15	18	24	28	41	46	47	47	Approvisionnement total
Supply per capita	40	45	55	58	77	85	85	83	Approvisionnement par habitant
Colombia									**Colombie**
Primary Energy production	1 882	2 419	3 046	3 335	4 486	5 109	5 282	5 700	Production d'énergie primaire
Net imports	*-891	-1 267	-1 987	-2 139	*-3 089	*-3 698	*-3 822	-4 160	Importations nettes
Changes in stocks	7	48	-43	16	6	-2	0	26	Variations des stocks
Total supply	983	1 104	1 103	1 180	1 390	1 412	1 460	1 513	Approvisionnement total
Supply per capita	30	30	28	27	30	30	31	32	Approvisionnement par habitant

18

Production, trade and supply of energy *(continue)*
Petajoules and gigajoules per capita
Production, commerce et fourniture d'énergie *(suite)*
pétajoules et gigajoules par habitant

Region, country or area	1990	1995	2000	2005	2010	2011	2012	2013	Région, pays ou zone
Comoros									**Comores**
Primary Energy production	1	2	2	2	2	3	3	3	Production d'énergie primaire
Net imports *	1	2	1	2	2	3	2	3	Importations nettes *
Total supply	2	*3	3	4	4	*5	*5	*5	Approvisionnement total
Supply per capita	5	*6	6	6	6	*7	*7	*7	Approvisionnement par habitant
Congo									**Congo**
Primary Energy production	361	419	609	563	724	712	677	619	Production d'énergie primaire
Net imports	*-325	*-375	-569	-519	-649	-584	-575	-522	Importations nettes
Changes in stocks	0	0	...	0	6	35	3	-4	Variations des stocks
Total supply	36	44	40	45	69	93	99	102	Approvisionnement total
Supply per capita	15	16	13	13	17	22	23	23	Approvisionnement par habitant
Cook Islands									**Îles Cook**
Net imports *	1	1	1	1	1	1	1	1	Importations nettes *
Total supply	*0	*0	*0	1	*1	*1	*1	*1	Approvisionnement total
Supply per capita	*17	*17	*22	45	*49	*49	*49	*48	Approvisionnement par habitant
Costa Rica									**Costa Rica**
Primary Energy production	30	46	67	93	104	105	106	124	Production d'énergie primaire
Net imports	40	65	70	85	99	102	99	98	Importations nettes
Changes in stocks	1	0	-1	0	0	1	-2	0	Variations des stocks
Total supply	69	110	137	177	203	205	207	222	Approvisionnement total
Supply per capita	22	32	35	41	45	45	44	47	Approvisionnement par habitant
Côte d'Ivoire									**Côte d'Ivoire**
Primary Energy production	179	178	242	451	467	514	523	527	Production d'énergie primaire
Net imports	*36	*75	*28	-27	-39	-27	3	18	Importations nettes
Changes in stocks	*1	*0	*-8	*5	3	-1	-2	-2	Variations des stocks
Total supply	215	253	278	418	425	487	528	547	Approvisionnement total
Supply per capita	17	17	17	23	21	24	25	25	Approvisionnement par habitant
Croatia									**Croatie**
Primary Energy production	...	174	149	159	176	158	144	152	Production d'énergie primaire
Net imports	...	116	170	216	185	191	180	168	Importations nettes
Changes in stocks	...	-2	-4	4	5	-2	-6	-2	Variations des stocks
Total supply	...	293	324	371	356	351	330	321	Approvisionnement total
Supply per capita	...	63	72	84	82	82	77	75	Approvisionnement par habitant
Cuba									**Cuba**
Primary Energy production	*213	170	241	205	200	206	206	*218	Production d'énergie primaire
Net imports	*412	286	232	*207	*366	*335	*336	*367	Importations nettes
Changes in stocks	-5	4	10	...	...	...	...	...	Variations des stocks
Total supply	630	452	462	412	566	541	542	*585	Approvisionnement total
Supply per capita	60	41	42	37	50	48	48	*51	Approvisionnement par habitant
Curaçao									**Curaçao**
Net imports	...	...	...	...	...	...	87	77	Importations nettes
Total supply	...	...	...	...	...	...	87	78	Approvisionnement total
Supply per capita	...	...	...	...	...	...	572	502	Approvisionnement par habitant
Cyprus									**Chypre**
Primary Energy production	0	0	0	0	4	4	4	5	Production d'énergie primaire
Net imports	57	78	88	94	104	91	91	78	Importations nettes
Changes in stocks	2	10	1	3	5	-5	1	1	Variations des stocks
Total supply	54	69	87	92	103	100	94	82	Approvisionnement total
Supply per capita	70	80	93	89	94	90	83	71	Approvisionnement par habitant
Czech Republic									**République tchèque**
Primary Energy production	...	1 356	1 281	1 374	1 321	1 343	1 363	1 258	Production d'énergie primaire
Net imports	...	367	411	527	469	491	443	478	Importations nettes
Changes in stocks	...	-30	-49	16	-71	47	28	-16	Variations des stocks
Total supply	...	1 753	1 741	1 885	1 860	1 788	1 779	1 751	Approvisionnement total
Supply per capita	...	170	170	184	177	170	169	166	Approvisionnement par habitant
Dem. P. R. Korea									**R. p. dém. de Corée**
Primary Energy production	*2 307	*2 414	787	923	872	843	849	1 009	Production d'énergie primaire
Net imports	*256	*249	41	-28	-77	-251	-246	-397	Importations nettes
Changes in stocks	...	...	2	...	...	...	...	...	Variations des stocks
Total supply	*2 563	*2 663	826	895	795	592	603	612	Approvisionnement total
Supply per capita	*127	*122	36	38	32	24	24	25	Approvisionnement par habitant

Region, country or area	1990	1995	2000	2005	2010	2011	2012	2013	Région, pays ou zone
Dem. Rep. of the Congo									**Rép. dém. du Congo**
Primary Energy production	490	616	677	866	855	901	877	906	Production d'énergie primaire
Net imports	*-13	-33	-43	-41	-23	-18	-17	-18	Importations nettes
Changes in stocks	0	0	-1	0	0	0	0	0	Variations des stocks
Total supply	477	584	635	825	831	883	860	888	Approvisionnement total
Supply per capita	13	13	13	14	13	13	12	12	Approvisionnement par habitant
Denmark [3]									**Danemark** [3]
Primary Energy production	420	649	1 149	1 298	968	855	784	699	Production d'énergie primaire
Net imports	298	217	-383	-498	-204	-113	-82	35	Importations nettes
Changes in stocks	-7	57	-19	14	-47	-7	-17	8	Variations des stocks
Total supply	725	811	785	786	812	749	720	725	Approvisionnement total
Supply per capita	141	155	147	145	146	134	128	129	Approvisionnement par habitant
Djibouti									**Djibouti**
Primary Energy production	2	2	3	3	3	3	3	3	Production d'énergie primaire
Net imports *	5	5	5	6	7	4	7	7	Importations nettes *
Changes in stocks	0	...	...	...	*0	*-2	0	0	Variations des stocks
Total supply	*6	*7	8	*9	11	10	*11	*11	Approvisionnement total
Supply per capita	*12	*11	10	*11	13	12	*13	*13	Approvisionnement par habitant
Dominica									**Dominique**
Primary Energy production	0	0	0	0	0	0	0	0	Production d'énergie primaire
Net imports *	1	1	1	2	2	2	2	2	Importations nettes *
Changes in stocks *	...	0	...	...	...	...	...	...	Variations des stocks *
Total supply	*1	1	2	2	*2	*2	*2	*2	Approvisionnement total
Supply per capita	*14	19	24	26	*29	*28	*29	*28	Approvisionnement par habitant
Dominican Republic									**Rép. dominicaine**
Primary Energy production	22	24	19	24	36	36	38	43	Production d'énergie primaire
Net imports	122	212	259	*239	274	280	291	295	Importations nettes
Changes in stocks	1	0	0	2	-2	1	0	0	Variations des stocks
Total supply	143	236	278	262	312	316	329	339	Approvisionnement total
Supply per capita	20	30	32	28	31	31	32	33	Approvisionnement par habitant
Ecuador									**Équateur**
Primary Energy production	684	938	990	1 281	1 185	1 226	1 246	1 293	Production d'énergie primaire
Net imports	-428	*-577	-612	-772	-617	-621	-660	-672	Importations nettes
Changes in stocks	0	...	20	4	-43	-28	-60	-59	Variations des stocks
Total supply	256	362	358	506	611	633	646	680	Approvisionnement total
Supply per capita	25	32	29	38	41	42	42	43	Approvisionnement par habitant
Egypt									**Égypte**
Primary Energy production	2 376	2 684	2 773	3 383	3 694	3 718	3 670	3 445	Production d'énergie primaire
Net imports	*-1 084	*-709	-449	-645	-511	-186	-224	-23	Importations nettes
Changes in stocks	*21	369	36	-4	-60	-6	-5	0	Variations des stocks
Total supply	1 271	1 605	2 288	2 742	3 243	3 538	3 451	3 422	Approvisionnement total
Supply per capita	22	26	34	37	40	42	40	39	Approvisionnement par habitant
El Salvador									**El Salvador**
Primary Energy production	60	70	89	104	95	94	96	95	Production d'énergie primaire
Net imports	32	67	77	85	81	80	86	76	Importations nettes
Changes in stocks	-1	2	0	0	-1	-5	-7	-6	Variations des stocks
Total supply	92	135	166	189	176	180	188	176	Approvisionnement total
Supply per capita	17	24	28	31	29	30	31	29	Approvisionnement par habitant
Equatorial Guinea									**Guinée équatoriale**
Primary Energy production	4	34	285	822	849	*925	*889	*848	Production d'énergie primaire
Net imports	*2	*-28	-276	*-770	*-781	*-842	*-801	*-776	Importations nettes
Total supply	6	6	9	52	68	*84	*88	*73	Approvisionnement total
Supply per capita	15	13	18	86	94	*111	*114	*91	Approvisionnement par habitant
Eritrea									**Érythrée**
Primary Energy production	...	31	21	21	24	25	26	27	Production d'énergie primaire
Net imports	...	13	9	10	7	7	7	7	Importations nettes
Changes in stocks	...	...	0	-2	0	0	0	0	Variations des stocks
Total supply	...	43	30	32	31	32	33	34	Approvisionnement total
Supply per capita	...	14	8	7	7	7	7	7	Approvisionnement par habitant

18

Production, trade and supply of energy *(continue)*
Petajoules and gigajoules per capita
Production, commerce et fourniture d'énergie *(suite)*
pétajoules et gigajoules par habitant

Region, country or area	1990	1995	2000	2005	2010	2011	2012	2013	Région, pays ou zone
Estonia									**Estonie**
Primary Energy production	...	139	129	163	205	211	213	237	Production d'énergie primaire
Net imports	...	74	66	57	30	27	32	21	Importations nettes
Changes in stocks	...	-3	-1	0	-2	0	12	1	Variations des stocks
Total supply	...	217	196	220	236	238	233	258	Approvisionnement total
Supply per capita	...	151	143	164	177	179	176	195	Approvisionnement par habitant
Ethiopia									**Éthiopie**
Primary Energy production	...	885	977	1 070	1 212	1 236	1 259	1 285	Production d'énergie primaire
Net imports	...	31	43	60	82	89	94	111	Importations nettes
Changes in stocks	...	*-2	-2	-3	1	-2	0	1	Variations des stocks
Total supply	...	918	1 022	1 132	1 293	1 327	1 352	1 394	Approvisionnement total
Supply per capita	...	16	16	15	15	15	15	15	Approvisionnement par habitant
Falkland Is. (Malvinas)									**Îles Falkland (Malvinas)**
Primary Energy production	0	*0	*0	*0	*0	*0	*0	*0	Production d'énergie primaire
Net imports	0	*0	*0	*1	*1	*1	*1	*1	Importations nettes
Total supply	0	*1	*1	*1	*1	*1	*1	*1	Approvisionnement total
Supply per capita	249	*221	*178	*235	*261	*255	*259	*258	Approvisionnement par habitant
Faroe Islands									**Îles Féroé**
Primary Energy production	0	0	0	0	0	0	0	0	Production d'énergie primaire
Net imports	9	10	*10	*11	*9	*8	*8	*8	Importations nettes
Total supply	9	10	*10	*11	*9	*8	*9	*9	Approvisionnement total
Supply per capita	197	227	*216	*219	*184	*168	*177	*179	Approvisionnement par habitant
Fiji									**Fidji**
Primary Energy production	10	13	11	9	6	7	6	8	Production d'énergie primaire
Net imports	9	*12	*11	*18	*23	*23	*23	*24	Importations nettes
Changes in stocks	-1	...	...	*0	...	...	...	...	Variations des stocks
Total supply	21	*25	22	28	28	29	28	31	Approvisionnement total
Supply per capita	28	*32	27	34	32	33	32	36	Approvisionnement par habitant
Finland									**Finlande**
Primary Energy production	459	507	622	695	726	714	718	755	Production d'énergie primaire
Net imports	713	636	727	767	722	767	642	658	Importations nettes
Changes in stocks	36	-26	1	26	-77	16	-54	37	Variations des stocks
Total supply	1 135	1 170	1 348	1 436	1 525	1 465	1 414	1 374	Approvisionnement total
Supply per capita	228	229	261	274	284	272	261	252	Approvisionnement par habitant
France [4]									**France** [4]
Primary Energy production	4 651	5 314	5 429	5 692	5 642	5 648	5 587	5 659	Production d'énergie primaire
Net imports	4 801	4 620	5 258	5 689	5 197	4 943	4 901	4 905	Importations nettes
Changes in stocks	71	10	148	49	-109	58	-71	-32	Variations des stocks
Total supply	9 381	9 924	10 540	11 332	10 948	10 534	10 559	10 597	Approvisionnement total
Supply per capita	165	171	178	186	174	166	166	166	Approvisionnement par habitant
French Guiana									**Guyane française**
Primary Energy production	0	0	1	*3	*3	*3	3	3	Production d'énergie primaire
Net imports *	11	12	9	8	10	10	9	10	Importations nettes *
Total supply	12	13	*10	11	*13	*13	*13	*13	Approvisionnement total
Supply per capita	100	90	*63	55	*54	*54	*52	*52	Approvisionnement par habitant
French Polynesia									**Polynésie française**
Primary Energy production	0	0	1	1	1	*1	*1	*1	Production d'énergie primaire
Net imports *	7	6	8	11	12	11	11	11	Importations nettes *
Total supply	*8	7	9	12	13	12	12	12	Approvisionnement total
Supply per capita	*38	32	37	48	49	46	44	44	Approvisionnement par habitant
Gabon									**Gabon**
Primary Energy production	609	822	630	622	602	603	585	569	Production d'énergie primaire
Net imports	-509	-765	-572	-545	-528	-512	-492	-470	Importations nettes
Changes in stocks	50	2	-4	5	-15	0	1	0	Variations des stocks
Total supply	50	56	62	72	88	91	94	99	Approvisionnement total
Supply per capita	54	52	50	52	57	58	58	60	Approvisionnement par habitant
Gambia									**Gambie**
Primary Energy production	4	5	6	6	6	6	7	7	Production d'énergie primaire
Net imports *	3	3	4	5	7	6	7	7	Importations nettes *
Changes in stocks	0	...	...	...	...	...	...	...	Variations des stocks
Total supply	7	8	9	10	*13	*13	*13	*14	Approvisionnement total
Supply per capita	7	7	7	7	*8	*7	*7	*7	Approvisionnement par habitant

Production, trade and supply of energy *(continue)*
Petajoules and gigajoules per capita
Production, commerce et fourniture d'énergie *(suite)*
pétajoules et gigajoules par habitant

Region, country or area	1990	1995	2000	2005	2010	2011	2012	2013	Région, pays ou zone
Georgia									**Géorgie**
Primary Energy production	...	22	28	53	55	47	46	59	Production d'énergie primaire
Net imports	...	37	65	82	76	102	109	103	Importations nettes
Changes in stocks	...	0	0	0	0	0	0	0	Variations des stocks
Total supply	...	59	93	135	131	149	155	163	Approvisionnement total
Supply per capita	...	12	20	30	31	35	38	40	Approvisionnement par habitant
Germany									**Allemagne**
Primary Energy production	...	6 050	5 645	5 704	5 375	5 145	5 124	5 032	Production d'énergie primaire
Net imports	...	7 849	8 207	8 354	8 084	7 870	7 867	8 202	Importations nettes
Changes in stocks	...	-137	-183	48	-192	54	-8	-27	Variations des stocks
Total supply	...	14 037	14 034	14 009	13 650	12 960	12 998	13 261	Approvisionnement total
Supply per capita	...	171	170	170	170	161	162	165	Approvisionnement par habitant
Ghana									**Ghana**
Primary Energy production	302	409	192	164	148	299	347	351	Production d'énergie primaire
Net imports	*49	*66	79	87	120	-16	-24	-30	Importations nettes
Changes in stocks	*1	...	...	...	0	0	0	5	Variations des stocks
Total supply	350	475	271	251	267	283	323	316	Approvisionnement total
Supply per capita	24	28	14	12	11	11	13	12	Approvisionnement par habitant
Gibraltar									**Gibraltar**
Net imports	3	5	6	7	8	7	8	8	Importations nettes
Changes in stocks	0	...	...	...	...	...	...	...	Variations des stocks
Total supply	2	4	5	6	7	7	7	8	Approvisionnement total
Supply per capita	90	162	197	221	239	231	232	245	Approvisionnement par habitant
Greece									**Grèce**
Primary Energy production	385	391	419	432	396	403	437	390	Production d'énergie primaire
Net imports	506	581	741	831	762	684	706	573	Importations nettes
Changes in stocks	-10	10	12	-18	-11	-44	18	-30	Variations des stocks
Total supply	902	963	1 148	1 280	1 169	1 131	1 124	993	Approvisionnement total
Supply per capita	89	90	104	115	105	101	101	90	Approvisionnement par habitant
Greenland									**Groenland**
Primary Energy production	...	...	1	1	1	*1	1	1	Production d'énergie primaire
Net imports	8	*7	*8	9	10	10	8	*8	Importations nettes
Changes in stocks	...	...	...	*1	1	-1	0	*0	Variations des stocks
Total supply	8	*7	*8	9	10	11	9	9	Approvisionnement total
Supply per capita	140	*128	*146	165	186	199	164	166	Approvisionnement par habitant
Grenada									**Grenade**
Primary Energy production	0	0	0	0	0	0	0	0	Production d'énergie primaire
Net imports	*2	*2	*3	3	4	4	4	5	Importations nettes
Changes in stocks	0	0	0	0	0	0	0	0	Variations des stocks
Total supply	2	2	3	3	4	4	4	5	Approvisionnement total
Supply per capita	17	23	29	33	38	36	39	43	Approvisionnement par habitant
Guadeloupe									**Guadeloupe**
Primary Energy production	1	*2	*1	*2	*2	*3	*3	*4	Production d'énergie primaire
Net imports	*17	*21	*23	26	*28	*28	*28	*28	Importations nettes
Total supply	18	22	*24	28	*31	*31	*31	*33	Approvisionnement total
Supply per capita	47	55	*57	63	*67	*68	*68	*70	Approvisionnement par habitant
Guatemala									**Guatemala**
Primary Energy production	143	178	229	256	279	302	309	310	Production d'énergie primaire
Net imports	52	75	75	117	111	117	122	122	Importations nettes
Changes in stocks	0	*1	-4	7	0	1	4	-20	Variations des stocks
Total supply	195	252	307	365	391	419	427	452	Approvisionnement total
Supply per capita	22	25	27	29	27	28	28	29	Approvisionnement par habitant
Guernsey									**Guernesey**
Net imports	...	...	...	1	1	1	1	0	Importations nettes
Total supply	...	...	...	1	1	1	1	0	Approvisionnement total
Supply per capita	...	...	...	17	14	18	18	6	Approvisionnement par habitant
Guinea									**Guinée**
Primary Energy production	91	110	107	109	112	112	113	113	Production d'énergie primaire
Net imports *	14	15	16	17	36	38	36	32	Importations nettes *
Changes in stocks	0	...	...	...	...	...	...	...	Variations des stocks
Total supply	106	126	123	126	148	149	149	145	Approvisionnement total
Supply per capita	18	17	15	14	13	13	13	12	Approvisionnement par habitant

Production, trade and supply of energy *(continue)*
Petajoules and gigajoules per capita
Production, commerce et fourniture d'énergie *(suite)*
pétajoules et gigajoules par habitant

Region, country or area	1990	1995	2000	2005	2010	2011	2012	2013	Région, pays ou zone
Guinea-Bissau									**Guinée-Bissau**
Primary Energy production	17	18	20	22	24	24	24	25	Production d'énergie primaire
Net imports *	3	4	3	4	5	5	5	5	Importations nettes *
Total supply	20	21	23	26	28	28	29	29	Approvisionnement total
Supply per capita	20	19	19	19	17	17	17	17	Approvisionnement par habitant
Guyana									**Guyana**
Primary Energy production	9	8	9	9	8	10	8	7	Production d'énergie primaire
Net imports	*16	*21	22	20	24	25	27	27	Importations nettes
Changes in stocks	...	0	0	0	0	0	0	0	Variations des stocks
Total supply	24	29	32	29	32	34	36	34	Approvisionnement total
Supply per capita	34	39	43	39	43	46	47	45	Approvisionnement par habitant
Haiti									**Haïti**
Primary Energy production	51	58	65	115	131	135	140	140	Production d'énergie primaire
Net imports	*11	*13	20	28	28	29	31	31	Importations nettes
Changes in stocks	-1	...	...	...	...	...	...	-1	Variations des stocks
Total supply	64	71	84	142	159	164	170	172	Approvisionnement total
Supply per capita	9	9	10	15	16	16	17	16	Approvisionnement par habitant
Honduras									**Honduras**
Primary Energy production	91	92	96	77	93	101	104	106	Production d'énergie primaire
Net imports	32	51	65	89	98	105	111	112	Importations nettes
Changes in stocks	0	1	5	-4	0	-5	0	0	Variations des stocks
Total supply	122	142	157	170	191	212	214	217	Approvisionnement total
Supply per capita	25	25	25	25	25	28	28	28	Approvisionnement par habitant
Hungary									**Hongrie**
Primary Energy production	615	583	486	434	461	451	442	426	Production d'énergie primaire
Net imports	594	521	573	729	627	543	517	499	Importations nettes
Changes in stocks	-1	11	10	8	10	-55	-29	-22	Variations des stocks
Total supply	1 211	1 093	1 049	1 155	1 078	1 049	988	947	Approvisionnement total
Supply per capita	117	106	103	115	108	105	99	95	Approvisionnement par habitant
Iceland									**Islande**
Primary Energy production	59	63	109	127	252	260	283	286	Production d'énergie primaire
Net imports	28	27	31	31	26	25	26	26	Importations nettes
Changes in stocks	0	0	2	1	-1	0	0	1	Variations des stocks
Total supply	87	90	139	158	278	285	308	312	Approvisionnement total
Supply per capita	341	338	494	534	875	889	952	957	Approvisionnement par habitant
India									**Inde**
Primary Energy production	9 218	10 985	12 084	18 293	22 531	22 918	23 218	22 401	Production d'énergie primaire
Net imports	1 196	*2 010	3 890	4 689	6 558	7 719	9 283	10 690	Importations nettes
Changes in stocks	45	*-41	-155	194	252	199	-28	-103	Variations des stocks
Total supply	10 370	13 035	16 130	22 787	28 836	30 438	32 528	33 195	Approvisionnement total
Supply per capita	12	14	15	20	23	24	26	26	Approvisionnement par habitant
Indonesia									**Indonésie**
Primary Energy production	6 735[5]	8 320[5]	8 129[5]	11 351	16 854	17 651	18 412	19 754	Production d'énergie primaire
Net imports	*-2 753[5]	*-3 634[5]	*-3 577[5]	-4 243	-8 522	-7 793	-8 212	-10 937	Importations nettes
Changes in stocks	223[5]	100[5]	-417[5]	21	2	-187	-72	62	Variations des stocks
Total supply	3 759[5]	4 587[5]	4 970[5]	7 087	8 330	10 044	10 271	8 754	Approvisionnement total
Supply per capita	20[5]	23[5]	23[5]	31	34	41	41	35	Approvisionnement par habitant
Iran (Islamic Rep. of)									**Iran (Rép. islamique d')**
Primary Energy production	7 454	9 302	11 121	13 006	15 646	15 814	13 473	12 544	Production d'énergie primaire
Net imports	-4 628	-5 412	-5 787	*-5 682	-6 525	-6 737	-4 082	-3 401	Importations nettes
Changes in stocks	0	0	-52	103	80	-89	-124	-30	Variations des stocks
Total supply	2 825	3 890	5 386	7 221	9 040	9 167	9 515	9 172	Approvisionnement total
Supply per capita	51	65	82	104	122	122	125	119	Approvisionnement par habitant
Iraq									**Iraq**
Primary Energy production	4 411	1 265	5 482	4 147	5 274	5 813	6 426	6 522	Production d'énergie primaire
Net imports	*-3 679	-175	-4 487	-2 656	-3 835	-4 156	-4 547	-4 460	Importations nettes
Changes in stocks	*169	0	0	3	1	0	7	-16	Variations des stocks
Total supply	563	1 089	995	1 488	1 437	1 656	1 872	2 077	Approvisionnement total
Supply per capita	32	54	42	54	47	52	57	61	Approvisionnement par habitant

Production, trade and supply of energy *(continue)*
Petajoules and gigajoules per capita
Production, commerce et fourniture d'énergie *(suite)*
pétajoules et gigajoules par habitant

Region, country or area	1990	1995	2000	2005	2010	2011	2012	2013	Région, pays ou zone
Ireland									**Irlande**
Primary Energy production	145	172	90	69	78	70	54	95	Production d'énergie primaire
Net imports	281	303	482	544	522	494	466	485	Importations nettes
Changes in stocks	10	28	-2	2	-10	12	-32	33	Variations des stocks
Total supply	416	446	573	610	610	552	551	548	Approvisionnement total
Supply per capita	118	124	151	147	132	119	118	117	Approvisionnement par habitant
Isle of Man									**Île de Man**
Primary Energy production *	...	...	...	0	0	0	0	0	Production d'énergie primaire *
Net imports	...	...	...	0	0	0	1	*0	Importations nettes
Total supply	...	...	...	*0	0	0	1	*0	Approvisionnement total
Supply per capita	...	...	...	*6	2	2	11	*4	Approvisionnement par habitant
Israel									**Israël**
Primary Energy production	18	23	27	87	162	197	136	270	Production d'énergie primaire
Net imports	450	625	719	685	804	730	886	755	Importations nettes
Changes in stocks	-12	-1	-17	5	0	-37	11	27	Variations des stocks
Total supply	480	649	763	767	966	965	1 012	997	Approvisionnement total
Supply per capita	107	122	127	116	130	128	132	128	Approvisionnement par habitant
Italy [6]									**Italie** [6]
Primary Energy production	1 072	1 235	1 183	1 165	1 237	1 306	1 466	1 542	Production d'énergie primaire
Net imports	5 375	5 481	6 239	6 504	6 000	5 715	5 344	4 965	Importations nettes
Changes in stocks	80	3	190	-70	69	-20	6	-23	Variations des stocks
Total supply	6 366	6 714	7 230	7 738	7 168	7 042	6 804	6 530	Approvisionnement total
Supply per capita	112	118	127	132	120	118	114	109	Approvisionnement par habitant
Jamaica									**Jamaïque**
Primary Energy production	13	9	11	15	6	7	7	7	Production d'énergie primaire
Net imports	*97	*118	144	139	95	102	97	*102	Importations nettes
Changes in stocks	-2	-3	7	1	-2	0	1	2	Variations des stocks
Total supply	113	130	148	151	103	108	103	108	Approvisionnement total
Supply per capita	48	53	57	56	38	39	37	39	Approvisionnement par habitant
Japan									**Japon**
Primary Energy production	3 121	4 086	4 387	4 172	4 130	2 144	1 189	1 172	Production d'énergie primaire
Net imports	15 344	16 665	17 456	17 704	16 742	17 217	17 766	17 798	Importations nettes
Changes in stocks	146	56	165	101	-5	42	50	-15	Variations des stocks
Total supply	18 318	20 696	21 678	21 774	20 877	19 319	18 904	18 984	Approvisionnement total
Supply per capita	150	166	172	172	164	152	149	149	Approvisionnement par habitant
Jersey									**Jersey**
Primary Energy production	...	...	...	*0	*0	1	*1	*1	Production d'énergie primaire
Net imports	...	1	1	2	2	3	*3	*3	Importations nettes
Total supply	...	1	1	3	3	3	*3	*3	Approvisionnement total
Supply per capita	...	9	17	28	27	33	*33	*34	Approvisionnement par habitant
Jordan									**Jordanie**
Primary Energy production	2	10	10	10	9	9	8	7	Production d'énergie primaire
Net imports	136	160	190	285	300	285	327	*307	Importations nettes
Changes in stocks	4	-2	-5	7	5	-12	7	-10	Variations des stocks
Total supply	133	172	205	289	303	305	329	325	Approvisionnement total
Supply per capita	39	39	42	54	46	45	47	45	Approvisionnement par habitant
Kazakhstan									**Kazakhstan**
Primary Energy production	...	2 663	3 367	5 131	6 770	6 961	7 030	7 198	Production d'énergie primaire
Net imports	...	-477	-1 819	-2 771	*-3 346	*-3 528	*-3 677	-3 570	Importations nettes
Changes in stocks	...	...	-12	8	88	-74	65	80	Variations des stocks
Total supply	...	2 185	1 560	2 352	3 335	3 507	3 287	3 548	Approvisionnement total
Supply per capita	...	137	104	155	204	212	195	208	Approvisionnement par habitant
Kenya									**Kenya**
Primary Energy production	217	239	248	338	*650	*662	*690	*719	Production d'énergie primaire
Net imports	*71	98	141	109	137	160	150	154	Importations nettes
Changes in stocks	0	...	...	4	-5	-1	0	0	Variations des stocks
Total supply	289	337	388	443	*793	*823	*841	*873	Approvisionnement total
Supply per capita	12	12	12	12	*20	*20	*20	*20	Approvisionnement par habitant
Kiribati									**Kiribati**
Primary Energy production	0	0	0	0	0	0	0	0	Production d'énergie primaire
Net imports	*0	*0	0	*1	*1	*1	*1	*1	Importations nettes
Total supply	*0	*0	1	*1	*1	*1	*1	*1	Approvisionnement total
Supply per capita	*4	*4	6	*10	*9	*8	*8	*8	Approvisionnement par habitant

18 Production, trade and supply of energy *(continue)*
Petajoules and gigajoules per capita
Production, commerce et fourniture d'énergie *(suite)*
pétajoules et gigajoules par habitant

Region, country or area	1990	1995	2000	2005	2010	2011	2012	2013	Région, pays ou zone
Kuwait[7]									**Koweït[7]**
Primary Energy production	2 799	4 802	4 748	6 080	5 557	6 366	7 136	7 113	Production d'énergie primaire
Net imports	*-2 123	-3 968	-3 941	-4 915	-4 229	-4 996	-5 573	-5 659	Importations nettes
Changes in stocks	*-36	-1	-17	31	-22	-18	-5	-21	Variations des stocks
Total supply	711	835	824	1 133	1 350	1 388	1 568	1 475	Approvisionnement total
Supply per capita	341	513	424	500	441	429	459	410	Approvisionnement par habitant
Kyrgyzstan									**Kirghizistan**
Primary Energy production	...	53	60	61	53	68	73	74	Production d'énergie primaire
Net imports	...	51	38	51	63	69	104	96	Importations nettes
Changes in stocks	...	3	-2	0	1	-3	4	2	Variations des stocks
Total supply	...	101	101	112	116	139	173	167	Approvisionnement total
Supply per capita	...	22	20	22	21	25	31	29	Approvisionnement par habitant
Lao People's Dem. Rep.									**Rép. dém. pop. lao**
Primary Energy production	54	57	76	80	79	78	82	85	Production d'énergie primaire
Net imports *	1	3	-7	-5	-4	-5	-5	1	Importations nettes *
Total supply	55	59	68	74	75	73	77	85	Approvisionnement total
Supply per capita	13	12	13	13	12	11	12	13	Approvisionnement par habitant
Latvia									**Lettonie**
Primary Energy production	...	60	59	78	83	87	98	90	Production d'énergie primaire
Net imports	...	138	94	116	78	102	99	96	Importations nettes
Changes in stocks	...	1	-4	5	-28	10	11	3	Variations des stocks
Total supply	...	197	157	190	189	179	185	183	Approvisionnement total
Supply per capita	...	79	66	82	91	87	91	91	Approvisionnement par habitant
Lebanon									**Liban**
Primary Energy production	5	8	7	10	9	9	9	10	Production d'énergie primaire
Net imports	*108	163	198	197	253	251	283	279	Importations nettes
Changes in stocks	...	...	-1	...	...	...	...	...	Variations des stocks
Total supply	*113	171	207	206	262	260	293	289	Approvisionnement total
Supply per capita	*38	49	55	51	60	57	59	55	Approvisionnement par habitant
Lesotho									**Lesotho**
Primary Energy production	12	14	20	20	21	21	21	21	Production d'énergie primaire
Net imports	1	1	0	*14	26	*25	*26	*26	Importations nettes
Total supply	13	15	20	34	47	*46	*46	*48	Approvisionnement total
Supply per capita	8	8	10	17	23	*23	*23	*23	Approvisionnement par habitant
Liberia									**Libéria**
Primary Energy production	32	25	43	53	64	66	69	71	Production d'énergie primaire
Net imports *	7	5	5	9	11	12	14	12	Importations nettes *
Total supply	38	30	49	63	75	79	82	83	Approvisionnement total
Supply per capita	18	14	17	20	19	19	20	19	Approvisionnement par habitant
Libya									**Libye**
Primary Energy production	3 020	3 190	3 117	4 062	4 295	1 353	3 590	2 554	Production d'énergie primaire
Net imports	-2 676	-2 538	-2 433	-3 335	-3 360	-635	-2 877	-1 848	Importations nettes
Changes in stocks	-124	...	...	...	84	153	0	0	Variations des stocks
Total supply	468	653	685	727	851	564	712	707	Approvisionnement total
Supply per capita	108	137	131	126	136	90	113	113	Approvisionnement par habitant
Liechtenstein									**Liechtenstein**
Primary Energy production	...	...	...	...	1	1	1	1	Production d'énergie primaire
Net imports	...	...	...	...	2	2	2	2	Importations nettes
Total supply	...	...	...	...	3	3	3	3	Approvisionnement total
Supply per capita	...	...	...	...	81	77	78	79	Approvisionnement par habitant
Lithuania									**Lituanie**
Primary Energy production	...	157	139	170	64	64	65	69	Production d'énergie primaire
Net imports	...	234	175	196	225	232	230	210	Importations nettes
Changes in stocks	...	14	8	4	3	0	-4	-2	Variations des stocks
Total supply	...	376	306	362	286	296	299	282	Approvisionnement total
Supply per capita	...	104	87	106	92	97	99	95	Approvisionnement par habitant
Luxembourg									**Luxembourg**
Primary Energy production	1	2	3	5	5	5	5	6	Production d'énergie primaire
Net imports	143	130	140	179	173	171	168	162	Importations nettes
Changes in stocks	0	-1	2	-1	-1	0	1	0	Variations des stocks
Total supply	143	133	142	185	178	176	173	168	Approvisionnement total
Supply per capita	376	327	325	405	351	339	325	307	Approvisionnement par habitant

Production, trade and supply of energy *(continue)*
Petajoules and gigajoules per capita
Production, commerce et fourniture d'énergie *(suite)*
pétajoules et gigajoules par habitant

Region, country or area	1990	1995	2000	2005	2010	2011	2012	2013	Région, pays ou zone
Madagascar									**Madagascar**
Primary Energy production	72	89	92	104	124	124	124	127	Production d'énergie primaire
Net imports *	13	18	26	24	25	29	35	39	Importations nettes *
Changes in stocks	*0	0	...	0	-1	-1	1	0	Variations des stocks
Total supply	85	107	118	127	151	154	158	166	Approvisionnement total
Supply per capita	8	8	8	7	7	7	7	7	Approvisionnement par habitant
Malawi									**Malawi**
Primary Energy production	55	54	76	80	85	86	86	87	Production d'énergie primaire
Net imports *	8	9	10	11	14	14	14	14	Importations nettes *
Total supply	63	63	86	91	99	100	100	101	Approvisionnement total
Supply per capita	7	6	8	7	7	7	6	6	Approvisionnement par habitant
Malaysia									**Malaisie**
Primary Energy production	1 794	2 791	3 082	3 770	3 450	3 394	3 369	3 529	Production d'énergie primaire
Net imports	-1 019	*-1 018	-1 116	-1 120	*-505	-272	-298	-57	Importations nettes
Changes in stocks	3	*5	8	18	*-9	54	-76	54	Variations des stocks
Total supply	773	1 768	1 958	2 632	2 955	3 068	3 146	3 417	Approvisionnement total
Supply per capita	42	85	84	101	105	107	108	116	Approvisionnement par habitant
Maldives									**Maldives**
Primary Energy production	0	0	0	0	0	0	0	0	Production d'énergie primaire
Net imports *	2	4	6	9	12	13	15	15	Importations nettes *
Total supply	*2	4	6	9	13	14	15	15	Approvisionnement total
Supply per capita	*9	14	24	30	39	40	45	43	Approvisionnement par habitant
Mali									**Mali**
Primary Energy production	37	42	45	49	52	53	53	53	Production d'énergie primaire
Net imports *	5	6	11	12	12	14	13	13	Importations nettes *
Total supply	43	48	57	60	64	66	66	66	Approvisionnement total
Supply per capita	5	5	5	5	4	4	4	4	Approvisionnement par habitant
Malta									**Malte**
Primary Energy production	...	...	...	0	0	0	0	0	Production d'énergie primaire
Net imports	*28	*30	28	37	34	36	39	35	Importations nettes
Changes in stocks	0	1	1	0	-1	2	1	4	Variations des stocks
Total supply	28	29	28	37	35	35	37	31	Approvisionnement total
Supply per capita	76	74	70	91	85	84	89	75	Approvisionnement par habitant
Marshall Islands *									**Îles Marshall ***
Net imports	1	1	1	1	1	1	1	1	Importations nettes
Total supply	1	1	1	1	1	1	1	1	Approvisionnement total
Supply per capita	14	19	20	23	28	28	28	28	Approvisionnement par habitant
Martinique									**Martinique**
Primary Energy production	0	0	*0	*0	*1	*1	*1	*1	Production d'énergie primaire
Net imports *	27	26	28	33	30	32	32	33	Importations nettes *
Changes in stocks	0	...	...	...	...	...	...	...	Variations des stocks
Total supply	27	*27	*28	33	31	33	33	*33	Approvisionnement total
Supply per capita	75	*73	*72	83	78	84	84	*84	Approvisionnement par habitant
Mauritania									**Mauritanie**
Primary Energy production	11	12	13	15	34	33	31	31	Production d'énergie primaire
Net imports	*13	*15	16	21	7	24	18	*20	Importations nettes
Changes in stocks	0	...	...	0	*-3	*11	*-1	0	Variations des stocks
Total supply	*23	*26	29	36	45	47	51	*51	Approvisionnement total
Supply per capita	*11	*12	11	12	12	13	13	*13	Approvisionnement par habitant
Mauritius									**Maurice**
Primary Energy production	13	15	13	12	11	11	10	9	Production d'énergie primaire
Net imports	21	28	34	44	49	50	52	51	Importations nettes
Changes in stocks	1	2	-3	-1	-2	-1	-1	2	Variations des stocks
Total supply	33	41	49	57	63	62	63	58	Approvisionnement total
Supply per capita	31	36	41	45	50	49	50	46	Approvisionnement par habitant
Mexico									**Mexique**
Primary Energy production	8 167	8 455	9 325	10 305	8 952	8 990	8 861	8 849	Production d'énergie primaire
Net imports	-2 996	-3 005	-3 173	-3 300	-1 659	-1 335	-954	-923	Importations nettes
Changes in stocks	21	-10	68	7	98	128	158	-18	Variations des stocks
Total supply	5 150	5 460	6 085	6 997	7 196	7 526	7 749	7 944	Approvisionnement total
Supply per capita	61	59	61	66	61	63	63	64	Approvisionnement par habitant

18
Production, trade and supply of energy *(continue)*
Petajoules and gigajoules per capita
Production, commerce et fourniture d'énergie *(suite)*
pétajoules et gigajoules par habitant

Region, country or area	1990	1995	2000	2005	2010	2011	2012	2013	Région, pays ou zone
Micronesia (Fed. States of)									**Micronésie (États féd. de)**
Primary Energy production	...	...	0	0	0	0	0	0	Production d'énergie primaire
Net imports *	...	...	2	2	2	2	2	2	Importations nettes *
Total supply	...	...	*2	2	2	*2	*2	*2	Approvisionnement total
Supply per capita	...	...	*18	16	17	*18	*20	*21	Approvisionnement par habitant
Mongolia									**Mongolie**
Primary Energy production	82	72	66	138	666	843	794	808	Production d'énergie primaire
Net imports	33	*17	21	-35	-451	-575	*-411	*-331	Importations nettes
Changes in stocks	...	...	...	-1	*-69	*-49	*9	*2	Variations des stocks
Total supply	116	89	87	104	284	316	373	475	Approvisionnement total
Supply per capita	53	39	36	41	105	115	133	166	Approvisionnement par habitant
Montenegro									**Monténégro**
Primary Energy production	...	...	...	27	37	29	29	32	Production d'énergie primaire
Net imports	...	...	...	17	14	16	13	10	Importations nettes
Total supply	...	...	...	44	51	45	42	42	Approvisionnement total
Supply per capita	...	...	...	71	82	73	68	67	Approvisionnement par habitant
Montserrat									**Montserrat**
Net imports *	0	1	0	1	1	1	1	1	Importations nettes *
Changes in stocks	...	...	...	...	0	0	*0	*0	Variations des stocks
Total supply	*0	*0	*0	*1	1	1	1	1	Approvisionnement total
Supply per capita	*41	*48	*75	*93	180	121	122	153	Approvisionnement par habitant
Morocco									**Maroc**
Primary Energy production	83	85	73	72	81	78	77	82	Production d'énergie primaire
Net imports	267	*354	401	530	*686	*696	*760	761	Importations nettes
Changes in stocks	-2	10	-2	0	10	-4	10	47	Variations des stocks
Total supply	352	428	477	603	758	779	828	796	Approvisionnement total
Supply per capita	14	16	17	20	23	24	25	23	Approvisionnement par habitant
Mozambique									**Mozambique**
Primary Energy production	235	248	302	430	516	548	671	734	Production d'énergie primaire
Net imports	*13	16	-11	-66	-95	-98	-182	-195	Importations nettes
Changes in stocks	0	0	0	-1	0	11	42	70	Variations des stocks
Total supply	248	264	292	364	421	440	447	469	Approvisionnement total
Supply per capita	18	17	16	18	17	18	17	18	Approvisionnement par habitant
Myanmar									**Myanmar**
Primary Energy production	446	464	648	927	960	985	989	972	Production d'énergie primaire
Net imports	6	27	-113	-307	-305	-298	-304	-290	Importations nettes
Changes in stocks	2	-7	-3	1	2	-8	2	1	Variations des stocks
Total supply	449	498	538	619	653	695	681	682	Approvisionnement total
Supply per capita	11	12	12	13	13	13	13	13	Approvisionnement par habitant
Namibia									**Namibie**
Primary Energy production	...	10	12	13	13	14	14	*15	Production d'énergie primaire
Net imports	*0	25	28	40	51	*46	*52	*49	Importations nettes
Changes in stocks	...	...	...	...	0	0	0	*0	Variations des stocks
Total supply	*0	36	40	52	65	61	66	63	Approvisionnement total
Supply per capita	*0	22	21	25	29	27	29	27	Approvisionnement par habitant
Nauru									**Nauru**
Net imports *	2	2	2	1	1	1	1	1	Importations nettes *
Changes in stocks	...	...	...	...	*0	*0	*0	0	Variations des stocks
Total supply	*2	*1	*1	*1	1	*1	*1	*1	Approvisionnement total
Supply per capita	*188	*148	*119	*84	58	*55	*55	*60	Approvisionnement par habitant
Nepal									**Népal**
Primary Energy production	146	271	310	349	391	342	341	386	Production d'énergie primaire
Net imports	8	25	39	39	62	*63	*65	*72	Importations nettes
Changes in stocks	-1	-1	...	...	...	...	...	...	Variations des stocks
Total supply	154	297	349	388	452	405	405	458	Approvisionnement total
Supply per capita	8	14	14	14	17	15	15	16	Approvisionnement par habitant
Netherlands									**Pays-Bas**
Primary Energy production	2 534	2 791	2 409	2 604	2 924	2 695	2 711	2 908	Production d'énergie primaire
Net imports	198	35	710	673	546	422	513	325	Importations nettes
Changes in stocks	8	-98	105	43	16	-86	-26	34	Variations des stocks
Total supply	2 723	2 923	3 014	3 235	3 454	3 203	3 250	3 199	Approvisionnement total
Supply per capita	183	190	190	198	208	192	194	190	Approvisionnement par habitant

18 Production, trade and supply of energy *(continue)*
Petajoules and gigajoules per capita
Production, commerce et fourniture d'énergie *(suite)*
pétajoules et gigajoules par habitant

Region, country or area	1990	1995	2000	2005	2010	2011	2012	2013	Région, pays ou zone
Netherlands Antilles [former]									**Antilles néerlandaises [anc.]**
Primary Energy production	0	0	0	0	0	*0	...	...	Production d'énergie primaire
Net imports	*71	*94	83	84	68	87	...	...	Importations nettes
Changes in stocks *	-11	...	...	...	...	...	...	...	Variations des stocks *
Total supply	82	*95	83	84	69	87	...	...	Approvisionnement total
Supply per capita	432	*500	464	452	334	410	...	...	Approvisionnement par habitant
New Caledonia									**Nouvelle-Calédonie**
Primary Energy production	2	1	2	1	1	2	2	2	Production d'énergie primaire
Net imports	*20	*26	*28	*36	*49	*48	*45	51	Importations nettes
Changes in stocks	...	...	...	...	3	2	-1	2	Variations des stocks
Total supply	21	27	30	37	46	48	48	51	Approvisionnement total
Supply per capita	125	140	141	161	186	191	189	200	Approvisionnement par habitant
New Zealand									**Nouvelle-Zélande**
Primary Energy production	501	546	630	573	771	742	739	739	Production d'énergie primaire
Net imports	58	82	106	167	77	97	111	143	Importations nettes
Changes in stocks	1	-3	-12	-8	12	5	-22	6	Variations des stocks
Total supply	557	631	749	749	835	835	872	876	Approvisionnement total
Supply per capita	164	172	194	181	191	190	196	196	Approvisionnement par habitant
Nicaragua									**Nicaragua**
Primary Energy production	59	66	57	62	66	66	78	89	Production d'énergie primaire
Net imports	28	37	49	57	56	67	60	59	Importations nettes
Changes in stocks	0	0	-1	0	-3	3	0	-1	Variations des stocks
Total supply	86	103	107	119	125	129	138	149	Approvisionnement total
Supply per capita	21	22	21	22	22	22	24	25	Approvisionnement par habitant
Niger									**Niger**
Primary Energy production	54	65	74	*70	56	62	88	101	Production d'énergie primaire
Net imports	*7	*9	*7	*7	13	13	-4	-13	Importations nettes
Changes in stocks	0	0	0	0	0	2	0	1	Variations des stocks
Total supply	62	73	81	*77	70	73	84	87	Approvisionnement total
Supply per capita	8	8	7	*6	4	4	5	5	Approvisionnement par habitant
Nigeria									**Nigéria**
Primary Energy production	5 511	7 010	8 257	9 819	10 598	10 817	11 105	10 668	Production d'énergie primaire
Net imports	*-3 364	*-3 788	-4 497	-5 392	-5 615	-5 486	-5 604	-5 091	Importations nettes
Changes in stocks	*-108	*-11	4	-25	-35	20	-105	-12	Variations des stocks
Total supply	2 255	3 232	3 756	4 453	5 019	5 310	5 607	5 589	Approvisionnement total
Supply per capita	23	29	30	32	31	32	33	32	Approvisionnement par habitant
Niue									**Nioué**
Primary Energy production	0	0	0	0	0	0	0	0	Production d'énergie primaire
Net imports *	0	0	0	0	0	0	0	0	Importations nettes *
Total supply *	0	0	0	0	0	0	0	0	Approvisionnement total *
Supply per capita *	26	34	44	56	64	65	65	65	Approvisionnement par habitant *
Norway [8]									**Norvège [8]**
Primary Energy production	4 977	7 543	9 340	9 372	8 760	8 307	8 561	8 133	Production d'énergie primaire
Net imports	-4 024	-6 559	-8 250	-8 240	-7 334	-7 117	-7 302	-6 747	Importations nettes
Changes in stocks	77	20	15	15	-1	25	10	-2	Variations des stocks
Total supply	876	964	1 075	1 117	1 426	1 164	1 249	1 388	Approvisionnement total
Supply per capita	207	221	239	242	292	235	249	273	Approvisionnement par habitant
Oman									**Oman**
Primary Energy production	1 531	1 931	2 359	2 351	2 847	2 934	3 014	3 151	Production d'énergie primaire
Net imports	*-1 392	*-1 710	-2 079	-1 906	-2 022	-1 980	-2 027	-2 141	Importations nettes
Changes in stocks	-14	-15	-41	0	-7	-2	2	-4	Variations des stocks
Total supply	153	236	322	445	832	956	985	1 015	Approvisionnement total
Supply per capita	82	106	142	183	283	298	278	260	Approvisionnement par habitant
Other non-specified areas									**Autres zones non-spécifiées**
Primary Energy production	461	465	502	528	538	562	551	561	Production d'énergie primaire
Net imports	1 601	2 260	3 150	3 747	4 112	3 977	3 901	3 939	Importations nettes
Changes in stocks	114	104	243	37	-1	-25	6	-35	Variations des stocks
Total supply	1 947	2 621	3 410	4 238	4 651	4 565	4 446	4 534	Approvisionnement total
Supply per capita	95	123	153	186	200	196	191	194	Approvisionnement par habitant

18 Production, trade and supply of energy *(continue)*
Petajoules and gigajoules per capita
Production, commerce et fourniture d'énergie *(suite)*
pétajoules et gigajoules par habitant

Region, country or area	1990	1995	2000	2005	2010	2011	2012	2013	Région, pays ou zone
Pakistan									**Pakistan**
Primary Energy production	906	1 192	1 403	2 020	2 113	2 205	*2 220	2 009	Production d'énergie primaire
Net imports	385	529	680	633	843	810	*816	*862	Importations nettes
Changes in stocks	...	...	0	10	2	0	*-5	3	Variations des stocks
Total supply	1 290	1 722	2 082	2 642	2 954	3 015	*3 043	2 868	Approvisionnement total
Supply per capita	12	14	14	17	17	17	*17	16	Approvisionnement par habitant
Palau *									**Palaos ***
Primary Energy production	...	0	0	0	0	0	0	0	Production d'énergie primaire
Net imports	...	1	1	2	3	3	3	3	Importations nettes
Total supply	...	2	2	3	3	3	3	3	Approvisionnement total
Supply per capita	...	95	90	138	152	154	153	152	Approvisionnement par habitant
Panama									**Panama**
Primary Energy production	24	22	28	32	26	27	31	31	Production d'énergie primaire
Net imports	42	41	75	88	97	103	*112	129	Importations nettes
Changes in stocks	4	1	-1	0	-20	-23	-10	1	Variations des stocks
Total supply	62	62	104	120	143	152	152	158	Approvisionnement total
Supply per capita	26	23	35	37	40	41	41	42	Approvisionnement par habitant
Papua New Guinea									**Papouasie-Nvl-Guinée**
Primary Energy production	251	263	220	170	80	66	97	99	Production d'énergie primaire
Net imports *	-163	-171	-121	-48	47	68	34	44	Importations nettes *
Changes in stocks	0	*3	*0	0	2	0	0	-2	Variations des stocks
Total supply	88	89	99	122	126	134	130	144	Approvisionnement total
Supply per capita	21	19	18	20	18	19	18	20	Approvisionnement par habitant
Paraguay									**Paraguay**
Primary Energy production	*145	194	276	288	327	337	343	335	Production d'énergie primaire
Net imports	*-59	-80	-122	-109	-98	-104	-105	-112	Importations nettes
Changes in stocks	*1	-4	2	-1	-4	-5	2	-5	Variations des stocks
Total supply	*84	118	153	180	234	236	236	228	Approvisionnement total
Supply per capita	*20	25	29	31	38	38	37	35	Approvisionnement par habitant
Peru									**Pérou**
Primary Energy production	420	408	419	455	810	930	920	982	Production d'énergie primaire
Net imports	-28	60	130	108	-22	-119	-119	-162	Importations nettes
Changes in stocks	-3	23	7	-3	-12	-15	-55	-38	Variations des stocks
Total supply	395	444	542	565	800	826	856	858	Approvisionnement total
Supply per capita	18	19	21	20	27	28	28	28	Approvisionnement par habitant
Philippines									**Philippines**
Primary Energy production	540	533	695	762	865	882	969	935	Production d'énergie primaire
Net imports	512	759	860	*697	718	707	772	819	Importations nettes
Changes in stocks	13	28	4	-10	11	8	14	-27	Variations des stocks
Total supply	1 039	1 264	1 551	1 469	1 572	1 581	1 727	1 781	Approvisionnement total
Supply per capita	17	18	20	17	17	17	18	18	Approvisionnement par habitant
Poland									**Pologne**
Primary Energy production	4 350	4 161	3 318	3 280	2 808	2 855	2 986	2 969	Production d'énergie primaire
Net imports	99	58	438	686	1 351	1 452	1 295	1 090	Importations nettes
Changes in stocks	43	-36	-24	82	-93	45	171	-61	Variations des stocks
Total supply	4 406	4 256	3 780	3 886	4 251	4 261	4 110	4 121	Approvisionnement total
Supply per capita	116	111	99	102	110	110	106	107	Approvisionnement par habitant
Portugal [9]									**Portugal [9]**
Primary Energy production	142	139	161	151	242	231	203	241	Production d'énergie primaire
Net imports	596	714	864	969	717	723	694	642	Importations nettes
Changes in stocks	19	12	0	16	-15	5	0	-16	Variations des stocks
Total supply	720	840	1 026	1 104	975	949	897	899	Approvisionnement total
Supply per capita	73	83	99	105	92	90	85	86	Approvisionnement par habitant
Puerto Rico									**Porto Rico**
Primary Energy production	*1	0	1	0	0	1	0	1	Production d'énergie primaire
Net imports	...	...	12	24	27	26	48	55	Importations nettes
Total supply	*1	0	13	24	27	27	48	56	Approvisionnement total
Supply per capita	*0	0	3	6	7	7	13	15	Approvisionnement par habitant

18 Production, trade and supply of energy *(continue)*
Petajoules and gigajoules per capita
Production, commerce et fourniture d'énergie *(suite)*
pétajoules et gigajoules par habitant

Region, country or area	1990	1995	2000	2005	2010	2011	2012	2013	Région, pays ou zone
Qatar									**Qatar**
Primary Energy production	1 009	1 363	2 449	3 718	7 428	8 802	9 184	9 271	Production d'énergie primaire
Net imports	-832	-811	-1 921	-2 991	-6 261	-7 479	-7 618	-7 860	Importations nettes
Changes in stocks	-22	0	-56	-1	3	-7	-13	15	Variations des stocks
Total supply	199	553	584	728	1 165	1 331	1 579	1 396	Approvisionnement total
Supply per capita	420	1 102	988	886	660	699	783	664	Approvisionnement par habitant
Republic of Korea									**République de Corée**
Primary Energy production	941	878	1 420	1 776	1 855	1 941	1 908	1 799	Production d'énergie primaire
Net imports	2 840	5 319	6 505	6 810	8 703	8 976	9 034	9 257	Importations nettes
Changes in stocks	-69	115	71	-179	117	44	-49	49	Variations des stocks
Total supply	3 851	6 083	7 854	8 764	10 441	10 874	10 992	11 007	Approvisionnement total
Supply per capita	90	136	171	186	213	220	222	221	Approvisionnement par habitant
Republic of Moldova									**République de Moldova**
Primary Energy production	...	1	3	4	8	10	10	12	Production d'énergie primaire
Net imports	...	*184	*64	*85	80	83	80	82	Importations nettes
Changes in stocks	...	2	0	0	-2	1	0	1	Variations des stocks
Total supply	...	183	66	89	91	93	91	93	Approvisionnement total
Supply per capita	...	42	16	24	22	23	22	23	Approvisionnement par habitant
Réunion									**Réunion**
Primary Energy production	7	8	7	6	8	7	8	*8	Production d'énergie primaire
Net imports *	20	27	36	44	50	53	50	50	Importations nettes *
Changes in stocks	0	0	1	1	0	1	-2	*-2	Variations des stocks
Total supply	28	35	42	48	57	59	58	*59	Approvisionnement total
Supply per capita	46	53	56	61	69	71	69	*70	Approvisionnement par habitant
Romania									**Roumanie**
Primary Energy production	1 598	1 340	1 194	1 173	1 155	1 162	1 142	1 090	Production d'énergie primaire
Net imports	771	591	331	446	311	315	328	240	Importations nettes
Changes in stocks	20	2	-8	-7	-7	-31	2	-8	Variations des stocks
Total supply	2 349	1 930	1 532	1 626	1 474	1 508	1 469	1 339	Approvisionnement total
Supply per capita	101	85	69	75	73	75	74	68	Approvisionnement par habitant
Russian Federation									**Fédération de Russie**
Primary Energy production	...	40 589	41 030	50 506	53 679	54 534	55 216	56 266	Production d'énergie primaire
Net imports	...	-13 397	-14 842	-22 870	-24 553	-24 266	-24 029	-25 332	Importations nettes
Changes in stocks	...	472	195	298	147	-83	69	225	Variations des stocks
Total supply	...	26 720	25 994	27 338	28 980	30 350	31 118	30 709	Approvisionnement total
Supply per capita	...	180	177	190	202	212	217	214	Approvisionnement par habitant
Rwanda									**Rwanda**
Primary Energy production	28	48	46	63	76	*78	*81	*83	Production d'énergie primaire
Net imports *	8	7	8	7	8	9	10	11	Importations nettes *
Changes in stocks	...	...	...	0	0	0	0	0	Variations des stocks
Total supply	35	54	54	71	84	*88	*91	*95	Approvisionnement total
Supply per capita	5	10	7	8	8	*8	*8	*9	Approvisionnement par habitant
Saint Helena									**Sainte-Hélène**
Primary Energy production	*0	0	0	0	0	0	0	0	Production d'énergie primaire
Net imports	0	*0	0	0	0	0	0	0	Importations nettes
Total supply	0	*0	0	0	0	0	0	0	Approvisionnement total
Supply per capita	17	*27	36	33	38	37	41	40	Approvisionnement par habitant
Saint Kitts and Nevis									**Saint-Kitts-et-Nevis**
Primary Energy production	1	1	1	1	...	...	...	...	Production d'énergie primaire
Net imports *	2	2	3	3	4	4	4	4	Importations nettes *
Changes in stocks	0	0	...	...	...	...	...	...	Variations des stocks
Total supply *	2	3	3	4	4	4	4	4	Approvisionnement total *
Supply per capita *	59	68	67	78	70	71	73	73	Approvisionnement par habitant *
Saint Lucia									**Sainte-Lucie**
Primary Energy production	0	0	0	0	0	0	0	0	Production d'énergie primaire
Net imports	2	4	*5	*6	*6	*6	*6	*6	Importations nettes
Changes in stocks	0	...	...	...	...	...	...	...	Variations des stocks
Total supply	2	5	5	*5	6	6	6	6	Approvisionnement total
Supply per capita	18	31	31	*33	33	33	33	33	Approvisionnement par habitant

18

Production, trade and supply of energy *(continue)*
Petajoules and gigajoules per capita
Production, commerce et fourniture d'énergie *(suite)*
pétajoules et gigajoules par habitant

Region, country or area	1990	1995	2000	2005	2010	2011	2012	2013	Région, pays ou zone
Saint Pierre and Miquelon									**Saint-Pierre-et-Miquelon**
Primary Energy production	...	...	*0	*0	0	0	0	*0	Production d'énergie primaire
Net imports	2	1	*1	*1	*1	*1	*1	*1	Importations nettes
Total supply	1	1	*1	*1	*1	*1	*1	*1	Approvisionnement total
Supply per capita	206	164	*125	*148	*159	*160	*158	*162	Approvisionnement par habitant
Saint Vincent-Grenadines									**Saint-Vincent-Grenadines**
Primary Energy production	0	0	0	0	*0	*0	*0	*0	Production d'énergie primaire
Net imports	*1	*2	2	*3	*3	*3	*4	*3	Importations nettes
Total supply	*1	*2	2	*3	3	3	4	3	Approvisionnement total
Supply per capita	*12	*19	21	*31	31	28	34	28	Approvisionnement par habitant
Samoa									**Samoa**
Primary Energy production	1	1	1	1	1	1	1	1	Production d'énergie primaire
Net imports	*2	*2	*2	*2	*3	3	*4	3	Importations nettes
Changes in stocks	...	...	...	*0	0	0	*0	*0	Variations des stocks
Total supply	*3	*3	*3	*3	*4	4	*4	4	Approvisionnement total
Supply per capita	*16	*16	*16	*18	*21	22	*23	22	Approvisionnement par habitant
Sao Tome and Principe									**Sao Tomé-et-Principe**
Primary Energy production	1	1	1	1	1	1	1	1	Production d'énergie primaire
Net imports	*1	*1	*1	*1	2	*2	*2	*2	Importations nettes
Total supply	*1	1	*2	*2	2	*2	*3	*3	Approvisionnement total
Supply per capita	*12	12	*12	*13	14	*14	*15	*14	Approvisionnement par habitant
Saudi Arabia [7]									**Arabie saoudite** [7]
Primary Energy production	16 769	19 932	20 284	24 162	22 115	24 664	26 010	25 573	Production d'énergie primaire
Net imports	*-13 260	-16 427	-15 797	-18 116	-14 731	-17 013	-17 932	-17 702	Importations nettes
Changes in stocks	*172	-12	0	0	-319	257	-243	-114	Variations des stocks
Total supply	3 336	3 516	4 487	6 047	7 703	7 395	8 320	7 985	Approvisionnement total
Supply per capita	207	190	224	252	274	257	282	264	Approvisionnement par habitant
Senegal									**Sénégal**
Primary Energy production	45	49	43	46	86	90	92	75	Production d'énergie primaire
Net imports	*41	*43	46	65	74	77	74	78	Importations nettes
Changes in stocks	0	0	1	3	-4	-4	-2	-2	Variations des stocks
Total supply	*86	*92	89	109	165	172	168	154	Approvisionnement total
Supply per capita	*12	*11	9	10	13	13	12	11	Approvisionnement par habitant
Serbia [10]									**Serbie** [10]
Primary Energy production	...	...	...	431	440	465	449	473	Production d'énergie primaire
Net imports	...	...	...	243	212	203	166	145	Importations nettes
Changes in stocks	...	...	...	2	4	-5	11	0	Variations des stocks
Total supply	...	...	...	672	648	673	604	618	Approvisionnement total
Supply per capita	...	...	...	68	72	75	67	69	Approvisionnement par habitant
Seychelles									**Seychelles**
Primary Energy production	0	0	0	0	0	0	0	0	Production d'énergie primaire
Net imports	1	*4	*9	*10	*9	*8	*10	*9	Importations nettes
Changes in stocks	*0	0	*0	...	...	...	...	...	Variations des stocks
Total supply	2	*3	8	10	10	8	10	9	Approvisionnement total
Supply per capita	31	*39	102	115	103	88	103	94	Approvisionnement par habitant
Sierra Leone									**Sierra Leone**
Primary Energy production	43	42	49	50	52	52	52	52	Production d'énergie primaire
Net imports *	7	5	6	7	9	11	13	15	Importations nettes *
Changes in stocks	0	...	...	...	...	...	...	...	Variations des stocks
Total supply	50	47	55	56	60	63	65	67	Approvisionnement total
Supply per capita	13	12	13	11	10	11	11	11	Approvisionnement par habitant
Singapore									**Singapour**
Primary Energy production	...	...	...	...	...	0	0	0	Production d'énergie primaire
Net imports	493	641	783	*774	*857	*766	*986	*1 086	Importations nettes
Changes in stocks	-163	2	27	-31	-31	-36	-69	-34	Variations des stocks
Total supply	656	640	756	805	888	801	1 054	1 120	Approvisionnement total
Supply per capita	218	184	193	189	175	154	199	207	Approvisionnement par habitant
Sint Maarten (Dutch part) *									**Saint-Martin (partie néerl.)** *
Net imports	...	...	...	...	...	...	12	11	Importations nettes
Total supply	...	...	...	...	...	...	12	12	Approvisionnement total
Supply per capita	...	...	...	...	...	...	333	327	Approvisionnement par habitant

18 Production, trade and supply of energy *(continue)*
Petajoules and gigajoules per capita
Production, commerce et fourniture d'énergie *(suite)*
pétajoules et gigajoules par habitant

Region, country or area	1990	1995	2000	2005	2010	2011	2012	2013	Région, pays ou zone
Slovakia									**Slovaquie**
Primary Energy production	...	210	263	265	250	259	261	269	Production d'énergie primaire
Net imports	...	514	485	516	475	470	422	435	Importations nettes
Changes in stocks	...	-22	4	4	-12	13	-4	-7	Variations des stocks
Total supply	...	746	744	778	737	717	688	711	Approvisionnement total
Supply per capita	...	139	138	144	136	132	127	131	Approvisionnement par habitant
Slovenia									**Slovénie**
Primary Energy production	...	124	129	146	153	158	148	149	Production d'énergie primaire
Net imports	...	129	141	160	150	147	151	133	Importations nettes
Changes in stocks	...	-1	2	0	0	0	4	-5	Variations des stocks
Total supply	...	254	269	306	302	305	295	287	Approvisionnement total
Supply per capita	...	129	136	153	147	148	143	139	Approvisionnement par habitant
Solomon Islands *									**Îles Salomon ***
Primary Energy production	3	3	3	3	3	3	3	3	Production d'énergie primaire
Net imports	2	2	2	3	3	3	3	3	Importations nettes
Total supply	5	5	5	6	6	6	6	6	Approvisionnement total
Supply per capita	17	15	13	12	11	11	11	11	Approvisionnement par habitant
Somalia									**Somalie**
Primary Energy production	58	69	85	103	124	128	128	125	Production d'énergie primaire
Net imports	0	0	*6	*4	*1	*-1	*1	*1	Importations nettes
Total supply	58	69	90	107	126	127	129	126	Approvisionnement total
Supply per capita	9	11	12	13	13	13	13	12	Approvisionnement par habitant
South Africa									**Afrique du Sud**
Primary Energy production	4 431	5 329	6 175	6 652	6 914	6 885	6 998	6 999	Production d'énergie primaire
Net imports	*-511	-1 036	-1 171	-1 183	-596	-512	-687	-640	Importations nettes
Changes in stocks	0	-68	-41	0	1	7	6	18	Variations des stocks
Total supply	3 921	4 360	5 044	5 468	6 316	6 366	6 304	6 341	Approvisionnement total
Supply per capita	107	105	113	114	122	122	119	119	Approvisionnement par habitant
South Sudan									**Soudan du sud**
Primary Energy production	...	...	...	...	...	...	74	217	Production d'énergie primaire
Net imports	...	...	...	...	...	...	-47	-190	Importations nettes
Total supply	...	...	...	...	...	...	27	28	Approvisionnement total
Supply per capita	...	...	...	...	...	...	2	2	Approvisionnement par habitant
Spain [11]									**Espagne [11]**
Primary Energy production	1 442	1 313	1 314	1 256	1 419	1 422	1 463	1 508	Production d'énergie primaire
Net imports	2 308	2 949	3 812	4 738	3 991	3 883	3 690	3 289	Importations nettes
Changes in stocks	6	69	52	78	73	-25	-142	-125	Variations des stocks
Total supply	3 744	4 191	5 074	5 915	5 337	5 330	5 295	4 923	Approvisionnement total
Supply per capita	96	106	126	136	115	114	114	106	Approvisionnement par habitant
Sri Lanka									**Sri Lanka**
Primary Energy production	135	146	156	163	184	180	173	184	Production d'énergie primaire
Net imports	54	80	144	161	170	195	217	188	Importations nettes
Changes in stocks	2	-1	4	1	-7	-4	8	-16	Variations des stocks
Total supply	187	225	296	324	360	381	382	388	Approvisionnement total
Supply per capita	11	12	16	16	18	19	19	19	Approvisionnement par habitant
State of Palestine									**État de Palestine**
Primary Energy production	...	...	...	8	9	9	8	8	Production d'énergie primaire
Net imports	...	...	20	49	45	50	50	53	Importations nettes
Changes in stocks	...	...	...	0	...	...	...	...	Variations des stocks
Total supply	...	...	20	58	54	59	58	61	Approvisionnement total
Supply per capita	...	...	6	16	13	14	13	14	Approvisionnement par habitant
Sudan									**Soudan**
Primary Energy production	...	...	...	...	...	...	616	672	Production d'énergie primaire
Net imports	...	...	...	...	...	...	-31	*-64	Importations nettes
Total supply	...	...	...	...	...	...	585	608	Approvisionnement total
Supply per capita	...	...	...	...	...	...	16	16	Approvisionnement par habitant
Sudan [former]									**Soudan [anc.]**
Primary Energy production	379	448	851	1 145	1 473	1 467	...	...	Production d'énergie primaire
Net imports	*73	*59	-287	-480	-762	-752	...	...	Importations nettes
Changes in stocks	*1	1	8	12	...	...	...	...	Variations des stocks
Total supply	451	506	556	654	711	716	...	...	Approvisionnement total
Supply per capita	17	17	16	17	16	15	...	...	Approvisionnement par habitant

18

Production, trade and supply of energy *(continue)*
Petajoules and gigajoules per capita
Production, commerce et fourniture d'énergie *(suite)*
pétajoules et gigajoules par habitant

Region, country or area	1990	1995	2000	2005	2010	2011	2012	2013	Région, pays ou zone
Suriname									**Suriname**
Primary Energy production	15	17	35	32	43	44	45	42	Production d'énergie primaire
Net imports	16	18	3	-4	0	-7	-3	-8	Importations nettes
Changes in stocks	0	0	...	...	0	0	0	0	Variations des stocks
Total supply	30	35	38	28	43	37	41	34	Approvisionnement total
Supply per capita	75	81	82	55	83	70	78	65	Approvisionnement par habitant
Swaziland									**Swaziland**
Primary Energy production	20	19	34	36	33	34	37	38	Production d'énergie primaire
Net imports *	...	2	7	3	14	12	11	11	Importations nettes *
Changes in stocks	0	...	...	*0	*0	0	0	0	Variations des stocks
Total supply	20	21	41	40	46	47	48	49	Approvisionnement total
Supply per capita	23	22	39	36	39	39	39	39	Approvisionnement par habitant
Sweden									**Suède**
Primary Energy production	1 235	1 312	1 257	1 430	1 364	1 375	1 494	1 450	Production d'énergie primaire
Net imports	726	747	724	739	714	687	542	594	Importations nettes
Changes in stocks	-5	-29	9	32	-33	-6	-42	0	Variations des stocks
Total supply	1 967	2 089	1 971	2 137	2 112	2 068	2 079	2 044	Approvisionnement total
Supply per capita	230	237	222	237	225	219	218	212	Approvisionnement par habitant
Switzerland [12]									**Suisse** [12]
Primary Energy production	428	468	500	458	526	513	530	539	Production d'énergie primaire
Net imports	584	521	524	628	564	535	545	569	Importations nettes
Changes in stocks	-5	-14	-18	6	-2	-8	7	-5	Variations des stocks
Total supply	1 017	1 003	1 041	1 080	1 091	1 056	1 068	1 113	Approvisionnement total
Supply per capita	152	142	145	145	139	133	133	136	Approvisionnement par habitant
Syrian Arab Republic									**Rép. arabe syrienne**
Primary Energy production	1 036	1 403	1 417	1 171	1 165	1 056	567	316	Production d'énergie primaire
Net imports	*-554	-813	-690	-431	-233	-174	59	226	Importations nettes
Changes in stocks	6	*5	...	1	20	45	0	0	Variations des stocks
Total supply	476	585	729	740	911	837	627	542	Approvisionnement total
Supply per capita	39	41	46	40	44	41	31	28	Approvisionnement par habitant
Tajikistan									**Tadjikistan**
Primary Energy production	...	56	53	65	64	64	70	72	Production d'énergie primaire
Net imports	...	40	38	33	29	25	30	34	Importations nettes
Total supply	...	95	91	98	93	89	99	106	Approvisionnement total
Supply per capita	...	17	15	15	12	12	12	13	Approvisionnement par habitant
Thailand									**Thaïlande**
Primary Energy production	1 089	1 322	1 694	2 117	2 934	*2 956	*3 130	3 036	Production d'énergie primaire
Net imports	*699	*1 289	*1 308	*1 869	*2 027	*1 957	*2 130	2 140	Importations nettes
Changes in stocks	4	-25	-67	-54	34	*-19	*59	-161	Variations des stocks
Total supply	1 784	2 635	3 069	4 040	4 927	4 931	5 200	5 336	Approvisionnement total
Supply per capita	31	44	49	61	74	74	77	79	Approvisionnement par habitant
TFYR of Macedonia									**ex-R.Y. de Macédoine**
Primary Energy production	...	112	118	108	66	71	63	59	Production d'énergie primaire
Net imports	...	43	48	52	53	59	61	55	Importations nettes
Changes in stocks	...	5	-4	-4	-1	1	-2	-3	Variations des stocks
Total supply	...	151	169	163	119	130	125	117	Approvisionnement total
Supply per capita	...	77	84	80	58	63	60	56	Approvisionnement par habitant
Timor-Leste									**Timor-Leste**
Primary Energy production	...	...	...	201	186	178	169	168	Production d'énergie primaire
Net imports *	...	...	...	-196	-182	-173	-164	-161	Importations nettes *
Total supply *	...	...	...	4	4	5	5	7	Approvisionnement total *
Supply per capita *	...	...	...	4	4	4	5	6	Approvisionnement par habitant *
Togo									**Togo**
Primary Energy production	77	94	74	84	99	102	105	108	Production d'énergie primaire
Net imports	*11	*15	*16	15	29	26	24	25	Importations nettes
Changes in stocks	0	0	0	-1	-1	-3	-2	-2	Variations des stocks
Total supply	87	108	89	99	130	131	131	134	Approvisionnement total
Supply per capita	24	26	19	18	20	20	19	19	Approvisionnement par habitant
Tonga									**Tonga**
Primary Energy production	0	0	0	0	0	0	0	0	Production d'énergie primaire
Net imports	1	1	1	*2	*2	*2	*3	*3	Importations nettes
Total supply	1	1	1	*2	2	2	2	3	Approvisionnement total
Supply per capita	12	14	15	*16	16	15	24	28	Approvisionnement par habitant

18 Production, trade and supply of energy *(continue)*
Petajoules and gigajoules per capita
Production, commerce et fourniture d'énergie *(suite)*
pétajoules et gigajoules par habitant

Region, country or area	1990	1995	2000	2005	2010	2011	2012	2013	Région, pays ou zone
Trinidad and Tobago									**Trinité-et-Tobago**
Primary Energy production	517	540	752	1 318	1 785	1 696	1 669	1 673	Production d'énergie primaire
Net imports	-262	*-236	-336	-800	-953	-871	-861	-863	Importations nettes
Changes in stocks	5	-16	7	-1	-13	-6	0	-13	Variations des stocks
Total supply	249	320	411	519	844	830	808	824	Approvisionnement total
Supply per capita	205	254	318	395	636	622	602	611	Approvisionnement par habitant
Tunisia									**Tunisie**
Primary Energy production	222	205	253	274	341	312	309	298	Production d'énergie primaire
Net imports	*-27	*6	47	71	79	96	115	135	Importations nettes
Changes in stocks	2	-9	7	-4	-7	3	1	2	Variations des stocks
Total supply	192	220	293	349	427	405	422	430	Approvisionnement total
Supply per capita	23	25	31	35	40	38	39	39	Approvisionnement par habitant
Turkey									**Turquie**
Primary Energy production	1 087	1 112	1 085	1 004	1 352	1 345	1 282	1 352	Production d'énergie primaire
Net imports	1 139	1 532	2 114	2 517	3 039	3 316	3 694	3 568	Importations nettes
Changes in stocks	35	45	-3	-28	-28	-55	57	23	Variations des stocks
Total supply	2 191	2 598	3 202	3 549	4 419	4 716	4 918	4 897	Approvisionnement total
Supply per capita	40	44	50	52	61	64	66	64	Approvisionnement par habitant
Turkmenistan									**Turkménistan**
Primary Energy production	...	1 376	1 928	2 584	1 982	2 737	2 853	3 210	Production d'énergie primaire
Net imports	...	-802	-1 303	-1 779	-1 030	-1 699	-1 780	-2 108	Importations nettes
Total supply	...	574	624	805	951	1 037	1 073	1 102	Approvisionnement total
Supply per capita	...	137	139	169	189	203	207	210	Approvisionnement par habitant
Turks and Caicos Islands									**Îles Turques-et-Caïques**
Primary Energy production	0	0	0	0	0	0	0	0	Production d'énergie primaire
Net imports	0	*0	*0	*2	*3	*3	*3	*3	Importations nettes
Total supply	0	*0	*0	*2	*3	*3	*3	*3	Approvisionnement total
Supply per capita	0	*13	*11	*57	*86	*84	*86	*85	Approvisionnement par habitant
Uganda									**Ouganda**
Primary Energy production	271	300	321	350	379	396	398	398	Production d'énergie primaire
Net imports *	9	12	17	26	46	49	46	54	Importations nettes *
Changes in stocks	0	...	...	...	...	...	...	...	Variations des stocks
Total supply	281	312	338	376	425	445	443	452	Approvisionnement total
Supply per capita	16	15	14	13	13	13	13	12	Approvisionnement par habitant
Ukraine									**Ukraine**
Primary Energy production	...	3 661	3 231	3 328	3 238	3 513	3 554	3 578	Production d'énergie primaire
Net imports	...	3 411	2 343	2 458	1 762	1 976	1 605	*1 320	Importations nettes
Changes in stocks	...	0	0	-86	-490	265	16	54	Variations des stocks
Total supply	...	7 073	5 574	5 872	5 491	5 225	5 142	4 844	Approvisionnement total
Supply per capita	...	138	114	125	120	115	113	107	Approvisionnement par habitant
United Arab Emirates									**Émirats arabes unis**
Primary Energy production	5 177	5 661	6 571	7 293	7 541	8 150	8 465	8 792	Production d'énergie primaire
Net imports	*-4 375	*-4 533	-4 853	-5 450	*-4 955	*-5 544	*-5 669	*-6 082	Importations nettes
Changes in stocks	12	21	...	...	...	...	...	...	Variations des stocks
Total supply	791	1 108	1 719	1 842	2 587	2 606	2 795	2 710	Approvisionnement total
Supply per capita	437	472	567	453	311	298	312	300	Approvisionnement par habitant
United Kingdom [13]									**Royaume-Uni** [13]
Primary Energy production	8 595	10 588	11 247	8 483	6 198	5 395	4 859	4 557	Production d'énergie primaire
Net imports	-151	-1 928	-2 169	736	1 965	2 434	3 045	3 383	Importations nettes
Changes in stocks	-84	-265	-138	0	-265	37	-78	20	Variations des stocks
Total supply	8 527	8 925	9 216	9 219	8 428	7 792	7 982	7 920	Approvisionnement total
Supply per capita	149	154	157	153	134	123	126	124	Approvisionnement par habitant
United Rep. of Tanzania									**Rép.-Unie de Tanzanie**
Primary Energy production	178	434	543	665	789	814	843	889	Production d'énergie primaire
Net imports	*29	*44	30	57	64	72	89	105	Importations nettes
Changes in stocks	0	...	...	...	...	...	...	...	Variations des stocks
Total supply	206	477	572	722	854	887	932	994	Approvisionnement total
Supply per capita	8	16	17	19	19	19	19	20	Approvisionnement par habitant

18

Production, trade and supply of energy *(continue)*
Petajoules and gigajoules per capita
Production, commerce et fourniture d'énergie *(suite)*
pétajoules et gigajoules par habitant

Region, country or area	1990	1995	2000	2005	2010	2011	2012	2013	Région, pays ou zone
United States [14]									**États-Unis** [14]
Primary Energy production	68 648	68 870	69 580	68 293	72 211	74 785	75 626	78 449	Production d'énergie primaire
Net imports	12 528	16 273	23 218	28 307	19 895	16 709	13 332	10 896	Importations nettes
Changes in stocks	1 606	-764	-1 980	-16	-263	135	159	-1 494	Variations des stocks
Total supply	79 570	85 906	94 779	96 616	92 368	91 359	88 798	90 838	Approvisionnement total
Supply per capita	314	323	336	326	298	292	282	286	Approvisionnement par habitant
Uruguay									**Uruguay**
Primary Energy production	51	45	46	45	89	84	82	94	Production d'énergie primaire
Net imports	44	66	70	75	85	99	126	94	Importations nettes
Changes in stocks	-2	4	-2	-5	1	-6	8	-7	Variations des stocks
Total supply	96	107	119	126	173	189	201	196	Approvisionnement total
Supply per capita	31	33	36	38	51	56	59	58	Approvisionnement par habitant
Uzbekistan									**Ouzbékistan**
Primary Energy production	...	1 984	2 307	2 365	2 309	2 400	2 382	2 267	Production d'énergie primaire
Net imports	...	-170	-191	-394	-499	-417	-354	-469	Importations nettes
Changes in stocks	...	88	...	...	...	...	...	...	Variations des stocks
Total supply	...	1 726	2 116	1 970	1 809	1 983	2 027	1 798	Approvisionnement total
Supply per capita	...	75	85	76	65	70	71	62	Approvisionnement par habitant
Vanuatu									**Vanuatu**
Primary Energy production	0	0	1	1	1	1	1	1	Production d'énergie primaire
Net imports *	1	1	1	1	2	2	2	2	Importations nettes *
Changes in stocks *	...	...	...	...	0	0	0	0	Variations des stocks *
Total supply	1	1	*2	2	3	3	3	3	Approvisionnement total
Supply per capita	8	7	*11	8	11	12	11	10	Approvisionnement par habitant
Venezuela (Boliv. Rep. of)									**Venezuela (Rép. boliv. du)**
Primary Energy production	5 953	7 962	8 965	8 283	8 139	8 184	8 013	7 716	Production d'énergie primaire
Net imports	-4 027	-5 772	-6 320	-5 883	*-4 983	*-5 193	-4 787	-4 807	Importations nettes
Changes in stocks	-18	24	15	-336	-54	-2	22	39	Variations des stocks
Total supply	1 944	2 167	2 629	2 737	3 211	2 993	3 204	2 871	Approvisionnement total
Supply per capita	99	98	108	103	111	102	107	95	Approvisionnement par habitant
Viet Nam									**Viet Nam**
Primary Energy production	787	1 136	1 733	2 612	2 747	2 806	2 856	2 887	Production d'énergie primaire
Net imports	-3	-207	-421	-795	-370	-252	-260	-307	Importations nettes
Changes in stocks	-6	20	50	70	5	-26	-72	26	Variations des stocks
Total supply	791	908	1 262	1 747	2 372	2 580	2 667	2 554	Approvisionnement total
Supply per capita	12	12	16	21	27	29	30	28	Approvisionnement par habitant
Wallis and Futuna Islands									**Îles Wallis-et-Futuna**
Net imports	...	...	...	0	0	0	0	0	Importations nettes
Total supply	...	...	...	0	0	0	0	0	Approvisionnement total
Supply per capita	...	...	...	26	27	28	26	26	Approvisionnement par habitant
Yemen									**Yémen**
Primary Energy production	...	743	909	844	803	765	618	759	Production d'énergie primaire
Net imports	...	*-611	-711	-529	-477	-490	-369	-416	Importations nettes
Changes in stocks	...	-21	0	40	...	...	...	...	Variations des stocks
Total supply	...	153	197	274	326	275	250	344	Approvisionnement total
Supply per capita	...	10	11	13	14	11	10	13	Approvisionnement par habitant
Zambia									**Zambie**
Primary Energy production	207	224	248	280	322	332	344	362	Production d'énergie primaire
Net imports	*14	*19	13	25	29	32	38	38	Importations nettes
Changes in stocks	...	-1	1	0	0	0	-1	-1	Variations des stocks
Total supply	222	244	260	305	351	364	382	400	Approvisionnement total
Supply per capita	28	27	25	27	25	25	26	26	Approvisionnement par habitant
Zimbabwe									**Zimbabwe**
Primary Energy production	372	362	379	379	361	372	391	422	Production d'énergie primaire
Net imports	*28	64	53	36	26	45	51	59	Importations nettes
Changes in stocks	0	5	-5	1	...	...	...	...	Variations des stocks
Total supply	400	422	438	414	386	417	441	480	Approvisionnement total
Supply per capita	38	36	35	33	28	29	30	32	Approvisionnement par habitant

Source:
United Nations Statistics Division, New York, *Energy Statistics Yearbook 2013.*

Source:
Organisation des Nations Unies, Division de statistique, New York, *Annuaire des statistiques de l'Energie 2013.*

18

Production, trade and supply of energy *(continue)*
Petajoules and gigajoules per capita
Production, commerce et fourniture d'énergie *(suite)*
pétajoules et gigajoules par habitant

1	Data exclude overseas territories.	1	Les données ne comprennent pas les territoires d'outre-mer.
2	For statistical purposes, the data for China do not include those for the Hong Kong Special Administrative Region (Hong Kong SAR), Macao Special Administrative Region (Macao SAR) and Taiwan Province of China.	2	Pour la présentation des statistiques, les données pour la Chine ne comprennent pas la Région Administrative Spéciale de Hong Kong (Hong Kong RAS), la Région Administrative Spéciale de Macao (Macao RAS) et la province de Taiwan.
3	Data exclude Greenland and the Danish Faroes.	3	Les données ne comprennent pas le Groenland et les îles Féroé danoises.
4	Including Monaco.	4	Y compris Monaco.
5	Data include Timor-Leste.	5	Les données comprennent Timor-Leste.
6	Data include San Marino and the Holy See.	6	Les données comprennent Saint-Marin et le Saint-Siège.
7	The data for crude oil production include 50 per cent of the output of the Neutral Zone.	7	Les données relatives à la production de pétrole brut comprennent 50 pour cent de la production de la Zone Neutre.
8	Including Svalbard and Jan Mayen Islands.	8	Y compris îles Svalbard et Jan Mayen.
9	Data includes the Azores and Madeira.	9	Les données comprennent Azores et Madère.
10	Excluding Kosovo.	10	Non compris Kosovo.
11	Data include the Canary Islands.	11	Les données comprennent les Iles Canaries.
12	Including Liechtenstein.	12	Y compris Liechtenstein.
13	Shipments of coal and oil to Jersey, Guernsey and the isle of Man from the United Kingdom are not classed as exports. Supplies of coal and oil to these islands are, therefore, included as part of UK supply. Exports of natural gas to the Isle of Man included with the exports to Ireland.	13	Les livraisons de charbon et de pétrole du Royaume-Uni à Jersey, Guernesey et à l'île de Man ne sont pas considérées comme des exportations ; l'approvisionnement de charbon et pétrole à ces îles fait donc partie de l'approvisionnement du Royaume-Uni. Les exportations de gaz naturel vers l'île de Man sont inclues dans les exportations vers l'Irlande.
14	Includes the 50 states and the District of Columbia. Oil statistics as well as coal trade statistics also include Puerto Rico, Guam, the U.S. Virgin Islands, American Samoa, Johnston Atoll, Midway Islands, Wake Island and the Northern Mariana Islands.	14	Comprend les 50 États et le District de Columbia. Les statistiques pétrolières et les statistiques du commerce de charbon incluent également Puerto Rico, Guam, les îles Vierges américaines, les Samoa américaines, Johnston Atoll, les îles Midway, l'île de Wake et les îles Mariannes septentrionale.

Country or area [&]	2004	2010	2012	2013	2014	2015	2016	Pays ou zone [&]
Afghanistan								**Afghanistan**
Vertebrates	31	31	32	32	32	31	35	Vertébrés
Invertebrates	1	1	1	1	2	2	2	Invertébrés
Plants	1	2	3	3	3	5	5	Plantes
Total	33	34	36	36	37	38	42	Total
Albania								**Albanie**
Vertebrates	33	53	54	54	53	54	63	Vertébrés
Invertebrates	4	47	55	55	56	58	62	Invertébrés
Plants	0	0	0	0	0	0	0	Plantes
Total	37	100	109	109	109	112	125	Total
Algeria								**Algérie**
Vertebrates	36	69	72	72	71	71	78	Vertébrés
Invertebrates	12	21	22	22	25	26	29	Invertébrés
Plants	2	15	16	17	17	17	18	Plantes
Total	50	105	110	111	113	114	125	Total
American Samoa								**Samoa américaines**
Vertebrates	18	21	22	24	25	25	27	Vertébrés
Invertebrates	5	57	57	64	64	64	64	Invertébrés
Plants	1	1	1	1	1	1	1	Plantes
Total	24	79	80	89	90	90	92	Total
Andorra								**Andorre**
Vertebrates	1	4	4	4	4	4	5	Vertébrés
Invertebrates	4	4	7	8	7	7	7	Invertébrés
Plants	0	0	0	0	0	0	0	Plantes
Total	5	8	11	12	11	11	12	Total
Angola								**Angola**
Vertebrates	44	77	83	85	87	87	99	Vertébrés
Invertebrates	6	7	7	7	9	9	9	Invertébrés
Plants	26	33	33	34	34	34	34	Plantes
Total	76	117	123	126	130	130	142	Total
Anguilla								**Anguilla**
Vertebrates	15	20	25	28	29	37	38	Vertébrés
Invertebrates	0	10	10	10	10	10	10	Invertébrés
Plants	3	3	3	3	3	4	4	Plantes
Total	18	33	38	41	42	51	52	Total
Antarctica								**Antarctique**
Vertebrates	8	6	6	6	6	6	6	Vertébrés
Invertebrates	0	0	0	0	0	0	0	Invertébrés
Plants	0	0	0	0	0	0	0	Plantes
Total	8	6	6	6	6	6	6	Total
Antigua and Barbuda								**Antigua-et-Barbuda**
Vertebrates	18	23	27	29	30	38	38	Vertébrés
Invertebrates	0	11	11	11	11	11	11	Invertébrés
Plants	4	4	4	4	4	4	4	Plantes
Total	22	38	42	44	45	53	53	Total
Argentina								**Argentine**
Vertebrates	134	157	159	160	160	160	172	Vertébrés
Invertebrates	10	12	13	13	13	13	14	Invertébrés
Plants	42	44	45	69	70	70	70	Plantes
Total	186	213	217	242	243	243	256	Total
Armenia								**Arménie**
Vertebrates	27	29	32	32	31	31	34	Vertébrés
Invertebrates	7	6	7	7	9	9	9	Invertébrés
Plants	1	1	1	1	71	71	71	Plantes
Total	35	36	40	40	111	111	114	Total

Country or area [&]	2004	2010	2012	2013	2014	2015	2016	Pays ou zone [&]
Aruba								**Aruba**
Vertebrates	17	20	20	20	19	25	28	Vertébrés
Invertebrates	1	1	1	2	2	2	2	Invertébrés
Plants	0	1	1	2	2	2	2	Plantes
Total	18	22	22	24	23	29	32	Total
Australia [1]								**Australie** [1]
Vertebrates	282	297	300	301	304	304	319	Vertébrés
Invertebrates	283	489	483	505	511	514	514	Invertébrés
Plants	56	67	86	90	91	91	# 93	Plantes
Total	621	853	869	896	906	909	926	Total
Austria								**Autriche**
Vertebrates	20	23	24	24	24	24	27	Vertébrés
Invertebrates	44	55	65	70	70	69	69	Invertébrés
Plants	3	4	13	13	13	13	# 21	Plantes
Total	67	82	102	107	107	106	117	Total
Azerbaijan								**Azerbaïdjan**
Vertebrates	32	41	42	42	43	43	47	Vertébrés
Invertebrates	6	4	6	6	7	7	7	Invertébrés
Plants	0	0	0	0	42	42	42	Plantes
Total	38	45	48	48	92	92	96	Total
Bahamas								**Bahamas**
Vertebrates	36	44	49	51	52	61	65	Vertébrés
Invertebrates	1	11	11	12	12	12	12	Invertébrés
Plants	5	7	8	8	8	8	8	Plantes
Total	42	62	68	71	72	81	85	Total
Bahrain								**Bahreïn**
Vertebrates	18	19	18	19	20	19	23	Vertébrés
Invertebrates	0	13	14	14	13	13	13	Invertébrés
Plants	0	0	0	0	0	0	0	Plantes
Total	18	32	32	33	33	32	36	Total
Bangladesh								**Bangladesh**
Vertebrates	73	104	105	106	108	109	120	Vertébrés
Invertebrates	0	2	2	7	7	7	7	Invertébrés
Plants	12	16	17	17	17	21	21	Plantes
Total	85	122	124	130	132	137	148	Total
Barbados								**Barbade**
Vertebrates	18	24	29	30	31	37	39	Vertébrés
Invertebrates	0	10	10	11	11	11	11	Invertébrés
Plants	2	2	2	3	3	3	3	Plantes
Total	20	36	41	44	45	51	53	Total
Belarus								**Bélarus**
Vertebrates	10	10	12	12	12	11	14	Vertébrés
Invertebrates	8	6	9	9	9	9	9	Invertébrés
Plants	0	0	1	1	1	1	1	Plantes
Total	18	16	22	22	22	21	24	Total
Belgium								**Belgique**
Vertebrates	25	15	17	17	17	16	23	Vertébrés
Invertebrates	11	11	14	14	14	14	14	Invertébrés
Plants	0	1	0	0	0	0	# 1	Plantes
Total	36	27	31	31	31	30	38	Total
Belize								**Belize**
Vertebrates	36	48	55	58	60	70	70	Vertébrés
Invertebrates	1	12	12	12	12	12	12	Invertébrés
Plants	30	32	32	33	33	35	36	Plantes
Total	67	92	99	103	105	117	118	Total
Benin								**Bénin**
Vertebrates	17	47	51	53	55	55	66	Vertébrés
Invertebrates	0	1	1	1	3	3	3	Invertébrés
Plants	14	14	14	15	16	16	17	Plantes
Total	31	62	66	69	74	74	86	Total

Country or area &	2004	2010	2012	2013	2014	2015	2016	Pays ou zone &
Bermuda								**Bermudes**
Vertebrates	18	18	23	24	25	28	35	Vertébrés
Invertebrates	25	28	28	28	28	28	28	Invertébrés
Plants	4	4	4	4	7	8	8	Plantes
Total	47	50	55	56	60	64	71	Total
Bhutan								**Bhoutan**
Vertebrates	40	50	51	52	52	52	54	Vertébrés
Invertebrates	1	1	1	1	1	1	1	Invertébrés
Plants	7	8	9	9	12	18	18	Plantes
Total	48	59	61	62	65	71	73	Total
Bolivia (Plurinational State of)								**Bolivie (État plurinational de)**
Vertebrates	79	90	110	111	114	114	124	Vertébrés
Invertebrates	1	1	3	3	3	3	3	Invertébrés
Plants	70	72	73	98	99	99	104	Plantes
Total	150	163	186	212	216	216	231	Total
Bonaire, Sint Eustatius and Saba								**Bonaire, Saint-Eustache et Saba**
Vertebrates	...	...	1	2	32	41	42	Vertébrés
Invertebrates	...	...	0	0	11	11	11	Invertébrés
Plants	...	...	0	0	3	3	3	Plantes
Total	...	...	1	2	46	55	56	Total
Bosnia and Herzegovina								**Bosnie-Herzégovine**
Vertebrates	29	44	44	45	46	48	51	Vertébrés
Invertebrates	10	22	31	34	36	36	36	Invertébrés
Plants	1	1	1	1	1	1	1	Plantes
Total	40	67	76	80	83	85	88	Total
Botswana								**Botswana**
Vertebrates	15	18	19	20	22	22	24	Vertébrés
Invertebrates	0	0	0	0	0	0	0	Invertébrés
Plants	0	0	1	2	2	2	2	Plantes
Total	15	18	20	22	24	24	26	Total
Bouvet Island								**Île Bouvet**
Vertebrates	1	2	2	3	3	3	3	Vertébrés
Invertebrates	0	0	0	0	0	0	0	Invertébrés
Plants	0	0	0	0	0	0	0	Plantes
Total	1	2	2	3	3	3	3	Total
Brazil								**Brésil**
Vertebrates	282[2]	341[2]	377	379	395	396	397	Vertébrés
Invertebrates	34	45	45	54	54	54	55	Invertébrés
Plants	381	387	404	504	516	516	521	Plantes
Total	697	773	826	937	965	966	973	Total
British Indian Ocean Terr.								**Terr. brit. de l'océan Indien**
Vertebrates	6	10	11	11	11	11	14	Vertébrés
Invertebrates	0	65	65	69	69	69	69	Invertébrés
Plants	1	1	1	1	1	1	1	Plantes
Total	7	76	77	81	81	81	84	Total
British Virgin Islands								**Îles Vierges britanniques**
Vertebrates	20	23	28	32	33	41	41	Vertébrés
Invertebrates	0	10	10	10	10	10	10	Invertébrés
Plants	10	10	10	10	10	10	10	Plantes
Total	30	43	48	52	53	61	61	Total
Brunei Darussalam								**Brunéi Darussalam**
Vertebrates	49	70	74	75	75	77	80	Vertébrés
Invertebrates	0	1	1	8	8	8	8	Invertébrés
Plants	99	99	99	104	104	104	104	Plantes
Total	148	170	174	187	187	189	192	Total
Bulgaria								**Bulgarie**
Vertebrates	35	39	42	42	43	42	49	Vertébrés
Invertebrates	9	27	35	36	37	37	38	Invertébrés
Plants	0	0	6	6	6	6	6	Plantes
Total	44	66	83	84	86	85	93	Total

Country or area [&]	2004	2010	2012	2013	2014	2015	2016	Pays ou zone [&]
Burkina Faso								**Burkina Faso**
Vertebrates	9	20	23	24	25	25	27	Vertébrés
Invertebrates	0	1	1	1	1	1	1	Invertébrés
Plants	2	3	3	3	3	3	3	Plantes
Total	11	24	27	28	29	29	31	Total
Burundi								**Burundi**
Vertebrates	22	43	46	46	46	45	44	Vertébrés
Invertebrates	4	7	7	7	7	7	7	Invertébrés
Plants	2	2	3	6	7	8	8	Plantes
Total	28	52	56	59	60	60	59	Total
Cabo Verde								**Cabo Verde**
Vertebrates	21	28	32	36	35	35	49	Vertébrés
Invertebrates	0	0	12	12	13	13	13	Invertébrés
Plants	2	3	3	3	3	3	3	Plantes
Total	23	31	47	51	51	51	65	Total
Cambodia								**Cambodge**
Vertebrates	72	107	124	125	128	130	137	Vertébrés
Invertebrates	0	67	68	77	77	77	79	Invertébrés
Plants	31	30	33	32	32	36	36	Plantes
Total	103	204	225	234	237	243	252	Total
Cameroon								**Cameroun**
Vertebrates	146	222	232	235	236	237	255	Vertébrés
Invertebrates	4	24	24	25	27	27	27	Invertébrés
Plants	334	378	376	379	425	433	490	Plantes
Total	484	624	632	639	688	697	772	Total
Canada								**Canada**
Vertebrates	62	63	68	69	69	68	79	Vertébrés
Invertebrates	11	12	14	15	18	22	25	Invertébrés
Plants	1	2	2	2	6	7	# 14	Plantes
Total	74	77	84	86	93	97	118	Total
Cayman Islands								**Îles Caïmanes**
Vertebrates	16	21	25	27	28	38	38	Vertébrés
Invertebrates	1	11	11	11	11	11	11	Invertébrés
Plants	2	2	2	2	22	22	22	Plantes
Total	19	34	38	40	61	71	71	Total
Central African Rep.								**Rép. centrafricaine**
Vertebrates	15	19	25	26	31	32	33	Vertébrés
Invertebrates	0	0	0	0	0	0	0	Invertébrés
Plants	15	17	17	18	22	22	24	Plantes
Total	30	36	42	44	53	54	57	Total
Chad								**Tchad**
Vertebrates	18	24	26	27	29	29	31	Vertébrés
Invertebrates	1	4	4	4	4	4	4	Invertébrés
Plants	2	2	3	4	5	5	6	Plantes
Total	21	30	33	35	38	38	41	Total
Chile								**Chili**
Vertebrates	83	95	95	96	98	98	111	Vertébrés
Invertebrates	0	9	10	12	12	12	13	Invertébrés
Plants	40	41	41	71	72	72	72	Plantes
Total	123	145	146	179	182	182	196	Total
China [3]								**Chine** [3]
Vertebrates	326	374	408	407	416	420	428	Vertébrés
Invertebrates	4	32	42	76	76	76	76	Invertébrés
Plants	443	453	461	475	503	544	# 575	Plantes
Total	773	859	911	958	995	1 040	1 079	Total
China, Hong Kong SAR								**Chine, Hong Kong RAS**
Vertebrates	32	37	44	44	45	45	46	Vertébrés
Invertebrates	1	6	6	8	8	8	8	Invertébrés
Plants	6	6	6	6	6	7	9	Plantes
Total	39	49	56	58	59	60	63	Total

Country or area [&]	2004	2010	2012	2013	2014	2015	2016	Pays ou zone [&]
China, Macao SAR								**Chine, Macao RAS**
Vertebrates	5	9	9	9	10	10	10	Vertébrés
Invertebrates	0	0	0	1	1	1	1	Invertébrés
Plants	0	0	0	0	0	0	0	Plantes
Total	5	9	9	10	11	11	11	Total
Christmas Island								**Île Christmas**
Vertebrates	12	16	17	17	18	18	20	Vertébrés
Invertebrates	0	16	16	18	18	18	18	Invertébrés
Plants	1	1	1	1	1	1	1	Plantes
Total	13	33	34	36	37	37	39	Total
Cocos (Keeling) Islands								**Îles des Cocos (Keeling)**
Vertebrates	5	11	12	12	12	11	12	Vertébrés
Invertebrates	0	17	17	20	20	20	20	Invertébrés
Plants	0	0	0	0	0	0	0	Plantes
Total	5	28	29	32	32	31	32	Total
Colombia								**Colombie**
Vertebrates	371	424	454	459	463	473	518	Vertébrés
Invertebrates	0	30	30	33	33	33	52	Invertébrés
Plants	222	227	226	238	245	245	257	Plantes
Total	593	681	710	730	741	751	827	Total
Comoros								**Comores**
Vertebrates	18	21	24	25	26	26	29	Vertébrés
Invertebrates	4	63	63	72	73	73	73	Invertébrés
Plants	5	5	5	7	7	7	7	Plantes
Total	27	89	92	104	106	106	109	Total
Congo								**Congo**
Vertebrates	29	61	62	66	70	71	81	Vertébrés
Invertebrates	1	5	5	5	7	7	7	Invertébrés
Plants	35	37	37	38	41	41	45	Plantes
Total	65	103	104	109	118	119	133	Total
Cook Islands								**Îles Cook**
Vertebrates	22	27	29	30	30	30	31	Vertébrés
Invertebrates	0	25	25	32	32	32	32	Invertébrés
Plants	1	1	1	1	11	11	11	Plantes
Total	23	53	55	63	73	73	74	Total
Costa Rica								**Costa Rica**
Vertebrates	112	142	151	156	156	162	166	Vertébrés
Invertebrates	9	27	28	30	30	30	31	Invertébrés
Plants	110	116	117	129	131	131	140	Plantes
Total	231	285	296	315	317	323	337	Total
Côte d'Ivoire								**Côte d'Ivoire**
Vertebrates	61	100	108	108	113	113	127	Vertébrés
Invertebrates	1	4	4	4	6	6	6	Invertébrés
Plants	105	106	104	105	107	107	112	Plantes
Total	167	210	216	217	226	226	245	Total
Croatia								**Croatie**
Vertebrates	46	77	83	85	84	84	93	Vertébrés
Invertebrates	11	21	59	63	66	67	68	Invertébrés
Plants	0	3	7	8	8	8	# 9	Plantes
Total	57	101	149	156	158	159	170	Total
Cuba								**Cuba**
Vertebrates	106	123	131	132	133	138	138	Vertébrés
Invertebrates	3	15	15	23	23	23	23	Invertébrés
Plants	163	166	166	176	176	176	179	Plantes
Total	272	304	312	331	332	337	340	Total
Curaçao								**Curaçao**
Vertebrates	...	...	11	12	29	36	36	Vertébrés
Invertebrates	...	...	0	0	11	11	11	Invertébrés
Plants	...	...	0	0	2	2	2	Plantes
Total	...	...	11	12	42	49	49	Total

Country or area [&]	2004	2010	2012	2013	2014	2015	2016	Pays ou zone [&]
Cyprus								**Chypre**
Vertebrates	24	31	33	34	33	34	41	Vertébrés
Invertebrates	0	4	5	5	7	8	12	Invertébrés
Plants	1	8	18	18	18	18	18	Plantes
Total	25	43	56	57	58	60	71	Total
Czech Republic								**République tchèque**
Vertebrates	22	10	11	11	11	11	14	Vertébrés
Invertebrates	19	19	23	24	25	24	24	Invertébrés
Plants	4	4	11	11	10	10	# 15	Plantes
Total	45	33	45	46	46	45	53	Total
Dem. P. R. Korea								**R. p. dém. de Corée**
Vertebrates	40	44	50	51	52	53	57	Vertébrés
Invertebrates	1	2	2	3	3	3	3	Invertébrés
Plants	3	6	8	8	8	8	17	Plantes
Total	44	52	60	62	63	64	77	Total
Dem. Rep. of the Congo								**Rép. dém. du Congo**
Vertebrates	84	162	166	168	172	171	183	Vertébrés
Invertebrates	22	51	51	51	53	53	53	Invertébrés
Plants	65	83	89	95	107	109	113	Plantes
Total	171	296	306	314	332	333	349	Total
Denmark								**Danemark**
Vertebrates	21	18	21	20	20	20	27	Vertébrés
Invertebrates	11	12	15	15	15	15	15	Invertébrés
Plants	3	3	2	2	1	1	# 4	Plantes
Total	35	33	38	37	36	36	46	Total
Djibouti								**Djibouti**
Vertebrates	19	29	34	33	34	34	38	Vertébrés
Invertebrates	0	50	50	57	57	57	57	Invertébrés
Plants	2	2	2	3	3	3	3	Plantes
Total	21	81	86	93	94	94	98	Total
Dominica								**Dominique**
Vertebrates	22	27	31	32	33	40	42	Vertébrés
Invertebrates	0	11	11	11	11	11	11	Invertébrés
Plants	11	10	10	10	11	11	11	Plantes
Total	33	48	52	53	55	62	64	Total
Dominican Republic								**Rép. dominicaine**
Vertebrates	72	80	86	89	90	95	95	Vertébrés
Invertebrates	2	16	16	16	16	16	16	Invertébrés
Plants	30	30	32	38	41	42	42	Plantes
Total	104	126	134	143	147	153	153	Total
Ecuador								**Équateur**
Vertebrates	288	356	383	393	394	395	417	Vertébrés
Invertebrates	48	62	62	65	65	65	70	Invertébrés
Plants	1 815	1 837	1 837	1 843	1 840	1 848	# 1 866	Plantes
Total	2 151	2 255	2 282	2 301	2 299	2 308	2 353	Total
Egypt								**Égypte**
Vertebrates	43	73	79	80	82	83	96	Vertébrés
Invertebrates	1	46	47	53	54	55	56	Invertébrés
Plants	2	2	2	2	3	3	3	Plantes
Total	46	121	128	135	139	141	155	Total
El Salvador								**El Salvador**
Vertebrates	23	39	42	44	47	47	47	Vertébrés
Invertebrates	1	6	6	7	7	7	10	Invertébrés
Plants	25	27	26	29	29	29	29	Plantes
Total	49	72	74	80	83	83	86	Total
Equatorial Guinea								**Guinée équatoriale**
Vertebrates	38	60	62	64	69	71	85	Vertébrés
Invertebrates	2	2	2	3	5	5	5	Invertébrés
Plants	61	68	67	68	77	79	88	Plantes
Total	101	130	131	135	151	155	178	Total

Country or area &	2004	2010	2012	2013	2014	2015	2016	Pays ou zone &
Eritrea								**Érythrée**
Vertebrates	31	44	49	51	51	51	58	Vertébrés
Invertebrates	0	50	50	58	58	58	58	Invertébrés
Plants	3	3	4	4	4	4	4	Plantes
Total	34	97	103	113	113	113	120	Total
Estonia								**Estonie**
Vertebrates	8	8	11	11	11	11	14	Vertébrés
Invertebrates	4	3	6	6	6	6	6	Invertébrés
Plants	0	0	0	0	0	0	# 2	Plantes
Total	12	11	17	17	17	17	22	Total
Ethiopia								**Éthiopie**
Vertebrates	65	79	83	86	89	89	92	Vertébrés
Invertebrates	6	15	15	15	15	15	15	Invertébrés
Plants	22	26	25	40	40	41	41	Plantes
Total	93	120	123	141	144	145	148	Total
Falkland Is. (Malvinas)								**Îles Falkland (Malvinas)**
Vertebrates	21	18	19	18	18	18	18	Vertébrés
Invertebrates	0	0	0	0	0	0	0	Invertébrés
Plants	5	5	5	5	5	5	5	Plantes
Total	26	23	24	23	23	23	23	Total
Faroe Islands								**Îles Féroé**
Vertebrates	11	13	13	13	13	14	20	Vertébrés
Invertebrates	0	0	0	0	0	0	0	Invertébrés
Plants	0	0	0	0	0	0	# 1	Plantes
Total	11	13	13	13	13	14	21	Total
Fiji								**Fidji**
Vertebrates	33	37	42	48	48	48	52	Vertébrés
Invertebrates	2	90	155	165	165	165	165	Invertébrés
Plants	66	65	65	65	65	65	65	Plantes
Total	101	192	262	278	278	278	282	Total
Finland								**Finlande**
Vertebrates	14	10	13	13	13	13	17	Vertébrés
Invertebrates	10	7	9	10	10	10	10	Invertébrés
Plants	1	1	2	2	2	2	# 8	Plantes
Total	25	18	24	25	25	25	35	Total
France								**France**
Vertebrates	53	62	66	70	70	70	80	Vertébrés
Invertebrates	65	91	121	127	132	133	137	Invertébrés
Plants	2	15	32	32	32	33	# 43	Plantes
Total	120	168	219	229	234	236	260	Total
French Guiana								**Guyane française**
Vertebrates	33	40	49	50	50	51	55	Vertébrés
Invertebrates	0	0	0	0	0	0	0	Invertébrés
Plants	16	16	16	16	16	16	18	Plantes
Total	49	56	65	66	66	67	73	Total
French Polynesia								**Polynésie française**
Vertebrates	46	54	62	61	63	63	63	Vertébrés
Invertebrates	29	59	60	65	65	65	65	Invertébrés
Plants	47	47	47	47	47	47	47	Plantes
Total	122	160	169	173	175	175	175	Total
French S. and Antarc. Terr.								**Terr. austr. et ant. françaises**
Vertebrates	17	21	22	22	22	22	24	Vertébrés
Invertebrates	0	0	0	0	0	0	0	Invertébrés
Plants	0	0	0	0	0	0	0	Plantes
Total	17	21	22	22	22	22	24	Total
Gabon								**Gabon**
Vertebrates	31	84	86	89	92	93	105	Vertébrés
Invertebrates	1	0	0	1	3	3	3	Invertébrés
Plants	107	120	119	120	132	135	162	Plantes
Total	139	204	205	210	227	231	270	Total

Country or area [&]	2004	2010	2012	2013	2014	2015	2016	Pays ou zone [&]
Gambia								**Gambie**
Vertebrates	17	39	45	47	48	48	60	Vertébrés
Invertebrates	0	0	0	0	2	2	2	Invertébrés
Plants	4	4	4	5	5	5	5	Plantes
Total	21	43	49	52	55	55	67	Total
Georgia								**Géorgie**
Vertebrates	33	37	38	38	39	38	42	Vertébrés
Invertebrates	10	9	12	13	15	15	15	Invertébrés
Plants	0	0	0	0	61	61	# 62	Plantes
Total	43	46	50	51	115	114	119	Total
Germany								**Allemagne**
Vertebrates	35	33	35	35	35	33	39	Vertébrés
Invertebrates	31	34	55	58	58	57	57	Invertébrés
Plants	12	12	17	17	17	17	# 24	Plantes
Total	78	79	107	110	110	107	120	Total
Ghana								**Ghana**
Vertebrates	43	83	94	95	99	99	112	Vertébrés
Invertebrates	0	1	3	3	5	5	5	Invertébrés
Plants	117	118	116	117	119	119	119	Plantes
Total	160	202	213	215	223	223	236	Total
Gibraltar								**Gibraltar**
Vertebrates	16	20	21	20	20	21	28	Vertébrés
Invertebrates	2	2	3	3	4	5	5	Invertébrés
Plants	0	0	0	0	0	0	0	Plantes
Total	18	22	24	23	24	26	33	Total
Greece								**Grèce**
Vertebrates	62	107	110	111	111	111	120	Vertébrés
Invertebrates	11	36	105	113	117	122	181	Invertébrés
Plants	2	13	57	57	58	58	# 60	Plantes
Total	75	156	272	281	286	291	361	Total
Greenland								**Groenland**
Vertebrates	11	13	15	15	16	16	21	Vertébrés
Invertebrates	0	0	0	0	0	0	0	Invertébrés
Plants	1	1	1	1	1	1	1	Plantes
Total	12	14	16	16	17	17	22	Total
Grenada								**Grenade**
Vertebrates	20	24	30	31	32	38	40	Vertébrés
Invertebrates	0	10	10	10	10	10	10	Invertébrés
Plants	3	3	3	3	3	3	3	Plantes
Total	23	37	43	44	45	51	53	Total
Guadeloupe [4]								**Guadeloupe** [4]
Vertebrates	25	30	34	36	37	45	45	Vertébrés
Invertebrates	1	16	16	16	16	16	16	Invertébrés
Plants	7	8	8	8	9	9	9	Plantes
Total	33	54	58	60	62	70	70	Total
Guam								**Guam**
Vertebrates	16	24	27	29	31	31	34	Vertébrés
Invertebrates	5	6	6	13	60	60	60	Invertébrés
Plants	3	4	4	4	4	4	4	Plantes
Total	24	34	37	46	95	95	98	Total
Guatemala								**Guatemala**
Vertebrates	115	140	151	162	168	176	176	Vertébrés
Invertebrates	8	8	8	9	9	9	13	Invertébrés
Plants	85	82	83	93	94	97	102	Plantes
Total	208	230	242	264	271	282	291	Total
Guernsey								**Guernesey**
Vertebrates	...	2	2	2	2	2	5	Vertébrés
Invertebrates	...	0	0	0	0	0	0	Invertébrés
Plants	...	0	0	0	0	0	0	Plantes
Total	...	2	2	2	2	2	5	Total

Threatened species *(continued)*
Number by taxonomic group

Espèces menacées *(suite)*
Nombre par groupe taxonomique

Country or area &	2004	2010	2012	2013	2014	2015	2016	Pays ou zone &
Guinea								**Guinée**
Vertebrates	42	107	113	119	122	122	132	Vertébrés
Invertebrates	3	5	5	5	7	7	7	Invertébrés
Plants	22	22	22	28	34	34	44	Plantes
Total	67	134	140	152	163	163	183	Total
Guinea-Bissau								**Guinée-Bissau**
Vertebrates	17	48	55	56	59	59	69	Vertébrés
Invertebrates	1	0	0	0	2	2	2	Invertébrés
Plants	4	4	4	5	5	5	5	Plantes
Total	22	52	59	61	66	66	76	Total
Guyana								**Guyana**
Vertebrates	41	46	60	62	62	63	65	Vertébrés
Invertebrates	1	1	1	1	1	1	1	Invertébrés
Plants	23	22	22	23	23	23	26	Plantes
Total	65	69	83	86	86	87	92	Total
Haiti								**Haïti**
Vertebrates	86	94	100	104	106	113	113	Vertébrés
Invertebrates	2	14	14	14	14	14	14	Invertébrés
Plants	28	29	32	38	39	42	42	Plantes
Total	116	137	146	156	159	169	169	Total
Heard Is. and McDonald Is.								**Île Heard-et-Îles MacDonald**
Vertebrates	12	12	12	12	12	12	12	Vertébrés
Invertebrates	0	0	0	0	0	0	0	Invertébrés
Plants	0	0	0	0	0	0	0	Plantes
Total	12	12	12	12	12	12	12	Total
Holy See								**Saint-Siège**
Vertebrates	...	1	1	1	1	1	1	Vertébrés
Invertebrates	...	0	0	0	0	0	0	Invertébrés
Plants	...	0	0	0	0	0	0	Plantes
Total	...	1	1	1	1	1	1	Total
Honduras								**Honduras**
Vertebrates	93	110	122	139	147	156	156	Vertébrés
Invertebrates	2	17	17	18	18	18	21	Invertébrés
Plants	111	113	113	118	119	120	123	Plantes
Total	206	240	252	275	284	294	300	Total
Hungary								**Hongrie**
Vertebrates	25	20	22	22	22	22	25	Vertébrés
Invertebrates	25	26	33	34	36	35	35	Invertébrés
Plants	1	1	9	10	10	10	# 12	Plantes
Total	51	47	64	66	68	67	72	Total
Iceland								**Islande**
Vertebrates	15	17	19	19	19	21	26	Vertébrés
Invertebrates	0	0	0	0	0	0	0	Invertébrés
Plants	0	0	0	0	0	0	0	Plantes
Total	15	17	19	19	19	21	26	Total
India								**Inde**
Vertebrates	283	390	491	514	521	520	528	Vertébrés
Invertebrates	23	113	123	134	135	135	135	Invertébrés
Plants	246	255	321	325	332	384	388	Plantes
Total	552	758	935	973	988	1 039	1 051	Total
Indonesia								**Indonésie**
Vertebrates	419	503	513	515	529	530	540	Vertébrés
Invertebrates	31	246	248	287	288	290	290	Invertébrés
Plants	383	393	393	404	408	426	427	Plantes
Total	833	1 142	1 154	1 206	1 225	1 246	1 257	Total
Iran (Islamic Rep. of)								**Iran (Rép. islamique d')**
Vertebrates	65	82	85	87	94	94	105	Vertébrés
Invertebrates	3	19	21	23	24	24	24	Invertébrés
Plants	1	1	2	2	3	3	3	Plantes
Total	69	102	108	112	121	121	132	Total

Country or area [&]	2004	2010	2012	2013	2014	2015	2016	Pays ou zone [&]
Iraq								**Iraq**
Vertebrates	33	45	44	45	51	51	54	Vertébrés
Invertebrates	2	15	16	16	17	17	17	Invertébrés
Plants	0	0	1	1	1	1	1	Plantes
Total	35	60	61	62	69	69	72	Total
Ireland								**Irlande**
Vertebrates	18	24	29	30	31	32	40	Vertébrés
Invertebrates	3	2	3	3	6	6	6	Invertébrés
Plants	1	1	1	1	1	1	# 3	Plantes
Total	22	27	33	34	38	39	49	Total
Isle of Man								**Île de Man**
Vertebrates	...	2	2	3	3	3	3	Vertébrés
Invertebrates	...	0	0	0	0	0	0	Invertébrés
Plants	...	0	0	0	0	0	0	Plantes
Total	...	2	2	3	3	3	3	Total
Israel								**Israël**
Vertebrates	47	73	77	76	78	78	89	Vertébrés
Invertebrates	10	58	60	67	73	74	74	Invertébrés
Plants	0	0	0	0	0	0	9	Plantes
Total	57	131	137	143	151	152	172	Total
Italy								**Italie**
Vertebrates	53	70	76	77	76	76	85	Vertébrés
Invertebrates	58	77	120	130	134	136	141	Invertébrés
Plants	3	27	65	66	66	67	# 79	Plantes
Total	114	174	261	273	276	279	305	Total
Jamaica								**Jamaïque**
Vertebrates	54	58	63	62	63	69	71	Vertébrés
Invertebrates	5	15	15	15	15	15	15	Invertébrés
Plants	208	209	209	212	214	214	214	Plantes
Total	267	282	287	289	292	298	300	Total
Japan								**Japon**
Vertebrates	148	158	165	164	171	171	180	Vertébrés
Invertebrates	45	157	164	170	170	170	171	Invertébrés
Plants	12	15	16	17	21	23	# 46	Plantes
Total	205	330	345	351	362	364	397	Total
Jersey								**Jersey**
Vertebrates	...	2	2	2	2	2	5	Vertébrés
Invertebrates	...	0	0	0	1	1	1	Invertébrés
Plants	...	0	0	0	0	0	0	Plantes
Total	...	2	2	2	3	3	6	Total
Jordan								**Jordanie**
Vertebrates	27	41	42	42	41	40	47	Vertébrés
Invertebrates	3	48	50	56	61	61	61	Invertébrés
Plants	0	1	1	1	1	1	5	Plantes
Total	30	90	93	99	103	102	113	Total
Kazakhstan								**Kazakhstan**
Vertebrates	48	53	55	55	55	55	57	Vertébrés
Invertebrates	4	4	6	6	7	7	7	Invertébrés
Plants	1	16	17	15	15	16	16	Plantes
Total	53	73	78	76	77	78	80	Total
Kenya								**Kenya**
Vertebrates	99	137	143	149	157	157	162	Vertébrés
Invertebrates	27	72	72	83	84	84	84	Invertébrés
Plants	103	129	131	187	187	222	222	Plantes
Total	229	338	346	419	428	463	468	Total
Kiribati								**Kiribati**
Vertebrates	10	17	19	20	19	19	21	Vertébrés
Invertebrates	1	73	73	81	81	81	81	Invertébrés
Plants	0	0	0	0	0	0	0	Plantes
Total	11	90	92	101	100	100	102	Total

Country or area [&]	2004	2010	2012	2013	2014	2015	2016	Pays ou zone [&]
Kuwait								**Koweït**
Vertebrates	20	28	28	28	29	29	36	Vertébrés
Invertebrates	0	13	14	14	13	13	13	Invertébrés
Plants	0	0	0	0	0	0	0	Plantes
Total	20	41	42	42	42	42	49	Total
Kyrgyzstan								**Kirghizistan**
Vertebrates	12	23	23	23	23	23	26	Vertébrés
Invertebrates	3	3	3	3	4	4	4	Invertébrés
Plants	1	14	14	14	14	14	14	Plantes
Total	16	40	40	40	41	41	44	Total
Lao People's Dem. Rep.								**Rép. dém. pop. lao**
Vertebrates	72	107	144	144	147	148	153	Vertébrés
Invertebrates	0	3	21	21	21	21	21	Invertébrés
Plants	19	22	30	32	32	41	41	Plantes
Total	91	132	195	197	200	210	215	Total
Latvia								**Lettonie**
Vertebrates	15	9	13	13	13	13	16	Vertébrés
Invertebrates	8	9	12	12	12	12	12	Invertébrés
Plants	0	0	0	0	0	0	# 1	Plantes
Total	23	18	25	25	25	25	29	Total
Lebanon								**Liban**
Vertebrates	25	44	47	47	47	48	58	Vertébrés
Invertebrates	1	5	10	10	15	16	16	Invertébrés
Plants	0	1	1	2	2	5	10	Plantes
Total	26	50	58	59	64	69	84	Total
Lesotho								**Lesotho**
Vertebrates	11	10	10	10	10	10	11	Vertébrés
Invertebrates	1	2	2	2	3	3	3	Invertébrés
Plants	1	4	4	4	4	4	4	Plantes
Total	13	16	16	16	17	17	18	Total
Liberia								**Libéria**
Vertebrates	45	91	93	94	98	98	108	Vertébrés
Invertebrates	3	9	9	9	11	11	11	Invertébrés
Plants	46	47	47	48	49	49	52	Plantes
Total	94	147	149	151	158	158	171	Total
Libya								**Libye**
Vertebrates	24	42	45	46	45	49	55	Vertébrés
Invertebrates	0	0	0	0	1	2	4	Invertébrés
Plants	1	2	3	3	3	3	3	Plantes
Total	25	44	48	49	49	54	62	Total
Liechtenstein								**Liechtenstein**
Vertebrates	3	0	0	0	0	0	2	Vertébrés
Invertebrates	5	2	4	4	4	4	4	Invertébrés
Plants	0	0	0	0	0	0	0	Plantes
Total	8	2	4	4	4	4	6	Total
Lithuania								**Lituanie**
Vertebrates	12	12	15	15	15	14	17	Vertébrés
Invertebrates	5	5	7	7	7	7	7	Invertébrés
Plants	0	0	1	1	1	1	# 2	Plantes
Total	17	17	23	23	23	22	26	Total
Luxembourg								**Luxembourg**
Vertebrates	6	1	2	2	2	2	4	Vertébrés
Invertebrates	4	4	7	7	7	7	7	Invertébrés
Plants	0	0	0	0	0	0	0	Plantes
Total	10	5	9	9	9	9	11	Total
Madagascar								**Madagascar**
Vertebrates	222	283	391	392	445	446	527	Vertébrés
Invertebrates	32	100	100	108	110	110	110	Invertébrés
Plants	276	280	365	373	374	409	607	Plantes
Total	530	663	856	873	929	965	1 244	Total

Country or area &	2004	2010	2012	2013	2014	2015	2016	Pays ou zone &
Malawi								**Malawi**
Vertebrates	25	127	129	131	132	130	135	Vertébrés
Invertebrates	11	17	16	16	16	16	16	Invertébrés
Plants	14	14	18	23	23	25	24	Plantes
Total	50	158	163	170	171	171	175	Total
Malaysia								**Malaisie**
Vertebrates	190	246	258	262	268	269	282	Vertébrés
Invertebrates	19	242	243	259	262	262	262	Invertébrés
Plants	683	692	695	705	706	721	721	Plantes
Total	892	1 180	1 196	1 226	1 236	1 252	1 265	Total
Maldives								**Maldives**
Vertebrates	12	20	23	23	23	23	29	Vertébrés
Invertebrates	0	39	39	46	46	46	46	Invertébrés
Plants	0	0	0	0	0	0	0	Plantes
Total	12	59	62	69	69	69	75	Total
Mali								**Mali**
Vertebrates	19	23	28	30	31	31	34	Vertébrés
Invertebrates	0	0	0	0	0	0	0	Invertébrés
Plants	6	6	7	7	8	8	8	Plantes
Total	25	29	35	37	39	39	42	Total
Malta								**Malte**
Vertebrates	22	20	23	23	23	23	30	Vertébrés
Invertebrates	3	3	3	3	4	4	5	Invertébrés
Plants	0	3	4	4	4	4	4	Plantes
Total	25	26	30	30	31	31	39	Total
Marshall Islands								**Îles Marshall**
Vertebrates	12	17	20	21	22	22	25	Vertébrés
Invertebrates	1	67	67	73	73	73	73	Invertébrés
Plants	0	0	0	0	0	0	0	Plantes
Total	13	84	87	94	95	95	98	Total
Martinique								**Martinique**
Vertebrates	20	22	25	27	28	34	33	Vertébrés
Invertebrates	1	1	1	2	2	2	2	Invertébrés
Plants	8	8	8	8	9	9	9	Plantes
Total	29	31	34	37	39	45	44	Total
Mauritania								**Mauritanie**
Vertebrates	25	57	64	65	66	66	80	Vertébrés
Invertebrates	1	1	1	1	3	3	3	Invertébrés
Plants	0	0	0	0	0	0	0	Plantes
Total	26	58	65	66	69	69	83	Total
Mauritius								**Maurice**
Vertebrates	28	36	39	39	41	41	45	Vertébrés
Invertebrates	32	98	97	107	118	118	118	Invertébrés
Plants	87	88	88	91	91	90	90	Plantes
Total	147	222	224	237	250	249	253	Total
Mayotte								**Mayotte**
Vertebrates	6	9	14	15	15	15	18	Vertébrés
Invertebrates	1	60	60	69	69	69	69	Invertébrés
Plants	0	0	0	0	0	0	0	Plantes
Total	7	69	74	84	84	84	87	Total
Mexico								**Mexique**
Vertebrates	446	609	620	622	628	639	648	Vertébrés
Invertebrates	41	79	80	90	92	93	102	Invertébrés
Plants	261	255	259	362	371	377	402	Plantes
Total	748	943	959	1 074	1 091	1 109	1 152	Total
Micronesia (Fed. States of)								**Micronésie (États féd. de)**
Vertebrates	22	35	42	43	43	44	48	Vertébrés
Invertebrates	4	108	108	115	115	115	115	Invertébrés
Plants	4	5	5	5	5	4	4	Plantes
Total	30	148	155	163	163	163	167	Total

Country or area &	2004	2010	2012	2013	2014	2015	2016	Pays ou zone &
Monaco								**Monaco**
Vertebrates	9	11	13	13	13	13	17	Vertébrés
Invertebrates	0	0	1	1	2	3	3	Invertébrés
Plants	0	0	0	0	0	0	0	Plantes
Total	9	11	14	14	15	16	20	Total
Mongolia								**Mongolie**
Vertebrates	36	33	33	33	33	33	37	Vertébrés
Invertebrates	3	3	3	3	3	3	3	Invertébrés
Plants	0	0	0	0	0	0	0	Plantes
Total	39	36	36	36	36	36	40	Total
Montenegro								**Monténégro**
Vertebrates	...	44	46	48	47	49	57	Vertébrés
Invertebrates	...	28	33	34	34	34	35	Invertébrés
Plants	...	0	2	2	2	2	2	Plantes
Total	...	72	81	84	83	85	94	Total
Montserrat								**Montserrat**
Vertebrates	18	21	25	28	29	37	36	Vertébrés
Invertebrates	0	11	11	11	11	11	11	Invertébrés
Plants	3	3	5	5	5	6	6	Plantes
Total	21	35	41	44	45	54	53	Total
Morocco								**Maroc**
Vertebrates	40	86	86	87	89	90	101	Vertébrés
Invertebrates	8	40	41	42	46	52	60	Invertébrés
Plants	2	31	33	34	34	34	# 38	Plantes
Total	50	157	160	163	169	176	199	Total
Mozambique								**Mozambique**
Vertebrates	64	98	103	105	117	117	127	Vertébrés
Invertebrates	5	59	58	67	67	67	67	Invertébrés
Plants	46	52	53	58	77	84	84	Plantes
Total	115	209	214	230	261	268	278	Total
Myanmar								**Myanmar**
Vertebrates	107	143	159	159	161	163	178	Vertébrés
Invertebrates	2	64	66	77	77	77	77	Invertébrés
Plants	38	42	46	46	47	61	61	Plantes
Total	147	249	271	282	285	301	316	Total
Namibia								**Namibie**
Vertebrates	44	66	71	72	74	73	81	Vertébrés
Invertebrates	1	0	0	1	4	4	4	Invertébrés
Plants	24	26	26	27	27	28	28	Plantes
Total	69	92	97	100	105	105	113	Total
Nauru								**Nauru**
Vertebrates	5	12	12	12	12	12	14	Vertébrés
Invertebrates	0	62	62	68	68	68	68	Invertébrés
Plants	0	0	0	0	0	0	0	Plantes
Total	5	74	74	80	80	80	82	Total
Nepal								**Népal**
Vertebrates	69	83	83	83	85	85	87	Vertébrés
Invertebrates	1	3	3	3	3	3	3	Invertébrés
Plants	7	7	9	9	12	17	17	Plantes
Total	77	93	95	95	100	105	107	Total
Netherlands								**Pays-Bas**
Vertebrates	27	18	20	20	20	19	26	Vertébrés
Invertebrates	7	6	10	10	10	10	10	Invertébrés
Plants	0	0	0	0	0	0	# 3	Plantes
Total	34	24	30	30	30	29	39	Total
Netherlands Antilles [former]								**Antilles néerlandaises [anc.]**
Vertebrates	26	24	28	31	...	...	...	Vertébrés
Invertebrates	0	11	11	11	...	...	...	Invertébrés
Plants	2	3	3	3	...	...	...	Plantes
Total	28	38	42	45	...	...	...	Total

Country or area &	2004	2010	2012	2013	2014	2015	2016	Pays ou zone &
New Caledonia								**Nouvelle-Calédonie**
Vertebrates	34	61	107	108	109	109	114	Vertébrés
Invertebrates	11	97	113	124	125	125	125	Invertébrés
Plants	217	257	257	258	259	259	286	Plantes
Total	262	415	477	490	493	493	525	Total
New Zealand								**Nouvelle-Zélande**
Vertebrates	114	117	119	118	130	130	128	Vertébrés
Invertebrates	14	15	15	42	46	46	46	Invertébrés
Plants	21	21	20	21	21	21	# 23	Plantes
Total	149	153	154	181	197	197	197	Total
Nicaragua								**Nicaragua**
Vertebrates	49	61	69	73	73	79	78	Vertébrés
Invertebrates	2	17	17	18	18	18	20	Invertébrés
Plants	39	43	42	43	43	44	46	Plantes
Total	90	121	128	134	134	141	144	Total
Niger								**Niger**
Vertebrates	12	22	25	27	27	27	29	Vertébrés
Invertebrates	1	2	2	1	1	1	1	Invertébrés
Plants	2	2	2	3	3	3	3	Plantes
Total	15	26	29	31	31	31	33	Total
Nigeria								**Nigéria**
Vertebrates	61	113	120	126	127	127	145	Vertébrés
Invertebrates	1	12	12	15	17	17	17	Invertébrés
Plants	170	172	168	168	188	189	197	Plantes
Total	232	297	300	309	332	333	359	Total
Niue								**Nioué**
Vertebrates	12	20	21	20	20	20	22	Vertébrés
Invertebrates	0	23	23	30	30	30	30	Invertébrés
Plants	0	0	0	0	0	0	0	Plantes
Total	12	43	44	50	50	50	52	Total
Norfolk Island								**Île Norfolk**
Vertebrates	21	19	17	17	17	17	18	Vertébrés
Invertebrates	12	21	21	23	23	23	23	Invertébrés
Plants	1	1	1	1	1	1	1	Plantes
Total	34	41	39	41	41	41	42	Total
Northern Mariana Islands								**Îles Mariannes du Nord**
Vertebrates	22	29	33	35	37	37	39	Vertébrés
Invertebrates	2	51	51	57	57	57	57	Invertébrés
Plants	4	5	5	5	5	5	5	Plantes
Total	28	85	89	97	99	99	101	Total
Norway								**Norvège**
Vertebrates	22	27	30	30	30	30	38	Vertébrés
Invertebrates	9	7	10	11	10	10	10	Invertébrés
Plants	2	2	3	3	3	4	# 14	Plantes
Total	33	36	43	44	43	44	62	Total
Oman								**Oman**
Vertebrates	48	47	52	53	54	53	62	Vertébrés
Invertebrates	1	26	27	32	31	31	31	Invertébrés
Plants	6	6	6	6	6	6	6	Plantes
Total	55	79	85	91	91	90	99	Total
Other non-specified areas								**Autres zones non-spécifiées**
Vertebrates	79	104	113	110	115	114	123	Vertébrés
Invertebrates	0	122	122	128	128	128	128	Invertébrés
Plants	78	78	77	79	82	83	# 85	Plantes
Total	157	304	312	317	325	325	336	Total
Pakistan								**Pakistan**
Vertebrates	70	92	97	97	100	99	110	Vertébrés
Invertebrates	0	15	15	18	18	18	18	Invertébrés
Plants	2	2	4	4	5	12	12	Plantes
Total	72	109	116	119	123	129	140	Total

Country or area &	2004	2010	2012	2013	2014	2015	2016	Pays ou zone &
Palau								**Palaos**
Vertebrates	13	22	25	26	27	27	31	Vertébrés
Invertebrates	5	102	137	146	146	146	146	Invertébrés
Plants	3	4	4	4	4	4	4	Plantes
Total	21	128	166	176	177	177	181	Total
Panama								**Panama**
Vertebrates	113	125	132	135	138	148	151	Vertébrés
Invertebrates	2	20	20	22	22	22	22	Invertébrés
Plants	195	202	200	202	203	203	208	Plantes
Total	310	347	352	359	363	373	381	Total
Papua New Guinea								**Papouasie-Nvl-Guinée**
Vertebrates	141	139	143	142	145	146	155	Vertébrés
Invertebrates	12	171	171	181	181	181	181	Invertébrés
Plants	142	143	143	145	145	151	152	Plantes
Total	295	453	457	468	471	478	488	Total
Paraguay								**Paraguay**
Vertebrates	40	38	39	39	39	39	40	Vertébrés
Invertebrates	0	0	0	0	0	0	0	Invertébrés
Plants	10	10	11	19	19	19	19	Plantes
Total	50	48	50	58	58	58	59	Total
Peru								**Pérou**
Vertebrates	232	274	307	310	316	317	351	Vertébrés
Invertebrates	2	3	7	8	8	8	8	Invertébrés
Plants	274	274	275	318	318	318	334	Plantes
Total	508	551	589	636	642	643	693	Total
Philippines								**Philippines**
Vertebrates	225	262	271	271	291	291	302	Vertébrés
Invertebrates	19	213	213	237	237	237	237	Invertébrés
Plants	212	222	224	229	233	239	239	Plantes
Total	456	697	708	737	761	767	778	Total
Pitcairn								**Pitcairn**
Vertebrates	15	20	21	20	20	20	21	Vertébrés
Invertebrates	5	15	15	16	16	16	16	Invertébrés
Plants	7	7	7	7	7	7	7	Plantes
Total	27	42	43	43	43	43	44	Total
Poland								**Pologne**
Vertebrates	27	17	20	20	20	19	23	Vertébrés
Invertebrates	15	16	22	23	23	22	22	Invertébrés
Plants	4	4	11	11	10	10	# 13	Plantes
Total	46	37	53	54	53	51	58	Total
Portugal								**Portugal**
Vertebrates	51	71	78	79	79	81	92	Vertébrés
Invertebrates	82	79	88	89	94	94	97	Invertébrés
Plants	15	21	81	81	81	81	# 83	Plantes
Total	148	171	247	249	254	256	272	Total
Puerto Rico								**Porto Rico**
Vertebrates	44	49	55	61	62	70	69	Vertébrés
Invertebrates	1	1	1	0	0	0	0	Invertébrés
Plants	52	53	53	54	57	57	57	Plantes
Total	97	103	109	115	119	127	126	Total
Qatar								**Qatar**
Vertebrates	12	19	20	21	22	22	26	Vertébrés
Invertebrates	0	13	14	14	13	13	13	Invertébrés
Plants	0	0	0	0	0	0	0	Plantes
Total	12	32	34	35	35	35	39	Total
Republic of Korea								**République de Corée**
Vertebrates	54	58	60	59	64	65	72	Vertébrés
Invertebrates	1	3	3	4	4	4	5	Invertébrés
Plants	0	3	6	6	7	7	# 32	Plantes
Total	55	64	69	69	75	76	109	Total

Country or area [&]	2004	2010	2012	2013	2014	2015	2016	Pays ou zone [&]
Republic of Moldova								**République de Moldova**
Vertebrates	22	24	22	22	22	22	26	Vertébrés
Invertebrates	5	3	6	6	5	5	5	Invertébrés
Plants	0	0	2	2	2	2	2	Plantes
Total	27	27	30	30	29	29	33	Total
Réunion								**Réunion**
Vertebrates	18	16	19	21	22	22	24	Vertébrés
Invertebrates	16	73	73	85	87	87	87	Invertébrés
Plants	14	15	15	17	17	17	17	Plantes
Total	48	104	107	123	126	126	128	Total
Romania								**Roumanie**
Vertebrates	40	39	42	42	43	42	49	Vertébrés
Invertebrates	22	24	35	38	39	38	38	Invertébrés
Plants	1	1	5	6	5	5	5	Plantes
Total	63	64	82	86	87	85	92	Total
Russian Federation								**Fédération de Russie**
Vertebrates	114	93	158	76	126	126	134	Vertébrés
Invertebrates	30	25	32	36	37	36	36	Invertébrés
Plants	7	8	11	12	54	55	# 60	Plantes
Total	151	126	201	124	217	217	230	Total
Rwanda								**Rwanda**
Vertebrates	30	49	51	50	53	51	51	Vertébrés
Invertebrates	4	2	2	2	2	2	2	Invertébrés
Plants	3	4	5	6	6	8	8	Plantes
Total	37	55	58	58	61	61	61	Total
Saint Barthélemy								**Saint-Barthélemy**
Vertebrates	...	4	6	6	7	16	18	Vertébrés
Invertebrates	...	11	11	11	11	11	11	Invertébrés
Plants	...	2	2	2	2	2	2	Plantes
Total	...	17	19	19	20	29	31	Total
Saint Helena [5]								**Sainte-Hélène [5]**
Vertebrates	32	31	31	34	37	37	39	Vertébrés
Invertebrates	2	2	2	2	13	13	15	Invertébrés
Plants	26	27	30	30	30	30	44	Plantes
Total	60	60	63	66	80	80	98	Total
Saint Kitts and Nevis								**Saint-Kitts-et-Nevis**
Vertebrates	17	24	28	30	32	39	39	Vertébrés
Invertebrates	0	10	10	10	10	10	10	Invertébrés
Plants	2	2	2	2	2	2	2	Plantes
Total	19	36	40	42	44	51	51	Total
Saint Lucia								**Sainte-Lucie**
Vertebrates	23	29	33	34	35	41	43	Vertébrés
Invertebrates	0	11	11	11	11	11	11	Invertébrés
Plants	6	6	6	6	6	6	6	Plantes
Total	29	46	50	51	52	58	60	Total
Saint Martin (French part)								**Saint-Martin (partie française)**
Vertebrates	...	4	9	10	30	38	37	Vertébrés
Invertebrates	...	11	11	11	10	10	10	Invertébrés
Plants	...	2	2	2	3	3	3	Plantes
Total	...	17	22	23	43	51	50	Total
Saint Pierre and Miquelon								**Saint-Pierre-et-Miquelon**
Vertebrates	2	4	6	6	7	7	11	Vertébrés
Invertebrates	0	0	0	0	0	0	0	Invertébrés
Plants	0	0	0	0	0	0	0	Plantes
Total	2	4	6	6	7	7	11	Total
Saint Vincent-Grenadines								**Saint-Vincent-Grenadines**
Vertebrates	20	24	31	32	33	39	41	Vertébrés
Invertebrates	0	10	10	10	10	10	10	Invertébrés
Plants	4	4	4	4	5	5	5	Plantes
Total	24	38	45	46	48	54	56	Total

Threatened species *(continued)*
Number by taxonomic group

Espèces menacées *(suite)*
Nombre par groupe taxonomique

Country or area &	2004	2010	2012	2013	2014	2015	2016	Pays ou zone &
Samoa								**Samoa**
Vertebrates	15	23	24	26	26	26	29	Vertébrés
Invertebrates	1	53	53	62	62	62	62	Invertébrés
Plants	2	2	2	2	2	2	2	Plantes
Total	18	78	79	90	90	90	93	Total
San Marino								**Saint-Marin**
Vertebrates	...	0	...	...	0	0	0	Vertébrés
Invertebrates	...	0	...	...	1	1	1	Invertébrés
Plants	...	0	...	...	0	0	0	Plantes
Total	...	0	...	...	1	1	1	Total
Sao Tome and Principe								**Sao Tomé-et-Principe**
Vertebrates	24	33	39	39	38	38	51	Vertébrés
Invertebrates	2	2	2	3	5	5	5	Invertébrés
Plants	35	35	35	36	38	38	38	Plantes
Total	61	70	76	78	81	81	94	Total
Saudi Arabia								**Arabie saoudite**
Vertebrates	37	47	51	52	54	57	66	Vertébrés
Invertebrates	1	53	55	60	59	59	59	Invertébrés
Plants	3	3	3	3	3	3	3	Plantes
Total	41	103	109	115	116	119	128	Total
Senegal								**Sénégal**
Vertebrates	40	72	80	82	83	82	95	Vertébrés
Invertebrates	0	1	10	11	13	13	13	Invertébrés
Plants	7	9	9	10	11	11	12	Plantes
Total	47	82	99	103	107	106	120	Total
Serbia								**Serbie**
Vertebrates	...	29	29	29	29	29	35	Vertébrés
Invertebrates	...	16	20	23	23	22	24	Invertébrés
Plants	...	1	4	5	5	5	# 6	Plantes
Total	...	46	53	57	57	56	65	Total
Serbia and Montenegro [former]								**Serbie-et-Monténégro [anc.]**
Vertebrates	42	...	...	...	...	...	...	Vertébrés
Invertebrates	19	...	...	...	...	...	...	Invertébrés
Plants	1	...	...	...	...	...	...	Plantes
Total	62	...	...	...	...	...	...	Total
Seychelles								**Seychelles**
Vertebrates	35	45	49	52	54	54	55	Vertébrés
Invertebrates	4	100	147	158	319	319	319	Invertébrés
Plants	45	45	61	62	62	62	62	Plantes
Total	84	190	257	272	435	435	436	Total
Sierra Leone								**Sierra Leone**
Vertebrates	35	77	82	86	90	92	103	Vertébrés
Invertebrates	4	6	6	6	8	8	8	Invertébrés
Plants	47	48	48	54	58	58	65	Plantes
Total	86	131	136	146	156	158	176	Total
Singapore								**Singapour**
Vertebrates	30	58	56	56	56	56	63	Vertébrés
Invertebrates	1	162	162	173	173	173	173	Invertébrés
Plants	54	57	57	58	58	58	58	Plantes
Total	85	277	275	287	287	287	294	Total
Sint Maarten (Dutch part)								**Saint-Martin (partie néerlandaise)**
Vertebrates	...	...	2	3	29	37	37	Vertébrés
Invertebrates	...	...	0	0	10	10	10	Invertébrés
Plants	...	...	0	0	2	2	2	Plantes
Total	...	...	2	3	41	49	49	Total
Slovakia								**Slovaquie**
Vertebrates	27	15	16	16	16	16	19	Vertébrés
Invertebrates	19	17	21	22	23	22	22	Invertébrés
Plants	2	2	7	8	7	7	# 10	Plantes
Total	48	34	44	46	46	45	51	Total

Country or area &	2004	2010	2012	2013	2014	2015	2016	Pays ou zone &
Slovenia								**Slovénie**
Vertebrates	32	36	40	42	41	42	50	Vertébrés
Invertebrates	42	59	73	74	75	76	77	Invertébrés
Plants	0	0	7	7	7	7	# 11	Plantes
Total	74	95	120	123	123	125	138	Total
Solomon Islands								**Îles Salomon**
Vertebrates	52	63	66	66	70	70	75	Vertébrés
Invertebrates	6	141	141	151	151	151	151	Invertébrés
Plants	16	16	16	17	17	17	17	Plantes
Total	74	220	223	234	238	238	243	Total
Somalia								**Somalie**
Vertebrates	46	56	61	60	62	60	66	Vertébrés
Invertebrates	1	51	51	62	62	62	62	Invertébrés
Plants	17	21	23	42	42	43	43	Plantes
Total	64	128	135	164	166	165	171	Total
South Africa								**Afrique du Sud**
Vertebrates	155	185	192	194	210	209	219	Vertébrés
Invertebrates	127	159	158	166	202	202	202	Invertébrés
Plants	75	97	99	101	116	116	116	Plantes
Total	357	441	449	461	528	527	537	Total
South Georgia and Sandwich Is.								**Géorgie du S. et Îles Sandwich**
Vertebrates	12	10	10	9	9	9	9	Vertébrés
Invertebrates	0	0	0	0	0	0	0	Invertébrés
Plants	0	0	0	0	0	0	0	Plantes
Total	12	10	10	9	9	9	9	Total
South Sudan								**Soudan du sud**
Vertebrates	...	...	14	15	27	26	30	Vertébrés
Invertebrates	...	...	0	0	0	0	0	Invertébrés
Plants	...	...	0	2	15	16	15	Plantes
Total	...	...	14	17	42	42	45	Total
Spain								**Espagne**
Vertebrates	76	118	123	124	122	123	135	Vertébrés
Invertebrates	63	67	196	211	216	215	236	Invertébrés
Plants	14	55	213	213	214	213	# 221	Plantes
Total	153	240	532	548	552	551	592	Total
Sri Lanka								**Sri Lanka**
Vertebrates	112	149	155	155	159	159	167	Vertébrés
Invertebrates	2	120	120	130	130	130	130	Invertébrés
Plants	280	283	286	286	287	291	291	Plantes
Total	394	552	561	571	576	580	588	Total
State of Palestine								**État de Palestine**
Vertebrates	4	16	18	18	18	20	23	Vertébrés
Invertebrates	0	2	2	2	4	4	4	Invertébrés
Plants	0	0	0	0	0	0	3	Plantes
Total	4	18	20	20	22	24	30	Total
Sudan								**Soudan**
Vertebrates	36	49	55	58	56	57	63	Vertébrés
Invertebrates	2	45	45	50	50	50	50	Invertébrés
Plants	17	18	17	19	16	16	16	Plantes
Total	55	112	117	127	122	123	129	Total
Suriname								**Suriname**
Vertebrates	32	38	47	48	48	49	55	Vertébrés
Invertebrates	0	1	1	1	1	1	1	Invertébrés
Plants	27	26	26	26	26	26	27	Plantes
Total	59	65	74	75	75	76	83	Total
Svalbard and Jan Mayen Is.								**Svalbard et îles Jan Mayen**
Vertebrates	9	3	4	5	5	5	7	Vertébrés
Invertebrates	0	0	0	0	0	0	0	Invertébrés
Plants	0	0	0	0	0	0	0	Plantes
Total	9	3	4	5	5	5	7	Total

Threatened species *(continued)*
Number by taxonomic group

Espèces menacées *(suite)*
Nombre par groupe taxonomique

Country or area [&]	2004	2010	2012	2013	2014	2015	2016	Pays ou zone [&]
Swaziland								**Swaziland**
Vertebrates	12	18	21	22	23	23	25	Vertébrés
Invertebrates	0	0	0	0	0	0	0	Invertébrés
Plants	11	11	11	11	11	11	11	Plantes
Total	23	29	32	33	34	34	36	Total
Sweden								**Suède**
Vertebrates	20	15	17	17	17	16	24	Vertébrés
Invertebrates	13	11	14	15	15	15	15	Invertébrés
Plants	3	3	5	5	4	5	# 14	Plantes
Total	36	29	36	37	36	36	53	Total
Switzerland								**Suisse**
Vertebrates	17	14	15	15	15	15	18	Vertébrés
Invertebrates	30	28	38	43	43	43	43	Invertébrés
Plants	2	3	4	4	4	4	# 10	Plantes
Total	49	45	57	62	62	62	71	Total
Syrian Arab Republic								**Rép. arabe syrienne**
Vertebrates	26	68	72	72	84	84	94	Vertébrés
Invertebrates	3	7	8	9	19	20	20	Invertébrés
Plants	0	3	3	4	4	4	13	Plantes
Total	29	78	83	85	107	108	127	Total
Tajikistan								**Tadjikistan**
Vertebrates	20	24	27	27	27	26	29	Vertébrés
Invertebrates	2	2	2	2	3	3	3	Invertébrés
Plants	2	14	13	12	12	12	12	Plantes
Total	24	40	42	41	42	41	44	Total
Thailand								**Thaïlande**
Vertebrates	136	201	231	231	233	234	245	Vertébrés
Invertebrates	1	185	200	211	211	211	211	Invertébrés
Plants	84	91	126	131	133	150	150	Plantes
Total	221	477	557	573	577	595	606	Total
TFYR of Macedonia								**ex-R.Y. de Macédoine**
Vertebrates	24	31	31	31	31	31	34	Vertébrés
Invertebrates	5	59	67	68	69	69	70	Invertébrés
Plants	0	0	0	0	0	0	0	Plantes
Total	29	90	98	99	100	100	104	Total
Timor-Leste								**Timor-Leste**
Vertebrates	11	18	18	19	19	19	20	Vertébrés
Invertebrates	0	0	0	1	1	1	1	Invertébrés
Plants	0	0	1	1	1	1	1	Plantes
Total	11	18	19	21	21	21	22	Total
Togo								**Togo**
Vertebrates	22	43	48	49	50	51	63	Vertébrés
Invertebrates	0	1	1	1	3	3	3	Invertébrés
Plants	10	10	11	12	12	12	12	Plantes
Total	32	54	60	62	65	66	78	Total
Tokelau								**Tokélaou**
Vertebrates	6	10	11	11	11	11	14	Vertébrés
Invertebrates	0	31	31	35	35	35	35	Invertébrés
Plants	0	0	0	0	0	0	0	Plantes
Total	6	41	42	46	46	46	49	Total
Tonga								**Tonga**
Vertebrates	11	19	21	23	23	23	26	Vertébrés
Invertebrates	2	35	37	47	47	47	47	Invertébrés
Plants	3	4	4	4	4	4	4	Plantes
Total	16	58	62	74	74	74	77	Total
Trinidad and Tobago								**Trinité-et-Tobago**
Vertebrates	32	37	45	44	45	53	56	Vertébrés
Invertebrates	0	10	10	10	10	10	10	Invertébrés
Plants	1	1	1	1	2	2	2	Plantes
Total	33	48	56	55	57	65	68	Total

Country or area &	2004	2010	2012	2013	2014	2015	2016	Pays ou zone &
Tunisia								**Tunisie**
Vertebrates	31	57	61	62	61	61	70	Vertébrés
Invertebrates	5	11	11	11	14	15	17	Invertébrés
Plants	0	7	7	7	7	7	7	Plantes
Total	36	75	79	80	82	83	94	Total
Turkey								**Turquie**
Vertebrates	76	130	134	134	191	191	199	Vertébrés
Invertebrates	13	15	36	38	70	76	78	Invertébrés
Plants	3	5	9	10	103	103	# 105	Plantes
Total	92	150	179	182	364	370	382	Total
Turkmenistan								**Turkménistan**
Vertebrates	35	37	38	38	38	38	42	Vertébrés
Invertebrates	5	5	6	6	7	7	7	Invertébrés
Plants	0	3	4	4	4	4	4	Plantes
Total	40	45	48	48	49	49	53	Total
Turks and Caicos Islands								**Îles Turques-et-Caïques**
Vertebrates	18	22	26	28	29	37	37	Vertébrés
Invertebrates	0	10	10	10	10	10	10	Invertébrés
Plants	2	2	7	7	9	9	9	Plantes
Total	20	34	43	45	48	56	56	Total
Tuvalu								**Tuvalu**
Vertebrates	7	14	15	15	15	15	17	Vertébrés
Invertebrates	1	71	71	78	78	78	78	Invertébrés
Plants	0	0	0	0	0	0	0	Plantes
Total	8	85	86	93	93	93	95	Total
Uganda								**Ouganda**
Vertebrates	77	110	114	116	121	117	116	Vertébrés
Invertebrates	19	15	15	19	19	19	19	Invertébrés
Plants	38	41	40	48	49	52	52	Plantes
Total	134	166	169	183	189	188	187	Total
Ukraine								**Ukraine**
Vertebrates	40	45	47	47	48	47	52	Vertébrés
Invertebrates	14	15	21	24	24	23	23	Invertébrés
Plants	1	1	17	17	16	17	# 19	Plantes
Total	55	61	85	88	88	87	94	Total
United Arab Emirates								**Émirats arabes unis**
Vertebrates	23	32	32	34	34	33	41	Vertébrés
Invertebrates	0	16	17	16	15	15	15	Invertébrés
Plants	0	0	0	0	0	0	0	Plantes
Total	23	48	49	50	49	48	56	Total
United Kingdom								**Royaume-Uni**
Vertebrates	32	48	52	53	53	53	61	Vertébrés
Invertebrates	10	11	15	15	17	18	18	Invertébrés
Plants	13	14	15	15	15	16	# 22	Plantes
Total	55	73	82	83	85	87	101	Total
United Rep. of Tanzania								**Rép.-Unie de Tanzanie**
Vertebrates	144	313	328	332	346	346	355	Vertébrés
Invertebrates	33	80	80	107	129	129	129	Invertébrés
Plants	239	298	302	475	504	602	602	Plantes
Total	416	691	710	914	979	1 077	1 086	Total
United States								**États-Unis**
Vertebrates	342	376	392	441	438	444	454	Vertébrés
Invertebrates	561	531	555	566	570	575	578	Invertébrés
Plants	240	245	256	271	279	280	# 447	Plantes
Total	1 143	1 152	1 203	1 278	1 287	1 299	1 479	Total
U.S. Minor Outlying islands								**Îles mineures éloignées des É-U**
Vertebrates	15	23	25	24	24	24	25	Vertébrés
Invertebrates	0	44	44	47	47	47	47	Invertébrés
Plants	0	0	0	0	0	0	0	Plantes
Total	15	67	69	71	71	71	72	Total

Country or area [&]	2004	2010	2012	2013	2014	2015	2016	Pays ou zone [&]
United States Virgin Is.								**Îles Vierges américaines**
Vertebrates	22	21	25	31	32	41	42	Vertébrés
Invertebrates	0	0	0	0	0	0	0	Invertébrés
Plants	9	12	12	12	12	12	12	Plantes
Total	31	33	37	43	44	53	54	Total
Uruguay								**Uruguay**
Vertebrates	48	78	80	80	79	79	82	Vertébrés
Invertebrates	1	1	1	2	2	2	2	Invertébrés
Plants	1	1	1	22	22	22	22	Plantes
Total	50	80	82	104	103	103	106	Total
Uzbekistan								**Ouzbékistan**
Vertebrates	29	34	35	35	35	34	39	Vertébrés
Invertebrates	1	1	2	2	3	3	3	Invertébrés
Plants	1	15	17	17	17	17	17	Plantes
Total	31	50	54	54	55	54	59	Total
Vanuatu								**Vanuatu**
Vertebrates	19	32	35	35	35	35	37	Vertébrés
Invertebrates	0	79	82	92	92	92	92	Invertébrés
Plants	10	10	10	10	10	10	10	Plantes
Total	29	121	127	137	137	137	139	Total
Venezuela (Boliv. Rep. of)								**Venezuela (Rép. boliv. du)**
Vertebrates	151	179	197	202	202	209	212	Vertébrés
Invertebrates	1	21	21	26	26	26	26	Invertébrés
Plants	67	70	70	77	77	77	82	Plantes
Total	219	270	288	305	305	312	320	Total
Viet Nam								**Viet Nam**
Vertebrates	144	186	229	229	235	239	260	Vertébrés
Invertebrates	0	92	114	125	126	127	152	Invertébrés
Plants	145	146	169	169	177	199	204	Plantes
Total	289	424	512	523	538	565	616	Total
Wallis and Futuna Islands								**Îles Wallis-et-Futuna**
Vertebrates	12	16	21	22	22	22	24	Vertébrés
Invertebrates	0	57	58	65	65	65	65	Invertébrés
Plants	1	1	1	1	1	1	1	Plantes
Total	13	74	80	88	88	88	90	Total
Western Sahara								**Sahara occidental**
Vertebrates	18	38	37	36	37	36	44	Vertébrés
Invertebrates	1	1	1	1	3	3	3	Invertébrés
Plants	0	0	0	0	0	0	0	Plantes
Total	19	39	38	37	40	39	47	Total
Yemen								**Yémen**
Vertebrates	34	48	55	55	55	57	67	Vertébrés
Invertebrates	2	62	63	69	68	68	68	Invertébrés
Plants	159	159	160	162	162	162	162	Plantes
Total	195	269	278	286	285	287	297	Total
Zambia								**Zambie**
Vertebrates	24	44	45	47	51	49	51	Vertébrés
Invertebrates	7	14	14	14	14	14	14	Invertébrés
Plants	8	9	11	14	14	20	20	Plantes
Total	39	67	70	75	79	83	85	Total
Zimbabwe								**Zimbabwe**
Vertebrates	24	34	36	36	38	38	39	Vertébrés
Invertebrates	2	5	5	5	5	5	5	Invertébrés
Plants	17	16	16	17	17	17	17	Plantes
Total	43	55	57	58	60	60	61	Total
Areas n.e.s								**Zones n.s.a**
Vertebrates	1	1	1	1	1	1	3	Vertébrés
Invertebrates	0	0	0	1	1	1	1	Invertébrés
Plants	0	0	0	0	0	0	0	Plantes
Total	1	1	1	2	2	2	4	Total

Source:

World Conservation Union (IUCN) / Species Survival Commission (SSC), Gland, Switzerland and Cambridge, United Kingdom, IUCN Red List of Threatened Species publication, last accessed July 2016.

Source:

Union mondiale pour la nature (UICN) / Commission de la sauvegarde des espèces, Gland, Suisse, et Cambridge, Royaume-Uni, La liste rouge des espèces menacées de l'UICN, dernier accès juillet 2016.

& Vertebrates consists of mammals, birds, reptiles, amphibians and fish.
Invertebrates consists of molluscs and other invertebrates.
Plants consists of plants, and since 2016, fungi and protists.

Reptiles, fishes, molluscs, other invertebrates, plants, fungi & protists:
please note that for these groups, there are still many species that have
not yet been assessed for the IUCN Red List and therefore their status is
not known (i.e. these groups have not yet been completely assessed).
Therefore the figures presented below for these groups should be
interpreted as the number of species known to be threatened within those
species that have been assessed to date, and not as the overall total
number of threatened species for each group.

1   Excluding overseas territories.
2   The figures for Amphibians displayed here are those that were
    agreed at the GAA Brazil workshop in April 2003; the "consistent
    Red List Categories" were not yet accepted by the Brazilian experts.

3   For statistical purposes, the data for China do not include those for
    the Hong Kong Special Administrative Region (Hong Kong SAR),
    Macao Special Administrative Region (Macao SAR) and Taiwan
    Province of China.
4   Excluding the north islands, Saint Barthélémy and Saint Martin
    (French part).
5   Including Ascension and Tristan da Cunha.

& Les vertébrés se composent de mammifères, oiseaux, reptiles, amphibiens
et poissons;
   les invertébrés se composent de mollusques et autres invertébrés.
   A partir de 2016, les plantes incluent aussi les champignons et les
protistes.

Veuillez noter que beaucoup d'espèces tels que les reptiles, les poissons, les
mollusques et autres invertébrés, les plantes, les champignons et les protistes
n'ont pas été encore évaluées dans le cadre de la Liste rouge de l'UICN ,
donc leur statut est inconnu pour le moment (c.-à-d. ces groupes ne sont que
partiellement évalués). En conséquent, les données présentées ci-dessous
pour chaque groupe doivent être interprétées comme le nombre d'espèces
connues et menacées parmi les espèces évaluées à ce jour, et non comme le
nombre total d'espèces menacées dans chaque groupe.

1   Non compris les départements d'outre-mer.
2   Les chiffres concernant les amphibiens sont ceux qui ont été
    convenus lors de l'atelier de l'Évaluation mondiale des amphibiens
    du Brésil en avril 2003 ; les "catégories conformes à la Liste rouge"
    n'ont pas encore été acceptées par les experts brésiliens.
3   Pour la présentation des statistiques, les données pour la Chine ne
    comprennent pas la Région Administrative Spéciale de Hong Kong
    (Hong Kong RAS), la Région Administrative Spéciale de Macao
    (Macao RAS) et la province de Taiwan.
4   Les îles du Nord, Saint-Barthélémy et Saint-Martin (partie français),
    sont exclues.
5   Y compris Ascension et Tristan da Cunha.

20

CO$_2$ emission estimates
From fossil fuel combustion, cement production and gas flared (thousand metric tons of carbon dioxide) and per capita

Estimation des émissions de CO$_2$
Dues à la combustion de combustibles fossiles, à la production de ciment et au gaz brûlés à la torchère (milliers de tonnes de dioxyde de carbone) et par habitant

Country or area[&]	1975	1985	1995	2005	2010	2011	2012	2013	Pays ou zone[&]
Total, all countries or areas									**Total, pays ou zones**
Thousand metric tons	**16 854**	**19 864**	**23 263**	**29 428**	**33 505**	**34 866**	**35 464**	**35 849**	**Milliers de tonnes**
Metric tons per capita	**4.1**	**4.1**	**4.1**	**4.5**	**4.8**	**5.0**	**5.0**	**5.0**	**Tonnes par habitant**
Afghanistan									**Afghanistan**
Thousand metric tons	2 127	3 509	1 243	1 327	8 471	12 251	20 455	21 269	Milliers de tonnes
Metric tons per capita	0.1	0.3	0.1	<0.0	0.3	0.4	0.7	0.7	Tonnes par habitant
Albania									**Albanie**
Thousand metric tons	4 595	7 880	2 087	4 254	4 598	5 240	4 712	4 815	Milliers de tonnes
Metric tons per capita	1.9	2.7	0.7	1.4	1.6	1.8	1.7	1.7	Tonnes par habitant
Algeria									**Algérie**
Thousand metric tons	32 031	72 786	95 346	107 128	119 178	121 187	129 988	134 216	Milliers de tonnes
Metric tons per capita	2.0	3.3	3.4	3.3	3.3	3.3	3.5	3.5	Tonnes par habitant
Andorra									**Andorre**
Thousand metric tons	...	...	407	576	517	491	491	491	Milliers de tonnes
Metric tons per capita	...	...	6.3	7.4	6.2	6.0	6.2	6.5	Tonnes par habitant
Angola									**Angola**
Thousand metric tons	4 415	4 701	10 975	19 156	29 057	30 341	33 399	32 464	Milliers de tonnes
Metric tons per capita	0.7	0.5	0.9	1.2	1.4	1.4	1.5	1.4	Tonnes par habitant
Anguilla									**Anguilla**
Thousand metric tons	...	...	...	128	150	143	143	136	Milliers de tonnes
Metric tons per capita	...	...	...	9.5	10.9	10.1	10.0	9.5	Tonnes par habitant
Antarctic Fisheries									**Pêcheries antarctiques**
Thousand metric tons	...	...	4	7	...	...	...	...	Milliers de tonnes
Metric tons per capita	...	...	3.1	6.1	...	...	...	...	Tonnes par habitant
Antigua and Barbuda									**Antigua-et-Barbuda**
Thousand metric tons	708	249	323	411	524	513	524	524	Milliers de tonnes
Metric tons per capita	10.2	3.8	4.7	4.9	6.0	5.8	5.9	5.8	Tonnes par habitant
Argentina									**Argentine**
Thousand metric tons	94 931	100 597	127 964	162 111	187 919	191 634	192 360	189 819	Milliers de tonnes
Metric tons per capita	3.6	3.3	3.7	4.2	4.5	4.6	4.6	4.5	Tonnes par habitant
Armenia									**Arménie**
Thousand metric tons	...	...	3 410	4 353	4 217	4 917	5 695	5 497	Milliers de tonnes
Metric tons per capita	...	...	1.1	1.4	1.4	1.7	1.9	1.8	Tonnes par habitant
Aruba									**Aruba**
Thousand metric tons	...	...	1 610	2 497	2 457	2 439	1 302	876	Milliers de tonnes
Metric tons per capita	...	...	20.0	24.7	24.2	23.9	12.7	8.5	Tonnes par habitant
Australia									**Australie**
Thousand metric tons	175 884	241 230	281 860	350 173	372 798	376 711	375 457	377 906	Milliers de tonnes
Metric tons per capita	12.7	15.3	15.5	17.2	16.8	16.7	16.4	16.2	Tonnes par habitant
Austria									**Autriche**
Thousand metric tons	54 393	54 701	59 783	74 216	67 832	65 258	62 247	62 409	Milliers de tonnes
Metric tons per capita	7.2	7.2	7.5	9.0	8.1	7.7	7.4	7.4	Tonnes par habitant
Azerbaijan									**Azerbaïdjan**
Thousand metric tons	...	...	33 333	34 338	30 678	33 447	35 555	35 643	Milliers de tonnes
Metric tons per capita	...	...	4.3	4.0	3.4	3.6	3.8	3.7	Tonnes par habitant
Bahamas									**Bahamas**
Thousand metric tons	8 254	1 511	1 683	1 742	2 464	1 907	1 918	3 110	Milliers de tonnes
Metric tons per capita	43.7	6.5	6.0	5.5	6.8	5.2	5.1	8.2	Tonnes par habitant
Bahrain									**Bahreïn**
Thousand metric tons	5 754	10 194	14 818	19 208	29 138	29 365	29 087	31 958	Milliers de tonnes
Metric tons per capita	21.7	24.5	26.5	26.5	23.1	22.5	21.8	23.7	Tonnes par habitant
Bangladesh									**Bangladesh**
Thousand metric tons	4 870	10 235	22 816	39 479	59 992	63 571	67 480	68 951	Milliers de tonnes
Metric tons per capita	0.1	0.1	0.2	0.3	0.4	0.4	0.4	0.4	Tonnes par habitant
Barbados									**Barbade**
Thousand metric tons	568	847	829	1 353	1 478	1 529	1 470	1 448	Milliers de tonnes
Metric tons per capita	2.3	3.3	3.2	5.0	5.3	5.5	5.2	5.1	Tonnes par habitant
Belarus									**Bélarus**
Thousand metric tons	...	...	60 773	59 167	62 658	63 655	63 223	63 769	Milliers de tonnes
Metric tons per capita	...	...	5.9	6.0	6.6	6.7	6.7	6.7	Tonnes par habitant

20

CO$_2$ emission estimates *(continued)*
From fossil fuel combustion, cement production and gas flared (thousand metric tons of carbon dioxide) and per capita

Estimation des émissions de CO$_2$ *(suite)*
Dues à la combustion de combustibles fossiles, à la production de ciment et au gaz brûlés à la torchère (milliers de tonnes de dioxyde de carbone) et par habitant

Country or area[&]	1975	1985	1995	2005	2010	2011	2012	2013	Pays ou zone[&]
Belgium									**Belgique**
Thousand metric tons	122 100	104 473	112 328	108 510	107 751	98 917	92 196	93 619	Milliers de tonnes
Metric tons per capita	12.5	10.6	11.1	10.4	9.9	9.0	8.3	8.4	Tonnes par habitant
Belize									**Belize**
Thousand metric tons	176	191	378	414	539	598	480	517	Milliers de tonnes
Metric tons per capita	1.3	1.1	1.7	1.5	1.7	1.8	1.4	1.5	Tonnes par habitant
Benin									**Bénin**
Thousand metric tons	444	744	1 327	2 395	5 090	5 306	5 512	5 798	Milliers de tonnes
Metric tons per capita	0.1	0.2	0.2	0.3	0.6	0.6	0.6	0.6	Tonnes par habitant
Bermuda									**Bermudes**
Thousand metric tons	458	451	447	396	572	396	444	363	Milliers de tonnes
Metric tons per capita	8.5	7.8	7.3	6.2	8.9	6.2	7.0	5.8	Tonnes par habitant
Bhutan									**Bhoutan**
Thousand metric tons	4	62	249	396	488	733	818	884	Milliers de tonnes
Metric tons per capita	0	0.1	0.5	0.6	0.7	1.0	1.1	1.2	Tonnes par habitant
Bolivia (Plurin. State of)									**Bolivie (État plurin. de)**
Thousand metric tons	4 059	4 125	9 461	12 185	15 203	16 146	18 793	19 703	Milliers de tonnes
Metric tons per capita	0.8	0.7	1.3	1.3	1.5	1.6	1.8	1.9	Tonnes par habitant
Bonaire, St Eust. and Saba									**Bonaire, St.-Eust. et Saba**
Thousand metric tons	...	...	...	...	...	...	312	323	Milliers de tonnes
Metric tons per capita	...	...	...	...	...	...	13.6	13.6	Tonnes par habitant
Bosnia and Herzegovina									**Bosnie-Herzégovine**
Thousand metric tons	...	...	3 429	16 204	21 276	23 887	22 259	21 907	Milliers de tonnes
Metric tons per capita	...	...	1.0	4.3	5.5	6.2	5.8	5.7	Tonnes par habitant
Botswana									**Botswana**
Thousand metric tons	187	1 159	3 047	4 096	4 796	4 400	4 448	5 423	Milliers de tonnes
Metric tons per capita	0.2	1.0	1.9	2.2	2.3	2.1	2.1	2.5	Tonnes par habitant
Brazil									**Brésil**
Thousand metric tons	151 165	181 249	258 347	347 309	419 754	439 413	470 029	503 677	Milliers de tonnes
Metric tons per capita	1.4	1.3	1.6	1.9	2.1	2.2	2.3	2.5	Tonnes par habitant
British Virgin Islands									**Îles Vierges britanniques**
Thousand metric tons	26	48	84	132	172	176	176	176	Milliers de tonnes
Metric tons per capita	2.3	3.7	4.7	6.0	6.3	6.3	6.1	6.0	Tonnes par habitant
Brunei Darussalam									**Brunéi Darussalam**
Thousand metric tons	7 081	2 596	4 789	5 005	8 203	9 696	9 652	7 785	Milliers de tonnes
Metric tons per capita	45.2	11.8	16.5	13.8	20.9	24.3	23.8	18.9	Tonnes par habitant
Bulgaria									**Bulgarie**
Thousand metric tons	73 061	89 541	58 005	47 917	44 118	49 343	44 873	39 563	Milliers de tonnes
Metric tons per capita	8.4	10.0	6.9	6.2	5.9	6.7	6.2	5.5	Tonnes par habitant
Burkina Faso									**Burkina Faso**
Thousand metric tons	220	477	627	1 126	1 962	2 211	2 629	3 058	Milliers de tonnes
Metric tons per capita	<0.0	0.1	0.1	0.1	0.1	0.1	0.1	0.2	Tonnes par habitant
Burundi									**Burundi**
Thousand metric tons	77	231	323	154	213	242	282	293	Milliers de tonnes
Metric tons per capita	<0.0	<0.0	<0.0	<0.0	<0.0	<0.0	<0.0	<0.0	Tonnes par habitant
Cabo Verde									**Cabo Verde**
Thousand metric tons	77	84	121	345	480	532	495	444	Milliers de tonnes
Metric tons per capita	0.3	0.3	0.3	0.7	1.0	1.1	1.0	0.9	Tonnes par habitant
Cambodia									**Cambodge**
Thousand metric tons	73	418	1 551	2 776	5 013	5 207	5 456	5 574	Milliers de tonnes
Metric tons per capita	0.0	<0.0	0.1	0.2	0.4	0.4	0.4	0.4	Tonnes par habitant
Cameroon									**Cameroun**
Thousand metric tons	1 162	6 476	4 364	3 696	6 780	5 768	6 128	6 813	Milliers de tonnes
Metric tons per capita	0.1	0.6	0.3	0.2	0.3	0.3	0.3	0.3	Tonnes par habitant
Canada									**Canada**
Thousand metric tons	397 143	422 079	467 704	551 759	492 588	497 330	481 602	475 735	Milliers de tonnes
Metric tons per capita	17.2	16.3	16.0	17.1	14.4	14.4	13.8	13.5	Tonnes par habitant
Cayman Islands									**Îles Caïmanes**
Thousand metric tons	59	194	367	480	565	594	543	539	Milliers de tonnes
Metric tons per capita	4.6	9.6	11.1	9.2	10.2	10.5	9.5	9.2	Tonnes par habitant

CO$_2$ emission estimates *(continued)*
From fossil fuel combustion, cement production and gas flared (thousand metric tons of carbon dioxide) and per capita
Estimation des émissions de CO$_2$ *(suite)*
Dues à la combustion de combustibles fossiles, à la production de ciment et au gaz brûlés à la torchère (milliers de tonnes de dioxyde de carbone) et par habitant

Country or area[&]	1975	1985	1995	2005	2010	2011	2012	2013	Pays ou zone[&]
Central African Rep.									**Rép. centrafricaine**
Thousand metric tons	103	161	235	235	264	279	293	297	Milliers de tonnes
Metric tons per capita	<0.0	0.1	0.1	0.1	0.1	0.1	0.1	0.1	Tonnes par habitant
Chad									**Tchad**
Thousand metric tons	183	180	103	400	517	539	543	609	Milliers de tonnes
Metric tons per capita	<0.0	<0.0	0.0	<0.0	<0.0	<0.0	<0.0	<0.0	Tonnes par habitant
Chile									**Chili**
Thousand metric tons	23 014	21 503	41 745	61 818	72 251	79 244	80 975	83 171	Milliers de tonnes
Metric tons per capita	2.2	1.8	2.9	3.8	4.3	4.6	4.7	4.7	Tonnes par habitant
China[1]									**Chine**[1]
Thousand metric tons	1 145 607	1 966 553	3 320 285	5 790 017	8 767 878	9 724 591	10 020 745	10 249 463	Milliers de tonnes
Metric tons per capita	1.2	1.9	2.8	4.4	6.5	7.2	7.4	7.5	Tonnes par habitant
China, Hong Kong SAR									**Chine, Hong Kong RAS**
Thousand metric tons	11 019	23 014	31 470	43 872	40 759	43 795	43 447	44 994	Milliers de tonnes
Metric tons per capita	2.5	4.3	5.1	6.5	5.8	6.2	6.1	6.3	Tonnes par habitant
China, Macao SAR									**Chine, Macao RAS**
Thousand metric tons	297	733	1 243	1 837	1 984	2 358	2 274	2 167	Milliers de tonnes
Metric tons per capita	1.2	2.5	3.1	3.8	3.7	4.3	4.1	3.8	Tonnes par habitant
Colombia									**Colombie**
Thousand metric tons	35 896	48 379	59 614	60 946	76 164	76 530	79 875	89 625	Milliers de tonnes
Metric tons per capita	1.5	1.6	1.7	1.4	1.7	1.7	1.7	1.9	Tonnes par habitant
Comoros									**Comores**
Thousand metric tons	33	48	66	103	128	150	147	161	Milliers de tonnes
Metric tons per capita	0.1	0.1	0.1	0.1	0.2	0.2	0.2	0.2	Tonnes par habitant
Congo									**Congo**
Thousand metric tons	1 100	1 280	1 566	983	1 925	2 233	2 384	2 483	Milliers de tonnes
Metric tons per capita	0.7	0.6	0.6	0.3	0.5	0.6	0.6	0.6	Tonnes par habitant
Cook Islands									**Îles Cook**
Thousand metric tons	15	22	22	62	70	70	70	70	Milliers de tonnes
Metric tons per capita	0.8	1.2	1.2	3.2	3.5	3.4	3.4	3.4	Tonnes par habitant
Costa Rica									**Costa Rica**
Thousand metric tons	2 039	2 266	4 866	6 868	7 569	7 741	7 767	7 616	Milliers de tonnes
Metric tons per capita	1.0	0.8	1.4	1.6	1.7	1.7	1.7	1.6	Tonnes par habitant
Côte d'Ivoire									**Côte d'Ivoire**
Thousand metric tons	3 990	7 327	7 132	7 825	6 164	6 399	8 390	8 988	Milliers de tonnes
Metric tons per capita	0.6	0.7	0.5	0.4	0.3	0.3	0.4	0.4	Tonnes par habitant
Croatia									**Croatie**
Thousand metric tons	...	...	17 052	22 724	20 444	20 058	18 163	17 712	Milliers de tonnes
Metric tons per capita	...	...	3.7	5.1	4.7	4.7	4.3	4.1	Tonnes par habitant
Cuba									**Cuba**
Thousand metric tons	27 066	32 578	25 709	26 006	38 375	35 988	36 157	39 340	Milliers de tonnes
Metric tons per capita	2.9	3.2	2.3	2.3	3.4	3.2	3.2	3.4	Tonnes par habitant
Curaçao									**Curaçao**
Thousand metric tons	...	...	...	...	...	...	6 029	5 255	Milliers de tonnes
Metric tons per capita	...	...	...	...	...	...	39.5	34.0	Tonnes par habitant
Cyprus									**Chypre**
Thousand metric tons	1 980	3 102	5 383	7 503	7 708	7 426	6 920	5 948	Milliers de tonnes
Metric tons per capita	3.0	4.4	6.3	7.3	7.0	6.6	6.1	5.2	Tonnes par habitant
Czech Republic									**République tchèque**
Thousand metric tons	...	...	123 831	120 109	111 579	106 908	101 037	98 661	Milliers de tonnes
Metric tons per capita	...	...	12.0	11.7	10.6	10.2	9.6	9.4	Tonnes par habitant
Dem. P. R. Korea									**R. p. dém. de Corée**
Thousand metric tons	107 740	144 898	236 676	75 573	66 600	48 133	49 112	50 091	Milliers de tonnes
Metric tons per capita	6.7	7.7	10.9	3.2	2.7	1.9	2.0	2.0	Tonnes par habitant
Dem. Rep. of the Congo									**Rép. dém. du Congo**
Thousand metric tons	3 238	3 641	2 123	1 500	1 991	2 468	2 560	2 780	Milliers de tonnes
Metric tons per capita	0.1	0.1	<0.0	<0.0	<0.0	<0.0	<0.0	<0.0	Tonnes par habitant
Denmark									**Danemark**
Thousand metric tons	55 779	60 073	57 172	47 095	46 641	40 645	36 406	38 067	Milliers de tonnes
Metric tons per capita	11.0	11.7	10.9	8.7	8.4	7.3	6.5	6.8	Tonnes par habitant

CO$_2$ emission estimates *(continued)*
From fossil fuel combustion, cement production and gas flared (thousand metric tons of carbon dioxide) and per capita
Estimation des émissions de CO$_2$ *(suite)*
Dues à la combustion de combustibles fossiles, à la production de ciment et au gaz brûlés à la torchère (milliers de tonnes de dioxyde de carbone) et par habitant

Country or area[&]	1975	1985	1995	2005	2010	2011	2012	2013	Pays ou zone[&]
Djibouti									**Djibouti**
Thousand metric tons	198	359	304	414	517	473	517	609	Milliers de tonnes
Metric tons per capita	0.9	0.9	0.5	0.5	0.6	0.6	0.6	0.7	Tonnes par habitant
Dominica									**Dominique**
Thousand metric tons	29	48	81	117	139	128	136	132	Milliers de tonnes
Metric tons per capita	0.4	0.7	1.1	1.7	1.9	1.8	1.9	1.8	Tonnes par habitant
Dominican Republic									**Rép. dominicaine**
Thousand metric tons	6 340	7 294	15 885	18 639	21 650	22 083	22 746	22 072	Milliers de tonnes
Metric tons per capita	1.2	1.1	2.0	2.0	2.2	2.2	2.2	2.2	Tonnes par habitant
Ecuador									**Équateur**
Thousand metric tons	7 363	19 431	22 842	30 264	38 074	39 640	40 473	43 527	Milliers de tonnes
Metric tons per capita	1.1	2.1	2.0	2.2	2.6	2.6	2.6	2.8	Tonnes par habitant
Egypt									**Égypte**
Thousand metric tons	31 100	63 938	95 723	167 208	202 715	217 163	217 068	213 012	Milliers de tonnes
Metric tons per capita	0.8	1.2	1.5	2.2	2.5	2.6	2.5	2.4	Tonnes par habitant
El Salvador									**El Salvador**
Thousand metric tons	2 109	1 988	5 027	6 454	6 351	6 648	7 198	6 359	Milliers de tonnes
Metric tons per capita	0.5	0.4	0.9	1.1	1.1	1.1	1.2	1.0	Tonnes par habitant
Equatorial Guinea									**Guinée équatoriale**
Thousand metric tons	62	66	132	4 712	4 679	6 769	5 831	5 412	Milliers de tonnes
Metric tons per capita	0.3	0.2	0.3	7.7	6.4	9.0	7.5	6.8	Tonnes par habitant
Eritrea									**Érythrée**
Thousand metric tons	...	...	935	766	513	594	660	667	Milliers de tonnes
Metric tons per capita	...	...	0.3	0.2	0.1	0.1	0.1	0.1	Tonnes par habitant
Estonia									**Estonie**
Thousand metric tons	...	...	17 521	16 755	18 108	18 606	17 624	19 915	Milliers de tonnes
Metric tons per capita	...	...	12.2	12.4	13.6	14.0	13.3	15.1	Tonnes par habitant
Ethiopia									**Éthiopie**
Thousand metric tons	1 210	1 815	2 563	5 119	6 586	7 726	8 562	10 634	Milliers de tonnes
Metric tons per capita	<0.0	<0.0	<0.0	0.1	0.1	0.1	0.1	0.1	Tonnes par habitant
Falkland Is. (Malvinas)									**Îles Falkland (Malvinas)**
Thousand metric tons	15	29	40	51	55	55	55	55	Milliers de tonnes
Metric tons per capita	6.7	15.5	16.9	17.5	19.1	19.1	19.1	19.0	Tonnes par habitant
Faroe Islands									**Îles Féroé**
Thousand metric tons	348	517	682	722	631	568	590	594	Milliers de tonnes
Metric tons per capita	8.7	11.4	15.7	15.0	12.9	11.7	12.2	12.3	Tonnes par habitant
Fiji									**Fidji**
Thousand metric tons	623	579	869	1 364	1 599	1 591	1 595	1 709	Milliers de tonnes
Metric tons per capita	1.1	0.8	1.1	1.7	1.9	1.8	1.8	1.9	Tonnes par habitant
Finland									**Finlande**
Thousand metric tons	46 102	49 666	52 713	54 646	61 910	56 637	48 826	46 300	Milliers de tonnes
Metric tons per capita	9.8	10.1	10.3	10.4	11.6	10.5	9.0	8.5	Tonnes par habitant
France [2]									**France [2]**
Thousand metric tons	446 919	400 880	349 117	385 171	352 769	331 537	332 956	333 191	Milliers de tonnes
Metric tons per capita	8.5	7.3	6.0	6.3	5.6	5.2	5.2	5.2	Tonnes par habitant
French Guiana									**Guyane française**
Thousand metric tons	117	367	909	590	744	759	722	759	Milliers de tonnes
Metric tons per capita	2.1	4.2	6.6	2.9	3.2	3.2	2.9	3.0	Tonnes par habitant
French Polynesia									**Polynésie française**
Thousand metric tons	213	587	429	821	858	832	814	821	Milliers de tonnes
Metric tons per capita	1.7	3.4	2.0	3.2	3.2	3.1	3.0	3.0	Tonnes par habitant
Gabon									**Gabon**
Thousand metric tons	5 368	6 318	4 624	4 888	4 536	4 470	4 580	4 756	Milliers de tonnes
Metric tons per capita	9.0	8.0	4.3	3.6	2.9	2.8	2.8	2.9	Tonnes par habitant
Gambia									**Gambie**
Thousand metric tons	99	172	216	323	473	436	473	491	Milliers de tonnes
Metric tons per capita	0.2	0.2	0.2	0.2	0.3	0.3	0.3	0.3	Tonnes par habitant
Georgia									**Géorgie**
Thousand metric tons	...	...	2 303	5 068	5 556	6 883	7 356	7 510	Milliers de tonnes
Metric tons per capita	...	...	0.4	1.1	1.3	1.7	1.8	1.8	Tonnes par habitant

CO$_2$ emission estimates *(continued)*
From fossil fuel combustion, cement production and gas flared (thousand metric tons of carbon dioxide) and per capita

Estimation des émissions de CO$_2$ *(suite)*
Dues à la combustion de combustibles fossiles, à la production de ciment et au gaz brûlés à la torchère (milliers de tonnes de dioxyde de carbone) et par habitant

Country or area[&]	1975	1985	1995	2005	2010	2011	2012	2013	Pays ou zone[&]
Germany									**Allemagne**
Thousand metric tons	...	...	864 110	797 180	758 537	732 120	739 300	757 313	Milliers de tonnes
Metric tons per capita	...	...	10.6	9.6	9.4	9.1	9.2	9.4	Tonnes par habitant
Ghana									**Ghana**
Thousand metric tons	2 747	3 326	5 277	6 993	9 956	9 831	11 877	14 620	Milliers de tonnes
Metric tons per capita	0.3	0.3	0.3	0.3	0.4	0.4	0.5	0.6	Tonnes par habitant
Gibraltar									**Gibraltar**
Thousand metric tons	88	44	275	400	466	455	462	491	Milliers de tonnes
Metric tons per capita	3.5	1.6	10.1	13.7	15.2	14.6	14.7	15.5	Tonnes par habitant
Greece									**Grèce**
Thousand metric tons	38 874	60 601	78 782	98 675	83 490	80 725	79 970	69 156	Milliers de tonnes
Metric tons per capita	4.3	6.1	7.4	8.8	7.5	7.2	7.2	6.3	Tonnes par habitant
Greenland									**Groenland**
Thousand metric tons	506	510	502	609	664	708	568	568	Milliers de tonnes
Metric tons per capita	10.2	9.6	9.1	10.6	11.7	12.6	10.1	10.1	Tonnes par habitant
Grenada									**Grenade**
Thousand metric tons	48	62	150	216	260	253	271	304	Milliers de tonnes
Metric tons per capita	0.5	0.6	1.5	2.1	2.5	2.4	2.6	2.9	Tonnes par habitant
Guadeloupe									**Guadeloupe**
Thousand metric tons	403	891	1 555	1 933	2 101	2 131	2 131	2 142	Milliers de tonnes
Metric tons per capita	1.2	2.5	3.9	4.3	4.6	4.7	4.6	4.6	Tonnes par habitant
Guatemala									**Guatemala**
Thousand metric tons	3 524	3 524	7 165	12 570	11 665	11 837	11 973	13 597	Milliers de tonnes
Metric tons per capita	0.6	0.4	0.7	1.0	0.8	0.8	0.8	0.9	Tonnes par habitant
Guinea									**Guinée**
Thousand metric tons	843	994	1 250	1 181	2 604	2 780	2 582	2 299	Milliers de tonnes
Metric tons per capita	0.2	0.2	0.2	0.1	0.2	0.3	0.2	0.2	Tonnes par habitant
Guinea-Bissau									**Guinée-Bissau**
Thousand metric tons	114	172	183	213	238	246	253	257	Milliers de tonnes
Metric tons per capita	0.1	0.2	0.1	0.1	0.1	0.1	0.1	0.1	Tonnes par habitant
Guyana									**Guyana**
Thousand metric tons	1 826	1 419	1 463	1 437	1 720	1 782	1 995	1 936	Milliers de tonnes
Metric tons per capita	2.5	1.9	2.0	1.9	2.3	2.3	2.6	2.5	Tonnes par habitant
Haiti									**Haïti**
Thousand metric tons	484	942	902	2 076	2 127	2 219	2 318	2 406	Milliers de tonnes
Metric tons per capita	0.1	0.1	0.1	0.2	0.2	0.2	0.2	0.2	Tonnes par habitant
Honduras									**Honduras**
Thousand metric tons	1 668	1 907	3 880	7 554	7 976	8 955	8 984	9 065	Milliers de tonnes
Metric tons per capita	0.6	0.4	0.7	1.1	1.1	1.2	1.2	1.1	Tonnes par habitant
Hungary									**Hongrie**
Thousand metric tons	75 390	86 105	60 370	58 019	50 216	47 763	44 521	41 441	Milliers de tonnes
Metric tons per capita	7.2	8.2	5.8	5.8	5.0	4.8	4.5	4.2	Tonnes par habitant
Iceland									**Islande**
Thousand metric tons	1 617	1 628	1 947	2 230	1 962	1 881	1 870	1 969	Milliers de tonnes
Metric tons per capita	7.4	6.7	7.3	7.5	6.2	5.9	5.8	6.1	Tonnes par habitant
India									**Inde**
Thousand metric tons	252 202	426 674	811 562	1 222 563	1 719 691	1 846 764	2 018 504	2 034 752	Milliers de tonnes
Metric tons per capita	0.4	0.6	0.8	1.1	1.4	1.5	1.6	1.6	Tonnes par habitant
Indonesia									**Indonésie**
Thousand metric tons	53 964	121 246	224 941	341 992	428 760	573 379	599 540	479 365	Milliers de tonnes
Metric tons per capita	0.4	0.7	1.1	1.5	1.8	2.3	2.4	1.9	Tonnes par habitant
Iran (Islamic Rep. of)									**Iran (Rép. islamique d')**
Thousand metric tons	139 350	160 769	273 338	468 991	605 216	619 166	643 863	616 976	Milliers de tonnes
Metric tons per capita	4.3	3.4	4.6	6.7	8.1	8.3	8.5	8.0	Tonnes par habitant
Iraq									**Iraq**
Thousand metric tons	33 157	44 675	77 902	113 523	112 196	133 930	153 013	167 813	Milliers de tonnes
Metric tons per capita	2.8	2.9	3.9	4.1	3.6	4.2	4.7	4.9	Tonnes par habitant
Ireland									**Irlande**
Thousand metric tons	22 053	26 725	32 970	43 538	40 242	35 827	35 720	34 965	Milliers de tonnes
Metric tons per capita	6.9	7.6	9.1	10.5	8.7	7.7	7.7	7.5	Tonnes par habitant

CO₂ emission estimates *(continued)*
From fossil fuel combustion, cement production and gas flared (thousand metric tons of carbon dioxide) and per capita

Estimation des émissions de CO₂ *(suite)*
Dues à la combustion de combustibles fossiles, à la production de ciment et au gaz brûlés à la torchère (milliers de tonnes de dioxyde de carbone) et par habitant

Country or area[&]	1975	1985	1995	2005	2010	2011	2012	2013	Pays ou zone[&]
Israel									**Israël**
Thousand metric tons	19 648	24 870	51 100	56 952	68 881	69 130	75 529	71 074	Milliers de tonnes
Metric tons per capita	5.9	6.1	9.6	8.6	9.3	9.1	9.8	9.1	Tonnes par habitant
Italy [3]									**Italie** [3]
Thousand metric tons	342 311	372 134	430 484	473 384	405 361	397 994	369 447	344 768	Milliers de tonnes
Metric tons per capita	6.2	6.6	7.6	8.1	6.8	6.7	6.2	5.8	Tonnes par habitant
Jamaica									**Jamaïque**
Thousand metric tons	8 188	5 046	9 179	10 499	7 257	7 723	7 327	7 726	Milliers de tonnes
Metric tons per capita	4.1	2.2	3.7	3.9	2.6	2.8	2.6	2.8	Tonnes par habitant
Japan									**Japon**
Thousand metric tons	870 073	915 397	1 183 447	1 239 255	1 171 841	1 191 056	1 229 574	1 243 384	Milliers de tonnes
Metric tons per capita	7.8	7.6	9.5	9.8	9.2	9.4	9.7	9.8	Tonnes par habitant
Jordan									**Jordanie**
Thousand metric tons	2 494	8 540	13 557	21 060	21 181	21 668	24 859	24 807	Milliers de tonnes
Metric tons per capita	1.2	3.0	3.1	4.0	3.3	3.2	3.6	3.4	Tonnes par habitant
Kazakhstan									**Kazakhstan**
Thousand metric tons	...	...	169 540	177 329	246 620	259 121	244 721	262 902	Milliers de tonnes
Metric tons per capita	...	...	10.6	11.7	15.1	15.7	14.6	15.4	Tonnes par habitant
Kenya									**Kenya**
Thousand metric tons	4 976	3 770	7 554	8 562	12 174	13 458	12 515	13 300	Milliers de tonnes
Metric tons per capita	0.4	0.2	0.3	0.3	0.3	0.3	0.3	0.3	Tonnes par habitant
Kiribati									**Kiribati**
Thousand metric tons	33	22	22	62	62	62	62	62	Milliers de tonnes
Metric tons per capita	0.7	0.3	0.3	0.7	0.6	0.6	0.6	0.6	Tonnes par habitant
Kuwait									**Koweït**
Thousand metric tons	16 828	29 160	54 822	71 547	89 625	91 030	101 147	97 960	Milliers de tonnes
Metric tons per capita	16.0	16.7	33.7	31.6	29.3	28.1	29.6	27.2	Tonnes par habitant
Kyrgyzstan									**Kirghizistan**
Thousand metric tons	...	...	4 529	5 390	6 384	7 657	10 121	9 842	Milliers de tonnes
Metric tons per capita	...	...	1.0	1.1	1.2	1.4	1.8	1.7	Tonnes par habitant
Lao People's Dem. Rep.									**Rép. dém. pop. lao**
Thousand metric tons	253	202	348	1 404	1 639	1 624	2 160	2 175	Milliers de tonnes
Metric tons per capita	0.1	0.1	0.1	0.3	0.3	0.3	0.3	0.3	Tonnes par habitant
Latvia									**Lettonie**
Thousand metric tons	...	...	9 443	7 506	8 075	7 301	7 063	7 081	Milliers de tonnes
Metric tons per capita	...	...	3.8	3.3	3.9	3.5	3.5	3.5	Tonnes par habitant
Lebanon									**Liban**
Thousand metric tons	6 432	8 056	13 579	16 208	20 055	20 451	22 640	22 581	Milliers de tonnes
Metric tons per capita	2.3	2.8	3.9	4.0	4.6	4.4	4.6	4.3	Tonnes par habitant
Lesotho									**Lesotho**
Thousand metric tons	...	...	...	1 419	2 259	2 200	2 222	2 296	Milliers de tonnes
Metric tons per capita	...	...	...	0.7	1.1	1.1	1.1	1.1	Tonnes par habitant
Liberia									**Libéria**
Thousand metric tons	1 485	722	330	730	796	895	1 030	957	Milliers de tonnes
Metric tons per capita	0.9	0.3	0.1	0.2	0.2	0.2	0.3	0.2	Tonnes par habitant
Libya									**Libye**
Thousand metric tons	11 580	31 419	46 021	52 108	61 349	39 391	52 398	51 041	Milliers de tonnes
Metric tons per capita	4.7	8.2	9.6	9.0	9.8	6.3	8.3	8.1	Tonnes par habitant
Liechtenstein									**Liechtenstein**
Thousand metric tons	...	...	...	...	55	48	48	51	Milliers de tonnes
Metric tons per capita	...	...	...	...	1.5	1.3	1.3	1.4	Tonnes par habitant
Lithuania									**Lituanie**
Thousand metric tons	...	...	16 204	13 902	13 469	13 788	13 832	12 640	Milliers de tonnes
Metric tons per capita	...	...	4.5	4.1	4.3	4.5	4.6	4.3	Tonnes par habitant
Luxembourg									**Luxembourg**
Thousand metric tons	11 844	9 168	8 317	11 547	10 968	10 939	10 664	10 161	Milliers de tonnes
Metric tons per capita	33.0	25.0	20.4	25.3	21.6	21.0	20.0	18.7	Tonnes par habitant
Madagascar									**Madagascar**
Thousand metric tons	1 694	1 056	1 265	1 742	1 958	2 314	2 666	3 077	Milliers de tonnes
Metric tons per capita	0.2	0.1	0.1	0.1	0.1	0.1	0.1	0.1	Tonnes par habitant

CO₂ emission estimates *(continued)*
From fossil fuel combustion, cement production and gas flared (thousand metric tons of carbon dioxide) and per capita
Estimation des émissions de CO₂ *(suite)*
Dues à la combustion de combustibles fossiles, à la production de ciment et au gaz brûlés à la torchère (milliers de tonnes de dioxyde de carbone) et par habitant

Country or area&	1975	1985	1995	2005	2010	2011	2012	2013	Pays ou zone&
Malawi									**Malawi**
Thousand metric tons	579	557	730	917	1 192	1 188	1 129	1 272	Milliers de tonnes
Metric tons per capita	0.1	0.1	0.1	0.1	0.1	0.1	0.1	0.1	Tonnes par habitant
Malaysia									**Malaisie**
Thousand metric tons	19 446	36 237	121 132	174 487	218 476	220 405	218 707	236 510	Milliers de tonnes
Metric tons per capita	1.6	2.3	5.8	6.7	7.8	7.7	7.6	8.0	Tonnes par habitant
Maldives									**Maldives**
Thousand metric tons	7	66	249	601	898	953	1 078	1 049	Milliers de tonnes
Metric tons per capita	<0.0	0.3	1.0	2.1	2.7	2.8	3.1	3.0	Tonnes par habitant
Mali									**Mali**
Thousand metric tons	341	407	469	898	964	1 045	994	1 027	Milliers de tonnes
Metric tons per capita	<0.0	<0.0	<0.0	0.1	0.1	0.1	0.1	0.1	Tonnes par habitant
Malta									**Malte**
Thousand metric tons	667	1 199	2 127	2 699	2 527	2 519	2 666	2 219	Milliers de tonnes
Metric tons per capita	2.2	3.4	5.5	6.6	6.1	6.1	6.4	5.3	Tonnes par habitant
Marshall Islands									**Îles Marshall**
Thousand metric tons	...	...	66	84	103	103	103	103	Milliers de tonnes
Metric tons per capita	...	...	1.3	1.7	2.0	2.0	2.0	2.0	Tonnes par habitant
Martinique									**Martinique**
Thousand metric tons	902	1 210	2 039	2 479	2 230	2 406	2 402	2 420	Milliers de tonnes
Metric tons per capita	2.8	3.6	5.5	6.2	5.6	6.1	6.1	6.1	Tonnes par habitant
Mauritania									**Mauritanie**
Thousand metric tons	510	656	1 093	1 588	2 237	2 395	2 655	2 648	Milliers de tonnes
Metric tons per capita	0.4	0.4	0.5	0.5	0.6	0.7	0.7	0.7	Tonnes par habitant
Mauritius									**Maurice**
Thousand metric tons	590	708	1 830	3 297	3 916	3 920	4 070	3 726	Milliers de tonnes
Metric tons per capita	0.7	0.7	1.6	2.6	3.2	3.1	3.2	2.9	Tonnes par habitant
Mexico									**Mexique**
Thousand metric tons	164 472	288 501	328 292	435 438	445 163	466 780	480 883	488 602	Milliers de tonnes
Metric tons per capita	2.7	3.8	3.6	4.1	3.7	3.9	3.9	4.0	Tonnes par habitant
Micronesia (Fed. States of)									**Micronésie (États féd. de)**
Thousand metric tons	...	...	...	121	121	128	143	147	Milliers de tonnes
Metric tons per capita	...	...	...	1.1	1.2	1.2	1.4	1.4	Tonnes par habitant
Mongolia									**Mongolie**
Thousand metric tons	4 067	9 028	7 921	8 562	24 661	27 583	32 728	41 591	Milliers de tonnes
Metric tons per capita	2.8	4.7	3.4	3.4	9.1	10.0	11.7	14.6	Tonnes par habitant
Montenegro									**Monténégro**
Thousand metric tons	...	...	...	...	2 582	2 571	2 336	2 248	Milliers de tonnes
Metric tons per capita	...	...	...	...	4.1	4.1	3.7	3.6	Tonnes par habitant
Montserrat									**Montserrat**
Thousand metric tons	11	26	33	37	66	40	44	51	Milliers de tonnes
Metric tons per capita	1.0	2.2	3.3	6.6	13.0	8.0	8.5	10.3	Tonnes par habitant
Morocco									**Maroc**
Thousand metric tons	11 111	17 866	30 374	45 771	55 958	57 686	62 724	58 558	Milliers de tonnes
Metric tons per capita	0.7	0.8	1.1	1.5	1.7	1.8	1.9	1.7	Tonnes par habitant
Mozambique									**Mozambique**
Thousand metric tons	2 886	1 148	1 115	1 822	2 736	3 223	3 121	4 019	Milliers de tonnes
Metric tons per capita	0.3	0.1	0.1	0.1	0.1	0.1	0.1	0.1	Tonnes par habitant
Myanmar									**Myanmar**
Thousand metric tons	4 587	6 711	6 960	11 599	12 515	14 298	12 934	12 603	Milliers de tonnes
Metric tons per capita	0.1	0.2	0.2	0.3	0.3	0.3	0.3	0.2	Tonnes par habitant
Namibia									**Namibie**
Thousand metric tons	...	...	1 632	2 310	3 099	2 824	3 374	2 948	Milliers de tonnes
Metric tons per capita	...	...	1.0	1.1	1.4	1.2	1.5	1.2	Tonnes par habitant
Nauru									**Nauru**
Thousand metric tons	103	125	106	62	44	40	40	44	Milliers de tonnes
Metric tons per capita	14.8	15.4	10.8	6.1	4.3	4.0	4.0	4.3	Tonnes par habitant
Nepal									**Népal**
Thousand metric tons	352	678	2 035	3 084	5 057	5 534	5 845	6 502	Milliers de tonnes
Metric tons per capita	<0.0	<0.0	0.1	0.1	0.2	0.2	0.2	0.2	Tonnes par habitant

CO$_2$ emission estimates *(continued)*
From fossil fuel combustion, cement production and gas flared (thousand metric tons of carbon dioxide) and per capita
Estimation des émissions de CO$_2$ *(suite)*
Dues à la combustion de combustibles fossiles, à la production de ciment et au gaz brûlés à la torchère (milliers de tonnes de dioxyde de carbone) et par habitant

Country or area[&]	1975	1985	1995	2005	2010	2011	2012	2013	Pays ou zone[&]
Netherlands									**Pays-Bas**
Thousand metric tons	153 020	147 949	168 304	170 255	181 649	172 400	169 379	169 973	Milliers de tonnes
Metric tons per capita	11.3	10.2	10.9	10.5	10.9	10.3	10.1	10.1	Tonnes par habitant
Netherlands Ant. [former]									**Antilles néerlandaises [anc.]**
Thousand metric tons	...	...	5 225	5 735	4 562	5 820	...	...	Milliers de tonnes
Metric tons per capita	...	...	27.5	30.8	22.0	27.3	...	...	Tonnes par habitant
New Caledonia									**Nouvelle-Calédonie**
Thousand metric tons	2 530	1 456	2 076	2 824	3 542	3 649	3 630	3 861	Milliers de tonnes
Metric tons per capita	19.7	9.4	10.9	12.2	14.4	14.6	14.3	15.0	Tonnes par habitant
New Zealand									**Nouvelle-Zélande**
Thousand metric tons	18 262	21 804	27 132	34 140	31 778	31 503	34 506	33 960	Milliers de tonnes
Metric tons per capita	5.9	6.7	7.4	8.3	7.3	7.2	7.8	7.6	Tonnes par habitant
Nicaragua									**Nicaragua**
Thousand metric tons	1 929	1 991	2 780	4 320	4 536	4 881	4 675	4 569	Milliers de tonnes
Metric tons per capita	0.7	0.6	0.6	0.8	0.8	0.8	0.8	0.8	Tonnes par habitant
Niger									**Niger**
Thousand metric tons	334	997	810	715	1 173	1 327	1 867	1 962	Milliers de tonnes
Metric tons per capita	0.1	0.1	0.1	0.1	0.1	0.1	0.1	0.1	Tonnes par habitant
Nigeria									**Nigéria**
Thousand metric tons	47 396	69 893	33 267	104 689	92 016	96 094	99 636	95 650	Milliers de tonnes
Metric tons per capita	0.7	0.8	0.3	0.7	0.6	0.6	0.6	0.6	Tonnes par habitant
Niue									**Nioué**
Thousand metric tons	4	4	7	11	11	11	11	11	Milliers de tonnes
Metric tons per capita	0.8	1.1	2.8	5.5	5.7	5.7	5.7	5.7	Tonnes par habitant
Norway									**Norvège**
Thousand metric tons	29 673	40 715	33 439	42 438	60 113	45 196	49 890	59 636	Milliers de tonnes
Metric tons per capita	7.4	9.8	7.7	9.2	12.3	9.1	9.9	11.7	Tonnes par habitant
Oman									**Oman**
Thousand metric tons	7 257	8 661	15 896	29 893	50 374	56 215	59 204	61 184	Milliers de tonnes
Metric tons per capita	8.1	5.6	7.1	12.3	17.1	17.5	16.7	15.7	Tonnes par habitant
Other non-specified areas									**Autres zones non-spécifiées**
Thousand metric tons	43 520	84 880	174 348	259 389	269 998	269 180	258 131	261 530	Milliers de tonnes
Metric tons per capita	2.7	4.4	8.2	11.4	11.6	11.6	11.1	11.2	Tonnes par habitant
Pakistan									**Pakistan**
Thousand metric tons	23 219	47 176	84 484	136 636	161 396	162 008	163 060	153 369	Milliers de tonnes
Metric tons per capita	0.3	0.5	0.7	0.8	1.0	0.9	0.9	0.8	Tonnes par habitant
Palau									**Palaos**
Thousand metric tons	194	161	106	191	216	224	224	224	Milliers de tonnes
Metric tons per capita	4.7	2.3	6.2	9.6	10.5	10.9	10.8	10.7	Tonnes par habitant
Panama									**Panama**
Thousand metric tons	3 663	2 629	2 985	6 839	9 164	10 099	10 084	10 363	Milliers de tonnes
Metric tons per capita	2.1	1.2	1.1	2.1	2.5	2.8	2.7	2.7	Tonnes par habitant
Papua New Guinea									**Papouasie-Nvl-Guinée**
Thousand metric tons	1 533	2 127	2 061	4 386	4 664	5 229	4 980	6 073	Milliers de tonnes
Metric tons per capita	0.6	0.6	0.4	0.7	0.7	0.7	0.7	0.8	Tonnes par habitant
Paraguay									**Paraguay**
Thousand metric tons	840	1 551	3 964	3 832	5 097	5 321	5 207	4 972	Milliers de tonnes
Metric tons per capita	0.3	0.4	0.8	0.7	0.8	0.8	0.8	0.8	Tonnes par habitant
Peru									**Pérou**
Thousand metric tons	21 965	19 512	23 883	37 136	57 594	49 633	54 521	57 154	Milliers de tonnes
Metric tons per capita	1.5	1.0	1.0	1.4	1.9	1.7	1.8	1.9	Tonnes par habitant
Philippines									**Philippines**
Thousand metric tons	32 526	28 049	60 711	74 832	84 920	85 584	91 319	98 239	Milliers de tonnes
Metric tons per capita	0.8	0.5	0.9	0.9	0.9	0.9	1.0	1.0	Tonnes par habitant
Poland									**Pologne**
Thousand metric tons	375 559	445 900	344 214	302 539	316 264	316 997	299 961	302 333	Milliers de tonnes
Metric tons per capita	11.0	12.0	8.9	7.9	8.2	8.2	7.8	7.8	Tonnes par habitant
Portugal									**Portugal**
Thousand metric tons	21 357	27 407	51 870	65 280	48 137	47 623	46 061	46 263	Milliers de tonnes
Metric tons per capita	2.3	2.8	5.1	6.2	4.5	4.5	4.4	4.4	Tonnes par habitant

CO$_2$ emission estimates *(continued)*
From fossil fuel combustion, cement production and gas flared (thousand metric tons of carbon dioxide) and per capita

Estimation des émissions de CO$_2$ *(suite)*
Dues à la combustion de combustibles fossiles, à la production de ciment et au gaz brûlés à la torchère (milliers de tonnes de dioxyde de carbone) et par habitant

Country or area[&]	1975	1985	1995	2005	2010	2011	2012	2013	Pays ou zone[&]
Qatar									**Qatar**
Thousand metric tons	10 957	12 405	31 789	50 957	72 618	80 436	94 125	85 023	Milliers de tonnes
Metric tons per capita	67.3	33.7	63.4	62.1	41.1	42.2	46.7	40.4	Tonnes par habitant
Republic of Korea									**République de Corée**
Thousand metric tons	81 829	178 334	374 771	462 922	566 717	589 401	584 080	592 499	Milliers de tonnes
Metric tons per capita	2.3	4.4	8.4	9.8	11.6	12.0	11.8	11.9	Tonnes par habitant
Republic of Moldova									**République de Moldova**
Thousand metric tons	...	...	11 192	4 895	4 932	5 038	4 925	4 976	Milliers de tonnes
Metric tons per capita	...	...	2.6	1.3	1.2	1.2	1.2	1.2	Tonnes par habitant
Réunion									**Réunion**
Thousand metric tons	484	994	1 830	3 447	4 169	4 272	4 250	4 316	Milliers de tonnes
Metric tons per capita	1.0	1.8	2.7	4.3	5.0	5.1	5.0	5.1	Tonnes par habitant
Romania									**Roumanie**
Thousand metric tons	162 257	194 344	127 102	96 457	79 288	84 873	81 723	70 736	Milliers de tonnes
Metric tons per capita	7.6	8.5	5.6	4.4	3.9	4.2	4.1	3.6	Tonnes par habitant
Russian Federation									**Fédération de Russie**
Thousand metric tons	...	...	1 631 474	1 615 090	1 675 027	1 768 073	1 835 609	1 789 074	Milliers de tonnes
Metric tons per capita	...	...	11.0	11.2	11.7	12.4	12.8	12.5	Tonnes par habitant
Rwanda									**Rwanda**
Thousand metric tons	176	616	462	528	590	664	737	799	Milliers de tonnes
Metric tons per capita	<0.0	0.1	0.1	0.1	0.1	0.1	0.1	0.1	Tonnes par habitant
Saint Helena									**Sainte-Hélène**
Thousand metric tons	...	4	11	11	11	11	11	11	Milliers de tonnes
Metric tons per capita	...	0.6	1.8	2.0	2.9	3.0	3.0	3.0	Tonnes par habitant
Saint Kitts and Nevis									**Saint-Kitts-et-Nevis**
Thousand metric tons	...	51	161	235	260	268	279	279	Milliers de tonnes
Metric tons per capita	...	1.2	3.8	4.8	5.0	5.0	5.2	5.1	Tonnes par habitant
Saint Lucia									**Sainte-Lucie**
Thousand metric tons	77	128	312	367	403	407	407	407	Milliers de tonnes
Metric tons per capita	0.7	1.0	2.1	2.2	2.3	2.3	2.2	2.2	Tonnes par habitant
Saint Pierre and Miquelon									**Saint-Pierre-et-Miquelon**
Thousand metric tons	33	33	70	66	70	70	70	73	Milliers de tonnes
Metric tons per capita	5.7	5.5	11.2	10.5	11.3	11.3	11.3	11.7	Tonnes par habitant
Saint Vincent-Grenadines									**Saint-Vincent-Grenadines**
Thousand metric tons	33	66	128	220	220	198	253	209	Milliers de tonnes
Metric tons per capita	0.4	0.6	1.2	2.0	2.0	1.8	2.3	1.9	Tonnes par habitant
Samoa									**Samoa**
Thousand metric tons	59	114	132	169	205	235	253	238	Milliers de tonnes
Metric tons per capita	0.4	0.7	0.8	1.0	1.1	1.2	1.4	1.2	Tonnes par habitant
Sao Tome and Principe									**Sao Tomé-et-Principe**
Thousand metric tons	18	55	48	77	99	103	114	114	Milliers de tonnes
Metric tons per capita	0.2	0.6	0.4	0.5	0.6	0.6	0.6	0.6	Tonnes par habitant
Saudi Arabia									**Arabie saoudite**
Thousand metric tons	83 336	172 653	236 254	397 642	520 553	500 729	565 998	541 429	Milliers de tonnes
Metric tons per capita	11.3	13.1	12.8	16.5	18.5	17.4	19.2	17.9	Tonnes par habitant
Senegal									**Sénégal**
Thousand metric tons	2 582	2 677	3 495	5 812	7 745	8 368	7 913	8 423	Milliers de tonnes
Metric tons per capita	0.6	0.4	0.4	0.6	0.6	0.6	0.6	0.6	Tonnes par habitant
Serbia									**Serbie**
Thousand metric tons	...	...	...	...	45 926	49 193	44 063	44 869	Milliers de tonnes
Metric tons per capita	...	...	...	...	5.1	5.5	4.9	5.0	Tonnes par habitant
Serbia and Monten. [former]									**Serbie-et-Monténégro [anc.]**
Thousand metric tons	...	...	40 579	60 172	...	...	...	...	Milliers de tonnes
Metric tons per capita	...	...	3.7	5.8	...	...	...	...	Tonnes par habitant
Seychelles									**Seychelles**
Thousand metric tons	59	150	198	689	689	598	704	645	Milliers de tonnes
Metric tons per capita	1.0	2.2	2.7	8.3	7.4	6.4	7.4	6.8	Tonnes par habitant
Sierra Leone									**Sierra Leone**
Thousand metric tons	524	664	400	546	726	898	1 030	1 192	Milliers de tonnes
Metric tons per capita	0.2	0.2	0.1	0.1	0.1	0.1	0.2	0.2	Tonnes par habitant

CO_2 emission estimates *(continued)*
From fossil fuel combustion, cement production and gas flared (thousand metric tons of carbon dioxide) and per capita
Estimation des émissions de CO_2 *(suite)*
Dues à la combustion de combustibles fossiles, à la production de ciment et au gaz brûlés à la torchère (milliers de tonnes de dioxyde de carbone)
et par habitant

Country or area[&]	1975	1985	1995	2005	2010	2011	2012	2013	Pays ou zone[&]
Singapore									**Singapour**
Thousand metric tons	24 543	33 417	47 110	30 359	43 945	38 327	54 627	50 557	Milliers de tonnes
Metric tons per capita	10.9	12.3	13.5	7.1	8.7	7.4	10.3	9.4	Tonnes par habitant
Sint Maarten (Dutch part)									**Saint-Martin (partie néerlandai**
Thousand metric tons	...	...	...	...	...	...	744	763	Milliers de tonnes
Metric tons per capita	...	...	...	...	...	...	21.2	20.9	Tonnes par habitant
Slovakia									**Slovaquie**
Thousand metric tons	...	...	42 115	39 358	36 241	34 525	32 765	33 678	Milliers de tonnes
Metric tons per capita	...	...	7.8	7.3	6.7	6.4	6.1	6.2	Tonnes par habitant
Slovenia									**Slovénie**
Thousand metric tons	...	...	14 404	15 867	15 423	15 192	14 928	14 441	Milliers de tonnes
Metric tons per capita	...	...	7.3	7.9	7.5	7.4	7.2	7.0	Tonnes par habitant
Solomon Islands									**Îles Salomon**
Thousand metric tons	59	150	143	161	194	198	198	198	Milliers de tonnes
Metric tons per capita	0.3	0.6	0.4	0.3	0.4	0.4	0.4	0.4	Tonnes par habitant
Somalia									**Somalie**
Thousand metric tons	502	858	11	594	612	576	638	623	Milliers de tonnes
Metric tons per capita	0.1	0.1	0.0	0.1	0.1	0.1	0.1	0.1	Tonnes par habitant
South Africa									**Afrique du Sud**
Thousand metric tons	185 202	324 214	360 363	416 916	474 099	475 038	472 071	471 239	Milliers de tonnes
Metric tons per capita	7.2	9.8	8.7	8.7	9.2	9.1	8.9	8.8	Tonnes par habitant
South Sudan									**Soudan du sud**
Thousand metric tons	...	...	...	...	...	...	1 331	1 448	Milliers de tonnes
Metric tons per capita	...	...	...	...	...	...	0.1	0.1	Tonnes par habitant
Spain									**Espagne**
Thousand metric tons	181 645	201 234	241 611	353 462	270 911	270 548	264 779	236 969	Milliers de tonnes
Metric tons per capita	5.1	5.2	6.1	8.1	5.8	5.8	5.7	5.1	Tonnes par habitant
Sri Lanka									**Sri Lanka**
Thousand metric tons	2 897	3 957	5 904	12 101	13 689	15 233	16 226	16 025	Milliers de tonnes
Metric tons per capita	0.2	0.3	0.3	0.6	0.7	0.7	0.8	0.8	Tonnes par habitant
State of Palestine									**État de Palestine**
Thousand metric tons	...	...	...	2 743	2 035	2 248	2 200	2 439	Milliers de tonnes
Metric tons per capita	...	...	...	0.8	0.5	0.6	0.5	0.6	Tonnes par habitant
Sudan									**Soudan**
Thousand metric tons	4 239	4 074	4 298	10 983	15 940	15 658	14 661	15 445	Milliers de tonnes
Metric tons per capita	0.3	0.2	0.1	0.3	0.4	0.3	0.4	0.4	Tonnes par habitant
Suriname									**Suriname**
Thousand metric tons	2 021	1 599	2 182	1 687	2 626	2 171	2 508	2 101	Milliers de tonnes
Metric tons per capita	5.5	4.3	5.0	3.4	5.1	4.1	4.7	3.9	Tonnes par habitant
Swaziland									**Swaziland**
Thousand metric tons	337	440	455	1 019	1 038	1 049	1 206	1 089	Milliers de tonnes
Metric tons per capita	0.7	0.6	0.5	0.9	0.9	0.9	1.0	0.9	Tonnes par habitant
Sweden									**Suède**
Thousand metric tons	80 824	62 394	55 155	51 562	52 024	51 734	47 048	44 327	Milliers de tonnes
Metric tons per capita	9.9	7.5	6.2	5.7	5.5	5.5	4.9	4.6	Tonnes par habitant
Switzerland									**Suisse**
Thousand metric tons	39 098	39 827	39 226	41 338	38 995	36 967	37 741	40 348	Milliers de tonnes
Metric tons per capita	6.1	6.1	5.5	5.5	5.0	4.7	4.7	5.0	Tonnes par habitant
Syrian Arab Republic									**Rép. arabe syrienne**
Thousand metric tons	11 184	29 651	41 664	50 627	61 606	56 908	43 821	36 065	Milliers de tonnes
Metric tons per capita	1.5	2.8	2.9	2.8	3.0	2.8	2.2	1.9	Tonnes par habitant
Tajikistan									**Tadjikistan**
Thousand metric tons	...	...	2 450	2 442	2 545	2 351	3 055	3 586	Milliers de tonnes
Metric tons per capita	...	...	0.4	0.4	0.3	0.3	0.4	0.4	Tonnes par habitant
Thailand									**Thaïlande**
Thousand metric tons	24 408	48 672	161 154	247 467	288 589	290 342	305 223	303 118	Milliers de tonnes
Metric tons per capita	0.6	0.9	2.7	3.7	4.3	4.3	4.5	4.5	Tonnes par habitant
TFYR of Macedonia									**ex-R.Y. de Macédoine**
Thousand metric tons	...	...	10 840	11 280	8 603	9 388	8 962	8 295	Milliers de tonnes
Metric tons per capita	...	...	5.5	5.5	4.2	4.5	4.3	4.0	Tonnes par habitant

20

CO$_2$ emission estimates *(continued)*
From fossil fuel combustion, cement production and gas flared (thousand metric tons of carbon dioxide) and per capita

Estimation des émissions de CO$_2$ *(suite)*
Dues à la combustion de combustibles fossiles, à la production de ciment et au gaz brûlés à la torchère (milliers de tonnes de dioxyde de carbone) et par habitant

Country or area[&]	1975	1985	1995	2005	2010	2011	2012	2013	Pays ou zone[&]
Timor-Leste									**Timor-Leste**
Thousand metric tons	...	...	...	176	235	246	293	440	Milliers de tonnes
Metric tons per capita	...	...	...	0.2	0.2	0.2	0.3	0.4	Tonnes par habitant
Togo									**Togo**
Thousand metric tons	312	550	953	1 338	2 611	2 409	2 164	2 230	Milliers de tonnes
Metric tons per capita	0.1	0.2	0.2	0.3	0.4	0.4	0.3	0.3	Tonnes par habitant
Tonga									**Tonga**
Thousand metric tons	33	48	95	114	117	103	180	209	Milliers de tonnes
Metric tons per capita	0.4	0.5	1.0	1.1	1.1	1.0	1.7	2.0	Tonnes par habitant
Trinidad and Tobago									**Trinité-et-Tobago**
Thousand metric tons	9 622	20 755	20 968	29 512	47 909	46 901	45 372	46 542	Milliers de tonnes
Metric tons per capita	9.5	17.7	16.6	22.4	36.1	35.1	33.8	34.5	Tonnes par habitant
Tunisia									**Tunisie**
Thousand metric tons	5 548	11 940	15 735	22 662	27 660	26 021	27 004	27 668	Milliers de tonnes
Metric tons per capita	1.0	1.6	1.8	2.3	2.6	2.4	2.5	2.5	Tonnes par habitant
Turkey									**Turquie**
Thousand metric tons	65 698	106 717	171 975	237 391	298 002	320 840	329 561	323 451	Milliers de tonnes
Metric tons per capita	1.7	2.2	2.9	3.5	4.1	4.4	4.4	4.3	Tonnes par habitant
Turkmenistan									**Turkménistan**
Thousand metric tons	...	...	34 000	48 338	57 290	62 592	64 994	66 893	Milliers de tonnes
Metric tons per capita	...	...	8.1	10.2	11.4	12.2	12.6	12.8	Tonnes par habitant
Turks and Caicos Islands									**Îles Turques-et-Caïques**
Thousand metric tons	...	...	11	121	191	191	198	198	Milliers de tonnes
Metric tons per capita	...	...	0.8	4.0	6.1	6.0	6.2	6.0	Tonnes par habitant
Uganda									**Ouganda**
Thousand metric tons	1 133	620	939	2 171	3 920	4 265	4 085	4 895	Milliers de tonnes
Metric tons per capita	0.1	<0.0	<0.0	0.1	0.1	0.1	0.1	0.1	Tonnes par habitant
Ukraine									**Ukraine**
Thousand metric tons	...	...	445 944	333 877	304 643	286 444	295 773	271 101	Milliers de tonnes
Metric tons per capita	...	...	8.7	7.1	6.7	6.3	6.5	6.0	Tonnes par habitant
United Arab Emirates									**Émirats arabes unis**
Thousand metric tons	31 070	49 926	70 641	116 149	160 813	159 584	172 356	169 122	Milliers de tonnes
Metric tons per capita	58.2	37.0	30.1	28.5	19.3	18.3	19.3	18.7	Tonnes par habitant
United Kingdom									**Royaume-Uni**
Thousand metric tons	603 643	559 856	538 118	542 580	493 608	447 935	467 198	457 473	Milliers de tonnes
Metric tons per capita	10.7	9.9	9.2	9.0	7.8	7.1	7.3	7.2	Tonnes par habitant
United Rep. of Tanzania									**Rép.-Unie de Tanzanie**
Thousand metric tons	2 285	2 358	3 553	5 504	7 107	8 093	9 545	10 752	Milliers de tonnes
Metric tons per capita	0.1	0.1	0.1	0.1	0.1	0.2	0.2	0.2	Tonnes par habitant
United States									**États-Unis**
Thousand metric tons	4 406 330	4 492 555	5 138 010	5 795 162	5 408 869	5 305 280	5 115 806	5 186 168	Milliers de tonnes
Metric tons per capita	19.8	18.4	19.0	19.3	17.2	16.8	16.1	16.1	Tonnes par habitant
Uruguay									**Uruguay**
Thousand metric tons	5 970	3 297	4 591	5 776	6 381	7 763	8 694	7 605	Milliers de tonnes
Metric tons per capita	2.1	1.1	1.4	1.7	1.9	2.3	2.6	2.2	Tonnes par habitant
Uzbekistan									**Ouzbékistan**
Thousand metric tons	...	...	100 080	112 878	104 168	113 684	115 815	103 226	Milliers de tonnes
Metric tons per capita	...	...	4.4	4.4	3.7	4.0	4.0	3.6	Tonnes par habitant
Vanuatu									**Vanuatu**
Thousand metric tons	55	121	66	59	121	132	114	106	Milliers de tonnes
Metric tons per capita	0.6	1.0	0.4	0.3	0.5	0.6	0.5	0.4	Tonnes par habitant
Venezuela (Boliv. Rep. of)									**Venezuela (Rép. boliv. du)**
Thousand metric tons	63 817	101 279	133 350	165 096	189 071	178 807	198 641	185 532	Milliers de tonnes
Metric tons per capita	5.0	5.9	6.1	6.2	6.5	6.1	6.6	6.1	Tonnes par habitant
Viet Nam									**Viet Nam**
Thousand metric tons	21 800	21 166	29 090	98 144	147 340	161 887	158 231	152 624	Milliers de tonnes
Metric tons per capita	0.4	0.4	0.4	1.2	1.7	1.8	1.8	1.7	Tonnes par habitant
Wallis and Futuna Islands									**Îles Wallis-et-Futuna**
Thousand metric tons	...	...	...	29	29	26	26	22	Milliers de tonnes
Metric tons per capita	...	...	...	1.9	2.1	1.8	1.8	1.6	Tonnes par habitant

20
CO$_2$ emission estimates *(continued)*
From fossil fuel combustion, cement production and gas flared (thousand metric tons of carbon dioxide) and per capita
Estimation des émissions de CO$_2$ *(suite)*
Dues à la combustion de combustibles fossiles, à la production de ciment et au gaz brûlés à la torchère (milliers de tonnes de dioxyde de carbone) et par habitant

Country or area[&]	1975	1985	1995	2005	2010	2011	2012	2013	Pays ou zone[&]
Yemen									**Yémen**
Thousand metric tons	...	...	11 250	20 044	23 432	19 666	18 291	25 346	Milliers de tonnes
Metric tons per capita	...	...	0.7	1.0	1.0	0.8	0.7	1.0	Tonnes par habitant
Zambia									**Zambie**
Thousand metric tons	4 081	2 754	2 171	2 288	2 710	2 934	3 480	3 825	Milliers de tonnes
Metric tons per capita	0.8	0.4	0.3	0.2	0.2	0.2	0.2	0.3	Tonnes par habitant
Zimbabwe									**Zimbabwe**
Thousand metric tons	8 320	10 264	15 130	10 770	9 267	11 628	12 934	13 781	Milliers de tonnes
Metric tons per capita	1.4	1.2	1.3	0.8	0.7	0.8	0.9	0.9	Tonnes par habitant

Source:
Carbon Dioxide Information Analysis Center (CDIAC) of the Oak Ridge National Laboratory, Oak Ridge, Tennessee, U.S.A., database on national CO$_2$ emission estimates, last accessed March 2016.

[&] These data are transformed from the original CDIAC data in metric tons of Carbon to metric tons of Carbon Dioxide by a factor of 3.667.

1   For statistical purposes, the data for China do not include those for the Hong Kong Special Administrative Region (Hong Kong SAR), Macao Special Administrative Region (Macao SAR) and Taiwan Province of China.
2   Including Monaco.
3   Including San Marino.

Source:
"Carbon Dioxide Information Analysis Center (CDIAC) of the Oak Ridge National Laboratory, Oak Ridge, Tennessee, U.S.A.", la base de données des estimations nationales des émissions de CO$_2$, dernier accès mars 2016.

[&] Les données originales fournies par CDIAC en tonnes métriques de carbone sont converties en tonnes métriques de dioxyde de carbone en multipliant par un facteur de 3,667.

1   Pour la présentation des statistiques, les données pour la Chine ne comprennent pas la Région Administrative Spéciale de Hong Kong (Hong Kong RAS), la Région Administrative Spéciale de Macao (Macao RAS) et la province de Taiwan.
2   Y compris Monaco.
3   Y compris Saint-Marin.

Patents
Resident filings (per million population), grants and patents in force

Brevets
Demandes émanant de résidents (par million d'habitants), délivrances et brevets en vigueur

Region, country or area	1985	1995	2005	2010	2012	2013	2014	Région, pays ou zone
Total, all countries or areas								**Total, tous pays ou zones**
Grants of patents	397 580	427 600	631 000	912 300	1 135 700	1 172 500	1 176 600	Brevets délivrés
Africa								**Afrique**
Grants of patents	9 180	6 403	4 200	9 000	10 200	13 500	14 000	Brevets délivrés
Northern America								**Amérique septentrionale**
Grants of patents	90 358	110 558	159 300	238 700	275 000	301 700	324 400	Brevets délivrés
Latin America and the Caribbean								**Amérique latine et Caraïbes**
Grants of patents	7 777	8 218	14 200	17 100	20 200	19 300	17 800	Brevets délivrés
Asia								**Asie**
Grants of patents	67 282	140 102	287 100	467 700	647 100	654 200	634 600	Brevets délivrés
Europe								**Europe**
Grants of patents	222 244	150 282	151 000	160 800	159 200	161 800	161 700	Brevets délivrés
Oceania								**Océanie**
Grants of patents	1 756	12 049	15 200	19 000	24 000	22 000	24 100	Brevets délivrés
Albania								**Albanie**
Resident filings (per mil. pop.)	...	...	...	...	...	...	4.0	Demandes de rés. (par mil. d'hab.)
Grants of patents	...	...	395	349	...	9	5	Brevets délivrés
Patents in force	...	...	...	349	...	4 322	...	Brevets en vigueur
Algeria								**Algérie**
Resident filings (per mil. pop.)	...	1.0	2.0	2.0	3.0	3.0	2.0	Demandes de rés. (par mil. d'hab.)
Grants of patents	...	118	443	1 076	352	5 127	5 372	Brevets délivrés
Patents in force	...	...	498	...	6 308	4 666	4 340	Brevets en vigueur
Argentina								**Argentine**
Resident filings (per mil. pop.)	...	19.0	27.0	14.0	18.0	16.0	12.0	Demandes de rés. (par mil. d'hab.)
Grants of patents	...	1 003	1 798	1 366	932	1 297	1 360	Brevets délivrés
Armenia								**Arménie**
Resident filings (per mil. pop.)	...	58.0	68.0	48.0	47.0	43.0	41.0	Demandes de rés. (par mil. d'hab.)
Grants of patents	...	52	126	124	117	99	108	Brevets délivrés
Patents in force	...	...	110	278	266	263	279	Brevets en vigueur
Australia								**Australie**
Resident filings (per mil. pop.)	...	99.0	125.0	109.0	116.0	132.0	85.0	Demandes de rés. (par mil. d'hab.)
Grants of patents	...	9 406	10 979	14 557	17 724	17 112	19 304	Brevets délivrés
Patents in force	...	...	96 403	96 293	112 176	122 811	128 407	Brevets en vigueur
Austria								**Autriche**
Resident filings (per mil. pop.)	300.0	217.0	404.0	497.0	491.0	490.0	475.0	Demandes de rés. (par mil. d'hab.)
Grants of patents	2 571	1 777	938	1 130	1 439	1 256	962	Brevets délivrés
Patents in force	...	...	10 126	10 066	10 715	110 202	118 494	Brevets en vigueur
Azerbaijan								**Azerbaïdjan**
Resident filings (per mil. pop.)	...	29.0	33.0	30.0	18.0	20.0	21.0	Demandes de rés. (par mil. d'hab.)
Grants of patents	...	9	195	126	111	78	97	Brevets délivrés
Patents in force	...	...	...	...	289	248	87	Brevets en vigueur
Bahamas								**Bahamas**
Resident filings (per mil. pop.)	13.0	...	...	...	8.0	3.0	5.0	Demandes de rés. (par mil. d'hab.)
Grants of patents	66	...	...	...	125	116	120	Brevets délivrés
Patents in force	...	...	...	...	...	...	1 536	Brevets en vigueur
Bahrain								**Bahreïn**
Resident filings (per mil. pop.)	...	...	...	...	8.0	...	5.0	Demandes de rés. (par mil. d'hab.)
Grants of patents	31	...	...	...	2	...	...	Brevets délivrés
Patents in force	...	...	...	...	2	123	117	Brevets en vigueur
Bangladesh								**Bangladesh**
Resident filings (per mil. pop.)	...	1.0	...	...	...	...	...	Demandes de rés. (par mil. d'hab.)
Grants of patents	118	80	182	92	153	134	121	Brevets délivrés
Patents in force	...	...	...	...	...	1 031	1 077	Brevets en vigueur
Barbados								**Barbade**
Resident filings (per mil. pop.)	...	...	...	...	...	11.0	3.0	Demandes de rés. (par mil. d'hab.)
Grants of patents	...	...	7	...	16	9	3	Brevets délivrés
Patents in force	...	...	49	57	...	...	...	Brevets en vigueur

Region, country or area	1985	1995	2005	2010	2012	2013	2014	Région, pays ou zone
Belarus								**Bélarus**
Resident filings (per mil. pop.)	...	61.0	121.0	197.0	189.0	167.0	82.0	Demandes de rés. (par mil. d'hab.)
Grants of patents	...	633	955	1 222	1 291	1 117	1 938	Brevets délivrés
Patents in force	...	...	...	4 444	...	4 478	5 176	Brevets en vigueur
Belgium								**Belgique**
Resident filings (per mil. pop.)	78.0	72.0	208.0	244.0	239.0	232.0	250.0	Demandes de rés. (par mil. d'hab.)
Grants of patents	1 976	1 216	708	532	795	745	373	Brevets délivrés
Patents in force	...	...	708	...	...	...	...	Brevets en vigueur
Belize								**Belize**
Grants of patents	...	...	...	...	4	10	28	Brevets délivrés
Patents in force	...	...	...	...	...	102	120	Brevets en vigueur
Bhutan								**Bhoutan**
Resident filings (per mil. pop.)	...	...	...	...	4.0	4.0	...	Demandes de rés. (par mil. d'hab.)
Grants of patents	...	...	...	...	2	...	...	Brevets délivrés
Patents in force	...	...	...	...	...	2	...	Brevets en vigueur
Bolivia (Plurinational State of)								**Bolivie (État plurinational de)**
Resident filings (per mil. pop.)	1.0	2.0	...	...	...	...	1.0	Demandes de rés. (par mil. d'hab.)
Grants of patents	62	47	...	...	...	...	97	Brevets délivrés
Patents in force	...	...	...	...	...	...	601	Brevets en vigueur
Bosnia and Herzegovina								**Bosnie-Herzégovine**
Resident filings (per mil. pop.)	...	...	17.0	15.0	1.0	2.0	11.0	Demandes de rés. (par mil. d'hab.)
Grants of patents	...	...	46	173	57	31	5	Brevets délivrés
Patents in force	...	...	120	716	734	583	503	Brevets en vigueur
Botswana								**Botswana**
Resident filings (per mil. pop.)	...	...	...	...	2.0	4.0	2.0	Demandes de rés. (par mil. d'hab.)
Grants of patents	...	...	...	...	1	3	...	Brevets délivrés
Patents in force	...	...	...	...	...	...	883	Brevets en vigueur
Brazil								**Brésil**
Resident filings (per mil. pop.)	14.0	17.0	22.0	22.0	24.0	25.0	23.0	Demandes de rés. (par mil. d'hab.)
Grants of patents	3 934	2 659	2 439	3 251	2 830	2 972	2 749	Brevets délivrés
Patents in force	...	...	32 571	40 022	39 592	35 517	24 976	Brevets en vigueur
Brunei Darussalam								**Brunéi Darussalam**
Resident filings (per mil. pop.)	...	...	...	...	49.0	48.0	61.0	Demandes de rés. (par mil. d'hab.)
Grants of patents	...	42	26	40	86	93	71	Brevets délivrés
Patents in force	...	...	...	...	...	119	...	Brevets en vigueur
Bulgaria								**Bulgarie**
Resident filings (per mil. pop.)	...	44.0	35.0	34.0	35.0	42.0	35.0	Demandes de rés. (par mil. d'hab.)
Grants of patents	130	375	313	251	101	125	72	Brevets délivrés
Patents in force	...	...	2 203	6 812	1 519	1 431	1 324	Brevets en vigueur
Burundi								**Burundi**
Grants of patents	...	1	...	...	...	...	...	Brevets délivrés
Canada								**Canada**
Resident filings (per mil. pop.)	81.0	83.0	160.0	134.0	135.0	130.0	118.0	Demandes de rés. (par mil. d'hab.)
Grants of patents	18 697	9 139	15 516	19 120	21 819	23 833	23 749	Brevets délivrés
Patents in force	...	...	125 110	133 355	144 363	153 781	161 442	Brevets en vigueur
Chile								**Chili**
Resident filings (per mil. pop.)	10.0	12.0	22.0	19.0	19.0	19.0	25.0	Demandes de rés. (par mil. d'hab.)
Grants of patents	448	133	311	1 020	770	898	1 168	Brevets délivrés
Patents in force	...	...	...	8 121	8 981	9 585	9 987	Brevets en vigueur
China [1]								**Chine** [1]
Resident filings (per mil. pop.)	4.0	8.0	72.0	219.0	396.0	519.0	587.0	Demandes de rés. (par mil. d'hab.)
Grants of patents	44	3 393	53 305	135 110	217 105	207 688	233 228	Brevets délivrés
Patents in force	...	...	182 396	564 760	875 385	1 033 908	1 196 497	Brevets en vigueur
China, Hong Kong SAR								**Chine, Hong Kong RAS**
Resident filings (per mil. pop.)	3.0	4.0	23.0	19.0	24.0	31.0	27.0	Demandes de rés. (par mil. d'hab.)
Grants of patents	1 030	1 960	6 518	5 353	5 035	6 564	5 932	Brevets délivrés
Patents in force	...	...	...	33 225	36 158	38 858	40 865	Brevets en vigueur
China, Macao SAR								**Chine, Macao RAS**
Resident filings (per mil. pop.)	...	...	6.0	7.0	9.0	11.0	3.0	Demandes de rés. (par mil. d'hab.)
Grants of patents	...	2	5	156	29	22	16	Brevets délivrés
Patents in force	...	...	12	377	435	442	451	Brevets en vigueur

Patents *(continued)*
Resident filings (per million population), grants and patents in force
Brevets *(suite)*
Demandes émanant de résidents (par million d'habitants), délivrances et brevets en vigueur

Region, country or area	1985	1995	2005	2010	2012	2013	2014	Région, pays ou zone
Colombia								**Colombie**
Resident filings (per mil. pop.)	2.0	4.0	2.0	3.0	4.0	5.0	5.0	Demandes de rés. (par mil. d'hab.)
Grants of patents	169	365	256	639	1 667	2 264	1 212	Brevets délivrés
Patents in force	...	...	...	...	4 172	5 967	6 710	Brevets en vigueur
Congo								**Congo**
Grants of patents	...	15	...	...	...	...	...	Brevets délivrés
Costa Rica								**Costa Rica**
Resident filings (per mil. pop.)	...	...	...	2.0	2.0	4.0	3.0	Demandes de rés. (par mil. d'hab.)
Grants of patents	...	...	...	45	65	106	114	Brevets délivrés
Patents in force	...	...	...	239	313	417	518	Brevets en vigueur
Côte d'Ivoire								**Côte d'Ivoire**
Resident filings (per mil. pop.)	...	...	...	...	3.0	...	...	Demandes de rés. (par mil. d'hab.)
Croatia								**Croatie**
Resident filings (per mil. pop.)	...	57.0	82.0	62.0	55.0	56.0	43.0	Demandes de rés. (par mil. d'hab.)
Grants of patents	...	25	140	82	155	159	90	Brevets délivrés
Patents in force	...	...	1 094	2 134	3 379	4 243	4 838	Brevets en vigueur
Cuba								**Cuba**
Resident filings (per mil. pop.)	...	10.0	9.0	...	3.0	2.0	2.0	Demandes de rés. (par mil. d'hab.)
Grants of patents	18	77	64	...	84	125	94	Brevets délivrés
Patents in force	...	...	653	...	1 417	972	927	Brevets en vigueur
Cyprus								**Chypre**
Resident filings (per mil. pop.)	...	...	53.0	35.0	46.0	40.0	44.0	Demandes de rés. (par mil. d'hab.)
Grants of patents	43	...	68	19	5	1	...	Brevets délivrés
Patents in force	...	...	3 521	333	246	82	149	Brevets en vigueur
Czech Republic								**République tchèque**
Resident filings (per mil. pop.)	...	61.0	65.0	99.0	96.0	108.0	102.0	Demandes de rés. (par mil. d'hab.)
Grants of patents	...	1 299	1 551	911	668	611	688	Brevets délivrés
Patents in force	...	...	10 165	9 633	8 608	7 780	7 157	Brevets en vigueur
Dem. P. R. Korea								**R. p. dém. de Corée**
Resident filings (per mil. pop.)	...	...	246.0	327.0	337.0	...	...	Demandes de rés. (par mil. d'hab.)
Grants of patents	...	...	3 583	6 290	6 550	...	...	Brevets délivrés
Denmark								**Danemark**
Resident filings (per mil. pop.)	167.0	236.0	523.0	625.0	539.0	583.0	596.0	Demandes de rés. (par mil. d'hab.)
Grants of patents	1 054	1 120	389	155	190	309	292	Brevets délivrés
Patents in force	...	...	56 978	47 732	47 085	51 277	51 345	Brevets en vigueur
Djibouti								**Djibouti**
Resident filings (per mil. pop.)	...	...	...	...	...	1.0	...	Demandes de rés. (par mil. d'hab.)
Dominican Republic								**Rép. dominicaine**
Resident filings (per mil. pop.)	...	...	...	...	2.0	1.0	1.0	Demandes de rés. (par mil. d'hab.)
Grants of patents	...	...	...	...	89	44	62	Brevets délivrés
Patents in force	...	...	...	...	201	229	294	Brevets en vigueur
Ecuador								**Équateur**
Resident filings (per mil. pop.)	...	1.0	1.0	...	...	...	...	Demandes de rés. (par mil. d'hab.)
Grants of patents	...	90	38	28	...	...	...	Brevets délivrés
Patents in force	...	...	38	199	...	...	...	Brevets en vigueur
Egypt								**Égypte**
Resident filings (per mil. pop.)	3.0	7.0	6.0	8.0	8.0	8.0	9.0	Demandes de rés. (par mil. d'hab.)
Grants of patents	298	346	147	321	634	465	415	Brevets délivrés
Patents in force	...	...	...	3 316	3 377	3 553	4 012	Brevets en vigueur
El Salvador								**El Salvador**
Resident filings (per mil. pop.)	3.0	1.0	...	...	...	...	...	Demandes de rés. (par mil. d'hab.)
Grants of patents	70	61	...	...	33	48	77	Brevets délivrés
Patents in force	...	...	...	...	...	...	1 642	Brevets en vigueur
Estonia								**Estonie**
Resident filings (per mil. pop.)	...	11.0	19.0	83.0	46.0	50.0	61.0	Demandes de rés. (par mil. d'hab.)
Grants of patents	...	...	163	120	116	78	38	Brevets délivrés
Patents in force	...	...	1 395	1 320	1 276	1 228	1 089	Brevets en vigueur
Ethiopia								**Éthiopie**
Grants of patents	...	...	9	...	...	...	...	Brevets délivrés
Fiji								**Fidji**
Grants of patents	15	...	...	...	...	...	...	Brevets délivrés

Region, country or area	1985	1995	2005	2010	2012	2013	2014	Région, pays ou zone
Finland								**Finlande**
Resident filings (per mil. pop.)	352.0	403.0	637.0	628.0	665.0	642.0	662.0	Demandes de rés. (par mil. d'hab.)
Grants of patents	2 160	2 347	1 757	923	836	711	787	Brevets délivrés
Patents in force	...	...	39 450	12 221	46 854	47 058	47 344	Brevets en vigueur
France								**France**
Resident filings (per mil. pop.)	212.0	209.0	354.0	373.0	372.0	372.0	379.0	Demandes de rés. (par mil. d'hab.)
Grants of patents	24 195	17 918	11 473	9 899	12 913	11 405	11 889	Brevets délivrés
Patents in force	...	...	343 568	435 915	490 941	500 114	510 490	Brevets en vigueur
Gambia								**Gambie**
Grants of patents	...	...	2	...	...	...	...	Brevets délivrés
Georgia								**Géorgie**
Resident filings (per mil. pop.)	...	61.0	52.0	41.0	31.0	25.0	24.0	Demandes de rés. (par mil. d'hab.)
Grants of patents	...	133	320	258	346	286	209	Brevets délivrés
Patents in force	...	...	1 040	1 044	1 501	2 050	1 486	Brevets en vigueur
Germany								**Allemagne**
Resident filings (per mil. pop.)	415.0	467.0	875.0	910.0	919.0	917.0	913.0	Demandes de rés. (par mil. d'hab.)
Grants of patents	19 500	16 000	17 063	13 678	11 332	13 858	15 030	Brevets délivrés
Patents in force	...	...	434 663	514 046	549 521	569 340	576 273	Brevets en vigueur
Greece								**Grèce**
Resident filings (per mil. pop.)	113.0	25.0	48.0	73.0	64.0	69.0	68.0	Demandes de rés. (par mil. d'hab.)
Grants of patents	3 294	350	320	479	291	282	316	Brevets délivrés
Patents in force	...	...	...	32 120	3 491	2 966	3 239	Brevets en vigueur
Guatemala								**Guatemala**
Resident filings (per mil. pop.)	9.0	3.0	1.0	...	...	...	1.0	Demandes de rés. (par mil. d'hab.)
Grants of patents	166	22	104	104	86	65	105	Brevets délivrés
Patents in force	...	...	...	590	681	746	840	Brevets en vigueur
Guyana								**Guyana**
Resident filings (per mil. pop.)	1.0	...	...	...	...	...	...	Demandes de rés. (par mil. d'hab.)
Grants of patents	22	...	...	...	...	...	...	Brevets délivrés
Patents in force	...	...	...	...	...	...	1 442	Brevets en vigueur
Haiti								**Haïti**
Resident filings (per mil. pop.)	1.0	...	...	...	...	...	...	Demandes de rés. (par mil. d'hab.)
Grants of patents	9	3	11	10	...	...	...	Brevets délivrés
Honduras								**Honduras**
Resident filings (per mil. pop.)	3.0	1.0	...	...	1.0	1.0	...	Demandes de rés. (par mil. d'hab.)
Grants of patents	19	...	85	81	136	132	94	Brevets délivrés
Patents in force	...	...	...	...	241	...	...	Brevets en vigueur
Hungary								**Hongrie**
Resident filings (per mil. pop.)	273.0	106.0	78.0	75.0	80.0	75.0	67.0	Demandes de rés. (par mil. d'hab.)
Grants of patents	2 095	1 910	1 126	65	477	1 351	376	Brevets délivrés
Patents in force	...	...	9 125	2 586	5 167	5 237	4 695	Brevets en vigueur
Iceland								**Islande**
Resident filings (per mil. pop.)	87.0	71.0	253.0	343.0	271.0	222.0	281.0	Demandes de rés. (par mil. d'hab.)
Grants of patents	21	11	101	139	47	43	54	Brevets délivrés
Patents in force	...	...	349	1 892	3 327	603	567	Brevets en vigueur
India								**Inde**
Resident filings (per mil. pop.)	1.0	2.0	4.0	7.0	8.0	9.0	9.0	Demandes de rés. (par mil. d'hab.)
Grants of patents	1 814	1 613	4 320	7 138	4 328	3 377	6 153	Brevets délivrés
Patents in force	...	...	16 419	47 224	42 991	45 103	49 272	Brevets en vigueur
Indonesia								**Indonésie**
Resident filings (per mil. pop.)	...	...	1.0	2.0	...	3.0	3.0	Demandes de rés. (par mil. d'hab.)
Patents in force	...	...	...	...	...	22 564	...	Brevets en vigueur
Iran (Islamic Rep. of)								**Iran (Rép. islamique d')**
Resident filings (per mil. pop.)	4.0	5.0	58.0	149.0	139.0	146.0	174.0	Demandes de rés. (par mil. d'hab.)
Grants of patents	339	166	2 890	5 372	5 681	3 476	3 060	Brevets délivrés
Patents in force	...	...	...	...	4 599	3 440	...	Brevets en vigueur
Iraq								**Iraq**
Resident filings (per mil. pop.)	19.0	4.0	...	...	...	...	...	Demandes de rés. (par mil. d'hab.)
Grants of patents	103	32	...	...	...	...	...	Brevets délivrés

21 Patents *(continued)*
Resident filings (per million population), grants and patents in force
Brevets *(suite)*
Demandes émanant de résidents (par million d'habitants), délivrances et brevets en vigueur

Region, country or area	1985	1995	2005	2010	2012	2013	2014 Région, pays ou zone
Ireland							**Irlande**
Resident filings (per mil. pop.)	205.0	233.0	264.0	274.0	231.0	192.0	191.0 Demandes de rés. (par mil. d'hab.)
Grants of patents	1 042	3 525	349	243	190	214	148 Brevets délivrés
Patents in force	...	...	...	79 040	96 583	108 218	111 109 Brevets en vigueur
Israel							**Israël**
Resident filings (per mil. pop.)	187.0	228.0	241.0	190.0	167.0	149.0	137.0 Demandes de rés. (par mil. d'hab.)
Grants of patents	1 636	2 029	2 269	2 293	3 386	3 697	3 984 Brevets délivrés
Patents in force	...	...	...	26 494	...	25 372	... Brevets en vigueur
Italy							**Italie**
Resident filings (per mil. pop.)	35.0	...	...	219.0	205.0	200.0	200.0 Demandes de rés. (par mil. d'hab.)
Grants of patents	...	9 164	5 534	16 106	5 625	8 114	7 795 Brevets délivrés
Patents in force	...	...	...	65 417	60 563	61 341	63 071 Brevets en vigueur
Jamaica							**Jamaïque**
Resident filings (per mil. pop.)	...	3.0	4.0	5.0	9.0	8.0	12.0 Demandes de rés. (par mil. d'hab.)
Grants of patents	...	4	...	...	20	36	28 Brevets délivrés
Patents in force	...	...	527	...	...	296	324 Brevets en vigueur
Japan							**Japon**
Resident filings (per mil. pop.)	2 272.0	2 661.0	2 880.0	2 265.0	2 250.0	2 134.0	2 092.0 Demandes de rés. (par mil. d'hab.)
Grants of patents	50 100	109 100	122 944	222 693	274 791	277 079	227 142 Brevets délivrés
Patents in force	...	...	1 123 055	1 423 432	1 694 435	1 838 177	1 920 490 Brevets en vigueur
Jordan							**Jordanie**
Resident filings (per mil. pop.)	...	...	9.0	7.0	8.0	5.0	6.0 Demandes de rés. (par mil. d'hab.)
Grants of patents	...	...	55	64	48	48	115 Brevets délivrés
Patents in force	...	...	...	312	346	317	377 Brevets en vigueur
Kazakhstan							**Kazakhstan**
Resident filings (per mil. pop.)	...	65.0	101.0	105.0	...	111.0	105.0 Demandes de rés. (par mil. d'hab.)
Grants of patents	...	1 281	...	1 868	...	1 500	1 504 Brevets délivrés
Patents in force	...	...	...	581	...	377	5 184 Brevets en vigueur
Kenya							**Kenya**
Resident filings (per mil. pop.)	...	...	1.0	2.0	3.0	3.0	3.0 Demandes de rés. (par mil. d'hab.)
Grants of patents	98	...	48	54	76	71	53 Brevets délivrés
Kiribati							**Kiribati**
Resident filings (per mil. pop.)	...	...	...	...	...	176.0	... Demandes de rés. (par mil. d'hab.)
Grants of patents	1	...	...	...	...	...	... Brevets délivrés
Kyrgyzstan							**Kirghizistan**
Resident filings (per mil. pop.)	...	26.0	...	26.0	20.0	20.0	23.0 Demandes de rés. (par mil. d'hab.)
Grants of patents	...	133	...	109	103	88	100 Brevets délivrés
Patents in force	...	...	...	112	341	348	375 Brevets en vigueur
Latvia							**Lettonie**
Resident filings (per mil. pop.)	...	85.0	53.0	101.0	107.0	152.0	56.0 Demandes de rés. (par mil. d'hab.)
Grants of patents	...	629	122	184	154	136	141 Brevets délivrés
Patents in force	...	...	4 012	5 680	6 833	6 329	6 763 Brevets en vigueur
Lebanon							**Liban**
Grants of patents	...	...	...	...	317	316	... Brevets délivrés
Lesotho							**Lesotho**
Resident filings (per mil. pop.)	...	5.0	...	...	...	...	... Demandes de rés. (par mil. d'hab.)
Grants of patents	...	7	...	...	...	...	... Brevets délivrés
Libya							**Libye**
Resident filings (per mil. pop.)	...	1.0	...	...	...	...	... Demandes de rés. (par mil. d'hab.)
Liechtenstein							**Liechtenstein**
Resident filings (per mil. pop.)	...	...	4 380.0	8 033.0	5 368.0	6 775.0	7 473.0 Demandes de rés. (par mil. d'hab.)
Lithuania							**Lituanie**
Resident filings (per mil. pop.)	...	29.0	21.0	38.0	43.0	47.0	50.0 Demandes de rés. (par mil. d'hab.)
Grants of patents	...	494	116	84	92	93	120 Brevets délivrés
Patents in force	...	...	768	642	599	519	520 Brevets en vigueur
Luxembourg							**Luxembourg**
Resident filings (per mil. pop.)	221.0	86.0	441.0	990.0	946.0	948.0	1 054.0 Demandes de rés. (par mil. d'hab.)
Grants of patents	418	...	29	87	112	...	152 Brevets délivrés
Patents in force	...	...	...	21 346	21 267	20 421	19 360 Brevets en vigueur

Region, country or area	1985	1995	2005	2010	2012	2013	2014	Région, pays ou zone
Madagascar								**Madagascar**
Resident filings (per mil. pop.)	...	2.0	...	...	...	...	...	Demandes de rés. (par mil. d'hab.)
Grants of patents	...	25	32	55	44	40	24	Brevets délivrés
Patents in force	...	...	249	387	439	514	390	Brevets en vigueur
Malawi								**Malawi**
Grants of patents	43	23	...	...	...	...	...	Brevets délivrés
Malaysia								**Malaisie**
Resident filings (per mil. pop.)	1.0	7.0	20.0	44.0	38.0	40.0	45.0	Demandes de rés. (par mil. d'hab.)
Grants of patents	1 150	1 753	2 508	2 160	2 460	2 660	2 705	Brevets délivrés
Patents in force	...	...	...	20 908	21 447	22 782	21 568	Brevets en vigueur
Malta								**Malte**
Resident filings (per mil. pop.)	...	30.0	...	104.0	74.0	132.0	150.0	Demandes de rés. (par mil. d'hab.)
Grants of patents	20	19	...	4	11	15	4	Brevets délivrés
Patents in force	...	...	...	832	615	560	490	Brevets en vigueur
Mauritius								**Maurice**
Resident filings (per mil. pop.)	4.0	3.0	...	...	3.0	2.0	...	Demandes de rés. (par mil. d'hab.)
Grants of patents	4	3	...	8	6	5	...	Brevets délivrés
Mexico								**Mexique**
Resident filings (per mil. pop.)	8.0	5.0	5.0	8.0	11.0	10.0	10.0	Demandes de rés. (par mil. d'hab.)
Grants of patents	977	3 538	8 098	9 399	12 358	10 368	9 819	Brevets délivrés
Patents in force	...	...	48 374	82 017	96 962	101 645	106 340	Brevets en vigueur
Monaco								**Monaco**
Resident filings (per mil. pop.)	521.0	423.0	503.0	543.0	612.0	714.0	735.0	Demandes de rés. (par mil. d'hab.)
Grants of patents	66	36	9	5	6	5	5	Brevets délivrés
Patents in force	...	...	37 483	53 859	42 838	41 976	53 893	Brevets en vigueur
Mongolia								**Mongolie**
Resident filings (per mil. pop.)	...	57.0	40.0	41.0	...	47.0	48.0	Demandes de rés. (par mil. d'hab.)
Grants of patents	5	117	197	96	...	212	216	Brevets délivrés
Patents in force	...	...	13 663	2 645	...	869	...	Brevets en vigueur
Montenegro								**Monténégro**
Resident filings (per mil. pop.)	...	...	...	37.0	60.0	37.0	21.0	Demandes de rés. (par mil. d'hab.)
Grants of patents	...	...	...	264	291	121	14	Brevets délivrés
Patents in force	...	...	...	264	858	1 448	1 933	Brevets en vigueur
Morocco								**Maroc**
Resident filings (per mil. pop.)	2.0	3.0	5.0	5.0	6.0	10.0	11.0	Demandes de rés. (par mil. d'hab.)
Grants of patents	313	354	556	808	979	937	...	Brevets délivrés
Patents in force	...	...	9 872	...	...	...	...	Brevets en vigueur
Mozambique								**Mozambique**
Grants of patents	...	...	14	...	...	...	...	Brevets délivrés
Nepal								**Népal**
Resident filings (per mil. pop.)	...	...	...	...	...	1.0	...	Demandes de rés. (par mil. d'hab.)
Grants of patents	1	1	3	...	2	1	...	Brevets délivrés
Patents in force	...	...	...	...	72	72	72	Brevets en vigueur
Netherlands								**Pays-Bas**
Resident filings (per mil. pop.)	134.0	137.0	614.0	511.0	444.0	485.0	543.0	Demandes de rés. (par mil. d'hab.)
Grants of patents	2 145	673	2 373	1 947	1 895	2 029	1 722	Brevets délivrés
Patents in force	...	...	14 091	...	12 753	12 704	12 518	Brevets en vigueur
New Zealand								**Nouvelle-Zélande**
Resident filings (per mil. pop.)	310.0	350.0	458.0	364.0	323.0	363.0	363.0	Demandes de rés. (par mil. d'hab.)
Grants of patents	1 732	2 641	4 189	4 347	6 152	4 752	4 677	Brevets délivrés
Patents in force	...	...	34 182	34 800	27 222	28 217	28 854	Brevets en vigueur
Nicaragua								**Nicaragua**
Resident filings (per mil. pop.)	1.0	...	...	...	1.0	...	...	Demandes de rés. (par mil. d'hab.)
Grants of patents	25	1	...	...	68	72	62	Brevets délivrés
Patents in force	...	...	...	...	260	328	387	Brevets en vigueur
Nigeria								**Nigéria**
Grants of patents	...	...	...	...	528	645	...	Brevets délivrés
Norway								**Norvège**
Resident filings (per mil. pop.)	222.0	259.0	247.0	334.0	312.0	317.0	318.0	Demandes de rés. (par mil. d'hab.)
Grants of patents	2 165	2 014	542	1 631	1 310	1 430	1 413	Brevets délivrés
Patents in force	...	...	...	12 755	15 396	19 297	21 882	Brevets en vigueur

Patents *(continued)*
Resident filings (per million population), grants and patents in force

Brevets *(suite)*
Demandes émanant de résidents (par million d'habitants), délivrances et brevets en vigueur

Region, country or area	1985	1995	2005	2010	2012	2013	2014	Région, pays ou zone
Pakistan								**Pakistan**
Resident filings (per mil. pop.)	...	...	1.0	1.0	1.0	1.0	1.0	Demandes de rés. (par mil. d'hab.)
Grants of patents	...	474	393	238	312	282	185	Brevets délivrés
Panama								**Panama**
Resident filings (per mil. pop.)	6.0	6.0	...	...	...	2.0	3.0	Demandes de rés. (par mil. d'hab.)
Grants of patents	72	80	228	378	325	266	166	Brevets délivrés
Patents in force	...	...	...	378	357	1 858	1 725	Brevets en vigueur
Papua New Guinea								**Papouasie-Nvl-Guinée**
Grants of patents	...	...	...	...	...	57	...	Brevets délivrés
Patents in force	...	...	...	...	...	42	...	Brevets en vigueur
Paraguay								**Paraguay**
Resident filings (per mil. pop.)	2.0	...	4.0	3.0	...	...	...	Demandes de rés. (par mil. d'hab.)
Grants of patents	8	...	...	...	...	...	...	Brevets délivrés
Peru								**Pérou**
Resident filings (per mil. pop.)	2.0	...	1.0	1.0	2.0	2.0	3.0	Demandes de rés. (par mil. d'hab.)
Grants of patents	148	...	388	365	431	287	332	Brevets délivrés
Patents in force	...	...	2 252	2 435	2 616	2 615	2 651	Brevets en vigueur
Philippines								**Philippines**
Resident filings (per mil. pop.)	2.0	2.0	2.0	2.0	2.0	2.0	3.0	Demandes de rés. (par mil. d'hab.)
Grants of patents	1 281	589	1 642	1 153	1 111	2 207	2 159	Brevets délivrés
Patents in force	...	...	...	52 527	...	...	...	Brevets en vigueur
Poland								**Pologne**
Resident filings (per mil. pop.)	138.0	67.0	56.0	90.0	126.0	121.0	116.0	Demandes de rés. (par mil. d'hab.)
Grants of patents	4 467	2 608	2 522	3 004	2 484	2 804	2 852	Brevets délivrés
Patents in force	...	...	14 578	30 021	41 242	47 610	53 183	Brevets en vigueur
Portugal								**Portugal**
Resident filings (per mil. pop.)	8.0	8.0	19.0	55.0	66.0	71.0	80.0	Demandes de rés. (par mil. d'hab.)
Grants of patents	960	960	231	140	112	130	97	Brevets délivrés
Patents in force	...	...	35 871	39 076	37 612	36 782	35 561	Brevets en vigueur
Qatar								**Qatar**
Resident filings (per mil. pop.)	...	...	...	...	3.0	7.0	4.0	Demandes de rés. (par mil. d'hab.)
Republic of Korea								**République de Corée**
Resident filings (per mil. pop.)	66.0	1 313.0	2 538.0	2 668.0	2 962.0	3 186.0	3 254.0	Demandes de rés. (par mil. d'hab.)
Grants of patents	2 268	12 512	73 512	68 843	113 467	127 330	129 786	Brevets délivrés
Patents in force	...	...	420 906	640 412	738 312	812 595	885 959	Brevets en vigueur
Republic of Moldova								**République de Moldova**
Resident filings (per mil. pop.)	...	73.0	105.0	42.0	26.0	19.0	19.0	Demandes de rés. (par mil. d'hab.)
Grants of patents	...	227	269	132	51	61	54	Brevets délivrés
Patents in force	...	...	1 108	1 018	635	471	384	Brevets en vigueur
Romania								**Roumanie**
Resident filings (per mil. pop.)	185.0	80.0	43.0	69.0	53.0	51.0	49.0	Demandes de rés. (par mil. d'hab.)
Grants of patents	2 786	1 860	759	447	384	451	356	Brevets délivrés
Patents in force	...	...	8 627	2 915	15 284	17 100	17 268	Brevets en vigueur
Russian Federation								**Fédération de Russie**
Resident filings (per mil. pop.)	...	118.0	166.0	203.0	204.0	203.0	169.0	Demandes de rés. (par mil. d'hab.)
Grants of patents	...	25 633	23 390	30 322	32 880	31 638	33 950	Brevets délivrés
Patents in force	...	...	123 089	181 904	181 515	194 248	208 320	Brevets en vigueur
Rwanda								**Rwanda**
Resident filings (per mil. pop.)	...	...	...	...	3.0	...	...	Demandes de rés. (par mil. d'hab.)
Grants of patents	1	...	3	...	24	...	...	Brevets délivrés
Patents in force	...	...	...	...	119	...	135	Brevets en vigueur
Saint Vincent-Grenadines								**Saint-Vincent-Grenadines**
Patents in force	...	...	...	...	...	28	...	Brevets en vigueur
Samoa								**Samoa**
Resident filings (per mil. pop.)	...	...	...	...	...	...	5.0	Demandes de rés. (par mil. d'hab.)
Grants of patents	4	2	...	...	126	...	...	Brevets délivrés
Patents in force	...	...	...	...	99	...	96	Brevets en vigueur
Saudi Arabia								**Arabie saoudite**
Resident filings (per mil. pop.)	...	2.0	6.0	12.0	...	25.0	32.0	Demandes de rés. (par mil. d'hab.)
Grants of patents	...	3	225	194	...	233	561	Brevets délivrés
Patents in force	...	...	...	...	...	1 988	2 338	Brevets en vigueur

Patents *(continued)*
Resident filings (per million population), grants and patents in force

Brevets *(suite)*
Demandes émanant de résidents (par million d'habitants), délivrances et brevets en vigueur

Region, country or area	1985	1995	2005	2010	2012	2013	2014	Région, pays ou zone
Serbia								**Serbie**
Resident filings (per mil. pop.)	...	...	50.0	40.0	27.0	29.0	30.0	Demandes de rés. (par mil. d'hab.)
Grants of patents	...	...	265	427	167	136	105	Brevets délivrés
Patents in force	...	...	...	1 477	2 303	2 644	2 964	Brevets en vigueur
Serbia and Montenegro [former]								**Serbie-et-Monténégro [anc.]**
Grants of patents	1 053	510	...	...	...	...	...	Brevets délivrés
Seychelles								**Seychelles**
Grants of patents	2	1	...	...	...	...	...	Brevets délivrés
Sierra Leone								**Sierra Leone**
Grants of patents	...	5	...	...	...	...	...	Brevets délivrés
Singapore								**Singapour**
Resident filings (per mil. pop.)	1.0	41.0	133.0	176.0	203.0	212.0	238.0	Demandes de rés. (par mil. d'hab.)
Grants of patents	416	1 750	7 530	4 442	5 633	5 575	5 538	Brevets délivrés
Patents in force	...	...	43 024	43 591	45 209	45 999	47 422	Brevets en vigueur
Slovakia								**Slovaquie**
Resident filings (per mil. pop.)	...	50.0	32.0	48.0	38.0	39.0	44.0	Demandes de rés. (par mil. d'hab.)
Grants of patents	...	381	560	376	161	115	94	Brevets délivrés
Patents in force	...	...	4 033	3 593	3 174	2 755	2 357	Brevets en vigueur
Slovenia								**Slovénie**
Resident filings (per mil. pop.)	...	158.0	215.0	282.0	...	...	...	Demandes de rés. (par mil. d'hab.)
Grants of patents	...	380	285	250	...	...	...	Brevets délivrés
Patents in force	...	...	5 201	1 485	...	...	...	Brevets en vigueur
Solomon Islands								**Îles Salomon**
Grants of patents	4	...	...	...	...	...	...	Brevets délivrés
Somalia								**Somalie**
Grants of patents	7	...	...	...	...	...	...	Brevets délivrés
South Africa								**Afrique du Sud**
Resident filings (per mil. pop.)	129.0	23.0	21.0	16.0	12.0	12.0	15.0	Demandes de rés. (par mil. d'hab.)
Grants of patents	6 768	5 113	1 831	5 331	6 205	4 756	5 065	Brevets délivrés
Patents in force	...	...	...	...	...	54 220	55 031	Brevets en vigueur
Spain								**Espagne**
Resident filings (per mil. pop.)	56.0	52.0	92.0	107.0	103.0	97.0	95.0	Demandes de rés. (par mil. d'hab.)
Grants of patents	9 115	686	2 769	2 773	2 720	3 004	3 235	Brevets délivrés
Patents in force	...	...	39 297	31 804	35 616	36 893	37 581	Brevets en vigueur
Sri Lanka								**Sri Lanka**
Resident filings (per mil. pop.)	2.0	4.0	8.0	11.0	...	16.0	...	Demandes de rés. (par mil. d'hab.)
Grants of patents	112	159	180	504	126	236	...	Brevets délivrés
Sudan								**Soudan**
Grants of patents	...	...	174	...	84	...	8	Brevets délivrés
Patents in force	...	...	...	...	27	...	...	Brevets en vigueur
Swaziland								**Swaziland**
Resident filings (per mil. pop.)	...	...	...	...	2.0	...	...	Demandes de rés. (par mil. d'hab.)
Grants of patents	30	...	...	...	3	...	...	Brevets délivrés
Patents in force	...	...	...	...	...	9	...	Brevets en vigueur
Sweden								**Suède**
Resident filings (per mil. pop.)	460.0	446.0	555.0	614.0	605.0	625.0	604.0	Demandes de rés. (par mil. d'hab.)
Grants of patents	5 681	1 541	1 911	1 380	999	685	588	Brevets délivrés
Patents in force	...	...	102 741	96 796	96 252	95 695	93 348	Brevets en vigueur
Switzerland								**Suisse**
Resident filings (per mil. pop.)	493.0	410.0	897.0	1 069.0	1 013.0	1 012.0	1 018.0	Demandes de rés. (par mil. d'hab.)
Grants of patents	6 421	1 303	...	741	455	534	677	Brevets délivrés
Patents in force	...	...	99 531	123 033	148 020	148 759	144 859	Brevets en vigueur
Syrian Arab Republic								**Rép. arabe syrienne**
Resident filings (per mil. pop.)	...	9.0	6.0	...	...	...	...	Demandes de rés. (par mil. d'hab.)
Grants of patents	...	71	72	...	...	...	...	Brevets délivrés
Tajikistan								**Tadjikistan**
Resident filings (per mil. pop.)	...	6.0	4.0	1.0	1.0	...	...	Demandes de rés. (par mil. d'hab.)
Grants of patents	...	47	...	3	1	2	...	Brevets délivrés
Patents in force	...	...	...	248	254	256	...	Brevets en vigueur

Region, country or area	1985	1995	2005	2010	2012	2013	2014	Région, pays ou zone
Thailand								**Thaïlande**
Resident filings (per mil. pop.)	1.0	2.0	14.0	18.0	15.0	23.0	15.0	Demandes de rés. (par mil. d'hab.)
Grants of patents	45	470	553	772	1 008	1 149	1 286	Brevets délivrés
Patents in force	...	...	...	10 201	11 065	11 211	11 623	Brevets en vigueur
TFYR of Macedonia								**ex-R.Y. de Macédoine**
Resident filings (per mil. pop.)	...	51.0	25.0	13.0	24.0	20.0	...	Demandes de rés. (par mil. d'hab.)
Grants of patents	...	163	373	406	520	378	...	Brevets délivrés
Trinidad and Tobago								**Trinité-et-Tobago**
Resident filings (per mil. pop.)	...	19.0	1.0	...	2.0	1.0	1.0	Demandes de rés. (par mil. d'hab.)
Grants of patents	...	87	...	...	57	34	39	Brevets délivrés
Tunisia								**Tunisie**
Resident filings (per mil. pop.)	2.0	3.0	6.0	11.0	14.0	10.0	13.0	Demandes de rés. (par mil. d'hab.)
Grants of patents	...	141	...	...	613	535	552	Brevets délivrés
Patents in force	...	...	...	...	...	3 685	...	Brevets en vigueur
Turkey								**Turquie**
Resident filings (per mil. pop.)	3.0	3.0	15.0	48.0	65.0	64.0	68.0	Demandes de rés. (par mil. d'hab.)
Grants of patents	385	763	823	...	1 004	1 211	1 276	Brevets délivrés
Patents in force	...	...	...	...	36 577	44 867	53 908	Brevets en vigueur
Uganda								**Ouganda**
Grants of patents	26	...	...	...	1	3	1	Brevets délivrés
Patents in force	...	...	...	...	...	30	26	Brevets en vigueur
Ukraine								**Ukraine**
Resident filings (per mil. pop.)	...	93.0	75.0	56.0	55.0	63.0	54.0	Demandes de rés. (par mil. d'hab.)
Grants of patents	...	1 350	3 719	3 874	3 405	3 635	3 319	Brevets délivrés
Patents in force	...	...	37 336	24 622	25 275	26 033	26 183	Brevets en vigueur
United Arab Emirates								**Émirats arabes unis**
Resident filings (per mil. pop.)	...	...	...	...	3.0	3.0	5.0	Demandes de rés. (par mil. d'hab.)
Grants of patents	...	...	...	...	40	63	110	Brevets délivrés
Patents in force	...	...	...	...	...	451	561	Brevets en vigueur
United Kingdom								**Royaume-Uni**
Resident filings (per mil. pop.)	348.0	321.0	372.0	333.0	316.0	305.0	309.0	Demandes de rés. (par mil. d'hab.)
Grants of patents	20 880	9 473	10 159	5 594	6 864	5 235	4 986	Brevets délivrés
Patents in force	...	...	...	424 209	459 447	469 941	498 904	Brevets en vigueur
United Rep. of Tanzania								**Rép.-Unie de Tanzanie**
Grants of patents	30	...	...	...	...	...	...	Brevets délivrés
United States								**États-Unis**
Resident filings (per mil. pop.)	268.0	466.0	703.0	782.0	856.0	909.0	894.0	Demandes de rés. (par mil. d'hab.)
Grants of patents	71 661	101 419	143 806	219 614	253 155	277 835	300 678	Brevets délivrés
Patents in force	...	...	1 683 968	2 017 318	2 239 231	2 387 502	2 527 750	Brevets en vigueur
Uruguay								**Uruguay**
Resident filings (per mil. pop.)	21.0	11.0	7.0	7.0	6.0	...	11.0	Demandes de rés. (par mil. d'hab.)
Grants of patents	196	36	...	29	22	...	31	Brevets délivrés
Patents in force	...	...	...	877	...	...	646	Brevets en vigueur
Uzbekistan								**Ouzbékistan**
Resident filings (per mil. pop.)	...	46.0	10.0	13.0	9.0	10.0	11.0	Demandes de rés. (par mil. d'hab.)
Grants of patents	...	1 233	407	192	175	184	179	Brevets délivrés
Patents in force	...	...	1 263	1 253	1 016	1 155	1 141	Brevets en vigueur
Venezuela (Boliv. Rep. of)								**Venezuela (Rép. boliv. du)**
Resident filings (per mil. pop.)	13.0	...	...	...	...	...	...	Demandes de rés. (par mil. d'hab.)
Grants of patents	351	...	...	...	...	...	...	Brevets délivrés
Viet Nam								**Viet Nam**
Resident filings (per mil. pop.)	...	...	2.0	4.0	4.0	5.0	5.0	Demandes de rés. (par mil. d'hab.)
Grants of patents	...	56	668	822	1 068	1 182	1 397	Brevets délivrés
Patents in force	...	...	...	9 103	11 524	10 615	14 593	Brevets en vigueur
Yemen								**Yémen**
Resident filings (per mil. pop.)	...	...	1.0	1.0	2.0	2.0	1.0	Demandes de rés. (par mil. d'hab.)
Grants of patents	...	...	...	...	51	62	20	Brevets délivrés
Patents in force	...	...	...	...	...	62	20	Brevets en vigueur
Zambia								**Zambie**
Resident filings (per mil. pop.)	...	...	...	...	...	1.0	1.0	Demandes de rés. (par mil. d'hab.)
Grants of patents	74	43	14	12	32	21	23	Brevets délivrés
Patents in force	...	...	2 695	3 858	4 384	4 122	4 161	Brevets en vigueur

Patents *(continued)*
Resident filings (per million population), grants and patents in force

Brevets *(suite)*
Demandes émanant de résidents (par million d'habitants), délivrances et brevets en vigueur

Region, country or area	1985	1995	2005	2010	2012	2013	2014 Région, pays ou zone
Zimbabwe							**Zimbabwe**
Resident filings (per mil. pop.)	4.0	5.0	...	...	...	...	... Demandes de rés. (par mil. d'hab.)
Grants of patents	212	105	...	...	...	...	... Brevets délivrés
OAPI [2]							**OAPI** [2]
Grants of patents	...	23	374	...	380	430	550 Brevets délivrés
Patents in force	...	...	...	...	3 120	...	... Brevets en vigueur
ARIPO [3]							**ARIPO** [3]
Grants of patents	1	64	164	111	205	271	254 Brevets délivrés
Patents in force	...	...	...	...	...	2 291	2 550 Brevets en vigueur
EAPO [4]							**EAPO** [4]
Grants of patents	...	...	1 201	1 802	1 541	1 581	1 600 Brevets délivrés
EPO [5]							**OEB** [5]
Grants of patents	15 117	41 609	53 258	58 108	65 665	66 696	64 608 Brevets délivrés
GCC [6]							**CCG** [6]
Grants of patents	...	...	105	362	358	553	503 Brevets délivrés
Patents in force	...	...	...	...	1 756	2 510	16 586 Brevets en vigueur

Source:
World Intellectual Property Organisation (WIPO), Geneva, WIPO statistics database, last accessed April 2016.

Source:
Organisation mondiale de la propriété intellectuelle (OMPI), Genève, la base de données statistiques de l'OMPI, dernier accès avril 2016

1 For statistical purposes, the data for China do not include those for the Hong Kong Special Administrative Region (Hong Kong SAR) and Macao Special Administrative Region (Macao SAR).

2 Members of the African Intellectual Property Organization (OAPI), which includes Benin, Burkina Faso, Cameroon, Central African Republic, Chad, Congo, Côte d'Ivoire, Equatorial Guinea, Gabon, Guinea, Guinea-Bissau, Mali, Mauritania, Niger, Senegal and Togo.

3 Members of the African Regional Intellectual Property Organization (ARIPO), which includes Botswana, Gambia, Ghana, Kenya, Lesotho, Liberia, Malawi, Mozambique, Namibia, Rwanda, Sao Tome and Principe, Sierra Leone, Somalia, Sudan, Swaziland, Tanzania, Uganda, Zambia and Zimbabwe.

4 Members of the Eurasian Patent Organization (EAPO), which includes Armenia, Azerbaijan, Belarus, Kazakhstan, Kyrgyzstan, Moldova, the Russian Federation, Tajikistan and Turkmenistan.

5 Members of the European Patent Organisation (EPO), which includes Albania, Austria, Belgium, Bulgaria, Croatia, Cyprus, Czech Republic, Denmark, Estonia, Finland, France, Germany, Greece, Hungary, Iceland, Ireland, Italy, Latvia, Liechtenstein, Lithuania, Luxembourg, Malta, Monaco, Netherlands, Norway, Poland, Portugal, Romania, San Marino, Serbia, Slovenia, Slovakia, Spain, Sweden, Switzerland, the Former Yugoslav Republic of Macedonia, Turkey and the United Kingdom.

6 Members of the Gulf Cooperation Council (GCC) Patent Office, which includes Bahrain, Kuwait, Oman, Qatar, Saudi Arabia and the United Arab Emirates.

1 Pour la présentation des statistiques, les données pour la Chine ne comprennent pas la Région Administrative Spéciale de Hong Kong (Hong Kong RAS) et la Région Administrative Spéciale de Macao (Macao RAS).

2 Les membres de l'Organisation africaine de la propriété intellectuelle (OAPI) incluent: Bénin, Burkina Faso, Cameroun, Congo, Côte d'Ivoire, Gabon, Guinée, Guinée-Bissau, Guinée équatoriale, Mali, Mauritanie, Niger, République centrafricaine, Sénégal, Tchad et Togo.

3 Les membres de l'Organisation régionale africaine de la propriété intellectuelle (ARIPO) incluent: Botswana, Gambie, Ghana, Kenya, Lesotho, Libéria, Malawi, Mozambique, Namibie, Ouganda, Rwanda, Sao Tomé-et-Principe, Sierra Leone, Somalie, Soudan, Swaziland, Tanzanie, Zambie et Zimbabwe.

4 Les membres de l'Organisation eurasienne de la propriété intellectuelle (EAPO) incluent: Arménie, Azerbaïdjan, Bélarus, Fédération de Russie, Kazakhstan, Kirghizistan, Moldavie, Tadjikistan et Turkménistan.

5 Les membres de l'Organisation européenne des brevets (OEB) incluent: Albanie, Allemagne, Autriche, Belgique, Bulgarie, Chypre, Croatie, Danemark, Espagne, Estonie, Finlande, France, Grèce, Hongrie, Irlande, Islande, Italie, L'ex-République yougoslave de Macédoine, Lettonie, Liechtenstein, Lituanie, Luxembourg, Malte, Monaco, Norvège, Pays-Bas, Pologne, Portugal, République tchèque, Roumanie, Royaume-Uni, Saint-Marin, Serbie, Slovaquie, Slovénie, Suède, Suisse et Turquie.

6 Les États suivants sont membres de l'Organisation de la propriété intellectuelle pour la Conseil de coopération du Golfe (CCG) incluent: Bahreïn, Arabie saoudite, Émirats arabes unis, Koweït, Oman et Qatar.

Cellular mobile telephone subscriptions
Number (thousands) and per 100 inhabitants

Abonnements aux services de téléphone cellulaire mobile
Nombre (milliers) et pour 100 habitants

Country or area	2000	2005	2010	2012	2013	2014	2015	Pays ou zone
Afghanistan								**Afghanistan**
Number (thousands)	0	1 200	* 10 216	15 340	16 807	18 407	# 19 709[1]	Nombre (en milliers)
Per 100 inhabitants	0	5	* 36	51	55	59	# 62[1]	Pour 100 habitants
Albania								**Albanie**
Number (thousands)	30	1 530	* 2 692	3 500	3 686	3 360	3 401	Nombre (en milliers)
Per 100 inhabitants	1	48	* 85	111	116	105	106	Pour 100 habitants
Algeria								**Algérie**
Number (thousands)	86	13 661	32 780	37 528	39 517	43 298	45 928	Nombre (en milliers)
Per 100 inhabitants	<0	40	88	98	101	108	113	Pour 100 habitants
American Samoa								**Samoa américaines**
Number (thousands)	2	...	...	...	...	...	...	Nombre (en milliers)
Per 100 inhabitants	3	...	...	...	...	...	...	Pour 100 habitants
Andorra								**Andorre**
Number (thousands)	24	65	66	64	64	66	71	Nombre (en milliers)
Per 100 inhabitants	36	79	84	82	81	83	88	Pour 100 habitants
Angola								**Angola**
Number (thousands)	26	1 611	* 9 403	12 785	13 285	14 053	13 885	Nombre (en milliers)
Per 100 inhabitants	<0	10	* 48	61	62	63	61	Pour 100 habitants
Anguilla								**Anguilla**
Number (thousands)	2	13	26	* 26	* 26	* 26	* 26	Nombre (en milliers)
Per 100 inhabitants	20	103	187	* 185	* 182	* 180	* 178	Pour 100 habitants
Antigua and Barbuda								**Antigua-et-Barbuda**
Number (thousands)	22	86	168	127[2]	114	120	* 126	Nombre (en milliers)
Per 100 inhabitants	28	104	193	143[2]	127	132	* 137	Pour 100 habitants
Argentina								**Argentine**
Number (thousands)	6 488	22 156	57 082	64 328	67 362	61 234	* 60 664	Nombre (en milliers)
Per 100 inhabitants	18	57	141	157	163	146	* 144	Pour 100 habitants
Armenia								**Arménie**
Number (thousands)	17	318	3 865	3 323	3 346	3 459	3 442	Nombre (en milliers)
Per 100 inhabitants	1	11	130	112	112	116	115	Pour 100 habitants
Aruba								**Aruba**
Number (thousands)	15	103	* 132	* 135	* 139	* 140	* 141	Nombre (en milliers)
Per 100 inhabitants	17	103	* 130	* 132	* 135	* 135	* 136	Pour 100 habitants
Ascension								**Ascension**
Number (thousands)	0	0	0	0	...	0	<0	Nombre (en milliers)
Australia								**Australie**
Number (thousands)	8 562	18 420	22 500	24 338	24 940	31 010	31 770[3]	Nombre (en milliers)
Per 100 inhabitants	44	90	100	106	107	131	133[3]	Pour 100 habitants
Austria								**Autriche**
Number (thousands)	6 117	8 665	12 241	13 588	13 272	12 953	13 471	Nombre (en milliers)
Per 100 inhabitants	76	105	146	161	156	152	157	Pour 100 habitants
Azerbaijan								**Azerbaïdjan**
Number (thousands)	420	2 242	9 100	10 125	10 130	10 553	10 697	Nombre (en milliers)
Per 100 inhabitants	5	26	100	109	108	111	111	Pour 100 habitants
Bahamas								**Bahamas**
Number (thousands)	32	228	428	* 300	* 287	315	311	Nombre (en milliers)
Per 100 inhabitants	11	69	119	* 81	* 76	82	80	Pour 100 habitants
Bahrain								**Bahreïn**
Number (thousands)	206	767	1 567	2 124	2 210	2 329	2 519	Nombre (en milliers)
Per 100 inhabitants	31	87	125	161	166	173	185	Pour 100 habitants
Bangladesh								**Bangladesh**
Number (thousands)	279	9 000	67 924	97 180	116 553	126 866	133 720[4]	Nombre (en milliers)
Per 100 inhabitants	<0	6	45	63	74	80	83[4]	Pour 100 habitants
Barbados								**Barbade**
Number (thousands)	28	206	350	349	308	305	335	Nombre (en milliers)
Per 100 inhabitants	11	75	125	123	108	107	116	Pour 100 habitants
Belarus								**Bélarus**
Number (thousands)	49	4 100	10 333	10 676	11 114	11 402	11 448	Nombre (en milliers)
Per 100 inhabitants	<0	42	109	114	119	123	124	Pour 100 habitants
Belgium								**Belgique**
Number (thousands)	5 629	9 605	12 154	12 313	12 315	12 735	12 938	Nombre (en milliers)
Per 100 inhabitants	55	91	111	111	111	114	116	Pour 100 habitants

Country or area	2000	2005	2010	2012	2013	2014	2015	Pays ou zone
Belize								**Belize**
Number (thousands)	17	* 96	194[5]	172[6]	175[5]	* 172	* 170	Nombre (en milliers)
Per 100 inhabitants	7	* 35	63[5]	53[6]	53[5]	* 51	* 49	Pour 100 habitants
Benin								**Bénin**
Number (thousands)	55	596	7 075	8 408	9 627	# 8 660[7]	9 318	Nombre (en milliers)
Per 100 inhabitants	1	7	74	84	93	# 82[7]	86	Pour 100 habitants
Bermuda								**Bermudes**
Number (thousands)	13	53	* 88	* 91	* 94	59	* 38	Nombre (en milliers)
Per 100 inhabitants	21	82	* 136	* 140	* 144	91	* 58	Pour 100 habitants
Bhutan								**Bhoutan**
Number (thousands)	0	36	394[8]	561[8]	544[8,9]	625[8]	676[8]	Nombre (en milliers)
Per 100 inhabitants	0	6	55[8]	76[8]	72[8,9]	82[8]	87[8]	Pour 100 habitants
Bolivia (Plurinational State of)								**Bolivie (État plurinational de)**
Number (thousands)	583	2 421	7 179	9 493[10]	10 426[10]	10 450	10 163	Nombre (en milliers)
Per 100 inhabitants	7	26	71	90[10]	98[10]	96	92	Pour 100 habitants
Bosnia and Herzegovina								**Bosnie-Herzégovine**
Number (thousands)	93	1 594	3 110	3 358	3 488	3 491	3 444	Nombre (en milliers)
Per 100 inhabitants	2	41	81	88	91	91	90	Pour 100 habitants
Botswana								**Botswana**
Number (thousands)	222	564	2 363	3 082[4]	3 247[4]	3 411	3 475	Nombre (en milliers)
Per 100 inhabitants	13	30	120	154[4]	161[4]	167	169	Pour 100 habitants
Brazil								**Brésil**
Number (thousands)	23 188	86 210	196 930	248 324	271 100	280 729	257 814	Nombre (en milliers)
Per 100 inhabitants	13	46	101	125	135	139	127	Pour 100 habitants
British Virgin Islands								**Îles Vierges britanniques**
Number (thousands)	...	...	48	49	* 53	47	* 42	Nombre (en milliers)
Per 100 inhabitants	...	...	175	173	* 188	165	* 146	Pour 100 habitants
Brunei Darussalam								**Brunéi Darussalam**
Number (thousands)	95	233	435	470	469	# 452[11]	463	Nombre (en milliers)
Per 100 inhabitants	29	63	109	114	112	# 107[11]	108	Pour 100 habitants
Bulgaria								**Bulgarie**
Number (thousands)	738	6 245	10 200	10 781	10 487[12]	9 487[12]	* 9 195	Nombre (en milliers)
Per 100 inhabitants	9	81	138	148	145[12]	132[12]	* 129	Pour 100 habitants
Burkina Faso								**Burkina Faso**
Number (thousands)	25	634	5 708	9 976	11 241	12 494	14 447	Nombre (en milliers)
Per 100 inhabitants	<0	5	37	61	66	72	81	Pour 100 habitants
Burundi								**Burundi**
Number (thousands)	16	153	* 1 678	2 247	2 537	3 193	4 998	Nombre (en milliers)
Per 100 inhabitants	<0	2	* 18	23	25	30	46	Pour 100 habitants
Cabo Verde								**Cabo Verde**
Number (thousands)	20	82	372	425	499	613	646	Nombre (en milliers)
Per 100 inhabitants	4	17	76	86	100	122	127	Pour 100 habitants
Cambodia								**Cambodge**
Number (thousands)	131	1 062	8 151	19 105	20 265	20 452	20 851	Nombre (en milliers)
Per 100 inhabitants	1	8	57	129	134	133	133	Pour 100 habitants
Cameroon								**Cameroun**
Number (thousands)	103	2 253	8 637	13 108	15 665	17 270	16 807[13]	Nombre (en milliers)
Per 100 inhabitants	1	12	42	60	70	76	72[13]	Pour 100 habitants
Canada								**Canada**
Number (thousands)	8 727	17 017	25 825	27 720[14]	28 360	28 789	* 29 390	Nombre (en milliers)
Per 100 inhabitants	28	53	76	80[14]	81	81	* 82	Pour 100 habitants
Cayman Islands								**Îles Caïmanes**
Number (thousands)	11	81[15]	101[15]	99[15]	98[15]	91[15]	93[15]	Nombre (en milliers)
Per 100 inhabitants	26	166[15]	181[15]	172[15]	168[15]	154[15]	155[15]	Pour 100 habitants
Central African Rep.								**Rép. centrafricaine**
Number (thousands)	5	100	979[1]	1 143[1]	* 1 360	1 156[1]	* 982	Nombre (en milliers)
Per 100 inhabitants	<0	3	23[1]	25[1]	* 29	25[1]	* 20	Pour 100 habitants
Chad								**Tchad**
Number (thousands)	6	210	2 875	4 402	4 561	5 252	5 466	Nombre (en milliers)
Per 100 inhabitants	<0	2	25	35	36	40	40	Pour 100 habitants
Chile								**Chili**
Number (thousands)	3 402	10 570	19 852	23 941	23 661	23 681	23 206	Nombre (en milliers)
Per 100 inhabitants	22	65	116	137	134	133	129	Pour 100 habitants
China								**Chine**
Number (thousands)	85 260	393 406	859 003	1 112 155	1 229 113	1 286 093	1 305 738	Nombre (en milliers)
Per 100 inhabitants	7	30	63	81	89	92	93	Pour 100 habitants

Cellular mobile telephone subscriptions *(continued)*
Number (thousands) and per 100 inhabitants

Abonnements aux services de téléphone cellulaire mobile *(suite)*
Nombre (milliers) et pour 100 habitants

Country or area	2000	2005	2010	2012	2013	2014	2015	Pays ou zone
China, Hong Kong SAR								**Chine, Hong Kong RAS**
Number (thousands)	5 447	8 544	13 794	16 388	17 098	16 959	16 736	Nombre (en milliers)
Per 100 inhabitants	80	124	196	229	237	234	229	Pour 100 habitants
China, Macao SAR								**Chine, Macao RAS**
Number (thousands)	141	533	1 122	1 613	1 722	1 856	1 896	Nombre (en milliers)
Per 100 inhabitants	33	114	210	290	304	323	324	Pour 100 habitants
Colombia								**Colombie**
Number (thousands)	2 257	21 850	44 478	49 066	50 295	55 330	57 327	Nombre (en milliers)
Per 100 inhabitants	6	51	96	103	104	113	116	Pour 100 habitants
Comoros								**Comores**
Number (thousands)	0	16	165	284	* 348	* 383	* 422	Nombre (en milliers)
Per 100 inhabitants	0	3	24	40	* 47	* 51	* 55	Pour 100 habitants
Congo								**Congo**
Number (thousands)	70	558	3 719	4 283	4 660	* 4 930	* 5 216	Nombre (en milliers)
Per 100 inhabitants	2	16	90	99	105	* 108	* 112	Pour 100 habitants
Costa Rica								**Costa Rica**
Number (thousands)	212	1 101	3 128	5 378	7 112	* 7 020	* 7 536	Nombre (en milliers)
Per 100 inhabitants	5	25	67	112	146	* 142	* 151	Pour 100 habitants
Côte d'Ivoire								**Côte d'Ivoire**
Number (thousands)	473	2 349	15 599	18 100	19 391	22 105	25 408	Nombre (en milliers)
Per 100 inhabitants	3	14	82	91	95	106	119	Pour 100 habitants
Croatia								**Croatie**
Number (thousands)	1 033	3 650	* 4 928[16]	4 971[16]	# 4 721[12]	4 461	4 416	Nombre (en milliers)
Per 100 inhabitants	23	83	* 114[16]	115[16]	# 110[12]	104	104	Pour 100 habitants
Cuba								**Cuba**
Number (thousands)	7	136	1 003	1 682	1 996	2 531	3 335	Nombre (en milliers)
Per 100 inhabitants	<0	1	9	15	18	22	30	Pour 100 habitants
Cyprus								**Chypre**
Number (thousands)	218	783	1 034	1 111	1 100	1 111	1 111	Nombre (en milliers)
Per 100 inhabitants	23	76	94	98	96	96	95	Pour 100 habitants
Czech Republic								**République tchèque**
Number (thousands)	4 346	11 776	12 934	13 522	13 719	13 913	* 13 925	Nombre (en milliers)
Per 100 inhabitants	42	115	123	127	128	130	* 129	Pour 100 habitants
Dem. P. R. Korea								**R. p. dém. de Corée**
Number (thousands)	0	0	432	* 1 700	* 2 420	* 2 800	* 3 240	Nombre (en milliers)
Per 100 inhabitants	0	0	2	* 7	* 10	* 11	* 13	Pour 100 habitants
Dem. Rep. of the Congo								**Rép. dém. du Congo**
Number (thousands)	15[17]	2 746[17]	11 820	20 093	28 232	37 103	37 753	Nombre (en milliers)
Per 100 inhabitants	<0[17]	5[17]	19	31	42	53	53	Pour 100 habitants
Denmark								**Danemark**
Number (thousands)	3 364	5 449	# 6 421[18]	7 293	7 031	7 160	7 266	Nombre (en milliers)
Per 100 inhabitants	63	101	# 116[18]	130	125	127	128	Pour 100 habitants
Djibouti								**Djibouti**
Number (thousands)	<0	44	166	212	244	287	* 312	Nombre (en milliers)
Per 100 inhabitants	<0	6	20	25	28	32	* 35	Pour 100 habitants
Dominica								**Dominique**
Number (thousands)	* 1	52	* 106	109	94	74	77	Nombre (en milliers)
Per 100 inhabitants	* 2	74	* 148	152	130	102	106	Pour 100 habitants
Dominican Republic								**Rép. dominicaine**
Number (thousands)	705	3 623	8 893	8 934	9 200	8 304[2]	8 797	Nombre (en milliers)
Per 100 inhabitants	8	39	89	87	88	79[2]	83	Pour 100 habitants
Ecuador								**Équateur**
Number (thousands)	482	6 246	14 781	16 457	16 626	16 606	12 888[2]	Nombre (en milliers)
Per 100 inhabitants	4	45	99	106	106	104	79[2]	Pour 100 habitants
Egypt								**Égypte**
Number (thousands)	1 360[3]	13 630	70 661	96 799	99 705	95 316	94 016	Nombre (en milliers)
Per 100 inhabitants	2	19	91	120	122	114	111	Pour 100 habitants
El Salvador								**El Salvador**
Number (thousands)	744	2 412	* 7 700[3]	* 8 649	8 992[16]	9 193	9 334	Nombre (en milliers)
Per 100 inhabitants	12	40	* 124[3]	* 137	142[16]	144	145	Pour 100 habitants
Equatorial Guinea								**Guinée équatoriale**
Number (thousands)	* 5	97	399	501	511	517	533	Nombre (en milliers)
Per 100 inhabitants	* 1	16	57	68	67	66	67	Pour 100 habitants
Eritrea								**Érythrée**
Number (thousands)	0	40	185	305	355	* 417	* 475	Nombre (en milliers)
Per 100 inhabitants	0	1	3	5	6	* 6	* 7	Pour 100 habitants

Country or area	2000	2005	2010	2012	2013	2014	2015	Pays ou zone
Estonia								**Estonie**
Number (thousands)	557	1 445	1 653[19]	2 071[19]	2 055	2 063[19]	1 904[19]	Nombre (en milliers)
Per 100 inhabitants	41	109	127[19]	160[19]	160	161[19]	149[19]	Pour 100 habitants
Ethiopia								**Éthiopie**
Number (thousands)	18	411	6 854	20 524	25 647	30 490	42 312[20]	Nombre (en milliers)
Per 100 inhabitants	<0	1	8	22	27	32	43[20]	Pour 100 habitants
Falkland Is. (Malvinas)								**Îles Falkland (Malvinas)**
Number (thousands)	0	1[21]	3	3	4	4	* 5	Nombre (en milliers)
Per 100 inhabitants	0	26[21]	108	114	119	137	* 164	Pour 100 habitants
Faroe Islands								**Îles Féroé**
Number (thousands)	17	42	59	59	* 60	61	* 63	Nombre (en milliers)
Per 100 inhabitants	37	86	120	119	* 121	124	* 127	Pour 100 habitants
Fiji								**Fidji**
Number (thousands)	55[22]	205[22]	698[3]	859	930	876[23]	966	Nombre (en milliers)
Per 100 inhabitants	7[22]	25[22]	81[3]	98	106	99[23]	108	Pour 100 habitants
Finland								**Finlande**
Number (thousands)	3 729	5 270	8 390[16]	9 320[16]	# 7 411[12]	7 603[12]	7 399[12]	Nombre (en milliers)
Per 100 inhabitants	72	100	156[16]	172[16]	# 137[12]	140[12]	136[12]	Pour 100 habitants
France								**France**
Number (thousands)	29 052	48 088	57 785	62 260	63 324	65 425	66 681	Nombre (en milliers)
Per 100 inhabitants	49	78	91	97	99	101	103	Pour 100 habitants
French Polynesia								**Polynésie française**
Number (thousands)	40	120	216	228	243	255	268	Nombre (en milliers)
Per 100 inhabitants	17	47	81	83	88	91	95	Pour 100 habitants
Gabon								**Gabon**
Number (thousands)	120[17]	737[17]	1 610	* 2 558	* 2 745	2 933	2 958	Nombre (en milliers)
Per 100 inhabitants	10[17]	53[17]	103	* 157	* 164	171	169	Pour 100 habitants
Gambia								**Gambie**
Number (thousands)	* 6	247	1 478	1 526	1 849	2 284	2 586	Nombre (en milliers)
Per 100 inhabitants	* <0	17	88	85	100	120	131	Pour 100 habitants
Georgia								**Géorgie**
Number (thousands)	195	1 174	# 3 978[10]	4 699	4 993	5 401	5 551	Nombre (en milliers)
Per 100 inhabitants	4	26	# 91[10]	108	115	125	129	Pour 100 habitants
Germany								**Allemagne**
Number (thousands)	48 202[17]	79 271[17]	# 88 400[1,12]	92 400[1,12]	100 034[1,12]	99 530[1,12]	96 360[1,12]	Nombre (en milliers)
Per 100 inhabitants	58[17]	95[17]	# 106[1,12]	112[1,12]	121[1,12]	120[1,12]	117[1,12]	Pour 100 habitants
Ghana								**Ghana**
Number (thousands)	130	2 875	17 437	25 618	28 026	30 361	35 008	Nombre (en milliers)
Per 100 inhabitants	1	13	72	101	108	115	130	Pour 100 habitants
Gibraltar								**Gibraltar**
Number (thousands)	6	* 20	* 30	35	...	38	...	Nombre (en milliers)
Per 100 inhabitants	20	* 69	* 103	119	...	130	...	Pour 100 habitants
Greece								**Grèce**
Number (thousands)	5 932	10 260	12 293	13 360	13 000	12 271	12 682[24]	Nombre (en milliers)
Per 100 inhabitants	54	93	111	120	117	110	114[24]	Pour 100 habitants
Greenland								**Groenland**
Number (thousands)	15	46	57	59	* 60	* 61	* 61	Nombre (en milliers)
Per 100 inhabitants	27	82	101	105	* 106	* 106	* 107	Pour 100 habitants
Grenada								**Grenade**
Number (thousands)	4	47	122	130	133	117	120	Nombre (en milliers)
Per 100 inhabitants	4	46	117	123	126	110	112	Pour 100 habitants
Guam								**Guam**
Number (thousands)	27	...	...	...	...	...	...	Nombre (en milliers)
Per 100 inhabitants	18	...	...	...	...	...	...	Pour 100 habitants
Guatemala								**Guatemala**
Number (thousands)	857	4 510	18 068	20 787	21 716	16 912[25]	18 121	Nombre (en milliers)
Per 100 inhabitants	8	36	126	138	140	107[25]	111	Pour 100 habitants
Guernsey								**Guernesey**
Number (thousands)	22	...	...	...	...	...	...	Nombre (en milliers)
Per 100 inhabitants	35	...	...	...	...	...	...	Pour 100 habitants
Guinea								**Guinée**
Number (thousands)	42	189	4 000	5 585	7 436	* 8 684	10 764	Nombre (en milliers)
Per 100 inhabitants	<0	2	37	49	63	* 72	87	Pour 100 habitants
Guinea-Bissau								**Guinée-Bissau**
Number (thousands)	0	99	677	1 049	940	1 108	1 238	Nombre (en milliers)
Per 100 inhabitants	0	7	43	63	55	63	69	Pour 100 habitants

Cellular mobile telephone subscriptions *(continued)*
Number (thousands) and per 100 inhabitants

Abonnements aux services de téléphone cellulaire mobile *(suite)*
Nombre (milliers) et pour 100 habitants

Country or area	2000	2005	2010	2012	2013	2014	2015	Pays ou zone
Guyana								**Guyana**
Number (thousands)	40	281	560	547	555	567	543	Nombre (en milliers)
Per 100 inhabitants	5	37	71	69	69	71	67	Pour 100 habitants
Haiti								**Haïti**
Number (thousands)	55	* 500	4 000	6 095	* 7 160	6 769	7 412	Nombre (en milliers)
Per 100 inhabitants	1	* 5	40	60	* 69	65	70	Pour 100 habitants
Honduras								**Honduras**
Number (thousands)	155	1 281	9 505	7 370[26]	7 767	7 725	8 048	Nombre (en milliers)
Per 100 inhabitants	2	19	125	93[26]	96	94	96	Pour 100 habitants
Hungary								**Hongrie**
Number (thousands)	3 076	9 320	12 012	11 579	11 590	11 726	11 786	Nombre (en milliers)
Per 100 inhabitants	30	92	120	116	116	118	119	Pour 100 habitants
Iceland								**Islande**
Number (thousands)	215	283	341	352	356	370	384	Nombre (en milliers)
Per 100 inhabitants	76	95	107	108	108	111	114	Pour 100 habitants
India								**Inde**
Number (thousands)	3 577	90 140	752 190[4]	864 721[2,4,27]	886 304[4,27]	944 009[4,27]	1 011 054[4]	Nombre (en milliers)
Per 100 inhabitants	<0	8	62[4]	70[2,4,27]	71[4,27]	74[4,27]	79[4]	Pour 100 habitants
Indonesia								**Indonésie**
Number (thousands)	3 669	46 910	211 290	281 964	313 227	325 583	* 338 426	Nombre (en milliers)
Per 100 inhabitants	2	21	88	114	125	129	* 132	Pour 100 habitants
Iran (Islamic Rep. of)								**Iran (Rép. islamique d')**
Number (thousands)	963	8 511[28]	54 052	58 158	65 246	68 891	74 219	Nombre (en milliers)
Per 100 inhabitants	1	12[28]	73	76	84	88	93	Pour 100 habitants
Iraq								**Iraq**
Number (thousands)	0	1 533	23 264	* 26 756	* 32 450	* 33 000	* 33 559	Nombre (en milliers)
Per 100 inhabitants	0	6	75	* 82	* 96	* 95	* 94	Pour 100 habitants
Ireland								**Irlande**
Number (thousands)	2 461	4 270	4 701	5 014[16]	4 881	4 914	4 902	Nombre (en milliers)
Per 100 inhabitants	65	103	105	110[16]	105	105	104	Pour 100 habitants
Israel								**Israël**
Number (thousands)	4 400	7 757	* 9 111	* 9 225	9 500	* 9 500	* 10 570	Nombre (en milliers)
Per 100 inhabitants	73	117	* 123	* 121	123	* 121	* 133	Pour 100 habitants
Italy								**Italie**
Number (thousands)	42 246	71 500	93 666	97 189[29]	96 863[29]	94 226	92 520	Nombre (en milliers)
Per 100 inhabitants	74	122	155	160[29]	159[29]	154	151	Pour 100 habitants
Jamaica								**Jamaïque**
Number (thousands)	367	1 981	3 182	2 715[26]	2 846	3 005	3 137	Nombre (en milliers)
Per 100 inhabitants	14	74	116	98[26]	102	107	112	Pour 100 habitants
Japan								**Japon**
Number (thousands)	66 784[30]	96 484[30]	123 287[30]	141 129[31]	147 888[31]	152 696[4,31]	158 591[4,31]	Nombre (en milliers)
Per 100 inhabitants	53[30]	76[30]	97[30]	111[31]	116[31]	120[4,31]	125[4,31]	Pour 100 habitants
Jersey								**Jersey**
Number (thousands)	45	...	...	...	...	...	...	Nombre (en milliers)
Per 100 inhabitants	52	...	...	...	...	...	...	Pour 100 habitants
Jordan								**Jordanie**
Number (thousands)	389	3 138	6 620	8 984	10 314	11 092	13 798	Nombre (en milliers)
Per 100 inhabitants	8	60	103	128	142	148	179	Pour 100 habitants
Kazakhstan								**Kazakhstan**
Number (thousands)	197	5 398	19 403	30 235	30 365	28 596	31 390	Nombre (en milliers)
Per 100 inhabitants	1	36	122	186	185	172	187	Pour 100 habitants
Kenya								**Kenya**
Number (thousands)	127[4]	4 612	24 969	30 732	31 830	33 633	37 716	Nombre (en milliers)
Per 100 inhabitants	<0[4]	13	61	71	72	74	81	Pour 100 habitants
Kiribati								**Kiribati**
Number (thousands)	<0	1	11	* 18	* 20	* 30	41	Nombre (en milliers)
Per 100 inhabitants	<0	1	11	* 18	* 20	* 29	39	Pour 100 habitants
Kuwait								**Koweït**
Number (thousands)	476	1 382[32]	3 979	* 5 100	* 6 410	7 600	* 8 305	Nombre (en milliers)
Per 100 inhabitants	25	60[32]	133	* 157	* 190	218	* 232	Pour 100 habitants
Kyrgyzstan								**Kirghizistan**
Number (thousands)	9	542	5 275	6 798	6 737[1]	7 563	7 579	Nombre (en milliers)
Per 100 inhabitants	<0	11	99	124	121[1]	134	133	Pour 100 habitants
Lao People's Dem. Rep.								**Rép. dém. pop. lao**
Number (thousands)	13	658	4 003	4 300[33]	4 613	4 619	3 727	Nombre (en milliers)
Per 100 inhabitants	<0	11	63	65[33]	68	67	53	Pour 100 habitants

Country or area	2000	2005	2010	2012	2013	2014	2015	Pays ou zone
Latvia								**Lettonie**
Number (thousands)	401	1 872	* 2 306	2 631	2 558	2 384	2 579[34]	Nombre (en milliers)
Per 100 inhabitants	17	84	* 110	128	125	117	127[34]	Pour 100 habitants
Lebanon								**Liban**
Number (thousands)	* 743	* 994	2 864	3 755	3 885	4 387	* 4 400	Nombre (en milliers)
Per 100 inhabitants	* 23	* 25	66	81	81	88	* 87	Pour 100 habitants
Lesotho								**Lesotho**
Number (thousands)	22	250	987	1 545	1 790	2 139	2 237	Nombre (en milliers)
Per 100 inhabitants	1	13	49	75	86	102	106	Pour 100 habitants
Liberia								**Libéria**
Number (thousands)	2	160	1 571	2 380[35]	2 551[36]	3 225	* 3 652	Nombre (en milliers)
Per 100 inhabitants	<0	5	40	57[35]	59[36]	73	* 81	Pour 100 habitants
Libya								**Libye**
Number (thousands)	40	2 000	* 10 900	* 9 587	* 10 235	* 10 076	* 9 918	Nombre (en milliers)
Per 100 inhabitants	1	36	* 180	* 156	* 165	* 161	* 157	Pour 100 habitants
Liechtenstein								**Liechtenstein**
Number (thousands)	* 10	* 28	* 36	36	38	41	41	Nombre (en milliers)
Per 100 inhabitants	* 30	* 79	* 98	98	104	109	109	Pour 100 habitants
Lithuania								**Lituanie**
Number (thousands)	524[1]	4 353[1]	4 891	4 997	4 566	4 268	4 184	Nombre (en milliers)
Per 100 inhabitants	15[1]	132[1]	159	165	151	142	140	Pour 100 habitants
Luxembourg								**Luxembourg**
Number (thousands)	303[17]	510	727	761	788	802	807	Nombre (en milliers)
Per 100 inhabitants	70[17]	111	143	145	149	149	149	Pour 100 habitants
Madagascar								**Madagascar**
Number (thousands)	63	510	7 712	8 779	8 461	9 714	* 11 152	Nombre (en milliers)
Per 100 inhabitants	<0	3	37	39	37	41	* 46	Pour 100 habitants
Malawi								**Malawi**
Number (thousands)	49	421	3 117	4 647	5 290	5 633	6 116	Nombre (en milliers)
Per 100 inhabitants	<0	3	21	29	32	33	35	Pour 100 habitants
Malaysia								**Malaisie**
Number (thousands)	5 122	19 545	33 859	41 325	43 005	44 929	44 111	Nombre (en milliers)
Per 100 inhabitants	22	76	120	141	145	149	144	Pour 100 habitants
Maldives								**Maldives**
Number (thousands)	8	204	494	561	625	666	740	Nombre (en milliers)
Per 100 inhabitants	3	68	152	166	181	189	207	Pour 100 habitants
Mali								**Mali**
Number (thousands)	10	762	7 440	14 613	19 749	23 506	22 699[33]	Nombre (en milliers)
Per 100 inhabitants	<0	6	53	98	129	149	140[33]	Pour 100 habitants
Malta								**Malte**
Number (thousands)	114	324	456	532	557	546	558	Nombre (en milliers)
Per 100 inhabitants	28	78	107	124	130	127	129	Pour 100 habitants
Marshall Islands								**Îles Marshall**
Number (thousands)	<0	1	...	...	...	16	* 16	Nombre (en milliers)
Per 100 inhabitants	1	1	...	...	...	29	* 29	Pour 100 habitants
Mauritania								**Mauritanie**
Number (thousands)	15[17]	746[17]	2 776[17]	4 025[17]	# 3 988[1]	3 753[1]	3 644[1]	Nombre (en milliers)
Per 100 inhabitants	1[17]	24[17]	77[17]	106[17]	# 103[1]	94[1]	89[1]	Pour 100 habitants
Mauritius								**Maurice**
Number (thousands)	180	657	1 191	1 486	1 534	1 652	1 762	Nombre (en milliers)
Per 100 inhabitants	15	54	97	120	123	132	141	Pour 100 habitants
Mayotte [1]								**Mayotte** [1]
Number (thousands)	0	...	...	...	...	...	...	Nombre (en milliers)
Per 100 inhabitants	0	...	...	...	...	...	...	Pour 100 habitants
Mexico								**Mexique**
Number (thousands)	14 078	47 129	91 383	* 100 727	106 748	104 798	* 106 831	Nombre (en milliers)
Per 100 inhabitants	14	43	78	* 83	87	85	* 85	Pour 100 habitants
Micronesia (Fed. States of)								**Micronésie (États féd. de)**
Number (thousands)	0	14	28	31	31	...	...	Nombre (en milliers)
Per 100 inhabitants	0	13	27	30	30	...	...	Pour 100 habitants
Monaco								**Monaco**
Number (thousands)	14	17	23	33	35	34	34	Nombre (en milliers)
Per 100 inhabitants	43	51	64	88	94	88	89	Pour 100 habitants
Mongolia								**Mongolie**
Number (thousands)	155	557	2 510	3 375	2 878[1]	3 027	3 068	Nombre (en milliers)
Per 100 inhabitants	6	22	93	121	101[1]	105	105	Pour 100 habitants

Cellular mobile telephone subscriptions *(continued)*
Number (thousands) and per 100 inhabitants

Abonnements aux services de téléphone cellulaire mobile *(suite)*
Nombre (milliers) et pour 100 habitants

Country or area	2000	2005	2010	2012	2013	2014	2015	Pays ou zone
Montenegro								**Monténégro**
Number (thousands)	...	543	* 1 170	* 991	* 994	1 013	1 008	Nombre (en milliers)
Per 100 inhabitants	...	88	* 189	* 160	* 160	163	162	Pour 100 habitants
Montserrat								**Montserrat**
Number (thousands)	<0	...	4	* 4	* 5	* 5	* 5	Nombre (en milliers)
Per 100 inhabitants	10	...	85	* 83	* 88	* 88	* 97	Pour 100 habitants
Morocco								**Maroc**
Number (thousands)	2 342	12 393	31 982	39 016	42 424	44 115[4]	43 080	Nombre (en milliers)
Per 100 inhabitants	8	41	101	120	129	132[4]	127	Pour 100 habitants
Mozambique								**Mozambique**
Number (thousands)	51	1 504	7 224	8 805	12 401	18 483	20 135	Nombre (en milliers)
Per 100 inhabitants	<0	7	30	35	48	70	74	Pour 100 habitants
Myanmar								**Myanmar**
Number (thousands)	13	129	* 594	3 730	6 832	29 029	41 529	Nombre (en milliers)
Per 100 inhabitants	<0	<0	* 1	7	13	54	77	Pour 100 habitants
Namibia								**Namibie**
Number (thousands)	82	449	1 950	2 147	2 728	2 671	2 443	Nombre (en milliers)
Per 100 inhabitants	4	22	90	95	118	114	102	Pour 100 habitants
Nauru								**Nauru**
Number (thousands)	* 1	...	6	7	...	...	...	Nombre (en milliers)
Per 100 inhabitants	* 12	...	62	68	...	...	...	Pour 100 habitants
Nepal								**Népal**
Number (thousands)	10	227	9 196	16 609[4]	21 362	23 021[4]	27 516[4]	Nombre (en milliers)
Per 100 inhabitants	<0	1	34	60[4]	77	82[4]	97[4]	Pour 100 habitants
Netherlands								**Pays-Bas**
Number (thousands)	10 755	15 834	19 179[2,34]	19 717	19 467[12]	19 562[12,37]	20 809	Nombre (en milliers)
Per 100 inhabitants	68	97	115[2,34]	118	116[12]	116[12,37]	124	Pour 100 habitants
New Caledonia								**Nouvelle-Calédonie**
Number (thousands)	50	134	221	* 231	* 241	* 243	* 246	Nombre (en milliers)
Per 100 inhabitants	24	59	90	* 91	* 94	* 94	* 93	Pour 100 habitants
New Zealand								**Nouvelle-Zélande**
Number (thousands)	1 542	3 530	4 710[38]	4 922[38]	4 766[38]	5 100[38]	5 600	Nombre (en milliers)
Per 100 inhabitants	40	85	108[38]	110[38]	106[38]	112[38]	122	Pour 100 habitants
Nicaragua								**Nicaragua**
Number (thousands)	90	1 119	3 962[17]	* 5 852[17]	* 6 809[17]	* 7 068[17]	7 264[17]	Nombre (en milliers)
Per 100 inhabitants	2	21	68[17]	* 98[17]	* 112[17]	* 115[17]	116[17]	Pour 100 habitants
Niger								**Niger**
Number (thousands)	2	324	3 669	5 396	* 7 006	* 8 236	* 8 959	Nombre (en milliers)
Per 100 inhabitants	<0	2	23	31	* 39	* 44	* 47	Pour 100 habitants
Nigeria								**Nigéria**
Number (thousands)	30	18 587	87 298	112 778	127 246	138 960	150 830	Nombre (en milliers)
Per 100 inhabitants	<0	13	55	67	73	78	82	Pour 100 habitants
Niue								**Nioué**
Number (thousands)	<0	...	...	...	...	...	...	Nombre (en milliers)
Per 100 inhabitants	22	...	...	...	...	...	...	Pour 100 habitants
Northern Mariana Islands								**Îles Mariannes du Nord**
Number (thousands)	3	...	...	...	...	...	...	Nombre (en milliers)
Per 100 inhabitants	4	...	...	...	...	...	...	Pour 100 habitants
Norway								**Norvège**
Number (thousands)	3 224	4 754	5 599	5 798	5 863	5 913	5 841[3]	Nombre (en milliers)
Per 100 inhabitants	72	103	114	116	116	116	114[3]	Pour 100 habitants
Oman								**Oman**
Number (thousands)	162	1 333	4 606	5 278	5 617	6 194	6 647	Nombre (en milliers)
Per 100 inhabitants	7	53	164	159	155	158	160	Pour 100 habitants
Pakistan								**Pakistan**
Number (thousands)	306	12 771	99 186	120 151	127 737	135 762[4]	125 900[39]	Nombre (en milliers)
Per 100 inhabitants	<0	8	57	67	70	73[4]	67[39]	Pour 100 habitants
Palau								**Palaos**
Number (thousands)	...	6	15	17	18	19	24	Nombre (en milliers)
Per 100 inhabitants	...	30	71	83	86	91	112	Pour 100 habitants
Panama								**Panama**
Number (thousands)	410	1 749	6 646	6 214	6 205	* 6 205	* 6 947	Nombre (en milliers)
Per 100 inhabitants	13	52	181	163	161	* 158	* 174	Pour 100 habitants
Papua New Guinea								**Papouasie-Nvl-Guinée**
Number (thousands)	9	75	1 909[40]	* 2 709	* 3 000	* 3 359	* 3 560	Nombre (en milliers)
Per 100 inhabitants	<0	1	28[40]	* 38	* 41	* 45	* 47	Pour 100 habitants

Country or area	2000	2005	2010	2012	2013	2014	2015	Pays ou zone
Paraguay								**Paraguay**
Number (thousands)	821	1 887	5 921	6 794	7 053	7 305	7 412	Nombre (en milliers)
Per 100 inhabitants	15	32	92	102	104	106	105	Pour 100 habitants
Peru								**Pérou**
Number (thousands)	1 274	5 583	29 115	29 388[16]	29 793[16]	31 880[12]	34 236	Nombre (en milliers)
Per 100 inhabitants	5	20	100	98[16]	98[16]	104[12]	110	Pour 100 habitants
Philippines								**Philippines**
Number (thousands)	6 454	34 779	83 150	101 978	102 824	111 326	* 120 255	Nombre (en milliers)
Per 100 inhabitants	8	41	89	105	105	111	* 118	Pour 100 habitants
Poland								**Pologne**
Number (thousands)	6 747	29 166	46 952[16]	54 086[16]	56 973[16]	# 56 905[12]	* 56 838	Nombre (en milliers)
Per 100 inhabitants	18	76	123[16]	142[16]	149[16]	# 149[12]	* 149	Pour 100 habitants
Portugal								**Portugal**
Number (thousands)	6 665	11 447	12 210[16]	# 11 918[12]	11 991[12]	11 896[12]	11 715[12]	Nombre (en milliers)
Per 100 inhabitants	65	109	115[16]	# 112[12]	113[12]	112[12]	110[12]	Pour 100 habitants
Puerto Rico								**Porto Rico**
Number (thousands)	1 318	1 993	2 934	3 050	3 085	3 209	3 205	Nombre (en milliers)
Per 100 inhabitants	35	53	79	83	84	87	87	Pour 100 habitants
Qatar								**Qatar**
Number (thousands)	121	717[17]	2 186[1]	2 601[1]	3 310[1]	3 306[1]	3 610	Nombre (en milliers)
Per 100 inhabitants	20	87[17]	125[1]	127[1]	153[1]	146[1]	154	Pour 100 habitants
Republic of Korea								**République de Corée**
Number (thousands)	26 816	38 342	50 767	53 624	54 681	57 290	58 935	Nombre (en milliers)
Per 100 inhabitants	58	82	105	109	111	116	118	Pour 100 habitants
Republic of Moldova								**République de Moldova**
Number (thousands)	139	1 090[17]	* 2 551	* 3 584	* 3 697	3 738	3 713	Nombre (en milliers)
Per 100 inhabitants	3	29[17]	* 71	* 102	* 106	108	108	Pour 100 habitants
Romania								**Roumanie**
Number (thousands)	2 499	13 354	24 360[41]	22 840[41]	22 910[41]	22 920[41]	23 120[41]	Nombre (en milliers)
Per 100 inhabitants	11	60	111[41]	105[41]	106[41]	106[41]	107[41]	Pour 100 habitants
Russian Federation								**Fédération de Russie**
Number (thousands)	3 263	120 000	237 689[17]	208 065[17]	218 300[17]	221 030	227 288	Nombre (en milliers)
Per 100 inhabitants	2	83	166[17]	145[17]	153[17]	155	160	Pour 100 habitants
Rwanda								**Rwanda**
Number (thousands)	39	223	3 549	5 691	6 689	7 747	8 760	Nombre (en milliers)
Per 100 inhabitants	<0	2	33	50	57	64	70	Pour 100 habitants
Saint Helena								**Sainte-Hélène**
Number (thousands)	0	0	0	0	0	0	1	Nombre (en milliers)
Per 100 inhabitants	0	0	0	0	0	0	31	Pour 100 habitants
Saint Kitts and Nevis								**Saint-Kitts-et-Nevis**
Number (thousands)	1	51[4]	80	76	* 77[42]	65	73	Nombre (en milliers)
Per 100 inhabitants	3	104[4]	153	142	* 142[42]	119	132	Pour 100 habitants
Saint Lucia								**Sainte-Lucie**
Number (thousands)	* 3	* 106	198	216	212	188[43]	188	Nombre (en milliers)
Per 100 inhabitants	* 2	* 64	112	119	116	103[43]	102	Pour 100 habitants
Saint Vincent-Grenadines								**Saint-Vincent-Grenadines**
Number (thousands)	2	71	132	127	125	115	113	Nombre (en milliers)
Per 100 inhabitants	2	65	121	116	115	105	104	Pour 100 habitants
Samoa								**Samoa**
Number (thousands)	3	24	* 90	100	100	107	113	Nombre (en milliers)
Per 100 inhabitants	1	13	* 48	53	52	56	59	Pour 100 habitants
San Marino								**Saint-Marin**
Number (thousands)	15	17	31	36	37	* 38	37	Nombre (en milliers)
Per 100 inhabitants	54	58	99	115	117	* 119	115	Pour 100 habitants
Sao Tome and Principe								**Sao Tomé-et-Principe**
Number (thousands)	0	12	103	122	125	* 129	* 132	Nombre (en milliers)
Per 100 inhabitants	0	8	58	65	65	* 65	* 65	Pour 100 habitants
Saudi Arabia								**Arabie saoudite**
Number (thousands)	1 376	14 164	51 564	53 000	53 104	52 735	52 796	Nombre (en milliers)
Per 100 inhabitants	7	57	189	187	184	180	177	Pour 100 habitants
Senegal								**Sénégal**
Number (thousands)	250	1 730	8 344	11 471	13 134	14 380	14 959	Nombre (en milliers)
Per 100 inhabitants	3	15	64	84	93	99	100	Pour 100 habitants
Serbia								**Serbie**
Number (thousands)	...	5 511[44]	9 915[44]	# 9 138[45]	9 199	9 345	9 156	Nombre (en milliers)
Per 100 inhabitants	...	67[44]	125[44]	# 118[45]	119	122	121	Pour 100 habitants

Cellular mobile telephone subscriptions *(continued)*
Number (thousands) and per 100 inhabitants

Abonnements aux services de téléphone cellulaire mobile *(suite)*
Nombre (milliers) et pour 100 habitants

Country or area	2000	2005	2010	2012	2013	2014	2015	Pays ou zone
Seychelles								**Seychelles**
Number (thousands)	26	59	118[46]	136	137	151	148	Nombre (en milliers)
Per 100 inhabitants	33	68	129[46]	148	147	162	158	Pour 100 habitants
Sierra Leone								**Sierra Leone**
Number (thousands)	12	...	2 000	* 2 210	* 4 000	* 4 757	* 5 657	Nombre (en milliers)
Per 100 inhabitants	<0	...	35	* 37	* 66	* 77	* 90	Pour 100 habitants
Singapore								**Singapour**
Number (thousands)	2 747	4 385	7 385[17]	8 068[17]	8 438[17]	8 104[47]	* 8 211[4]	Nombre (en milliers)
Per 100 inhabitants	70	98	145[17]	152[17]	156[17]	147[47]	* 146[4]	Pour 100 habitants
Slovakia								**Slovaquie**
Number (thousands)	1 244[17]	4 540[17]	5 925	6 094	6 208	6 378	6 676	Nombre (en milliers)
Per 100 inhabitants	23[17]	84[17]	109	112	114	117	122	Pour 100 habitants
Slovenia								**Slovénie**
Number (thousands)	1 216	1 759[1]	2 122	2 241	2 284	2 326	2 354	Nombre (en milliers)
Per 100 inhabitants	61	88[1]	103	108	110	112	113	Pour 100 habitants
Solomon Islands								**Îles Salomon**
Number (thousands)	1	6	116	302	323	377	425	Nombre (en milliers)
Per 100 inhabitants	<0	1	22	55	58	66	73	Pour 100 habitants
Somalia								**Somalie**
Number (thousands)	80	500	* 648	* 2 300	5 183	5 500	* 5 836	Nombre (en milliers)
Per 100 inhabitants	1	6	* 7	* 23	49	51	* 52	Pour 100 habitants
South Africa								**Afrique du Sud**
Number (thousands)	8 339	33 960	50 372	* 68 394	76 865	79 281	85 197	Nombre (en milliers)
Per 100 inhabitants	19	70	98	* 131	146	149	159	Pour 100 habitants
South Sudan								**Soudan du sud**
Number (thousands)	...	...	1 500[48]	2 300[35]	2 853[35]	2 876[35]	* 2 899	Nombre (en milliers)
Per 100 inhabitants	...	...	14[48]	21[35]	25[35]	25[35]	* 24	Pour 100 habitants
Spain								**Espagne**
Number (thousands)	24 265	42 694	51 389	50 665	50 159	50 806	50 926	Nombre (en milliers)
Per 100 inhabitants	60	98	111	108	107	108	108	Pour 100 habitants
Sri Lanka								**Sri Lanka**
Number (thousands)	430	3 362	17 359	19 333	20 315	22 123	24 385	Nombre (en milliers)
Per 100 inhabitants	2	17	84	92	96	103	113	Pour 100 habitants
State of Palestine								**État de Palestine**
Number (thousands)	7	568	2 604	3 135	3 190	3 198	3 531	Nombre (en milliers)
Per 100 inhabitants	<0	16	65	74	74	72	78	Pour 100 habitants
Sudan								**Soudan**
Number (thousands)	* 23[49]	* 1 828[49]	18 093	27 659	27 658	27 797	27 939	Nombre (en milliers)
Per 100 inhabitants	* <0[49]	* 5[49]	42	74	73	72	71	Pour 100 habitants
Suriname								**Suriname**
Number (thousands)	41	233	521	569	869	* 928	* 991	Nombre (en milliers)
Per 100 inhabitants	9	47	99	106	161	* 171	* 181	Pour 100 habitants
Swaziland								**Swaziland**
Number (thousands)	33	200	726	805	* 893	* 917	* 941	Nombre (en milliers)
Per 100 inhabitants	3	18	61	65	* 71	* 72	* 73	Pour 100 habitants
Sweden								**Suède**
Number (thousands)	6 372	9 104	10 992	11 848	12 014	12 313	12 639	Nombre (en milliers)
Per 100 inhabitants	72	101	117	125	126	128	130	Pour 100 habitants
Switzerland								**Suisse**
Number (thousands)	4 639	6 834	9 644	10 561	* 11 049	* 11 150	* 11 700	Nombre (en milliers)
Per 100 inhabitants	65	92	123	132	* 137	* 137	* 142	Pour 100 habitants
Syrian Arab Republic								**Rép. arabe syrienne**
Number (thousands)	* 30	2 950	11 696	12 980	12 291	14 040	13 904	Nombre (en milliers)
Per 100 inhabitants	* <0	16	54	59	56	64	62	Pour 100 habitants
Tajikistan								**Tadjikistan**
Number (thousands)	1	265	5 941	* 6 528	* 7 537	* 7 999	* 8 489	Nombre (en milliers)
Per 100 inhabitants	<0	4	78	* 82	* 92	* 95	* 99	Pour 100 habitants
Thailand								**Thaïlande**
Number (thousands)	3 056	30 460	71 726	85 012	93 849	97 096	84 797	Nombre (en milliers)
Per 100 inhabitants	5	46	108	127	140	144	126	Pour 100 habitants
TFYR of Macedonia								**ex-R.Y. de Macédoine**
Number (thousands)	116	1 131	2 153	2 235	2 237	2 224	2 223	Nombre (en milliers)
Per 100 inhabitants	6	54	102	106	106	106	105	Pour 100 habitants
Timor-Leste								**Timor-Leste**
Number (thousands)	...	33	473	* 621	* 650	1 376	1 377	Nombre (en milliers)
Per 100 inhabitants	...	3	44	* 56	* 57	119	117	Pour 100 habitants

Cellular mobile telephone subscriptions *(continued)*
Number (thousands) and per 100 inhabitants

Abonnements aux services de téléphone cellulaire mobile *(suite)*
Nombre (milliers) et pour 100 habitants

Country or area	2000	2005	2010	2012	2013	2014	2015	Pays ou zone
Togo								**Togo**
Number (thousands)	50	434	2 602[50]	3 312[50]	4 263[50]	4 516	4 657	Nombre (en milliers)
Per 100 inhabitants	1	8	41[50]	50[50]	63[50]	65	65	Pour 100 habitants
Tokelau								**Tokélaou**
Number (thousands)	0	0	...	...	...	...	...	Nombre (en milliers)
Per 100 inhabitants	0	0	...	...	...	...	...	Pour 100 habitants
Tonga								**Tonga**
Number (thousands)	* <0	* 30	* 54	* 56	* 58	* 68	* 70	Nombre (en milliers)
Per 100 inhabitants	* <0	* 30	* 52	* 53	* 55	* 64	* 66	Pour 100 habitants
Trinidad and Tobago								**Trinité-et-Tobago**
Number (thousands)	162	924	1 894	1 884	1 944	1 981	2 123	Nombre (en milliers)
Per 100 inhabitants	13	71	143	141	145	147	158	Pour 100 habitants
Tunisia								**Tunisie**
Number (thousands)	119	5 681	11 114	12 844	12 712	14 284	14 598	Nombre (en milliers)
Per 100 inhabitants	1	57	105	118	116	128	130	Pour 100 habitants
Turkey								**Turquie**
Number (thousands)	16 133	43 609	61 770	67 681	69 661	71 888	73 639	Nombre (en milliers)
Per 100 inhabitants	26	64	86	91	93	95	96	Pour 100 habitants
Turkmenistan								**Turkménistan**
Number (thousands)	8	* 105	3 198	* 5 900	* 6 125	* 7 206	* 7 842	Nombre (en milliers)
Per 100 inhabitants	<0	* 2	63	* 114	* 117	* 136	* 146	Pour 100 habitants
Tuvalu								**Tuvalu**
Number (thousands)	0	1	2	* 3	* 3	* 4	* 4	Nombre (en milliers)
Per 100 inhabitants	0	13	16	* 28	* 34	* 38	* 40	Pour 100 habitants
Uganda								**Ouganda**
Number (thousands)	127	1 315	12 828[4]	16 356[4]	18 069[4]	20 366[4]	20 220[4]	Nombre (en milliers)
Per 100 inhabitants	1	5	38[4]	45[4]	48[4]	52[4]	50[4]	Pour 100 habitants
Ukraine								**Ukraine**
Number (thousands)	819	30 014	53 929	59 344	62 459	61 170	60 720	Nombre (en milliers)
Per 100 inhabitants	2	64	117	130	138	144	144	Pour 100 habitants
United Arab Emirates								**Émirats arabes unis**
Number (thousands)	1 428	4 534	10 926	13 775	16 064	16 819	17 943	Nombre (en milliers)
Per 100 inhabitants	47	109	129	150	172	178	187	Pour 100 habitants
United Kingdom								**Royaume-Uni**
Number (thousands)	43 452	65 472	76 730	78 329	78 674	78 461	80 284	Nombre (en milliers)
Per 100 inhabitants	74	109	124	125	125	124	126	Pour 100 habitants
United Rep. of Tanzania								**Rép.-Unie de Tanzanie**
Number (thousands)	111	2 964	20 984	27 219	27 443	31 863	39 666	Nombre (en milliers)
Per 100 inhabitants	<0	8	47	57	56	63	76	Pour 100 habitants
United States								**États-Unis**
Number (thousands)	109 478	203 700	285 118	304 838	310 698	355 500	382 307	Nombre (en milliers)
Per 100 inhabitants	38	68	91	96	97	110	118	Pour 100 habitants
United States Virgin Is. *								**Îles Vierges américaines ***
Number (thousands)	35	80	...	...	...	...	...	Nombre (en milliers)
Per 100 inhabitants	32	75	...	...	...	...	...	Pour 100 habitants
Uruguay								**Uruguay**
Number (thousands)	411	1 155	4 437[16]	4 995[16]	5 268[16]	5 497[16]	5 495[16]	Nombre (en milliers)
Per 100 inhabitants	12	35	132[16]	147[16]	155[16]	161[16]	160[16]	Pour 100 habitants
Uzbekistan								**Ouzbékistan**
Number (thousands)	53	720	* 20 952	20 274[26]	* 21 500	21 639	21 783	Nombre (en milliers)
Per 100 inhabitants	<0	3	* 75	71[26]	* 74	74	73	Pour 100 habitants
Vanuatu								**Vanuatu**
Number (thousands)	<0	13	170	146[38]	127[38]	156	175	Nombre (en milliers)
Per 100 inhabitants	<0	6	72	59[38]	50[38]	60	66	Pour 100 habitants
Venezuela (Boliv. Rep. of)								**Venezuela (Rép. boliv. du)**
Number (thousands)	5 447	12 496	27 880	30 569	* 30 896	30 528	* 29 094	Nombre (en milliers)
Per 100 inhabitants	22	47	96	102	* 102	99	* 93	Pour 100 habitants
Viet Nam								**Viet Nam**
Number (thousands)	789	9 593	111 570	131 674	123 736	136 148	* 122 000	Nombre (en milliers)
Per 100 inhabitants	1	11	125	145	135	147	* 131	Pour 100 habitants
Wallis and Futuna Islands								**Îles Wallis-et-Futuna**
Number (thousands)	0	0	...	...	...	0	...	Nombre (en milliers)
Yemen								**Yémen**
Number (thousands)	32	2 278	11 085	13 900	* 16 845	* 17 100	* 17 359	Nombre (en milliers)
Per 100 inhabitants	<0	11	49	58	* 69	* 68	* 68	Pour 100 habitants

Country or area	2000	2005	2010	2012	2013	2014	2015	Pays ou zone
Zambia								**Zambie**
Number (thousands)	99	950	5 447	10 525	10 396[33]	10 115	11 558	Nombre (en milliers)
Per 100 inhabitants	1	8	41	75	72[33]	67	74	Pour 100 habitants
Zimbabwe								**Zimbabwe**
Number (thousands)	266	647	7 700	12 614	13 633	11 799	12 757	Nombre (en milliers)
Per 100 inhabitants	2	5	59	92	96	81	85	Pour 100 habitants

Source:

International Telecommunication Union (ITU), Geneva, the ITU database, last accessed July 2016.

Source:

Union internationale des télécommunications (UIT), Genève, la base de données de l'UIT, dernier accès juillet 2016.

1	Data refers to active subscriptions only.	1	Les données concernent les abonnements actifs.
2	Decrease due to operator updating their records to exclude inactive subscriptions.	2	Baisse due à l'exclusion des comptes inactifs par l'un des principaux opérateurs.
3	Data as at the end of June.	3	Données à la fin de juin.
4	Data as at the end of December.	4	Données à la fin de décembre.
5	Includes mobile GSM mobile base.	5	Y compris Mobile GSM Mobile Base.
6	Includes Mobile GSM and AMPS Post and Pre Mobile Base.	6	Y compris Mobile GSM et AMPS Post et Pre Mobile Base en décembre 2010.
7	Includes only active pre-paid subscriptions.	7	Ne comprend que les abonnements actifs et prépayés .
8	Both Bhutan Telecom and Tashi Cell provide mobile-cellular services.	8	Buthan Telecom and Tashi Cell sont les deux fournisseurs de téléphonie mobile.
9	Decrease in total number of subscribers is attributed mainly to foreign workers leaving the country.	9	Diminution du nombre total d'abonnés attribuable principalement au départ des travailleurs étrangers quittant le pays.
10	Data refers to subscriptions active in the last month (or 30 days).	10	Les données portent sur les abonnements actifs dans le dernier mois (ou 30 jours).
11	Previous year data refer to total number of configured SIMs instead of active subscriptions.	11	Les données de l'année précédente se réfèrent au nombre total de cartes SIMs configurées plutôt que les abonnements actifs.
12	Excludes data-only subscriptions.	12	À l'exclusion des abonnements aux services d'accès aux données.
13	Data as at the end of September.	13	Données à la fin de septembre.
14	Includes retail subscriptions.	14	Y compris les abonnements de détail.
15	Data refers to number of mobile handsets in operation.	15	Les données portent sur le nombre de téléphones portables en fonctionnement.
16	Includes data-only subscriptions.	16	Y compris les abonnements aux services d'accès aux données.
17	Includes inactive subscriptions.	17	Y compris les abonnés inactifs.
18	Excludes telemetry subscriptions.	18	À l'exclusion des abonnements aux services de télémétrie.
19	Excludes prepaid cards that are used to provide Travel SIM/WorldMobile service.	19	À l'exclusion des cartes SIM prépayées utilisées pour fournir des services de voyage World Mobile.
20	There is an ongoing Telecom Expansion Project (TEP), which resulted in about 12 Million new subscriber than the previous year.	20	Le Plan d'amélioration technologique (TEP) a donné lieu à environ 12 millions de nouveaux abonnés par rapport à l'année précédente.
21	GSM mobile was only launched on 12 December 2005.	21	Les services GSM n'ont été lancés que le 12 décembre 2005.
22	Year ending 31 March of the following year.	22	Année finissant le 31 Mars de l'année suivante.
23	Reduction in multiple sim usage per subscriber (of different network operators).	23	Baisse du nombre d'utilisations de cartes SIM multiples par abonné (de différents opérateurs réseau).
24	Four companies compete on the local market. The latest company entered the market at the 4 quarter of 2014. Data on the new entrant concern the 9th month of 2015.	24	Quatre entreprises se font concurrence sur le marché local. La dernière entreprise est rentrée sur le marché durant le quatrième trimestre 2014, mais les données sur ce nouveau concurrent se référent au neuvième mois de 2015.
25	New tax on numbering resources, which has prompted operators to return several numbers, either inactive ones or with low consumption.	25	Nouvelle taxe sur les ressources de numérotation, qui a incité les opérateurs à rendre plusieurs numéros inactifs ou sous-utilisés.
26	Decrease due to merger or closing of operators in the mobile market.	26	Baisse découlant de la fusion ou de la fermeture d'opérateurs de téléphonie mobile.
27	Including fixed wireless local loop (WLL) subscriptions.	27	Y compris les abonnements au réseau d'accès hertzien (WLL).
28	Data as at the end of October.	28	Données à la fin de octobre.
29	Includes Mobile Network Operators (MNO) and Mobile Visual Network Operators (MVNO).	29	Y compris les opérateurs de réseau mobile (ORM) et les opérateurs de réseaux virtuels mobiles (ORVM)
30	Including Personal Handy-phone System (PHS) subscriptions.	30	Y compris les abonnements au système PHS (Personal Handy System).
31	Including Personal Handy-phone System (PHS) and data-only subscriptions.	31	Y compris la téléphonie PHS et les abonnements aux services d'accès aux données.
32	Does not include data from all operators.	32	Ne comprend pas les données de tous les opérateurs.
33	Decrease due to implementation of subscriber registration.	33	Baisse due à la mise en place d'un système d'enregistrement des abonnés.
34	Data as at the end of July.	34	Données à la fin de juillet.

35	Data obtained from all four mobile GSM operators.
36	Data obtained from all three mobile GSM operators.
37	Data refers to the third quarter.
38	Data refers to subscriptions active in the last 3 months (or 90 days).
39	Figure is reported after a biometric re-verification of SIMs by all Cellular Mobile Operators.
40	Global System for Mobile Communications (GSM).
41	Data refers to pre-paid subscriptions active in the last 6 months (or 180 days).
42	Data as at the end of March.
43	Decrease due to change in operator's method of counting pre-paid subscriptions.
44	Includes inactive pre-paid subscriptions.
45	Data refers to pre-paid subscriptions active in the last 3 months (or 90 days).
46	Data as at end of January of the following year.
47	Decline was due to the regulatory controls on prepaid SIM cards which restricts each end user to hold no more than 3 prepaid cards.
48	Data obtained from five mobile operators.
49	Canar counted as fixed line.
50	Includes active subscriptions using CDMA (Code Division Multiple Access) and GSM (Global System for Mobiles).

35	Chiffre obtenu des quatre opérateurs de téléphonie mobile dans le pays.
36	Chiffre obtenu des trois opérateurs de téléphonie mobile dans le pays.
37	Troisième trimestre.
38	Les données portent sur les abonnements actifs au cours des trois derniers mois (90 jours).
39	Le chiffre est rapporté après une nouvelle vérification biométrique des cartes SIM par tous les opérateurs de téléphonie mobile cellulaire.
40	GSM.
41	Les données portent sur les abonnements prépayés actifs pendant les six derniers mois (180 jours).
42	Données à la fin de mars.
43	Baisse due aux modifications apportées à la méthode utilisée par l'opérateur pour calculer le nombre d'abonnements prépayés.
44	Y compris les abonnements prépayés inactifs.
45	Les données portent sur les abonnements prépayés actifs pendant le dernier mois (30 jours).
46	Année finissant le 31 javier de l'année suivante.
47	La baisse est due aux contrôles réglementaires sur les cartes SIM prépayées restreignant chaque utilisateur final à ne pas détenir plus de 3 cartes prépayées.
48	Chiffre obtenu des cinq opérateurs de téléphonie mobile dans le pays.
49	Canar comptabilisé comme ligne fixe.
50	Y compris les lignes mobiles GSM et AMRC actives.

Internet usage
Percentage of individuals per country

Utilisation d'Internet
Pourcentage de personnes par pays

Country or area	2000	2005	2010	2012	2013	2014	2015	Pays ou zone
Afghanistan	...	1.2	* 4.0	* 5.5	* 5.9	* 7.0	* 8.3	Afghanistan
Albania	0.1	6.0	45.0	* 54.7	* 57.2	* 60.1	* 63.3	Albanie
Algeria	0.5	5.8	12.5	* 15.2	* 16.5	* 25.0	* 38.2	Algérie
Andorra	10.5	37.6	81.0	* 86.4	* 94.0	* 95.9	* 96.9	Andorre
Angola	0.1	1.1	* 2.8	* 6.5	* 8.9	10.2	* 12.4	Angola
Anguilla	22.4	* 29.0	* 49.6	* 59.2	* 64.8	* 70.4	* 76.0	Anguilla
Antigua and Barbuda	6.5	* 27.0	* 47.0	* 58.0	* 63.4	* 64.0	* 65.2	Antigua-et-Barbuda
Argentina	7.0	17.7	* 45.0	* 55.8	* 59.9	* 64.7	* 69.4	Argentine
Armenia	1.3	5.3	* 25.0	* 37.5	41.9	54.6	* 58.2	Arménie
Aruba	15.4	* 25.4	* 62.0	* 74.0	* 78.9	* 83.8	* 88.7	Aruba
Ascension	34.9	...	35.0	* 41.0	...	...	...	Ascension
Australia	46.8	63.0[1]	* 76.0	79.0[1]	83.5[1]	* 84.0	84.6[1]	Australie
Austria	33.7[2]	58.0[3]	75.2[3]	80.0[4]	80.6[3]	81.0[3]	83.9[3]	Autriche
Azerbaijan	0.1	8.0	46.0[5]	* 54.2	73.0[5]	75.0[5]	77.0[5]	Azerbaïdjan
Bahamas *	8.0	25.0	43.0	71.7	72.0	76.9	78.0	Bahamas *
Bahrain	6.2	21.3	55.0	* 88.0[1]	90.0[1]	90.5[1]	93.5[1]	Bahreïn
Bangladesh	* 0.1	* 0.2	* 3.7	* 5.0	6.6	* 13.9	* 14.4	Bangladesh
Barbados	4.0	* 52.5	* 65.1	* 71.2	* 71.8	* 75.2	* 76.1	Barbade
Belarus	1.9	...	31.8[6]	46.9[6]	# 54.2[7]	59.0[7]	62.2[7]	Bélarus
Belgium	29.4	* 55.8	75.0	80.7[4]	82.2[3]	85.0[3]	85.1[3]	Belgique
Belize	6.0	* 17.0	28.2[8]	* 31.0	* 33.6	* 38.7	* 41.6	Belize
Benin	0.2	1.3	3.1	* 4.5	* 4.9	* 6.0	* 6.8	Bénin
Bermuda	42.9	65.4	* 84.2	* 91.3	* 95.3	* 96.8	* 98.3	Bermudes
Bhutan	0.4	3.8	* 13.6	* 15.6	* 22.4	* 30.3	* 39.8	Bhoutan
Bolivia (Plurinational State of)	* 1.4	5.2	* 22.4	35.3[8,9]	37.0[4,8]	34.6[4,8]	* 45.1	Bolivie (État plurinational de)
Bosnia and Herzegovina	1.1	21.3	* 42.8	* 52.8	* 57.8	* 60.8	* 65.1	Bosnie-Herzégovine
Botswana	2.9	* 3.3	6.0	* 11.5	* 15.0	* 18.5	* 27.5	Botswana
Brazil	2.9	21.0[4,10]	40.7[4,10]	48.6[10]	51.0[10]	54.6[10]	* 59.1	Brésil
British Virgin Islands	...	...	37.0	37.6	...	...	...	Îles Vierges britanniques
Brunei Darussalam	9.0	36.5	* 53.0	* 60.3	* 64.5	* 68.8	* 71.2	Brunéi Darussalam
Bulgaria	5.4	20.0[3]	46.2[3]	51.9[4]	53.1[3]	55.5[3]	56.7[3]	Bulgarie
Burkina Faso	0.1	0.5	* 2.4	* 3.7	* 9.1	* 9.4	* 11.4	Burkina Faso
Burundi	0.1	0.5	* 1.0	* 1.2	1.3[8]	* 1.4	* 4.9	Burundi
Cabo Verde	1.8	6.1	* 30.0	34.7	* 37.5	* 40.3	* 43.0	Cabo Verde
Cambodia	<0.0	0.3	1.3	4.9	* 6.8	* 14.0	* 19.0	Cambodge
Cameroon	0.3	1.4	* 4.3	...	...	16.2[1]	* 20.7	Cameroun
Canada	51.3	71.6[6]	# 80.3[6]	83.0[6,11]	* 85.8	* 87.1	* 88.5	Canada
Cayman Islands	...	38.0	* 66.0	* 69.7	* 71.4	* 74.1	* 77.0	Îles Caïmanes
Central African Rep.	0.1	* 0.3	* 2.0	* 3.0	* 3.5	* 4.0	* 4.6	Rép. centrafricaine
Chad	<0.0	0.4	* 1.7	* 2.1	* 2.3	* 2.5	* 2.7	Tchad
Chile	16.6[8]	* 31.2[8]	* 45.0[8]	* 55.1	58.0[8]	* 61.1	* 64.3	Chili
China [12]	1.8	8.5	34.3	* 42.3	* 45.8[7,13]	* 47.9[7,13]	* 50.3[7,13]	Chine [12]
China, Hong Kong SAR [10]	27.8[9]	56.9	72.0	72.9	74.2	79.9	84.9	Chine, Hong Kong RAS [10]
China, Macao SAR	* 13.6	* 34.9	55.2[14]	61.3[14]	65.8[14]	69.8[14]	77.6[14]	Chine, Macao RAS
Colombia	2.2	11.0	36.5[8]	49.0[8]	51.7[8]	52.6[8]	55.9[8]	Colombie
Comoros	0.3	* 2.0	* 5.1	* 6.0	* 6.5	* 7.0	* 7.5	Comores
Congo	<0.0	* 1.5	* 5.0	* 6.1	* 6.6	* 7.1	* 7.6	Congo
Costa Rica	5.8	22.1[8]	36.5[4,8]	47.5[4,8]	# 46.0[4]	* 53.0	59.8[4,8]	Costa Rica
Côte d'Ivoire	0.2	1.0	* 2.7	* 5.0	* 8.4	* 14.6	* 21.0	Côte d'Ivoire
Croatia	6.6	33.1[3]	56.6[3]	61.9[4]	66.7[3]	68.6[3]	69.8[3]	Croatie
Cuba	0.5[15]	9.7[16]	15.9[16]	* 21.2	27.9[7,16]	29.1[7]	* 31.1	Cuba
Cyprus	15.3	32.8[3]	53.0[3]	60.7[4]	65.5[3]	69.3[3]	71.7[3]	Chypre
Czech Republic	9.8	35.3[3]	68.8[3]	73.4[4]	74.1[3]	79.7[3]	81.3[3]	République tchèque
Dem. P. R. Korea	0.0[17]	0.0[17]	* 0.0	* 0.0	...	...	...	R. p. dém. de Corée
Dem. Rep. of the Congo *	<0.0	0.2	0.7	1.7	2.2	3.0	3.8	Rép. dém. du Congo *
Denmark	39.2[18]	82.7[3]	88.7[3]	92.3[4]	94.6[3]	96.0[3]	96.3[3]	Danemark
Djibouti	0.2	1.0	* 6.5	* 8.3	* 9.5	* 10.7	* 11.9	Djibouti
Dominica	8.8	* 38.5	47.5	* 49.8	* 51.0	57.5	* 67.6	Dominique
Dominican Republic	3.7	11.5	* 31.4	42.3[19]	45.9[19]	* 49.6	51.9[19]	Rép. dominicaine
Ecuador	1.5	* 6.0	29.0[8]	35.1[8]	40.3[8]	45.6[8]	48.9[8]	Équateur
Egypt	* 0.6	12.8	21.6[7]	26.4[7]	29.4[7]	33.9[7]	35.9	Égypte

Country or area	2000	2005	2010	2012	2013	2014	2015	Pays ou zone
El Salvador	1.2	* 4.2[10]	15.9[10]	20.3	23.1[9,10]	24.8[10]	* 26.9	El Salvador
Equatorial Guinea	0.1	1.1	6.0	* 13.9	* 16.4	* 18.9	* 21.3	Guinée équatoriale
Eritrea	0.1	…	* 0.6	* 0.8	* 0.9	* 1.0	* 1.1	Érythrée
Estonia	28.6	61.5[3]	74.1[3,4]	78.4[4]	79.4[3]	84.2[3]	88.4[3]	Estonie
Ethiopia	<0.0	0.2	* 0.8	* 2.9	* 4.6	* 7.7	* 11.6	Éthiopie
Falkland Is. (Malvinas)	58.6	* 84.0	95.8	* 96.9	* 96.9	* 97.6	* 98.3	Îles Falkland (Malvinas)
Faroe Islands	32.9	* 67.9	75.2	* 85.3	* 90.0	* 93.3	* 94.2	Îles Féroé
Fiji	1.5	8.5	* 20.0	* 33.7	* 37.1	* 41.8	* 46.3	Fidji
Finland	37.2[1,20]	74.5[3]	86.9[3]	89.9[4]	91.5[3]	92.4[3]	92.7[3]	Finlande
France	14.3[1,9]	42.9[21,22]	77.3[3]	81.4[4]	81.9[3]	83.8[3]	84.7[3]	France
French Polynesia	6.4	21.5	49.0	* 52.9	* 56.8	* 60.7	* 64.6	Polynésie française
Gabon	1.2	4.9	* 7.2	* 8.6	* 9.2	* 20.0	* 23.5	Gabon
Gambia	0.9	* 3.8	9.2	* 12.4	* 14.0	* 16.5	* 17.1	Gambie
Georgia	0.5	* 6.1	26.9	* 36.9	43.3[7,9]	* # 44.0[4,7]	45.2[4,7]	Géorgie
Germany	30.2	68.7[3]	82.0[3]	82.4[4]	84.2[3]	86.2[3]	87.6[3]	Allemagne
Ghana	0.2	1.8	# 7.8[19]	* 10.6	12.3	* 18.9	* 23.5	Ghana
Gibraltar	19.1	* 39.1	65.0	* 65.0	…	…	…	Gibraltar
Greece	9.1	24.0[3]	44.4[3]	55.1[4]	59.9[3]	63.2[3]	66.8[3]	Grèce
Greenland	31.7	57.7	63.0	* 64.9	* 65.8	* 66.7	* 67.6	Groenland
Grenada	4.1	* 20.5	* 27.0	* 32.0	* 35.0	51.6	* 53.8	Grenade
Guam	16.1	38.6	* 54.0	* 61.5	* 65.4	* 69.3	* 73.1	Guam
Guatemala	0.7	5.7	* 10.5	* 16.0	* 19.7	* 23.4	* 27.1	Guatemala
Guernsey	31.9	73.6	…	…	…	…	…	Guernesey
Guinea	0.1	0.5	* 1.0	* 1.5	* 1.6	* 1.7	* 4.7	Guinée
Guinea-Bissau	0.2	1.9	* 2.5	* 2.9	* 3.1	* 3.3	* 3.5	Guinée-Bissau
Guyana	6.6	…	29.9	* 33.0	* 35.0	* 37.4	* 38.2	Guyana
Haiti	0.2	* 6.4	* 8.4	* 9.8	* 10.6	* 11.4	* 12.2	Haïti
Honduras	1.2	* 6.5[8]	11.1	18.1	17.8	* 19.1	* 20.4	Honduras
Hungary	7.0	39.0[3]	65.0[3]	70.6[4]	72.6[3]	76.1[3]	72.8[3]	Hongrie
Iceland	44.5	87.0[3]	93.4[3,4]	96.2[4]	96.5[3]	98.2[3]	* 98.2	Islande
India	0.5	* 2.4	* 7.5	* 12.6	* 15.1	* 21.0	* 26.0	Inde
Indonesia	0.9	3.6	10.9	14.5[8]	14.9[8]	17.1[8]	22.0[8]	Indonésie
Iran (Islamic Rep. of)	0.9	* 8.1	15.9[7]	22.7[7]	30.0[7]	* 39.4	* 44.1	Iran (Rép. islamique d')
Iraq	…	* 0.9	2.5	* 7.1	* 9.2	13.2	* 17.2	Iraq
Ireland	17.9[1]	41.6[3]	69.9[3]	76.9[4]	78.2[3]	79.7[3]	80.1[3]	Irlande
Israel	20.9	25.2	67.5[23]	70.8[4,23]	70.3[4,23]	75.0[23]	* 78.9	Israël
Italy	23.1	35.0[3]	53.7[3]	55.8[4]	58.5[3]	62.0[3]	65.6[3]	Italie
Jamaica	3.1	* 12.8	27.7[2]	33.8[2]	37.1[2]	40.4[2]	* 43.2	Jamaïque
Japan	30.0[24]	66.9[7]	78.2[7]	79.5[7]	88.2[7]	89.1[7]	* 93.3	Japon
Jersey	9.2	31.3	…	* 41.0	…	…	…	Jersey
Jordan	2.6	12.9	27.2[8]	37.0[8]	41.4[8]	46.2[8]	* 53.4	Jordanie
Kazakhstan	0.7	3.0	31.6[3]	* 53.3	* 63.0	* 66.0	72.9[25]	Kazakhstan
Kenya	0.3	3.1	14.0	* 32.1	* 39.0	* 43.4	* 45.6	Kenya
Kiribati	* 1.8	* 4.0	9.1	* 10.7	* 11.5	* 12.3	* 13.0	Kiribati
Kuwait	6.7	25.9	* 61.4	* 70.5	* 75.5	* 78.7	* 82.1	Koweït
Kyrgyzstan	1.0	10.5	* 16.3	* 19.8	* 23.0	* 28.3	* 30.2	Kirghizistan
Lao People's Dem. Rep.	0.1	0.9	7.0	* 10.7	* 12.5	* 14.3	* 18.2	Rép. dém. pop. lao
Latvia	6.3	46.0[3]	68.4[3]	73.1[4]	75.2[3]	75.8[3]	79.2[3]	Lettonie
Lebanon	8.0	10.1[7]	* 43.7[1]	* 61.2	* 70.5	* 73.0	* 74.0	Liban
Lesotho	0.2	* 2.6	* 3.9	* 4.6	* 5.0	* 11.0	* 16.1	Lesotho
Liberia	<0.0	…	2.3	* 2.6	* 3.2	* 5.4	* 5.9	Libéria
Libya	0.2	* 3.9	* 14.0	…	* 16.5	* 17.8	* 19.0	Libye
Liechtenstein	36.5	63.4	* 80.0	* 89.4	* 93.8	* 95.2	* 96.6	Liechtenstein
Lithuania	6.4	36.2[3,9]	62.1[3,9]	67.2[4]	68.5[3]	72.1[3]	71.4[3]	Lituanie
Luxembourg	22.9	70.0[3]	90.6[3]	91.9[4]	93.8[3]	94.7[3]	97.3[3]	Luxembourg
Madagascar	0.2	* 0.6	* 1.7	* 2.3	* 3.0	* 3.7	* 4.2	Madagascar
Malawi	0.1	0.4	2.3	* 4.4	* 5.1	* 5.8	* 9.3	Malawi
Malaysia	21.4	48.6	56.3	65.8	57.1[1]	63.7[1]	71.1[1]	Malaisie
Maldives	* 2.2[26]	* 6.9[26]	26.5[1]	* 38.9	* 44.1	* 49.3	* 54.5	Maldives
Mali	0.1	0.5	* 2.0	* 2.8	* 3.5	* 7.0	* 10.3	Mali
Malta	13.1	41.2[3]	63.0[3]	68.2[4]	68.9[3]	73.2[3]	76.2[3]	Malte
Marshall Islands	1.5	3.9	* 7.0	* 12.5	* 14.0	* 16.8	* 19.3	Îles Marshall
Mauritania	0.2	0.7	* 4.0	* 5.0	* 6.2	* 10.7	* 15.2	Mauritanie

Country or area	2000	2005	2010	2012	2013	2014	2015	Pays ou zone
Mauritius	7.3	* 15.2	28.3[8]	35.4[8]	40.1[8]	44.8[8]	50.1[8]	Maurice
Mayotte	1.2	...	...	...	...	...	...	Mayotte
Mexico	5.1	* 17.2[27]	* 31.1[27]	39.8[7]	43.5[7]	44.4[7]	# 57.4[7]	Mexique
Micronesia (Fed. States of)	3.7	11.9	* 20.0	* 26.0	* 27.8	* 29.7	* 31.5	Micronésie (États féd. de)
Monaco	42.2	55.5	75.0	* 87.0	* 90.7	* 92.4	* 93.4	Monaco
Mongolia	1.3	...	10.2	16.4	17.7	19.9	21.4	Mongolie
Montenegro	...	* 27.1	* 37.5	56.8[3]	60.3[3]	* 61.0	* 64.6	Monténégro
Montserrat	...	...	35.0	...	...	...	...	Montserrat
Morocco	0.7	15.1[22,28]	* 52.0[25,29]	55.4[9,25]	56.0[9,30]	56.8[4,30]	57.1[4,8]	Maroc
Mozambique	0.1	* 0.9	4.2	* 4.8	* 5.4	* 5.9	* 9.0	Mozambique
Myanmar	...	0.1	0.3	* 1.4	* 1.8	* 11.5	* 21.8	Myanmar
Namibia	1.6	* 4.0	11.6	* 12.9	* 13.9	* 14.8	* 22.3	Namibie
Nepal	0.2	0.8	7.9[27]	* 11.1	* 13.3	* 15.4	* 17.6	Népal
Netherlands	44.0	81.0[3,9]	90.7[3,9]	92.9[4]	94.0[3]	93.2[3]	93.1[3]	Pays-Bas
New Caledonia	13.9	32.4	* 42.0	* 58.0	* 66.0	* 70.0	* 74.0	Nouvelle-Calédonie
New Zealand	* 47.4	* 62.7	* 80.5	82.0[1,9]	* 82.8	* 85.5	* 88.2	Nouvelle-Zélande
Nicaragua	1.0	2.6	* 10.0	* 13.5	* 15.5	* 17.6	* 19.7	Nicaragua
Niger	<0.0	* 0.2	* 0.8	* 1.4	* 1.7	* 2.0	* 2.2	Niger
Nigeria	0.1	* 3.5	* 24.0	32.8	38.0	* 42.7	* 47.4	Nigéria
Niue	26.5	51.7	* 77.0	...	...	...	...	Nioué
Norway	* 52.0	82.0[3]	93.4[3]	94.7[4]	95.1[3]	96.3[3]	96.8[3]	Norvège
Oman	* 3.5	6.7	35.8[8]	* 60.0	66.5[8,31]	* 70.2	* 74.2	Oman
Other non-specified areas	28.1	58.0	71.5	* 76.0	* 80.0	* 84.0	* 88.0	Autres zones non-spécifiées
Pakistan	...	6.3	* 8.0	* 10.0	* 10.9	* 13.8	* 18.0[10]	Pakistan
Panama	6.6	11.5	* 40.1	40.3[10,13]	44.0[10]	44.9[10]	51.2[10]	Panama
Papua New Guinea	0.8	1.7	1.3[10]	* 3.5	* 5.1	* 6.5	* 7.9	Papouasie-Nvl-Guinée
Paraguay	0.7	7.9[4,10]	19.8[4,10]	29.3[4,10]	36.9[4,10]	43.0[4,10]	* 44.4	Paraguay
Peru	* 3.1	* 17.1	34.8[7]	38.2[7]	39.2[7]	40.2[7]	40.9[7]	Pérou
Philippines	2.0	* 5.4	25.0	* 36.2	* 37.0	* 39.7	* 40.7	Philippines
Poland	7.3	38.8[3]	62.3[3]	62.3[4]	62.8[3]	66.6[3]	68.0[3]	Pologne
Portugal	* 16.4	35.0[3]	53.3[3]	60.3[4]	62.1[3]	64.6[3]	68.6[3]	Portugal
Puerto Rico	10.5	* 23.4	45.3[19]	69.0	69.0[32]	76.1[32]	* 79.5	Porto Rico
Qatar	4.9	24.7	69.0	69.3	* 85.3	* 91.5	92.9[1]	Qatar
Republic of Korea	44.7[14,22]	73.5[14]	83.7[14]	84.1[14]	84.8[14]	87.9[3]	89.9[3]	République de Corée
Republic of Moldova	1.3	14.6	* 32.3	* 43.4	* 45.0	* 46.6	* 49.8	République de Moldova
Romania	3.6	* 21.5[3]	39.9[3]	45.9[4]	49.8[3]	54.1[3]	55.8[3]	Roumanie
Russian Federation	2.0	15.2	43.0[3]	63.8	68.0[33]	70.5[9,33]	73.4[9,33]	Fédération de Russie
Rwanda	0.1	* 0.6	* 8.0	* 8.0	* 9.0	* 10.6	* 18.0	Rwanda
Saint Helena	5.9	15.9	24.9	* 37.6	...	...	...	Sainte-Hélène
Saint Kitts and Nevis	5.9	* 34.0	63.0	* 64.0	* 64.6	68.0	* 75.7	Saint-Kitts-et-Nevis
Saint Lucia	5.1	21.6	43.3	45.9[1,9]	* 46.2	* 50.0	* 52.4	Sainte-Lucie
Saint Vincent-Grenadines	* 3.2	* 9.2	* 33.7	* 40.0	* 43.5	47.4	* 51.8	Saint-Vincent-Grenadines
Samoa	0.6	3.4	* 7.0	* 12.9	* 15.3	* 21.2	* 25.4	Samoa
San Marino	48.8	50.3	...	...	...	...	...	Saint-Marin
Sao Tome and Principe	4.6	* 13.8	18.8	* 21.6	* 23.0	* 24.4	* 25.8	Sao Tomé-et-Principe
Saudi Arabia	2.2	12.7	41.0	* 54.0	* 60.5	64.7[28]	69.6[28]	Arabie saoudite
Senegal	0.4	4.8	* 8.0	* 10.8	* 13.1	* 17.7	* 21.7	Sénégal
Serbia	...	* 26.3[3]	40.9	* 48.1	53.5[3]	62.1[3]	65.3[3]	Serbie
Seychelles	* 7.4	25.4	* 41.0	* 47.1	* 50.4	* 54.3	* 58.1	Seychelles
Sierra Leone	0.1	0.2	* 0.6	* 1.3	* 1.7	* 2.1	* 2.5	Sierra Leone
Singapore	36.0	61.0[1]	* 71.0[5]	72.0[5]	80.9[5]	79.0	* 82.1	Singapour
Slovakia	9.4	55.2[3]	75.7[3,4]	76.7[4]	77.9[3]	80.0[3]	* 85.0	Slovaquie
Slovenia	15.1	46.8[4]	70.0[3]	68.4[4]	72.7[3]	71.6[3]	73.1[3]	Slovénie
Solomon Islands	0.5	0.8	* 5.0	* 7.0	* 8.0	* 9.0	* 10.0	Îles Salomon
Somalia	<0.0	* 1.1	...	* 1.4	* 1.5	* 1.6	* 1.8	Somalie
South Africa	5.3	7.5	* 24.0	* 41.0	* 46.5	* 49.0	* 51.9	Afrique du Sud
South Sudan *	...	...	7.0	...	14.1	15.9	17.9	Soudan du sud *
Spain	13.6[2,34]	47.9[3,9]	65.8[10]	69.8[4]	71.6[3]	76.2[3]	78.7[3]	Espagne
Sri Lanka	0.6	* 1.8	12.0	* 18.3	* 21.9	25.8	* 30.0	Sri Lanka
State of Palestine	1.1	* 16.0[10]	* 37.4	43.4	* 46.6	53.7[10]	* 57.4	État de Palestine
Sudan	<0.0	1.3	* 16.7	21.0[1]	* 22.7	* 24.6	* 26.6	Soudan
Suriname	2.5	6.4	31.6	* 34.7	* 37.4	* 40.1	* 42.8	Suriname
Swaziland	0.9	* 3.7	11.0	* 20.8	* 24.7	* 27.1	* 30.4	Swaziland

Country or area	2000	2005	2010	2012	2013	2014	2015	Pays ou zone
Sweden	45.7	84.8[3]	90.0[35]	93.2[4]	94.8[3]	92.5[3]	90.6[3]	Suède
Switzerland [2,13]	47.1	70.1	83.9	85.2	86.3	87.4	88.0	Suisse [2,13]
Syrian Arab Republic	0.2	* 5.6	20.7	* 24.3	* 26.2	* 28.1	* 30.0	Rép. arabe syrienne
Tajikistan	<0.0	0.3	* 11.6	* 14.5	* 16.0	* 17.5	* 19.0	Tadjikistan
Thailand	3.7	15.0	22.4	26.5[7]	28.9[7]	# 34.9[7]	39.3[7]	Thaïlande
TFYR of Macedonia	2.5	* 26.5[3]	51.9[36]	57.4[3,4]	65.2[3,4]	68.1[3,4]	70.4[3]	ex-R.Y. de Macédoine
Timor-Leste	...	0.1	* 0.2	* 2.3	* 8.0	* 11.3	* 13.4	Timor-Leste
Togo *	0.8	1.8	3.0	4.0	4.5	5.7	7.1	Togo *
Tonga	2.4	* 4.9	* 16.0	* 33.0	* 35.0	* 40.0	* 45.0	Tonga
Trinidad and Tobago	7.7	* 29.0	* 48.5	* 59.5	* 63.8	* 65.1	* 69.2	Trinité-et-Tobago
Tunisia	2.8	9.7	36.8	* 41.4	* 43.8	* 46.2	* 48.5	Tunisie
Turkey [3]	3.8	15.5[9]	39.8[9]	45.1[9]	46.3[9]	51.0[9]	53.7	Turquie [3]
Turkmenistan	0.1	* 1.0	* 3.0	* 7.2	* 9.6	* 12.2	* 15.0	Turkménistan
Tuvalu	5.2	...	* 25.0	* 35.0	* 37.0	* 39.2	* 42.7	Tuvalu
Uganda	0.2	1.7	12.5	* 14.7	* 16.2	* 17.7	* 19.2	Ouganda
Ukraine	0.7	* 3.7[22,37]	23.3	35.3	41.0	46.2	* 49.3	Ukraine
United Arab Emirates	23.6	* 40.0	68.0	85.0[4,36]	* 88.0	90.4[4,36]	* 91.2	Émirats arabes unis
United Kingdom	26.8[4,6]	70.0[3]	85.0[3]	87.5[4]	89.8[3]	91.6[3]	92.0[3]	Royaume-Uni
United Rep. of Tanzania	0.1	* 1.1	* 2.9	* 4.0	* 4.4	* 4.9	* 5.4	Rép.-Unie de Tanzanie
United States	* 43.1	* 68.0	71.7[14]	74.7[14]	71.4[14]	* 73.0	74.6[14]	États-Unis
United States Virgin Is.	13.8	* 27.3	* 31.2	* 40.5	* 45.3	* 50.1	* 54.8	Îles Vierges américaines
Uruguay	10.5	20.1	46.4[7]	54.5[7]	57.7[7]	61.5[7]	64.6[7]	Uruguay
Uzbekistan	0.5	3.3	* 15.9	* 23.6	* 26.8	* 35.5	* 42.8	Ouzbékistan
Vanuatu	2.1	5.1	8.0	* 10.6	* 11.3	* 18.8	* 22.4	Vanuatu
Venezuela (Boliv. Rep. of)	3.4	12.6	* 37.4	49.1	* 54.9	* 57.0	* 61.9	Venezuela (Rép. boliv. du)
Viet Nam	0.3	12.7	30.7	* 39.5	* 43.9	* 48.3	* 52.7	Viet Nam
Wallis and Futuna Islands	4.8	6.7	8.2	* 9.0	...	...	...	Îles Wallis-et-Futuna
Yemen	0.1	* 1.0	12.4	* 17.4	* 20.0	* 22.6	* 25.1	Yémen
Zambia	0.2	* 2.9	10.0	* 13.5	* 15.4	* 19.0	* 21.0	Zambie
Zimbabwe	* 0.4	* 2.4	* 6.4	* 12.0	* 15.5	16.3[14]	* 16.4	Zimbabwe

Source:

International Telecommunication Union (ITU), Geneva, the ITU database, last accessed July 2016.

Source:

Union internationale des télécommunications (UIT), Genève, la base de données de l'UIT, dernier accès juillet 2016.

1	Population age 15+.
2	Persons aged 14 years and over.
3	Population age 16-74.
4	Users in the last 3 months.
5	Population age 7+.
6	Persons aged 16 years and over.
7	Population age 6+.
8	Population age 5+.
9	Users in the last 12 months.
10	Population age 10+.
11	Residents of Canada excluding: Residents of the Yukon, Northwest Territories and Nunavut, Inmates of Institutions, Persons living on Indian Reserves, and Full time members of the Canadian Forces.
12	For statistical purposes, the data for China do not include those for the Hong Kong Special Administrative Region (Hong Kong SAR), Macao Special Administrative Region (Macao SAR) and Taiwan Province of China.
13	Users in the last 6 months.
14	Population age 3+.
15	Refers only to users with access to the international network.

1	Population âgée de plus de 15 ans.
2	Population âgée de plus de 14 ans.
3	Population âgée de 16 à 74 ans.
4	Utilisateurs au cours des 3 derniers mois.
5	Population âgée de plus de 7 ans.
6	Personnes âgées de 16 ans et plus.
7	Population âgée de plus de 6 ans.
8	Population âgée de plus de 5 ans.
9	Utilisateurs au cours des 12 derniers mois.
10	Population âgée de plus de 10 ans.
11	Les résidents du Canada a l'exclusion: des résidents du Yukon, des Territoires du Nord-Ouest et du Nunavut, des pensionnaires des établissements, des personnes vivant sur les réserves indiennes et les personnes à plein-temps dans les Forces canadiennes.
12	Pour la présentation des statistiques, les données pour la Chine ne comprennent pas la Région Administrative Spéciale de Hong Kong (Hong Kong RAS), la Région Administrative Spéciale de Macao (Macao RAS) et la province de Taiwan.
13	Utilisateurs au cours des 6 derniers mois.
14	Population âgée de plus de 3 ans.
15	Uniquement utilisateurs ayant accès au réseau international.

16	Including users of the international network and also those having access only to the Cuban network.	16	Y compris les utilisateurs du réseau international et ceux qui n'ont accès qu'au réseau cubain.
17	Commercially not available. Local Intranet available in country.	17	Non disponible commercialement. L'Intranet local est disponible dans le pays.
18	E-mail users.	18	Utilisateurs du courrier électronique
19	Population age 12+.	19	Population âgée de plus de 12 ans.
20	Has used at least one other Internet application besides e-mail in last 3 months.	20	Ayant au moins utilisé une application Internet autre que le courrier électronique au cours des trois derniers mois. Population âgée de plus de 15 ans.
21	Population age 11+	21	Population âgée de plus de 11 ans.
22	Users in the last month.	22	Utilisateurs au cours du mois passé.
23	Population age 20+.	23	Population âgée de plus de 20 ans.
24	PC-based only.	24	Sur ordinateur PC seulement.
25	Population age 6-74.	25	Population âgée de 6 à 74 ans.
26	Excluding mobile internet users.	26	Les utilisateurs d'Internet mobile non-inclus.
27	December.	27	Décembre.
28	Population age 12-65.	28	Population âgée 12 a 65 ans.
29	Living in electrified areas.	29	Vivant dans des zones électrifiées.
30	Population age 5 to 75.	30	Population âgée de 5 à 75.
31	Excluding population living in workers' camps.	31	Non compris la population vivant dans les camps de travailleurs.
32	Population age 18+.	32	Population âgée de plus de 18 ans.
33	Population age 15-72.	33	Population âgée de 15 à 72 ans.
34	November.	34	Novembre.
35	Population age 16-75.	35	Population âgée de 15 à 75 ans.
36	Population age 15-74.	36	Population âgée de 15 à 74 ans.
37	Population age 15-59.	37	Populationâgée de 15 à 59.

24

Tourist/visitor arrivals and tourism expenditure
Thousands arrivals and millions of US dollars

Arrivées de touristes/visiteurs et dépenses touristiques
Milliers d'arrivées et millions de dollars É.-U.

Country or area of destination	Series& Série&	1995	2005	2010	2012	2013	2014	Pays ou zone de destination
Afghanistan								**Afghanistan**
Tourism expenditure		...	...	167	168	154	91	Dépenses touristiques
Albania								**Albanie**
Tourist/visitor arrivals [1]	TF	...	...	2 191	3 156	2 857	3 341	Arrivées de touristes/visiteurs [1]
Tourism expenditure		70	880	1 780	1 623	1 670	1 849	Dépenses touristiques
Algeria								**Algérie**
Tourist/visitor arrivals [2]	VF	520	1 443	2 070	2 634	2 733	2 301	Arrivées de touristes/visiteurs [2]
Tourism expenditure		...	477	324	295	326	347	Dépenses touristiques
American Samoa	TF							**Samoa américaines**
Tourist/visitor arrivals		34	25	23	23	21	22	Arrivées de touristes/visiteurs
Andorra	TF							**Andorre**
Tourist/visitor arrivals		...	2 418	#1 808	2 238	2 328	2 363	Arrivées de touristes/visiteurs
Angola								**Angola**
Tourist/visitor arrivals	TF	9	210	425	528	650	595	Arrivées de touristes/visiteurs
Tourism expenditure		27	103	726	711	1 241	1 597	Dépenses touristiques
Anguilla								**Anguilla**
Tourist/visitor arrivals [1]	TF	39	62	62	65	69	71	Arrivées de touristes/visiteurs [1]
Tourism expenditure [3]		50	86	99	113	122	...	Dépenses touristiques [3]
Antigua and Barbuda								**Antigua-et-Barbuda**
Tourist/visitor arrivals [1]	TF	220	245[4]	230[4]	247[4]	243[4]	249[4]	Arrivées de touristes/visiteurs [1]
Tourism expenditure [3]		247	309	298	319	299	...	Dépenses touristiques [3]
Argentina								**Argentine**
Tourist/visitor arrivals	TF	2 289	3 823	5 325	5 587	#5 246	5 931	Arrivées de touristes/visiteurs
Tourism expenditure		2 550	3 209	5 629	5 538	4 918	5 218	Dépenses touristiques
Armenia								**Arménie**
Tourist/visitor arrivals	TF	12	319	684	963	1 084	1 204	Arrivées de touristes/visiteurs
Tourism expenditure		14	243	694	853	905	994	Dépenses touristiques
Aruba								**Aruba**
Tourist/visitor arrivals	TF	619	733	824	904	979	1 072	Arrivées de touristes/visiteurs
Tourism expenditure		554	...	1 254	1 412	1 511	1 632	Dépenses touristiques
Australia								**Australie**
Tourist/visitor arrivals [5]	VF	3 726	5 499	5 790	6 032	6 382	6 868	Arrivées de touristes/visiteurs [5]
Tourism expenditure		11 915	19 719	31 064	34 497	33 575	34 117	Dépenses touristiques
Austria								**Autriche**
Tourist/visitor arrivals	TCE	17 173	19 952	22 004	24 151	24 813	25 291	Arrivées de touristes/visiteurs
Tourism expenditure [3]		13 435	16 243	18 757	18 937	20 220	20 907	Dépenses touristiques [3]
Azerbaijan								**Azerbaïdjan**
Tourist/visitor arrivals	TF	...	693	1 280	1 986	2 130	2 160	Arrivées de touristes/visiteurs
Tourism expenditure		87	100	792	2 634	2 618	2 713	Dépenses touristiques
Bahamas								**Bahamas**
Tourist/visitor arrivals	TF	1 598	1 608	1 370	1 422	1 364	1 427	Arrivées de touristes/visiteurs
Tourism expenditure		1 356	2 081	2 159	2 333	2 305	2 470	Dépenses touristiques
Bahrain								**Bahreïn**
Tourist/visitor arrivals	VF	2 311	6 313	11 952	8 062	9 163	10 452	Arrivées de touristes/visiteurs
Tourism expenditure		593	1 603	2 163	1 742	1 865	1 915	Dépenses touristiques
Bangladesh								**Bangladesh**
Tourist/visitor arrivals	TF	156	208	303	125	148	125	Arrivées de touristes/visiteurs
Tourism expenditure		...	82	104	105	131	154	Dépenses touristiques
Barbados								**Barbade**
Tourist/visitor arrivals	TF	442	548	532	536	509	521	Arrivées de touristes/visiteurs
Tourism expenditure		630	1 081	1 074	947	992	...	Dépenses touristiques
Belarus								**Bélarus**
Tourist/visitor arrivals [6]	TF	161	91	119	119	137	137	Arrivées de touristes/visiteurs [6]
Tourism expenditure		28	346	665	986	1 156	1 230	Dépenses touristiques

24

Tourist/visitor arrivals and tourism expenditure *(continued)*
Thousands arrivals and millions of US dollars

Arrivées de touristes/visiteurs et dépenses touristiques *(suite)*
Milliers d'arrivées et millions de dollars É.-U.

Country or area of destination	Series& Série&	1995	2005	2010	2012	2013	2014	Pays ou zone de destination
Belgium								**Belgique**
Tourist/visitor arrivals	TCE	5 560	6 747	7 186	7 560	7 684	7 887	Arrivées de touristes/visiteurs
Tourism expenditure		...	10 881	...	13 711	14 429	15 302	Dépenses touristiques
Belize								**Belize**
Tourist/visitor arrivals	TF	131	237	242	277	294	321	Arrivées de touristes/visiteurs
Tourism expenditure [3]		78	214	264	298	351	380	Dépenses touristiques [3]
Benin								**Bénin**
Tourist/visitor arrivals	TF	138	176	199	220	231	242	Arrivées de touristes/visiteurs
Tourism expenditure		...	108	149	174	193	...	Dépenses touristiques
Bermuda								**Bermudes**
Tourist/visitor arrivals [4]	TF	387	270	232	232	236	224	Arrivées de touristes/visiteurs [4]
Tourism expenditure		...	...	...	448	447	420	Dépenses touristiques
Bhutan								**Bhoutan**
Tourist/visitor arrivals	TF	5	14	#41	105	116	134	Arrivées de touristes/visiteurs
Tourism expenditure		5	19	64	93	116	120	Dépenses touristiques
Bolivia (Plur. State of)								**Bolivie (État plur. de)**
Tourist/visitor arrivals	TF	284	524	679	798	798	871	Arrivées de touristes/visiteurs
Tourism expenditure		92	345	339	631	639	736	Dépenses touristiques
Bonaire								**Bonaire**
Tourist/visitor arrivals	TF	59	63	71	...	...	...	Arrivées de touristes/visiteurs
Tourism expenditure [3]		37	87	...	...	...	...	Dépenses touristiques [3]
Bosnia and Herzegovina								**Bosnie-Herzégovine**
Tourist/visitor arrivals	TCE	...	217	365	439	529	536	Arrivées de touristes/visiteurs
Tourism expenditure		...	557	662	686	752	779	Dépenses touristiques
Botswana								**Botswana**
Tourist/visitor arrivals	TF	521	1 474	1 973	1 614	1 544	...	Arrivées de touristes/visiteurs
Tourism expenditure		176	563	80	36	113	...	Dépenses touristiques
Brazil								**Brésil**
Tourist/visitor arrivals	TF	1 991	5 358	5 161	5 677	5 813	6 430	Arrivées de touristes/visiteurs
Tourism expenditure		1 085	4 168	5 963	6 890	7 014	7 403	Dépenses touristiques
British Virgin Islands								**Îles Vierges britanniques**
Tourist/visitor arrivals	TF	219	337	330	351	366	386	Arrivées de touristes/visiteurs
Tourism expenditure		211	412	389	398	421	459	Dépenses touristiques
Brunei Darussalam								**Brunéi Darussalam**
Tourist/visitor arrivals [4]	TF	...	126	214	209	225	201	Arrivées de touristes/visiteurs [4]
Tourism expenditure [3]		...	191	...	92	...	...	Dépenses touristiques [3]
Bulgaria								**Bulgarie**
Tourist/visitor arrivals	TF	3 466	4 837	6 047	6 541	6 898	7 311	Arrivées de touristes/visiteurs
Tourism expenditure		662	3 063	...	...	...	4 557	Dépenses touristiques
Burkina Faso								**Burkina Faso**
Tourist/visitor arrivals	THS	124	245	274	237	218	191	Arrivées de touristes/visiteurs
Tourism expenditure		...	46	105	...	...	...	Dépenses touristiques
Burundi								**Burundi**
Tourist/visitor arrivals [2]	TF	34	148	#142	...	...	...	Arrivées de touristes/visiteurs [2]
Tourism expenditure		2	2	2	3	3	...	Dépenses touristiques
Cabo Verde								**Cabo Verde**
Tourist/visitor arrivals	THS	28	198	336	482	503	494	Arrivées de touristes/visiteurs
Tourism expenditure		29	177	387	465	483	452	Dépenses touristiques
Cambodia								**Cambodge**
Tourist/visitor arrivals	TF	220[4]	1 422[7]	2 508[7]	3 584[7]	4 210[7]	4 503[7]	Arrivées de touristes/visiteurs
Tourism expenditure		71	929	1 671	2 663	2 895	3 220	Dépenses touristiques
Cameroon								**Cameroun**
Tourist/visitor arrivals	TF	...	...	569	812	912	...	Arrivées de touristes/visiteurs
Tourism expenditure		75	229	171	377	607	...	Dépenses touristiques
Canada								**Canada**
Tourist/visitor arrivals	TF	16 932	18 771	16 219	16 344	16 059	16 537	Arrivées de touristes/visiteurs
Tourism expenditure		9 176	15 887	18 438	20 696	20 941	...	Dépenses touristiques

Tourist/visitor arrivals and tourism expenditure *(continued)*
Thousands arrivals and millions of US dollars

Arrivées de touristes/visiteurs et dépenses touristiques *(suite)*
Milliers d'arrivées et millions de dollars É.-U.

Country or area of destination	Series& Série&	1995	2005	2010	2012	2013	2014	Pays ou zone de destination
Cayman Islands								**Îles Caïmanes**
Tourist/visitor arrivals [4]	TF	361	168	288	322	345	383	Arrivées de touristes/visiteurs [4]
Tourism expenditure		394	356	465	470	480	...	Dépenses touristiques
Central African Rep.								**Rép. centrafricaine**
Tourist/visitor arrivals [8]	TF	26	12	54	71	...	...	Arrivées de touristes/visiteurs [8]
Tourism expenditure		4	7	14	15	...	...	Dépenses touristiques
Chad								**Tchad**
Tourist/visitor arrivals	TF	...	...	71	86	100	122	Arrivées de touristes/visiteurs
Tourism expenditure		43	...	...	...	...	...	Dépenses touristiques
Chile								**Chili**
Tourist/visitor arrivals	TF	1 540	2 027	2 801[2]	3 554[2]	3 576[2]	3 674[2]	Arrivées de touristes/visiteurs
Tourism expenditure		1 186	1 682	2 422	3 114	3 144	3 134	Dépenses touristiques
China								**Chine**
Tourist/visitor arrivals [9]	TF	20 034	46 809	55 664	57 725	55 686	55 622	Arrivées de touristes/visiteurs [9]
Tourism expenditure [3]		8 730	29 296	45 814	50 028	51 664	56 913	Dépenses touristiques [3]
China, Hong Kong SAR								**Chine, Hong Kong RAS**
Tourist/visitor arrivals	TF	...	14 773	20 085	23 770	25 661	27 770	Arrivées de touristes/visiteurs
Tourism expenditure		...	13 588	27 208	37 098	42 426	46 031	Dépenses touristiques
China, Macao SAR								**Chine, Macao RAS**
Tourist/visitor arrivals *	TF	4 202	9 014	11 926[10]	13 577[10]	14 268[10]	14 566[10]	Arrivées de touristes/visiteurs *
Tourism expenditure		3 233	8 190	28 214	44 368	52 389	51 556	Dépenses touristiques
Colombia								**Colombie**
Tourist/visitor arrivals	TF	1 399	#933	1 405	2 175	2 288	2 565	Arrivées de touristes/visiteurs
Tourism expenditure		887	1 891	3 441	4 363	4 759	4 887	Dépenses touristiques
Comoros								**Comores**
Tourist/visitor arrivals	TF	23	26	15	...	...	...	Arrivées de touristes/visiteurs
Tourism expenditure		22	24	35	39	...	...	Dépenses touristiques
Congo								**Congo**
Tourist/visitor arrivals	TF	...	...	194[2]	256[2]	343[2]	*373	Arrivées de touristes/visiteurs
Tourism expenditure [3]		14	40	63	*73	*38	...	Dépenses touristiques [3]
Cook Islands								**Îles Cook**
Tourist/visitor arrivals	TF	48	88	104	122	121	121	Arrivées de touristes/visiteurs
Tourism expenditure		28	91	111	168	168	175	Dépenses touristiques
Costa Rica								**Costa Rica**
Tourist/visitor arrivals	TF	785	1 679	2 100	2 343	2 428	2 527	Arrivées de touristes/visiteurs
Tourism expenditure		763	1 810	2 179	2 732	2 684	2 954	Dépenses touristiques
Côte d'Ivoire								**Côte d'Ivoire**
Tourist/visitor arrivals	VF	...	...	252	289	380	471	Arrivées de touristes/visiteurs
Tourism expenditure		103	93	213	173	191	...	Dépenses touristiques
Croatia								**Croatie**
Tourist/visitor arrivals	TCE	1 485	#7 743	9 111	10 369	10 948	11 623	Arrivées de touristes/visiteurs
Tourism expenditure		...	7 625	8 299	8 912	9 715	10 079	Dépenses touristiques
Cuba								**Cuba**
Tourist/visitor arrivals [4]	TF	742	2 261	2 507	2 815	2 829	2 970	Arrivées de touristes/visiteurs [4]
Tourism expenditure		1 100	2 591	2 218	2 613	2 608	2 546	Dépenses touristiques
Curaçao								**Curaçao**
Tourist/visitor arrivals [4]	TF	224	222	342	421	441	455	Arrivées de touristes/visiteurs [4]
Tourism expenditure		175	244	438	676	778	820	Dépenses touristiques
Cyprus								**Chypre**
Tourist/visitor arrivals	TF	2 100	2 470	2 173	2 465	2 405	2 441	Arrivées de touristes/visiteurs
Tourism expenditure		2 018	2 644	2 371	2 696	3 015	...	Dépenses touristiques
Czech Republic								**République tchèque**
Tourist/visitor arrivals	TF	...	9 404	8 629	10 123	10 300	10 617	Arrivées de touristes/visiteurs
Tourism expenditure		...	5 772	8 068	8 174	7 792	7 611	Dépenses touristiques
Dem. Rep. of the Congo								**Rép. dém. du Congo**
Tourist/visitor arrivals	TF	35	61	81[4]	167[11]	191[11]	...	Arrivées de touristes/visiteurs
Tourism expenditure [3]		...	3	11	7	8	<0	Dépenses touristiques [3]

24 Tourist/visitor arrivals and tourism expenditure *(continued)*
Thousands arrivals and millions of US dollars

Arrivées de touristes/visiteurs et dépenses touristiques *(suite)*
Milliers d'arrivées et millions de dollars É.-U.

Country or area of destination	Series& Série&	1995	2005	2010	2012	2013	2014	Pays ou zone de destination
Denmark								**Danemark**
Tourist/visitor arrivals	TCE	...	9 587	9 425	#8 443	8 557	#10 267	Arrivées de touristes/visiteurs
Tourism expenditure [3]		3 691	5 293	5 704	6 135	6 490	7 002	Dépenses touristiques [3]
Djibouti								**Djibouti**
Tourist/visitor arrivals	THS	21	30	51	60	63	...	Arrivées de touristes/visiteurs
Tourism expenditure [3]		5	7	18	21	22	...	Dépenses touristiques [3]
Dominica								**Dominique**
Tourist/visitor arrivals	TF	60	79	77	79	78	82	Arrivées de touristes/visiteurs
Tourism expenditure [3]		42	57	94	78	102	126	Dépenses touristiques [3]
Dominican Republic								**Rép. dominicaine**
Tourist/visitor arrivals [2,4]	TF	1 776[12]	3 691	4 125	4 563	4 690	5 141	Arrivées de touristes/visiteurs [2,4]
Tourism expenditure [3]		1 571	3 518	4 163	4 687	5 064	5 637	Dépenses touristiques [3]
Ecuador								**Équateur**
Tourist/visitor arrivals [1]	VF	440	860	1 047	1 272	1 364	1 557	Arrivées de touristes/visiteurs [1]
Tourism expenditure		315	488	786	1 039	1 251	1 487	Dépenses touristiques
Egypt								**Égypte**
Tourist/visitor arrivals	TF	2 871	8 244	14 051	11 196	9 174	9 628	Arrivées de touristes/visiteurs
Tourism expenditure		2 954	7 206	13 633	10 823	6 747	7 979	Dépenses touristiques
El Salvador								**El Salvador**
Tourist/visitor arrivals	TF	235	1 127	1 150	1 255	1 283	1 345	Arrivées de touristes/visiteurs
Tourism expenditure		152	656	646	900	1 054	1 285	Dépenses touristiques
Equatorial Guinea								**Guinée équatoriale**
Tourism expenditure		1	...	...	...	...	...	Dépenses touristiques
Eritrea								**Érythrée**
Tourist/visitor arrivals [2]	VF	315	83	84	...	...	...	Arrivées de touristes/visiteurs [2]
Tourism expenditure		58	66	...	...	...	...	Dépenses touristiques
Estonia								**Estonie**
Tourist/visitor arrivals	TF	530	#1 917	#2 372	2 744	2 873	2 918	Arrivées de touristes/visiteurs
Tourism expenditure		452	1 229	...	...	2 022	2 231	Dépenses touristiques
Ethiopia								**Éthiopie**
Tourist/visitor arrivals	TF	103[13]	227[2,14]	468[2,14]	597[2,14]	681[2,14]	770[2,14]	Arrivées de touristes/visiteurs
Tourism expenditure		177	533	1 434	1 980	...	...	Dépenses touristiques
Fiji								**Fidji**
Tourist/visitor arrivals [1]	TF	318	545	632	661	658	693	Arrivées de touristes/visiteurs [1]
Tourism expenditure		369	722	825	989	971	...	Dépenses touristiques
Finland								**Finlande**
Tourist/visitor arrivals	TCE	1 779	#2 080	2 319	2 778	2 797	2 731	Arrivées de touristes/visiteurs
Tourism expenditure		2 383	3 069	4 510	5 415	...	...	Dépenses touristiques
France								**France**
Tourist/visitor arrivals	TF	60 033	#74 988	76 647	81 980	83 634	83 767	Arrivées de touristes/visiteurs
Tourism expenditure		31 295	52 139	56 187	64 001	66 049	66 803	Dépenses touristiques
French Guiana								**Guyane française**
Tourist/visitor arrivals	TF	...	95	189	187	180	185	Arrivées de touristes/visiteurs
Tourism expenditure		...	44	...	...	...	...	Dépenses touristiques
French Polynesia								**Polynésie française**
Tourist/visitor arrivals [1]	TF	172	208	154	169	164	181	Arrivées de touristes/visiteurs [1]
Tourism expenditure [3]		...	530	405	438	458	...	Dépenses touristiques [3]
Gabon								**Gabon**
Tourist/visitor arrivals [15]	TF	125	269	...	...	...	...	Arrivées de touristes/visiteurs [15]
Tourism expenditure		94	13	...	...	...	...	Dépenses touristiques
Gambia								**Gambie**
Tourist/visitor arrivals [16]	TF	45	108	91	157	171	156	Arrivées de touristes/visiteurs [16]
Tourism expenditure		...	59	80	99	...	...	Dépenses touristiques
Georgia								**Géorgie**
Tourist/visitor arrivals	TF	...	...	1 067	1 790	2 065	2 229	Arrivées de touristes/visiteurs
Tourism expenditure		...	287	737	1 565	1 916	1 972	Dépenses touristiques

Tourist/visitor arrivals and tourism expenditure *(continued)*
Thousands arrivals and millions of US dollars

Arrivées de touristes/visiteurs et dépenses touristiques *(suite)*
Milliers d'arrivées et millions de dollars É.-U.

Country or area of destination	Series& Série&	1995	2005	2010	2012	2013	2014	Pays ou zone de destination
Germany								**Allemagne**
Tourist/visitor arrivals	TCE	14 847	21 500	26 875	30 411	31 545	32 999	Arrivées de touristes/visiteurs
Tourism expenditure		24 052	40 531	49 128	51 646	55 312	55 924	Dépenses touristiques
Ghana								**Ghana**
Tourist/visitor arrivals [2]	TF	286	429	931	903	994	1 093	Arrivées de touristes/visiteurs [2]
Tourism expenditure		30	867	706	1 154	1 010	1 027	Dépenses touristiques
Greece								**Grèce**
Tourist/visitor arrivals	TF	10 130	14 765	15 007	15 518	17 920	22 033	Arrivées de touristes/visiteurs
Tourism expenditure		4 182	13 453	13 858	14 671	17 436	19 481	Dépenses touristiques
Grenada								**Grenade**
Tourist/visitor arrivals	TF	108	99	110	116	113	134	Arrivées de touristes/visiteurs
Tourism expenditure [3]		76	71	112	122	120	...	Dépenses touristiques [3]
Guadeloupe								**Guadeloupe**
Tourist/visitor arrivals [4,17]	TF	640[18]	372	392	325	487[19]	...	Arrivées de touristes/visiteurs [4,17]
Tourism expenditure		458	306	510	...	671	...	Dépenses touristiques
Guam	TF							**Guam**
Tourist/visitor arrivals		1 362	1 228	1 197	1 304	1 334	1 343	Arrivées de touristes/visiteurs
Guatemala								**Guatemala**
Tourist/visitor arrivals	TF	...	...	1 219	1 305	1 331	1 455	Arrivées de touristes/visiteurs
Tourism expenditure [3]		213	791	1 378	1 419	1 479	1 564	Dépenses touristiques [3]
Guinea								**Guinée**
Tourist/visitor arrivals [20]	TF	...	45	12	96	56	33	Arrivées de touristes/visiteurs [20]
Tourism expenditure		1	...	2	2	...	...	Dépenses touristiques
Guinea-Bissau								**Guinée-Bissau**
Tourist/visitor arrivals [4]	TF	...	5	...	...	...	...	Arrivées de touristes/visiteurs [4]
Tourism expenditure		...	...	14	7	...	...	Dépenses touristiques
Guyana								**Guyana**
Tourist/visitor arrivals	TF	106	117[21]	152[21]	177[21]	158[21]	206[21]	Arrivées de touristes/visiteurs
Tourism expenditure [3]		33	35	80	64	77	79	Dépenses touristiques [3]
Haiti								**Haïti**
Tourist/visitor arrivals [4]	TF	145	112	255[2]	349[2]	420[2]	465[2]	Arrivées de touristes/visiteurs [4]
Tourism expenditure [3]		90	80	383	447	546	578	Dépenses touristiques [3]
Honduras								**Honduras**
Tourist/visitor arrivals	TF	271	673	863	895	863	868	Arrivées de touristes/visiteurs
Tourism expenditure		85	465	627	684	618	642	Dépenses touristiques
Hungary								**Hongrie**
Tourist/visitor arrivals	TF	...	9 979	9 510	10 353	10 624	12 140	Arrivées de touristes/visiteurs
Tourism expenditure		2 938	4 761	6 595	6 149	6 671	7 479	Dépenses touristiques
Iceland								**Islande**
Tourist/visitor arrivals	TF	190	374	489	673	807	998	Arrivées de touristes/visiteurs
Tourism expenditure [3]		186	413	562	865	1 076	1 367	Dépenses touristiques [3]
India								**Inde**
Tourist/visitor arrivals [1]	TF	2 124	3 919	5 776	6 578	6 968	7 679	Arrivées de touristes/visiteurs [1]
Tourism expenditure		...	7 659	...	18 340	19 042	20 756	Dépenses touristiques
Indonesia								**Indonésie**
Tourist/visitor arrivals	TF	4 324	5 002	7 003	8 044	8 802	9 435	Arrivées de touristes/visiteurs
Tourism expenditure		...	5 094	7 618	9 463	10 302	11 567	Dépenses touristiques
Iran (Islamic Rep. of)								**Iran (Rép. islamique d')**
Tourist/visitor arrivals	VF	568	1 889	2 938	3 834	4 769	4 967	Arrivées de touristes/visiteurs
Tourism expenditure		205	1 025	2 631	2 483	3 212	3 676	Dépenses touristiques
Iraq								**Iraq**
Tourist/visitor arrivals	VF	61	...	1 518	1 111	892	...	Arrivées de touristes/visiteurs
Tourism expenditure		...	186	1 736	1 640	...	...	Dépenses touristiques
Ireland								**Irlande**
Tourist/visitor arrivals [22]	TF	4 818	7 333	#7 134	7 550	8 260	8 813	Arrivées de touristes/visiteurs [22]
Tourism expenditure		2 698	6 780	8 187	9 064	9 538	11 093	Dépenses touristiques

24

Tourist/visitor arrivals and tourism expenditure *(continued)*
Thousands arrivals and millions of US dollars

Arrivées de touristes/visiteurs et dépenses touristiques *(suite)*
Milliers d'arrivées et millions de dollars É.-U.

Country or area of destination	Series& Série&	1995	2005	2010	2012	2013	2014	Pays ou zone de destination
Israel								**Israël**
Tourist/visitor arrivals [1]	TF	2 215	1 903	2 803	2 886	2 962	2 927	Arrivées de touristes/visiteurs [1]
Tourism expenditure [23]		3 491	3 427	5 824	6 180	6 452	6 439	Dépenses touristiques [23]
Italy								**Italie**
Tourist/visitor arrivals [24]	TF	31 052	36 513	43 626	46 360	47 704	48 576	Arrivées de touristes/visiteurs [24]
Tourism expenditure [3]		28 731	35 319	38 438	40 960	43 829	45 547	Dépenses touristiques [3]
Jamaica								**Jamaïque**
Tourist/visitor arrivals [25]	TF	1 147	1 479	1 922	1 986	2 008	2 080	Arrivées de touristes/visiteurs [25]
Tourism expenditure [3]		1 069	1 545	2 001	2 069	2 074	2 255	Dépenses touristiques [3]
Japan								**Japon**
Tourist/visitor arrivals [1]	VF	3 345	6 728	8 611	8 358	10 364	13 413	Arrivées de touristes/visiteurs [1]
Tourism expenditure		4 894	15 554	15 356	16 197	16 865	20 790	Dépenses touristiques
Jordan								**Jordanie**
Tourist/visitor arrivals	TF	1 075	#2 987[2]	4 207[2]	4 162[2]	3 945[2]	3 990[2]	Arrivées de touristes/visiteurs
Tourism expenditure		973	1 759	4 390	5 123	5 145	5 537	Dépenses touristiques
Kazakhstan								**Kazakhstan**
Tourist/visitor arrivals	TF	...	3 143	2 991	4 437	4 926	4 560	Arrivées de touristes/visiteurs
Tourism expenditure		155	801	1 236	1 572	1 601	1 555	Dépenses touristiques
Kenya								**Kenya**
Tourist/visitor arrivals	TF	918	1 399	1 470	1 619	1 434	1 261	Arrivées de touristes/visiteurs
Tourism expenditure		785	969	1 620	2 004	1 829	1 833	Dépenses touristiques
Kiribati								**Kiribati**
Tourist/visitor arrivals [26]	TF	4	4	5	5	6	...	Arrivées de touristes/visiteurs [26]
Tourism expenditure		2	3	5	5	4	4	Dépenses touristiques
Kuwait								**Koweït**
Tourist/visitor arrivals	VF	1 443	3 474	5 208	5 729	6 217	6 528	Arrivées de touristes/visiteurs
Tourism expenditure		307	413	574	780	619	615	Dépenses touristiques
Kyrgyzstan								**Kirghizistan**
Tourist/visitor arrivals	VF	...	...	855	2 406	3 076	2 849	Arrivées de touristes/visiteurs
Tourism expenditure		...	94	212	486	585	468	Dépenses touristiques
Lao People's Dem. Rep.								**Rép. dém. pop. lao**
Tourist/visitor arrivals	TF	60	672	1 670	2 291	2 700	3 164	Arrivées de touristes/visiteurs
Tourism expenditure		52	143	385	461	613	...	Dépenses touristiques
Latvia								**Lettonie**
Tourist/visitor arrivals [27]	TF	539	1 116	1 373	1 435	1 536	1 843	Arrivées de touristes/visiteurs [27]
Tourism expenditure		37	446	...	...	1 190	1 244	Dépenses touristiques
Lebanon								**Liban**
Tourist/visitor arrivals [28]	TF	450	1 140	2 168	1 366	1 274	1 355	Arrivées de touristes/visiteurs [28]
Tourism expenditure		710	5 969	8 026	6 853	6 412	6 576	Dépenses touristiques
Lesotho								**Lesotho**
Tourist/visitor arrivals	TF	87	...	414	317	320	...	Arrivées de touristes/visiteurs
Tourism expenditure [3]		27	27	23	22	17	17	Dépenses touristiques [3]
Liberia [3]								**Libéria [3]**
Tourism expenditure		...	67	12	...	...	91	Dépenses touristiques
Libya								**Libye**
Tourist/visitor arrivals	THS	...	81	...	...	...	...	Arrivées de touristes/visiteurs
Tourism expenditure		4	301	170	...	...	...	Dépenses touristiques
Liechtenstein	TCE							**Liechtenstein**
Tourist/visitor arrivals		...	...	64	62[29]	60[29]	61[29]	Arrivées de touristes/visiteurs
Lithuania								**Lituanie**
Tourist/visitor arrivals	TF	650	2 000	1 507	1 900	2 012	2 063	Arrivées de touristes/visiteurs
Tourism expenditure [3]		77	920	958	1 317	1 374	1 383	Dépenses touristiques [3]
Luxembourg								**Luxembourg**
Tourist/visitor arrivals	TCE	768	913	805	950	945	1 038	Arrivées de touristes/visiteurs
Tourism expenditure		...	...	4 559	5 273	5 652	5 488	Dépenses touristiques
Madagascar								**Madagascar**
Tourist/visitor arrivals [30]	TF	75	277	196	256	196	222	Arrivées de touristes/visiteurs [30]
Tourism expenditure		106	190	319	563	578	...	Dépenses touristiques

24

Tourist/visitor arrivals and tourism expenditure *(continued)*
Thousands arrivals and millions of US dollars

Arrivées de touristes/visiteurs et dépenses touristiques *(suite)*
Milliers d'arrivées et millions de dollars É.-U.

Country or area of destination	Series& Série&	1995	2005	2010	2012	2013	2014	Pays ou zone de destination
Malawi								**Malawi**
Tourist/visitor arrivals [31]	TF	192	438	746	770	795	...	Arrivées de touristes/visiteurs [31]
Tourism expenditure		22	48	45	35	33	36	Dépenses touristiques
Malaysia								**Malaisie**
Tourist/visitor arrivals [32]	TF	7 469	16 431	24 577	25 033	25 715	27 437	Arrivées de touristes/visiteurs [32]
Tourism expenditure [3]		3 969	8 846	18 152	20 251	21 500	22 600	Dépenses touristiques [3]
Maldives								**Maldives**
Tourist/visitor arrivals [4]	TF	315	395	792	958	1 125	1 205	Arrivées de touristes/visiteurs [4]
Tourism expenditure		...	...	...	2 032	2 424	2 772	Dépenses touristiques
Mali								**Mali**
Tourist/visitor arrivals	TF	...	...	169	134	142	168	Arrivées de touristes/visiteurs
Tourism expenditure		26	149	208	144	178	...	Dépenses touristiques
Malta								**Malte**
Tourist/visitor arrivals	TF	1 116	1 171[33]	1 339[33]	1 443[33]	1 582[33]	1 690[33]	Arrivées de touristes/visiteurs
Tourism expenditure [3]		656	755	1 066	1 260	1 402	1 521	Dépenses touristiques [3]
Marshall Islands								**Îles Marshall**
Tourist/visitor arrivals	TF	6[4]	9[35]	5[4]	5[4]	...	...	Arrivées de touristes/visiteurs
Tourism expenditure		3[34]	4	4	4	4	5	Dépenses touristiques
Martinique								**Martinique**
Tourist/visitor arrivals	TF	457	484	478	488	490	490	Arrivées de touristes/visiteurs
Tourism expenditure		384	280	472	462	484	483	Dépenses touristiques
Mauritania								**Mauritanie**
Tourism expenditure		...	...	...	50	50	42	Dépenses touristiques
Mauritius								**Maurice**
Tourist/visitor arrivals	TF	422	761	935	965	993	1 039	Arrivées de touristes/visiteurs
Tourism expenditure		616	1 189	1 585	1 778	1 593	1 719	Dépenses touristiques
Mexico								**Mexique**
Tourist/visitor arrivals [2]	TF	20 241	21 915	23 290	23 403	24 151	29 346	Arrivées de touristes/visiteurs [2]
Tourism expenditure		6 847	12 801	12 628	13 320	14 311	16 607	Dépenses touristiques
Micronesia (Fed. States of)								**Micronésie (États féd. de)**
Tourist/visitor arrivals [36]	TF	...	19	45	38	42	35	Arrivées de touristes/visiteurs [36]
Tourism expenditure [3]		...	...	24	22	24	...	Dépenses touristiques [3]
Monaco	THS							**Monaco**
Tourist/visitor arrivals		233	286	279	292	328	329	Arrivées de touristes/visiteurs
Mongolia								**Mongolie**
Tourist/visitor arrivals [37]	TF	108	338	456	476	418	393	Arrivées de touristes/visiteurs [37]
Tourism expenditure		33	203	288	480	228	215	Dépenses touristiques
Montenegro								**Monténégro**
Tourist/visitor arrivals	TCE	...	272	1 088	1 264	1 324	1 350	Arrivées de touristes/visiteurs
Tourism expenditure		...	...	765	860	929	959	Dépenses touristiques
Montserrat								**Montserrat**
Tourist/visitor arrivals	TF	18	10	6	7	7	9	Arrivées de touristes/visiteurs
Tourism expenditure [3]		17	9	6	7	8	8	Dépenses touristiques [3]
Morocco								**Maroc**
Tourist/visitor arrivals [2]	TF	2 602	5 843	9 288	9 375	10 046	10 283	Arrivées de touristes/visiteurs [2]
Tourism expenditure		1 469	5 426	8 176	8 491	8 201	...	Dépenses touristiques
Mozambique								**Mozambique**
Tourist/visitor arrivals	TF	...	578[38]	#1 718[39]	2 113[39]	1 886[39]	1 661[39]	Arrivées de touristes/visiteurs
Tourism expenditure		...	138	135	224	228	225	Dépenses touristiques
Myanmar								**Myanmar**
Tourist/visitor arrivals	TF	194	660	792	1 059	2 044	3 081	Arrivées de touristes/visiteurs
Tourism expenditure		169	83	91	550	964	1 613	Dépenses touristiques
Namibia								**Namibie**
Tourist/visitor arrivals	TF	272	778	984	1 079	1 176	...	Arrivées de touristes/visiteurs
Tourism expenditure		...	363	560	598	524	517	Dépenses touristiques
Nepal								**Népal**
Tourist/visitor arrivals [40]	TF	363	375	603	803	798	790	Arrivées de touristes/visiteurs [40]
Tourism expenditure		232	160	378	379	460	502	Dépenses touristiques

24

Tourist/visitor arrivals and tourism expenditure *(continued)*
Thousands arrivals and millions of US dollars

Arrivées de touristes/visiteurs et dépenses touristiques *(suite)*
Milliers d'arrivées et millions de dollars É.-U.

Country or area of destination	Series& Série&	1995	2005	2010	2012	2013	2014	Pays ou zone de destination
Netherlands								**Pays-Bas**
Tourist/visitor arrivals	TCE	6 574	10 012	10 883	11 680	12 783	13 925	Arrivées de touristes/visiteurs
Tourism expenditure [3]		6 578	8 770	11 653	12 261	13 751	14 682	Dépenses touristiques [3]
New Caledonia								**Nouvelle-Calédonie**
Tourist/visitor arrivals [2]	TF	86	101	99	112	108	107	Arrivées de touristes/visiteurs [2]
Tourism expenditure [3]		108	149	129	165	168	...	Dépenses touristiques [3]
New Zealand								**Nouvelle-Zélande**
Tourist/visitor arrivals	TF	...	2 353	2 435	2 473	2 629	2 772	Arrivées de touristes/visiteurs
Tourism expenditure [3]		2 318	6 486	6 523	7 142	7 496	8 400	Dépenses touristiques [3]
Nicaragua								**Nicaragua**
Tourist/visitor arrivals	TF	281	712[2]	1 011[2]	1 180[2]	1 229[2]	1 330[2]	Arrivées de touristes/visiteurs
Tourism expenditure [3]		50	206	313	421	417	445	Dépenses touristiques [3]
Niger								**Niger**
Tourist/visitor arrivals	TF	35	58	74	94	123	135	Arrivées de touristes/visiteurs
Tourism expenditure		...	44	106	51	59	...	Dépenses touristiques
Nigeria								**Nigéria**
Tourist/visitor arrivals	TF	656	1 010	1 555	486	600	...	Arrivées de touristes/visiteurs
Tourism expenditure		47	139	736	639	616	601	Dépenses touristiques
Niue								**Nioué**
Tourist/visitor arrivals [41]	TF	2	3	6	5	7	...	Arrivées de touristes/visiteurs [41]
Tourism expenditure		2	1	2	...	...	...	Dépenses touristiques
Northern Mariana Islands								**Îles Mariannes du Nord**
Tourist/visitor arrivals [4]	TF	669	498	375	401	439	460	Arrivées de touristes/visiteurs [4]
Tourism expenditure		655	...	...	...	...	...	Dépenses touristiques
Norway								**Norvège**
Tourist/visitor arrivals	TF	2 880[42]	3 824	4 767	#4 538[18]	4 778[18]	4 855[18]	Arrivées de touristes/visiteurs
Tourism expenditure		2 730	4 243	5 299	6 785	6 554	6 490	Dépenses touristiques
Oman								**Oman**
Tourist/visitor arrivals	TF	...	891	1 441	1 241	1 392	1 519	Arrivées de touristes/visiteurs
Tourism expenditure		...	627	1 256	1 781	1 913	1 949	Dépenses touristiques
Other non-specified areas								**Autres zones non-spécifiées**
Tourist/visitor arrivals [2]	VF	2 332	3 378	5 567	7 311	8 016	9 910	Arrivées de touristes/visiteurs [2]
Tourism expenditure		3 985	5 740	10 387	14 115	14 782	17 423	Dépenses touristiques
Pakistan								**Pakistan**
Tourist/visitor arrivals	TF	378	798	907	966	...	...	Arrivées de touristes/visiteurs
Tourism expenditure		582	828	998	1 014	938	971	Dépenses touristiques
Palau								**Palaos**
Tourist/visitor arrivals [43]	TF	53	81	86	119	105	141	Arrivées de touristes/visiteurs [43]
Tourism expenditure [44]		...	63	76	109	117	131	Dépenses touristiques [44]
Panama								**Panama**
Tourist/visitor arrivals	TF	345	702	1 324	1 606	1 658	1 745	Arrivées de touristes/visiteurs
Tourism expenditure		372	1 108	2 621	4 534	5 119	5 748	Dépenses touristiques
Papua New Guinea								**Papouasie-Nvl-Guinée**
Tourist/visitor arrivals	TF	42	69	140	168	174	182	Arrivées de touristes/visiteurs
Tourism expenditure		...	9	2	...	4	3	Dépenses touristiques
Paraguay								**Paraguay**
Tourist/visitor arrivals [5]	TF	438	#341[45]	465[45]	579[45]	610[45]	649[45]	Arrivées de touristes/visiteurs [5]
Tourism expenditure		162	96	243	291	299	314	Dépenses touristiques
Peru								**Pérou**
Tourist/visitor arrivals	TF	479	#1 571[2,46]	2 299[2,46]	2 846[2,46]	3 164[2,46]	*3 215[2,46]	Arrivées de touristes/visiteurs
Tourism expenditure		521	1 438	2 475	3 074	3 925	3 831	Dépenses touristiques
Philippines								**Philippines**
Tourist/visitor arrivals [2]	TF	1 760	2 623	3 520	4 273	4 681	4 833	Arrivées de touristes/visiteurs [2]
Tourism expenditure		1 141	2 863	3 441	4 963	5 599	6 052	Dépenses touristiques
Poland								**Pologne**
Tourist/visitor arrivals	TF	19 215	15 200	12 470	14 840	15 800	16 000	Arrivées de touristes/visiteurs
Tourism expenditure		6 927	7 128	10 037	11 888	12 432	12 311	Dépenses touristiques

24

Tourist/visitor arrivals and tourism expenditure *(continued)*
Thousands arrivals and millions of US dollars

Arrivées de touristes/visiteurs et dépenses touristiques *(suite)*
Milliers d'arrivées et millions de dollars É.-U.

Country or area of destination	Series& Série&	1995	2005	2010	2012	2013	2014	Pays ou zone de destination
Portugal								**Portugal**
Tourist/visitor arrivals	TCE	4 572	5 769	6 756	7 503	8 097	9 092	Arrivées de touristes/visiteurs
Tourism expenditure		5 646	9 042	12 985	14 582	16 210	17 784	Dépenses touristiques
Puerto Rico								**Porto Rico**
Tourist/visitor arrivals [30]	TF	3 131	3 686	3 186	3 069	3 200	3 246	Arrivées de touristes/visiteurs [30]
Tourism expenditure [47]		1 828	3 239	3 211	3 193	3 334	3 438	Dépenses touristiques [47]
Qatar								**Qatar**
Tourist/visitor arrivals	TF	...	...	1 700	2 346	2 611	2 826	Arrivées de touristes/visiteurs
Tourism expenditure		...	...	...	7 220	8 452	10 576	Dépenses touristiques
Republic of Korea								**République de Corée**
Tourist/visitor arrivals [48]	VF	3 753	6 023	8 798	11 140	12 176	14 202	Arrivées de touristes/visiteurs [48]
Tourism expenditure		6 670	8 290	14 367	18 851	19 644	23 008	Dépenses touristiques
Republic of Moldova								**République de Moldova**
Tourist/visitor arrivals [49]	TCE	...	67	64	89	96	94	Arrivées de touristes/visiteurs [49]
Tourism expenditure		71	138	222	279	318	313	Dépenses touristiques
Réunion								**Réunion**
Tourist/visitor arrivals	TF	304	409	420	447	416	406	Arrivées de touristes/visiteurs
Tourism expenditure		216	384	392	404	403	387	Dépenses touristiques
Romania								**Roumanie**
Tourist/visitor arrivals	VF	5 445	5 839	7 498	7 937	8 019	8 442	Arrivées de touristes/visiteurs
Tourism expenditure		689	1 325	1 631	1 901	2 048	2 225	Dépenses touristiques
Russian Federation								**Fédération de Russie**
Tourist/visitor arrivals	VF	10 290	22 201	22 281	28 177	30 792	32 421	Arrivées de touristes/visiteurs
Tourism expenditure		...	7 805	13 239	17 876	20 198	19 451	Dépenses touristiques
Rwanda								**Rwanda**
Tourist/visitor arrivals	TF	...	...	504	815	864	926	Arrivées de touristes/visiteurs
Tourism expenditure		4	67	224	337	351	...	Dépenses touristiques
Saba	TF							**Saba**
Tourist/visitor arrivals		10	12	12	...	...	...	Arrivées de touristes/visiteurs
Saint Kitts and Nevis								**Saint-Kitts-et-Nevis**
Tourist/visitor arrivals [30]	TF	79	141	98	104	107	113	Arrivées de touristes/visiteurs [30]
Tourism expenditure [3]		63	121	90	95	101	110	Dépenses touristiques [3]
Saint Lucia								**Sainte-Lucie**
Tourist/visitor arrivals [1]	TF	231	318	306	307	319	338	Arrivées de touristes/visiteurs [1]
Tourism expenditure [3]		230	382	309	337	354	...	Dépenses touristiques [3]
Saint Vincent-Grenadines								**St.-Vincent-Grenadines**
Tourist/visitor arrivals [30]	TF	60	96	72	74	72	71	Arrivées de touristes/visiteurs [30]
Tourism expenditure [3]		53	104	86	94	92	...	Dépenses touristiques [3]
Samoa								**Samoa**
Tourist/visitor arrivals	TF	68	102	122	126	116	120	Arrivées de touristes/visiteurs
Tourism expenditure		36	74	124	148	137	146	Dépenses touristiques
San Marino [50]	THS							**Saint-Marin** [50]
Tourist/visitor arrivals		28	50	120	139	71	75	Arrivées de touristes/visiteurs
Sao Tome and Principe								**Sao Tomé-et-Principe**
Tourist/visitor arrivals	TF	6	16	8	...	...	...	Arrivées de touristes/visiteurs
Tourism expenditure		...	...	...	...	31	56	Dépenses touristiques
Saudi Arabia								**Arabie saoudite**
Tourist/visitor arrivals	TF	3 325	8 037	10 850	16 332	15 772	18 259	Arrivées de touristes/visiteurs
Tourism expenditure		...	...	7 536	8 400	8 690	9 263	Dépenses touristiques
Senegal								**Sénégal**
Tourist/visitor arrivals	TF	...	769	*900	*962	*1 063	*836	Arrivées de touristes/visiteurs
Tourism expenditure [3]		168	242	453	407	439	...	Dépenses touristiques [3]
Serbia								**Serbie**
Tourist/visitor arrivals	TCE	...	453	683	810	922	1 029	Arrivées de touristes/visiteurs
Tourism expenditure		...	308	950	1 080	1 221	1 352	Dépenses touristiques
Seychelles								**Seychelles**
Tourist/visitor arrivals	TF	121	129	175	208	230	233	Arrivées de touristes/visiteurs
Tourism expenditure		224	269	352	429	484	481	Dépenses touristiques

24

Tourist/visitor arrivals and tourism expenditure *(continued)*
Thousands arrivals and millions of US dollars

Arrivées de touristes/visiteurs et dépenses touristiques *(suite)*
Milliers d'arrivées et millions de dollars É.-U.

Country or area of destination	Series& Série&	1995	2005	2010	2012	2013	2014	Pays ou zone de destination
Sierra Leone								**Sierra Leone**
Tourist/visitor arrivals [4]	TF	14	40	39	60	81	44	Arrivées de touristes/visiteurs [4]
Tourism expenditure [3]		57	64	26	47	66	35	Dépenses touristiques [3]
Singapore								**Singapour**
Tourist/visitor arrivals	TF	6 070	7 079	9 161	11 098	11 899	11 864	Arrivées de touristes/visiteurs
Tourism expenditure [3]		7 611	6 209	14 178	18 939	19 301	19 203	Dépenses touristiques [3]
Sint Eustatius [51]	TF							**Saint-Eustache** [51]
Tourist/visitor arrivals		9	10	11	...	...	...	Arrivées de touristes/visiteurs
Sint Maarten (Dutch part)								**Saint-Martin (partie néerl.)**
Tourist/visitor arrivals [4,52]	TF	460	468	443	457	467	500	Arrivées de touristes/visiteurs [4,52]
Tourism expenditure		...	...	681	854	871	930	Dépenses touristiques
Slovakia								**Slovaquie**
Tourist/visitor arrivals	TF	...	#6 184	5 415	6 235	...	...	Arrivées de touristes/visiteurs
Tourism expenditure		630	1 282	2 335	2 365	...	2 619	Dépenses touristiques
Slovenia								**Slovénie**
Tourist/visitor arrivals	TCE	732	1 555	1 869	2 156	2 259	2 411	Arrivées de touristes/visiteurs
Tourism expenditure		1 128	1 894	2 721	2 737	2 895	2 939	Dépenses touristiques
Solomon Islands								**Îles Salomon**
Tourist/visitor arrivals	TF	12	9	21	24	24	20	Arrivées de touristes/visiteurs
Tourism expenditure		17	6	51	65	71	64	Dépenses touristiques
South Africa								**Afrique du Sud**
Tourist/visitor arrivals	TF	4 488[53]	7 369[53]	#8 074	9 188	9 537	#9 549[54]	Arrivées de touristes/visiteurs
Tourism expenditure		2 654	8 629	10 309	11 202	10 468	10 484	Dépenses touristiques
Spain								**Espagne**
Tourist/visitor arrivals	TF	34 920	55 914	52 677	57 464	60 675	64 995	Arrivées de touristes/visiteurs
Tourism expenditure [3]		25 368	49 565	54 305	57 877	62 584	65 100	Dépenses touristiques [3]
Sri Lanka								**Sri Lanka**
Tourist/visitor arrivals [1]	TF	403	549	654	1 006	1 275	1 527	Arrivées de touristes/visiteurs [1]
Tourism expenditure		367	729	1 044	1 756	2 506	3 278	Dépenses touristiques
State of Palestine								**État de Palestine**
Tourist/visitor arrivals	THS	...	88	522	490[55]	545[55]	556[55]	Arrivées de touristes/visiteurs
Tourism expenditure [3]		255	52	409	469	524	543	Dépenses touristiques [3]
Sudan								**Soudan**
Tourist/visitor arrivals [2]	TF	...	...	...	575	591	684	Arrivées de touristes/visiteurs [2]
Tourism expenditure [3]		...	...	...	772	773	967	Dépenses touristiques [3]
Sudan [former]								**Soudan [anc.]**
Tourist/visitor arrivals	TF	29	246[2]	495[2]	...	...	...	Arrivées de touristes/visiteurs
Tourism expenditure [3]		8	114	82	...	...	...	Dépenses touristiques [3]
Suriname								**Suriname**
Tourist/visitor arrivals	TF	43[56]	161	205	240	249	252	Arrivées de touristes/visiteurs
Tourism expenditure		52	96	69	79	92	103	Dépenses touristiques
Swaziland								**Swaziland**
Tourist/visitor arrivals	TF	300[57]	837	868	888	968	...	Arrivées de touristes/visiteurs
Tourism expenditure		54	77	51	30	13	16	Dépenses touristiques
Sweden								**Suède**
Tourist/visitor arrivals	TF	...	...	...	12 372	10 980	10 522	Arrivées de touristes/visiteurs
Tourism expenditure [3]		3 471	6 665	8 653	10 608	11 535	12 696	Dépenses touristiques [3]
Switzerland								**Suisse**
Tourist/visitor arrivals [58]	THS	6 946	7 229	8 628	8 566	8 967	9 158	Arrivées de touristes/visiteurs [58]
Tourism expenditure		11 354	11 949	17 617	19 613	20 329	21 006	Dépenses touristiques
Syrian Arab Republic								**Rép. arabe syrienne**
Tourist/visitor arrivals	TCE	815	3 571[2]	#8 546[2,59]	...	...	...	Arrivées de touristes/visiteurs
Tourism expenditure		...	2 035	6 308	...	...	...	Dépenses touristiques
Tajikistan								**Tadjikistan**
Tourist/visitor arrivals	VF	...	...	160	244	208	213	Arrivées de touristes/visiteurs
Tourism expenditure		...	9	32	60	49	107	Dépenses touristiques

24

Tourist/visitor arrivals and tourism expenditure *(continued)*
Thousands arrivals and millions of US dollars

Arrivées de touristes/visiteurs et dépenses touristiques *(suite)*
Milliers d'arrivées et millions de dollars É.-U.

Country or area of destination	Series& Série&	1995	2005	2010	2012	2013	2014	Pays ou zone de destination
Thailand								**Thaïlande**
Tourist/visitor arrivals	TF	6 952[2]	11 567[2]	15 936	22 354	26 547	24 810	Arrivées de touristes/visiteurs
Tourism expenditure		9 257	12 102	23 809	37 769	45 740	42 063	Dépenses touristiques
TFYR of Macedonia								**ex-R.Y. de Macédoine**
Tourist/visitor arrivals	TCE	147	197	262	351	400	425	Arrivées de touristes/visiteurs
Tourism expenditure		...	116	199	237	270	298	Dépenses touristiques
Timor-Leste								**Timor-Leste**
Tourist/visitor arrivals [60]	TF	...	...	40	58	79	60	Arrivées de touristes/visiteurs [60]
Tourism expenditure [3]		...	...	24	21	29	35	Dépenses touristiques [3]
Togo								**Togo**
Tourist/visitor arrivals	THS	53	81	202	235	327	282	Arrivées de touristes/visiteurs
Tourism expenditure		...	27	105	206	233	...	Dépenses touristiques
Tonga								**Tonga**
Tourist/visitor arrivals [4]	TF	29	42	47	47	48	50	Arrivées de touristes/visiteurs [4]
Tourism expenditure		...	15	28	46	48	...	Dépenses touristiques
Trinidad and Tobago								**Trinité-et-Tobago**
Tourist/visitor arrivals [4]	TF	260	463	388	455	434	412	Arrivées de touristes/visiteurs [4]
Tourism expenditure		232	593	630	...	...		Dépenses touristiques
Tunisia								**Tunisie**
Tourist/visitor arrivals	TF	4 120[1]	6 378[1]	7 828	6 999	7 352	7 163	Arrivées de touristes/visiteurs
Tourism expenditure		1 838	2 800	3 477	2 931	2 863	3 042	Dépenses touristiques
Turkey								**Turquie**
Tourist/visitor arrivals	TF	7 083	20 273	31 364[62]	35 698[62]	37 795[62]	39 811[62]	Arrivées de touristes/visiteurs
Tourism expenditure [61]		...	20 760	26 318	31 566	35 037	37 371	Dépenses touristiques [61]
Turkmenistan	TF							**Turkménistan**
Tourist/visitor arrivals		218	12	...	...	...	...	Arrivées de touristes/visiteurs
Turks and Caicos Islands								**Îles Turques-et-Caïques**
Tourist/visitor arrivals	TF	79	176	281	292	291	357	Arrivées de touristes/visiteurs
Tourism expenditure		53	...	...	...	...	...	Dépenses touristiques
Tuvalu								**Tuvalu**
Tourist/visitor arrivals	TF	1	1	2	1	1	1	Arrivées de touristes/visiteurs
Tourism expenditure [3]		...	1	2	3	2	...	Dépenses touristiques [3]
Uganda								**Ouganda**
Tourist/visitor arrivals	TF	160	468	946	1 197	1 206	1 266	Arrivées de touristes/visiteurs
Tourism expenditure		...	382	802	1 157	1 355	811	Dépenses touristiques
Ukraine								**Ukraine**
Tourist/visitor arrivals	TF	3 716	17 631	21 203	23 013	24 671	12 712	Arrivées de touristes/visiteurs
Tourism expenditure		...	3 542	4 696	5 988	5 931	2 263	Dépenses touristiques
United Arab Emirates								**Émirats arabes unis**
Tourist/visitor arrivals [63,64]	THS	2 315	7 126	...	...	...	...	Arrivées de touristes/visiteurs [63,64]
Tourism expenditure		632[63,65]	3 218	8 577	10 924	12 389	13 969	Dépenses touristiques
United Kingdom								**Royaume-Uni**
Tourist/visitor arrivals	TF	21 719	28 039	28 295	29 282	31 063	32 613	Arrivées de touristes/visiteurs
Tourism expenditure		27 577	39 647	40 216	47 052	53 522	62 830	Dépenses touristiques
United Rep. of Tanzania								**Rép.-Unie de Tanzanie**
Tourist/visitor arrivals	TF	285	590	754	1 043	1 063	1 113	Arrivées de touristes/visiteurs
Tourism expenditure		...	835	1 279	1 754	1 912	2 043	Dépenses touristiques
United States								**États-Unis**
Tourist/visitor arrivals	TF	43 318	49 206	60 010	66 657	69 995	#75 011	Arrivées de touristes/visiteurs
Tourism expenditure		93 743	122 077	167 996	200 997	214 542	220 757	Dépenses touristiques
United States Virgin Is.								**Îles Vierges américaines**
Tourist/visitor arrivals	TF	454	593	590	580	570	602	Arrivées de touristes/visiteurs
Tourism expenditure		822	1 432	1 013	1 153	1 232	...	Dépenses touristiques
Uruguay								**Uruguay**
Tourist/visitor arrivals	TF	2 022	1 808	2 353	2 695	2 683	2 682	Arrivées de touristes/visiteurs
Tourism expenditure		725	699	1 669	2 219	2 015	1 861	Dépenses touristiques

24

Tourist/visitor arrivals and tourism expenditure *(continued)*
Thousands arrivals and millions of US dollars

Arrivées de touristes/visiteurs et dépenses touristiques *(suite)*
Milliers d'arrivées et millions de dollars É.-U.

Country or area of destination	Series& Série&	1995	2005	2010	2012	2013	2014	Pays ou zone de destination
Uzbekistan								**Ouzbékistan**
Tourist/visitor arrivals	TF	92	242	975	...	1 969	...	Arrivées de touristes/visiteurs
Tourism expenditure [3]		...	28	121	...	...	...	Dépenses touristiques [3]
Vanuatu								**Vanuatu**
Tourist/visitor arrivals	TF	44	62	97	108	110	109	Arrivées de touristes/visiteurs
Tourism expenditure		...	104	242	268	314	284	Dépenses touristiques
Venezuela (Boliv. Rep. of)								**Venezuela (Rép. boliv. du)**
Tourist/visitor arrivals	TF	700	706	526	988	986	857	Arrivées de touristes/visiteurs
Tourism expenditure		995	722	794	904	926	...	Dépenses touristiques
Viet Nam								**Viet Nam**
Tourist/visitor arrivals	VF	1 351	3 477	5 050	6 848	7 572	7 874	Arrivées de touristes/visiteurs
Tourism expenditure		...	2 300	4 450	6 850	7 250	7 330	Dépenses touristiques
Yemen								**Yémen**
Tourist/visitor arrivals	TF	61	336	1 025[2]	874[2]	990[2]	...	Arrivées de touristes/visiteurs
Tourism expenditure		...	...	1 291	1 005	1 097	1 199	Dépenses touristiques
Zambia								**Zambie**
Tourist/visitor arrivals	TF	163	669	815	859	915	947	Arrivées de touristes/visiteurs
Tourism expenditure [3]		...	447	492	518	552	642	Dépenses touristiques [3]
Zimbabwe								**Zimbabwe**
Tourist/visitor arrivals	VF	1 416	1 559	2 239	1 794	1 833	1 905	Arrivées de touristes/visiteurs
Tourism expenditure		145	99	634	749	856	827	Dépenses touristiques

Source:
World Tourism Organization (UNWTO), Madrid, UNWTO statistics database, last accessed April 2016.

The majority of the expenditure data have been provided to the WTO by the International Monetary Fund (IMF).

& Series (by order of priority, see Annex II):

TF: Arrivals of non-resident tourists at national borders.
VF: Arrivals of non-resident visitors at national borders.
TCE: Arrivals of non-resident tourists in all types of accommodation establishments.
THS: Arrivals of non-resident tourists in hotels and similar establishments.

1 Excluding nationals residing abroad.
2 Including nationals residing abroad.
3 Excluding passenger transport.
4 Arrivals by air.
5 Excluding nationals residing abroad and crew members.

6 Package tour only.
7 Arrivals by all means of transport.
8 Arrivals by air to Bangui only.
9 For statistical purposes, the data for China do not include those for the Hong Kong Special Administrative Region (Hong Kong SAR), Macao Special Administrative Region (Macao SAR) and Taiwan Province of China.
10 Does not include other non-residents namely workers, students, etc.

11 The arrivals data relate only to three border posts (N'Djili airport in Kinshasa, the Luano airport in Lubumbashi, and the land border-crossing of Kasumbalesa in Katanga province).

12 Excluding the passengers at Herrera airport.
13 Arrivals to Bole airport only.

Source:
Organisation mondiale du tourisme (OMT), Madrid, la base de données de l'OMT, dernier accès Avril 2016.

La majorité des données sur les dépenses touristiques sont celles que le Fonds monétaire international (FMI) a fournies à l'Organisation mondiale du tourisme (OMT).

& Série (par ordre de priorite, voir annexe II):

TF : Arrivées de touristes non résidents aux frontières nationales.
VF: Arrivées de visiteurs non résidents aux frontières nationales.
TCE: Arrivées de touristes non résidents dans tous les types d'établissements d'hébergement touristique.
THS: Arrivées de touristes non résidents dans les hôtels et établissements assimilés.

1 A l'exclusion des nationaux résidant à l'étranger.
2 Y compris les nationaux du pays résidant à l'étranger.
3 Non compris le transport de passagers.
4 Arrivées par voie aérienne
5 A l'exclusion des nationaux du pays résidant à l'étranger et des membres des équipages.
6 Tourisme organisé.
7 Arrivées par tous moyens de transport confondus.
8 Arrivées par voie aérienne à Bangui uniquement.
9 Pour la présentation des statistiques, les données pour la Chine ne comprennent pas la Région Administrative Spéciale de Hong Kong (Hong Kong RAS), la Région Administrative Spéciale de Macao (Macao RAS) et la province de Taiwan.
10 Ne comprend pas d'autres catégories de non-résidents tels que les travailleurs, les étudiants, etc.
11 Les données des entrées ne concernent que 3 postes frontaliers : l'aéroport de N'Djili à Kinshasa, l'aéroport de la Luano à Lubumbashi et le poste frontière terrestre situé à Kasumbalesa dans la province du Katanga.
12 A l'exclusion des passagers à l'aéroport de Herrera.
13 Arrivées à l'aéroport de Bole uniquement.

24

Tourist/visitor arrivals and tourism expenditure *(continued)*
Thousands arrivals and millions of US dollars

Arrivées de touristes/visiteurs et dépenses touristiques *(suite)*
Milliers d'arrivées et millions de dollars É.-U.

14	Arrivals through all ports of entry.	14	Arrivées à travers tous les ports d'entrée.
15	Arrivals of non-resident tourists at Libreville airport.	15	Arrivées de touristes non résidents à l'aéroport de Libreville.
16	Charter tourists only.	16	Arrivées en vols à la demande seulement.
17	Excluding the north islands (Saint Martin and Saint Barthelemy).	17	Les îles du Nord (Saint Martin et Saint Barthélemy) sont exclues.
18	Non-resident tourists staying in all types of accommodation establishments.	18	Arrivées de touristes non résidents dans tous les types d'établissements d'hébergement touristique.
19	Including residents and non-residents.	19	Y compris les résidents et les non-résidents.
20	Arrivals by air at Conakry airport.	20	Arrivées par voie aérienne à l'aéroport de Conakry.
21	Arrivals to Timehri airport only.	21	Arrivées à l'aéroport de Timehri seulement.
22	Including tourists from Northern Ireland.	22	Y compris touristes à Irlande du Nord.
23	Including the expenditures of foreign workers in Israel.	23	Y compris les dépenses des travailleurs étrangers en Israël.
24	Excluding seasonal and border workers.	24	A l'exclusion des travailleurs saisonniers et frontaliers.
25	Arrivals of non-resident tourists by air. Including nationals residing abroad. E/D cards.	25	Arrivées de touristes non résidents par air. Y compris les nationaux qui résident à l'étranger. Cartes d'embarquement et de débarquement.
26	Air arrivals. Tarawa and Christmas Island.	26	Arrivées par voie aérienne. Tarawa et Ile Christmas.
27	Non-resident departures. Survey of persons crossing the state border.	27	Départs de non-résidents. Enquête menée auprès de personnes franchissant la frontière de l'État.
28	Excluding nationals residing abroad, Syrian nationals and Palestinians.	28	À l'exclusion des nationaux résidant à l'étranger, Syriens et Palestiniens.
29	Excluding long term tourists on campgrounds and in holiday flats.	29	A l'exclusion des touristes à long terme en camping ou dans des appartements de vacances.
30	Arrivals of non-resident tourists by air.	30	Arrivées de touristes non résidents par voie aérienne.
31	Departures.	31	Départs.
32	Including Singapore residents crossing the frontier by road through Johore Causeway.	32	Y compris les résidents de Singapour traversant la frontière par voie terrestre à travers le Johore Causeway.
33	Departures by air and by sea.	33	Départs par voies aérienne et maritime.
34	Fiscal years (October 1 to September 30).	34	Années fiscales (du 1er octobre au 30 septembre).
35	Air and sea arrivals.	35	Arrivées par voie aérienne et maritime.
36	Arrivals in the States of Kosrae, Chuuk, Pohnpei and Yap; excluding FSM citizens.	36	Arrivées dans les États de Kosrae, Chuuk, Pohnpei et Yap; non compris les citoyens FSM.
37	Excluding diplomats and foreign residents in Mongolia.	37	Sont exclus les diplomates et les étrangers qui résident en Mongolie.
38	The data correspond only to 12 border posts.	38	les données ne couvrent que 12 postes frontière.
39	The data of all the border posts of the country are used.	39	Les données de l'ensemble des postes frontières du pays sont utilisées.
40	Including arrivals from India.	40	Y compris les arrivées à Inde.
41	Including Niueans residing usually in New Zealand.	41	Y compris les nationaux de Niue résidant habituellement en Nouvelle-Zélande.
42	Non-resident tourists staying in registered hotels.	42	Non résidents touristes dans les hôtels enregistrés.
43	Air arrivals (Palau International Airport).	43	Arrivées par voie aérienne (Aéroport international de Palau).
44	Fiscal years ending September 30.	44	Exercices terminant le 30 Septembre.
45	E/D cards in the "Silvio Petirossi" airport and passenger counts at the national border crossings.	45	Cartes d'embarquement et de débarquement à l'aéroport Silvio Petirossi et comptages des passagers lors du franchissement des frontières nationales.
46	Including tourists with identity document other than a passport.	46	Nouvelle série estimée comprenant les touristes avec une pièce d'identité autre qu'un passeport.
47	Fiscal years (July-June).	47	Année fiscale de juillet à juin.
48	Including nationals residing abroad and crew members.	48	Y compris les nationaux résidant à l'étranger et membres des équipages.
49	Excluding the left side of the river Nistru and the municipality of Bender.	49	La rive gauche de la rivière Nistru et la municipalité de Bender sont exclues.
50	Including Italian tourists.	50	Y compris les touristes italiens.
51	Excluding Netherlands Antillean residents.	51	A l'exclusion des résidents des Antilles Néerlandaises.
52	Including arrivals to Saint Martin (the French side of the island).	52	Y compris les arrivées à Saint-Martin (la partie française de l'île).
53	Excluding arrivals for work and contract workers.	53	À l'exclusion des arrivées par travail et les travailleurs contractuels.
54	Excluding transit.	54	A l'exclusion des personnes en transit.
55	West Bank only.	55	Cisjordanie seulement.
56	Arrivals at Zanderij Airport.	56	Arrivées à l'aéroport de Zanderij.
57	Arrivals in hotels only.	57	Arrivées dans les hôtels uniquement.
58	Including health establishments.	58	Y compris les établissements de cure.
59	Including Iraqi nationals.	59	Y compris les nationaux iraquiens.
60	Arrivals by air at Dili Airport.	60	Arrivées par voie aérienne à l'aéroport de Dili.
61	Including expenditure of the nationals residing abroad.	61	Y compris dépenses des nationaux résidant à l'étranger.
62	Turkish citizens resident abroad are included.	62	Citoyens turcs résidant à l'étranger sont inclus.
63	Including domestic tourism and nationals residing abroad.	63	Y compris le tourisme intérieur et les nationaux du pays résidant à

24

Tourist/visitor arrivals and tourism expenditure *(continued)*
Thousands arrivals and millions of US dollars

Arrivées de touristes/visiteurs et dépenses touristiques *(suite)*
Milliers d'arrivées et millions de dollars É.-U.

				l'étranger.
64	Arrivals at hotels only.		64	Arrivées dans les hôtels uniquement.
65	Hotel revenues.		65	Recettes des hôtels.

25

Net disbursements of official development assistance to recipients
Total, bilateral and multilateral aid (millions of US dollars); and as a percentage of Gross National Income (GNI)

Décaissements nets d'aide publique au développement aux bénéficiaires
Total, bilatérale et multilatérale d'aide (millions de dollars É.-U.); et en pourcentage du Revenu National Brut (RNB)

Region, country or area[&]	1975	1985	1995	2005	2010	2012	2013	2014	Région, pays ou zone[&]
World[1]									**Monde**[1]
Bilateral	**15 196**	**26 356**	**48 142**	**83 033**	**94 207**	**96 003**	**111 066**	**119 123**	**Bilatérale**
Multilateral	**3 597**	**5 920**	**11 004**	**25 620**	**37 133**	**36 735**	**39 735**	**41 952**	**Multilatérale**
Total	**18 793**	**32 277**	**59 145**	**108 652**	**131 340**	**132 738**	**150 800**	**161 075**	**Total**
% of GNI	**1.7**	**1.3**	**1.0**	**1.2**	**0.7**	**0.6**	**0.6**	**0.6**	**% du RNB**
Africa[1]									**Afrique**[1]
Bilateral	5 678	10 085	16 798	24 096	29 983	33 576	38 686	35 037	Bilatérale
Multilateral	1 105	2 247	5 015	11 740	17 732	17 555	18 030	19 156	Multilatérale
Total	6 783	12 332	21 814	35 836	47 716	51 132	56 715	54 193	Total
% of GNI	4.4	4.0	4.6	3.8	2.6	2.3	2.5	2.3	% du RNB
Northern Africa[1,2]									**Afrique septentrionale**[1,2]
Bilateral	3 069	2 858	2 698	1 688	1 837	3 214	7 611	5 579	Bilatérale
Multilateral	179	142	277	980	823	1 526	1 173	1 774	Multilatérale
Total	3 248	3 000	2 975	2 667	2 660	4 740	8 784	7 354	Total
% of GNI	8.2	2.7	1.9	0.8	0.5	0.7	1.3	1.2	% du RNB
Sub-Saharan Africa[1,3]									**Afrique subsaharienne**[1,3]
Bilateral	2 557	7 017	13 747	21 884	27 058	28 927	29 743	27 707	Bilatérale
Multilateral	926	1 968	4 594	10 532	16 425	15 491	16 271	16 614	Multilatérale
Total	3 483	8 985	18 341	32 415	43 483	44 418	46 014	44 321	Total
% of GNI	3.0	4.6	5.8	5.1	3.4	2.9	2.9	2.7	% du RNB
Americas[1]									**Amériques**[1]
Bilateral	777	3 015	5 450	4 602	8 078	7 921	7 969	7 809	Bilatérale
Multilateral	597	398	935	2 106	2 854	2 145	2 263	2 140	Multilatérale
Total	1 375	3 413	6 384	6 708	10 932	10 065	10 232	9 949	Total
% of GNI	0.4	0.5	0.4	0.3	0.2	0.2	0.2	0.2	% du RNB
America, North[1]									**Amérique du Nord**[1]
Bilateral	254	1 989	2 971	2 260	4 923	3 603	3 352	3 190	Bilatérale
Multilateral	280	203	513	1 011	1 886	1 103	1 264	1 249	Multilatérale
Total	534	2 193	3 484	3 271	6 809	4 706	4 616	4 439	Total
% of GNI	0.5	0.9	0.8	0.3	0.5	0.3	0.3	0.3	% du RNB
America, South[1]									**Amérique du Sud**[1]
Bilateral	435	903	2 252	2 063	1 922	3 384	3 174	3 513	Bilatérale
Multilateral	317	150	344	792	858	875	750	699	Multilatérale
Total	752	1 052	2 597	2 855	2 780	4 259	3 924	4 212	Total
% of GNI	0.3	0.2	0.2	0.2	0.1	0.1	0.1	0.1	% du RNB
Asia[1]									**Asie**[1]
Bilateral	6 393	9 033	14 158	39 056	26 879	24 155	33 395	41 937	Bilatérale
Multilateral	1 813	2 747	4 415	7 579	9 896	9 267	11 156	11 847	Multilatérale
Total	8 206	11 780	18 573	46 635	36 774	33 422	44 551	53 785	Total
% of GNI	1.5	0.9	0.6	1.0	0.4	0.2	0.3	0.3	% du RNB
Eastern and South-Eastern Asia[1,4]									**Asie orientale et du Sud-Est**[1,4]
Bilateral	1 495	2 448	7 361	6 613	4 974	3 466	3 068	3 432	Bilatérale
Multilateral	371	498	1 411	1 779	2 515	2 667	2 790	2 742	Multilatérale
Total	1 866	2 946	8 771	8 391	7 489	6 133	5 858	6 173	Total
% of GNI	0.7	0.4	0.4	0.3	0.1	0.1	0.1	0.1	% du RNB
South-central Asia[1,5]									**Asie centrale et du Sud**[1,5]
Bilateral	2 457	2 678	4 035	7 260	13 209	12 431	14 177	12 571	Bilatérale
Multilateral	1 308	1 998	2 409	4 395	5 499	5 196	6 451	7 184	Multilatérale
Total	3 765	4 676	6 444	11 655	18 708	17 627	20 628	19 754	Total
% of GNI	2.7	1.6	1.2	1.0	0.8	0.7	0.7	0.6	% du RNB
Western Asia[1,6]									**Asie occidentale**[1,6]
Bilateral	2 407	3 856	2 447	24 330	7 777	7 487	15 133	23 343	Bilatérale
Multilateral	134	181	504	1 182	1 709	1 249	1 771	1 738	Multilatérale
Total	2 541	4 037	2 951	25 512	9 486	8 736	16 904	25 081	Total
% of GNI	1.9	1.1	0.7	3.7	3.2	1.0	2.0	8.4	% du RNB

25

Net disbursements of official development assistance to recipients *(continued)*
Total, bilateral and multilateral aid (millions of US dollars); and as a percentage of Gross National Income (GNI)

Décaissements nets d'aide publique au développement aux bénéficiaires *(suite)*
Total, bilatérale et multilatérale d'aide (millions de dollars É.-U.); et en pourcentage du Revenu National Brut (RNB)

Region, country or area[&]	1975	1985	1995	2005	2010	2012	2013	2014	Région, pays ou zone[&]
Europe[1,7]									**Europe**[1,7]
Bilateral	96	383	1 750	2 457	3 105	3 735	2 761	3 733	Bilatérale
Multilateral	58	28	534	1 605	2 787	4 342	4 676	4 880	Multilatérale
Total	154	410	2 285	4 062	5 892	8 076	7 437	8 613	Total
% of GNI	0.3	0.6	1.0	0.6	0.6	0.7	0.6	0.8	% du RNB
Oceania[1]									**Océanie**[1]
Bilateral	589	859	1 804	819	1 599	1 886	1 827	1 582	Bilatérale
Multilateral	24	38	62	341	269	294	322	281	Multilatérale
Total	613	897	1 865	1 161	1 868	2 179	2 149	1 863	Total
% of GNI	17.1	14.5	12.3	11.7	12.1	9.6	9.5	27.5	% du RNB
Areas not specified									**Zones non spécifiées**
Bilateral	...	2 982	8 182	12 003	24 564	24 731	26 428	29 024	Bilatérale
Multilateral	...	462	42	2 248	3 595	3 132	3 287	3 647	Multilatérale
Total	1 662	3 445	8 224	14 251	28 159	27 863	29 715	32 672	Total
Afghanistan									**Afghanistan**
Bilateral	41	7	129	2 295	5 759	6 066	4 718	4 258	Bilatérale
Multilateral	29	9	84	542	713	601	545	565	Multilatérale
Total	69	16	213	2 838	6 472	6 667	5 262	4 823	Total
% of GNI	2.9	...	...	45.1	40.5	32.3	25.7	23.0	% du RNB
Albania									**Albanie**
Bilateral	...	...	118	196	253	232	171	166	Bilatérale
Multilateral	...	...	63	123	112	118	99	114	Multilatérale
Total	...	...	181	319	364	350	269	280	Total
% of GNI	...	...	7.3	3.7	3.1	2.9	2.1	2.1	% du RNB
Algeria									**Algérie**
Bilateral	213	161	270	251	128	93	139	100	Bilatérale
Multilateral	23	11	24	96	71	51	62	58	Multilatérale
Total	235	172	294	347	198	145	201	158	Total
% of GNI	1.5	0.3	0.7	0.4	0.1	0.1	0.1	0.1	% du RNB
Angola									**Angola**
Bilateral	4	72	304	216	148	146	149	96	Bilatérale
Multilateral	1	19	113	198	89	96	134	135	Multilatérale
Total	5	90	416	415	237	241	283	231	Total
% of GNI	...	1.4	11.3	1.7	0.3	0.2	0.3	0.2	% du RNB
Anguilla									**Anguilla**
Bilateral	2	2	3	4	-1	2	4	...	Bilatérale
Multilateral	<0	1	1	<0	9	3	4	...	Multilatérale
Total	2	3	3	4	8	5	8	...	Total
Antigua and Barbuda									**Antigua-et-Barbuda**
Bilateral	2	2	2	7	5	-<0	<0	1	Bilatérale
Multilateral	1	1	<0	1	14	3	2	2	Multilatérale
Total	2	3	2	8	19	2	2	2	Total
% of GNI	...	1.5	0.5	0.8	1.7	0.2	0.2	0.2	% du RNB
Argentina									**Argentine**
Bilateral	3	30	126	64	91	77	-5	12	Bilatérale
Multilateral	23	9	17	32	30	102	37	37	Multilatérale
Total	26	39	143	96	121	180	33	49	Total
% of GNI	0.1	0.1	0.1	0.1	<0.0	<0.0	<0.0	<0.0	% du RNB
Armenia									**Arménie**
Bilateral	...	...	111	103	231	150	125	128	Bilatérale
Multilateral	...	...	107	67	111	121	154	137	Multilatérale
Total	...	...	218	170	342	271	279	265	Total
% of GNI	...	...	14.8	3.3	3.5	2.6	2.5	2.3	% du RNB
Aruba									**Aruba**
Bilateral	...	...	23	...	...	...	...	...	Bilatérale
Multilateral	...	...	3	...	...	...	...	...	Multilatérale
Total	...	12	26	...	...	...	...	...	Total
Azerbaijan									**Azerbaïdjan**
Bilateral	...	...	67	125	80	222	-186	132	Bilatérale
Multilateral	...	...	53	91	80	63	112	83	Multilatérale
Total	...	...	120	217	160	285	-73	215	Total
% of GNI	...	...	3.9	1.9	0.3	0.5	-0.1	0.3	% du RNB

Net disbursements of official development assistance to recipients *(continued)*
Total, bilateral and multilateral aid (millions of US dollars); and as a percentage of Gross National Income (GNI)

Décaissements nets d'aide publique au développement aux bénéficiaires *(suite)*
Total, bilatérale et multilatérale d'aide (millions de dollars É.-U.); et en pourcentage du Revenu National Brut (RNB)

Region, country or area[&]	1975	1985	1995	2005	2010	2012	2013	2014	Région, pays ou zone[&]
Bahamas									**Bahamas**
Bilateral	<0	<0	3	...	...	...	...	...	Bilatérale
Multilateral	1	1	2	...	...	...	...	...	Multilatérale
Total	1	1	4	...	...	...	...	...	Total
Bahrain									**Bahreïn**
Bilateral	25	73	100	...	...	...	...	...	Bilatérale
Multilateral	1	1	<0	...	...	...	...	...	Multilatérale
Total	26	74	100	...	...	...	...	...	Total
% of GNI	...	2.2	1.7	...	...	...	...	...	% du RNB
Bangladesh									**Bangladesh**
Bilateral	777	619	907	588	582	1 097	1 398	1 341	Bilatérale
Multilateral	295	508	375	731	822	1 052	1 231	1 077	Multilatérale
Total	1 072	1 127	1 282	1 319	1 404	2 148	2 629	2 418	Total
% of GNI	5.5	5.0	3.3	1.8	1.1	1.5	1.6	1.3	% du RNB
Barbados									**Barbade**
Bilateral	3	6	-2	-5	-1	...	...	...	Bilatérale
Multilateral	3	1	1	3	18	...	...	...	Multilatérale
Total	5	7	-1	-2	16	...	...	...	Total
% of GNI	1.3	0.6	-0.1	-0.1	0.4	...	...	...	% du RNB
Belarus									**Bélarus**
Bilateral	...	...	...	40	96	59	71	77	Bilatérale
Multilateral	...	...	...	18	42	44	35	43	Multilatérale
Total	...	...	...	58	139	103	106	120	Total
% of GNI	...	...	...	0.2	0.3	0.2	0.2	0.2	% du RNB
Belize									**Belize**
Bilateral	8	21	12	6	9	7	28	11	Bilatérale
Multilateral	1	1	7	6	16	18	21	25	Multilatérale
Total	9	22	18	12	25	25	50	36	Total
% of GNI	8.2	11.0	3.0	1.2	2.0	1.7	3.3	...	% du RNB
Benin									**Bénin**
Bilateral	36	62	228	173	389	331	329	312	Bilatérale
Multilateral	18	32	52	174	301	178	331	289	Multilatérale
Total	54	94	280	347	690	509	660	600	Total
% of GNI	8.0	9.2	13.2	8.0	10.6	6.8	8.0	6.9	% du RNB
Bermuda									**Bermudes**
Bilateral	...	1	-2	...	...	...	...	...	Bilatérale
Multilateral	...	<0	0	...	...	...	...	...	Multilatérale
Total	<0	1	-2	...	...	...	...	...	Total
Bhutan									**Bhoutan**
Bilateral	<0	9	62	52	87	89	83	67	Bilatérale
Multilateral	2	14	10	38	44	72	51	63	Multilatérale
Total	2	23	71	90	130	161	134	130	Total
% of GNI	...	17.0	26.8	11.2	8.7	9.6	8.1	7.4	% du RNB
Bolivia (Plurinational State of)									**Bolivie (État plurinational de)**
Bilateral	27	166	558	486	567	525	536	559	Bilatérale
Multilateral	29	30	153	157	148	133	164	113	Multilatérale
Total	56	196	711	643	715	658	700	672	Total
% of GNI	2.2	4.0	11.0	7.0	3.8	2.6	2.4	2.1	% du RNB
Bosnia and Herzegovina									**Bosnie-Herzégovine**
Bilateral	...	...	900	318	287	309	220	280	Bilatérale
Multilateral	...	...	66	231	222	262	302	352	Multilatérale
Total	...	...	966	549	508	571	522	632	Total
% of GNI	...	...	59.4	4.8	3.0	3.4	2.9	3.4	% du RNB
Botswana									**Botswana**
Bilateral	39	71	71	8	94	53	73	51	Bilatérale
Multilateral	12	24	19	40	60	20	35	49	Multilatérale
Total	51	96	90	48	153	73	107	100	Total
% of GNI	14.4	9.4	1.9	0.5	1.3	0.5	0.7	0.6	% du RNB

Net disbursements of official development assistance to recipients *(continued)*
Total, bilateral and multilateral aid (millions of US dollars); and as a percentage of Gross National Income (GNI)

Décaissements nets d'aide publique au développement aux bénéficiaires *(suite)*
Total, bilatérale et multilatérale d'aide (millions de dollars É.-U.); et en pourcentage du Revenu National Brut (RNB)

Region, country or area[&]	1975	1985	1995	2005	2010	2012	2013	2014	Région, pays ou zone[&]
Brazil									**Brésil**
Bilateral	101	91	237	139	362	1 100	1 032	830	Bilatérale
Multilateral	66	31	34	104	88	181	113	82	Multilatérale
Total	166	122	271	243	450	1 281	1 145	912	Total
% of GNI	0.1	0.1	<0.0	<0.0	<0.0	0.1	0.1	<0.0	% du RNB
British Virgin Islands									**Îles Vierges britanniques**
Bilateral	2	2	<0	...	...	...	...	...	Bilatérale
Multilateral	<0	<0	1	...	...	...	...	...	Multilatérale
Total	2	2	1	...	...	...	...	...	Total
Brunei Darussalam									**Brunéi Darussalam**
Bilateral	<0	1	4	...	...	...	...	...	Bilatérale
Multilateral	0	<0	<0	...	...	...	...	...	Multilatérale
Total	<0	1	4	...	...	...	...	...	Total
% of GNI	...	...	0.1	...	...	...	...	...	% du RNB
Burkina Faso									**Burkina Faso**
Bilateral	62	134	350	377	553	711	616	672	Bilatérale
Multilateral	27	54	140	316	491	431	428	448	Multilatérale
Total	89	188	490	693	1 044	1 142	1 044	1 120	Total
% of GNI	9.5	12.2	20.7	12.7	11.4	10.7	9.5	9.0	% du RNB
Burundi									**Burundi**
Bilateral	33	91	147	162	281	283	259	250	Bilatérale
Multilateral	15	46	140	202	349	238	297	252	Multilatérale
Total	47	137	287	364	629	521	556	502	Total
% of GNI	11.5	12.1	29.0	33.1	31.2	21.2	20.5	16.2	% du RNB
Cabo Verde									**Cabo Verde**
Bilateral	4	42	91	111	249	216	213	182	Bilatérale
Multilateral	5	22	25	51	79	29	32	48	Multilatérale
Total	9	64	116	162	328	246	245	230	Total
% of GNI	...	48.0	23.9	17.3	20.6	14.6	13.8	12.9	% du RNB
Cambodia									**Cambodge**
Bilateral	78	9	434	371	523	623	581	556	Bilatérale
Multilateral	4	5	117	165	210	184	224	243	Multilatérale
Total	82	14	551	536	733	807	805	799	Total
% of GNI	...	...	16.3	8.9	6.9	6.0	5.6	5.0	% du RNB
Cameroon									**Cameroun**
Bilateral	72	134	378	276	303	335	430	511	Bilatérale
Multilateral	37	18	64	138	237	262	318	342	Multilatérale
Total	110	152	443	414	541	596	748	852	Total
% of GNI	4.5	1.9	5.4	2.6	2.3	2.3	2.6	2.6	% du RNB
Cayman Islands									**Îles Caïmanes**
Bilateral	1	-<0	-1	...	...	...	...	...	Bilatérale
Multilateral	1	<0	<0	...	...	...	...	...	Multilatérale
Total	2	<0	-1	...	...	...	...	...	Total
Central African Rep.									**Rép. centrafricaine**
Bilateral	39	73	126	37	78	65	111	289	Bilatérale
Multilateral	16	31	42	52	183	162	91	321	Multilatérale
Total	55	104	168	89	261	227	202	610	Total
% of GNI	14.5	12.1	15.3	6.6	13.1	10.4	13.0	34.1	% du RNB
Chad									**Tchad**
Bilateral	33	118	165	164	276	274	216	165	Bilatérale
Multilateral	34	61	70	221	215	198	242	224	Multilatérale
Total	67	179	235	385	491	472	458	388	Total
% of GNI	7.7	17.4	16.3	6.8	4.8	3.9	3.7	2.9	% du RNB
Chile									**Chili**
Bilateral	96	37	147	85	154	111	54	194	Bilatérale
Multilateral	31	3	9	83	44	14	26	47	Multilatérale
Total	127	40	157	167	197	125	79	241	Total
% of GNI	1.8	0.3	0.2	0.2	0.1	0.1	<0.0	0.1	% du RNB

Net disbursements of official development assistance to recipients *(continued)*
Total, bilateral and multilateral aid (millions of US dollars); and as a percentage of Gross National Income (GNI)

Décaissements nets d'aide publique au développement aux bénéficiaires *(suite)*
Total, bilatérale et multilatérale d'aide (millions de dollars É.-U.); et en pourcentage du Revenu National Brut (RNB)

Region, country or area[&]	1975	1985	1995	2005	2010	2012	2013	2014	Région, pays ou zone[&]
China[8]									**Chine**[8]
Bilateral	...	697	2 550	1 486	318	-608	-1 022	-1 255	Bilatérale
Multilateral	...	242	921	329	327	415	350	295	Multilatérale
Total	...	939	3 471	1 814	645	-193	-672	-960	Total
% of GNI	...	0.3	0.5	0.1	<0.0	0.0	-<0.0	-<0.0	% du RNB
China, Hong Kong SAR									**Chine, Hong Kong RAS**
Bilateral	-1	18	14	...	...	...	...	...	Bilatérale
Multilateral	<0	2	4	...	...	...	...	...	Multilatérale
Total	-1	20	18	...	...	...	...	...	Total
% of GNI	-<0.0	0.1	<0.0	...	...	...	...	...	% du RNB
China, Macao SAR									**Chine, Macao RAS**
Bilateral	<0	<0	-4	...	...	...	...	...	Bilatérale
Multilateral	0	<0	0	...	...	...	...	...	Multilatérale
Total	<0	<0	-4	...	...	...	...	...	Total
% of GNI	...	<0.0	-0.1	...	...	...	...	...	% du RNB
Colombia									**Colombie**
Bilateral	51	45	157	537	577	693	765	1 110	Bilatérale
Multilateral	32	17	12	83	98	70	89	111	Multilatérale
Total	83	61	169	621	675	763	854	1 221	Total
% of GNI	0.7	0.2	0.2	0.4	0.3	0.2	0.2	0.3	% du RNB
Comoros									**Comores**
Bilateral	19	32	27	5	20	5	-28	25	Bilatérale
Multilateral	2	15	15	18	47	64	107	49	Multilatérale
Total	22	47	42	23	67	69	79	74	Total
% of GNI	35.9	41.7	17.9	6.0	12.7	12.1	12.8	11.5	% du RNB
Congo									**Congo**
Bilateral	43	55	115	1 350	1 048	67	82	47	Bilatérale
Multilateral	13	14	10	76	267	73	68	59	Multilatérale
Total	56	69	125	1 425	1 315	139	151	106	Total
% of GNI	7.6	3.4	10.2	35.4	14.6	1.3	1.4	0.9	% du RNB
Cook Islands									**Îles Cook**
Bilateral	5	9	12	7	13	18	15	23	Bilatérale
Multilateral	<0	1	1	1	1	3	1	4	Multilatérale
Total	6	10	13	8	13	21	16	27	Total
Costa Rica									**Costa Rica**
Bilateral	16	265	16	13	78	16	15	20	Bilatérale
Multilateral	14	14	13	13	16	13	18	33	Multilatérale
Total	30	279	30	26	94	28	33	54	Total
% of GNI	1.6	6.2	0.3	0.1	0.3	0.1	0.1	0.1	% du RNB
Côte d'Ivoire									**Côte d'Ivoire**
Bilateral	82	110	948	16	477	1 075	844	414	Bilatérale
Multilateral	18	7	264	75	368	1 560	427	508	Multilatérale
Total	100	117	1 212	91	845	2 635	1 272	922	Total
% of GNI	2.7	1.9	12.1	0.6	3.5	10.1	4.3	2.8	% du RNB
Croatia									**Croatie**
Bilateral	...	...	48	58	11	...	...	...	Bilatérale
Multilateral	...	...	5	65	121	...	...	...	Multilatérale
Total	...	...	53	123	132	...	...	...	Total
% of GNI	...	...	0.2	0.3	0.2	...	...	...	% du RNB
Cuba									**Cuba**
Bilateral	11	31	46	65	78	55	66	229	Bilatérale
Multilateral	9	11	17	23	51	33	35	33	Multilatérale
Total	20	42	63	88	129	88	101	262	Total
% of GNI	0.2	0.2	0.2	0.2	0.2	...	...	...	% du RNB
Cyprus									**Chypre**
Bilateral	18	31	15	...	...	...	...	...	Bilatérale
Multilateral	13	6	7	...	...	...	...	...	Multilatérale
Total	31	37	21	...	...	...	...	...	Total
% of GNI	6.3	1.5	0.2	...	...	...	...	...	% du RNB

Net disbursements of official development assistance to recipients *(continued)*
Total, bilateral and multilateral aid (millions of US dollars); and as a percentage of Gross National Income (GNI)

Décaissements nets d'aide publique au développement aux bénéficiaires *(suite)*
Total, bilatérale et multilatérale d'aide (millions de dollars É.-U.); et en pourcentage du Revenu National Brut (RNB)

Region, country or area[&]	1975	1985	1995	2005	2010	2012	2013	2014	Région, pays ou zone[&]
Dem. P. R. Korea									**R. p. dém. de Corée**
Bilateral	...	...	3	33	28	62	68	109	Bilatérale
Multilateral	...	...	11	54	50	36	41	45	Multilatérale
Total	...	6	13	88	79	98	109	153	Total
Dem. Rep. of the Congo									**Rép. dém. du Congo**
Bilateral	219	223	136	1 094	1 039	1 902	1 445	1 313	Bilatérale
Multilateral	39	83	59	788	2 442	945	1 139	1 085	Multilatérale
Total	258	305	195	1 882	3 481	2 847	2 584	2 398	Total
% of GNI	2.5	4.5	4.0	16.4	17.8	11.0	9.5	8.3	% du RNB
Djibouti									**Djibouti**
Bilateral	34	71	94	45	104	114	101	111	Bilatérale
Multilateral	<0	11	12	29	27	34	47	52	Multilatérale
Total	34	81	105	74	131	149	148	163	Total
% of GNI	...	...	20.5	9.6	...	...	...	...	% du RNB
Dominica									**Dominique**
Bilateral	6	13	18	13	3	16	8	10	Bilatérale
Multilateral	2	4	7	8	30	9	12	6	Multilatérale
Total	8	17	25	21	32	26	20	16	Total
% of GNI	...	17.8	11.9	6.2	6.7	5.2	4.0	3.0	% du RNB
Dominican Republic									**Rép. dominicaine**
Bilateral	13	195	106	13	59	187	78	93	Bilatérale
Multilateral	17	12	13	67	116	74	69	74	Multilatérale
Total	30	207	119	81	175	261	147	167	Total
% of GNI	0.8	4.4	0.8	0.3	0.3	0.5	0.3	0.3	% du RNB
Ecuador									**Équateur**
Bilateral	21	116	196	152	98	97	102	118	Bilatérale
Multilateral	49	19	27	74	49	51	45	43	Multilatérale
Total	69	135	223	226	147	148	147	160	Total
% of GNI	0.9	0.8	1.0	0.6	0.2	0.2	0.2	0.2	% du RNB
Egypt									**Égypte**
Bilateral	2 421	1 700	1 875	720	402	1 239	5 386	3 218	Bilatérale
Multilateral	100	82	155	316	187	568	122	314	Multilatérale
Total	2 521	1 783	2 030	1 036	589	1 807	5 508	3 532	Total
% of GNI	22.5	5.7	3.4	1.2	0.3	0.7	2.1	1.3	% du RNB
El Salvador									**El Salvador**
Bilateral	4	327	277	132	202	186	120	72	Bilatérale
Multilateral	34	18	19	72	78	34	49	26	Multilatérale
Total	37	345	296	204	280	220	169	98	Total
% of GNI	2.0	9.4	3.2	1.2	1.3	1.0	0.7	0.4	% du RNB
Equatorial Guinea									**Guinée équatoriale**
Bilateral	1	9	23	18	75	9	1	-3	Bilatérale
Multilateral	1	8	10	20	10	5	3	4	Multilatérale
Total	2	17	33	38	85	14	4	1	Total
% of GNI	2.1	29.5	28.3	0.9	0.9	0.1	<0.0	<0.0	% du RNB
Eritrea									**Érythrée**
Bilateral	...	...	119	229	47	69	9	1	Bilatérale
Multilateral	...	...	29	121	114	65	72	83	Multilatérale
Total	2	...	148	349	161	134	81	83	Total
% of GNI	...	...	25.3	32.1	7.7	4.4	2.4	2.2	% du RNB
Ethiopia									**Éthiopie**
Bilateral	73	541	653	1 204	2 220	2 158	2 272	2 194	Bilatérale
Multilateral	59	177	224	724	1 232	1 063	1 613	1 391	Multilatérale
Total	132	718	876	1 928	3 453	3 221	3 885	3 585	Total
% of GNI	...	7.6	11.5	15.6	11.6	7.5	8.2	6.6	% du RNB
Fiji									**Fidji**
Bilateral	18	28	41	32	57	86	69	73	Bilatérale
Multilateral	2	3	4	34	19	22	22	19	Multilatérale
Total	19	32	44	66	76	107	91	92	Total
% of GNI	2.9	2.8	2.3	2.2	2.5	2.9	2.4	2.3	% du RNB

25

Net disbursements of official development assistance to recipients *(continued)*
Total, bilateral and multilateral aid (millions of US dollars); and as a percentage of Gross National Income (GNI)

Décaissements nets d'aide publique au développement aux bénéficiaires *(suite)*
Total, bilatérale et multilatérale d'aide (millions de dollars É.-U.); et en pourcentage du Revenu National Brut (RNB)

Region, country or area[&]	1975	1985	1995	2005	2010	2012	2013	2014	Région, pays ou zone[&]
French Polynesia									**Polynésie française**
Bilateral	71	171	448	...	...	...	...	...	Bilatérale
Multilateral	<0	1	3	...	...	...	...	...	Multilatérale
Total	72	172	451	...	...	...	...	...	Total
% of GNI	10.4	11.4	11.3	...	...	...	...	...	% du RNB
Gabon									**Gabon**
Bilateral	58	55	140	11	80	50	68	92	Bilatérale
Multilateral	3	6	4	50	24	23	21	19	Multilatérale
Total	60	61	144	60	104	73	90	111	Total
% of GNI	3.0	2.0	3.3	0.7	0.8	0.5	0.6	0.7	% du RNB
Gambia									**Gambie**
Bilateral	4	33	25	7	50	67	52	28	Bilatérale
Multilateral	4	15	20	53	70	72	64	71	Multilatérale
Total	8	48	45	60	120	139	115	100	Total
% of GNI	7.0	19.1	5.9	10.0	13.0	15.7	13.3	12.8	% du RNB
Georgia									**Géorgie**
Bilateral	...	...	104	182	358	405	366	307	Bilatérale
Multilateral	...	...	106	110	270	255	281	255	Multilatérale
Total	...	...	209	292	628	660	647	563	Total
% of GNI	...	...	8.1	4.5	5.5	4.2	4.1	3.4	% du RNB
Ghana									**Ghana**
Bilateral	99	105	361	658	1 125	1 156	798	645	Bilatérale
Multilateral	25	89	288	493	565	642	532	482	Multilatérale
Total	124	194	648	1 151	1 690	1 799	1 330	1 126	Total
% of GNI	4.5	4.4	10.2	10.9	5.3	4.5	2.8	3.1	% du RNB
Gibraltar									**Gibraltar**
Bilateral	...	...	<0	...	...	...	...	...	Bilatérale
Multilateral	...	...	0	...	...	...	...	...	Multilatérale
Total	3	29	<0	...	...	...	...	...	Total
Grenada									**Grenade**
Bilateral	2	32	8	33	10	1	2	14	Bilatérale
Multilateral	1	2	3	20	24	7	9	26	Multilatérale
Total	3	34	11	53	34	8	11	39	Total
% of GNI	...	...	4.1	7.9	4.6	1.0	1.4	4.6	% du RNB
Guatemala									**Guatemala**
Bilateral	25	71	181	191	322	247	410	185	Bilatérale
Multilateral	15	12	27	65	67	57	86	92	Multilatérale
Total	40	83	208	257	389	303	495	277	Total
% of GNI	1.1	0.9	1.4	1.0	1.0	0.6	0.9	0.5	% du RNB
Guinea									**Guinée**
Bilateral	20	74	305	94	44	-690	192	236	Bilatérale
Multilateral	4	39	111	104	174	1 030	281	325	Multilatérale
Total	24	113	416	198	218	340	473	561	Total
% of GNI	...	...	11.5	7.5	5.1	6.5	8.2	9.2	% du RNB
Guinea-Bissau									**Guinée-Bissau**
Bilateral	17	34	95	7	-36	24	45	15	Bilatérale
Multilateral	4	22	23	59	161	55	59	94	Multilatérale
Total	21	56	118	66	125	79	104	109	Total
% of GNI	18.7	35.3	49.8	11.5	14.8	8.2	11.0	10.6	% du RNB
Guyana									**Guyana**
Bilateral	7	21	59	80	105	67	53	123	Bilatérale
Multilateral	3	6	27	70	62	47	48	36	Multilatérale
Total	10	27	86	150	167	114	102	159	Total
% of GNI	2.1	7.0	15.1	19.3	7.3	4.0	3.4	4.9	% du RNB
Haiti									**Haïti**
Bilateral	18	120	603	260	2 336	1 026	902	812	Bilatérale
Multilateral	38	29	120	165	700	246	249	272	Multilatérale
Total	56	149	722	426	3 036	1 272	1 152	1 084	Total
% of GNI	...	...	...	9.7	45.7	16.0	13.6	12.4	% du RNB

25 Net disbursements of official development assistance to recipients *(continued)*
Total, bilateral and multilateral aid (millions of US dollars); and as a percentage of Gross National Income (GNI)

Décaissements nets d'aide publique au développement aux bénéficiaires *(suite)*
Total, bilatérale et multilatérale d'aide (millions de dollars É.-U.); et en pourcentage du Revenu National Brut (RNB)

Region, country or area[&]	1975	1985	1995	2005	2010	2012	2013	2014	Région, pays ou zone[&]
Honduras									**Honduras**
Bilateral	24	245	301	507	435	400	426	455	Bilatérale
Multilateral	26	25	101	183	198	168	201	149	Multilatérale
Total	51	270	402	690	633	568	627	604	Total
% of GNI	4.6	7.8	11.0	7.5	4.2	3.3	3.7	3.3	% du RNB
India									**Inde**
Bilateral	879	844	868	669	1 619	772	1 130	1 110	Bilatérale
Multilateral	726	743	862	1 207	1 192	896	1 306	1 874	Multilatérale
Total	1 605	1 587	1 729	1 876	2 812	1 667	2 435	2 984	Total
% of GNI	1.6	0.7	0.5	0.2	0.2	0.1	0.1	0.2	% du RNB
Indonesia									**Indonésie**
Bilateral	455	514	1 251	2 253	967	-201	-173	-575	Bilatérale
Multilateral	232	83	50	281	423	265	238	187	Multilatérale
Total	687	597	1 301	2 534	1 390	65	65	-388	Total
% of GNI	2.2	0.7	0.7	0.9	0.2	<0.0	<0.0	-0.1	% du RNB
Iran (Islamic Rep. of)									**Iran (Rép. islamique d')**
Bilateral	-12	2	161	69	81	118	102	48	Bilatérale
Multilateral	13	9	25	40	40	31	27	33	Multilatérale
Total	2	11	187	109	121	149	129	81	Total
% of GNI	0.0	<0.0	0.2	0.1	...	<0.0	<0.0	...	% du RNB
Iraq									**Iraq**
Bilateral	101	19	273	21 999	2 040	1 180	1 405	1 233	Bilatérale
Multilateral	9	1	60	58	138	121	136	137	Multilatérale
Total	110	20	333	22 057	2 178	1 301	1 541	1 370	Total
% of GNI	...	...	...	43.5	1.6	0.6	0.7	0.6	% du RNB
Israel									**Israël**
Bilateral	466	1 978	334	...	...	...	...	...	Bilatérale
Multilateral	2	<0	2	...	...	...	...	...	Multilatérale
Total	467	1 978	336	...	...	...	...	...	Total
% of GNI	3.9	8.5	0.4	...	...	...	...	...	% du RNB
Jamaica									**Jamaïque**
Bilateral	16	161	81	-9	-9	-8	-14	24	Bilatérale
Multilateral	9	7	27	48	150	34	84	69	Multilatérale
Total	24	169	108	40	141	27	70	92	Total
% of GNI	0.8	9.2	1.9	0.4	1.1	0.2	0.5	...	% du RNB
Jordan									**Jordanie**
Bilateral	471	598	461	591	688	958	1 121	2 452	Bilatérale
Multilateral	21	9	78	117	263	199	282	247	Multilatérale
Total	492	607	539	708	951	1 157	1 403	2 699	Total
% of GNI	35.0	12.4	8.4	5.5	3.6	3.8	4.2	7.6	% du RNB
Kazakhstan									**Kazakhstan**
Bilateral	...	...	54	198	154	84	43	46	Bilatérale
Multilateral	...	...	11	31	72	45	45	42	Multilatérale
Total	...	...	65	229	226	129	88	88	Total
% of GNI	...	...	0.3	0.4	0.2	0.1	<0.0	0.1	% du RNB
Kenya									**Kenya**
Bilateral	99	366	505	478	1 157	1 928	2 181	1 553	Bilatérale
Multilateral	26	61	227	281	468	725	1 130	1 112	Multilatérale
Total	125	427	732	759	1 625	2 653	3 312	2 665	Total
% of GNI	3.9	7.2	8.4	4.1	4.1	5.3	6.1	4.4	% du RNB
Kiribati									**Kiribati**
Bilateral	6	11	14	20	21	61	52	59	Bilatérale
Multilateral	<0	1	2	8	2	4	12	20	Multilatérale
Total	6	12	15	28	23	65	65	79	Total
% of GNI	...	38.8	16.9	17.4	10.8	24.7	23.1	35.2	% du RNB
Kosovo									**Kosovo**
Bilateral	...	...	...	...	215	412	377	361	Bilatérale
Multilateral	...	...	...	...	312	155	192	219	Multilatérale
Total	...	...	...	...	528	567	569	580	Total
% of GNI	...	...	...	...	8.9	8.5	7.9	7.8	% du RNB

25

Net disbursements of official development assistance to recipients *(continued)*
Total, bilateral and multilateral aid (millions of US dollars); and as a percentage of Gross National Income (GNI)

Décaissements nets d'aide publique au développement aux bénéficiaires *(suite)*
Total, bilatérale et multilatérale d'aide (millions de dollars É.-U.); et en pourcentage du Revenu National Brut (RNB)

Region, country or area[&]	1975	1985	1995	2005	2010	2012	2013	2014	Région, pays ou zone[&]
Kuwait									**Koweït**
Bilateral	-2	3	4	...	...	...	...	...	Bilatérale
Multilateral	2	1	<0	...	...	...	...	...	Multilatérale
Total	<0	4	4	...	...	...	...	...	Total
% of GNI	0.0	<0.0	<0.0	...	...	...	...	...	% du RNB
Kyrgyzstan									**Kirghizistan**
Bilateral	...	...	191	177	255	260	381	454	Bilatérale
Multilateral	...	...	94	91	127	212	156	170	Multilatérale
Total	...	...	285	268	382	472	536	624	Total
% of GNI	...	...	17.5	11.3	8.5	7.3	7.6	8.7	% du RNB
Lao People's Dem. Rep.									**Rép. dém. pop. lao**
Bilateral	26	19	252	165	285	258	283	347	Bilatérale
Multilateral	13	21	55	137	128	151	138	126	Multilatérale
Total	38	40	307	302	414	409	421	472	Total
% of GNI	...	1.7	17.5	11.3	6.2	4.7	4.0	4.2	% du RNB
Lebanon									**Liban**
Bilateral	7	65	147	141	322	589	376	583	Bilatérale
Multilateral	7	17	44	89	126	122	245	237	Multilatérale
Total	14	82	191	230	448	712	621	820	Total
% of GNI	...	...	1.6	1.1	1.2	1.7	1.4	1.8	% du RNB
Lesotho									**Lesotho**
Bilateral	12	65	81	26	92	188	199	49	Bilatérale
Multilateral	16	28	31	42	164	86	121	55	Multilatérale
Total	28	92	113	67	256	274	320	104	Total
% of GNI	11.4	19.2	9.4	3.6	9.9	9.6	11.2	4.0	% du RNB
Liberia									**Libéria**
Bilateral	8	72	51	119	956	361	318	446	Bilatérale
Multilateral	12	19	72	104	457	205	217	299	Multilatérale
Total	20	90	123	222	1 413	566	535	744	Total
% of GNI	3.3	11.2	...	56.5	126.9	35.8	32.5	43.9	% du RNB
Libya									**Libye**
Bilateral	-6	3	4	17	2	50	71	173	Bilatérale
Multilateral	8	2	2	6	7	37	57	38	Multilatérale
Total	2	5	6	24	9	87	129	210	Total
% of GNI	...	...	...	0.1	<0.0	0.1	0.2	0.5	% du RNB
Madagascar									**Madagascar**
Bilateral	43	118	192	514	259	207	257	258	Bilatérale
Multilateral	39	67	107	399	214	168	242	326	Multilatérale
Total	82	185	299	913	472	375	499	583	Total
% of GNI	3.6	6.7	10.0	18.4	5.5	3.9	4.9	5.7	% du RNB
Malawi									**Malawi**
Bilateral	43	72	285	301	558	773	716	532	Bilatérale
Multilateral	20	41	149	272	456	395	413	398	Multilatérale
Total	63	112	434	573	1 015	1 169	1 130	930	Total
% of GNI	10.1	10.4	32.2	21.1	19.2	28.5	30.4	22.8	% du RNB
Malaysia									**Malaisie**
Bilateral	88	219	104	14	-25	5	-138	-2	Bilatérale
Multilateral	11	10	5	12	27	10	18	14	Multilatérale
Total	99	229	108	26	2	15	-120	12	Total
% of GNI	1.0	0.8	0.1	<0.0	0.0	<0.0	-<0.0	0.0	% du RNB
Maldives									**Maldives**
Bilateral	2	6	51	49	71	23	8	6	Bilatérale
Multilateral	1	3	7	27	40	33	14	19	Multilatérale
Total	3	9	58	76	111	57	21	25	Total
% of GNI	...	7.7	15.2	7.0	5.5	2.5	0.9	1.0	% du RNB
Mali									**Mali**
Bilateral	67	308	401	392	767	790	872	771	Bilatérale
Multilateral	76	70	139	330	322	204	526	462	Multilatérale
Total	144	379	540	721	1 089	994	1 398	1 234	Total
% of GNI	17.6	29.4	22.3	14.2	12.1	10.1	13.1	10.7	% du RNB

25

Net disbursements of official development assistance to recipients *(continued)*
Total, bilateral and multilateral aid (millions of US dollars); and as a percentage of Gross National Income (GNI)

Décaissements nets d'aide publique au développement aux bénéficiaires *(suite)*
Total, bilatérale et multilatérale d'aide (millions de dollars É.-U.); et en pourcentage du Revenu National Brut (RNB)

Region, country or area[&]	1975	1985	1995	2005	2010	2012	2013	2014	Région, pays ou zone[&]
Malta									**Malte**
Bilateral	33	18	7	...	...	...	...	...	Bilatérale
Multilateral	1	<0	2	...	...	...	...	...	Multilatérale
Total	34	18	9	...	...	...	...	...	Total
% of GNI	6.4	1.5	0.3	...	...	...	...	...	% du RNB
Marshall Islands									**Îles Marshall**
Bilateral	...	...	36	55	25	83	88	55	Bilatérale
Multilateral	...	...	2	2	7	1	6	1	Multilatérale
Total	...	...	39	57	32	84	94	56	Total
% of GNI	...	...	25.4	31.7	16.3	39.3	41.4	...	% du RNB
Mauritania									**Mauritanie**
Bilateral	68	179	163	91	274	291	182	172	Bilatérale
Multilateral	15	28	67	97	97	117	112	85	Multilatérale
Total	83	208	230	189	370	408	293	257	Total
% of GNI	18.7	32.9	17.1	8.4	8.7	8.8	6.0	5.4	% du RNB
Mauritius									**Maurice**
Bilateral	15	24	15	13	41	105	69	23	Bilatérale
Multilateral	13	3	8	22	84	73	80	26	Multilatérale
Total	28	27	23	35	125	178	148	49	Total
% of GNI	...	2.6	0.6	0.6	1.3	1.6	1.2	0.4	% du RNB
Mayotte									**Mayotte**
Bilateral	...	20	107	201	603	...	...	...	Bilatérale
Multilateral	...	<0	1	<0	1	...	...	...	Multilatérale
Total	...	21	108	201	604	...	...	...	Total
Mexico									**Mexique**
Bilateral	-6	121	360	151	414	334	490	648	Bilatérale
Multilateral	62	23	24	29	56	82	71	158	Multilatérale
Total	56	144	384	180	470	417	561	807	Total
% of GNI	0.1	0.1	0.1	<0.0	0.1	<0.0	0.1	0.1	% du RNB
Micronesia									**Micronésie**
Bilateral	...	...	74	104	62	142	141	114	Bilatérale
Multilateral	...	...	3	3	1	2	3	3	Multilatérale
Total	...	...	77	107	63	143	143	116	Total
% of GNI	...	...	33.0	41.2	20.8	41.8	41.7	...	% du RNB
Mongolia									**Mongolie**
Bilateral	...	...	184	159	220	353	358	237	Bilatérale
Multilateral	...	...	25	61	81	91	69	77	Multilatérale
Total	...	5	209	220	301	444	427	315	Total
% of GNI	...	0.2	14.7	8.9	4.6	3.9	3.6	2.8	% du RNB
Montenegro									**Monténégro**
Bilateral	...	...	...	4	54	53	47	21	Bilatérale
Multilateral	...	...	...	0	26	50	71	80	Multilatérale
Total	...	...	...	4	80	104	118	102	Total
% of GNI	...	...	...	0.2	2.0	2.5	2.6	2.2	% du RNB
Montserrat									**Montserrat**
Bilateral	4	2	9	27	16	36	50	35	Bilatérale
Multilateral	<0	<0	1	1	10	5	5	5	Multilatérale
Total	4	2	9	28	26	40	55	40	Total
Morocco									**Maroc**
Bilateral	260	836	447	394	712	1 099	1 512	1 339	Bilatérale
Multilateral	10	30	54	338	278	366	492	908	Multilatérale
Total	270	866	501	732	990	1 465	2 004	2 247	Total
% of GNI	3.0	6.4	1.4	1.3	1.1	1.6	2.0	2.2	% du RNB
Mozambique									**Mozambique**
Bilateral	14	242	798	814	1 430	1 582	1 753	1 481	Bilatérale
Multilateral	6	54	264	483	512	492	562	622	Multilatérale
Total	20	296	1 062	1 297	1 941	2 074	2 315	2 103	Total
% of GNI	...	6.8	51.4	20.9	19.7	13.9	14.9	12.9	% du RNB

Net disbursements of official development assistance to recipients *(continued)*
Total, bilateral and multilateral aid (millions of US dollars); and as a percentage of Gross National Income (GNI)

Décaissements nets d'aide publique au développement aux bénéficiaires *(suite)*
Total, bilatérale et multilatérale d'aide (millions de dollars É.-U.); et en pourcentage du Revenu National Brut (RNB)

Region, country or area[&]	1975	1985	1995	2005	2010	2012	2013	2014	Région, pays ou zone[&]
Myanmar									**Myanmar**
Bilateral	17	237	109	34	212	359	3 124	1 105	Bilatérale
Multilateral	40	107	42	111	143	146	811	275	Multilatérale
Total	58	344	150	145	355	504	3 935	1 380	Total
% of GNI	...	...	...	...	...	0.7	6.9	2.2	% du RNB
Namibia									**Namibie**
Bilateral	...	...	170	68	202	180	187	160	Bilatérale
Multilateral	...	...	20	58	55	75	74	67	Multilatérale
Total	...	5	190	125	256	255	261	227	Total
% of GNI	...	0.4	4.6	1.8	2.4	2.0	2.0	1.7	% du RNB
Nauru									**Nauru**
Bilateral	...	...	3	9	27	33	28	20	Bilatérale
Multilateral	...	...	<0	<0	1	3	1	2	Multilatérale
Total	<0	<0	3	9	28	36	29	22	Total
Nepal									**Népal**
Bilateral	14	105	303	285	447	450	467	499	Bilatérale
Multilateral	29	126	126	139	370	317	404	381	Multilatérale
Total	44	231	429	424	818	767	871	880	Total
% of GNI	2.8	8.8	9.7	5.2	5.1	4.0	4.5	4.4	% du RNB
Netherlands Antilles [former]									**Antilles néerlandaises [anc.]**
Bilateral	28	64	97	...	...	...	...	...	Bilatérale
Multilateral	5	1	2	...	...	...	...	...	Multilatérale
Total	33	65	98	...	...	...	...	...	Total
New Caledonia									**Nouvelle-Calédonie**
Bilateral	65	145	448	...	...	...	...	...	Bilatérale
Multilateral	<0	<0	3	...	...	...	...	...	Multilatérale
Total	65	145	451	...	...	...	...	...	Total
% of GNI	7.9	17.0	12.4	...	...	...	...	...	% du RNB
Nicaragua									**Nicaragua**
Bilateral	17	91	585	593	541	411	353	313	Bilatérale
Multilateral	25	17	65	171	121	120	143	118	Multilatérale
Total	41	108	649	763	662	532	497	430	Total
% of GNI	2.8	4.3	17.2	12.4	7.8	5.2	4.7	3.7	% du RNB
Niger									**Niger**
Bilateral	106	226	211	269	400	543	430	399	Bilatérale
Multilateral	31	72	61	254	339	347	367	519	Multilatérale
Total	137	298	272	522	739	890	797	918	Total
% of GNI	13.2	21.3	14.9	15.4	13.0	13.1	10.6	11.4	% du RNB
Nigeria									**Nigéria**
Bilateral	65	19	88	5 961	1 077	1 068	1 310	1 236	Bilatérale
Multilateral	16	13	122	448	980	844	1 206	1 240	Multilatérale
Total	81	32	211	6 409	2 058	1 912	2 515	2 476	Total
% of GNI	0.3	0.1	0.8	6.5	0.6	0.4	0.5	0.5	% du RNB
Niue									**Nioué**
Bilateral	...	...	8	19	14	20	18	13	Bilatérale
Multilateral	...	...	<0	2	1	<0	<0	1	Multilatérale
Total	2	4	8	21	15	20	18	14	Total
Northern Mariana Islands									**Îles Mariannes du Nord**
Bilateral	81	158	<0	...	...	...	...	...	Bilatérale
Multilateral	<0	1	-1	...	...	...	...	...	Multilatérale
Total	81	159	-1	...	...	...	...	...	Total
Oman									**Oman**
Bilateral	138	75	74	15	-24	...	...	...	Bilatérale
Multilateral	2	2	1	4	2	...	...	...	Multilatérale
Total	140	78	75	19	-22	...	...	...	Total
% of GNI	7.8	0.8	0.6	0.1	-<0.0	...	...	...	% du RNB
Other non-specified areas									**Autres zones non-spécifiées**
Total	-20	-10	<0	...	...	...	...	...	Total
% of GNI	-0.1	-<0.0	0.0	...	...	...	...	...	% du RNB

Net disbursements of official development assistance to recipients *(continued)*
Total, bilateral and multilateral aid (millions of US dollars); and as a percentage of Gross National Income (GNI)

Décaissements nets d'aide publique au développement aux bénéficiaires *(suite)*
Total, bilatérale et multilatérale d'aide (millions de dollars É.-U.); et en pourcentage du Revenu National Brut (RNB)

Region, country or area[&]	1975	1985	1995	2005	2010	2012	2013	2014	Région, pays ou zone[&]
Pakistan									**Pakistan**
Bilateral	628	405	473	884	2 180	1 239	1 458	2 002	Bilatérale
Multilateral	119	363	349	731	840	777	733	1 610	Multilatérale
Total	747	767	821	1 615	3 020	2 016	2 191	3 612	Total
% of GNI	6.5	2.3	1.4	1.4	1.6	0.9	0.9	1.4	% du RNB
Palau									**Palaos**
Bilateral	...	...	142	23	28	15	34	23	Bilatérale
Multilateral	...	...	0	<0	1	<0	1	1	Multilatérale
Total	...	...	142	24	28	16	35	23	Total
% of GNI	...	...	145.2	12.8	15.9	7.5	16.1	9.6	% du RNB
Panama									**Panama**
Bilateral	18	61	32	4	112	35	-9	-215	Bilatérale
Multilateral	15	8	8	23	14	16	16	18	Multilatérale
Total	33	69	40	27	126	51	7	-196	Total
% of GNI	1.7	1.3	0.5	0.2	0.4	0.1	<0.0	-0.5	% du RNB
Papua New Guinea									**Papouasie-Nvl-Guinée**
Bilateral	294	241	342	213	417	533	532	467	Bilatérale
Multilateral	11	16	28	54	94	135	125	110	Multilatérale
Total	305	257	370	267	511	669	657	577	Total
% of GNI	23.8	11.0	8.5	5.9	5.5	4.5	4.5	...	% du RNB
Paraguay									**Paraguay**
Bilateral	11	42	130	38	69	71	90	33	Bilatérale
Multilateral	26	8	10	13	51	32	38	27	Multilatérale
Total	37	50	139	51	120	103	129	60	Total
% of GNI	...	...	1.7	0.7	0.7	0.5	0.5	0.2	% du RNB
Peru									**Pérou**
Bilateral	52	291	340	361	-368	333	303	268	Bilatérale
Multilateral	24	21	31	89	69	52	57	57	Multilatérale
Total	77	312	371	450	-300	385	360	325	Total
% of GNI	0.5	1.9	0.8	0.7	-0.2	0.2	0.2	0.2	% du RNB
Philippines									**Philippines**
Bilateral	150	422	843	490	378	-104	89	526	Bilatérale
Multilateral	27	37	60	77	151	102	101	149	Multilatérale
Total	177	459	902	567	530	-3	190	676	Total
% of GNI	1.2	1.6	1.2	0.4	0.2	0.0	0.1	0.2	% du RNB
Qatar									**Qatar**
Bilateral	-1	1	4	...	...	...	...	...	Bilatérale
Multilateral	1	1	-<0	...	...	...	...	...	Multilatérale
Total	1	2	4	...	...	...	...	...	Total
% of GNI	<0.0	<0.0	0.1	...	...	...	...	...	% du RNB
Republic of Korea									**République de Corée**
Bilateral	208	-13	54	...	...	...	...	...	Bilatérale
Multilateral	40	4	3	...	...	...	...	...	Multilatérale
Total	248	-9	57	...	...	...	...	...	Total
% of GNI	1.2	-<0.0	<0.0	...	...	...	...	...	% du RNB
Republic of Moldova									**République de Moldova**
Bilateral	...	...	...	94	212	256	193	325	Bilatérale
Multilateral	...	...	...	75	260	218	154	192	Multilatérale
Total	...	...	...	169	472	473	347	517	Total
% of GNI	...	...	...	5.1	7.5	5.8	3.9	5.9	% du RNB
Rwanda									**Rwanda**
Bilateral	54	123	403	276	601	510	659	517	Bilatérale
Multilateral	36	54	291	302	431	369	427	517	Multilatérale
Total	90	177	695	577	1 031	879	1 086	1 034	Total
% of GNI	15.9	10.4	53.5	22.6	18.2	12.3	14.7	13.3	% du RNB
Saint Helena									**Sainte-Hélène**
Bilateral	...	...	12	22	...	168	133	126	Bilatérale
Multilateral	...	...	<0	<0	...	0	6	5	Multilatérale
Total	3	12	13	23	54	168	139	131	Total

Net disbursements of official development assistance to recipients *(continued)*
Total, bilateral and multilateral aid (millions of US dollars); and as a percentage of Gross National Income (GNI)

Décaissements nets d'aide publique au développement aux bénéficiaires *(suite)*
Total, bilatérale et multilatérale d'aide (millions de dollars É.-U.); et en pourcentage du Revenu National Brut (RNB)

Region, country or area[&]	1975	1985	1995	2005	2010	2012	2013	2014	Région, pays ou zone[&]
Saint Kitts and Nevis									**Saint-Kitts-et-Nevis**
Bilateral	1	3	2	1	-2	7	7	...	Bilatérale
Multilateral	<0	1	2	2	13	15	23	...	Multilatérale
Total	2	4	4	3	11	22	29	...	Total
% of GNI	...	5.8	1.8	0.5	1.7	3.1	3.9	...	% du RNB
Saint Lucia									**Sainte-Lucie**
Bilateral	8	4	31	4	<0	7	3	7	Bilatérale
Multilateral	1	3	17	6	41	20	21	11	Multilatérale
Total	9	7	48	10	41	27	24	18	Total
% of GNI	...	3.8	9.3	1.2	3.4	2.1	1.9	1.4	% du RNB
Saint Vincent-Grenadines									**Saint-Vincent-Grenadines**
Bilateral	6	3	32	4	-<0	-1	-2	1	Bilatérale
Multilateral	<0	3	15	4	17	9	9	8	Multilatérale
Total	6	5	48	8	17	9	8	9	Total
% of GNI	18.1	5.0	18.7	1.5	2.5	1.2	1.1	1.3	% du RNB
Samoa									**Samoa**
Bilateral	4	14	38	24	98	80	72	63	Bilatérale
Multilateral	9	6	5	20	50	40	47	30	Multilatérale
Total	13	19	43	44	148	121	118	93	Total
% of GNI	...	...	...	9.9	23.2	15.7	15.5	12.0	% du RNB
Sao Tome and Principe									**Sao Tomé-et-Principe**
Bilateral	<0	6	72	9	35	33	26	16	Bilatérale
Multilateral	1	6	12	23	14	16	26	23	Multilatérale
Total	1	12	84	32	49	49	52	39	Total
% of GNI	...	...	...	27.0	24.5	18.7	16.8	11.6	% du RNB
Saudi Arabia									**Arabie saoudite**
Bilateral	-6	16	17	23	...	...	...	...	Bilatérale
Multilateral	9	12	<0	2	...	...	...	...	Multilatérale
Total	3	27	17	25	...	...	...	...	Total
% of GNI	<0.0	<0.0	<0.0	<0.0	...	...	...	...	% du RNB
Senegal									**Sénégal**
Bilateral	103	249	489	426	654	765	695	820	Bilatérale
Multilateral	36	42	163	272	283	308	297	287	Multilatérale
Total	139	290	652	698	937	1 073	992	1 107	Total
% of GNI	6.5	10.3	13.8	8.2	7.3	7.7	6.8	7.2	% du RNB
Serbia									**Serbie**
Bilateral	...	...	92	836	299	429	204	13	Bilatérale
Multilateral	...	...	3	231	361	660	576	358	Multilatérale
Total	...	...	95	1 066	660	1 089	780	371	Total
% of GNI	...	...	...	4.1	1.7	2.8	1.8	0.9	% du RNB
Seychelles									**Seychelles**
Bilateral	7	19	11	11	45	20	15	-2	Bilatérale
Multilateral	<0	3	2	6	11	15	11	12	Multilatérale
Total	7	22	13	17	56	36	25	10	Total
% of GNI	15.6	13.5	2.6	1.8	6.1	3.3	1.9	0.7	% du RNB
Sierra Leone									**Sierra Leone**
Bilateral	7	43	136	131	221	223	256	584	Bilatérale
Multilateral	10	20	76	209	237	217	191	327	Multilatérale
Total	17	64	212	340	458	440	447	911	Total
% of GNI	2.5	7.7	26.0	21.4	17.6	12.8	9.9	20.9	% du RNB
Singapore									**Singapour**
Bilateral	8	21	16	...	...	...	...	...	Bilatérale
Multilateral	5	2	1	...	...	...	...	...	Multilatérale
Total	12	23	17	...	...	...	...	...	Total
% of GNI	0.2	0.1	<0.0	...	...	...	...	...	% du RNB
Slovenia									**Slovénie**
Bilateral	...	...	40	...	...	...	...	...	Bilatérale
Multilateral	...	...	12	...	...	...	...	...	Multilatérale
Total	...	...	53	...	...	...	...	...	Total
% of GNI	...	...	0.2	...	...	...	...	...	% du RNB

Net disbursements of official development assistance to recipients *(continued)*
Total, bilateral and multilateral aid (millions of US dollars); and as a percentage of Gross National Income (GNI)

Décaissements nets d'aide publique au développement aux bénéficiaires *(suite)*
Total, bilatérale et multilatérale d'aide (millions de dollars É.-U.); et en pourcentage du Revenu National Brut (RNB)

Region, country or area[&]	1975	1985	1995	2005	2010	2012	2013	2014	Région, pays ou zone[&]
Solomon Islands									**Îles Salomon**
Bilateral	22	16	42	167	301	275	263	179	Bilatérale
Multilateral	1	5	6	32	40	30	26	19	Multilatérale
Total	22	21	48	198	340	305	288	199	Total
% of GNI	45.5	13.5	14.9	47.7	68.5	32.8	27.4	18.1	% du RNB
Somalia									**Somalie**
Bilateral	110	249	134	117	309	804	882	905	Bilatérale
Multilateral	55	102	54	123	197	186	172	204	Multilatérale
Total	165	351	188	240	506	990	1 054	1 109	Total
% of GNI	23.3	42.4	...	...	...	...	...	...	% du RNB
South Africa									**Afrique du Sud**
Bilateral	...	...	358	431	791	745	1 042	683	Bilatérale
Multilateral	...	...	28	259	236	321	254	387	Multilatérale
Total	...	...	386	690	1 027	1 066	1 295	1 070	Total
% of GNI	...	...	0.3	0.3	0.3	0.3	0.4	0.3	% du RNB
South Sudan									**Soudan du sud**
Bilateral	...	...	...	...	...	1 050	1 140	1 676	Bilatérale
Multilateral	...	...	...	...	...	137	260	288	Multilatérale
Total	...	...	...	...	...	1 187	1 400	1 964	Total
% of GNI	...	...	...	...	...	12.0	13.0	20.0	% du RNB
Sri Lanka									**Sri Lanka**
Bilateral	81	343	412	898	287	238	159	229	Bilatérale
Multilateral	67	125	142	263	296	250	242	259	Multilatérale
Total	148	468	554	1 161	583	489	401	488	Total
% of GNI	4.4	7.9	4.3	4.8	1.2	0.8	0.6	0.7	% du RNB
State of Palestine									**État de Palestine**
Bilateral	...	...	374	640	1 818	1 566	2 110	1 976	Bilatérale
Multilateral	...	...	140	376	695	439	492	510	Multilatérale
Total	...	...	514	1 016	2 513	2 005	2 601	2 487	Total
% of GNI	...	...	13.8	19.6	26.4	16.5	19.1	...	% du RNB
Sudan									**Soudan**
Bilateral	227	972	183	1 449	1 559	1 077	1 260	668	Bilatérale
Multilateral	67	153	54	376	469	289	243	204	Multilatérale
Total	294	1 125	237	1 826	2 028	1 366	1 503	872	Total
% of GNI	6.2	9.3	1.8	7.4	3.4	2.3	2.4	1.2	% du RNB
Suriname									**Suriname**
Bilateral	50	9	74	31	80	32	22	10	Bilatérale
Multilateral	3	2	3	13	25	8	8	2	Multilatérale
Total	53	11	77	44	104	40	30	13	Total
% of GNI	11.6	1.2	11.1	2.5	2.4	0.8	0.6	...	% du RNB
Swaziland									**Swaziland**
Bilateral	5	18	49	-1	21	56	47	42	Bilatérale
Multilateral	10	6	9	47	70	32	69	44	Multilatérale
Total	14	24	58	47	91	88	116	86	Total
% of GNI	...	...	3.2	1.7	2.8	2.6	3.7	2.7	% du RNB
Syrian Arab Republic									**Rép. arabe syrienne**
Bilateral	803	591	306	-4	-27	1 594	3 457	4 016	Bilatérale
Multilateral	25	19	50	74	162	78	181	182	Multilatérale
Total	827	610	356	70	135	1 672	3 638	4 198	Total
% of GNI	11.7	3.6	3.1	0.3	...	...	...	...	% du RNB
Tajikistan									**Tadjikistan**
Bilateral	...	...	49	142	255	260	242	227	Bilatérale
Multilateral	...	...	16	110	178	133	148	129	Multilatérale
Total	...	...	65	252	433	393	389	356	Total
% of GNI	...	...	5.5	11.3	7.8	5.2	4.6	3.9	% du RNB
Thailand									**Thaïlande**
Bilateral	67	403	812	-240	-123	-207	-56	255	Bilatérale
Multilateral	18	55	25	72	111	72	82	96	Multilatérale
Total	85	457	837	-168	-12	-135	26	351	Total
% of GNI	0.6	1.2	0.5	-0.1	0.0	-<0.0	<0.0	0.1	% du RNB

25

Net disbursements of official development assistance to recipients *(continued)*
Total, bilateral and multilateral aid (millions of US dollars); and as a percentage of Gross National Income (GNI)

Décaissements nets d'aide publique au développement aux bénéficiaires *(suite)*
Total, bilatérale et multilatérale d'aide (millions de dollars É.-U.); et en pourcentage du Revenu National Brut (RNB)

Region, country or area[&]	1975	1985	1995	2005	2010	2012	2013	2014	Région, pays ou zone[&]
TFYR of Macedonia									**ex-R.Y. de Macédoine**
Bilateral	...	...	34	154	126	77	96	61	Bilatérale
Multilateral	...	...	45	74	67	73	102	150	Multilatérale
Total	...	...	79	227	193	150	197	211	Total
% of GNI	...	...	1.8	3.7	2.1	1.6	1.9	1.9	% du RNB
Timor-Leste									**Timor-Leste**
Bilateral	...	...	<0	149	252	223	213	196	Bilatérale
Multilateral	...	...	0	35	40	59	44	51	Multilatérale
Total	<0	...	<0	185	291	282	257	247	Total
% of GNI	...	...	...	22.2	8.8	6.0	4.8	6.4	% du RNB
Togo									**Togo**
Bilateral	31	74	156	42	206	145	109	78	Bilatérale
Multilateral	11	37	35	41	198	96	115	130	Multilatérale
Total	42	111	191	83	404	241	224	208	Total
% of GNI	6.8	15.3	15.1	4.0	14.6	7.3	6.1	5.1	% du RNB
Tokelau									**Tokélaou**
Bilateral	...	2	3	16	14	19	24	19	Bilatérale
Multilateral	...	<0	<0	<0	<0	<0	<0	<0	Multilatérale
Total	<0	2	4	16	15	19	24	19	Total
Tonga									**Tonga**
Bilateral	3	11	37	23	59	68	60	61	Bilatérale
Multilateral	1	2	2	9	11	10	22	19	Multilatérale
Total	3	13	39	32	70	78	81	80	Total
% of GNI	...	21.2	18.9	12.4	18.8	16.6	18.2	18.0	% du RNB
Trinidad and Tobago									**Trinité-et-Tobago**
Bilateral	3	3	14	-11	2	...	...	...	Bilatérale
Multilateral	3	4	11	9	2	...	...	...	Multilatérale
Total	5	7	25	-2	4	...	...	...	Total
% of GNI	0.2	0.1	0.5	-<0.0	<0.0	...	...	...	% du RNB
Tunisia									**Tunisie**
Bilateral	176	144	48	235	427	589	313	509	Bilatérale
Multilateral	38	17	27	128	124	428	398	412	Multilatérale
Total	214	160	74	362	550	1 017	710	921	Total
% of GNI	5.1	2.0	0.4	1.2	1.3	2.4	1.6	...	% du RNB
Turkey									**Turquie**
Bilateral	19	168	296	-61	713	1 016	657	992	Bilatérale
Multilateral	40	12	16	457	334	2 094	2 186	2 450	Multilatérale
Total	58	180	313	396	1 047	3 110	2 843	3 442	Total
% of GNI	0.1	0.3	0.2	0.1	0.1	0.4	0.4	0.4	% du RNB
Turkmenistan									**Turkménistan**
Bilateral	...	...	27	21	31	29	26	20	Bilatérale
Multilateral	...	...	4	10	15	9	10	15	Multilatérale
Total	...	...	31	30	46	38	36	34	Total
% of GNI	...	...	1.2	0.4	0.2	0.1	0.1	0.1	% du RNB
Turks and Caicos Islands									**Îles Turques-et-Caïques**
Bilateral	3	5	5	3	...	...	...	...	Bilatérale
Multilateral	<0	1	1	2	...	...	...	...	Multilatérale
Total	3	6	6	5	...	...	...	...	Total
Tuvalu									**Tuvalu**
Bilateral	...	3	7	6	13	22	21	26	Bilatérale
Multilateral	...	<0	1	4	<0	2	5	9	Multilatérale
Total	<0	3	8	9	13	24	27	34	Total
% of GNI	...	...	...	24.9	26.1	42.2	48.7	...	% du RNB
Uganda									**Ouganda**
Bilateral	37	84	570	696	1 110	1 010	1 055	1 090	Bilatérale
Multilateral	15	96	263	497	578	632	646	543	Multilatérale
Total	52	179	833	1 192	1 688	1 642	1 701	1 633	Total
% of GNI	2.2	5.2	14.6	13.6	9.1	7.1	7.1	6.3	% du RNB

25

Net disbursements of official development assistance to recipients *(continued)*
Total, bilateral and multilateral aid (millions of US dollars); and as a percentage of Gross National Income (GNI)

Décaissements nets d'aide publique au développement aux bénéficiaires *(suite)*
Total, bilatérale et multilatérale d'aide (millions de dollars É.-U.); et en pourcentage du Revenu National Brut (RNB)

Region, country or area[&]	1975	1985	1995	2005	2010	2012	2013	2014	Région, pays ou zone[&]
Ukraine									**Ukraine**
Bilateral	...	...	...	256	437	522	424	950	Bilatérale
Multilateral	...	...	...	156	214	245	360	454	Multilatérale
Total	...	...	...	412	651	768	783	1 404	Total
% of GNI	...	...	...	0.5	0.5	0.4	0.4	1.1	% du RNB
United Arab Emirates									**Émirats arabes unis**
Bilateral	3	3	6	...	...	...	...	...	Bilatérale
Multilateral	1	1	-<0	...	...	...	...	...	Multilatérale
Total	4	4	5	...	...	...	...	...	Total
United Rep. of Tanzania									**Rép.-Unie de Tanzanie**
Bilateral	237	400	613	829	1 919	2 017	2 249	1 570	Bilatérale
Multilateral	57	77	258	670	1 037	807	1 182	1 078	Multilatérale
Total	293	477	871	1 499	2 957	2 823	3 431	2 648	Total
% of GNI	...	...	17.0	10.8	9.9	7.6	8.0	5.5	% du RNB
Uruguay									**Uruguay**
Bilateral	-1	3	59	-2	30	-2	15	75	Bilatérale
Multilateral	13	2	6	17	17	21	21	14	Multilatérale
Total	12	5	66	14	47	19	36	89	Total
% of GNI	0.4	0.1	0.4	0.1	0.1	<0.0	0.1	0.2	% du RNB
Uzbekistan									**Ouzbékistan**
Bilateral	...	...	76	129	151	98	156	155	Bilatérale
Multilateral	...	...	8	41	81	157	137	169	Multilatérale
Total	...	...	84	170	232	255	293	324	Total
% of GNI	...	...	0.6	1.2	0.6	0.5	0.5	0.5	% du RNB
Vanuatu									**Vanuatu**
Bilateral	12	20	43	29	105	96	85	96	Bilatérale
Multilateral	<0	2	2	11	3	5	5	3	Multilatérale
Total	12	22	46	39	108	101	91	98	Total
% of GNI	...	17.7	21.0	10.7	15.9	13.6	11.8	...	% du RNB
Venezuela (Boliv. Rep. of)									**Venezuela (Rép. boliv. du)**
Bilateral	-<0	9	35	14	32	30	22	25	Bilatérale
Multilateral	19	2	8	36	20	18	13	16	Multilatérale
Total	18	11	43	50	53	48	35	41	Total
% of GNI	0.1	<0.0	0.1	<0.0	<0.0	<0.0	<0.0	<0.0	% du RNB
Viet Nam									**Viet Nam**
Bilateral	367	115	724	1 389	2 001	2 840	2 678	2 831	Bilatérale
Multilateral	22	38	111	525	938	1 274	1 406	1 387	Multilatérale
Total	390	153	835	1 913	2 939	4 114	4 083	4 218	Total
% of GNI	...	...	4.1	3.4	2.6	2.8	2.5	2.4	% du RNB
Wallis and Futuna Islands									**Îles Wallis-et-Futuna**
Bilateral	2	<0	1	72	123	116	103	99	Bilatérale
Multilateral	<0	<0	<0	<0	5	4	2	<0	Multilatérale
Total	2	<0	1	72	127	120	106	100	Total
Yemen									**Yémen**
Bilateral	211	342	109	128	418	507	733	852	Bilatérale
Multilateral	40	89	63	169	249	203	306	313	Multilatérale
Total	251	431	171	297	667	709	1 039	1 164	Total
% of GNI	...	...	4.2	2.0	2.3	2.3	3.0	...	% du RNB
Zambia									**Zambie**
Bilateral	68	244	1 765	695	652	677	773	730	Bilatérale
Multilateral	18	75	266	478	266	279	369	264	Multilatérale
Total	86	319	2 031	1 172	917	955	1 142	995	Total
% of GNI	3.8	16.1	57.0	15.2	4.9	3.9	4.5	3.9	% du RNB
Zimbabwe									**Zimbabwe**
Bilateral	4	218	436	161	478	704	565	496	Bilatérale
Multilateral	<0	18	53	211	234	295	260	262	Multilatérale
Total	4	235	489	373	713	999	824	758	Total
% of GNI	0.1	4.3	7.2	6.8	7.7	8.7	6.6	6.0	% du RNB

25

Net disbursements of official development assistance to recipients *(continued)*
Total, bilateral and multilateral aid (millions of US dollars); and as a percentage of Gross National Income (GNI)

Décaissements nets d'aide publique au développement aux bénéficiaires *(suite)*
Total, bilatérale et multilatérale d'aide (millions de dollars É.-U.); et en pourcentage du Revenu National Brut (RNB)

Source:

Organization for Economic Co-operation and Development (OECD), Paris, the OECD Development Assistance Committee database, last accessed March 2016.

& Official Development Assistance (ODA) is defined as those flows to developing countries and multilateral institutions provided by official agencies, including state and local governments, or by their executive agencies, each transaction of which meets the following tests: i) it is administered with the promotion of the economic development and welfare of developing countries as its main objective; and ii) it is concessional in character and conveys a grant element of at least 25 per cent.

1   Includes regional aid disbursements in addition to the disbursements made to individual countries and areas.
2   Excludes Sudan.
3   Includes Sudan.
4   Excludes Myanmar.
5   Includes Azerbaijan, Armenia, Georgia and Myanmar. Excludes Iran (Islamic Rep. of).
6   Excludes Azerbaijan, Armenia, Georgia, Turkey and Cyprus. Includes Iran (Islamic Rep. of).
7   Includes Turkey and Cyprus.
8   For statistical purposes, the data for China do not include those for the Hong Kong Special Administrative Region (Hong Kong SAR) and Macao Special Administrative Region (Macao SAR).

Source:

Organisation de coopération et de développement économiques (OCDE), Paris, la base de données du comité d'aide au développement de l'OCDE, dernier accès mars 2016.

& On entend par Aide publique au développement (APD), l'ensemble des flux financiers à destination des pays en développement et des institutions multilatérales par des organismes publics (y compris les autorités étatiques et locales) ou par leurs organes exécutifs satisfaisant les critères suivants : a) elle est administrée dans le but premier de promouvoir le développement économique et social des pays en développement ; b) elle est accordée à des conditions de faveur et comporte une élément de subvention d'au moins 25 pour cent.

1   Comprend les versements d'aides régionales en plus des versements effectués aux aux différents pays et régions.
2   Exclut le Soudan.
3   Comprend le Soudan.
4   Non compris Myanmar.
5   Y compris l'Azerbaïdjan, l'Arménie, la Géorgie et le Myanmar. Non compris la Rép. Islamique d 'Iran.
6   Non compris l'Azerbaïdjan, l'Arménie, la Géorgie, la Turquie et Chypre. Y compris la Rép. Islamique d 'Iran.
7   Y compris la Turquie et Chypre.
8   Pour la présentation des statistiques, les données pour la Chine ne comprennent pas la Région Administrative Spéciale de Hong Kong (Hong Kong RAS) et la Région Administrative Spéciale de Macao (Macao RAS).

Net disbursements of official development assistance from donors
Millions of US dollars and as a percentage of gross national income (GNI)

Décaissements nets d'aide publique au développement par des donateurs
Millions de dollares É.-U. et en pourcentage du revenu national brut (RNB)

Country or area	1975	1985	1995	2005	2010	2012	2013	2014	Pays ou zone
Total									**Total**
$ millions	20 183	33 855	65 424	120 771	147 644	150 614	167 175	178 345	$millions
DAC total									**CAD total**
$ millions	13 286	28 774	58 896	108 296	128 369	126 911	134 832	137 222	$millions
% of GNI	0.34	0.33	0.26	0.32	0.31	0.28	0.30	0.30	% du RNB
Australia[1]									**Australie**[1]
$ millions	552	749	1 194	1 680	3 826	5 403	4 846	4 382	$millions
% of GNI	0.65	0.48	0.34	0.25	0.32	0.36	0.33	0.31	% du RNB
Austria[1]									**Autriche**[1]
$ millions	79	248	620	1 573	1 208	1 106	1 171	1 235	$millions
% of GNI	0.21	0.38	0.27	0.52	0.32	0.28	0.27	0.28	% du RNB
Belgium[1]									**Belgique**[1]
$ millions	378	440	1 034	1 963	3 004	2 315	2 300	2 448	$millions
% of GNI	0.60	0.55	0.38	0.53	0.64	0.47	0.45	0.46	% du RNB
Bulgaria									**Bulgarie**
$ millions	...	...	...	...	40	40	50	49	$millions
% of GNI	...	...	...	...	0.09	0.08	0.10	0.09	% du RNB
Canada[1]									**Canada**[1]
$ millions	880	1 631	2 067	3 756	5 214	5 650	4 947	4 240	$millions
% of GNI	0.54	0.49	0.38	0.34	0.34	0.32	0.27	0.24	% du RNB
Croatia									**Croatie**
$ millions	...	...	...	...	...	21	45	72	$millions
% of GNI	...	...	...	...	...	0.04	0.08	0.13	% du RNB
Cyprus[2]									**Chypre**[2]
$ millions	...	...	...	15	51	25	20	...	$millions
% of GNI	...	...	...	0.09	0.23	0.11	0.10	...	% du RNB
Czech Republic[1]									**République tchèque**[1]
$ millions	...	...	...	135	228	220	211	212	$millions
% of GNI	...	...	...	0.11	0.13	0.12	0.11	0.11	% du RNB
Denmark[1]									**Danemark**[1]
$ millions	205	440	1 623	2 109	2 871	2 693	2 927	3 003	$millions
% of GNI	0.55	0.80	0.96	0.81	0.91	0.83	0.85	0.86	% du RNB
Estonia									**Estonie**
$ millions	...	...	...	10	19	23	31	38	$millions
% of GNI	...	...	...	0.08	0.10	0.11	0.13	0.14	% du RNB
European Union (EU)[3]									**Union européenne (UE)**[3]
$ millions	722	1 510	5 398	9 390	12 747	17 479	15 959	16 451	$millions
Finland[1]									**Finlande**[1]
$ millions	48	211	388	902	1 333	1 320	1 435	1 635	$millions
% of GNI	0.17	0.40	0.31	0.46	0.55	0.53	0.54	0.59	% du RNB
France[1]									**France**[1]
$ millions	1 493	3 134	8 443	10 026	12 915	12 028	11 339	10 620	$millions
% of GNI	0.44	0.61	0.55	0.47	0.50	0.45	0.41	0.37	% du RNB
Germany[1]									**Allemagne**[1]
$ millions	1 689	2 942	7 524	10 082	12 985	12 939	14 228	16 566	$millions
% of GNI	0.40	0.47	0.31	0.36	0.39	0.37	0.38	0.42	% du RNB
Greece[1]									**Grèce**[1]
$ millions	...	...	...	384	508	327	239	247	$millions
% of GNI	...	...	...	0.17	0.17	0.13	0.10	0.11	% du RNB
Hungary									**Hongrie**
$ millions	29	84	...	100	114	118	128	144	$millions
% of GNI	...	...	...	0.11	0.09	0.10	0.10	0.11	% du RNB
Iceland[1]									**Islande**[1]
$ millions	...	...	...	27	29	26	35	37	$millions
% of GNI	...	...	...	0.18	0.29	0.22	0.25	0.22	% du RNB
Ireland[1]									**Irlande**[1]
$ millions	8	39	153	719	895	808	846	816	$millions
% of GNI	0.09	0.24	0.29	0.42	0.52	0.47	0.46	0.38	% du RNB

Net disbursements of official development assistance from donors *(continued)*
Millions of US dollars and as a percentage of gross national income (GNI)

Décaissements nets d'aide publique au développement par des donateurs *(suite)*
Millions de dollares É.-U. et en pourcentage du revenu national brut (RNB)

Country or area	1975	1985	1995	2005	2010	2012	2013	2014	Pays ou zone
Israel									**Israël**
$ millions	...	...	...	95	145	181	202	200	$millions
% of GNI	...	...	...	0.07	0.07	0.07	0.07	0.07	% du RNB
Italy [1]									**Italie** [1]
$ millions	182	1 098	1 623	5 091	2 996	2 737	3 430	4 009	$millions
% of GNI	0.10	0.26	0.15	0.29	0.15	0.14	0.17	0.19	% du RNB
Japan [1]									**Japon** [1]
$ millions	1 148	3 797	14 489	13 126	11 058	10 605	11 582	9 266	$millions
% of GNI	0.23	0.29	0.27	0.28	0.20	0.17	0.23	0.19	% du RNB
Kazakhstan									**Kazakhstan**
$ millions	...	...	...	...	...	...	8	33	$millions
% of GNI	...	...	...	...	...	...	0.00	0.02	% du RNB
Kuwait									**Koweït**
$ millions	859	647	384	218	232	180	231	277	$millions
Latvia									**Lettonie**
$ millions	...	...	...	11	16	21	24	25	$millions
% of GNI	...	...	...	0.07	0.06	0.07	0.08	0.08	% du RNB
Liechtenstein									**Liechtenstein**
$ millions	...	...	...	...	27	29	28	27	$millions
% of GNI	...	...	...	...	0.62	0.75	0.65	...	% du RNB
Lithuania									**Lituanie**
$ millions	...	...	...	16	37	52	50	46	$millions
% of GNI	...	...	...	0.06	0.10	0.13	0.11	0.10	% du RNB
Luxembourg [1]									**Luxembourg** [1]
$ millions	...	8	65	256	403	399	429	423	$millions
% of GNI	...	0.17	0.36	0.79	1.05	1.00	1.00	1.06	% du RNB
Malta									**Malte**
$ millions	...	...	...	...	14	19	18	20	$millions
% of GNI	...	...	...	...	0.18	0.23	0.20	0.20	% du RNB
Netherlands [1]									**Pays-Bas** [1]
$ millions	608	1 136	3 226	5 115	6 357	5 523	5 435	5 573	$millions
% of GNI	0.74	0.91	0.81	0.82	0.81	0.71	0.67	0.64	% du RNB
New Zealand [1]									**Nouvelle-Zélande** [1]
$ millions	66	54	123	274	342	449	457	506	$millions
% of GNI	0.52	0.25	0.23	0.27	0.26	0.28	0.26	0.27	% du RNB
Norway [1]									**Norvège** [1]
$ millions	184	574	1 244	2 794	4 372	4 753	5 581	5 086	$millions
% of GNI	0.65	1.01	0.86	0.94	1.05	0.93	1.07	1.00	% du RNB
Other non-specified areas									**Autres zones non-spécifiées**
$ millions	...	...	92	483	381	305	272	274	$millions
% of GNI	...	...	...	0.14	0.10	0.06	0.05	0.05	% du RNB
Poland [1]									**Pologne** [1]
$ millions	32	19	...	205	378	421	487	452	$millions
% of GNI	...	...	...	0.07	0.08	0.09	0.10	0.09	% du RNB
Portugal [1]									**Portugal** [1]
$ millions	...	10	258	377	649	581	488	430	$millions
% of GNI	...	0.05	0.25	0.21	0.29	0.28	0.23	0.19	% du RNB
Republic of Korea [1]									**République de Corée** [1]
$ millions	...	...	116	752	1 174	1 597	1 755	1 857	$millions
% of GNI	...	...	0.02	0.10	0.12	0.14	0.13	0.13	% du RNB
Romania									**Roumanie**
$ millions	...	...	...	...	114	142	134	214	$millions
% of GNI	...	...	...	...	0.07	0.09	0.07	0.11	% du RNB
Russian Federation [4]									**Fédération de Russie** [4]
$ millions	...	...	...	...	472	465	714	876	$millions
% of GNI	...	...	...	...	0.03	0.02	0.03	0.05	% du RNB
Saudi Arabia [5]									**Arabie saoudite** [5]
$ millions	2 569	2 443	305	1 026	3 480	1 299	5 683	13 634	$millions
Slovakia [1]									**Slovaquie** [1]
$ millions	...	...	...	57	74	80	86	83	$millions
% of GNI	...	...	...	...	0.09	0.09	0.09	0.09	% du RNB

26

Net disbursements of official development assistance from donors *(continued)*
Millions of US dollars and as a percentage of gross national income (GNI)

Décaissements nets d'aide publique au développement par des donateurs *(suite)*
Millions de dollares É.-U. et en pourcentage du revenu national brut (RNB)

Country or area	1975	1985	1995	2005	2010	2012	2013	2014	Pays ou zone
Slovenia[1]									**Slovénie**[1]
$ millions	...	...	...	35	59	58	62	62	$millions
% of GNI	...	...	...	0.11	0.13	0.13	0.13	0.12	% du RNB
Spain[1]									**Espagne**[1]
$ millions	...	169	1 348	3 018	5 949	2 037	2 348	1 877	$millions
% of GNI	...	0.10	0.24	0.27	0.43	0.16	0.17	0.13	% du RNB
Sweden[1]									**Suède**[1]
$ millions	566	840	1 704	3 362	4 533	5 240	5 827	6 233	$millions
% of GNI	0.78	0.86	0.77	0.94	0.97	0.97	1.01	1.09	% du RNB
Switzerland[1]									**Suisse**[1]
$ millions	104	303	1 084	1 772	2 300	3 052	3 200	3 522	$millions
% of GNI	0.18	0.31	0.33	0.42	0.39	0.47	0.46	0.50	% du RNB
Thailand									**Thaïlande**
$ millions	...	...	...	...	4	11	36	69	$millions
% of GNI	...	...	...	...	0.00	0.00	0.01	0.02	% du RNB
Timor-Leste									**Timor-Leste**
$ millions	...	...	...	...	...	...	...	3	$millions
Turkey									**Turquie**
$ millions	...	...	107	601	967	2 533	3 308	3 591	$millions
% of GNI	...	...	0.06	0.17	0.13	0.32	0.40	0.45	% du RNB
United Arab Emirates[6]									**Émirats arabes unis**[6]
$ millions	1 998	366	243	510	414	759	5 402	5 080	$millions
% of GNI	...	...	...	...	0.14	0.20	1.34	1.26	% du RNB
United Kingdom[1]									**Royaume-Uni**[1]
$ millions	904	1 530	3 202	10 772	13 053	13 891	17 871	19 306	$millions
% of GNI	0.38	0.33	0.29	0.47	0.57	0.56	0.70	0.70	% du RNB
United States[1]									**États-Unis**[1]
$ millions	4 161	9 403	7 367	27 935	29 656	30 652	31 267	33 096	$millions
% of GNI	0.27	0.24	0.10	0.23	0.20	0.19	0.18	0.19	% du RNB
Areas not specified									**Zones non spécifiées**
$ millions	721	32	...	...	...	...	...	...	$millions

Source:
Organization for Economic Co-operation and Development (OECD), Paris, the OECD Development Assistance Committee database, last accessed April 2016.

Source:
Organisation de coopération et de développement économiques (OCDE), Paris, la base de données du comité d'aide au développement de l'OCDE, dernier accès avril 2016.

1   Development Assistance Committee member (OECD)
2   For government controlled areas.
3   Refers to European Union institutions.
4   Some of the debt relief reported by Russia from 2014 onwards may correspond to the credits included in these estimates. Therefore, the statistics currently published on ODA by Russia and the estimates from the previous Chairman's reports should not be used at the same time.
5   Saudi Arabia's reporting to the OECD on its development co-operation programme consists of aggregate figures on humanitarian and development assistance by region, multilateral aid and loan disbursements and repayments by the Saudi Fund for Development.

6   The United Arab Emirates began to report at activity level in 2009. These data represent flows from all government agencies. Commitments are set equal to disbursements for agencies other than the Abu Dhabi Fund for Development.

1   Le Comité d'aide au développement (l'OCDE)
2   Pour les zones contrôlées par le gouvernement.
3   Se référant aux institutions de l'Union Européenne.
4   L'allégement de la dette déclarée par la Russie à partir de 2014 correspond aux crédits inclus dans ces estimations. Les statistiques de la la Russie concernant l'APD et les estimations des rapports précédents du Président ne peuvent donc pas être utilisées conjointement.
5   La déclaration de l'Arabie saoudite auprès de l'OCDE relatif au programme de coopération au développement est constituée de données globales sur l'aide humanitaire, l'aide au développement par région, l'aide multilatérale, et les déboursements et les remboursements de prêts par les Fonds saoudien pour le développement.
6   Les Émirats arabes unis ont commencé à présenter au niveau d"activité en 2009. Ces données représentent les flux de tous les organismes gouvernementaux. Les engagements sont égaux aux décaissements pour les organismes autres que le Fond d'Abu Dhabi pour le développement.

Annex I - Country and area nomenclature, regional and other groupings

The *Statistical Yearbook* lists countries or areas based on the United Nations Standard Country Codes (Series M, No. 49), prepared by the Statistics Division of the United Nations Secretariat and first issued in 1970[1]. The list of countries or areas; the composition of geographical regions and economic, trade and other groupings used within this yearbook (as at 31 July 2016) are presented in this Annex. The names of countries or areas refer to their short form used in day-to-day operations of the United Nations and not necessarily to their official name as used in formal documents[2]. As an aid to statistical data processing, a unique standard three-digit numerical code is assigned to each country or area and to each geographical region and grouping of countries or areas. For reference, these codes, where applicable, are presented to the left of each country or area listed in this Annex. These codes range from 000 to 899, inclusive.

A.   Changes in country or area names (since 31 July 2006)

The geographical extent of a country or area, or the composition of a geographical region, may change over time and users should take such changes into account in interpreting the tables in this Yearbook. A change in the name of a country or area while its geographical coverage remains the same is not usually accompanied by a change in its numerical code. Changes in numerical codes and names in the general period covered by the statistics (since 31 July 2006) in the *Yearbook*, are shown below;

Numerical code	Country or area (added or changed)	Date of change	Numerical code	Country or area (Name changes with no change in code)	Date of change
728	South Sudan	2011	132	Cabo Verde, *previously Cape Verde*	2013
729	Sudan	2011	275	State of Palestine, *previously Occupied Palestinian Territory*	2013
531	Curaçao	2010			
534	Sint Maarten (Dutch part)	2010	434	Libya, *previously Libyan Arab Jamahiriya*	2011
535	Bonaire, Sint Eustatius and Saba	2010	068	Bolivia (Plurinational State of), *previously Bolivia*	2009
652	Saint Barthélemy	2007			
663	Saint Martin (French part)	2007	498	Republic of Moldova, *previously Moldova (since 2007)*	2008
499	Montenegro	2006			
688	Serbia	2006	862	Venezuela (Bolivarian Republic of), *previously Venezuela*	2006

B.   Regional groupings

The scheme of regional groupings given on the next page is based mainly on 6 continents; these continental regions, except Antarctica, are further subdivided into 22 sub-regions and 2 intermediary regions (Sub-Saharan Africa and Latin America and the Caribbean) that are drawn as to obtain greater homogeneity in sizes of population, demographic circumstances and accuracy of demographic statistics. This nomenclature is widely used in international statistics and is followed to the greatest extent possible in the present *Yearbook* in order to promote consistency and facilitate comparability and analysis. However, it is by no means universal in international statistical compilation, even at the level of continental regions, and variations in international statistical sources and methods dictate many unavoidable differences in particular fields in the present *Yearbook*. General differences are indicated in the footnotes to the classification presented below. More detailed differences are given in the footnotes and technical notes to individual tables.

Neither is there international standardization in the use of the terms "developed" and "developing" countries, areas or regions. These terms are used in the present publication to refer to regional groupings generally considered as "developed": these are Northern America (numerical code 021), Europe (150), Japan (392) and Australia and New Zealand (053). These designations are intended for statistical convenience and do not necessarily express a judgement about the stage reached by a particular country or area in the development process. Differences from this usage are indicated in the notes to individual tables.

---

[1] Four revisions of this document were published in 1975, 1982, 1996 and 1999 (ST/ESA/STAT/SER.M/49/Rev.4). Further revisions are now published on the United Nations Statistics Division website under the M49 section.

[2] A listing in the 6 official languages of the United Nations is contained in Terminology Bulletin No. 347/Rev.1 or the UNTERM website, prepared by the Department of General Assembly Affairs and Conference Services of the United Nations Secretariat.

001	**World**
002	**Africa**
202	**Sub-Saharan Africa**	**015**	**Northern Africa**

202 Sub-Saharan Africa

014 Eastern Africa
- 086 British Indian Ocean Territory
- 108 Burundi
- 174 Comoros
- 262 Djibouti
- 232 Eritrea
- 231 Ethiopia
- 260 French Southern and Antarctic Territories
- 404 Kenya
- 450 Madagascar
- 454 Malawi
- 480 Mauritius
- 175 Mayotte
- 508 Mozambique
- 638 Réunion
- 646 Rwanda
- 690 Seychelles
- 706 Somalia
- 728 South Sudan note 736
- 800 Uganda
- 834 United Republic of Tanzania note /a
- 894 Zambia
- 716 Zimbabwe

017 Middle Africa
- 024 Angola
- 120 Cameroon
- 140 Central African Republic
- 148 Chad
- 178 Congo
- 180 Democratic Republic of the Congo
- 226 Equatorial Guinea
- 266 Gabon
- 678 Sao Tome and Principe

018 Southern Africa
- 072 Botswana
- 426 Lesotho
- 516 Namibia
- 710 South Africa
- 748 Swaziland

011 Western Africa
- 204 Benin
- 854 Burkina Faso
- 132 Cabo Verde
- 384 Côte d'Ivoire
- 270 Gambia
- 288 Ghana
- 324 Guinea
- 624 Guinea-Bissau
- 430 Liberia
- 466 Mali
- 478 Mauritania
- 562 Niger
- 566 Nigeria
- 654 Saint Helena note /b
- 686 Senegal
- 694 Sierra Leone
- 768 Togo

015 Northern Africa
- 012 Algeria
- 818 Egypt
- 434 Libya
- 504 Morocco
- 729 Sudan note 736
- 788 Tunisia
- 732 Western Sahara

Note
- 736 *Sudan [former]*
- /a Includes
- --- *Zanzibar*
- /b Includes
- --- *Ascension*
- --- *Tristan da Cunha*

019	**Americas**
419	**Latin America and the Caribbean**	**021**	**Northern America** note 003

419 Latin America and the Caribbean

029 Caribbean
- 660 Anguilla
- 028 Antigua and Barbuda
- 533 Aruba
- 044 Bahamas
- 052 Barbados
- 535 Bonaire, Sint Eustatius and Saba note 530
- 092 British Virgin Islands
- 136 Cayman Islands
- 192 Cuba
- 531 Curaçao note 530
- 212 Dominica
- 214 Dominican Republic
- 308 Grenada
- 312 Guadeloupe
- 332 Haiti
- 388 Jamaica
- 474 Martinique
- 500 Montserrat
- 630 Puerto Rico
- 652 Saint Barthélemy
- 659 Saint Kitts and Nevis
- 662 Saint Lucia
- 663 Saint Martin (French part)
- 670 Saint Vincent and the Grenadines
- 534 Sint Maarten (Dutch part) note 530
- 780 Trinidad and Tobago
- 796 Turks and Caicos Islands
- 850 United States Virgin Islands

013 Central America
- 084 Belize
- 188 Costa Rica
- 222 El Salvador
- 320 Guatemala
- 340 Honduras
- 484 Mexico
- 558 Nicaragua
- 591 Panama

005 South America
- 032 Argentina
- 068 Bolivia (Plurinational State of)
- 076 Brazil
- 152 Chile
- 170 Colombia
- 218 Ecuador
- 238 Falkland Islands (Malvinas)
- 254 French Guiana
- 328 Guyana
- 600 Paraguay
- 604 Peru
- 740 Suriname
- 239 South Georgia and the South Sandwich Islands
- 858 Uruguay
- 862 Venezuela (Bolivarian Republic of)

021 Northern America note 003
- 060 Bermuda
- 124 Canada
- 304 Greenland
- 666 Saint Pierre and Miquelon
- 840 United States of America

Note
- 003 *The continent of* **North America** *comprises Northern America, Caribbean and Central America.*

Tables may list
- 530 *Netherlands Antilles [former]*

010	**Antarctica**

142   **Asia**

143	**Central Asia**	496	Mongolia	**034**	**Southern Asia**	196	Cyprus
398	Kazakhstan	410	Republic of Korea	004	Afghanistan	268	Georgia
417	Kyrgyzstan			050	Bangladesh	368	Iraq
762	Tajikistan	**035**	**South-eastern Asia**	064	Bhutan	376	Israel
795	Turkmenistan	096	Brunei Darussalam	356	India	400	Jordan
860	Uzbekistan	116	Cambodia	364	Iran (Islamic	414	Kuwait
		360	Indonesia		Republic of)	422	Lebanon
030	**Eastern Asia**	418	Lao People's	462	Maldives	512	Oman
156	China		Democratic Republic	524	Nepal	634	Qatar
344	China, Hong Kong Special	458	Malaysia	586	Pakistan	682	Saudi Arabia
	Administrative Region	104	Myanmar	144	Sri Lanka	275	State of Palestine
446	China, Macao Special	608	Philippines			760	Syrian Arab Republic
	Administrative Region	702	Singapore	**145**	**Western Asia**	792	Turkey
408	Democratic People's	764	Thailand	051	Armenia	784	United Arab Emirates
	Republic of Korea	626	Timor-Leste	031	Azerbaijan	887	Yemen
392	Japan	704	Viet Nam	048	Bahrain		

150   **Europe**

151	**Eastern Europe**	831	Guernsey note 830	**039**	**Southern Europe**	**155**	**Western Europe**
112	Belarus	352	Iceland	008	Albania	040	Austria
100	Bulgaria	372	Ireland	020	Andorra	056	Belgium
203	Czech Republic	833	Isle of Man	070	Bosnia and Herzegovina	250	France
348	Hungary	832	Jersey note 830	191	Croatia	276	Germany
616	Poland	428	Latvia	292	Gibraltar	438	Liechtenstein
498	Republic of Moldova	440	Lithuania	300	Greece	442	Luxembourg
642	Romania	578	Norway note 074	336	Holy See	492	Monaco
643	Russian Federation	744	Svalbard and Jan	380	Italy	528	Netherlands
703	Slovakia		Mayen Islands	470	Malta	756	Switzerland
804	Ukraine	752	Sweden	499	Montenegro note 891		
		826	United Kingdom of	620	Portugal	**Note**	**Tables may list**
154	**Northern Europe**		Great Britain and	674	San Marino	*830*	*Channel Islands*
248	Åland Islands		Northern Ireland	688	Serbia note 891 & /c	*074*	*Bouvet Island*
208	Denmark			705	Slovenia	*891*	*Serbia and*
233	Estonia			724	Spain		*Montenegro [former]*
234	Faroe Islands			807	The former Yugoslav	*/c*	*Kosovo*
246	Finland				Republic of Macedonia		

009   **Oceania**

053	**Australia and New Zealand**	**057**	**Micronesia**	**061**	**Polynesia**
036	Australia	316	Guam	016	American Samoa
162	Christmas Island	296	Kiribati	184	Cook Islands
166	Cocos (Keeling) Islands	584	Marshall Islands	258	French Polynesia
334	Heard Island and McDonald Islands	583	Micronesia (Federated States of)	570	Niue
554	New Zealand	520	Nauru	612	Pitcairn
574	Norfolk Island	580	Northern Mariana Islands	882	Samoa
		585	Palau	772	Tokelau
054	**Melanesia**	581	United States minor outlying islands	776	Tonga
242	Fiji			798	Tuvalu
540	New Caledonia			876	Wallis and Futuna Islands
598	Papua New Guinea				
090	Solomon Islands				
548	Vanuatu				

Other groupings

Following is a list of other groupings and their compositions presented in the *Yearbook*. These groupings are organized mainly around economic and trade interests in regional associations.

063	**Andean Common Market (ANCOM)**						
068	Bolivia (Plurinational State of)	170	Colombia	218	Ecuador	604	Peru

066	**Asia-Pacific Economic Cooperation (APEC)**						
036	Australia	344	China, Hong Kong Special	554	New Zealand	702	Singapore
096	Brunei Darussalam		Administrative Region	598	Papua New Guinea	158	Taiwan Province of China
		360	Indonesia	604	Peru	764	Thailand
124	Canada	392	Japan	608	Philippines	840	United States of America
152	Chile	458	Malaysia	410	Republic of Korea	704	Viet Nam
156	China	484	Mexico	643	Russian Federation		

073	**Association of Southeast Asian Nations (ASEAN)**								
096	Brunei Darussalam	360	Indonesia	458	Malaysia	608	Philippines	764	Thailand
116	Cambodia	418	Lao PDR	104	Myanmar	702	Singapore	704	Viet Nam

130	**Caribbean Community and Common Market (CARICOM)**						
028	Antigua and Barbuda	084	Belize	332	Haiti	662	Saint Lucia
044	Bahamas (member of the Community only)	212	Dominica	388	Jamaica	670	Saint Vincent and the Grenadines
		308	Grenada	500	Montserrat	740	Suriname
052	Barbados	328	Guyana	659	Saint Kitts and Nevis	780	Trinidad and Tobago

395	**Central American Common Market (CACM)**								
188	Costa Rica	222	El Salvador	320	Guatemala	340	Honduras	558	Nicaragua

171	**Common Market for Eastern and Southern Africa (COMESA)**								
108	Burundi	262	Djibouti	404	Kenya	480	Mauritius	748	Swaziland
174	Comoros	818	Egypt	434	Libya	646	Rwanda	800	Uganda
180	Democratic Republic of the Congo	232	Eritrea	450	Madagascar	690	Seychelles	894	Zambia
		231	Ethiopia	454	Malawi	729	Sudan	716	Zimbabwe

172	**Commonwealth of Independent States (CIS)[3]**							
051	Armenia	398	Kazakhstan	643	Russian Federation	804	Ukraine
031	Azerbaijan	417	Kyrgyzstan	762	Tajikistan	860	Uzbekistan
112	Belarus	498	Republic of Moldova	795	Turkmenistan		

692	**Economic and Monetary Community of Central Africa (EMCCA)**										
120	Cameroon	140	Central African Republic	148	Chad	178	Congo	226	Equatorial Guinea	266	Gabon

892	**Economic Community of West African States (ECOWAS)**								
204	Benin	384	Côte d'Ivoire	324	Guinea	466	Mali	686	Senegal
854	Burkina Faso	270	Gambia	624	Guinea-Bissau	562	Niger	694	Sierra Leone
132	Cabo Verde	288	Ghana	430	Liberia	566	Nigeria	768	Togo

098	**Euro Area**								
040	Austria	246	Finland	372	Ireland	528	Netherlands	724	Spain
056	Belgium	250	France	380	Italy	620	Portugal		
196	Cyprus	276	Germany	442	Luxembourg	703	Slovakia		
233	Estonia	300	Greece	470	Malta	705	Slovenia		

197	**European Free Trade Association (EFTA)**							
352	Iceland	438	Liechtenstein	578	Norway	756	Switzerland

097	**European Union (EU)**								
040	Austria	208	Denmark	348	Hungary	470	Malta	705	Slovenia
056	Belgium	233	Estonia	372	Ireland	528	Netherlands	724	Spain
100	Bulgaria	246	Finland	380	Italy	616	Poland	752	Sweden

[3] Georgia (numerical code 268) can be listed as part of this group; for example, in International Merchandise Trade statistics.

097	European Union (EU) *(Continued)*								
191	Croatia	250	France	428	Latvia	620	Portugal	826	United Kingdom
196	Cyprus	276	Germany	440	Lithuania	642	Romania		
203	Czech Republic	300	Greece	442	Luxembourg	703	Slovakia		

095	Latin American Integration Association (LAIA)						
032	Argentina	170	Colombia	591	Panama	862	Venezuela (Bolivarian Republic of)
068	Bolivia (Plurinational State of)	192	Cuba	600	Paraguay		
076	Brazil	218	Ecuador	604	Peru		
152	Chile	484	Mexico	858	Uruguay		

199	Least developed countries (LDCs)						
004	Afghanistan	262	Djibouti	450	Madagascar	706	Somalia
024	Angola	226	Equatorial Guinea	454	Malawi	728	South Sudan
050	Bangladesh	232	Eritrea	466	Mali	729	Sudan
204	Benin	231	Ethiopia	478	Mauritania	626	Timor-Leste
064	Bhutan	270	Gambia	508	Mozambique	768	Togo
854	Burkina Faso	324	Guinea	104	Myanmar	798	Tuvalu
108	Burundi	624	Guinea-Bissau	524	Nepal	800	Uganda
116	Cambodia	332	Haiti	562	Niger	834	United Republic of Tanzania
140	Central African Republic	296	Kiribati	646	Rwanda		
148	Chad	418	Lao People's Democratic Republic	678	Sao Tome and Principe	548	Vanuatu
174	Comoros			686	Senegal	887	Yemen
180	Democratic Republic of the Congo	426	Lesotho	694	Sierra Leone	894	Zambia
		430	Liberia	090	Solomon Islands		

071	North American Free Trade Agreement (NAFTA)				
124	Canada	484	Mexico	840	United States of America

198	Organisation for Economic Co-operation and Development (OECD)						
036	Australia	250	France	442	Luxembourg	705	Slovenia
040	Austria	276	Germany	484	Mexico	724	Spain
056	Belgium	300	Greece	528	Netherlands	752	Sweden
124	Canada	348	Hungary	554	New Zealand	756	Switzerland
152	Chile	352	Iceland	578	Norway	792	Turkey
203	Czech Republic	372	Ireland	616	Poland	826	United Kingdom of Great Britain and Northern Ireland
208	Denmark	376	Israel	620	Portugal		
233	Estonia	380	Italy	410	Republic of Korea		
246	Finland	392	Japan	703	Slovakia	840	United States of America

399	Organization of the Petroleum Exporting Countries (OPEC)						
012	Algeria	364	Iran (Islamic Republic of)	566	Nigeria	862	Venezuela (Bolivarian Republic of)
024	Angola	368	Iraq	634	Qatar		
218	Ecuador	414	Kuwait	682	Saudi Arabia		
360	Indonesia	434	Libya	784	United Arab Emirates		

711	Southern African Customs Union (SACU)						
072	Botswana	426	Lesotho	516	Namibia	710	South Africa	748	Swaziland

069	Southern Common Market (MERCOSUR)						
032	Argentina	076	Brazil	858	Uruguay		
068	Bolivia (Plurinational State of)	600	Paraguay	862	Venezuela (Bolivarian Republic of)		

Annexe I - Nomenclature des pays ou zones, groupements régionaux et autres groupements

L'Annuaire statistique répertorie de pays ou de zones sur la base de l'opuscule United Nations Standard Country Codes (Série M, n° 49) rédigée par la Division de statistique du Secrétariat de l'Organisation des Nations Unies et publiée en 1970[1]. La liste des

---

[1] Quatre révisions de ce document ont été publiés en 1975, 1982, 1996 et 1999 (ST/ESA/STAT/SER.M/49/Rev.4). D'autres révisions sont maintenant publiés sur le site internet de la Division de statistique des Nations Unies.

pays ou zones; la composition des régions géographiques et économiques, commerciales et d'autres groupements utilisés dans cet annuaire (au 31 juillet 2016) sont présentées dans la présente annexe. On a eu recours à la forme brève des noms des pays et des zones usitées au cours des activités courantes des Nations Unies, et pas nécessairement aux désignations officielles utilisées dans les documents officiels[2]. Afin de faciliter le traitement des données statistiques, un code numérique unique et standard à trois chiffres est attribué à chaque pays ou zone et à chaque région géographique et groupement de pays ou de zones. Pour référence, ces codes, le cas échéant, sont présentés à la gauche de chaque pays ou zone visées dans cette annexe. Ces codes s'échelonnent entre 000 et 899.

A. Changements dans le nom des pays ou zones (depuis 31 juillet 2006)

L'étendue géographique d'un pays ou d'une zone ou la composition d'une région géographique peuvent varier au fil du temps et les utilisateurs devront tenir compte de telles modifications lorsqu'ils ont recours aux codes indiques dans la présente publication et qu'ils établissent des rapports entre les données portant sur différentes périodes. La modification du nom d'un pays ou d'une zone alors que son espace géographique demeure le même n'est pas habituellement accompagnée d'un changement de son code numérique. Changements dans les codes numériques et noms dans la période générale couverts par les statistiques (depuis le 31 juillet 2006) dans l'Annuaire, sont présentés ci-dessous;

Code numérique	Pays ou region (nouveaux codes et codes modifies)	Date de la modification	Code numérique	Pays ou region (Changements de nom sans modification du code)	Date de la modification
728	Soudan du Sud	2011	132	Cabo Verde, *ex-Cap-Vert*	2013
729	Soudan	2011	434	État de Palestine, *ex-Territoire palestinien occupé*	2013
531	Curaçao	2010			
534	Saint-Martin (partie néerlandaise)	2010	275	Libye, *ex-Jamahiriya arabe libyenne*	2011
535	Bonaire, Saint-Eustache et Saba	2010			
652	Saint-Barthélemy	2007	068	Bolivie (État plurinational de), *ex-Bolivie*	2009
663	Saint-Martin (partie française)	2007			
499	Monténégro	2006	498	République de Moldova, *ex-Moldova (depuis 2007)*	2008
688	Serbie	2006			
			862	Venezuela (République bolivarienne du), *ex-Venezuela*	2006

B.   Groupements régionaux

Le système des groupements régionaux à la page suivante est principalement basé sur les 6 continents; ces régions continentales, à l'exception de l'Antarctique, sont subdivisées en 22 sous-régions et 2 régions intermédiaires (Afrique subsaharienne et l'Amérique latine et Caraïbes) afin d'obtenir une homogénéité accrue les effectifs de population, les situations démographiques et la précision des statistiques démographiques. Cette nomenclature est couramment utilisée aux fins des statistiques internationales et a été appliquée autant qu'il a été possible dans le présent *Annuaire* en vue de renforcer la cohérence et de faciliter la comparaison et l'analyse. Son utilisation pour l'établissement des statistiques internationales n'est cependant rien moins qu'universelle, même au niveau des régions continentales, et les variations que présentent les sources et méthodes statistiques internationales entraîent inévitablement de nombreuses différences dans certains domaines de cet *Annuaire*. Les différences d'ordre général sont indiquées dans les notes figurant au bas de la classification présentée ci-dessous. Les différences plus spécifiques sont mentionnées dans les notes techniques et notes de bas de page accompagnant les divers tableaux.

L'application des expressions "développés" et "en développement" aux pays, zones ou régions n'est pas non plus normalisée à l'échelle internationale. Ces expressions sont utilisées dans la présente publication en référence aux groupements régionaux généralement considérés comme "développés", à savoir l'Amérique septentrionale (code numérique 021), l'Europe (150), le Japon (392) et l'Australie et la Nouvelle-Zélande (053). Ces appellations sont employées pour des raisons de commodité statistique et n'expriment pas nécessairement un jugement sur le stade de développement atteint par tel ou tel pays ou zone. Les cas différant de cet usage sont signalés dans les notes accompagnant les tableaux concernés.

---

[2] Le Bulletin terminologique No 347/Rev.1 ou UNTERM website, intitulé "Noms de pays" établi par le Département des affaires de l'Assemblée générale et des services de conférence du Secrétariat de l'Organisation des Nations Unies, répertorie les noms des États Membres dans les six langues de l'Organisation.

001	**Monde**					

002	**Afrique**

202	**Afrique subsaharienne**		015	**Afrique septentrionale**

014 Afrique orientale

108	Burundi
174	Comores
262	Djibouti
232	Érythrée
231	Éthiopie
404	Kenya
450	Madagascar
454	Malawi
480	Maurice
175	Mayotte
508	Mozambique
800	Ouganda
834	République-Unie de Tanzanie note /a
638	Réunion
646	Rwanda
690	Seychelles
706	Somalie
728	Soudan du Sud note 736
260	Terres australes et antarctiques françaises

086	Territoire britannique de l'océan Indien
894	Zambie
716	Zimbabwe

017 Afrique centrale

024	Angola
120	Cameroun
178	Congo
266	Gabon
226	Guinée équatoriale
140	République centrafricaine
180	République démocratique du Congo
678	Sao Tomé-et-Principe
148	Tchad

018 Afrique australe

710	Afrique du Sud
072	Botswana
426	Lesotho

| 516 | Namibie |
| 748 | Swaziland |

011 Afrique occidentale

204	Bénin
854	Burkina Faso
132	Cabo Verde
384	Côte d'Ivoire
270	Gambie
288	Ghana
324	Guinée
624	Guinée-Bissau
430	Libéria
466	Mali
478	Mauritanie
562	Niger
566	Nigéria
654	Sainte-Hélène note /b
686	Sénégal
694	Sierra Leone
768	Togo

015 Afrique septentrionale

012	Algérie
818	Égypte
434	Libye
504	Maroc
732	Sahara occidental
729	Soudan note 736
788	Tunisie

Note	**Certains tableaux peuvent**
736	*Soudan [anc.]*
/a	Incluent
---	*Zanzibar*
/b	Incluent
---	*Ascension*
---	*Tristan da Cunha*

019	**Amériques**

419	**Amérique latine et Caraïbes**		021	**Amérique septentrionale** note 003

029 Caraïbes

660	Anguilla
028	Antigua-et-Barbuda
533	Aruba
044	Bahamas
052	Barbade
535	Bonaire, Saint-Eustache et Saba note 530
192	Cuba
531	Curaçao note 530
212	Dominique
308	Grenade
312	Guadeloupe
332	Haïti
136	Îles Caïmanes
796	Îles Turques-et-Caïques
850	Îles Vierges américaines
092	Îles Vierges britanniques
388	Jamaïque
474	Martinique
500	Montserrat
630	Porto Rico
214	République dominicaine

652	Saint-Barthélemy
659	Saint-Kitts-et-Nevis
662	Sainte-Lucie
663	Saint-Martin (partie française)
534	Saint-Martin note 530 (partie néerlandaise)
670	Saint-Vincent-et-les Grenadines
780	Trinité-et-Tobago

013 Amérique centrale

084	Belize
188	Costa Rica
222	El Salvador
320	Guatemala
340	Honduras
484	Mexique
558	Nicaragua
591	Panama

005 Amérique du Sud

032	Argentine
068	Bolivie (État plurinational de)
076	Brésil
152	Chili
170	Colombie
218	Équateur
239	Géorgie du Sud-et-les Îles Sandwich du Sud
328	Guyana
254	Guyane française
238	Îles Falkland (Malvinas)
600	Paraguay
604	Pérou
740	Suriname
858	Uruguay
862	Venezuela (République bolivarienne du)

021 Amérique septentrionale note 003

060	Bermudes
124	Canada
840	États-Unis d'Amérique
304	Groenland
666	Saint-Pierre-et-Miquelon

| **Note** | |
| *003* | *Le continent de **l'Amérique du Nord** comprend l'Amérique septentrionale, les Caraïbes et l'Amérique centrale* |

| **Certains tableaux peuvent** | |
| *530* | *Antilles néerlandaises [anc.]* |

010	**Antarctique**

142	**Asie**						

142   Asie

143 Asie centrale		**035 Asie du Sud-Est**		**034 Asie méridionale**		784	Émirats arabes unis
398	Kazakhstan	096	Brunéi Darussalam	004	Afghanistan	275	État de Palestine
417	Kirghizistan	116	Cambodge	050	Bangladesh	268	Géorgie
860	Ouzbékistan	360	Indonésie	064	Bhoutan	368	Iraq
762	Tadjikistan	458	Malaisie	356	Inde	376	Israël
795	Turkménistan	104	Myanmar	364	Iran (République	400	Jordanie
		608	Philippines		islamique d')	414	Koweït
030	**Asie orientale**	418	République	462	Maldives	422	Liban
156	Chine		démocratique	524	Népal	512	Oman
344	Chine, région administrative		populaire lao	586	Pakistan	634	Qatar
	spéciale de Hong Kong	702	Singapour	144	Sri Lanka	760	République arabe
446	Chine, région administrative	764	Thaïlande				syrienne
	spéciale de Macao	626	Timor-Leste	**145**	**Asie occidentale**	792	Turquie
392	Japon	704	Viet Nam	682	Arabie saoudite	887	Yémen
496	Mongolie			051	Arménie		
410	République de Corée			031	Azerbaïdjan		
408	République populaire			048	Bahreïn		
	démocratique de Corée			196	Chypre		

150	**Europe**					

150   Europe

151 Europe orientale		831	Guernesey [note 830]	**039 Europe méridionale**		**155 Europe occidentale**	
112	Bélarus	833	Île de Man	008	Albanie	276	Allemagne
100	Bulgarie	248	Îles d'Åland	020	Andorre	040	Autriche
643	Fédération de Russie	234	Îles Féroé	070	Bosnie-Herzégovine	056	Belgique
348	Hongrie	744	Îles Svalbard-et-Jan	191	Croatie	250	France
616	Pologne		Mayen	724	Espagne	438	Liechtenstein
498	République de	372	Irlande	807	Ex-République	442	Luxembourg
	Moldova	352	Islande		yougoslave de Macédoine	492	Monaco
203	République tchèque	832	Jersey [note 830]	292	Gibraltar	528	Pays-Bas
642	Roumanie	428	Lettonie	300	Grèce	756	Suisse
703	Slovaquie	440	Lituanie	380	Italie		
804	Ukraine	578	Norvège [note 074]	470	Malte	**Note**	**Certains tableaux**
		826	Royaume-Uni de	499	Monténégro [note 891]		**peuvent énumérer**
154	**Europe**		Grande-Bretagne et	620	Portugal	830	*Îles Anglo-Normandes*
	septentrionale		d'Irlande du Nord	674	Saint-Marin	074	*Île Bouvet*
208	Danemark	752	Suède	336	Saint-Siège	891	*Serbie-et-Monténégro*
233	Estonie			688	Serbie [note 891 & /c]		*[anc.]*
246	Finlande			705	Slovénie	/c	Kosovo

009	**Océanie**					

009   Océanie

053 Australie et Nouvelle-Zélande		598	Papouasie-Nouvelle-Guinée	**061 Polynésie**	
036	Australie	548	Vanuatu	184	Îles Cook
162	Île Christmas			876	Îles Wallis-et-Futuna
334	Île Heard-et-Îles MacDonald	**057**	**Micronésie**	570	Nioué
574	Île Norfolk	316	Guam	612	Pitcairn
166	Îles des Cocos (Keeling)	580	Îles Mariannes du Nord	258	Polynésie française
554	Nouvelle-Zélande	584	Îles Marshall	882	Samoa
		581	Îles mineures éloignées des États-Unis	016	Samoa américaines
054	**Mélanésie**	296	Kiribati	772	Tokelau
242	Fidji	583	Micronésie (États fédérés de)	776	Tonga
090	Îles Salomon	520	Nauru	798	Tuvalu
540	Nouvelle-Calédonie	585	Palaos		

C. Autres groupements

On trouvera ci-après une liste des autres groupements et de leur composition, présentée dans l'*Annuaire*. Ces groupements correspondent essentiellement à des intérêts économiques et commerciaux d'après les associations régionales.

071 — Accord de libre-échange nord-américain (ALENA)

124	Canada	840	États-Unis d'Amérique	484	Mexique

073 — Association des nations de l'Asie du Sud-Est (ANASE)

096	Brunéi Darussalam	360	Indonésie	104	Myanmar	418	République démocratique populaire lao	764	Thaïlande
116	Cambodge	458	Malaisie	608	Philippines	702	Singapour	704	Viet Nam

097 — Association européenne de libre-échange (AELE)

352	Islande	438	Liechtenstein	578	Norvège	756	Suisse

095 — Association latino-américaine d'intégration (ALAI)

032	Argentine	170	Colombie	591	Panama	862	Venezuela (République bolivarienne du)
068	Bolivie (État plurinational de)	192	Cuba	600	Paraguay		
076	Brésil	218	Équateur	604	Pérou		
152	Chili	484	Mexique	858	Uruguay		

130 — Communauté des Caraïbes et Marché commun des Caraïbes (CARICOM)

028	Antigua-et-Barbuda	084	Belize	332	Haïti	662	Sainte-Lucie
044	Bahamas (membre de la communauté seulement)	212	Dominique	388	Jamaïque	670	Saint-Vincent-et-les Grenadines
		308	Grenade	500	Montserrat	740	Suriname
052	Barbade	328	Guyana	659	Saint-Kitts-et-Nevis	780	Trinité-et-Tobago

172 — Communauté d'Etats indépendants (CEI)[3]

051	Arménie	643	Fédération de Russie	417	Kirghizistan	762	Tadjikistan
031	Azerbaïdjan	268	Géorgie	860	Ouzbékistan	795	Turkménistan
112	Bélarus	398	Kazakhstan	498	République de Moldova	804	Ukraine

392 — Communauté économique des Etats de l'Afrique de l'Ouest (CEDEAO)

204	Bénin	384	Côte d'Ivoire	324	Guinée	466	Mali	686	Sénégal
854	Burkina Faso	270	Gambie	624	Guinée-Bissau	562	Niger	694	Sierra Leone
132	Cabo Verde	288	Ghana	430	Libéria	566	Nigéria	768	Togo

592 — Communauté économique et monétaire des Etats de l'Afrique Centrale (CEMAC)

120	Cameroun	178	Congo	266	Gabon	226	Guinée équatoriale	140	République centrafricaine	148	Tchad

966 — Coopération économique Asie-Pacifique (CEAP)

036	Australie	840	États-Unis d'Amérique	598	Papouasie-Nouvelle-Guinée	704	Viet Nam
096	Brunéi Darussalam	643	Fédération de Russie	604	Pérou		
124	Canada	360	Indonésie	608	Philippines		
152	Chili	392	Japon	158	Province chinoise de Taïwan		
156	Chine	458	Malaisie	410	République de Corée		
344	Chine, région administrative spéciale de Hong Kong	484	Mexique	702	Singapour		
		554	Nouvelle-Zélande	764	Thaïlande		

963 — Marché commun andin (ANCOM)

068	Bolivie (État plurinational de)	170	Colombie	218	Équateur	604	Pérou

395 — Marché commun centraméricain (MCCA)

188	Costa Rica	222	El Salvador	320	Guatemala	340	Honduras	558	Nicaragua

471 — Marché commun de l'Afrique de l'Est et de l'Afrique australe (COMESA)

108	Burundi	232	Érythrée	450	Madagascar	180	République démocratique du Congo	729	Soudan
174	Comores	231	Éthiopie	454	Malawi			748	Swaziland
262	Djibouti	404	Kenya	480	Maurice	646	Rwanda	894	Zambie
818	Égypte	434	Libye	800	Ouganda	690	Seychelles	716	Zimbabwe

---

[3] La Géorgie (code numérique 268) peut figurer dans ce groupe; par exemple, dans le cadre des statistiques du commerce international de marchandises.

069   Marché commun du Sud (MERCOSUR)

032	Argentine	076	Brésil	858	Uruguay
068	Bolivie (État plurinational de)	600	Paraguay	862	Venezuela (République bolivarienne du)

198   Organisation de coopération et de développement économiques (OCDE)

276	Allemagne	840	États-Unis d'Amérique	392	Japon	203	République tchèque
036	Australie	246	Finlande	442	Luxembourg	826	Royaume-Uni de Grande-Bretagne et d'Irlande du Nord
040	Autriche	250	France	484	Mexique		
056	Belgique	300	Grèce	578	Norvège		
124	Canada	348	Hongrie	554	Nouvelle-Zélande	703	Slovaquie
152	Chili	372	Irlande	528	Pays-Bas	705	Slovénie
208	Danemark	376	Israël	616	Pologne	752	Suède
724	Espagne	352	Islande	620	Portugal	756	Suisse
233	Estonie	380	Italie	410	République de Corée	792	Turquie

399   Organisation des pays exportateurs de pétrole (OPEP)

012	Algérie	218	Équateur	434	Libye	862	Venezuela (République bolivarienne du)
024	Angola	360	Indonésie	414	Koweït		
682	Arabie saoudite	364	Iran (République islamique d')	566	Nigéria		
784	Émirats arabes unis	368	Iraq	634	Qatar		

199   Pays les moins avancés (PMA)

004	Afghanistan	270	Gambie	478	Mauritanie	686	Sénégal
024	Angola	324	Guinée	508	Mozambique	694	Sierra Leone
050	Bangladesh	226	Guinée équatoriale	104	Myanmar	706	Somalie
204	Bénin	624	Guinée-Bissau	524	Népal	728	Soudan du Sud
064	Bhoutan	332	Haïti	562	Niger	729	Soudan
854	Burkina Faso	090	Îles Salomon	800	Ouganda	148	Tchad
108	Burundi	296	Kiribati	140	République centrafricaine	626	Timor-Leste
116	Cambodge	426	Lesotho	180	République démocratique du Congo	768	Togo
174	Comores	430	Libéria	418	République démocratique populaire lao	798	Tuvalu
262	Djibouti	450	Madagascar	834	République-Unie de Tanzanie	548	Vanuatu
232	Érythrée	454	Malawi	646	Rwanda	887	Yémen
231	Éthiopie	466	Mali	678	Sao Tomé-et-Principe	894	Zambie

711   Union douanière d'Afrique australe

710	Afrique du Sud	072	Botswana	426	Lesotho	516	Namibie	748	Swaziland

097   Union européenne (UE)

276	Allemagne	208	Danemark	348	Hongrie	470	Malte	826	Royaume-Uni de Grande-Bretagne et d'Irlande du Nord
040	Autriche	724	Espagne	372	Irlande	528	Pays-Bas		
056	Belgique	233	Estonie	380	Italie	616	Pologne		
100	Bulgarie	246	Finlande	428	Lettonie	620	Portugal	703	Slovaquie
191	Croatie	250	France	440	Lituanie	203	République tchèque	705	Slovénie
196	Chypre	300	Grèce	442	Luxembourg	642	Roumanie	752	Suède

098   Zone euro

276	Allemagne	724	Espagne	300	Grèce	470	Malte	705	Slovénie
040	Autriche	233	Estonie	372	Irlande	528	Pays-Bas		
056	Belgique	246	Finlande	380	Italie	620	Portugal		
196	Chypre	250	France	442	Luxembourg	703	Slovaquie		

Annex II: Technical notes

Chapter I: World summary

Table 1: World statistics – selected series

These world aggregates are obtained from other tables in this *Yearbook*, where available, and are compiled from statistical publications and databases of the United Nations and the specialized agencies and other institutions. The technical notes of the relevant table in this *Yearbook* should be consulted for detailed information on definition, source, compilation and coverage.

Chapter II: Population and migration

Table 2: Population, surface area and density

"Mid-year population estimates" and "density" are taken from the estimates and projections prepared by the United Nations Population Division, published in *World Population Prospects: The 2015 Revision*. "Surface area" are obtained from the *Demographic Yearbook*, through this source only official national data are reported.

"Mid-year population estimates" is the de facto population in a country, area or region as of 1 July of the year indicated. Figures are presented in millions.

"Density" is the population per square Kilometre.

"Surface area" estimates include inland waters.

Table 3: International migrants and refugees

"International migrant stock" are taken from the estimates and projections prepared by the United Nations Population Division, published in *International migrant stock: The 2015 Revision*. "Refugees and others of concern to UNHCR" are obtained from the United Nations High Commissioner for Refugees, published in the Population Statistics database.

"International migrant stock" represents the number of persons born in a country other than that in which they live. When information on country of birth was not recorded, data on the number of persons having foreign citizenship was used instead. In the absence of any empirical data, estimates were imputed. Data refer to mid-2015. Figures for international migrant stock as a percentage of the population are the outcome of dividing the estimated international migrant stock by the estimated total population and multiplying the result by 100.

"Refugees" include individuals recognised under the 1951 Convention relating to the Status of Refugees; its 1967 Protocol; the 1969 OAU Convention Governing the Specific Aspects of Refugee Problems in Africa; those recognised in accordance with the UNHCR Statute; individuals granted complementary forms of protection; or those enjoying temporary protection. Since 2007, the refugee population also includes people in a refugee-like situation.

"Asylum-seekers" are individuals who have sought international protection and whose claims for refugee status have not yet been determined, irrespective of when they may have been lodged.

"Other" represents the following 5 categories:

Internally displaced persons (IDPs) are people or groups of individuals who have been forced to leave their homes or places of habitual residence, in particular as a result of, or in order to avoid the effects of armed conflict, situations of generalised violence, violations of human rights, or natural or man-made disasters, and who have not crossed an international border. For the purposes of UNHCR's statistics, this population only includes conflict-generated IDPs to whom the Office extends protection and/or assistance. Since 2007, the IDP population also includes people in an IDP-like situation. For global IDP estimates, see www.internal-displacement.org.

Returned refugees are former refugees who have returned to their country of origin spontaneously or in an organised fashion but are yet to be fully integrated. Such return would normally only take place in conditions of safety and dignity.

Returned IDPs refer to those IDPs who were beneficiaries of UNHCR's protection and assistance activities and who returned to their areas of origin or habitual residence during the year.

Stateless persons are defined under international law as persons who are not considered as nationals by any State under the operation of its law. In other words, they do not possess the nationality of any State. UNHCR statistics refer to persons who fall under the agency's statelessness mandate because they are stateless according to this international definition, but data from some countries may also include persons with undetermined nationality.

Others of concern refers to individuals who do not necessarily fall directly into any of the groups above, but to whom UNHCR extends its protection and/or assistance services, based on humanitarian or other special grounds.

Chapter III: Gender

Table 4: Proportion of seats held by women in national parliament

The table shows the percentage of seats held by women members in single or lower chambers of national parliaments as at January/ February each year (see table footnotes for specific details). National parliaments can be bicameral or unicameral. This table covers the single chamber in unicameral parliaments and the lower chamber in bicameral parliaments. It does not cover the upper chamber of bi-cameral parliaments. Seats are usually won by members in general parliamentary elections. Seats may also be filled by nomination, appointment, indirect election, rotation of members and by-election. The proportion of seats held by women in national parliament is derived by dividing the total number of seats occupied by women by the total number of seats in parliament. There is no weighting or normalising of statistics. The source for this table is the Inter-Parliamentary Union (IPU), see www.ipu.org for further information.

Chapter IV: Education

Data in Tables 5 and 6 are presented using the 2011 revision of UNESCO's International Standard Classification of Education (ISCED). Data are presented in the tables based on the three main levels of educations defined as follows;

"Primary education" (ISCED level 1) programmes are typically designed to provide students with fundamental skills in reading, writing and mathematics (i.e. literacy and numeracy) and establish a solid foundation for learning and understanding core areas of knowledge, personal and social development, in preparation for lower secondary education. It focuses on learning at a basic level of complexity with little, if any, specialisation.

"Secondary education" (ISCED level 2 and 3) is divided into two different stages, i.e. lower secondary and upper secondary. Lower secondary education programmes are typically designed to build on the learning outcomes from primary. Usually, they aim to lay the foundation for lifelong learning and human development upon which education systems may then expand further educational opportunities. Upper secondary education programmes are typically designed to complete secondary education in preparation for tertiary education or provide skills relevant to employment, or both. Programmes at this level offer students more varied, specialised and in-depth instruction than programmes at lower secondary. They are more differentiated, with an increased range of options and streams available. Teachers are often highly qualified in the subjects or fields of specialisation they teach, particularly in the higher grades.

Tertiary education (ISCED levels 5-8) builds on secondary education, providing learning activities in specialised fields of education. It aims at learning at a high level of complexity and specialisation. Tertiary education includes what is commonly understood as academic education but also includes advanced vocational or professional education. It comprises ISCED levels 5, 6, 7 and 8, which are labelled as short-cycle tertiary education, Bachelor's or equivalent level, Master's or equivalent level, and doctoral or equivalent level, respectively. The content of programmes at the tertiary level is more complex and advanced than in lower ISCED levels.

For more information about the International Standard Classification of Education (ISCED) 2011 please refer to: http://www.uis.unesco.org/Education/Documents/isced-2011-en.pdf

Table 5: Education at the primary, secondary and tertiary levels

The table shows the number of students enrolled as well as the gross enrolment ratio which is the number of students enrolled, regardless of age, expressed as a percentage of the eligible official school-age population corresponding to the same level of education in a given school year. Enrolment is measured at the beginning of the school or academic year. The gross enrolment ratio at each level will include all pupils whatever their ages, whereas the population is limited to the range of official school ages. Therefore, for countries with almost universal education among the school-age population, the gross enrolment ratio can exceed 100 if the actual age distribution of pupils extends beyond the official school ages.

Table 6: Public expenditure on education

Public expenditure on education consists of current and capital expenditures on education by local, regional and national governments, including municipalities. Household contributions are excluded. Current expenditure on education includes expenditure for goods and services consumed within the current year and which would need to be renewed if needed the following year. It includes expenditure on: staff salaries and benefits; contracted or purchased services; other resources including books and teaching materials; welfare services; and other current expenditure such as subsidies to students and households, furniture and equipment, minor repairs, fuel, telecommunications, travel, insurance and rents. Capital expenditure on education includes expenditure for assets that last longer than one year. It includes expenditure for construction, renovation and major repairs of buildings and the purchase of heavy equipment or vehicles.

Chapter V: Health

Table 7: Health personnel

The table shows four main categories of health personnel (out of 9 categories available in the source); Physicians which includes generalist medical practitioners and specialist medical practitioners; Nursing and midwifery personnel which includes nursing professionals, midwifery professionals, nursing associate professionals and midwifery associate professionals. Traditional midwives are not included here; Dentistry personnel includes dentists, dental assistants, dental technicians and related occupations; and Pharmaceutical personnel which includes pharmacists, pharmaceutical assistants, pharmaceutical technicians and related occupations.

The data are obtained from the World Health Organisation's (WHO) Global Health Workforce Statistics database which are compiled from several sources such as national population censuses, labour force and employment surveys, national statistical products and routine administrative information systems. As a result, considerable variability remains across countries in the coverage, quality and reference year of the original data. In general, the denominator data for health workforce density (i.e. national population estimates) were obtained from the United Nations Population Division's *World Population Prospects* publication. In some cases, the official report provided only workforce density indicators, from which estimates of the stock were then calculated.

The classification of health workers used is based on criteria for vocational education and training, regulation of health professions, and activities and tasks of jobs, i.e. a framework for categorizing key workforce variables according to shared characteristics. The WHO framework largely draws on the latest revisions to the internationally standardized classification systems of the International Labour Organization (International Standard Classification of Occupations), United Nations Educational, Scientific and Cultural Organization (International Standard Classification of Education), and the United Nations Statistics Division (International Standard Industrial Classification of All Economic Activities). Depending on the nature of each country's situation and the means of measurement, data are available for up to 9 categories of health workers in the aggregated set, and up to 18 categories in the disaggregated set. The latter essentially reflects attempts to better distinguish some subgroups of the workforce according to assumed differences in skill level and skill specialization.

Table 8: Expenditure on health

Total expenditure on health is the sum of all outlays for health maintenance, restoration or enhancement paid for in cash or supplied in kind. It is the sum of General Government Expenditure on Health and Private Expenditure on Health. General government expenditure on health is the sum of health outlays paid for in cash or supplied in kind by government entities, such as the Ministry of Health, other ministries, parastatal organizations or social security agencies (without double counting government transfers to social security and extra budgetary funds). It includes all expenditure made by these entities, regardless of the source, so includes any donor funding passing through them. It includes transfer payments to households to offset medical care costs and extra budgetary funds to finance health services and goods. It includes current and capital expenditure. More information on the definition, methodology, sources and limitations of the data can be found on the Global Health Expenditure Database (see http://apps.who.int/nha/database/DocumentationCentre/Index/fr)

Chapter VI: Crime

Table 9: Intentional homicides and other crimes

"Intentional homicides" and "other crimes" are taken from the United Nations Office on Drugs and Crime, published in their statistics database.

"Intentional Homicide" means unlawful death purposefully inflicted on a person by another person. Data on intentional homicide should also include serious assault leading to death and death as a result of a terrorist attack. It should exclude attempted homicide, manslaughter, death due to legal intervention, justifiable homicide in self-defence and death due to armed conflict.

"Assault" means physical attack against the body of another person resulting in serious bodily injury, excluding indecent/sexual assault, threats and slapping/punching. 'Assault' leading to death should also be excluded.

"Kidnapping" means unlawfully detaining a person or persons against their will (including through the use of force, threat, fraud or enticement) for the purpose of demanding for their liberation an illicit gain or any other economic gain or other material benefit, or in order to oblige someone to do or not to do something. "Kidnapping" excludes disputes over child custody.

"Theft" means depriving a person or organisation of property without force with the intent to keep it . "Theft" excludes Burglary, housebreaking, Robbery, and Theft of a Motor Vehicle, which are recorded separately.

Total "Sexual violence" means rape and sexual assault, including Sexual Offences against Children.

Chapter VII: National accounts

The National Accounts Main Aggregates Database (available at http://unstats.un.org/unsd/snaama) presents national accounts data for more than 200 countries and areas of the world. It is the basis for the publication of National Account Statistics: Analysis of Main Aggregates (AMA), a publication prepared by the Statistics Division of the Department for Economic and Social Affairs of the United Nations Secretariat with the generous co-operation of national statistical offices. The database is updated in December of each year with newly available national accounts data for all countries and areas.

The National Accounts Main Aggregates Database is based on the data obtained from the United Nations National Accounts Questionnaire (NAQ) introduced in October 1999, which in turn is based on the System of National Accounts 1993 (1993 SNA). The data are supplemented with estimates prepared by the Statistics Division. The updated SNA, called the System of National Accounts 2008 (2008 SNA) was finalised in September 2009. As of 2015, 63 countries and territories (European Union Member States, Albania, Argentina, Australia, Brazil, Brunei Darussalam, Canada, China, Hong Kong SAR, China, Macao SAR, Dominican Republic, Ecuador, India, Indonesia, Israel, Kenya, Mexico, Mongolia, New Zealand, Nigeria, Pakistan, Peru, Philippines, Republic of Korea, Serbia, Singapore, South Africa, Swaziland, Timor-Leste, Uganda, Ukraine, the United States of America and Zambia) have started submitting data according to the 2008 SNA.

Every effort has been made to present the estimates of the various countries or areas in a form designed to facilitate international comparability. To this end, important differences in concept, scope, coverage and classification have been described in the footnotes for individual countries. Such differences should be taken into account to avoid misleading comparisons. Data contained in the tables relate to the calendar year for which they are shown, except in several cases. These special cases are posted on the National Accounts Main Aggregates Database website (http://unstats.un.org/unsd/snaama/notes.asp). The figures shown are the most recent estimates and revisions available at the time of compilation. In general, figures for the most recent year are to be regarded as provisional. The sums of the components in the tables may not necessarily add up to totals shown because of rounding.

Table 10: Gross domestic product and gross domestic product per capita

This table shows gross domestic product (GDP) and GDP per capita in US dollars at current prices, GDP at constant 2005 prices and the corresponding real rates of growth. The tables are intended to facilitate international comparisons of levels of income generated in production. Official data and estimates of total and per capita GDP at current prices have been converted to US dollars, while total GDP at constant prices are converted to 2005 prices before conversion to US dollars using the 2005 exchange rates. The conversion methodology to US dollars is described in the document on the methodology for the National Accounts Main Aggregates Database http://unstats.un.org/unsd/snaama/methodology.pdf). For inter-country comparisons over time, it would be more appropriate to use the growth rate in the table based on constant price data, which are more indicative of inter-country and intra-grouping comparisons of trends in total GDP. The growth rate shown in the table is computed as geometric mean of annual growth rates expressed as percentages for the years.

Table 11: Gross value added by kind of economic activity

This table presents the shares of the components of gross value added at current prices by kind of economic activity.

Sector	Comprises of (in terms of ISIC 3):
Agriculture	Agriculture, hunting, forestry and fishing (ISIC A-B)
Industry	Mining and quarrying, Manufacturing, Electricity, gas and water supply (ISIC C-E) Construction (ISIC F)
Services	Wholesale and retail trade; repair of motor vehicles, motorcycles and personal and household goods, Hotels and restaurants (ISIC G-H) Transport, storage and communications (ISIC I) Other activities which includes financial intermediation, real estate, renting and business activities, public administration and defense; compulsory social security, education, health and social work, other community, social and personal service activities, private households with employed persons (ISIC J-P).

Chapter VIII: Finance

Detailed information and current figures relating to table 12 are contained in International Financial Statistics, published by the International Monetary Fund (see also http://elibrary-data.imf.org) and in the United Nations Monthly Bulletin of Statistics.

Table 12: Balance of payments summary

A balance of payments can be broadly described as the record of an economy's international economic transactions. It shows (a) transactions in goods, services and income between an economy and the rest of the world, (b) changes of ownership and other changes in that economy's monetary gold, special drawing rights (SDRs) and claims on and liabilities to the rest of the world, and (c) unrequited transfers and counterpart entries needed to balance in the accounting sense any entries for the foregoing transactions and changes which are not mutually offsetting. The balance of payments data are presented on the basis of the methodology and presentation of the sixth edition of the Balance of Payments Manual (BPM6), published by the International Monetary Fund in November 2013. The BPM6 incorporates several major changes to take account of developments in international trade and finance over the years, and to better harmonize the Fund's balance of payments methodology with the methodology of the 2008 System of National Accounts (SNA). The detailed definitions concerning the content of the basic categories of the balance of payments are given in the Balance of Payments Manual (sixth edition)

Brief explanatory notes are given below to clarify the scope of the major items.

Current account is a record of all transactions in the balance of payments covering the exports and imports of goods and services, payments of income, and current transfers between residents of a country and non-residents.

Capital account, n.i.e. refers mainly to capital transfers linked to the acquisition of a fixed asset other than transactions relating to debt forgiveness plus the disposal of nonproduced, nonfinancial assets, and to capital transfers linked to the disposal of fixed assets by the donor or to the financing of capital formation by the recipient, plus the acquisition of nonproduced, nonfinancial assets.

Financial account, n.i.e. is the net sum of the balance of direct investment, portfolio investment, and other investment transactions.

Reserves and related items is the sum of transactions in reserve assets, LCFARs, exceptional financing, and use of Fund credit and loans.

Chapter IX: Labour market

A comparable and comprehensive collection of data on labour force and related topics are available from the International Labour Organisation's (ILO) *Key Indicators of the Labour Market (KILM)* publication, which is updated every 2 years. More timely information is contained in the ILO's ILOSTAT data repository (see www.ilo.org/ilostat) which publishes data as it is received from the countries either on an annual, quarterly or monthly basis but does not include all the consistency checks nor include all the sources used by KILM. For various reasons, national definitions of employment and unemployment often differ from the recommended international standard definitions and thereby limit international comparability. Inter-country comparisons are also complicated by a variety of types of data collection systems used to obtain information on employed and unemployed persons. The ILOSTAT website provides a comprehensive description of the methodology underlying the labour series.

Table 13: Labour force and unemployment

Labour force participation rate is calculated by expressing the number of persons in the labour force as a percentage of the working-age population. The labour force is the sum of the number of persons employed and the number of unemployed (see ILO's current International Recommendations on Labour Statistics). The working-age population is the population above a certain age, prescribed for the measurement of economic characteristics. The data refer to the age group of 15 years and over and are based on ILO's modelled estimates, unless otherwise stated in a footnote.

Unemployment" is defined to include persons above a certain age who, during a specified period of time were:

(a)   "Without work", i.e. were not in paid employment or self-employment;

(b)   "Currently available for work", i.e. were available for paid employment or self-employment during the reference period; and

(c)   "Seeking work", i.e. had taken specific steps in a specified period to find paid employment or self-employment

Persons not considered to be unemployed include:

(a)   Persons intending to establish their own business or farm, but who had not yet arranged to do so and who were not seeking work for pay or profit;

(b)   Former unpaid family workers not at work and not seeking work for pay or profit.

The series generally represent the total number of persons wholly unemployed or temporarily laid-off. Percentage figures, where given, are calculated by comparing the number of unemployed to the total members of that group of the labour force on which the unemployment data are based.

Table 14: Employment by economic activity

The employment table presents the percentage distribution of employed persons by economic activity, according to International Standard Industry Classification (ISIC) version 4.

Chapter X: Price and production indices

Table 15: Consumer price indices

A consumer price index is usually estimated as a series of summary measures of the period-to-period proportional change in the prices of a fixed set of consumer goods and services of constant quantity and characteristics, acquired, used or paid for by the reference population. Each summary measure is constructed as a weighted average of a large number of elementary aggregate indices. Each of the elementary aggregate indices is estimated using a sample of prices for a defined set of goods and services obtained in, or by residents of, a specific region from a given set of outlets or other sources of consumption goods and services. The table presents the general consumer price index for all groups of consumption items combined, and the food index including non-alcoholic beverages only. Where alcoholic beverages and/or tobacco are included, this is indicated in footnotes.

Table 16: Agricultural production

"Agriculture" relates to the production of all crops and livestock products. The "Food Index" includes those commodities which are considered edible and contain nutrients. The index numbers of agricultural output and food production are calculated by the Laspeyres formula with the base year period 2004-2006. The latter is provided in order to diminish the impact of annual fluctuations in agricultural output during base years on the indices for the period. Production quantities of each commodity are weighted by 2004-2006 average national producer prices and summed for each year. The index numbers are based on production data for a calendar year. These may differ in some instances from those actually produced and published by the individual countries themselves due to variations in concepts, coverage, weights and methods of calculation. Efforts have been made to estimate these methodological differences to achieve a better international comparability of data. Detailed data on agricultural production are published by FAO in its *Statistical Yearbook*.

Chapter XI: International merchandise trade

The *International Trade Statistics Yearbook* (ITSY) provides an overview of the latest trends of trade in goods and services of most countries and areas in the world, a publication prepared by the Statistics Division of the Department for Economic and Social Affairs of the United Nations Secretariat. The yearbook, see http://comtrade.un.org/pb/, is released in two volumes; Volume I is compiled earlier in the year to present an advanced overview of international merchandise trade from the previous year, Volume II, generally released six months later, contains detailed tables showing international trade for individual commodities and 11 world trade tables covering trade values and indices. Volume II also contains updated versions of world trade tables. The table in this yearbook are also updated monthly in the United Nations Monthly Bulletin of Statistics and on the trade statistics website, see http://unstats.un.org/unsd/trade/data/tables.asp#annual.

The statistics in this Yearbook have been compiled by national statistical authorities largely consistent with the United Nations recommended International Merchandise Trade Statistics, Concepts and Definitions 2010 (IMTS 2010). Depending on what parts of the economic territory are included in the statistical territory, the trade data-compilation system adopted by a country (its trade system) may be referred to as general or special.

General trade system	The statistical territory coincides with the economic territory. Consequently, it is recommended that the statistical territory of a country applying the general trade system comprises all applicable territorial elements. In this case, imports include goods entering the free circulation area, premises for inward processing, industrial free zones, premises for customs warehousing or commercial free zones and exports include goods leaving those territorial elements
Special trade system	(strict definition) The statistical territory comprises only a particular part of the economic territory, so that certain flows of goods which are in the scope of IMTS 2010 are not included in either import or export statistics of the compiling country. The strict definition of the special trade system is in use when the statistical territory comprises only the free circulation area, that is, the part within which goods "may be disposed of without customs restriction". Consequently, in such a case, imports include only goods entering the free circulation area of a compiling country and exports include only goods leaving the free circulation area of a compiling country
	(relaxed definition) (a) goods that enter a country for, or leave it after, inward processing, as well as (b) goods that enter or leave an industrial free zone, are also recorded and included in international merchandise trade statistics

Generally, all countries report their detailed merchandise trade data according to the Harmonized Commodity Description and Coding System (HS) and the data correspond and are then presented by Standard International Trade Classifications (SITC, Rev.3). Data refer to calendar years; however, for those countries which report according to some other reference year, the data are presented in the year which covers the majority of the reference year used by the country.

FOB-type values include the transaction value of the goods and the value of services performed to deliver goods to the border of the exporting country. CIF-type values include the transaction value of the goods, the value of services performed to deliver goods to the border of the exporting country and the value of the services performed to deliver the goods from the border of the exporting country to the border of the importing country. Therefore, data for the statistical value of imported goods are presented as a CIF-type value and the statistical value of exported goods as an FOB-type value.

Conversion of values from national currencies into United States dollars is done by means of currency conversion factors based on official exchange rates. Values in currencies subject to fluctuation are converted into United States dollars using weighted average exchange rates specially calculated for this purpose. The weighted average exchange rate for a given currency for a given year is the component monthly factors, furnished by the International Monetary Fund in its International Financial Statistics publication, weighted by the value of the relevant trade in each month; a monthly factor is the exchange rate (or the simple average rate) in effect during that month. These factors are applied to total imports and exports and to the trade in individual commodities with individual countries.

Table 17: Total imports, exports and balance of trade

Figures on the total imports and exports of countries (or areas) presented in this table are mainly taken from International Financial Statistics published monthly by the International Monetary Fund (IMF) but also from other sources such as national publications and websites and the United Nations *Monthly Bulletin of Statistics* Questionnaire, see the *International Trade Statistics Yearbook* for further details. Estimates for missing data are made in order to arrive to regional totals but are otherwise not shown. The estimation process is automated using quarterly year-on-year growth rates for the extrapolation of missing quarterly data (unless quarterly data can be estimated using available monthly data within the quarter). The conversion factors applied to data in this table are published quarterly in the United Nations *Monthly Bulletin of Statistics* and are also available on the United Nations trade statistics website: http://unstats.un.org/unsd/trade/data/tables.asp#annual.

Chapter XII: Energy

The Energy Statistics Yearbook (available at http://unstats.un.org/unsd/energy/yearbook) is a comprehensive collection of international energy statistics for over 220 countries and areas. The yearbook is prepared by the Statistics Division of the Department for Economic and Social Affairs of the United Nations Secretariat. The yearbook is produced every year with newly available data on energy production, trade, stock changes, bunkers and consumption for all countries and areas, and a historical series back to 1950 are available. The data are compiled primarily from the annual energy questionnaire distributed by the United Nations Statistics Division and supplemented by official national statistical publications, as well as publications from international and regional organizations. Where official data are not available or are inconsistent, estimates are made by the Statistics Division based on governmental, professional or commercial materials.

The period to which the data refer is the calendar year, with the exception of the data of the following countries which refer to the fiscal year: Afghanistan and Iran (Islamic Rep. of) – beginning 21 March of the year stated; Australia, Bangladesh, Bhutan, Egypt (for the latter two, electricity only), Nepal - ending June of the year stated; Pakistan - starting July of the year stated; India, Myanmar and New Zealand – beginning April of the year stated. Data on a per capita basis use population data from the United Nations Population Division as a denominator.

Table 18: Production, trade and supply of energy

Data are presented in petajoules (gigajoules per capita), to which the individual energy commodities are converted in the interests of international uniformity and comparability. To convert from original units to joules, the data in original units (metric tons, metric tons of oil equivalent, kilowatt hours, cubic metres) are multiplied by conversion factors. For a list of the relevant conversion factors and a detailed description of methods, see the Energy Statistics Yearbook.

Included in the production of commercial primary energy for solids are hard coal, lignite, peat and oil shale; liquids are comprised of crude petroleum, natural gas liquids, other hydrocarbons, additives and oxygenates, and liquid biofuels; gas comprises natural gas and primary steam/heat; and electricity is comprised of primary electricity generation from hydro, nuclear, geothermal, wind, tide, wave and solar sources.

Net imports (imports less exports and bunkers) and changes in stocks, refer to all primary and secondary forms of energy (including feedstocks). Within net imports; bunkers refer to bunkers of aviation gasoline, jet fuel and of hard coal, gasdiesel oil and residual fuel oil. International trade of energy commodities is based on the "general trade" system, that is, all goods entering and leaving the national boundary of a country are recorded as imports and exports.

Included in the consumption of energy are primary forms of solid fuels, net imports and changes in stocks of secondary fuels; liquids which is energy use of oil products includes feedstocks and refinery gas, and direct use of crude petroleum; gases include the consumption of natural gas and primary heat, net imports and changes in stocks of manufactured gases; and electricity which is primary electricity production and net imports of electricity. Consumption for some of the petroleum products is negative due to the exclusion of inter-product transfers from the calculations. Negative consumption of electricity is due to negligible primary electricity production as compared to net exports. More generally, negative consumption can represent a residual or statistical difference between production and exports when a particular product is mainly exported.

Chapter XIII: Environment

Table 19: Threatened species

Data on the number of threatened species in each group of animals and plants are compiled by the World Conservation Union (IUCN)/ Species Survival Commission (SSC) and published in the IUCN Red List of Threatened Species. The list provides a catalogue of those species that are considered globally threatened. The number of threatened species for any particular country will change between years for a number of reasons, including:

- New information being available to refine the assessment (e.g., confirmation that the species occurs or does not occur in a particular country, confirmation that the species is or is not threatened, etc.)

- Taxonomic changes (e.g., what was previously recognised as one species is now split into several separate species, or has now been merged with another species).

- Corrections (e.g., the previous assessment may have missed a particular country out of its country occurrence list or included a specific country by mistake).

- Genuine status changes (e.g., a species may have genuinely deteriorated or improved in status and therefore has moved into or out of the threatened categories).

The categories used in the Red List are as follows; extinct, extinct in the wild, critically endangered; endangered, vulnerable, near threatened and data deficient.

Table 20: CO_2 emissions estimates

The source of the data presented on the emissions of carbon dioxide (CO2) is the Carbon Dioxide Information Analysis Centre (CDIAC) of the Oak Ridge National Laboratory in the USA, see http://cdiac.ornl.gov/. The CDIAC estimates of CO2 emissions are derived primarily from United Nations energy statistics on the consumption of liquid and solid fuels and gas consumption and flaring, and from cement production estimates from the Bureau of Mines of the U.S. Department of Interior. The emissions presented in the table are in units of 1,000 metric tons of CO_2; to convert CO_2 into carbon, divide the data by 3.667. Full details of the procedures for calculating emissions are given in Global, Regional, and National Annual CO_2 Emissions Estimates from Fossil Fuel Burning, Hydraulic Cement Production, and Gas Flaring and on the CDIAC web site. Relative to other industrial sources for which CO2 emissions are estimated, statistics on gas flaring activities are sparse and sporadic. In countries where gas flaring activities account for a considerable proportion of the total CO_2 emissions, the sporadic nature of gas flaring statistics may produce spurious or misleading trends in national CO_2 emissions over the period covered by the table.

Chapter XIV: Science and technology

Table 21: Patents

A patent is granted by a national patent office or by a regional office that does the work for a number of countries, such as the European Patent Office and the African Regional Intellectual Property Organisation. Under such regional systems, an applicant requests protection for the invention in one or more countries, and each country decides as to whether to offer patent protection within its borders. The World Intellectual Property Organisation (WIPO) administered Patent Cooperation Treaty (PCT) provides for the filling of a single international patent application which has the same effect as national applications filed in the designated countries. Data include resident intensity, patents granted and patents in force. Patent intensity is presented as the resident patent fillings per million population, where as resident Intellectual Property (IP) filling refers to an application filed by an applicant at its national IP office. IP grant (registration) data are based on the same concept. In force refers to a patent or other form of IP protection that is currently valid. Country of origin is used to catgorise IP data by resident (domestic) and non-resident (foreign). The residence of the first-named applicant (or inventor) recorded in the IP document (e.g. patent or trademark application) is used to classify IP data by country of origin. The data are compiled and published by the WIPO.

Chapter XV: Communication

The statistics included in Tables 22 and 23 were obtained from the statistics database (see www.itu.int) and the *Yearbook of Statistics*, Telecommunication Services of the International Telecommunication Union.

Table 22: Cellular mobile telephone subscriptions

The number of mobile cellular telephone subscriptions (as well as the number of subscriptions per 100 inhabitants) refers to portable telephones subscribing to an automatic public mobile telephone service using cellular technology, which provides access to the Public Switched Telephone Network (PSTN). Users of both post-paid subscriptions and pre-paid accounts are included. The number of subscriptions per 100 inhabitants is calculated by dividing the number of subscriptions by the population and multiplying by 100.

Table 23: Internet usage

The table shows percentage of individuals using the internet and replaces the statistics shown in previous yearbooks such as the "Number (thousands) of fixed (wired) internet subscriptions" and "fixed (wired) internet subscriptions per 100 inhabitants". Besides capturing the use of the Internet, this indicator is able to measure changes in Internet access and use. In countries where many people access the Internet at work, at school, at cybercafés or other public locations, increases in public access serve to increase the number of users despite limited numbers of Internet subscriptions and of households with Internet access. Developing countries especially tend to have many Internet users per Internet subscriptions, reflecting that home access is not the primary location of access.

Chapter XVI: International tourism and transport

The data on international tourism have been supplied by the World Tourism Organization (UNWTO) from detailed tourism information published in either the *Yearbook of Tourism Statistics* or *Compendium of Tourism Statistics,* see www.unwto.org/statistics for further information. For statistical purposes, the term "international visitor" describes "any person who travels to a country other than that in which he/she has his/her usual residence but outside his/her usual environment for a period not exceeding 12 months and whose main purpose of visit is other than the exercise of an activity remunerated from within the country visited". There are four series presented in the UNWTO *yearbook* and *compendium*, but only one series is selected to be presented in this yearbook, generally based on the following priority order to best describe an "international visitor";

Order	Series code	Series name
1	TF	*Arrivals of non-resident tourists at national borders* are visitors who stay at least one night in a collective or private accommodation in the country visited (excludes same-day visitors)
2	VF	*Arrivals of non-resident visitors at national borders* are visitors as defined in series "TF" as well as same-day visitors who do not spend the night in a collective or private accommodation in the country visited
3	TCE	*Arrivals of non-resident tourists in all types of accommodation establishments*
4	THS	*Arrivals of non-resident tourists in hotels and similar establishments*

The figures do not include immigrants, residents in a frontier zone, persons domiciled in one country or area and working in an adjoining country or area, members of the armed forces and diplomats and consular representatives when they travel from their country of origin to the country in which they are stationed and vice-versa. The figures also exclude persons in transit who do not formally enter the country through passport control, such as air transit passengers who remain for a short period in a designated area of the air terminal or ship passengers who are not permitted to disembark. This category includes passengers transferred directly between airports or other terminals. Other passengers in transit through a country are classified as visitors.

Table 24: Tourist/visitor arrivals and tourism expenditure

Data on arrivals of non-resident (or international) visitors may be obtained from different sources. In some cases data are obtained from border statistics derived from administrative records (police, immigration, traffic counts and other types of controls), border surveys and registrations at accommodation establishments. Totals correspond to the total number of arrivals from the regions indicated in the table. When a person visits the same country several times a year, an equal number of arrivals is recorded. Likewise, if a person visits several countries during the course of a single trip, his/her arrival in each country is recorded separately. Consequently, arrivals cannot be assumed to be equal to the number of persons travelling.

Expenditure associated with tourism activity of visitors has been traditionally identified with the travel item of the Balance of Payments (BOP): in the case of inbound tourism, those expenditures in the country of reference associated with non-resident visitors are registered as "credits" in the BOP and refer to "travel receipts". The new conceptual framework approved by the United Nations Statistical Commission in relation to the measurement of tourism macroeconomic activity (the so-called Tourism Satellite Account) considers that "tourism industries and products" includes transport of passengers. Consequently, a better estimate of tourism-related expenditures by resident and non-resident visitors in an international scenario would be, in terms of the BOP, the value of the travel item plus that of the passenger transport item. Nevertheless, users should be aware that BOP estimates include, in addition to expenditures associated with visitors, those related to other types of individuals. The data published should allow international comparability and therefore correspond to those published by the International Monetary Fund and provided by the Central Banks, any exceptions are listed within the *Compendium of Tourism Statistics* and the *Yearbook of Tourism Statistics,* see www.unwto.org/statistics for further information.

Chapter XVII: Development assistance

Table 25: Net disbursements of official development assistance to recipients

The table presents estimates of flows of financial resources to individual recipients either directly (bilaterally) or through multilateral institutions (multilaterally). The multilateral institutions include the World Bank Group, regional banks, financial institutions of the European Union and a number of United Nations institutions, programmes and trust funds. The source of data is the Development Assistance Committee (DAC) of OECD to which member countries reported data on their flow of resources to developing countries and territories, countries and territories in transition, and multilateral institutions. Additional information on definitions, methods and sources can be found in OECD's *Geographical Distribution of Financial Flows to Developing Countries* publication, also see http://stats.oecd.org/ for further information.

Table 26: Net disbursements of official development assistance from donors

The table presents the development assistance expenditures of donor countries. This table includes donors' contributions to multilateral agencies; therefore, the overall totals differ from those in table 25, which include disbursements by multilateral agencies.

***

Annexe II: Notes techniques

Chapitre I : Aperçu mondial

Tableau 1 : Statistiques mondiales – séries principales

Ces séries d'agrégats mondiaux sont obtenues à partir d'autres tableaux figurant dans le présent *Annuaire*, lorsque c'est possible, et sont établies à partir de publications et de bases de données statistiques des Nations Unies et des organismes spécialisés ainsi que d'autres institutions. Pour davantage d'information sur les définitions, les sources, les méthodes de compilation et la couverture des données, il convient de se référer aux notes techniques du tableau correspondant dans le présent *Annuaire*.

Chapitre II : Population et migration

Tableau 2 : Population, superficie et densité

Les données concernant les estimations démographiques de milieu d'année et de la densité proviennent des estimations et des projections préparées par la Division de la population de l'Organisation des Nations Unies, qui sont publiées dans *Perspectives de la population mondiale : Révision de 2015*. Les données concernant la superficie sont extraites de l'*Annuaire démographique*, qui ne contient que les données nationales officielles.

Les estimations démographiques de milieu d'année correspondent à la population de facto d'un pays, d'une zone ou d'une région au 1er juillet de l'année indiquée. Les chiffres sont donnés en millions.

La densité est exprimée en nombre d'habitants par kilomètre carré.

Les estimations de la superficie comprennent les eaux intérieures.

Tableau 3 : Migrants internationaux et réfugiés

Le stock international de migrants provient d'estimations et de projections préparées par la Division de la population de l'Organisation des Nations Unies, publiées dans *International migrant stock: The 2015 Revision*. Les données sur les « réfugiés et autres personnes relevant de la compétence du HCR » ont été obtenues auprès du Haut-Commissaire des Nations Unies pour les réfugiés, et sont publiées dans la base de données de statistiques démographiques.

Le stock international de migrants représente le nombre de personnes nées dans un autre pays que celui dans lequel elles vivent. Lorsque les informations concernant le pays d'origine font défaut, on a utilisé les données sur le nombre de personnes de nationalité étrangère. Et à défaut de données empiriques, on a eu recours à des estimations. Les données font référence au milieu de l'année 2015. Le stock international de migrants en pourcentage de la population est obtenu en divisant le stock international de migrants estimé par la population totale estimée et en multipliant le résultat par 100.

Les réfugiés sont les personnes reconnues comme telles au sens de la Convention de 1951 relative au statut des réfugiés, de son protocole de 1967 ou de la Convention de l'OUA de 1969 régissant les aspects propres aux problèmes des réfugiés en Afrique; celles reconnues comme réfugiés conformément au Statut du HCR; les personnes qui bénéficient d'une forme de protection complémentaire ou jouissent d'une protection temporaire. Depuis 2007, la population des réfugiés inclut également les personnes dont la situation est assimilable à celle des réfugiés.

Les demandeurs d'asile sont des personnes qui ont déposé une demande de protection internationale et qui n'ont pas encore obtenu le statut de réfugié, quelle que soit la date à laquelle la demande a été présentée.

Les « autres personnes relevant de la compétence du HCR » se composent des cinq catégories suivantes :

Les personnes déplacées sont des personnes ou groupes de personnes qui ont été forcés ou contraints de fuir ou de quitter leurs foyers ou leur lieu de résidence habituel, notamment en raison d'un conflit armé, de situations de violence généralisée, de violations des droits de l'homme ou de catastrophes naturelles ou provoquées par l'homme, ou pour en éviter les effets, et qui n'ont pas franchi les frontières internationalement reconnues d'un État. Aux fins des statistiques du HCR, cette population ne comprend que les personnes déplacées en raison d'un conflit qui bénéficient de la protection et/ou de l'assistance du HCR. Depuis 2007, la population de personnes déplacées inclut également les personnes dont la situation est assimilable à celle des personnes déplacées. On trouvera des estimations de la population mondiale des personnes déplacées sur le site www.internal-displacement.org.

Les réfugiés de retour sont d'anciens réfugiés rentrés dans leur pays d'origine, soit spontanément, soit de façon organisée, mais qui ne sont pas encore pleinement intégrés. Ces retours ne se font normalement que lorsque leur sécurité et leur dignité peuvent être garanties.

Les déplacés de retour sont des personnes déplacée qui bénéficiaient des activités de protection et d'assistance du HCR et qui sont revenues à leur lieu d'origine ou de résidence habituel au cours de l'année.

Les apatrides sont définis par le droit international comme des personnes qu'aucun État ne considère comme ses ressortissants par application de sa législation. En d'autres termes, les apatrides ne possèdent la nationalité d'aucun État. Les statistiques du HCR incluent les personnes qui relèvent de sa compétence en vertu de cette définition, mais les données de certains pays peuvent aussi inclure des personnes de nationalité indéterminée.

Les autres personnes relevant de la compétence du HCR sont des personnes qui ne relèvent pas directement d'une des catégories ci-dessus, mais auxquelles le HCR assure protection et/ou assistance pour des raisons humanitaires ou d'autre motifs particuliers.

Chapitre III : La situation des femmes

Tableau 4 : Proportion de sièges occupés par des femmes au parlement national

Ce tableau indique le pourcentage des sièges des chambres uniques ou basses des parlements nationaux occupés par des femmes, en janvier ou février de chaque année (voir les notes du tableau pour plus de détails). Les parlements nationaux peuvent être bicaméraux ou unicaméraux. Ce tableau porte sur la chambre unique des parlements unicaméraux et sur la chambre basse des parlements bicaméraux. Il ne porte pas sur la chambre haute des parlements bicaméraux. Les sièges sont habituellement attribués aux membres à l'issue d'élections parlementaires générales. Certains sièges peuvent aussi être pourvus à l'issue de nominations, de désignations, d'élections indirectes, de roulement des membres et d'élections partielles. La proportion d'élues est obtenue en divisant le nombre total de sièges occupés par des femmes par le nombre total de sièges que compte le parlement. Les statistiques ne sont ni pondérées ni normalisées. La source de ce tableau est l'Union interparlementaire, voir www.ipu.org pour plus d'informations

Chapitre IV : Éducation

Les données des tableaux 5 et 6 sont présentées, conformément à la Classification internationale type de l'éducation de l'UNESCO (CITE, révision de 2011). Les données sont présentées dans les tableaux sur la base des trois principaux niveaux d'éducation, présentés comme suit :

L'enseignement primaire (CITE niveau 1) désigne les programmes éducatifs habituellement conçus pour apporter aux élèves les compétences fondamentales en lecture, écriture et en mathématiques (c'est-à-dire en calcul) afin d'établir une base solide pour la compréhension et l'apprentissage des principaux domaines de la connaissance et favoriser le développement personnel et social dans le but de les préparer à l'entrée dans le premier cycle de l'enseignement secondaire. Il privilégie l'enseignement à un niveau de complexité élémentaire avec peu ou pas de spécialisation.

L'enseignement secondaire (niveaux 2 et 3 de la CITE) est divisé en deux parties : le premier et le second cycles du secondaire. Les programmes du premier cycle de l'enseignement secondaire sont généralement conçus de manière à renforcer les acquis scolaires du primaire. L'objectif consiste habituellement à établir la base d'un apprentissage et d'un développement humain valables pour toute la vie et que les systèmes éducatifs pourront ensuite enrichir par de nouvelles possibilités d'éducation. Les programmes du deuxième cycle de l'enseignement secondaire visent en général à achever l'enseignement secondaire et à préparer à l'entrée dans l'enseignement supérieur et/ou à enseigner des compétences utiles à l'exercice d'un emploi. Les programmes de ce niveau offrent aux élèves un enseignement plus varié, spécialisé et approfondi que les programmes du premier cycle. Ils sont davantage différenciés et proposent un éventail plus large d'options et de filières. Les enseignants sont souvent hautement qualifiés dans les matières ou domaines spécialisés qu'ils enseignent, en particulier dans les classes supérieures.

L'enseignement supérieur (niveaux 5 à 8 de la CITE) se fonde sur les acquis de l'enseignement secondaire et offre des activités d'apprentissage dans des domaines d'éducation spécialisés. Il vise à transmettre des connaissances très spécialisées et d'un niveau de complexité élevé. L'enseignement supérieur inclut ce que l'on qualifie habituellement d'enseignement académique mais il comprend aussi l'enseignement technique ou professionnel avancé. Il comprend les niveaux 5, 6, 7 et 8 de la CITE, appelés respectivement enseignement supérieur de cycle court, enseignement du niveau de la licence ou équivalent, niveau master ou équivalent, et niveau doctorat ou équivalent. Le contenu des programmes de l'enseignement supérieur est plus complexe et plus avancé que celui des niveaux inférieurs de la CITE.

On trouvera davantage d'informations sur la Classification internationale type de l'éducation (CITE) 2011 sur le site : http://www.uis.unesco.org/Education/Documents/isced-2011-fr.pdf.

Tableau 5 : Enseignement primaire, secondaire et supérieur

Le tableau montre le nombre d'élèves scolarisés ainsi que le taux de scolarisation brut qui est le nombre d'étudiants inscrits, quel que soit leur âge, exprimé en pourcentage de la population d'âge scolaire officiellement admissible correspondant au même niveau d'enseignement pour une année scolaire donnée. Les inscriptions sont mesurées au début de l'année scolaire ou universitaire. Le taux de scolarisation brut pour chaque niveau comprend tous les élèves, quel que soit leur âge, tandis que la population générale considérée ne comprend que ceux dont l'âge correspond à l'âge scolaire officiel. De ce fait, pour les pays dont la population d'âge scolaire est quasi totalement scolarisée, le taux de scolarisation brut peut dépasser 100 si la répartition des âges effectifs des élèves s'étend au-delà des âges scolaires officiels.

Tableau 6 : Dépenses publiques afférentes à l'éducation

Les dépenses publiques afférentes à l'éducation consistent en dépenses courantes et dépenses en capital engagées par l'administration aux niveaux local, régional et national, y compris les municipalités. Les contributions des ménages sont exclues. Les dépenses d'éducation courantes comprennent les dépenses en biens et en services consommés dans l'année en cours et qui devront être renouvelées au besoin l'année suivante. Elles comprennent les dépenses au titre des salaires et avantages du personnel; des services achetés ou obtenus par contrat; d'autres ressources, notamment de manuels et autres matériels d'enseignement; des services sociaux; et d'autres dépenses courantes telles que les subventions aux étudiants et aux ménages, l'ameublement et le matériel, les petites réparations, le combustible, les télécommunications, les voyages, l'assurance et les loyers. Les dépenses en capital pour l'éducation consistent en achats de biens dont la durée dépasse une année. Elles comprennent les dépenses de construction, de rénovation et de grosses réparations de bâtiments ainsi que l'achat de matériel lourd et de véhicules.

Chapitre V : Santé

Tableau 7 : Le personnel de santé

Le tableau présente quatre grandes catégories de personnel de santé (sur les 9 catégories disponibles à la source) : la catégorie des médecins, qui comprend les médecins généralistes et les spécialistes; celle des infirmiers et sages-femmes comprend les infirmiers et sages-femmes qualifiés, les infirmiers auxiliaires professionnels et les sages-femmes auxiliaires professionnelles. Les accoucheuses traditionnelles ne sont pas incluses; le personnel de dentisterie comprend les dentistes, les assistants dentaires, les techniciens dentaires et les professions associées; et le personnel du secteur pharmaceutique comprend les pharmaciens, les pharmaciens assistants, les préparateurs en pharmacie et les professions associées.

Les données sont extraites des statistiques mondiales des personnels de santé de la base de données de l'Organisation mondiale de la Santé (OMS), qui sont établies à partir de plusieurs sources comme les recensements de population nationaux, les enquêtes sur la population active et l'emploi, les productions statistiques nationales et les données régulières des administrations. Il existe de ce fait d'un pays à l'autre une variabilité considérable dans la couverture, la qualité et l'année de référence des données. Généralement, les données du dénominateur pour le calcul de la densité du personnel de santé (c'est-à-dire les estimations de la population nationale) sont extraites des *Perspectives de la population*

mondiale publiées par la Division de la population de l'Organisation des Nations Unies. Parfois le rapport officiel ne fournit que les indicateurs de la densité du personnel de santé, à partir desquels le stock est ensuite estimé.

La classification des personnels du secteur de la santé utilisée repose sur les critères de l'enseignement et formation techniques et professionnels, la réglementation des professions de santé et les activités et tâches des postes, c'est-à-dire sur un cadre de catégorisation des principales variables des personnels selon des caractéristiques communes. Le cadre de l'OMS repose en grande partie sur les révisions les plus récentes des systèmes de classification internationaux normalisés de l'Organisation internationale du Travail (Classification internationale type des professions), de l'Organisation des Nations Unies pour l'éducation, la science et la culture (Classification internationale type de l'éducation) et de la Division de statistique de l'ONU (Classification internationale type, par industrie, de toutes les branches d'activité économique). En fonction de la situation propre à chaque pays et des moyens de mesure, les données disponibles dans l'ensemble agrégé décrivent jusqu'à 9 catégories de personnels de santé, et jusqu'à 18 catégories dans l'ensemble désagrégé. Ce dernier reflète essentiellement une tentative de mieux distinguer certains sous-groupes des personnels du secteur de la santé en fonction de différences supposées dans les niveaux de compétence et de spécialisation.

Tableau 8 : Dépenses de santé

Le montant total des dépenses de santé représente la somme de tous les frais encourus pour maintenir, rétablir ou améliorer l'état de santé, qu'il s'agisse de versements en espèces ou de services fournis en nature. C'est la somme des dépenses de santé des administrations publiques et des dépenses de santé privées. Les dépenses de santé des administrations publiques sont la somme des frais de santé engagés par des organismes publics, comme le Ministère de la santé, d'autres ministères, des organismes parapublics ou les caisses de sécurité sociale (sans double comptabilisation des transferts publics aux administrations de sécurité sociale et aux fonds extrabudgétaires), qu'il s'agisse de versements en espèces ou de services fournis en nature. Elle comprend toutes les dépenses effectuées par ces organismes, quelle qu'en soit la source, y compris donc le financement des bailleurs de fonds canalisé par eux. Elle inclut les paiements de transfert aux ménages en compensation du coût des soins médicaux et le financement des services et produits de santé par des fonds extrabudgétaires. Elle comprend les dépenses courantes et en capital. On trouvera davantage d'informations sur la définition, la méthodologie, les sources et les limitations des données sur le site de la base de données mondiale des dépenses de santé (voir http://apps.who.int/nha/database/DocumentationCentre/Index/fr).

Chapitre VI : Criminalité

Tableau 9 : Homicides intentionnels et autres crimes

Les données des « homicides intentionnels » et des « autres crimes » proviennent des données recueillies dans la base de données statistiques de l'Office des Nations Unies contre la drogue et le crime.

L'homicide intentionnel est défini comme la mort illégale d'une personne causée par une autre ayant l'intention de tuer ou de blesser gravement. Les données de l'homicide intentionnel doivent aussi inclure les violences suivies de mort et la mort résultant d'une attaque terroriste. Elles excluent la tentative d'homicide intentionnel, l'homicide involontaire, la mort causée par une intervention légale, l'homicide justifiable en état de légitime défense et la mort causée par un conflit armé.

L'agression est une atteinte à l'intégrité physique d'une autre personne entraînant des dommages corporels graves, à l'exclusion des actes préjudiciables à caractère sexuel, des menaces et des gifles/coups de poing. Les agressions graves ayant entraîné la mort sont également exclues.

L'enlèvement désigne la détention et soustraction illégales d'une ou de plusieurs personnes contre leur volonté (y compris par le recours à la force, aux menaces, à la fraude ou à l'incitation) aux fins d'exiger pour leur libération un gain illicite ou un autre avantage économique ou matériel, ou pour contraindre une personne à suivre ou à ne pas suivre une ligne de conduite. L'enlèvement exclut les différends relatifs à la garde d'un enfant.

Le vol est l'appropriation ou l'obtention illégale d'un bien dans l'intention d'en priver une personne ou une organisation de manière permanente sans son consentement et sans recours à la force. Le vol exclut le cambriolage, l'entrée avec effraction, le vol qualifié et le vol de véhicule motorisé, qui sont comptabilisés séparément.

La violence sexuelle désigne le viol et l'agression sexuelle, y compris les agressions sexuelles contre les enfants.

Chapitre VII : Comptes nationaux

La base de données des principaux agrégats des comptes nationaux (consultable sur le site http://unstats.un.org/unsd/snaama) présente les données des comptes nationaux de plus de 200 pays et régions du monde. Elle constitue la base de l'analyse des principaux agrégats des statistiques de la comptabilité nationale (*National Account Statistics: Analysis of Main Aggregates*), une publication préparée par la Division de statistique du Département des affaires économiques et sociales du Secrétariat des Nations Unies avec le généreux concours des offices nationaux de la statistique. La base de données est mise à jour chaque année au mois de décembre avec les données nouvellement disponibles des comptes nationaux de tous les pays et régions.

La base de données des principaux agrégats des comptes nationaux repose sur les données extraites du Questionnaire sur la comptabilité nationale des Nations Unies (NAQ) introduit en octobre 1999, qui lui-même repose sur le Système de comptes nationaux 1993 (SCN 1993). Les données sont complétées par des estimations préparées par la Division de statistique. Le SCN actualisé, appelé Système de comptes nationaux 2008 (SCN 2008) a été finalisé en septembre 2009. En 2015, 63 pays et territoires (États membres de l'Union européenne, Albanie, Argentine, Australie, Brésil, Brunéi Darussalam, Canada, Chine, RAS de Hong Kong, Chine, RAS de Macao, République dominicaine, Équateur, Inde, Indonésie, Israël, Kenya, Mexique, Mongolie, Nouvelle-Zélande, Nigéria, Pakistan, Pérou, Philippines, République de Corée, Serbie, Singapour,

Afrique du Sud, Swaziland, Timor-Leste, Ouganda, Ukraine, États-Unis d'Amérique et Zambie) ont commencé de soumettre des données conformément au SCN 2008.

Tout est mis en œuvre pour présenter les estimations des divers pays ou régions sous une forme conçue pour faciliter la comparabilité internationale. À cette fin, les différences importantes entre les concepts, la portée, la couverture et la classification sont décrites dans les notes de chaque pays. Ces différences doivent être prises en compte afin d'éviter les comparaisons fallacieuses. Les données contenues dans les tableaux se rapportent à l'année civile pour laquelle elles sont présentées, sauf exceptions. Ces exceptions sont affichées sur le site Web de la base de données des principaux agrégats des comptes nationaux (http://unstats.un.org/unsd/snaama/notes.asp). Les chiffres présentés sont les estimations et révisions les plus récentes disponibles au moment de leur établissement. En général, les chiffres de l'année la plus récente doivent être considérés comme provisoires. Les sommes des composantes des tableaux ne correspondent pas nécessairement aux totaux indiqués en raison des arrondis.

Tableau 10 : Produit intérieur brut et produit intérieur brut par habitant

Ce tableau présente le produit intérieur brut (PIB) et le PIB par habitant en dollars des États-Unis aux prix courants, le PIB à prix constants de 2005 et les taux de croissance réels correspondants. Les tableaux sont destinés à faciliter les comparaisons internationales des niveaux de revenu générés par la production. Les données et les estimations officielles du PIB total et du PIB par habitant aux prix courants sont converties en dollars des États-Unis, tandis que celles du PIB total à prix constants sont converties aux prix de 2005 avant conversion en dollars aux taux de change en vigueur en 2005. La méthode de conversion en dollars des États-Unis est décrite dans le document sur la méthodologie de la base de données des principaux agrégats des comptes nationaux (http://unstats.un.org/unsd/snaama/methodology.pdf). Pour les comparaisons entre pays sur la durée, il est plus approprié d'utiliser les taux de croissance à prix constants du tableau, qui représentent mieux les tendances du PIB total dans les comparaisons entre pays et entre groupes de pays. Le taux de croissance indiqué dans le tableau est calculé comme la moyenne géométrique des taux de croissance annuelle exprimés en pourcentages pour les années.

Tableau 11 : Valeur ajoutée brute par type d'activité économique

Ce tableau présente les parts des composantes de la valeur ajoutée brute aux prix courants par type d'activité économique.

Secteur	Composé de (selon la nomenclature CITI 3)
Agriculture	Agriculture, chasse, sylviculture et pêches (CITI A-B)
Industrie	Activités extractives, activités de fabrication, production et distribution d'électricité, de gaz et d'eau (CITI C-E) Construction (CITI F)
Services	Commerce de gros et de détail; réparation de véhicules automobiles, de motocycles et de biens personnels et domestiques, hôtels et restaurants (CITI G-H) Transports, entreposage et communications (CITI I) Autres activités, y compris intermédiation financière, immobilier, location et activités de services aux entreprises, administration publique et défense, sécurité sociale obligatoire, éducation, santé et action sociale, autres activités de services collectifs, sociaux et personnels, ménages privés employant du personnel domestique (CITI J-P).

Chapitre VIII : Finances

Des informations détaillées et les chiffres courants concernant le tableau 12 figurent dans *Statistiques financières internationales*, une publication du Fonds monétaire international (voir aussi http://elibrary-data.imf.org) et dans le *Bulletin mensuel de statistique des Nations Unies*.

Tableau 12 : Résumé de la balance des paiements

La balance des paiements peut être décrite d'une façon générale comme l'enregistrement des transactions économiques internationales d'une économie. Elle présente : a) les transactions en biens, services et revenus entre une économie et le reste du monde, b) les changements de propriété et les autres changements des avoirs de cette économie en or monétaire, droits de tirage spéciaux (DTS) et en créances et engagements envers le reste du monde, et c) les transferts sans contrepartie et les écritures de contrepartie nécessaires pour équilibrer au sens comptable les écritures au titre des transactions et changements susmentionnés qui ne s'annulent pas mutuellement. Les données de la balance des paiements sont présentées sur la base de la méthodologie et de la présentation de la sixième édition du *Manuel de la balance des paiements* (MBP6), publiée par le Fonds monétaire international en novembre 2013. Le MBP6 incorpore plusieurs modifications majeures pour tenir compte des évolutions du commerce international et de la finance internationale au fil des années, et de mieux harmoniser la méthodologie de la balance des paiements du Fonds avec celle du Système de comptes nationaux 2008 (SCN 2008). Les définitions détaillées concernant le contenu des catégories fondamentales de la balance des paiements sont données dans le *Manuel de la balance des paiements* (sixième édition).

De brèves notes explicatives sont données ci-dessous pour clarifier la portée des principaux éléments.

Le compte des transactions courantes enregistre toutes les transactions de la balance des paiements couvrant les exportations et les importations de biens et de services, les revenus et les transferts courants entre les résidents d'un pays et des non-résidents.

Le compte de capital, n.i.a. porte principalement sur les transferts de capital liés à l'acquisition d'un actif fixe autres que les transactions relatives à l'annulation de la dette, plus la cession d'actifs non financiers non produits, et les transferts de capital liés à la cession d'actifs fixes par le donateur ou au financement de la formation de capital par le récipiendaire, plus l'acquisition d'actifs non financiers non produits.

Le compte d'opérations financières, n.i.a. représente le solde net de l'investissement direct, de l'investissement de portefeuille et des autres transactions d'investissement.

Les réserves sont la somme des transactions sur actifs de réserve, les engagements constituant des avoirs de réserve pour les autorités étrangères, le financement exceptionnel et l'utilisation des crédits du FMI.

Chapitre IX : Marché du travail

Une collection complète de données comparables sur la population active et les sujets connexes est disponible sous la forme des *Indicateurs clé du marché du travail* (KILM) publiés par l'Organisation internationale du Travail (OIT), et mis à jour tous les deux ans. Des données plus contemporaines sont accessibles sur ILOSTAT (www.ilo.org/ilostat), le dépôt de données de l'OIT, qui publie des données annuellement, trimestriellement ou mensuellement à mesure de leur communication par les pays mais sans procéder à toutes les vérifications de leur cohérence ni inclure toutes les sources utilisées par les KILM. Pour diverses raisons, les définitions nationales de l'emploi et du chômage diffèrent souvent des définitions internationales normalisées recommandées, ce qui limite la comparabilité internationale. Les comparaisons entre pays se trouvent en outre compliquées par la diversité des systèmes de collecte de données utilisés pour recueillir des informations sur les personnes employées et les chômeurs. Le site Web d'ILOSTAT offre une description complète de la méthodologie employée pour établir les séries sur la main-d'œuvre.

Tableau 13 : Population active et chômage

Le taux de participation à la population active est calculé en exprimant le nombre de personnes de la population active sous forme de pourcentage de la population en âge de travailler. La population active est la somme du nombre de personnes employées et du nombre de personnes sans emploi (voir les Recommandations internationales en vigueur sur les statistiques du travail de l'OIT). La population en âge de travailler est la population d'âge supérieur à un certain seuil, prescrit pour la mesure des caractéristiques économiques. Les données concernent le groupe des personnes de 15 ans et plus et reposent sur les estimations modélisées de l'OIT, sauf indication contraire en note de bas de page.

La définition du chômage inclut les personnes en âge de travailler qui, au cours d'une certaine période, étaient :

a) « Sans emploi », c'est-à-dire sans emploi rémunéré ou indépendant;

b) « Disponibles », c'est-à-dire libres de contracter un emploi rémunéré ou de pratiquer un emploi indépendant au cours de la période de référence; et

c) « À la recherche d'un emploi », c'est-à-dire qui ont pris des mesures déterminées pour trouver un emploi rémunéré ou indépendant au cours de la période spécifiée

Ne sont pas considérées comme sans emploi :

a) Les personnes qui ont l'intention d'établir leur propre activité ou exploitation agricole, mais n'ont pas encore pris les dispositions nécessaires à cet effet et ne recherchent pas un emploi en vue d'une rémunération ou d'un profit;

b) Les anciens travailleurs familiaux non rémunérés qui n'ont pas d'emploi et ne sont pas à la recherche d'un emploi pour rémunération ou profit.

Les séries représentent généralement le nombre total des personnes au chômage complet ou temporairement mises à pied. Les données exprimées en pourcentages, lorsqu'elles figurent dans le tableau, sont calculées en comparant le nombre des chômeurs au total des membres du groupe de la population active sur lequel les données du chômage sont basées.

Tableau 14 : Emploi par activité économique

Le tableau de l'emploi présente, exprimée en pourcentages, la répartition des personnes employées par activité économique, conformément à la Classification internationale type, par industrie, de toutes les branches d'activité économique (CITI) version 4.

Chapitre X : Indices des prix et de la production

Tableau 15 : Indices des prix à la consommation

Un indice des prix à la consommation est généralement estimé sous forme d'une série de mesures synthétiques des variations relatives, d'une période à l'autre, des prix d'un ensemble fixe de biens et de services de consommation d'une quantité et de caractéristiques constantes, acquis, utilisés ou payés par la population de référence. Chaque mesure synthétique est construite comme la moyenne pondérée d'un grand nombre d'indices d'agrégats élémentaires. L'indice de chaque agrégat élémentaire est estimé au moyen d'un échantillon de prix pour un ensemble défini de biens et de services obtenus dans une région donnée ou par ses résidents auprès d'un ensemble donné de points de vente ou d'autres sources de biens et de services de consommation. Le tableau présente les indices généraux des prix à la consommation pour tous les groupes d'articles de consommation combinés, et un indice des prix des produits alimentaires ne comprenant que les boissons non alcoolisées. Lorsque les prix des boissons alcoolisées et/ou du tabac sont inclus, cela est indiqué en note.

Tableau 16 : Indices de la production agricole

L'agriculture désigne la production de tous les produits de culture et d'élevage. L'indice des produits alimentaires comprend les produits qui sont considérés comme comestibles et qui contiennent des éléments nutritifs. Les indices de la production agricole et de la production alimentaire sont calculés par la formule de Laspeyres avec comme période de base les années 2004-2006. Le choix d'une période de base de plusieurs années permet de réduire l'incidence des fluctuations annuelles de la production agricole sur les indices de cette période. Les quantités de chaque denrée produites sont pondérées par la moyenne nationale des prix à la production pour la période 2004-2006 et sommées pour chaque année. Les indices reposent sur les données de la production d'une année civile. Ceux-ci peuvent différer dans certains cas de ceux effectivement produits et publiés par les pays eux-mêmes en raison de différences dans les concepts, la couverture, les coefficients de pondération et les méthodes de calcul. On s'efforce d'estimer ces différences méthodologiques pour obtenir une meilleure comparabilité internationale des données. Des données détaillées sur la production agricole sont publiées par la FAO dans son *Annuaire statistique*.

Chapitre XI : Commerce international des marchandises

Préparé par la Division de statistique du Département des affaires économiques et sociales du Secrétariat des Nations Unies, l'*Annuaire statistique du commerce international* (ITSY) offre un aperçu des tendances récentes du commerce de biens et de services de la plupart des pays et régions du monde. L'*Annuaire*, voir http://comtrade.un.org/pb/, est publié en deux volumes; le Volume I est établi plus tôt dans l'année pour présenter un aperçu préliminaire du commerce international de marchandises de l'année précédente; le Volume II, publié généralement six mois plus tard, contient des tableaux détaillés qui présentent le commerce international par produit et 11 tableaux du commerce mondial couvrant les valeurs commerciales et les indices. Le Volume II contient également des versions actualisées des tableaux du commerce mondial. Les tableaux de cet *Annuaire* sont également mis à jour mensuellement dans le Bulletin mensuel de statistique des Nations Unies et sur le site des statistiques du commerce, voir http://unstats.un.org/unsd/trade/data/tables.asp#annual.

Les statistiques présentées dans cet *Annuaire* sont établies par les autorités statistiques nationales de façon largement conforme aux concepts et définitions des statistiques du commerce international de marchandises recommandés par les Nations Unies en 2010 (IMTS 2010). En fonction des parties du territoire économique incluses dans le territoire statistique, le système d'établissement des données du commerce adopté par un pays (son système de commerce) sera appelé général ou spécial.

Système de commerce général	Le territoire statistique coïncide avec le territoire économique. Il est donc recommandé que le territoire statistique d'un pays qui applique le système de commerce général englobe tous les éléments territoriaux applicables. Dans ce cas, les importations comprennent les biens qui entrent dans la zone de libre circulation, les installations de perfectionnement actif, les zones franches industrielles, les entrepôts sous douane ou les zones franches commerciales, et les exportations comprennent les biens qui quittent ces éléments territoriaux.
Système de commerce spécial	(définition stricte) Le territoire statistique ne comprend qu'une partie spécifique du territoire économique, de sorte que certains flux de marchandises auxquels les recommandations IMTS 2010 sont applicables ne sont inclus ni dans les statistiques d'importation ni dans les statistiques d'exportation du pays déclarant. La définition stricte du système de commerce spécial est utilisée lorsque le territoire statistique ne comprend que la zone de libre circulation, c'est-à-dire la partie dans laquelle les marchandises « peuvent être écoulées sans restriction douanière ». Par conséquent, dans ce cas, les importations ne comprennent que les marchandises qui entrent dans la zone de libre circulation du pays déclarant et les exportations ne comprennent que les marchandises qui quittent la zone de libre circulation du pays déclarant .
	(définition assouplie) a) les marchandises qui entrent dans un pays aux fins du perfectionnement actif ou en ressortent après, ainsi que b) les marchandises qui entrent dans une zone franche industrielle ou en sortent, sont aussi enregistrées et incluses dans les statistiques du commerce international de marchandises.

Tous les pays communiquent en général leurs données détaillées du commerce de marchandises conformément au Système harmonisé de désignation et de codification des marchandises (SH) et les données correspondent à la Classification type pour le commerce international (CTCI, rév. 3), dans laquelle elles sont ensuite présentées. Les données se réfèrent à des années civiles; cependant, pour les pays qui communiquent leurs données selon une autre année de référence, les données sont présentées dans l'année qui couvre la plus grande partie de l'année de référence utilisée par le pays.

Les valeurs FOB comprennent la valeur transactionnelle des marchandises et la valeur des services fournis pour livrer les marchandises à la frontière du pays exportateur. Les valeurs CIF comprennent la valeur transactionnelle des marchandises, la valeur des services fournis pour livrer les marchandises à la frontière du pays exportateur et la valeur des services fournis pour livrer les marchandises depuis la frontière du pays exportateur jusqu'à la frontière du pays importateur. De ce fait, les données de la valeur statistique des marchandises importées sont présentées dans le format CIF et celles de la valeur statistique des marchandises exportées dans le format FOB.

La conversion des valeurs libellées en monnaies nationales en dollars des États-Unis s'effectue au moyen de facteurs de conversion de devises fondés sur les taux de change officiels. Les valeurs libellées en monnaies sujettes à des fluctuations sont converties en dollars des États-Unis au moyen de taux de change moyens pondérés calculés spécialement à cette fin. Le taux de change moyen pondéré d'une monnaie donnée pour une année donnée est donné par les facteurs composants mensuels, fournis par le Fonds monétaire international dans sa publication *Statistiques financières internationales*, pondérés par la valeur du commerce concerné pour chaque mois; les facteurs mensuels sont les taux de change (ou les taux moyens simples) en vigueur au cours de ce mois. Ces facteurs sont appliqués aux importations et aux exportations totales et au commerce de chaque produit avec chaque pays.

Tableau 17 : Total des importations, des exportations et balance commerciale

Les données des importations totales et des exportations totales des pays (ou régions) présentées dans ce tableau proviennent principalement des *Statistiques financières internationales* publiées mensuellement par le Fonds monétaire international (FMI) mais aussi d'autres sources comme les publications et les sites web nationaux et le Questionnaire du *Bulletin mensuel de statistique* des Nations Unies, voir l'*Annuaire statistique du commerce international* pour plus de détails. Les données manquantes sont estimées afin de parvenir à des totaux régionaux mais ne sont pas présentées. La procédure d'estimation est automatisée au moyen des taux de croissance en glissement annuel trimestriels pour l'extrapolation des données trimestrielles manquantes (sauf si les données trimestrielles peuvent être estimées au moyen des données mensuelles disponibles au cours du trimestre). Les facteurs de conversion appliqués aux données de ce tableau sont publiés trimestriellement dans le *Bulletin mensuel de statistique* des Nations Unies et sont aussi disponibles sur le site Web des statistiques du commerce de l'ONU : http://unstats.un.org/unsd/trade/data/tables.asp#annual.

Chapitre XII : Énergie

L'*Annuaire statistique de l'énergie* (consultable sur le site http://unstats.un.org/unsd/energy/yearbook) est une collection complète de statistiques internationales de l'énergie qui couvre plus de 220 pays et régions. L'*Annuaire* est préparé par la Division de statistique du Département des affaires économiques et sociales du Secrétariat des Nations Unies. L'*Annuaire* est produit chaque année avec les nouvelles données disponibles sur la production d'énergie, le commerce, les variations de stocks, les réserves et la consommation de tous les pays et régions, et une série historique remontant à 1950 est disponible. Les données sont établies essentiellement à partir du questionnaire annuel sur l'énergie diffusé par la Division de statistique des Nations Unies et complétées par des publications statistiques officielles nationales, ainsi que des publications des organisations internationales et régionales. Lorsque les données officielles ne sont pas disponibles ou ne sont pas cohérentes, la Division de statistique établit des estimations en s'appuyant sur des éléments d'origine gouvernementale, professionnelle ou commerciale.

Les données se réfèrent à l'année civile, sauf celles des pays suivants qui se rapportent à l'exercice budgétaire : Afghanistan et Iran (République islamique d') – commençant le 21 mars de l'année indiquée; Australie, Bangladesh, Bhoutan, Égypte (électricité seulement pour ces deux derniers pays), Népal – finissant en juin de l'année indiquée; Pakistan – commençant en juillet de l'année indiquée; Inde, Myanmar et Nouvelle-Zélande – commençant en avril de l'année indiquée. Les données exprimées par habitant utilisent comme dénominateur les données démographiques de la Division de la population de l'Organisation des Nations Unies.

Tableau 18 : Production, commerce et fourniture d'énergie

Les données sont présentées en pétajoules (ou en gigajoules par habitant), unités auxquelles chaque produit énergétique est converti afin d'assurer l'uniformité et la comparabilité internationales des données. Pour convertir en joules les unités originelles, celles-ci (tonnes métriques, tonnes métriques d'équivalent pétrole, kilowattheures, mètres cubes) sont multipliées par des facteurs de conversion. Voir l'*Annuaire statistique de l'énergie* pour une liste des facteurs de conversion et une description détaillée des méthodes utilisées.

Les combustibles solides inclus dans la production commerciale d'énergie primaire sont l'anthracite, la lignite, la tourbe et le schiste bitumineux; les liquides comprennent le pétrole brut, les condensats de gaz naturel, les autres hydrocarbures, additifs et oxygénats, et les biocarburants liquides; les gaz comprennent le gaz naturel et la vapeur/chaleur primaire; et l'électricité comprend la production primaire d'électricité d'origine hydraulique, nucléaire, géothermique, éolienne, marémotrice, houlomotrice et solaire.

Les importations nettes (importations moins exportations et bunkers) et les variations des stocks font référence à toutes les formes d'énergie primaire et secondaire (y compris les produits intermédiaires). Dans les importations nettes, les bunkers font référence aux bunkers d'essence aviation, de carburéacteur et d'anthracite, de gazole/carburant diesel et de fiouls résiduels. Le commerce international de produits énergétiques repose sur le système du « commerce général », c'est-à-dire que toutes les marchandises entrant sur le territoire national d'un pays ou en sortant sont enregistrées comme importations ou exportations.

La consommation d'énergie comprend les formes primaires des combustibles solides, les importations nettes et les variations des stocks des combustibles secondaires; les liquides comprennent les produits pétroliers utilisés à des fins de production d'énergie y compris les produits intermédiaires, le gaz de raffinerie et le pétrole brut utilisé directement; les gaz comprennent la consommation de gaz naturel et de chaleur primaire, les importations nettes et les variations des stocks de gaz manufacturés; et l'électricité comprend la production primaire d'électricité et les importations nettes d'électricité. La consommation de certains produits pétroliers est négative en raison de l'exclusion des calculs des transferts entre produits. La consommation négative d'électricité est due à une production d'électricité primaire négligeable par rapport aux exportations nettes. Plus généralement, une consommation négative peut représenter une différence résiduelle ou statistique entre la production et les exportations lorsqu'un produit donné est principalement exporté.

Chapitre XIII : Environnement

Tableau 19 : Espèces menacées

Les données relatives au nombre d'espèces menacées dans chaque groupe d'animaux et de plantes sont établies par l'Union internationale pour la conservation de la nature et de ses ressources (UICN)/ Commission de sauvegarde des espèces (CSE) et publiées dans la liste rouge des espèces menacées de l'UICN. Cette liste fournit le catalogue des espèces considérées comme menacées au niveau mondial. Le nombre d'espèces menacées dans un pays donné change au cours des années pour diverses raisons, entre autres :

- La disponibilité d'informations nouvelles permet de raffiner l'évaluation (de confirmer par exemple que l'espèce est ou n'est pas présente dans un pays donné, que l'espèce est ou n'est pas menacée, etc.);

- L'évolution de la taxonomie (par exemple ce qu'on reconnaissait auparavant comme une certaine espèce est à présent réparti entre plusieurs espèces distinctes, ou fusionné avec une autre espèce);

- Les corrections (par exemple l'évaluation précédente peut avoir omis d'inscrire un pays donné dans la liste appropriée ou avoir inclus un pays donné par erreur);

- Changement de statut réel (par exemple la situation d'une espèce peut s'être réellement détériorée ou améliorée et l'espèce avoir par conséquent été incorporée aux catégories menacées ou en avoir été retirée).

Les catégories utilisées dans la Liste rouge sont les suivantes : espèce éteinte, éteinte à l'état sauvage, gravement menacée d'extinction; menacée d'extinction, vulnérable, quasi menacée et données insuffisantes.

Tableau 20 : Estimations des émissions de CO_2

Les données sur les émissions de dioxyde de carbone (CO_2) proviennent du « Carbon Dioxide Information Analysis Centre » (CDIAC) du laboratoire national d'Oak Ridge aux États-Unis, voir http://cdiac.ornl.gov/. Les estimations des émissions de CO_2 du CDIAC sont calculées essentiellement à partir de statistiques de l'énergie des Nations Unies sur la consommation de combustibles liquides et solides et la consommation et le torchage de gaz, ainsi que des estimations de la production de ciment provenant du Bureau des Mines du Département de l'intérieur des États-Unis. Les émissions sont présentées dans le tableau en unités de 1 000 tonnes métriques de CO_2; pour convertir le CO_2 en carbone, il faut diviser les chiffres par 3,667. Tous les détails des procédures de calcul des émissions sont données dans "Global, Regional, and National Annual CO_2 Emissions Estimates from Fossil Fuel Burning, Hydraulic Cement Production, and Gas Flaring" ainsi que sur le site Web du CDIAC. Par rapport aux autres sources industrielles dont les émissions de CO_2 sont estimées, les données relatives aux activités de torchage des gaz sont rares et sporadiques. Dans les pays où les activités de torchage des gaz représentent une proportion considérable des émissions totales de CO_2, la nature sporadique des données de torchage des gaz peut aboutir à l'apparition de tendances aberrantes ou fallacieuses des émissions nationales de CO_2 sur la période couverte par le tableau.

Chapitre XIV : Science et technologie

Tableau 21 : Brevets

Les brevets sont accordés par un office national des brevets ou un office régional qui accomplit cette tâche pour de nombreux pays, comme l'Office européen des brevets et l'African Regional Intellectual Property Organisation. Dans le cadre de ces systèmes régionaux, le requérant demande que son invention soit protégée dans un ou plusieurs pays, et il appartient à chaque pays d'accorder ou non la protection d'un brevet sur son territoire. Le Traité de coopération en matière de brevets (PCT), administré par l'Organisation mondiale de la propriété intellectuelle (OMPI) prévoit le dépôt d'une demande de brevet international unique dotée de la même validité que des demandes nationales déposées dans les pays désignés. Les données comprennent l'intensité de l'activité des résidents, les brevets délivrés et les brevets en vigueur. L'intensité du dépôt de brevets est présentée comme le nombre de demandes de brevet déposées par des résidents par million d'habitants, tandis que le dépôt d'une demande de propriété intellectuelle (PI) par des résidents fait référence à une demande déposée par un requérant auprès de son office national de la propriété intellectuelle. Les statistiques des droits de propriété intellectuelle (enregistrement) reposent sur le même concept. L'expression « en vigueur » désigne un brevet ou une autre forme de protection de la PI en cours de validité. Le pays d'origine sert à catégoriser les données de PI en PI résidente (intérieure) et non résidente (étrangère). La résidence du premier demandeur cité (ou inventeur) enregistré dans le document de PI (par exemple une demande de brevet ou de dépôt de marque) sert à classer les données de PI par pays d'origine. Les données sont établies et publiées par l'OMPI.

Chapitre XV : Communications

Les statistiques présentées aux tableaux 22 et 23 sont extraites de la base de données statistiques (voir www.itu.int) et de *l'Annuaire statistique* des Services de télécommunications de l'Union internationale des télécommunications.

Tableau 22 : Abonnements aux services de téléphone cellulaire mobile

Le nombre d'abonnements aux services de téléphone cellulaire mobile (ainsi que le nombre d'abonnements pour 100 habitants) fait référence aux utilisateurs de téléphones portables abonnés à un service automatique de téléphonie mobile public utilisant une technologie cellulaire, qui permet d'accéder au Réseau téléphonique public commuté (RTPC). Sont pris en compte aussi bien les abonnements prépayés que post-payés. Le nombre d'abonnements pour 100 habitants est calculé en divisant le nombre d'abonnements par le chiffre de la population et en multipliant le résultat par 100.

Tableau 23 : Utilisation d'Internet

Le tableau indique le pourcentage de personnes qui utilisent Internet et remplace les statistiques présentées dans les annuaires précédents comme le « Nombre (en milliers) d'abonnements à l'Internet fixe (filaire) » et le « Nombre d'abonnements à l'Internet fixe (filaire) pour 100 habitants ». Cet indicateur, outre qu'il saisit l'usage d'Internet, permet de mesurer les évolutions de l'accès à Internet et de son utilisation. Dans les pays où de nombreuses personnes accèdent à Internet au travail, à l'école, dans les cybercafés ou d'autres lieux publics, la multiplication des accès publics accroît le nombre d'utilisateurs malgré le nombre limité des abonnements et des foyers qui ont accès à Internet. Les pays en développement en particulier comptent souvent de nombreux utilisateurs par abonnement à Internet, ce qui traduit le fait que le foyer n'est pas le lieu principal d'accès.

Chapitre XVI : Tourisme et transport internationaux

Les données sur le tourisme international sont fournies par l'Organisation mondiale du tourisme (OMT) qui publie des renseignements détaillés sur le tourisme dans l'*Annuaire des statistiques du tourisme* ou le *Compendium des statistiques du tourisme*, voir www.unwto.org/statistics pour des informations plus complètes. Aux fins de l'établissement des statistiques, l'expression « visiteur international » décrit « toute personne qui se rend dans un pays autre que celui dans lequel il ou elle a son lieu de résidence habituelle, mais différent de son environnement habituel, pour une période de 12 mois au maximum, dans un but principal autre que celui d'y exercer une activité rémunérée ». L'*Annuaire* et le *Compendium* de l'OMT comportent quatre séries distinctes, mais une seule est retenue pour inclusion dans le présent *Annuaire*, en fonction en général de l'ordre de priorité suivant afin de décrire au mieux le « visiteur international »;

Ordre	Code de la série	Nom de la série
1	TF	*Arrivées de touristes non résidents aux frontières nationales* désigne les visiteurs qui passent au moins une nuit dans un logement collectif ou privé dans le pays visité (exclut les visiteurs d'un jour)
2	VF	*Arrivées de visiteurs non résidents aux frontières nationales* désigne les visiteurs définis dans la série « TF » ainsi que les visiteurs d'un jour qui ne passent pas la nuit dans un logement collectif ou privé dans le pays visité
3	TCE	*Arrivées de touristes non résidents dans tous les types d'établissements d'hébergement*
4	THS	*Arrivées de touristes non résidents dans les hôtels et établissements similaires*

Ces chiffres ne comprennent pas les immigrants, les résidents frontaliers, les personnes domiciliées dans un pays ou une région et qui travaillent dans un pays ou une région limitrophe, les membres des forces armées et les diplomates et les représentants consulaires lorsqu'ils se rendent de leur pays d'origine au pays où ils sont en poste et vice-versa. Ne sont pas inclus non plus les voyageurs en transit qui n'entrent pas formellement dans le pays en faisant viser leur passeport, comme les passagers d'un vol en escale qui demeurent pendant un court laps de temps dans une zone distincte d'une aérogare ou les passagers d'un navire qui ne sont pas autorisés à débarquer. Cette catégorie inclut les passagers transférés directement d'une aérogare à une autre ou à un autre terminal. Les autres passagers en transit dans un pays sont classés comme visiteurs.

Tableau 24 : Arrivées de touristes/visiteurs et dépenses touristiques

Les données relatives aux arrivées de visiteurs non résidents (ou internationaux) peuvent être obtenues de différentes sources. Dans certains cas, elles proviennent des statistiques frontalières tirées des registres administratifs (contrôles de police, de l'immigration, comptages du trafic routier et autres types de contrôle), des enquêtes statistiques aux frontières et des enregistrements d'établissements d'hébergement. Les totaux correspondent au nombre total d'arrivées depuis les régions indiquées dans le tableau. Lorsqu'une personne visite le même pays plusieurs fois dans l'année, un même nombre d'arrivées est enregistré. De même, si une personne visite plusieurs pays au cours d'un même voyage, son arrivée dans chaque pays est enregistrée séparément. On ne peut donc pas supposer que le nombre des arrivées soit égal au nombre de personnes qui voyagent.

Les dépenses associées à l'activité touristique des visiteurs sont traditionnellement identifiées au poste « Voyages » de la balance des paiements (BDP) : dans le cas du tourisme dans le pays récepteur, les dépenses dans le pays de référence associées aux visiteurs non résidents sont enregistrées comme « crédits » dans la BDP et il s'agit de « recettes au titre des voyages ». Le nouveau cadre conceptuel approuvé par la Commission de statistique des Nations Unies concernant la mesure de l'activité touristique à l'échelle macroéconomique (le compte satellite du tourisme) considère que la notion « industries et produits touristiques » inclut le transport de passagers. Par conséquent, une meilleure estimation des dépenses liées au tourisme par des visiteurs résidents et non résidents dans un scénario international serait, du point de vue de la BDP, la somme des valeurs du poste « Voyages » et du poste « Transport de passagers ». Néanmoins, les utilisateurs doivent être conscients de ce que les estimations de la BDP comprennent, outre les dépenses associées aux visiteurs, celles liées à d'autres types d'individus. Les données publiées doivent permettre la comparabilité internationale et donc correspondre à celles publiées par le Fonds monétaire international et fournies par les banques centrales, les exceptions sont listées dans le *Compendium des statistiques du tourisme* et l'*Annuaire des statistiques du tourisme*, voir www.unwto.org/statistics pour des informations plus complètes.

Chapitre XVII : Aide au développement

Tableau 25 : Décaissements nets d'aide publique au développement aux bénéficiaires

Le tableau présente des estimations des flux de ressources financières à destination des pays bénéficiaires soit directement (aide bilatérale) soit par l'intermédiaire d'institutions multilatérales (aide multilatérale). Les institutions multilatérales comprennent le Groupe de la Banque mondiale, des banques régionales, les institutions financières de l'Union européenne et un certain nombre d'institutions, de programmes et de fonds d'affectation spéciale des Nations Unies. Les données ont été obtenues auprès du Comité d'aide au développement (CAD) de l'OCDE, auquel les pays membres communiquent des données sur les flux de ressources qu'ils mettent à la disposition de pays et territoires en développement, de pays et territoires en transition, et des institutions multilatérales. On trouvera davantage d'informations sur les définitions, les méthodes et les sources dans la publication de l'OCDE intitulée *Répartition géographique des ressources financières allouées aux pays en développement*, ainsi que sur le site http://stats.oecd.org/.

Tableau 26 : Décaissements nets d'aide publique au développement par des donateurs

Le tableau présente les dépenses d'aide au développement des pays donateurs. Ce tableau inclut les contributions des donateurs aux institutions multilatérales; les totaux diffèrent donc de ceux du tableau 25, qui incluent les décaissements des institutions multilatérales.

* * *

Annex III

Conversion coefficients and factors

The metric system of weights and measures is employed in the *Statistical Yearbook*. In this system, the relationship between units of volume and capacity is: 1 litre = 1 cubic decimetre (dm^3) exactly (as decided by the 12th International Conference of Weights and Measures, New Delhi, November 1964).

Section A shows the equivalents of the basic metric, British imperial and United States units of measurements. According to an agreement between the national standards institutions of English-speaking nations, the British and United States units of length, area and volume are now identical, and based on the yard = 0.9144 metre exactly. The weight measures in both systems are based on the pound = 0.45359237 kilogram exactly (Weights and Measures Act 1963 (London), and *Federal Register announcement of 1 July 1959: Refinement of Values for the Yard and Pound* (Washington D.C.)).

Section B shows various derived or conventional conversion coefficients and equivalents.

Section C shows other conversion coefficients or factors which have been utilized in the compilation of certain tables in the *Statistical Yearbook*. Some of these are only of an approximate character and have been employed solely to obtain a reasonable measure of international comparability in the tables.

For a comprehensive survey of international and national systems of weights and measures and of units' weights for a large number of commodities in different countries, see *World Weights and Measures*.

A. Equivalents of metric, British imperial and United States units of measure

Annexe III

Coefficients et facteurs de conversion

L'*Annuaire statistique* utilise le système métrique pour les poids et mesures. La relation entre unités métriques de volume et de capacité est: 1 litre = 1 décimètre cube (dm^3) exactement (comme fut décidé à la Conférence internationale des poids et mesures, New Delhi, novembre 1964).

La section A fournit les principaux équivalents des systèmes de mesure métrique, britannique et américain. Suivant un accord entre les institutions de normalisation nationales des pays de langue anglaise, les mesures britanniques et américaines de longueur, superficie et volume sont désormais identiques, et sont basées sur le yard = 0.9144 mètre exactement. Les mesures de poids se rapportent, dans les deux systèmes, à la livre (pound) = 0.45359237 kilogramme exactement (*Weights and Measures Act 1963* (Londres), et *Federal Register Announcement of 1 July 1959: Refinement of Values for the Yard and Pound* (Washington, D.C.)).

La section B fournit divers coefficients et facteurs de conversion conventionnels ou dérivés.

La section C fournit d'autres coefficients ou facteurs de conversion utilisés dans l'élaboration de certains tableaux de l'*Annuaire statistique*. Certains coefficients ou facteurs de conversion ne sont que des approximations et ont été utilisés uniquement pour obtenir un degré raisonnable de comparabilité sur le plan international.

Pour une étude d'ensemble des systèmes internationaux et nationaux de poids et mesures, et d'unités de poids pour un grand nombre de produits dans différents pays, voir *World Weights and Measures*.

A. Equivalents des unités métriques, britanniques et des Etats-Unis

Metric units Unités métriques	British imperial and US equivalents Equivalents en mesures britanniques et des Etats-Unis		British imperial and US units Unités britanniques et des Etats-Unis	Metric equivalents Equivalents en mesures métriques
Length — Longueur				
1 centimetre – centimètre (cm)	0.3937008	inch	1 inch	2.540 cm
1 metre – mètre (m)	3.280840	feet	1 foot	30.480 cm
	1.093613	yard	1 yard	0.9144 m
1 kilometre – kilomètre (km)	0.6213712	mile	1 mile	1609.344 m
	0.5399568	international nautical mile	1 international nautical mile	1852.000 m
Area — Superficie				
1 square centimetre – (cm^2)	0.1550003	square inch	1 square inch	6.45160 cm^2
1 square metre – (m^2)	10.763910	square feet	1 square foot	9.290304 dm^2
	1.195990	square yards	1 square yard	0.83612736 m^2
1 hectare – (ha)	2.471054	acres	1 acre	0.4046856 ha
1 square kilometre – (km^2)	0.3861022	square mile	1 square mile	2.589988 km^2
Volume				
1 cubic centimetre – (cm^3)	0.06102374	cubic inch	1 cubic inch	16.38706 cm^3
1 cubic metre – (m^3)	35.31467	cubic feet	1 cubic foot	28.316847 dm^3
	1.307951	cubic yards	1 cubic yard	0.76455486 m^3
Capacity — Capacité				
1 litre (l)	0.8798766	British imperial quart	1 British imperial quart	1.136523 l
	1.056688	U.S. liquid quart	1 U.S. liquid quart	0.9463529 l
	0.908083	U.S. dry quart	1 U.S. dry quart	1.1012208 l
1 hectolitre (hl)	21.99692	British imperial gallons	1 British imperial gallon	4.546092 l
	26.417200	U.S. gallons	1 U.S. gallon	3.785412 l
	2.749614	British imperial bushels	1 imperial bushel	36.368735 l
	2.837760	U.S. bushels	1 U.S. bushel	35.239067 l

Metric units Unités métriques	British imperial and US equivalents Equivalents en mesures britanniques et des Etats-Unis		British imperial and US units Unités britanniques et des Etats-Unis	Metric equivalents Equivalents en mesures métriques
Weight or mass — Poids				
1 kilogram (kg)	35.27396	av. ounces	1 av. ounce	28.349523 g
	32.15075	troy ounces	1 troy ounce	31.10348 g
	2.204623	av. pounds	1 av. pound	453.59237 g
			1 cental (100 lb.)	45.359237 kg
			1 hundredweight (112 lb.)	50.802345 kg
1 ton – tonne (t)	1.1023113	short tons	1 short ton (2 000 lb.)	0.9071847 t
	0.9842065	long tons	1 long ton (2 240 lb.)	1.0160469 t

B. Various conventional or derived coefficients

Air transport

1 passenger-mile = 1.609344 passenger kilometre
1 short ton-mile = 1.459972 tonne-kilometre
1 long ton-mile = 1.635169 tonne kilometre

Electric energy

1 Kilowatt (kW) = 1.34102 British horsepower (hp)
  1.35962 cheval vapeur (cv)

C. Other coefficients or conversion factors employed in *Statistical Yearbook* tables

Roundwood

Equivalent in solid volume without bark.

Sugar

1 metric ton raw sugar = 0.9 metric ton refined sugar
For the United States and its possessions:
1 metric ton refined sugar = 1.07 metric tons raw sugar

Energy

1 metric ton peat = .325 metric ton of coal oil equivalent
1 ton oil equivalent = .4186 GJ or 11.63 MWh

B. Divers coefficients conventionnels ou dérivés

Transport aérien

1 voyageur (passager) – kilomètre = 0.621371 passenger-mile
1 tonne-kilomètre = 0.684945 short ton-mile
  0.611558 long ton-mile

Energie électrique

1 British horsepower (hp) = 0.7457 kW
  1 cheval vapeur (cv) = 0.735499 kW

C. Autres coefficients ou facteurs de conversion utilisés dans les tableaux de l'*Annuaire statistique*

Bois rond

Equivalences en volume solide sans écorce.

Sucre

1 tonne métrique de sucre brut = 0.9 tonne métrique de sucre raffiné
Pour les États-Unis et leurs possessions:
1 tonne métrique de sucre raffiné = 1.07 tonne métrique de sucre brut

Energie

1 tonne métrique d'équivalent charbon = 3.08 tonnes métrique de tourbe
1 GJ = 2.39 tonne d'équivalent pétrol or 1 MWh = .086 tonne métrique d'équivalent pétrol

Annex IV - Tables added, omitted and discontinued

A. Tables added

The present issue of the *Statistical Yearbook* includes the following tables which were not presented in the previous issue:

Table 3	International migrants and refugees
Table 9	Intentional homicides and other crimes
Table 12	Balance of payments summary
Table 13	Labour force and unemployment (previously Unemployment)
Table 18	Production, trade and supply of energy (previously Production, trade and consumption of commercial energy)

B. Tables omitted

The following tables which were presented in previous issues are not presented in the present issue. They will be updated in future issues of the *Yearbook* when new data become available:

- Civil aviation: scheduled airline traffic
- Gross domestic expenditure on research and development
- Index of industrial production
- Land
- Personnel in research and development (R & D)
- Water supply and sanitation coverage
- Population and rates of growth in urban areas and capital cities
- Ratio of girls to boys in primary, secondary and tertiary levels
- Selected indicators of life expectancy, childbearing, age dependency ratio and mortality
- Teaching staff at the primary, secondary and tertiary level

C. Tables discontinued

The following tables have been discontinued:

- Aluminium, unwrought
- Beer
- Cement
- Cigarettes
- Cereals
- Cinema infrastructure
- Energy commodities
- Exchange rates
- External debt stocks and flows, long term
- External debt stocks, (long term) public and publicly guaranteed
- Fertilizers
- Food supply
- Gross domestic product by type of expenditure
- International reserves, excluding gold
- Manufactured goods exports
- Meat and fish
- Outbound tourism
- Paper and paperboard
- Pig iron and crude steel
- Producer price indices
- Rates of discount of central banks
- Raw sugar
- Roundwood
- Share of women in wage employment in the non-agricultural sector
- Short-term interest rates
- Socio-economic development assistance through the United Nations system
- Trade indices of volume and unit value/price
- Tourist/visitor arrivals by region of origin
- Wages in manufacturing
- Woven cotton and wool

Annexe IV - Tableaux ajoutés, supprimés et discontinués

A. Tableaux ajoutés

Dans ce numéro de l'Annuaire statistique, les tableaux suivants n'ont pas été présentés dans le numéro antérieur, et ont été ajoutés:

Tableau 3	Migrants internationaux et réfugiés
Tableau 9	Homicides intentionnels et autres crimes
Tableau 12	Résumé de la balance des paiements
Tableau 13	Population active et chômage (précédemment Chômage)
Tableau 18	Production, commerce et fourniture d'énergie (précédemment Production, commerce et consommation d'énergie commerciale)

B. Tableaux supprimés

Les tableaux suivants qui ont été repris dans les éditions antérieures n'ont pas été repris dans la présente édition. Ils seront actualisés dans les futures livraisons de l'Annuaire à mesure que des données nouvelles deviendront disponibles:

- Accès à l'eau et a l'assainissement
- Aviation civile: trafic aérien régulier
- Dépenses intérieures brutes de recherche et développement
- Indices de la production industrielle
- Personnel employé dans la recherche et le développement
- Personnel enseignant au niveau primaire, secondaire et supérieur
- Population et taux de croissance dans les zones urbaines et capitales
- Rapport filles/garçons dans l'enseignement primaire, secondaire et supérieur
- Sélection d'indicateurs de l'espérance de vie, de la maternité, du ratio de dépendance
- Terres

C. Tableaux discontinués

Les tableaux suivants ont été discontinués:

- Aluminium non travaille
- Arrivées de touristes/visiteurs par régions de provenance
- Assistance en matière de développement socioéconomique fournie par le système des Nations Unies
- Bière
- Bois rond
- Céréales
- Cigarettes
- Ciment
- Coton et laine tisses
- Cours des changes
- Dépenses imputées au produit intérieur brut
- Disponibilités alimentaires
- Engrais
- Exploitation cinématographique
- Exportations des produits manufactures
- Fonte et acier brut
- Importations et exportations totales, indices de valeur unitaire et de volume
- Indices des prix à la production
- Papiers et cartons
- Principaux biens de l'énergie
- Salaires dans les industries manufacturières
- Stocks et flux de la dette extérieure, à long terme
- Stock de la dette extérieur garantie par l'Etat (à long ter
- Sucre brut
- Réserves internationales, l'or non compris
- Taux d'escompte des banques centrales
- Taux d'intérêt à court terme
- Tourisme à l'étranger
- Proportion de femmes salariées dans le secteur non agri
- Viande et halieutique